MARINE MAMMAL BIOLOGY
An Evolutionary Approach

T0256836

Marine Mammal Biology

An Evolutionary Approach

EDITED BY

A. RUS HOELZEL

School of Biological and Biomedical Sciences
University of Durham
UK

Blackwell
Publishing

BLACKWELL PUBLISHING
350 Main Street, Malden, MA 02148-5020, USA
9600 Garsington Road, Oxford OX4 2DQ, UK
550 Swanston Street, Carlton, Victoria 3053, Australia

First published 2002 by Blackwell Science Ltd

7   2008

Library of Congress Cataloging-in-Publication Data has been applied for

ISBN 978-0-632-05232-5 (paperback)

A catalogue record for this title is available from the British Library.

Set in 9.5 on 12 pt Galliard
by Graphicraft Ltd, Hong Kong

The publisher's policy is to use permanent paper from mills that operate a sustainable
forestry policy, and which has been manufactured from pulp processed using acid-free and
elementary chlorine-free practices. Furthermore, the publisher ensures that the text paper
and cover board used have met acceptable environmental accreditation standards.

For further information on
Blackwell Publishing, visit our website:
www.blackwellpublishing.com

Contents

Contributors

Daryl J. Boness
Department of Conservation Biology
National Zoological Park
Smithsonian Institution
Washington
DC 20008
USA

W. Don Bowen
Marine Fish Division
Bedford Institute of Oceanography
1 Challenger Drive
PO Box 1006
Dartmouth
Nova Scotia B2Y 4A2
Canada

Ian L. Boyd
Sea Mammal Research Unit
Gatty Marine Laboratory
University of St Andrews
St Andrews
Fife KY16 8LB
UK

Phillip J. Clapham
Large Whale Biology Program
Protected Species Branch
Northeast Fisheries Science Center
166 Water Street
Woods Hole
MA 02543
USA

Richard C. Connor
Department of Biological Sciences
UMass-Dartmouth

North Dartmouth
MA 02747
USA

Guido Dehnhardt
Department of Zoology and Neurobiology
University of Bochum
D-44780 Bochum
Germany

Jim A. Estes
A-316 Earth and Marine Sciences Building
University of California
Santa Cruz
CA 95064
USA

Robert C. Fleischer
Department of Conservation Biology
National Zoological Park
Smithsonian Institution
Washington
DC 20008
USA

Ilya I. Glezer
Department of Cell Biology and Anatomical
  Sciences
Sophie Davis School for Biomedical Education
City University Medical School at City College of
  New York
138 Street at Convent Avenue
New York
NY 10031
USA

Simon D. Goldsworthy
Department of Zoology
Latrobe University
Victoria 3086
Australia

Philip S. Hammond
NERC Sea Mammal Research Unit
Gatty Marine Laboratory
University of St Andrews
St Andrews
Fife KY16 8LB
UK

John E. Heyning
Natural History Museum of Los Angeles County
900 Exposition Boulevard
Los Angeles
CA 90007
USA

A. Rus Hoelzel
School of Biological and Biomedical Sciences
University of Durham
South Road
Durham DH1 3LE
UK

David Kastak
Long Marine Laboratory
University of California
Santa Cruz
CA 95060
USA

Gina M. Lento
School of Biological Sciences
University of Auckland
Private Bag 92019
Auckland
New Zealand

Bernie J. McConnell
NERC Sea Mammal Research Unit
Gatty Marine Laboratory
University of St Andrews
St Andrews

Fife KY16 8LB
UK

Anthony R. Martin
British Antarctic Survey
High Cross
Madingley Road
Cambridge CB3 0ET
UK

Sarah L. Mesnick
National Marine Fisheries Service
Southwest Fisheries Science Center
La Jolla Laboratory
La Jolla
CA 92038
USA

Edward H. Miller
Biology Department
Memorial University of Newfoundland
St John's
Newfoundland A1B 3X9
Canada

Andy J. Read
Duke University Marine Laboratory
135 Duke Marine Lab Road
Beaufort
NC 20516
USA

Randall R. Reeves
27 Chandler Lane
Hudson
Quebec J0P 1H0
Canada

Peter J.H. Reijnders
Alterra-Marine and Coastal Zone Research
PO Box 167
1790 AD Den Burg
The Netherlands

Ronald J. Schusterman
Long Marine Laboratory
University of California
Santa Cruz

CA 95060
USA

Peter T. Stevick
NERC Sea Mammal Research Unit
Gatty Marine Laboratory
University of St Andrews
St Andrews
Fife KY16 8L
UK

Peter L. Tyack
Biology Department
Woods Hole Oceanographic Institution
Woods Hole
MA 02543
USA

Terrie M. Williams
Department of Evolutionary and Ecological Biology
Long Marine Laboratory Center for Ocean Health
University of California at Santa Cruz
Santa Cruz
CA 95064
USA

Graham A.J. Worthy
Department of Biology
University of Central Florida
4000 Central Florida Boulevard
Orlando
FL 32816-2368
USA

Preface

Whales, porpoises and dolphins (cetaceans), seals, sea lions, fur seals and walruses (pinnipeds), manatees and dugongs (sirenians), sea otters and polar bears (carnivores) are all classified as marine mammals. Although the groups of mammals that inhabit the marine environment are taxonomically diverse, the sea is a powerful and in some ways unifying influence. For this reason it is useful to explore the similarities and differences among the various mammalian species that have adapted to life in the oceans. This evolutionary approach is the theme of this book, and it is relevant to all aspects of marine mammal biology. At the same time, the evolutionary history of the terrestrial ancestors of each of these groups has an important influence on the life history of individual species in the marine environment. Cetaceans were the first to adapt to a wholly marine existence, with the earliest fossil species dating back to the Eocene. Their closest terrestrial relatives are ungulates, but fully aquatic species first appeared as early as 50–60 million years ago. Among marine mammals, cetaceans and sirenians show the most extreme physical adaptations to aquatic life (and represent the earliest radiations), but retain numerous characteristics common to all mammals. These include giving live birth to dependent young, homeothermy and the need to breath air; and such factors limit and shape the evolutionary potential of marine mammal species. The other major taxonomic group is the pinnipeds, which radiated more recently in the Oligocene. These species remain comparatively similar to their carnivore terrestrial relatives, and have amphibious life histories, returning to land to breed and moult. A key objective of this book was to consider all aspects of the biology of marine mammal species in light of their adaptation to aquatic habitats, and to use a comparative approach to assess the role of phylogenetic history and evolutionary potential. A further objective was to provide an inclusive, advanced text on the biology of a set of species that are of great interest for their diversity, behaviour and interactions with us.

The book begins with an overview of the various species and their global distribution, followed by a review of their evolutionary history. The next few chapters describe the physiological and anatomical adaptations that evolved to meet the challenges of the marine environment, including thermoregulation, osmoregulation, streamlining, the capacity for deep diving, sensory adaptations, and changes in neural anatomy and function. Further chapters build on this foundation with detailed treatments of energetics, vocal anatomy and behaviour, echolocation, feeding ecology, an ecological perspective on social behaviour, and problem solving and memory. A review of life history and reproductive strategies is integrated with chapters on patterns of movement and population genetics. Finally, our interactions with these species, and our influence on the fate of their populations are discussed in a chapter on conservation and management.

By far the most taxonomically diverse (and most studied) of the marine mammals are the cetaceans and pinnipeds, and much of this volume focuses on these species. However, the authors have created a truly comparative analysis, incorporating data on all species we know as marine mammals. This volume provides the background and reference base a student new to the subject would require, but the authors have also made an effort to highlight the discussions and analyses at the forefront of their respective disciplines. There are also frequent illustrations and extensive summary tables providing easily accessible details on aspects of marine mammal biology and life history.

I would like to thank the authors for their excellent contributions to this volume, and their patience and responsiveness during various rounds of revision. I also thank those who critically reviewed chapters: Ian Boyd, Greg Donovan, Peter Evans, Ewan Fordyce, Stefania Gaspari, Jonathan Gordon, Phil Hammond, Sam Ridgeway, Sean Twiss, Graham Worthy and Louise Wynen. I thank the staff at Blackwell Science, especially Ian Sherman, Sarah Shannon, Cee Brandson, Jane Andrew and Katrina McCallum for good advice and enthusiastic assistance. The authors of Chapter 8 would also like to thank Stacey Reese for technical assistance and Deborah Austin and Carrie Beck for helpful comments. The authors of Chapter 13 thank Colleen Reichmuth Kastak and Robert Gisiner.

A. Rus Hoelzel, 2001

CHAPTER 1

Diversity and Zoogeography

Anthony R. Martin and Randall R. Reeves

1.1 INTRODUCTION

An examination of marine mammal diversity—in terms of morphology, ecology, life history and geographical distribution—reveals much about these animals and the processes which brought about today's complement of species. It is important to bear in mind, however, that in studying the present-day fauna, we are only looking at a snapshot in geological time. For reasons that will be explored in this and later chapters, the array of organisms presently inhabiting the world's oceans, estuaries, lakes and rivers differs from that of a thousand years ago, and dramatically from that of 1 million and 10 million years ago. Similarly, were we to revisit the planet a thousand or a million years in the future we would probably see differences (e.g. in numbers of species and their geographical ranges) that would cause us to change any conclusions and predictions based on what we see now. The earth's marine mammal fauna is dynamic, changing slowly but continuously in normal circumstances, yet much more rapidly within the past few centuries due to the influence of our own species on the planet's aquatic ecosystems.

What, exactly, is a 'marine mammal'? The usual interpretation of this term, adopted here, is that it covers animals within the orders Cetacea and Sirenia, three families of Carnivora (the pinnipeds), plus two other carnivores—the sea otter and polar bear. However, one or more species from each of the three major groups live in fresh water, so the term *aquatic* mammal would be more correct. Regardless of the collective name used, there is no reason to exclude the freshwater species as they are derived from, and closely related to, animals living in marine habitats. In some cases (e.g. Saimaa and Baikal seals) their isolation in enclosed waters occurred in fairly recent geological time.

Rather more than 120 living species of marine mammals are currently recognized: 84 cetaceans, 36 pinnipeds, 4 sirenians, the sea otter and the polar bear. But specialists disagree on how many species there are, and the number varies as new genetic or morphological information throws further light on systematic relationships (see Chapter 2). Similarly, the number of races or subspecies is neither universally agreed nor stable (see Chapter 11). For consistency of approach, we have largely adopted the conclusions of Rice (1998), who provides an excellent review of the literature on this subject.

1.2 DIVERSITY

1.2.1 Introduction

Although marine mammals vary greatly in form and function, one over-riding factor has influenced and constrained their appearance: they live for much or all of their time in a much denser medium than that inhabited by terrestrial mammals. The most profound differences between living in water and air for mammals are: (i) the inability to exchange lung gases at all times; (ii) the increased rate of integumentary heat loss; (iii) relative weightlessness; (iv) greater resistance to movement; (v) changed characteristics of sound propagation; and (vi) low-light conditions in all but near-surface waters. The degree of adaptation to an aquatic lifestyle depends, not surprisingly, on the proportion of time spent in water.

Fig. 1.1 Representative mysticetes: 1, pygmy right; 2, right; 3, blue; 4, minke; and 5, humpback whales.

(a)

Fig. 1.2 Representative odontocetes. (a) 1, sperm whale; 2, Hubbs' beaked whale; 3, long-finned pilot whale; 4, narwhale; and 5, killer whale. (*continued*)

(b)

Fig. 1.2 (*cont'd*) (b) 1, Risso's dolphin; 2, hourglass dolphin; 3, southern right whale dolphin; 4, bottlenose dolphin; 5, Hector's dolphin; 6, boto; and 7, harbour porpoise.

Whales, dolphins and porpoises (collectively known as cetaceans) (Figs 1.1 and 1.2), which spend their entire lives in water and are unable to live on land, demonstrate extreme morphological adaptations. Cetaceans breathe through nostrils on the dorsal surface of their head, permitting rapid gas exchange without pausing at the water surface, and have breath-hold capabilities (more than an hour in some species) far exceeding those of land mammals. Their thick layer of insulating blubber also serves as a depot for fat storage, and heat loss is further reduced by vascular heat-exchange systems (Chapter 3). Some cetaceans grow to extraordinary size; this is possible because they do not need to support their own weight. Their smooth skin, rigid bodies, internal genitalia and lack of hind limbs all help reduce drag. Broad, horizontal tail flukes driven by strong musculature provide powerful and efficient

Fig. 1.3 Sirenians, polar bear and sea otter: 1, polar bear; 2, sea otter; 3, West Indian manatee; 4, Amazonian manatee; and 5, dugong.

propulsion through water. A dorsal fin provides rotational stability for most species. Finally, most cetaceans (the odontocetes at least) have sophistic-ated sound-processing systems, allowing them to interpret received sounds and propagate their own signals for echolocation (Chapter 6). The majority of cetaceans use sound, not light, as the primary means of gathering environmental information.

Sirenians (manatees and the dugong) (Fig. 1.3) are also obligate water dwellers and share many characteristics with the cetaceans, including the basic form of the body. Their distinct genealogy (distantly linked with elephants (Chapter 2)) and lifestyle (primarily herbivorous in warm, shallow waters) have, however, ensured that sirenians differ significantly from cetaceans. They have paired nostrils positioned

anteriorly on the head which precludes rapid gas exchange during locomotion, and very dense skeletal bones which help neutralize the buoyancy of the blubber. They have no echolocation ability and less sensitive hearing than cetaceans. Their dentition is appropriate for the mastication of plants, unlike all other marine mammals. The four living sirenians are of only moderate size (typically 2.5–4.5 m in length) although Steller's sea cow, which was rendered extinct by humans about 230 years ago, grew to 8 m and 3.5 t.

The amphibious lifestyle of seals and their allies (the pinnipeds) (Fig. 1.4) precludes some of the extreme aquatic adaptations seen in the Cetacea and Sirenia. Their carnivoran lineage has inevitably resulted in quite different features from the other

Fig. 1.4 Representative seals: 1, walrus; 2, southern elephant seal; 3, leopard seal; 4, ribbon seal; 5, grey seal; 6, ringed seal; 7, Mediterranean monk seal; and 8, South American sea lion.

two groups. The term 'pinniped' means fin-footed, and indeed these animals have hindlimbs modified to provide propulsion and forelimbs capable of assisting locomotion on land or ice as well as in water. Most also have dense fur and a flexible, streamlined body. The double nostrils are positioned anteriorly on the head as in sirenians, and the areas immediately below the nostrils are covered in typical carnivoran sensory bristles, varying in length, stiffness and density across species. The dentition of pinnipeds is more like that of other carnivores than that of cetaceans or sirenians. They have differentiated teeth, including incisors (except the walrus), canines and postcanines. Like other carnivores, pinnipeds eat flesh. Most consume fish or cephalopods primarily, but some specialize on zooplankton (e.g. crabeater seal), molluscs (walrus) or warm-blooded prey (leopard seal). Pinnipeds principally rely on vision and touch rather than sound to gather information about their environment, having relatively large eyes and no proven echolocation abilities. As a group they possess exceptional diving capability, exceeding that of cetaceans after taking into account the effects of body size (Schreer & Kovacs 1997). The largest pinnipeds are elephant seals (*Mirounga* spp.), with males reaching more than 4 m and 2.5 t (Ling & Bryden 1981; McGinnis & Schusterman 1981), but even these are small in comparison with Steller's sea cow and the great whales.

The two other carnivores considered as marine mammals, the sea otter and the polar bear (Fig. 1.3), are less adapted to an aquatic lifestyle. In fact, the polar bear has achieved greater specialization for life in a cold habitat above water than an aquatic one and should perhaps be viewed as a water-adapted terrestrial mammal. Although capable of sustained swimming, its hollow hair fibres provide too much buoyancy for prolonged diving and its long legs allow rapid terrestrial locomotion but relatively poor mobility in water. Polar bears are specialist predators of seals, but will hunt other animals (e.g. belugas and even humans) or scavenge when the availability of seals is low. Sea otters are perfectly at home in the sea, alternately diving for bottom-dwelling molluscs and crustaceans, then rafting for hours on the surface. Their hind feet are large and flipper-like, and their pelage provides an efficient barrier against heat loss.

1.2.2 Cetaceans

None of the groups considered in this volume has a greater diversity of morphological form than the order Cetacea. The largest living cetacean, and indeed the largest animal known to have lived on earth, is the blue whale (*Balaenoptera musculus*). With an adult body length of up to 33 m and a body mass of up to 190,000 kg, the largest blue whales have a mass some 3000 times greater than the vaquita (*Phocoena sinus*) or the Hector's dolphin (*Cephalorhynchus hectori*), which do not exceed 1.7 m and 60 kg. Between these extremes, an extraordinary range of body shapes and sizes reflects differing habitat and diet, social structure and behaviour. The order Cetacea comprises two very different living suborders, Mysticeti (baleen whales, Fig. 1.1) and Odontoceti (toothed whales, Fig. 1.2), discussed separately below. In the English language, cetaceans are divided into 'whales', 'dolphins' and 'porpoises' (Table 1.1). These terms broadly reflect body size (in decreasing order) rather than taxonomy, and can be confusing. For example, the killer whale (*Orcinus orca*) and the two pilot whales (*Globicephala* spp.) are actually large marine dolphins (Delphinidae).

1.2.2.1 Mysticetes

Fourteen mysticete species are currently recognized, ranging in body size from the blue whale down to the pygmy right whale (*Caperea marginata*), which reaches about 6.5 m and 3.5 t. Unusually for mammals, adult female baleen whales are larger than males, typically by about 5% in body length. Mys-

Latin name	English name(s)
Order CARNIVORA	
Family Otariidae. Fur seals and sea lions	
Arctocephalus pusillus	Tasmanian and Cape fur seals
Arctocephalus gazella	Antarctic fur seal, Kerguelen fur seal
Arctocephalus tropicalis	Subantarctic fur seal
Arctocephalus townsendi	Guadalupe fur seal
Arctocephalus philippii	Juan Fernández fur seal
Arctocephalus forsteri	South Australian and New Zealand fur seal, Australasian fur seal
Arctocephalus australis	South American fur seal
Arctocephalus galapagoensis	Galápagos fur seal
Callorhinus ursinus	Northern fur seal
Zalophus japonicus	Japanese sea lion
Zalophus californianus	California sea lion
Zalophus wollebaeki	Galápagos sea lion
Eumetopias jubatus	Northern sea lion, Steller's sea lion
Neophoca cinerea	Australian sea lion
Phocarctos hookeri	New Zealand sea lion, Hooker's sea lion
Otaria flavescens (= *O. byronia*)	South American sea lion

Table 1.1 Recent marine mammals. (Adapted from Rice 1998; see also International Whaling Commission 2001.)

(continued)

Table 1.1 (*cont'd*)

Latin name	English name(s)
Family Odobenidae. Walrus	
Odobenus rosmarus	Walrus
Family Phocidae. Seals	
Erignathus barbatus	Bearded seal
Phoca vitulina	Harbour seal
Phoca largha	Spotted seal, larga seal
Pusa hispida	Ringed seal
Pusa caspica	Caspian seal
Pusa sibirica	Baikal seal
Halichoerus grypus	Grey seal
Histriophoca fasciata	Ribbon seal
Pagophilus groenlandicus	Harp seal
Cystophora cristata	Hooded seal
Monachus tropicalis	Caribbean monk seal
Monachus monachus	Mediterranean monk seal
Monachus schauinslandi	Hawaiian monk seal
Mirounga leonina	Southern elephant seal
Mirounga angustirostris	Northern elephant seal
Leptonychotes weddellii	Weddell seal
Ommatophoca rossii	Ross seal
Lobodon carcinophaga	Crabeater seal
Hydrurga leptonyx	Leopard seal
Family Ursidae. Bears	
Ursus maritimus	Polar bear
Family Mustelidae. Weasels and otters	
Enhydra lutris	Sea otter
Order CETACEA	
Suborder Mysticeti	
Family Balaenidae. Right whales	
Eubalaena glacialis	North Atlantic right whale
Eubalaena japonica	North Pacific right whale
Eubalaena australis	Southern right whale
Balaena mysticetus	Bowhead whale
Family Neobalaenidae. Pygmy right whale	
Caperea marginata	Pygmy right whale
Family Balaenopteridae. Rorquals	
Megaptera novaeangliae	Humpback whale
Balaenoptera acutorostrata	Northern minke whale, common minke whale (includes 'dwarf')
Balaenoptera bonaerensis	Antarctic minke whale
Balaenoptera edeni	Pygmy Bryde's whale
Balaenoptera brydei	Bryde's whale
Balaenoptera borealis	Sei whale
Balaenoptera physalus	Fin whale
Balaenoptera musculus	Blue whale
Family Eschrichtiidae. Gray whale	
Eschrichtius robustus	Gray whale

(*continued on p. 8*)

Table 1.1 (*cont'd*)

Latin name	English name(s)
Suborder Odontoceti	
Family Physeteridae. Sperm whale	
Physeter macrocephalus	Sperm whale
Family Kogiidae. Pygmy, or short-headed, sperm whales	
Kogia breviceps	Pygmy sperm whale
Kogia sima	Dwarf sperm whale
Family Ziphiidae. Beaked whales	
Ziphius cavirostris	Cuvier's beaked whale, Goosebeak whale
Berardius arnuxii	Arnoux's beaked whale
Berardius bairdii	Baird's beaked whale
Tasmacetus sheperdi	Shepherd's beaked whale
Indopacetus pacificus	Indo-Pacific beaked whale, Longman's beaked whale
Hyperoodon ampullatus	Northern bottlenose whale
Hyperoodon planifrons	Southern bottlenose whale
Mesoplodon hectori	Hector's beaked whale
Mesoplodon mirus	True's beaked whale
Mesoplodon europaeus	Gervais' beaked whale
Mesoplodon bidens	Sowerby's beaked whale
Mesoplodon grayi	Gray's beaked whale
Mesoplodon peruvianus	Pygmy beaked whale, lesser beaked whale
Mesoplodon bowdoini	Andrews' beaked whale
Mesoplodon traversii (formerly *bahamondi*)	Spade-toothed whale (formerly bahamonde's beaked whale)
Mesoplodon carlhubbsi	Hubbs' beaked whale
Mesoplodon ginkgodens	Ginkgo-toothed whale
Mesoplodon stejnegeri	Stejneger's beaked whale
Mesoplodon layardii	Strap-toothed whale
Mesoplodon densirostris	Blainville's beaked whale, dense-beaked whale
Family Platanistidae. South Asian river dolphins	
Platanista gangetica	Ganges dolphin, Indus dolphin, susu, bhulan
Family Iniidae. Amazon river dolphin	
Inia geoffrensis	Boto, Amazon river dolphin
Family Lipotidae. Chinese river dolphin	
Lipotes vexillifer	Baiji, Yangtze dolphin, Chinese river dolphin
Family Pontoporiidae. La Plata dolphin	
Pontoporia blainvillei	Franciscana
Family Monodontidae. Beluga and narwhal	
Delphinapterus leucas	Beluga, white whale
Monodon monoceros	Narwhal
Family Delphinidae. Dolphins	
Cephalorhynchus commersonii	Commerson's dolphin
Cephalorhynchus eutropia	Chilean dolphin
Cephalorhynchus heavisidii	Heaviside's dolphin

(*continued*)

Table 1.1 (*cont'd*)

Latin name	English name(s)
Cephalorhynchus hectori	Hector's dolphin
Steno bredanensis	Rough-toothed dolphin
Sousa teuszi	Atlantic humpbacked dolphin
Sousa plumbea	Indian humpbacked dolphin
Sousa chinensis	Pacific humpbacked dolphin
Sotalia fluviatilis	Tucuxi
Tursiops truncatus	Common bottlenose dolphin
Tursiops aduncus	Indian Ocean bottlenose dolphin
Stenella attenuata	Pan-tropical spotted dolphin
Stenella frontalis	Atlantic spotted dolphin
Stenella longirostris	Spinner dolphin
Stenella clymene	Clymene dolphin
Stenella coeruleoalba	Striped dolphin
Delphinus delphis	Short-beaked common dolphin
Delphinus capensis	Long-beaked common dolphin
Lagenodelphis hosei	Fraser's dolphin
Lagenorhynchus albirostris	White-beaked dolphin
Lagenorhynchus acutus	Atlantic white-sided dolphin
Lagenorhynchus obliquidens	Pacific white-sided dolphin
Lagenorhynchus obscurus	Dusky dolphin
Lagenorhynchus australis	Peale's dolphin
Lagenorhynchus cruciger	Hourglass dolphin
Lissodelphis borealis	Northern right whale dolphin
Lissodelphis peronii	Southern right whale dolphin
Grampus griseus	Risso's dolphin
Peponocephala electra	Melon-headed whale
Feresa attenuata	Pygmy killer whale
Pseudorca crassidens	False killer whale
Orcinus orca	Killer whale
Globicephala melas	Long-finned pilot whale
Globicephala macrorhynchus	Short-finned pilot whale
Orcaella brevirostris	Irrawaddy dolphin
Family Phocoenidae. Porpoises	
Neophocaena phocaenoides	Finless porpoise
Phocoena phocoena	Harbour porpoise, common porpoise
Phocoena sinus	Vaquita
Phocoena spinipinnis	Burmeister's porpoise
Phocoena dioptrica	Spectacled porpoise
Phocoenoides dalli	Dall's porpoise
Order SIRENIA	
Family Trichechidae. Manatees	
Trichechus manatus	West Indian manatee, Caribbean manatee
Trichechus senegalensis	West African manatee
Trichechus inunguis	Amazonian manatee
Family Dugongidae. Dugong and sea cow	
Dugong dugon	Dugong
Hydrodamalis gigas	Steller's sea cow

ticetes are filter feeders, consuming vast numbers of small organisms. Their teeth have been replaced by triangular plates of baleen, closely packed and rooted in the rostrum (upper jaw), one row on each side of the mouth. Baleen is formed principally of keratin and is continuously growing, its length maintained by wear at the lower (distal) end. Baleen plates are slightly curved to provide lateral strength, and are semirigid. The inner edge of each plate has a fringe of fibres, and these combine to form a dense mat lining the inside of the mouth. Food-laden water passes in through the open mouth and out through the baleen, leaving the food organisms trapped on the fibres. The diameter of the fibres varies from species to species and determines the size of prey organism which can be efficiently filtered from the water. The finest fibres are found in right whales (Balaenidae), which consume amphipods, euphausiids and copepods as small as 5 mm long.

Among the mysticete families, Balaenopteridae has the most species. Balaenopterids (commonly known as rorquals) are recognizable by having pleats, or 'ventral grooves', extending from the chin to the chest or belly. The pleated tissue is elastic and effectively forms a bag during feeding, allowing the whale to take in huge volumes of water before forcing it out through the baleen (Fig. 1.5). The huge tongue is then used to wipe prey off the mat of baleen fibres before swallowing. Whales that use this feeding technique are known as gulp feeders (Gaskin 1982). The seven species in the genus *Balaenoptera* (blue, fin, sei, Bryde's (two species), northern minke and Antarctic minke whales) share a sleek body shape, permitting rapid movement through the water (Fig. 1.1). The humpback whale (*Megaptera novae-angliae*), the only extant member of its genus, has a stockier build and much longer flippers than the other balaenopterids.

The family Balaenidae (right whales) is a group of rotund, slow-swimming whales with huge mouths and no dorsal fins. The huge girth is due to an unusually thick blubber layer. Balaenids have the longest baleen of all mysticetes, up to 4 m, and their mouths are steeply arched to accommodate this feeding apparatus. Unlike the balaenopterids, right whales have no ventral grooves and feed by passing a continuous stream of water through the open mouth and out through the baleen during active swimming. Two basic types of right whale exist. The black right whales (genus *Eubalaena*, three species) have raised, roughened patches of skin on the head, known as callosities. The similar bowhead or Greenland right whale (*Balaena mysticetus*) has a white chin and no callosities.

Fig. 1.5 Blue whale with distended throat.

The family Neobalaenidae has one living species, the pygmy right whale. It is much smaller and sleeker than the balaenids and has a falcate dorsal fin. But the head shape, especially the arched jaw, is recognizably similar.

The final mysticete is the gray whale (*Eschrichtius robustus*), in some respects intermediate between the rorquals and right whales. Uniquely among baleen whales, the gray whale is primarily a bottom feeder, filtering small organisms from the upper layer of the sea bed. The tongue is used as a piston to suck food items into its mouth (Ray & Schevill 1974). Gray whales have short baleen, with coarse fibres.

1.2.2.2 Odontocetes

The suborder Odontoceti, the toothed whales, embraces at least 70 species of whales, dolphins and porpoises belonging to around 10 families spanning a great range of shapes and sizes. The number, size and shape of teeth are extremely variable both across and within families. In females of some species the teeth do not even erupt past the gumline. Odontocetes are principally consumers of fish and squid, but some species (e.g. belugas, *Delphinapterus leucas*; Vladykov 1946) also take benthic invertebrates and the killer whale includes warm-blooded prey in its diet (Dahlheim & Heyning 1999). All odontocetes have a single blowhole set to the left of the mid-line, reflecting an asymmetric skull, and probably all use echolocation. Most toothed whales are sexually dimorphic in body size, with males larger than females in all except some beaked whales, the *Cephalorhynchus* dolphins and the porpoises.

The largest odontocete is the sperm whale (*Physeter macrocephalus*) (Fig. 1.2). Sexual dimorphism is extreme in this species. Males reach an average adult size of 15 m and 45 t with a maximum of 18.5 m and 57 t, while females average some 11 m and 20 t (Rice 1989). The sperm whale is also exceptional in other ways. It dives deeper than any other whale (with the possible exception of some of the larger beaked whales), and its head is proportionally bigger than that of any other species. The head is a complex structure, essentially consisting of a reservoir of oils and waxes supported by a cradle of dense bone. The mandible is slim and underslung. The functions of the various elements of the sperm whale's head, especially the spermaceti organ, have long been matters of scientific conjecture and controversy. Principal hypotheses are that the spermaceti organ acts as an acoustic lens for focusing emitted sound, perhaps to immobilize prey (Norris & Møhl 1983), and that it is used in diving to alter the animal's buoyancy (Clarke 1978).

The two species most closely related to the sperm whale are, in contrast, among the smallest animals known as whales. The pygmy sperm whale (*Kogia breviceps*) and dwarf sperm whale (*K. sima*) form the family Kogiidae. Adult *K. breviceps* reach no more than 3.3 m and about 400 kg, while *K. sima* attain only about 2.7 m and 210 kg. Both have a narrow, underslung lower jaw, but otherwise their external appearance differs from the sperm whale's. They have an upright dorsal fin and proportionally both a smaller head and longer flippers.

The family Monodontidae comprises two living species: the narwhal (*Monodon monoceros*) and the white whale or beluga. These two small whales (maximum body length about 5 m in both species) share the unusual characteristic of having no dorsal fin. Furthermore, and uniquely among cetaceans, the male narwhal grows a long spiralled tusk. This extraordinary secondary sexual characteristic is actually a tooth, rooted in the upper left jaw, which emerges horizontally through the lip and grows to a maximum length of more than 3 m. The tusk angles slightly to the left of the whale's centre line, so forward motion produces an asymmetric lateral force. To counteract this force, and to help prevent the animal swimming in circles, narwhals have fused neck vertebrae. The tusk is normally the only tooth that erupts in male narwhals, but very rarely a tooth emerges on the right side too, resulting in the 'double-tuskers' often seen in museums. Females usually have no erupted teeth. In contrast, belugas of both sexes have a full complement of peg-like teeth.

The family of beaked whales, Ziphiidae, includes at least 20 species. These small to medium-sized whales share the characteristics of a distinctively narrow rostrum with the lower jaw extending at least to the tip of the upper, a shallow or non-existent notch between the tail flukes, a dorsal fin set well back on the body and two conspicuous throat grooves which converge anteriorly to form a 'V'. They also have three or four fused cervical vertebrae and extensive skull asymmetry. All but Shepherd's and Gray's beaked whales (*Tasmacetus shepherdi* and *Mesoplodon grayi*) have only two (genus *Mesoplodon* and usually *Hyperoodon* and *Ziphius*) or four (genus *Berardius*) non-vestigial teeth, located in the mandible. For most species, these erupt only in adult males. The number, shape and position of teeth are diagnostic for most species and may be the only characteristics by which they can be identified, short of molecular analyses.

Although beaked whales occur throughout the world's oceans, they are the least-known group of cetaceans. Most are dark (brown or grey) on the dorsal surface, becoming lighter ventrally. Adults, especially males, are often covered in pale, linear scars, sometimes in parallel pairs, suggesting that the teeth are used in aggressive encounters.

The Ziphiidae are currently split into six genera, one of which (*Mesoplodon*) contains at least 13 small (6 m or shorter) species. The smallest mesoplodont (*M. peruvianus*), at < 4 m, was described only recently (Reyes *et al.* 1991). Possibly, the most distinctively patterned beaked whale is the strap-toothed whale (*M. layardii*), which has an extensive light grey blaze on an otherwise black body. This species is also unique in that males possess a flattened tooth on each side of the lower jaw which curves over the rostrum with age, eventually preventing the animal from opening its mouth more than a few centimetres. As this extraordinary tooth development does not prevent the whale from feeding, prey are probably ingested by suction.

The genus *Berardius* is represented by two whales (*B. bairdii* and *B. arnuxii*) with a maximum body size of around 13 m and 10 m, respectively. Shepherd's beaked whale (*Tasmacetus shepherdi*) is unique among the group in having many teeth in both jaws (90 or more in some specimens) in addition to a pair of more typical 'tusk' teeth at the tip of the mandible in males (Mead 1989). Measured specimens have been up to 7 m in body length, but this is one of the rarest whales and few have yet been carefully examined.

By far the largest and most diverse odontocete family is Delphinidae, embracing some 35 species of dolphins in 17 genera (Rice 1998; LeDuc *et al.*

1999). Because some of these species are extremely abundant, most of the individual cetaceans alive in the world's oceans belong to this family. Body size in delphinids varies from the tiny *Cephalorhynchus* dolphins with a length of 1.7 m or less and a mass of 30–60 kg to the killer whale, males of which can reach a maximum of 9 m and 10 t. They also vary greatly in coloration, and the family includes the most strikingly marked cetaceans. Among these are the striped and spotted dolphins (*Stenella coeruleo-alba*, *S. attenuata* and *S. frontalis*), the common dolphins (genus *Delphinus*), Fraser's dolphin (*Lagenodelphis hosei*) and the *Lagenorhynchus* and *Cephalorhynchus* dolphins. They have a diversity of stripes, spots, swirls, blazes and patches in shades of black, white, grey and even tan or yellow (Mitchell 1970; Perrin 1975). At the other extreme are species with simple countershading of plain dark above and lighter below. Examples are the bottlenose dolphins (*Tursiops* spp.) (Fig. 1.2), tucuxi (*Sotalia fluviatilis*) and some races of the spinner dolphin (*Stenella longirostris*).

Six delphinids with superficially similar characteristics (body form and colour, reduced dentition in all but one, no beak and three or more fused neck vertebrae) have been collectively known as 'blackfish'. The group varies in size from a little over 2 m for the pygmy killer whale (*Feresa attenuata*) to the killer whale, in which the adult male has a greatly exaggerated dorsal fin and is twice the mass of the female. The black body coloration is broken in several species with white patches on the ventral surface, and the killer whale is immediately recognizable with its postocular blaze and entirely white chin and throat. The three 'killer' whales (*Orcinus*, *Pseudorca* and *Feresa*) and the two pilot whales—long-finned (*Globicephala melas*) and short-finned (*G. macrorhynchus*)—have thick, widely spaced teeth in both jaws, effective for grasping prey. The melon-headed whale (*Peponocephala electra*) differs from the rest of the group, having 20–25 pairs of small teeth in each jaw compared to 8–13 pairs for the other blackfish.

Most of the other delphinids share a 'classic' dolphin shape, with a distinct beak and prominent dorsal fin, two or more fused cervical vertebrae and 20 or more pairs of teeth in the upper jaw. None is more than 4 m long. The 'classic' dolphins embrace eight genera, of which four (*Steno, Sotalia, Tursiops* and *Lagenodelphis*) are represented by only one or two living species. *Lagenorhynchus* and *Stenella*, with six and at least five species, respectively, are the most diverse of these genera.

The four diminutive *Cephalorhynchus* species lack a well-defined beak and have characteristically rounded dorsal fins, but are otherwise perfectly dolphin-like. The highly gregarious northern (*Lissodelphis borealis*) and southern (*L. peronii*) right whale dolphins are extremely slim, and neither has any vestige of a dorsal fin. Both species are black above and white below with clear lines of demarcation between the two colours.

The Irrawaddy dolphin (*Orcaella brevirostris*) has external similarities to the beluga and has been placed by some systematists in the Monodontidae, but recent morphological and genetic evidence indicates that it is a delphinid and may be most closely related to the killer whale (see Chapter 2). *Orcaella* has a rounded head with no beak, a flexible neck allowing unusual mobility of the head and a low number of peg-like teeth in both upper and lower jaws. In common with only the beluga, this species can 'pucker' the lips and shoot a directed stream of water from the mouth; this ability may be useful for foraging in mud or bottom sediments.

The family Phocoenidae, the porpoises, comprises six species of very small cetaceans (all < 2.2 m) with small flippers and no beak. Five porpoises have small dorsal fins, but the finless porpoise (*Neophocaena phocaenoides*), as its name suggests, has no dorsal fin at all. Porpoises differ from other odontocetes in having laterally compressed or spatulate teeth which collectively form a cutting edge. There are three porpoise genera: *Phocoena, Phocoenoides* and *Neophocaena*. The harbour porpoise (*Phocoena phocoena*), vaquita (*P. sinus*), Burmeister's porpoise (*P. spinipinnis*) and finless porpoise are uniformly grey or black on the dorsal surface, fading to a lighter shade ventrally. The first three also have a dark eye patch and flipper stripe (axilla to lower lip), and some finless porpoises have white lips.

The remaining two porpoises—spectacled (*Phocoena dioptrica*) and Dall's (*Phocoenoides dalli*)—are strikingly marked with black dorsally and white beneath, though in *P. dalli* the head and neck are black too. Dall's porpoise differs from all other

porpoises in having a robust shape with a very 'deep' body.

The final odontocetes to consider are some of the most ancient—the river dolphins. The term 'river dolphin' applied in a taxonomic sense is, though, somewhat of a misnomer for two reasons. Firstly, one member of this group, the franciscana (*Pontoporia blainvillei*), does not live in rivers; secondly, some cetaceans which have populations living in rivers (e.g. *Sotalia, Orcaella* and *Neophocaena*) are excluded because they are taxonomically distant from these long-beaked species.

The river dolphins grow to an adult body size of no more than 2.5 m and 200 kg. They differ morphologically, behaviourally and physiologically from all other cetaceans. Characteristics common to the four species currently recognized include a very long, narrow forceps-like beak, large spatulate flippers, small eyes and flexible necks. The *Platanista* dolphins of the South Asian subcontinent are effectively blind due to the evolutionary loss of a crystalline lens. They have some 120 long, interlocking teeth for grasping fish and unique cranial processes thought to be related to echolocation.

The boto or Amazon river dolphin (*Inia geoffrensis*), also known as the pink dolphin because of its body colour in some parts of the range, is the only extant odontocete with differentiated dentition. Anterior teeth are simple, peg-like structures, but the posterior teeth have cusps on the inner face. These teeth indicate an unusual diet, and indeed the boto is capable of crushing both armoured fish and turtles (da Silva 1983). Its extremely flexible body, with broad, rotatable flippers and low dorsal fin, allows this species to swim and manoeuvre within an unusual habitat—the tangled vegetation and root systems of the Amazonian flooded forest.

1.2.3 Sirenians

The order Sirenia embraces two distinctly different groups of animals—the manatees (three extant species) and the dugong (one extant species) (Fig. 1.3). The manatees' structures are more generalized and adapted for slow movement in quiet and confined spaces, while the dugong's tail musculature and cetacean-like flukes provide for rapid acceleration and active swimming in exposed waters (Domning

1977; Anderson 1979). Manatees have two crescentic nostrils on top of the snout, a bristly muzzle with drooping jowls and a 'normal' mouth. In contrast, the dugong's face is dominated by a bizarre, downward-orientated rostral disk, which is really a modified and greatly expanded upper lip (Fig. 1.3). The male dugong has a pair of tusks that protrude a few centimetres just behind the corners of the facial disk. These incisors are worn into a chisel shape but are not known to be used as weapons (Nishiwaki & Marsh 1989).

Differences in facial structure reflect the differing diets of manatees and dugongs (Wells *et al.* 1999). Manatees are capable of feeding throughout the water column and reaching above the surface to crop emergent water plants and overhanging terrestrial ones, while dugongs are obligate bottom feeders. The forelimbs of manatees, in particular, are manoeuvrable enough to be used for stuffing food into the mouth, as well as for 'walking' along the bottom, sculling, turning and braking. Dugongs sometimes use their pectoral flippers as props while maintaining position on the sea floor.

1.2.4 Pinnipeds

As a group, the pinnipeds are less morphologically diverse than the cetaceans (Fig. 1.4). Body size ranges from the small lake seals in Eurasia (Baikal seal (*Pusa sibirica*) males reach no more than 1.5 m and 70 kg) to the elephant seals in which males reach 4 m and 2.5 t—thus a 2.5-fold difference in length and a 35-fold difference in mass, compared with 19- and 3000-fold differences, respectively, in cetaceans.

There are only three extant pinniped families, one of which, Odobenidae (walrus), is monotypic. The walrus (*Odobenus rosmarus*) is the only pinniped with long external tusks (Fig. 1.4). Both sexes have these highly modified upper canines, which are used for combat and threat displays, for anchorage or leverage while resting beside or clambering onto an ice floe or, in exceptional cases, to kill phocid prey (Lowry & Fay 1984). The walrus's mouth and face are uniquely adapted for foraging on clam beds. A highly enervated muzzle, covered with stiff bristles, is used to sense the substrate and detect food. A 'vacuum pump', with the tongue as the piston, is used to suck the siphons and feet of clams and other molluscs from their shells (Fay 1982).

Walruses have thick skin (up to 4 cm in adult males), an ample blubber layer (5–10 cm) and no external ears. Males can be over 3 m long and weigh 1200 kg; females are substantially smaller. The walrus's flipper structure permits the animal to move quite swiftly on land or ice when necessary, in a similar way to otariids (see below).

The Otariidae, or eared seals, are distinguished from other pinnipeds by their external ear flaps (pinnae); long, hairless or sparsely haired foreflippers with splayed digits and vestigial nails; and relatively large hind-flippers that can be rotated beneath the body. Their flipper structure allows otariids to 'walk', albeit awkwardly. Sexual dimorphism is pronounced in these polygynous seals, with males considerably larger than females in all species. Adult males develop a thick, robust neck region, and in many species they have a mane of longer hair reminiscent of a lion's. Their large size and powerful neck are essential for establishing and maintaining dominance on the breeding rookeries.

1.2.4.1 Otariids

The 16 otariid species are allocated to seven genera (Rice 1998). The only consistent feature that distinguishes fur seals from sea lions is the presence of a dense underfur, consisting of 30 or more secondary hairs associated with each primary hair, in the pelage of fur seals (Repenning *et al.* 1971; Warneke & Shaughnessy 1985). In addition, the sea lions generally have a broader, blunter nose than the fur seals.

The genus *Arctocephalus* embraces eight species, the southern fur seals. Morphological differences between the species are fairly subtle, especially in females and juveniles. The snout of *A. townsendi* males is markedly long and pointed, while those of *A. gazella* and *A. tropicalis* are almost blunt by comparison. Coloration is typically some shade of brown or grey, with countershading. The northern fur seal (*Callorhinus ursinus*) has proportionally longer flippers than its southern counterparts. In all fur seals, adult males have roughly five times the mass of adult females.

The sea lions are assigned to five genera, all but *Zalophus* being monospecific. Adult males are about

0.5 m longer (2.4 m vs 1.8 m) and three times more massive (300 kg vs 100 kg) than females. A striking feature of the male is its noticeably raised forehead, formed by the prominent saggital crest on the skull. Both sexes are chocolate brown, with little or no countershading.

The largest otariid is the northern (Steller's) sea lion (*Eumetopias jubatus*). Males grow to 3 m and can weigh 1 t, while females reach only about 2.2 m and 270 kg. The head and bulging neck of the bull are massive.

The three southern sea lions (South American *Otaria flavescens*; sometimes called *O. byronia*; plus the Australian, *Neophoca cinerea*, and New Zealand or Hooker's, *Phocarctos hookeri*, sea lions) are all roughly the same size as the California sea lion. Male *Otaria* have the most characteristic profile of any sea lion. Their heads are enormous, dwarfing the rest of the body, and the snout is broad and slightly upturned. The mane has very long hair and is somewhat lighter in colour than the dark brown body. Male Australian sea lions have a more complex colour pattern than the other sea lions (King 1983). The top of the head and the nape are white, in contrast to the chocolate brown body overall.

1.2.4.2 Phocids

The true or earless seals, the Phocidae, comprise the largest and most diverse of the pinniped families. Rice (1998) recognizes 13 modern phocid genera, all but four of them monospecific. Phocids are readily distinguished from otariids by their lack of pinnae (external ears); their shorter, haired fore-flippers with claws on all five digits; and their non-rotatable hind-flippers. They spend far less time in an 'upright' posture than the otariids when hauled out, and generally can only hunch or wriggle their way across the land or ice, in contrast to the ungainly walking of otariids and walruses. In most phocid species, males grow larger than females, but in some (e.g. the bearded seal (*Erignathus barbatus*), the Antarctic phocids and the monk seals) the reverse is true. Regardless of which sex grows larger, sexual dimorphism in size is much less pronounced in phocids than in otariids. Male phocids tend to be more vividly coloured than females, particularly in those species with relatively complex patterning

(e.g. ribbon and harp seals, *Histriophoca fasciata* and *Pagophilus groenlandicus*, respectively). Whereas otariids have scrotal testes and four mammary teats, phocids (and the walrus) have internal testes and either two (most species) or four (monk seals and the bearded seal) teats.

Within the Phocidae, there is considerable diversity in colour pattern and body size. Harbour, spotted, Caspian and Weddell seals (*Phoca vitulina*, *P. largha*, *Pusa caspica* and *Leptonychotes weddellii*, respectively) have spotted coats as adults. Ringed seals (*Pusa hispida*), as their name implies, have light rings on a grey background (Fig. 1.4). Adult ribbon seals have three dramatic, white ribbon-like bands (Fig. 1.4), and adult harp seals have a black face and a dark horseshoe-shaped saddle across the back and flanks. Baikal and monk seals are unspotted and fairly drab in comparison. Mediterranean monk seals (*Monachus monachus*) often have an extensive white ventral patch (Fig. 1.4), and bearded seals have a variable colour pattern, sometimes with muted streaking and blotches. The bearded seal is distinguished by its impressively long, white vibrissae (thus the name 'bearded') and squarish flippers. The smaller phocids are less than 2 m long and weigh no more than 150 kg, whereas the larger ones (the bearded, Weddell and monk seals) range to about 3 m and 400 kg or more.

Two northern phocids diverge from the typical seal body form. The Latin species name of the grey seal (*Halichoerus grypus*) means 'hook-nosed' and the larger males, in particular, have a strongly arched and elongate snout (a Roman nose) (Fig. 1.4). Coloration of grey seals is highly variable, from almost black in adult males to mainly cream in adult females, with dark or light irregular spotting. The hooded seal (*Cystophora cristata*) is so named because of the adult male's remarkable inflatable nasal sac. When inflated, the hood becomes a taut, bulging, bi-lobed protuberance that dominates the entire area from behind the eyes to well in front of the mouth. The seals may inflate it when disturbed, during the pupping and mating season or even while lying quietly on the ice. In addition to the hood or crest, the male often causes its nasal septum to extrude, forming a large red balloon-like structure. The pelts of adult hooded seals have dark spots and blotches on a silvery ground colour.

These seals are 2.5–2.7 m long and weigh up to about 400 kg.

Each of the three Antarctic phocids in addition to the Weddell seal has a distinctive morphology. The Ross seal (*Ommatophoca rossii*) is a large species (to 3 m and 210 kg) with the shortest hair of any phocid, and a streaky grey coloration pattern. It has huge eyes, probably reflecting the need to collect as much light as possible in this seal's normally dark world underneath the ice. The crabeater seal (*Lobodon carcinophaga*) is long and slim, with unique dentition. The postcanines have cusps with separate, well-defined lobes, and the upper and lower ones interlock to form a sieve. This sieve allows the seal to filter its main food, krill, from sea water in a manner reminiscent of the baleen whales. Crabeaters suck food into the mouth, then press the tongue against the palate and raise the lips to let the sea water escape (Klages & Cockcroft 1990). Crabeater seals are brownish to grey on the back, shading to silver on the belly.

The leopard seal (*Hydrurga leptonyx*) is almost reptilian in appearance, mainly owing to the large head, with a marked constriction in the neck, and huge gape (Fig. 1.4). The canine teeth are exceptionally long, the postcanines massive. In form, the latter resemble those of the crabeater seal except they have three rather than four or five cusps. Leopard seals are predators on penguins and seals although they also consume krill (probably straining them much like crabeater seals), squid and fish. Female leopard seals can grow to 3.6 m and 450 kg; males are smaller. The coloration is bipartite and spotted, with dark grey on the back and light silvery on the belly.

Elephant seals (Fig. 1.4) are the largest pinnipeds, and males of the southern species (*Mirounga leonina*) are substantially larger than their northern congeners (*M. angustirostris*), while the smaller females of the two species are roughly the same size (to 2–3 m and 600–900 kg). Although not closely related, elephant seals and walruses resemble one another in the texture and appearance of their brownish, sparsely haired skin. Elephant seals share with monk seals the unusual feature of having a 'catastrophic' moult, whereby the skin sloughs away in large patches rather than having hairs fall out individually. The elephant seal's most distinguishing feature, and the reason for its common name, is its inflatable proboscis, which is highly developed in adult males and is used for display during the breeding season.

1.3 ZOOGEOGRAPHY

Our ability to understand and interpret today's marine mammal distributions has increased substantially in recent decades due to advances in other fields of science, notably plate tectonics and global climate change. Changes in the shape, size and even existence of seas and oceans over geological time have profoundly influenced the occurrence and dispersal of marine animals.

Brief examination of the distribution of marine mammals living today demonstrates the critical importance of water temperature in defining their geographical limits. The influence of temperature may be directly on the animal itself, or may be indirect if the mammal is seeking temperature-sensitive prey resources. As we shall see below, the position of the land and its surrounding shelf (if any), and water temperature, are the two over-riding influences on marine mammal distribution. Because water temperature is correlated with latitude, species often occur within latitudinal bands—sometimes in one hemisphere, less often in both. Figure 1.6 illustrates some typical distributional patterns.

Despite progress in understanding the environmental history of our planet, we remain far from having a full comprehension of marine mammal zoogeography. Not only is much of the evidence of past distribution fragmentary, but we have little idea about the ranges of some species alive today. The ziphiid whales are a case in point. For only a handful of the 20 or so extant species in this family are we reasonably confident about their precise range, and for some we have only the locations of a few strandings, separated by huge expanses of ocean.

1.3.1 Cetaceans

1.3.1.1 Mysticetes

With few exceptions the baleen whales are migratory, occupying different parts of their range on a seasonal basis. Indeed, some mysticetes migrate further than any other mammals. The reasons for their migrations may be related to energy and

(a)

(b)

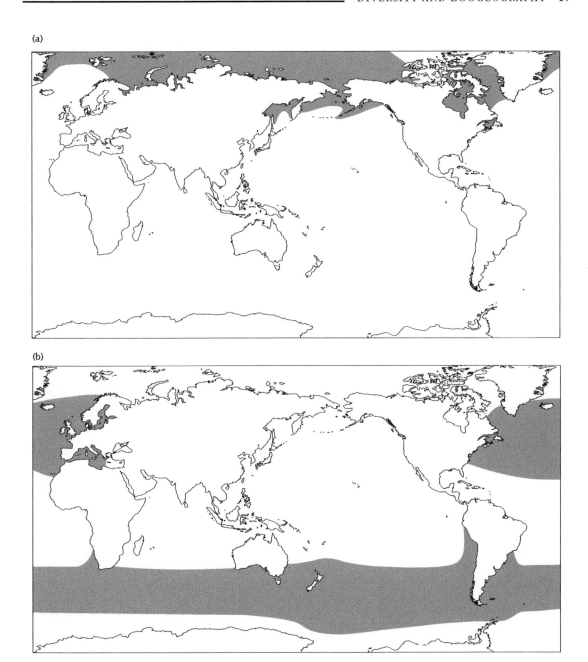

Fig. 1.6 Representative climate–zone distributions of cetaceans: (a) northern circumpolar—beluga; (b) antitropical—long-finned pilot whale. (*continued on p. 18*)

nutrition, or to the need for particular environmental conditions for reproduction. A population's winter and summer ranges can be separated by many thousands of kilometres (see Chapter 7).

The two types of balaenid whale have different distributions and migrational patterns. The bowhead is one of only three cetaceans with an exclusively Arctic range (the other two are monodontids); its migrations allow it to exploit rich high-latitude food resources during the short ice-free season. The morphologically similar right whales occupy temperate latitudes in both hemispheres. Their

(c)

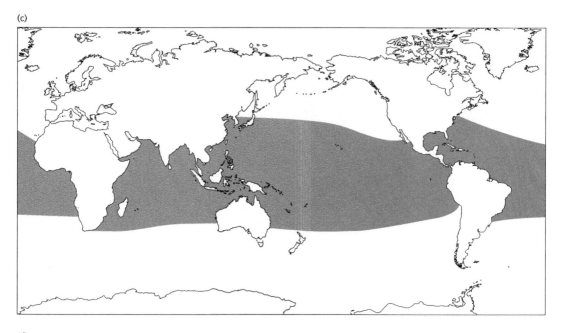

(d)

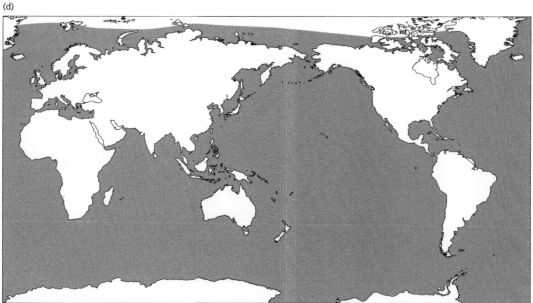

Fig. 1.6 (*cont'd*) (c) Pan-tropical—spinner dolphin; and (d) cosmopolitan—killer whale. This and all subsequent maps in Chapter 1 are intended to demonstrate patterns of distribution rather than to be precisely accurate in every detail. (Adapted from Jefferson *et al.* 1993.)

antitropical distribution ensures that northern and southern hemisphere populations remain separate, though they clearly have common ancestry (Schaeff *et al.* 1991; Rosenbaum *et al.* 2000). The pygmy right

whale is also a temperate species, but is confined to a narrow latitudinal band (roughly 30–52°S).

The gray whale is restricted to shelf waters by its bottom-feeding habits and coastal migration

(Swartz 1986). Gray whales are highly migrational, seeking sheltered warm-water lagoons in which to calve and nurse young during the winter after a summer spent feeding in high latitudes (Bering, Chukchi and Okhotsk Seas). They currently inhabit only the North Pacific but were also present in the North Atlantic as recently as the 1600s and possibly even the early 1800s (Mead & Mitchell 1984).

Some of the balaenopterid whales (blue, fin, sei and humpback at least) are cosmopolitan. Until recently it was assumed that they, like the gray whale, adhered to strict annual latitudinal migrations, but recent evidence indicates that some segments of populations, and indeed some populations, may not do so (e.g. fin whales in the Mediterranean Sea and possibly the Gulf of California; Bérubé *et al.* 1998). Although distribution maps may show a continuity across the equator, and some populations of humpbacks do in fact transgress the equator in their annual migrations (Stone *et al.* 1990), most northern and southern populations are 6 months out of phase so they rarely, if ever, come into contact (Mackintosh 1965). The northern Indian Ocean population of humpbacks is unusual in that it is denied access to high latitudes by the Asian land mass, and these whales are thought to be non-migratory, finding adequate food resources and breeding habitat in the same region (Reeves *et al.* 1991; Mikhalev 1997).

The Bryde's whales (*Balaenoptera edeni, B. brydei*) are the only balaenopterids with a narrow latitudinal range. Their combined distribution is pantropical with a limit at some 40° of latitude. It is probable that most Bryde's whales do not migrate large distances, but some might. Seasonal onshore–offshore or north–south movements are known to occur, perhaps due to food availability. Bryde's whales tagged near the equator in the western North Pacific were later killed on the pelagic whaling grounds centred at 25–30°N (Ohsumi 1980).

1.3.1.2 Odontocetes

All three sperm whales (*Physeter, Kogia sima* and *K. breviceps*) are creatures of deep water, normally coming close to coasts only where there is little or no continental shelf (e.g. volcanic islands like the Azores and some of the Lesser Antilles) or when dead or dying. Although *Physeter* is cosmopolitan

with a wide latitudinal range like that of many rorquals, the two smaller species occur primarily in tropical and warm temperate latitudes. *Physeter* is unique among cetaceans in having substantially different ranges for different segments of the population (Best 1979). Females and their offspring occur in matrilineal groups which are restricted to waters of 15°C or warmer and therefore lower latitudes (mostly < 45° of latitude). Males remain in these pods in their early years but eventually leave to join other subadult males in groups which venture polewards in both hemispheres. Adult males can exploit food resources at up to 78°N or 70°S.

The narwhal and many populations of the beluga are pagophilic and migratory, their movements dominated by the seasonal encroachment and recession of sea ice in the coldest waters of the northern hemisphere. Narwhals have a discontinuous circumpolar distribution, but the 30 or so beluga populations (International Whaling Commission 2000) form an almost continuous ring around the Arctic Ocean (Fig. 1.6). Belugas have been satellite-tracked in heavy pack ice to 80°N (Richard *et al.* 2001; Suydam *et al.* 2001), and are clearly as comfortable in these extreme conditions as are narwhals. Nevertheless, a few beluga populations live in semi-enclosed waters which are free of ice for much of the year, the southernmost being endemic to the lower St Lawrence River at only 47°N.

The beaked and bottlenose whales (ziphiids) are deep-water animals. Because they tend to be sparsely distributed, inconspicuous at the surface and difficult to identify at sea, there are few (in some cases no) records of live animals for a number of species. For this reason knowledge of distribution is often limited to the locations of beach-cast animals which may have drifted great distances from their normal range. Caution must therefore be used when trying to interpret ziphiid distributional patterns, which may well reflect those of their equally mysterious prey: deep-sea cephalopods.

Many ziphiids have antitropical distributions (Fig. 1.7), and in a few instances pairs of closely related species have almost mirror-image distributions each side of the equator, probably indicating common ancestry. In these cases, the cross-over may have occurred during a period of global cooling when seas near the equator were colder and the

(a)

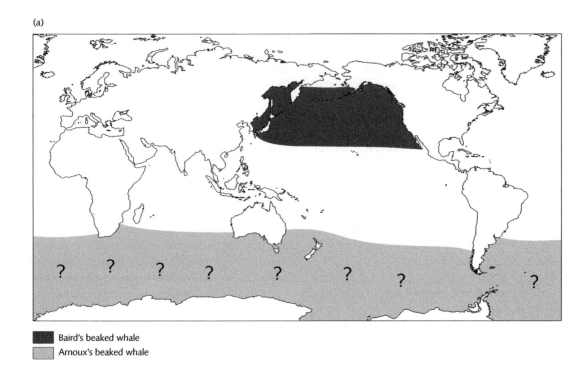

■ Baird's beaked whale
▨ Arnoux's beaked whale

(b)

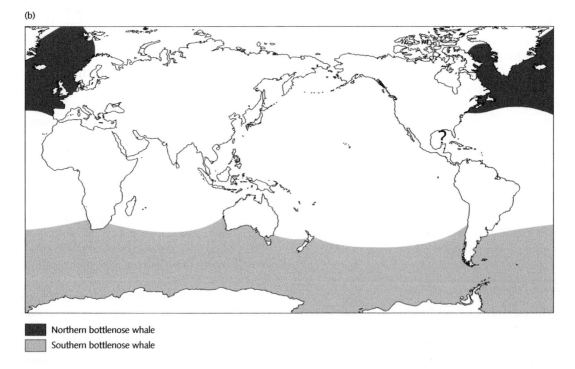

■ Northern bottlenose whale
▨ Southern bottlenose whale

Fig. 1.7 Representative ziphiid (beaked whale) distributions: (a) Baird's and Arnoux's beaked whales, the *Berardius* species pair; (b) northern and southern bottlenose whales, the *Hyperoodon* species pair. (*continued*)

(c)

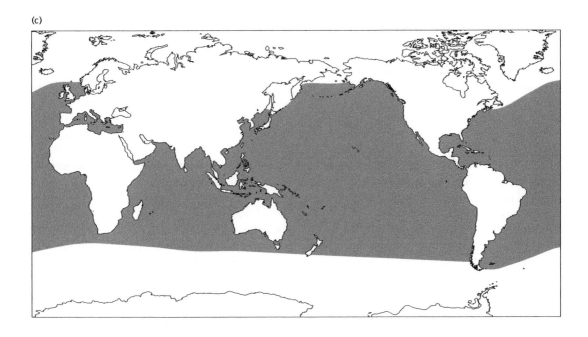

(d)

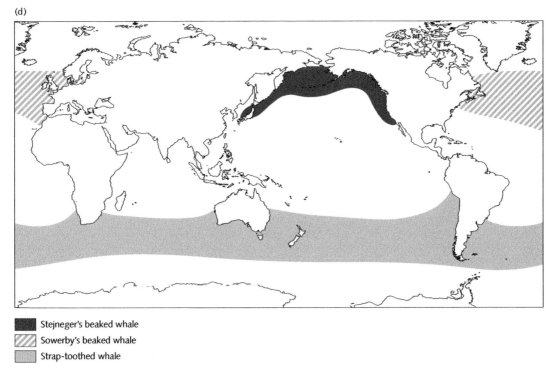

Stejneger's beaked whale
Sowerby's beaked whale
Strap-toothed whale

Fig. 1.7 (*cont'd*) (c) Cuvier's beaked whale; and (d) three mesoplodonts—strap-toothed, Sowerby's beaked and Stejneger's beaked whales. (Adapted from Jefferson *et al.* 1993.)

parent species occurred in lower latitudes (Davies 1963). The genera *Hyperoodon* and *Berardius* present interesting variations on this theme. Both have a circumpolar species in the Southern Hemisphere (*H. planifrons* and *B. arnuxii*), and another at temperate latitudes in one northern hemisphere ocean basin—*H. ampullatus* in the North Atlantic and *B. bairdii* in the North Pacific. Other ziphiids have ranges which differ from this antitropical pattern (Fig. 1.7). Blainville's and Cuvier's beaked whales are both relatively well known and widely distributed from tropical to cold-temperate climes, the latter occurring in all but polar waters.

The large family Delphinidae includes animals with a wide variety of distributional patterns: some coastal, some oceanic; some tropical, others antitropical; some almost global, others endemic to relatively small areas; most marine but some fresh water. A preference for deep, warm waters is characteristic of most of the *Stenella* species, *Delphinus*, *Steno* and *Lagenodelphis*. *Delphinus* and *Stenella coeruleoalba* can be found as far north as 60°N in the North Atlantic (presumably due to the warm

Gulf Stream), but elsewhere they occur mainly in lower latitudes. The range of Risso's dolphin is similar to that of *D. delphis*, but extends into somewhat higher latitudes (and thus slightly lower surface temperatures) in all oceans. Antitropical offshore ranges are found in several of the *Lagenorhynchus* species (*L. cruciger* and *L. acutus*) and the two species of *Lissodelphis*. These latter are clearly a species pair, occurring in similar habitats in both hemispheres, but land barriers have prevented movement of *L. borealis* into the North Atlantic.

Some delphinids are entirely coastal. The humpbacked dolphins (genus *Sousa*), Irrawaddy dolphin (*Orcaella*) and marine-form tucuxi are restricted to tropical or subtropical coasts, including the estuaries and lower ends of some rivers. The four diminutive *Cephalorhynchus* dolphins are also coastal, with nonoverlapping ranges in temperate waters of the southern hemisphere: one each side of South America, one off southwestern Africa and one around New Zealand (Fig. 1.8). The deep waters separating the southern continents have apparently acted as barriers to the exchange of animals between the population

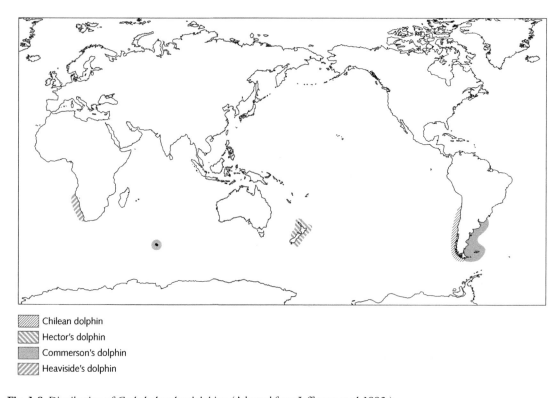

	Chilean dolphin
	Hector's dolphin
	Commerson's dolphin
	Heaviside's dolphin

Fig. 1.8 Distribution of *Cephalorhynchus* dolphins. (Adapted from Jefferson *et al.* 1993.)

centres. The substantial morphological differences between the *Cephalorhynchus* species can only have evolved in the absence of genetic interchange for a very long period of time. Intriguingly, though, Commerson's dolphins (*C. commersonii*) are also resident around the Kerguelen Islands in the southern Indian Ocean. The extreme isolation of this archipelago begs fascinating questions about the means and timing of colonization, and the degree of similarity between these Kerguelen dolphins and those in South America (Robineau 1983, 1986).

Of the six blackfish, four (false and pygmy killer whales, short-finned pilot whale and melon-headed whale) have warm-water (circumtropical and warm temperate) distributions. The long-finned pilot whale has an antitropical distribution (Southern Ocean and North Atlantic) and the killer whale has one of the most extensive ranges of all cetaceans, occurring from the Arctic Ocean to the Antarctic ice edge and most areas in between (Fig. 1.6).

The almost continuous low-latitude oceanic distribution of the short-finned pilot whale disguises some interesting complexities which may be more common in cetaceans as a whole than we realize. Two very different forms of this whale occur off eastern Japan, one to the north of the other (Kasuya *et al.* 1988; Wada 1988). These two populations remain geographically separate, apparently because of different habitat preferences (essentially defined by water temperature). The boundary between the populations moves latitudinally with seasonal changes in sea surface temperature.

Four of the six porpoises (*Phocoena phocoena, P. sinus, P. spinipinnis* and *Neophocaena phocaenoides*) are inshore species, venturing out of shelf waters very rarely. In fact, *N. phocaenoides* often occurs in estuaries and is commonly encountered more than 1000 km upstream from the ocean in the Chang-jiang (Yangtze River).

The distribution of *P. phocoena* is interesting on two counts. Firstly, it is the only small cetacean with an exclusively northern temperate/subarctic range which occurs in both the Atlantic and Pacific Oceans. Somehow this species has managed to pass around one or both of the great continents which isolate the two regions. Global climatic cycles have resulted in Arctic warming during quite recent millennia (witness the warming periods in the North American Arctic between 11 000 and

8500 years BP and again between about 5000 and 3000 years BP; Dyke & Morris 1990), so we can presume that porpoises traversed the Russian Arctic coast and/or the Northwest Passage during one or more warmer periods. The second point of interest is that the harbour porpoise is found in the Black Sea but is essentially absent from the Mediterranean (Read 1999). Gaskin (1982) speculated that isolation of the Black Sea population occurred during the Pleistocene, when one or more cooling events allowed porpoises to penetrate the Mediterranean. Subsequent warming may have rendered most of the Mediterranean inhospitable to this species.

It is striking that the four coastal porpoises have non-overlapping ranges (Fig. 1.9) which cumulatively cover a large proportion of the world's continental coastlines outside the polar regions. Three (harbour, Burmeister's and finless) have distributions covering thousands of kilometres, but the vaquita is restricted to a tiny area in the northern Gulf of California. The species is critically endangered and, although anthropogenic influences may finally force it to extinction (Rojas-Bracho & Taylor 1999), the vaquita had become vulnerable by retreating into this geographical cul-de-sac long before humans began adding to its problems. In contrast, the numerous Dall's porpoise, especially the form *Phocoenoides dalli dalli*, is comfortable both in nearshore and offshore waters.

The final species of the family, the spectacled porpoise, is the least known. This animal is very rarely seen at sea, and the small number of strandings provides an ambiguous indication of its habitat preference. There seems little doubt that the species is restricted to cold temperate and subantarctic waters; the paucity of sightings may indicate that it is either very rare or that it occurs offshore in areas with little observer coverage.

The extant river dolphins occupy five of the largest drainage systems in the tropics—those of the Indus, Ganges/Brahmaputra, Changjiang (Yangtze), Orinoco and Amazon; additionally the franciscana lives in the subtropical and warm temperate coastal waters and estuaries of eastern South America (Fig. 1.9c). Botos (*Inia geoffrensis*) living in the Amazon are subtly different to those in the Orinoco, but there is uncertainty about whether interchange of animals can occur between these two great watersheds (Best & da Silva 1989). The population living in Bolivia is

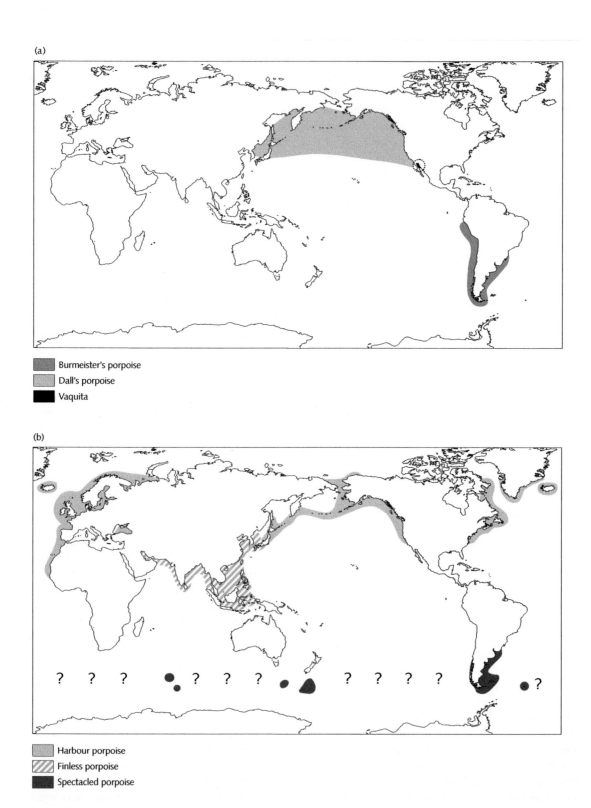

(a)

Burmeister's porpoise

Dall's porpoise

Vaquita

(b)

Harbour porpoise

Finless porpoise

Spectacled porpoise

Fig. 1.9 Porpoise and river dolphin distribution: (a) Burmeister's and Dall's porpoises, and vaquita; (b) harbour, spectacled and finless porpoises. (*continued*)

(c)

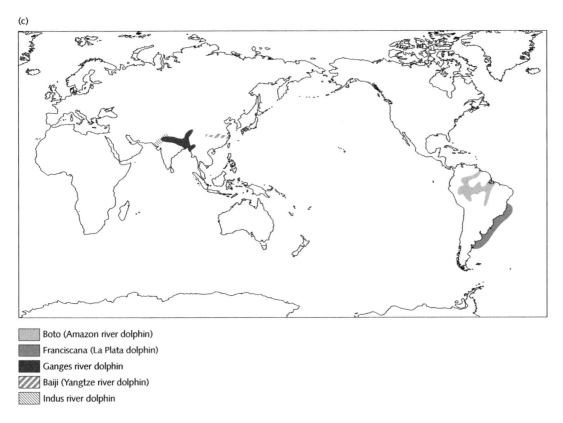

Boto (Amazon river dolphin)
Franciscana (La Plata dolphin)
Ganges river dolphin
Baiji (Yangtze river dolphin)
Indus river dolphin

Fig. 1.9 (*cont'd*) (c) Ganges, Indus and Yangtze river dolphins, boto and franciscana. (Adapted from Jefferson *et al.* 1993.)

certainly isolated geographically, however, and has been so long enough to have developed significant morphological differences from the Amazon and Orinoco botos (da Silva 1983). Therefore, while most distribution maps show *Inia* occurring in one contiguous area, at least two isolated populations exist and are evolving (or have already evolved) into distinct species (da Silva 1994).

1.3.2 Sirenians

The modern sirenians, apart from the extinct Steller's sea cow, are tropical and subtropical animals (Fig. 1.10). The dugongids (dugong and Steller's sea cow) are marine, while the trichechids (manatees) are closely tied to sources of fresh water. The Amazonian manatee is an obligate freshwater species and the other two manatees, West Indian and West African, occur in both sea water and fresh water. A

critical determinant of sirenian zoogeography is access to abundant aquatic plants. The dugong specializes on tropical seagrasses across the tropical and subtropical Indo-West Pacific. Pockets of dugongs occur (or did before being extirpated) around many South Pacific and Indian Ocean islands.

Although dugongs are able to forage on seagrass beds as deep as 20 m and they are known to travel to reefs hundreds of kilometres from shore, their dispersal is clearly limited by deep, wide expanses of ocean. They prefer sheltered waters no colder than 18°C and no deeper than about 6 m. The other modern dugongid, Steller's sea cow, is only known to have lived in Recent times in the shallow coastal waters of the Commander Islands, west of the Aleutians. Its ecological niche was clearly narrow, as it subsisted entirely on cold-tolerant kelps and was probably limited in its diving and swimming capabilities (Domning 1978).

(a)

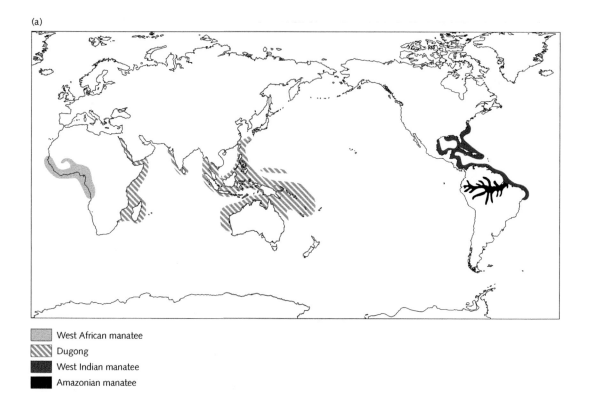

West African manatee
Dugong
West Indian manatee
Amazonian manatee

(b)

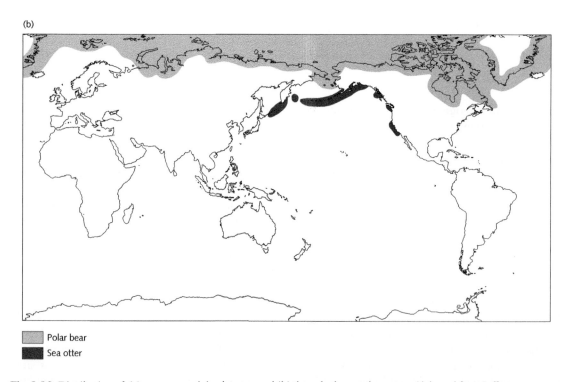

Polar bear
Sea otter

Fig. 1.10 Distribution of: (a) manatees and the dugong; and (b) the polar bear and sea otter. (Adapted from Jefferson *et al.* 1993.)

The West Indian manatee has a coastal range extending from the southeastern United States to Bahía, Brazil (Fig. 1.10a) (Lefebvre *et al.* 1989). The species' distribution is disjunct in many areas, due either to stretches of unsuitable habitat or extermination by people. Manatees may have 'island hopped' across the Lesser Antilles to maintain a connection between South America and the Greater Antilles. Alternatively, they could have crossed between Mexico's Yucatán Peninsula and Cuba, or followed the banks between Jamaica and the Caribbean coasts of Honduras and Guatemala (Lefebvre *et al.* 1989). Translocation of manatees in the 1960s led to their establishment in the Panama Canal, and this in turn has resulted in their dispersal westwards all the way to the Pacific Ocean (Montgomery *et al.* 1982).

The Amazonian manatee is entirely confined to fresh water. It inhabits all three types of Amazonian water (white, clear and black). Seasonal changes in water level influence manatee distribution. The animals move with rising water into the várzea and igapó where they gain access to new, tender aquatic macrophytes. As water levels decline, they find refuge either in perennial lake systems or in the main river channels. *Trichechus manatus* and *T. inunguis* occur in close proximity in the mouth of the Amazon in northern Brazil, but the latter seems to predominate in the area that is hydrologically and phytogeographically part of the Amazon basin (Domning 1981). Human-mediated mixing of the two species has occurred in Panama (Mou Sue *et al.* 1990) and possibly Guyana (Garcia-Rodriguez *et al.* 1998).

1.3.3 Pinnipeds

Although less aquatically adapted and less diverse in form than the cetaceans, the pinnipeds have been equally successful in terms of their numerical abundance and global distribution. They occur at all except the highest latitudes and have managed to establish a year-round presence in the ice zones of both poles. Moreover, seals have managed to persist in continental lake systems with no direct link to the sea. In general, pinnipeds are more successful than cetaceans at living permanently in higher latitudes, whereas cetaceans are far more abundant and diverse than pinnipeds in the tropics.

1.3.3.1 *Odobenids*

The walrus has a disjunct northern circumpolar distribution (Fig. 1.11). The three main populations (Pacific, Atlantic and central Russian Arctic) are separated by long stretches of year-round ice. In modern times, walruses have been associated with areas that are at least seasonally covered in sea ice, which the animals use as an alternative to land as a platform for resting and sunning. Migrations, where known, follow the seasonal pattern of ice-edge movement. Unlike some phocids (see below), the walrus does not maintain holes in sea ice. All populations have been subjected to hunting, often to the extent of wiping out the species from large regions (e.g. the Gulf of St Lawrence and the Scotian Shelf of eastern Canada), so the current distribution of the walrus is much reduced (see also Chapter 14).

1.3.3.2 *Otariids*

For the most part, otariids inhabit cool temperate and subpolar regions, although they also occur in low-latitude sites influenced by cold current systems (Figs 1.12 and 1.13). Most species form large concentrations at particular shore sites where pupping and mating take place. Otariids are entirely absent from the North Atlantic but are widespread and diverse in both the North Pacific and the Southern Ocean. In the South Atlantic they do not occur regularly north of Angola in the east, or southern Brazil in the west.

The highly adaptable genus *Arctocephalus* has an intriguing pattern of distribution (Bonner 1968; Repenning *et al.* 1971). The extreme reduction of many fur seal populations by commercial sealing has had a strong influence on their zoogeography. These secondary effects confound analyses of present-day status and distribution. Of the eight recognized species, six occur exclusively in the Southern Ocean, one (*A. galapagoensis*) in the eastern equatorial Pacific and one (*A. townsendi*) in the subtropical to temperate far-eastern North Pacific (Fig. 1.13). Some of the species breed within a single island group, while others have longitudinally extensive distributions that encompass numerous island groups. A widely accepted hypothesis is that primitive otariids entered the Southern Ocean from the North Pacific

(a)

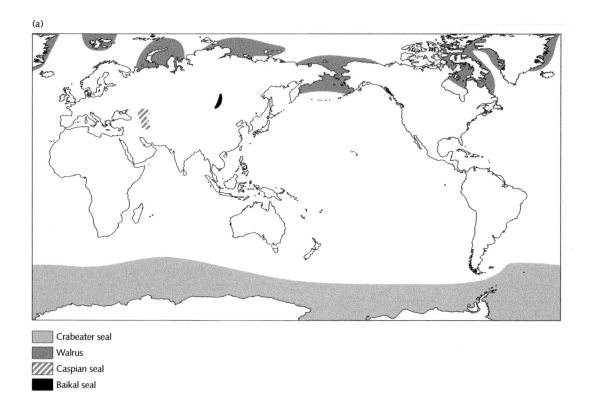

　　　Crabeater seal
　　　Walrus
　　　Caspian seal
　　　Baikal seal

(b)

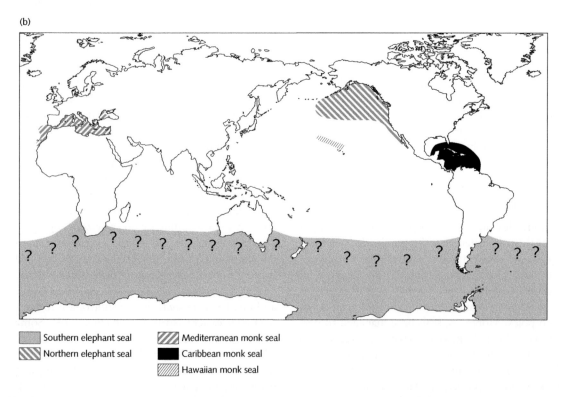

　　　Southern elephant seal　　　Mediterranean monk seal
　　　Northern elephant seal　　　Caribbean monk seal
　　　　　　　　　　　　　　　　Hawaiian monk seal

Fig. 1.11 Distribution of: (a) the walrus, and Baikal, Caspian and crabeater seals; (b) monk and elephant seals. (*continued*)

(c)

(d)

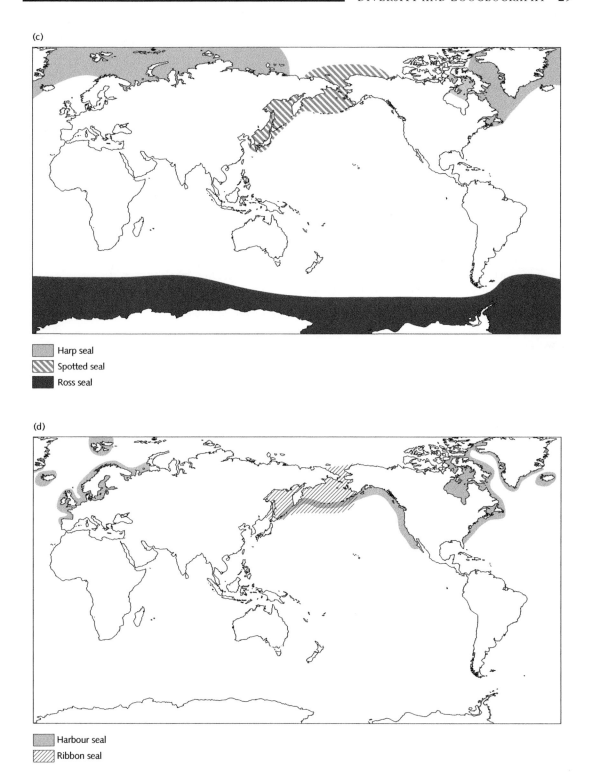

Harp seal
Spotted seal
Ross seal

Harbour seal
Ribbon seal

Fig. 1.11 (*cont'd*) (c) Harp, spotted, and Ross seals; (d) harbour and ribbon seals. (Adapted from Jefferson *et al.* 1993.)

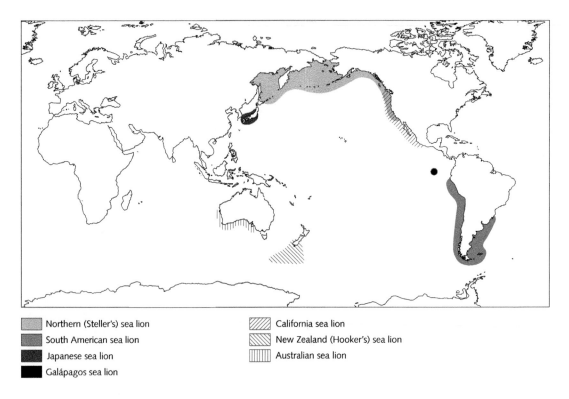

▨ Northern (Steller's) sea lion	▨ California sea lion
▨ South American sea lion	▨ New Zealand (Hooker's) sea lion
▨ Japanese sea lion	▨ Australian sea lion
▨ Galápagos sea lion	

Fig. 1.12 Distribution of the sea lions. (Adapted from Jefferson *et al.* 1993.)

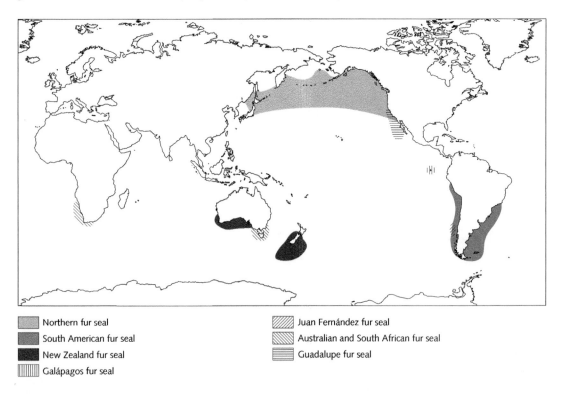

▨ Northern fur seal	▨ Juan Fernández fur seal
▨ South American fur seal	▨ Australian and South African fur seal
▨ New Zealand fur seal	▨ Guadalupe fur seal
▨ Galápagos fur seal	

Fig. 1.13 Distribution of some of the fur seals. (Adapted from Jefferson *et al.* 1993.)

and radiated to most of the circumantarctic islands in association with the prevailing West Wind Drift (Repenning *et al.* 1971). The core distributions of some species overlap and result in hybridization (Shaughnessy *et al.* 1988; Gales *et al.* 1992) (see Chapter 11). Vagrant subantarctic fur seals have been documented as far as 8000 km from the nearest part of their core distribution (Torres & Aguayo 1984), and such wide dispersal by individuals is typical of this and other fur seal species.

The northern, or North Pacific, fur seal (*Callorhinus ursinus*) has two main populations, often called 'herds'. During the pupping and breeding season, the larger herd hauls out on the Pribilof Islands in the eastern Bering Sea; the smaller, on the Commander Islands in the western Bering Sea. After leaving their northern rookeries in the autumn, these seals lead an entirely pelagic existence, moving along the continental shelf to southern California in the east and Japan in the west.

The *Zalophus* sea lions historically occurred in three discrete populations. The Asian population (Japanese sea lion) may be extinct or is very nearly so, having last been reliably reported from an island in the Sea of Japan in the 1950s. The Galápagos sea lion lives on the equator, hauling out on hot, dry beaches but feeding in the cool, nutrient-rich waters near the islands. Although this sea lion is sympatric with the Galápagos fur seal (*Arctocephalus galapagoensis*), the two species select different types of haul-out habitat, sea lions preferring wind-exposed sandy or rocky beaches and fur seals opting for rocky shores with steep, broken relief (Trillmich 1987). The two species also exhibit differences in diving behaviour and diet that doubtless limit competition.

The California sea lion is much more wide ranging and abundant than its congeners. The Pacific population inhabits coastal and inshore waters, as well as some offshore islands, from Cabo San Lucas at the southern tip of Baja California north to Vancouver Island. Only males migrate regularly north of California. A genetically isolated population occurs in the Gulf of California (Maldonado *et al.* 1995).

The northern, or Steller's, sea lion is another North Pacific endemic. Although their ranges overlap in the region from California to Vancouver Island, and indeed small numbers of northern sea lion pups have been born as far south as the California Channel Islands, the northern sea lion basically replaces the California sea lion as a breeding species to the north of California. It formerly bred as far north as the Pribilofs and is the only otariid that regularly hauls out on sea ice (Rice 1998).

Australian sea lions form small, widespread, temporally asynchronous breeding colonies along the west and south coasts of Australia (Gales & Costa 1997) (Fig. 1.12). Another regional endemic is the New Zealand, or Hooker's, sea lion whose limited range is centred in the subantarctic Auckland Islands, south of New Zealand. The South American sea lion has a much more extensive range and a larger total population (by at least an order of magnitude) than the other two southern hemisphere sea lions. Rookeries are scattered along the mainland coast and on islands (including the Falklands) from the tip of South America to northern Peru and southern Brazil.

1.3.3.3 *Phocids*

The phocids are the most widespread of the pinniped families, with representatives at most latitudes in both the northern and southern hemispheres (Fig. 1.11). The northern group, Phocinae, includes temperate region, subarctic and Arctic species, while the southern group, Monachinae, embraces the warm-water monk seals, the wide-ranging elephant seals and the Antarctic ice seals. Among the phocines, the bearded seal and the 'ubiquitous' ringed seal are non-migratory northern circumpolar species, remaining in Arctic regions all year round. The ringed seal's strong, sharp foreclaws are adapted for excavating sea ice, so this animal manages to maintain breathing holes in the thick land-fast ice as well as in the shifting pack ice. Bearded seals, in contrast, must remain near open water and they pup on the pack ice along leads and polynyas. The Baikal seal, a relict freshwater species confined to Lake Baikal in eastern Russia, is the only species other than the ringed seal to give birth in ice lairs. Another landlocked species, the Caspian seal, pups on the surface of the winter pack ice in the northern Caspian Sea and disperses widely throughout this saline inland sea at other seasons.

Most of the other phocines are also ice adapted and give birth on ice surfaces. Two North Atlantic endemics, the harp seal and the hooded seal, undertake long migrations between high and mid latitudes. They congregate in winter on floes in the southern margins of the ice zone, giving birth and mating in the late winter and early spring. Another North Atlantic endemic, the grey seal, occupies more temperate regions. It breeds in dense colonies, mainly on land but also on land-fast ice in eastern Canada and the Baltic Sea (Bonner 1981). The ribbon seal and spotted seal are endemic to the North Pacific (Fig. 1.11c). The spotted seal is closely related to the harbour seal (*Phoca vitulina*; known as the common seal in the UK) whose distribution follows the margins of both the North Atlantic and North Pacific, reaching warm temperate latitudes in the south and Arctic latitudes in the north (Fig. 1.11c). Harbour seals occur in coastal embayments and river mouths, sometimes ascending rivers to become established in freshwater lake systems.

The Monachinae, or southern phocids, include three groups: the monk seals, elephant seals and Antarctic ice seals. Monk seals are the only truly warm-water pinnipeds, as others that inhabit equatorial or low latitudes either migrate seasonally to richer high-latitude feeding grounds (northern elephant seals) or live in close proximity to cold, nutrient-rich current systems (e.g. Galápagos fur seals and sea lions). The monk seals are poorly adapted to the presence of man. At the time of Columbus, Caribbean monk seals (*Monachus tropicalis*) inhabited the coasts of many islands in the northern and western Caribbean Sea and Gulf of Mexico. They persisted on Mexico's Yucatán Peninsula until the late nineteenth century, and the last known colony was on Serranilla Bank between Jamaica and Honduras in the early 1950s. The modern distribution of Hawaiian monk seals (*M. schauinslandi*) has been centred in the remote and largely uninhabited northwestern Hawaiian Islands. Mediterranean monk seals (*M. monachus*) were historically widespread in the Mediterranean and Black Seas and along the northwest coast of Africa. In the twentieth century, however, they have been reduced to scattered localities in the Mediterranean Sea and eastern North Atlantic.

The four Antarctic monachines are all circumpolar and have overlapping but distinctive ranges that reflect their unique adaptations. Ross seals are among the most polar of the group. They haul out year-round in the heavy, consolidated pack ice to as far south as 78°S, but may forage as far away as the Antarctic Convergence. Crabeater seals are the southern equivalent of ringed seals—ubiquitous and exceedingly abundant. Weddell seals live mainly along the land-fast ice close to the Antarctic continent, with small disjunct groups on Signy Island, the South Orkneys and at South Georgia. The fourth member of this group is the leopard seal, whose distribution encompasses not only the entire Antarctic but also many subantarctic islands (e.g. Heard, Kerguelen, South Georgia).

The circumpolar southern elephant seal occurs mainly in subantarctic latitudes, centred between about 40°S and 60°S. There are three main breeding populations, at the Macquarie and Campbell Islands south of New Zealand, the Indian Ocean subantarctic island groups (Heard, Kerguelen, Crozet, Prince Edward and Marion), and the South Atlantic subantarctic islands (Bouvet, South Georgia and Falklands) together with the Antarctic Peninsula (Fig. 1.11b). Breeding colonies also occur in Tierra del Fuego and near Punta Norte, Argentina.

Northern elephant seals are endemic to the eastern North Pacific, where they breed almost exclusively on island and mainland beaches between Baja California and northern California. Like their southern counterparts, they are extraordinary divers and undertake lengthy feeding migrations away from the breeding grounds. Of particular interest is that both males and females make a double migration annually away from their terrestrial haul-out grounds, once after pupping and breeding and again after moulting. Between their periods on shore, the seals remain at sea for several months at a time and travel at least as far west as 173°W and as far north as the Aleutian Islands (Stewart & DeLong 1995).

1.3.4 Sea otter

The sea otter probably evolved in the North Pacific and has always been endemic to that single ocean basin. It formerly occupied virtually the entire Pacific rim, but its present-day distribution (Fig. 1.10b) is highly fragmented, largely as a result of past commercial exploitation. As obligate consumers of benthic invertebrates, sea otters are restricted to nearshore waters where such prey are abundant and accessible. Their limited capability for dispersal across the deep, wide passes separating oceanic islands has made recolonization a slow process. To some extent, the present-day distribution of sea otters is a result of re-introduction and supplementation efforts in North America (see Chapter 14).

1.3.5 Polar bear

The polar bear's distribution (Fig. 1.10b) closely tracks that of its primary prey, the northern phocid seals, especially the ringed seal. Polar bears live in close association with pack ice. They depend on the ice as a platform for hunting and as a means of transport. Adult male polar bears are true nomads, wandering thousands of kilometres across the ice in search of food and mates. Females are more conservative in light of their need for denning habitats, either on land or on relatively stable pack ice. On a circumpolar scale, the polar bear population is divided into about 19 relatively discrete populations (Derocher *et al.* 1998).

1.4 CONCLUSIONS

1.4.1 Cetaceans

Cetaceans occur in almost every part of the world's seas and oceans, having adapted both to the extreme cold of ice-covered polar waters and the unremitting heat of the tropics. They occur in rivers thousands of kilometres from the sea, and also in the centre of the Pacific Ocean thousands of kilometres from land. They have adjusted to life in water as shallow as 2 m (e.g. bottlenose dolphins in Florida) and as deep as thousands of metres (e.g. beaked whales). This is clearly a very successful and adaptive group of animals.

At a lower taxonomic level, there is huge variation between and even within species. At one end of the scale are species like the killer, sperm, blue, fin and humpback whales which have an almost global distribution. At the other, some species are restricted to a narrow range of water temperatures (e.g. bowhead, beluga and narwhal in cold waters; finless porpoise, spinner dolphin and tucuxi in the tropics). We also find considerable strategical differences in terms of population movements; some are resident in an area all year round (e.g. St Lawrence River belugas, Florida bottlenose dolphins, botos) while others make long migrations (e.g. most rorquals, gray whales and Beaufort Sea belugas).

The fact that the body surface area : volume ratio is inversely related to body size means that heat loss is proportionally higher in smaller animals. If this were a critical factor in cetacean distribution, we would expect to find that small cetaceans are excluded from polar regions, but this is not the case. Of the three species which remain in high (northern) latitudes all year round, completing their entire life cycle in or near sea ice, two (beluga and narwhal) are relatively small (5 m or less). At the other end of the globe, the hourglass dolphin (*Lagenorhynchus cruciger*), one of the smaller delphinids, occurs in Antarctic waters as cold as 1°C. Neither is it the case that the largest cetaceans must avoid tropical waters to avoid overheating. Bryde's whales and female sperm whales, for example, remain in low latitudes year-round, and the humpback whale routinely moves to tropical waters for the winter.

Cetaceans as a group, and some individual species, have wide tolerances of water temperature and depth, but other species have adapted to a particular habitat and are now constrained by that adaptation. Extreme examples are the monodontids, the river dolphins, the genus *Sousa* and the vaquita, all of which occur in tightly defined habitats and relatively small geographical areas. Most cetaceans have wider distributions than these, yet have not spread to all corners of the earth. They are limited by one or more factors, such as water temperature, depth or depth gradient, acting directly on them or indirectly through their prey. Even cosmopolitan species have patchy, or non-uniform, distributions reflecting preferences for particular habitats. A recent study of the widely distributed Risso's dolphin showed that its core habitat in the northern Gulf of Mexico was a relatively small area—the steep upper continental slope (Baumgartner 1997).

The two pilot whales provide an excellent example of temperature-defined distributions. They are very similar in size, shape and general morphology and have similar diets. Yet *Globicephala melas* is divided geographically into two cool-water populations (one in each hemisphere) by the tropical/subtropical belt inhabited by *G. macrorhynchus*. The zone of overlap between these two species is small; they obviously have very different habitat requirements. We mentioned above that a similar mechanism even separates two populations of *G. macrorhynchus* off Japan.

With regard to population movements we find, predictably, that seasonal migrations are more common in species subjected to greater seasonal climatic variation, i.e. those occurring at higher latitudes (see Chapter 7 for a detailed discussion). The best-known migrations are those of mysticetes, particularly where the animals follow a coastline or use coastal sites and can be seen easily; gray and humpback whales come to mind here, and both move many thousands of kilometres in each direction. Migrations among odontocetes are less well understood, mostly because of the difficulty of studying them. Recent satellite telemetry studies have shown large-scale seasonal movements in some beluga populations (Smith & Martin 1994; Richard *et al.* 2001), and we know that narwhals must undertake similar migrations to avoid being cut off by encroaching sea ice in the autumn. But direct evidence is lacking for most odontocetes, and the available information can be ambiguous. Furthermore, there are many examples of year-round occurrence of odontocetes in relatively high latitudes (e.g. pilot whales around the Faeroes; Peale's and Commerson's dolphins around the Falkland Islands), so there would appear to be no over-riding reason for odontocetes to move to lower latitudes for the winter. The diversity of migratory strategies within the Cetacea demonstrates how adaptation has allowed this group of mammals to survive and flourish in a great variety of climates and habitats (see Chapter 7).

1.4.2 Sirenians

Sirenians are constrained by their herbivorous diet (see Chapter 8), notwithstanding that they occasionally feed on sessile marine macroinvertebrates such as ascidians, sea cucumbers and sea pens (O'Shea *et al.* 1991; Preen 1995), and that manatees eat fish caught in gillnets (Powell 1978). Although much less widespread latitudinally than cetaceans and pinnipeds, the extant sirenians have an impressive record of dispersal. The dugong's marine adaptations have allowed it to establish far-flung outposts in areas separated by wide expanses of deep, inhospitable stretches of open ocean—for example, the populations near New Caledonia and Vanuatu in Melanesia, and Palau in Micronesia. For their part, some manatee populations live (or lived) mainly in marine conditions round oceanic islands (Lefebvre *et al.* 1989) while other populations are sufficiently freshwater-adapted to have penetrated into the hearts of the African and South American continents, to upstream of Timbuktu in the Niger River (Hatt 1934) and to Ecuador in the Amazon system (Timm *et al.* 1986). Thanks to human intervention, manatees have transgressed the isthmus of Central America via the Panama Canal (see above). In so doing, they have probably become the only marine mammals to have extended their range from one ocean basin to another in modern times, although a humpbacked dolphin has recently been documented in the Mediterranean Sea having presumably swum through that other marine shortcut—the Suez Canal (Kerem *et al.* 2001).

1.4.3 Pinnipeds

As with cetaceans, some endemic pinniped species have adapted to a narrow and well-defined ecological niche (e.g. Australian sea lion, two Galápagos otariids, ribbon seal), while other more generalist species thrive in a variety of conditions (e.g. harbour seal, California sea lion, leopard seal). Although the Antarctic phocids are all more or less circumpolar, only two of the Arctic pinnipeds (the ringed and bearded seals) are truly circumpolar, while the walrus is nearly so. The harbour seal's amphiboreal distribution is very much like that of the harbour porpoise. Unlike the porpoise, however, the harbour seal has a sister species, the spotted seal, living in the ice-strewn high latitudes of the North Pacific. There are no river seals as there are river dolphins, nor are there cetacean equivalents of the seal species and subspecies living in landlocked water bodies (Caspian, Baikal and Saimaa and Ladoga ringed seals).

In summary, the pinnipeds are diverse and widely distributed; indeed their success rivals that of other widespread, abundant carnivores such as the Canidae, Felidae, Mustelidae and Ursidae. Their amphibious existence has its constraints, requiring them to return to land or stable ice platforms for parturition, lactation and moulting. But this requirement can also be viewed as an advantage, for it allows the pinnipeds to conserve energy by resting in a thermally less stressful environment and to give birth and nurse their young in relative safety (out of the reach of sharks and killer whales at least). Pinnipeds have coevolved with cetaceans, managing to swim in the same waters and chase the same prey, even to dive to similar depths and migrate over equally huge distances. The extent to which these two groups of mammals have converged on similar strategies to exploit the riches of the earth's oceans, seas, lakes and rivers is nothing short of remarkable.

REFERENCES

Anderson, P.K. (1979) Dugong behavior: on being a marine mammalian grazer. *Biologist* **612**, 113–114.

Baumgartner, M.F. (1997) The distribution of Risso's dolphin (*Grampus griseus*) with respect to the physiography of the northern Gulf of Mexico. *Marine Mammal Science* **13**, 614–638.

Bérubé, M., Aguilar, A., Dendanto, D. *et al.* (1998) Population genetic structure of North Atlantic, Mediterranean Sea and Sea of Cortez fin whales, *Balaenoptera physalus* (Linnaeus 1758): analysis of mitochondrial and nuclear loci. *Molecular Ecology* **7**, 585–599.

Best, P.B. (1979) Social organization in sperm whales, *Physeter macrocephalus*. In: *Behavior of Marine Animals: Current Perspectives in Research, Vol. 3. Cetaceans* (H.E. Winn & B.L. Olla, eds), pp. 227–289. Plenum, New York.

Best, R.C. & da Silva, V.M.F. (1989) Biology, status and conservation of *Inia geoffrensis* in the Amazon and Orinoco river basins. In: *Biology and Conservation of the River Dolphins* (W.F. Perrin, R.L. Brownell Jr, K. Zhou & J. Liu, eds), pp. 23–34. Occasional Papers of the IUCN Species Survival Commission No. 3. International Union for Conservation of Nature and Natural Resources, Gland, Switzerland.

Bonner, W.N. (1968) *The Fur Seal of South Georgia*. British Antarctic Survey Scientific Report No. 56. British Antarctic Survey, London.

Bonner, W.N. (1981) Grey seal *Halichoerus grypus* Fabricius, 1791. In: *Handbook of Marine Mammals, Vol. 2. Seals*. (S.H. Ridgway & R.J. Harrison, eds), pp. 111–144. Academic Press, London.

Clarke, M.R. (1978) Buoyancy control as a function of the spermaceti organ in the sperm whale. *Journal of the Marine Biological Association of the UK* **58**, 27–71.

da Silva, V.M.F. (1983) *Ecologia alimentar dos Golfinhos de água doce da Amazônia*. MSc Thesis, University of Amazônia, Manaus, Brazil.

da Silva, V.M.F. (1994) *Aspects of the biology of the Amazonian dolphins genus Inia and Sotalia fluviatilis*. PhD Dissertation, University of Cambridge, UK.

Dahlheim, M.E. & Heyning, J.E. (1999) Killer whale *Orcinus orca* (Linnaeus, 1758). In: *Handbook of Marine Mammals, Vol. 6. The Second Book of Dolphins and the Porpoises* (S.H. Ridgway & R. Harrison, eds), pp. 281–322. Academic Press, London.

Davies, J.L. (1963) The antitropical factor in cetacean speciation. *Evolution* **17**, 107–116.

Derocher, A.E., Garner, G.W., Lunn, N.J. & Wiig, Ø., eds (1998) *Polar Bears: Proceedings of the Twelfth Working Meeting of the IUCN/SSC Polar Bear Specialist Group*. IUCN, Gland, Switzerland and Cambridge, UK.

Domning, D.P. (1977) Observations on the myology of *Dugong dugon* (Müller). *Smithsonian Contributions to Zoology* **226**, 57 pp.

Domning, D.P. (1978) *Sirenian Evolution in the North Pacific Ocean*. University of California Publications in Geological Sciences No. 118. University of California Press, Berkeley.

Domning, D.P. (1981) Distribution and status of manatees *Trichechus* spp. near the mouth of the Amazon River, Brazil. *Biological Conservation* **19**, 85–97.

Dyke, A.S. & Morris, T.F. (1990) Postglacial history of the bowhead whale and of driftwood penetration; implications for paleoclimate, central Canadian Arctic. *Geological Survey of Canada Paper* **89-24**, 19 pp.

Fay, F.H. (1982) *Ecology and Biology of the Pacific Walrus, Odobenus rosmarus divergens Illiger*. United States Department of the Interior, Fish and Wildlife Service, North American Fauna No. 74. US Government Printing Office, Washington, DC.

Gales, N.J. & Costa, D.P. (1997) The Australian sea lion, a review of an unusual life history. In: *Marine Mammal Research in the Southern Hemisphere, Vol. 1. Status, Ecology and Medicine* (M. Hindell & C. Kemper, eds), pp. 78–87. Surrey Beatty & Sons, Chipping Norton, UK.

Gales, N.J., Coughran, D.K. & Queale, L.F. (1992) Records of subantarctic fur seals *Arctocephalus tropicalis* in Australia. *Australian Mammalogy* **15**, 135–138.

Garcia-Rodriguez, A.I., Bowen, B.W., Domning, D. *et al.* (1998) Phylogeography of the West Indian manatee (*Trichechus manatus*): how many populations and how many taxa? *Molecular Ecology* **7**, 1137–1149.

Gaskin, D.E. (1982) *The Ecology of Whales and Dolphins*. Heinemann, London.

Hatt, R.T. (1934) The American Museum Congo Expedition manatee and other Recent manatees. *Bulletin of the American Museum of Natural History* **66** (IV), 533–566.

International Whaling Commission (2000) Report of the 1999 Annual Meeting of the International Whaling Commission Scientific Committee, Annex I: Report of the Standing Sub-Committee on small cetaceans. *Journal of Cetacean Research and Management* **2**, 235–263.

International Whaling Commission (2001) Annex U. Report of the Working Group on Nomenclature. *Journal of Cetacean Research and Management* **3** (Suppl.), 363–365.

Jefferson, T.A., Leatherwood, S. & Webber, M.A. (1993) *Marine Mammals of the World*. FAO Species Identification Guide. United Nations Environment Programme, Food and Agriculture Organization of the United Nations, Rome.

Kasuya, T., Miyashita, T. & Kasamatsu, F. (1988) Segregation of two forms of short-finned pilot whales off the Pacific coast of Japan. *Scientific Reports of the Whales Research Institute (Tokyo)* **39**, 77–90.

Kerem, D., Goffman, O. & Spanier, E. (2001) Sighting of a single humpback dolphin (*Sousa* sp.) along the Mediterranean coast of Israel. *Marine Mammal Science* **17**, 170–171.

King, J.E. (1983) *Seals of the World*, 2nd edn. British Museum (Natural History), London.

Klages, N.T.W. & Cockcroft, V.G. (1990) Feeding behaviour of a captive crabeater seal. *Polar Biology* **10**, 403–404.

LeDuc, R.G., Perrin, W.F. & Dizon, A.E. (1999) Phylogenetic relationships among the delphinid cetaceans based on full cytochrome *b* sequences. *Marine Mammal Science* **15**, 619–648.

Lefebvre, L.W., O'Shea, T.J., Rathbun, G.B. & Best, R.C. (1989) Distribution, status, and biogeography of the West Indian manatee. In: *Biogeography of the West Indies* (C.A. Woods, ed.), pp. 567–610. Sandhill Crane Press, Gainesville, FL.

Ling, J.K. & Bryden, M.M. (1981) Southern elephant seal *Mirounga leonina* Linnaeus, 1758. In: *Handbook of Marine Mammals, Vol. 2. Seals* (S.H. Ridgway & R.J. Harrison, eds), pp. 297–327. Academic Press, London.

Lowry, L.F. & Fay, F.H. (1984) Seal eating by walruses in the Bering and Chukchi seas. *Polar Biology* **3**, 11–18.

McGinnis, S.M. & Schusterman, R.J. (1981) Northern elephant seal *Mirounga angustirostris* Gill, 1866. In: *Handbook of Marine Mammals, Vol. 2. Seals* (S.H. Ridgway & R.J. Harrison, eds), pp. 329–349. Academic Press, London.

Mackintosh, N.A. (1965) *The Stocks of Whales*. Fishing News (Books), London.

Maldonado, J.E., Davila, F.O., Stewart, B.S., Geffen, E. & Wayne, R.K. (1995) Intraspecific genetic differentiation in California sea lions (*Zalophus californianus*) from southern California and the Gulf of California. *Marine Mammal Science* **11**, 46–58.

Mead, J.G. (1989) Shepherd's beaked whale *Tasmacetus shepherdi* Oliver, 1937. In: *Handbook of Marine Mammals, Vol. 4. River Dolphins and the Larger Toothed Whales* (S.H.

Ridgway & R. Harrison, eds), pp. 309–320. Academic Press, London.

Mead, J.G. & Mitchell, E.D. (1984) Atlantic gray whales. In: *The Gray Whale Eschrichtius robustus* (M.L. Jones, S.L. Swartz & S. Leatherwood, eds), pp. 33–53. Academic Press, Orlando.

Mikhalev, Y.A. (1997) Humpback whales *Megaptera novaeangliae* in the Arabian Sea. *Marine Ecology Progress Series* **149**, 13–21.

Mitchell, E. (1970) Pigmentation pattern evolution in delphinid cetaceans: an essay in adaptive coloration. *Canadian Journal of Zoology* **48**, 717–740.

Montgomery, G.G., Gale, N.B. & Murdoch Jr, W.P. (1982) Have manatee entered the eastern Pacific Ocean? *Mammalia* **46**, 257–258.

Mou Sue, L.L., Chen, D.H., Bonde, R.K. & O'Shea, T.J. (1990) Distribution and status of manatees (*Trichechus manatus*) in Panama. *Marine Mammal Science* **6**, 234–241.

Nishiwaki, N. & Marsh, H. (1989) Dugong *Dugong dugon* (Müller, 1776). In: *Handbook of Marine Mammals, Vol. 3. The Sirenians and Baleen Whales* (S.H. Ridgway & R. Harrison, eds), pp. 1–31. Academic Press, London.

Norris, K.S. & Møhl, B. (1983) Can odontocetes debilitate prey with sound? *American Naturalist* **122**, 85–104.

Ohsumi, S. (1980) Bryde's whales in the North Pacific in 1978. *Report of the International Whaling Commission* **30**, 315–318.

O'Shea, T.J., Rathbun, G.B., Bonde, R.K., Buergelt, C.D. & Odell, D.K. (1991) An epizootic of Florida manatees associated with a dinoflagellate bloom. *Marine Mammal Science* **7**, 165–179.

Perrin, W.F. (1975) *Variation of Spotted and Spinner Porpoise (Genus Stenella) in the Eastern Tropical Pacific and Hawaii.* Bulletin of the Scripps Institution of Oceanography No. 21. University of California San Diego, La Jolla, CA.

Powell, J.A. (1978) Evidence of carnivory in manatees (*Trichechus manatus*). *Journal of Mammalogy* **59**, 442.

Preen, A. (1995) Diet of dugongs: are they omnivores? *Journal of Mammalogy* **76**, 163–171.

Ray, G.C. & Schevill, W.E. (1974) Feeding of a captive gray whale (*Eschrichtius robustus*). *Marine Fisheries Review* **36** (4), 31–38.

Read, A.J. (1999) Harbour porpoise *Phocoena phocoena* (Linnaeus, 1758). In: *Handbook of Marine Mammals, Vol. 6. The Second Book of Dolphins and the Porpoises* (S.H. Ridgway & R. Harrison, eds), pp. 323–355. Academic Press, San Diego.

Reeves, R.R., Leatherwood, S. & Papastavrou, V. (1991) Possible stock affinities of humpback whales in the northern Indian Ocean. In: *Cetaceans and Cetacean Research in the Indian Ocean Sanctuary* (S. Leatherwood & G.P. Donovan, eds), pp. 259–269. Marine Mammal Technical Report No. 3. United Nations Environment Programme, Nairobi, Kenya.

Repenning, C.A., Peterson, R.S. & Hubbs, C.L. (1971) Contributions to the systematics of the southern fur seals, with particular reference to the Juan Fernández and Guadalupe species. In: *Antarctic Research Series, Vol. 18. Antarctic Pinnipedia* (W.H. Burt, ed.), pp. 1–34. American Geophysical Union, Washington, DC.

Reyes, J.C., Mead, J.G. & Van Waerebeek, K. (1991) A new species of beaked whale *Mesoplodon peruvianus*. *Marine Mammal Science* 7, 1–24.

Rice, D.W. (1989) Sperm whale *Physeter macrocephalus* Linnaeus, 1758. In: *Handbook of Marine Mammals, Vol. 4. River Dolphins and the Larger Toothed Whales* (S.H. Ridgway & R. Harrison, eds), pp. 177–233. Academic Press, London.

Rice, D.W. (1998) *Marine Mammals of the World, Systematics and Distribution*. Special Publication No. 4. Society for Marine Mammalogy, Lawrence, KA.

Richard, P.R., Martin, A.R. & Orr, J.R. (2001) Summer and autumn movements of belugas of the eastern Beaufort Sea stock. *Arctic* 54 (3), 223–236.

Robineau, D. (1983) Morphologie externe et pigmentation du dauphin de Commerson *Cephalorhynchus commersonii* (Lacepede, 1804), en particulier celui des Iles Kerguelen. *Canadian Journal of Zoology* 62, 2465–2475.

Robineau, D. (1986) Valeur adaptive des caracteres morphologiques distinctifs (taille et pigmentation) d'une population isolee d'un dauphin subantarctique, *Cephalorhynchus commersonii* (Lacepede, 1804). *Mammalia* 50, 357–368.

Rojas-Bracho, L. & Taylor, B.L. (1999) Risk factors affecting the vaquita (*Phocoena sinus*). *Marine Mammal Science* 15 (4), 974–989.

Rosenbaum, H.C., Brownell Jr, R.L., Brown, M.W. *et al.* (2000) Worldwide genetic differentiation of *Eubalaena*: questioning the number of right whale species. *Molecular Ecology* 9 (11), 1793–1802.

Schaeff, C., Kraus, S., Brown, M. *et al.* (1991) Preliminary analysis of mitochondrial DNA variation within and between the right whale species *Eubalaena glacialis* and *Eubalaena australis*. *Report of the International Whaling Commission, Special Issue* 13, 217–223.

Schreer, J.F. & Kovacs, K.M. (1997) Allometry of diving capacity in air-breathing vertebrates. *Canadian Journal of Zoology* 75, 339–358.

Shaughnessy, P.D., Shaughnessy, G.L. & Fletcher, L. (1988) Recovery of the fur seal population at Macquarie Island. *Papers and Proceedings of the Royal Society of Tasmania* 122, 177–187.

Smith, T.G. & Martin, A.R. (1994) Distribution and movements of belugas, *Delphinapterus leucas*, in the Canadian

High Arctic. *Canadian Journal of Fisheries and Aquatic Sciences* 51, 1653–1663.

Stewart, B.S. & DeLong, R.L. (1995) Double migrations of the northern elephant seal, *Mirounga angustirostris*. *Journal of Mammalogy* 76, 196–205.

Stone, G.S., Flórez-Gonzalez, L. & Katona, S. (1990) Whale migration record. *Nature* 346, 705.

Suydam, R.S., Lowry, L.F., Frost, K.J., O'Corry-Crowe, G.M. & Pikok Jr, D. (2001) Satellite tracking of eastern Chukchi Sea beluga whales into the Arctic Ocean. *Arctic* 54 (3) 237–243.

Swartz, S.L. (1986) Gray whale migratory, social and breeding behaviour. *Report of the International Whaling Commission, Special Issue* 8, 207–229.

Timm, R.M., Albuja, V.L. & Clauson, B.L. (1986) Ecology, distribution, harvest, and conservation of the Amazonian manatee, *Trichechus inunguis*, in Ecuador. *Biotropica* 18, 150–156.

Torres, D. & Aguayo, A. (1984) Presence of *Arctocephalus tropicalis* (Gray 1872) at the Juan Fernandez Archipelago, Chile. *Acta Zoological Fennica* 172, 133–134.

Trillmich, F. (1987) Galapagos fur seal, *Arctocephalus galapagoensis*. In: *Status, Biology, and Ecology of Fur Seals: Proceedings of an International Symposium and Workshop, Cambridge, England, 23–27 April 1984* (J.P. Croxall & R.L. Gentry, eds), pp. 23–27. NOAA Technical Report NMFS No. 51. US Department of Commerce, National Oceanic and Atmospheric Administration, National Marine Fisheries Service, Seattle, WA.

Vladykov, V.-D. (1946) Nourriture du Marsouin Blanc ou Béluga (*Delphinapterus leucas*) du fleuve Saint-Laurent. *Contribution de Départment des Pêcheries, Québec* 17, 130 pp.

Wada, S. (1988) Genetic differentiation between two forms of short-finned pilot whales off the Pacific coast of Japan. *Scientific Reports of the Whales Research Institute (Tokyo)* 39, 91–101.

Warneke, R.M. & Shaughnessy, P.D. (1985) *Arctocephalus pusillus*, the South African and Australian fur seal: taxonomy, evolution, biogeography, and life history. In: *Studies of Sea Mammals in South Latitudes* (J.K. Ling & M.M. Bryden, eds), pp. 53–77. South Australian Museum, Adelaide.

Wells, R.S., Boness, D.J. & Rathbun, G.B. (1999) Behavior. In: *Biology of Marine Mammals* (J.E. Reynolds III & S.A. Rommel, eds), pp. 324–422. Smithsonian Institution Press, Washington, DC.

CHAPTER 2

The Evolution of Marine Mammals

John E. Heyning and Gina M. Lento

2.1 INTRODUCTION

Over the past 50 million years, several groups of mammals have evolved secondary adaptations for an aquatic existence. A few of these groups, such as the hippo-like desmostylians, are extinct, but the majority are extant. With just over 120 modern species (Rice 1998), marine mammals are not taxonomically diverse compared to other marine animals, yet they are important components of many ecosystems. Interestingly, marine mammals have evolved from only two terrestrial groups, or clades, of mammals. The first is the Order Carnivora, which includes such well-known creatures as lions, tigers and bears. From this order arose the pinnipeds (seals,

sea lions, walrus) and sea otter. The other clade is the Ungulata, a group that includes all the modern orders of hooved animals. From these ranks, we now recognize the cetaceans (whales, dolphins, porpoises), the sirenians (dugongs, manatees) and the extinct desmostylians.

2.1.1 Systematics

Systematics is the study of defining evolutionary relationships among organisms, both extinct and extant. A phylogeny is a hypothesis of those evolutionary relationships, and is the bedrock of any evolutionary study (Boxes 2.1 and 2.2). It is upon a phylogenetic tree (which is simply a graphical representation of the evolutionary hypothesis) that

---

BOX 2.1 CLADISTICS AND PHENETICS: PHILOSOPHICAL APPROACHES TO PHYLOGENETIC RECONSTRUCTION

Most modern systematists use analytical methods based on a philosophical approach called **cladistics** (see Wiley 1981; Wiley *et al.* 1991 for reviews). The basic tenets of cladistics are quite simple and straightforward: organisms are deemed to be related based on shared derived characters (called **synapomorphies**). These characters are defined as having arisen in the common ancestor of the taxa in question. These characters are subsequently passed on to their descendant taxa. Groups of related taxa and their descendants are called **clades** regardless of taxonomic rank. A **character** can be any feature of an organism that is useful in phylogenetic analysis. Characters can be morphological or molecular, behavioural or physiological. Whereas characters are the feature, the **character state** is the specific condition expressed in the individual or taxa. For example, the character of limb number in animals is expressed in states of 0, 2, 4, 6, 8 or 10. In molecular terms, that character

of a given nucleotide position in a DNA sequence has four character states: G, A, T and C.

A second and slightly older philosphical approach to phylogenetics is called **phenetics**). In contrast to cladistics, phenetic analyses result in trees generated from indices of the overall similarity between organisms. This includes both shared ancestral (called **sympleisiomorphies**) and shared derived character states. Most phenetic techniques generate a similarity matrix. Many systematists do not consider the phenetic approach to be useful, as it is only weakly based in evolutionary theory and may obscure the phylogeny by pairing up distantly related taxa simply because they both have retained numerous ancestral characters. Although phenetics is a questionable technique for determining higher level evolutionary relationships, it is useful for identifying species or populations within species.

BOX 2.2 SOME USEFUL NOMENCLATURE

Monophyletic groups (clades) are those that include only the hypothetical ancestor and all of its descendants (e.g. below left, clade 1 + 2 or clade 1 + 2 + 3 + 4).
Paraphyletic groups are those which include the ancestor taxa, but not all the descendants (e.g. below centre, group 2 + 3 + 4). Paraphyletic groups often reflect a certain level of morphological organization or 'grade', and exclude some or all of the more derived

descendants. For example, in classic cetacean taxonomy, the suborder Archaeoceti is notably paraphyletic, as this group of fossil cetaceans includes the most primitive whales of a certain morphological grade, but not the descendant suborders Mysticeti and Odontoceti.
Polyphyletic groups contain collections of descendants from more than one ancestor but not the ancestors (e.g. below right, group 2 + 3).

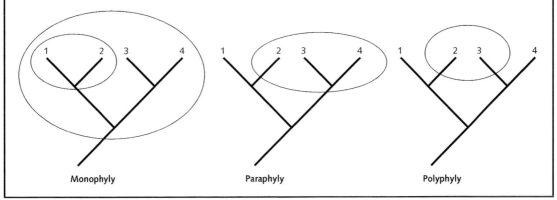

features such as behaviour, physiology or geographical distribution can be mapped. This is used to evaluate whether similar features in two organisms arose due to inheritance from a common ancestor or evolved via convergence. Thus, this chapter provides the basic framework for the following chapters. It should be noted from the outset that no proposed phylogeny can be proven. They are only hypotheses of relationships. Proof that a phylogeny is correct would require perfect knowledge of all past evolutionary events. Since that knowledge is unattainable, researchers today can only 'infer' past events from phylogenetic reconstructions of evolutionary relationships.

2.1.1.1 Molecular and morphological data

One of the seemingly great debates in modern systematics concerns the role of molecular versus morphological data in producing phylogenies. Much of this debate has centred on a few well-publicized conflicts between phylogenies based on morphological versus molecular data. Most systematists concur that neither molecular nor morphological datasets are inherently superior *per se* (e.g. Moritz & Hillis 1996, p. 5). Biomolecules such as DNA have the advantage of objective character state definition, a

virtually inexhaustible abundance of characters (considering each nucleotide as a separate character), and some theoretically predictable evolutionary behaviour that may be quantified. However, the accumulation of multiple nucleotide substitutions per site will eventually reach a point of saturation beyond which true phylogenetic signal is obscured. How long it takes to reach this point depends on the rate at which the locus in question is evolving (e.g. mitochondrial DNA, especially the hypervariable control region, evolves faster than ribosomal RNA genes or nuclear DNA). Morphological characters evolve more slowly than molecules over long periods of time, but suffer from the sometimes problematic subjective character definition, and are often not discriminatory enough to distinguish differences at the population level. It is clear though that both types of data are valuable in phylogenetic inference. Further, it is now understood by most researchers that some combination of data types may provide the best resource for resolving evolutionary histories (Kluge 1989; Kluge & Wolf 1993). The process by which different types of data are combined and analysed forms the basis for an entire branch of evolutionary systematics research (e.g. Pamilo & Nei 1988; Donoghue *et al.* 1989; Smith &

Littlewood 1994; Vrana *et al.* 1994; Huelsenbeck *et al.* 1996).

What is of greater importance is the proper collection of data and method of analysis. The methodology should be explicit, repeatable and based on evolutionary theory. Of equal importance is whether the specific dataset is appropriate for answering the question at hand. For example, hypervariable regions of the genome or subtle morphological differences, such as tooth or vertebral counts, may be useful for identifying species or resolving the relationships among closely related taxa. However, such data provides little or no resolution regarding the relationships among higher level taxa. It should be remembered that while no phylogeny can be proven, some phylogenies can be based on stronger evidence than others. Careful scrutiny for homologous characters, the choice of relevant outgroup(s), and increasing taxonomic and/or character sampling can sometimes help resolve some apparent conflicts among datasets.

2.1.1.2 Homology

One of the most critical aspects of evaluating the phylogenetic utility of characters is evaluating whether or not they are homologous. For morphologists, this should involve careful and objective comparisons regarding the details of anatomical structures and the position of the two anatomical features (i.e. character states) under question. Further evidence for homology may be obtained through developmental studies (i.e. embryology) and the fossil record. The question of homology is just as critical for molecular data. In DNA sequence data, positional homology is determined by the relative location of each nucleotide in a comparative alignment of several sequences. However, for some regions of DNA, such as hypervariable areas, positional homology can be confounded by nucleotide insertion/deletion events that result in gaps along the sequence of one or more of the taxa under examination.

For both morphological and molecular data, homoplasy (having a character state in common due to convergences or reversals rather than direct descent) can present problems. For example, homodont dentition may have arisen numerous times in parallel within toothed whales. In a similar way, a

cysteine ('C') at one particular nucleotide position is assumed to be homologous with a C at the same position in another taxon. However, with only a limited number of nucleotides (four), multiple substitutions at a single site are undetectable, and, theoretically, can add cryptic homoplasy to the sequence data.

2.1.1.3 The role of fossil taxa

The inclusion of fossil taxa in a phylogenetic analysis can have a significant impact on tree topology (e.g. Gauthier *et al.* 1988). Fossil taxa contribute most to the fuller resolution of phylogenetic trees when they 'fill in' the gaps of long divergent lineages. This is because fossil taxa help form a more complete morphological series which, in turn, helps elucidate character change over time and assists in the confirmation of homology. However, when fossil taxa are included, there are often substantially more missing entries in the data matrix because soft tissue characters and biomolecules typically do not fossilize and most fossil taxa are represented by incomplete osteological material. In phylogenetic analyses, data matrices with numerous empty cells can be problematic, often resulting in numerous polytomies (unresolved nodes of a phylogenetic tree) (Novacek 1992; Wilkinson 1995). The importance of fossil taxa in a phylogenetic analysis is therefore a trade-off between the completeness of that fossil and its temporal position in the clade and the amount of divergence within the clade (Huelsenbeck 1991). The stratigraphic position of a fossil provides evidence for the earliest known existence of a clade. However, such temporal data can not be used to place a clade in phylogenetic position as the fossil record is too incomplete (Heyning & Thacker 1999).

2.1.1.4 Effects of rapid evolutionary radiation

Regardless of the type of data examined, the occurrence of rapid adaptive radiations can impede phylogenetic resolution. Rapid radiations can affect phylogenetic reconstruction in two ways. If the event of rapid species diversification occurred in the distant geological past, the chance that subsequent evolutionary changes will obscure the original event is higher. Graphically this amounts to very

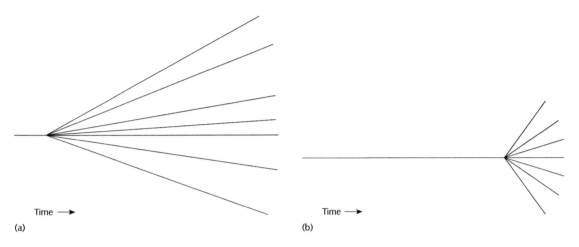

Fig. 2.1 Graphical representations of the effects of rapid radiation events on a phylogenetic tree. (a) An event of rapid species diversification which has occurred in the distant geological past shown here by very short internal branches between deep nodes giving rise to very long-terminal branches. (b) An event that has occurred in the recent geological past after which insufficient time has elapsed to allow the accumulation of discernible differences. This is represented here by a large polytomy including many terminal lineages, sometimes called a phylogenetic 'bush'.

short internal branches between deep nodes giving rise to very long terminal branches on a phylogenetic tree (Fig. 2.1a). Finding the right characters that might illuminate this distant event is difficult. Alternatively, if the rapid radiation event has occurred in the recent geological past, insufficient time will have elapsed to allow the accumulation of discernable differences, either morphological or molecular, between the taxa concerned. Graphically this is represented by a large polytomy including many terminal lineages, sometime called a phylogenetic bush (Fig. 2.1b).

2.2 ORDER CARNIVORA

Several groups of carnivores have taken to a partially marine existence, although all must come back to land, at least to give birth. The most diverse are the pinnipeds—the seals, sea lions and walruses. The order Carnivora is divided into two large groups— the Feliformia, which includes the viverrids (civets), the herpestids (mongooses), the hyaenids (hyenas) and the felids (cats), and the Caniformia, which includes the canids (dogs, jackals and foxes) and the Arctoidea (Fig. 2.2). All marine carnivores are within the arctoid lineage. The arctoids include the mustelids (skunks, ferrets, otters and weasels), the procyonids (raccoons), the ursids (bears), the phocids (true seals), the otariids (fur seals and sea lions, also called the eared seals) and the odobenids (walruses). However, as pinnipeds are deeply nested in the arctoid lineage, they should not be recognized as a distinct order or suborder of mammals.

Other marine carnivores include the polar bear, *Ursus maritimus*. Analysis of mtDNA, allozymes and karyotypes suggest that polar bears are a very recent evolutionary offshoot of the brown bear, *Ursus arctos*, and in particular of the brown bears from the islands off southeast Alaska (Rice 1998). Fossils suggest that the two diverged in the middle Pleistocene (Kurten 1964).

There are three species of marine carnivores in the weasel/otter family (Mustelidae). Along the northeast coast of North America, the extinct sea mink (*Mustela macrodon*) has been found primarily from Native American midden sites. The chungungo or marine otter, *Lutra felina*, is found in the coastal waters of South America. Lastly, the well-known sea otter, *Enhydra lutris*, is found in the coastal temperate and subarctic waters of the North Pacific. These two otters are closely related to the freshwater otters. The sea otter appears to be related to the late Miocene to early Pliocene *Enhydritherium*, found along both coasts of North America and Europe (Berta & Morgan 1985; Lambert 1997).

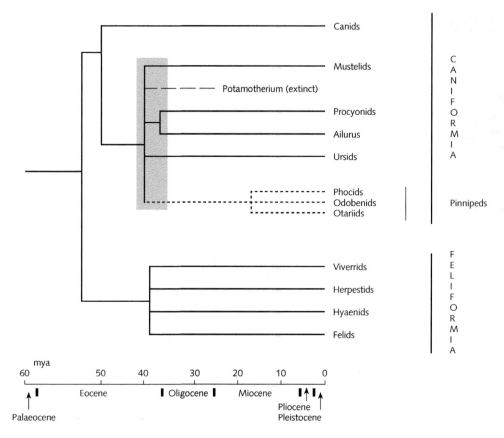

Fig. 2.2 A phylogenetic tree of carnivore evolution adapted from Wayne *et al.* (1991). The timescale below the tree indicates the times of divergence based on DNA–DNA hybridization data. The grey area refers to the polytomy involved in discerning the origin of the pinniped superfamily (diagonal striped lines) within the Arctoid carnivores. The dashed line shows the relative position of the fossil Potamotherium, which has played a key role in the support for a mustelid–phocid sister relationship. Note the polytomy among the feloids at approximately the same geological time as the canoids. A rapid radiation event has been proposed for both suborders (Wayne *et al.* 1991; Janczewski *et al.* 1992).

2.2.1 Pinnipeds

The pinnipeds (from the Latin meaning 'fin-footed') are a group of marine mammals originating from the arctoid branch of the Carnivore order. This group consists of the true seals, sea lions, fur seals, walruses, and their extinct ancestors. While their general affinities to the arctoid carnivores are universally accepted, the exact relationships of the pinnipeds to terrestrial carnivores or among the pinnipeds themselves have been under debate for more than 100 years (Mivart 1885; McLaren 1960).

The extant true seals, Phocidae, are distinguished from the eared seals by the lack of external ear flaps,

short forelimbs, short vibrissae (or whiskers), a pelvis which does not permit quadrupedal movement on land, an inflated tympanic bulla (which houses the middle ear), and locomotion in water by undulation of the hindlimbs (Fig. 2.3). They are divided into two basic groups, the Monachinae ('southern' seals) and the Phocinae ('northern' seals). The sea lions and fur seals (or 'eared' seals), Otariidae, can be discerned from the phocids by the presence of external ear flaps, elongate forelimbs resembling flippers, long vibrissae, a rotatable hind flipper that allows quadrupedal locomotion on land, and locomotion in water by vertical flapping of the fore flippers (Fig. 2.4). They are divided into two

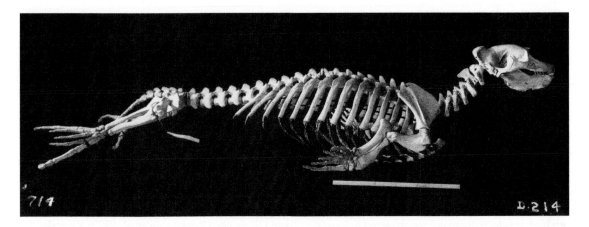

Fig. 2.3 Skeleton of a modern phocid, the elephant seal (*Mirounga angustirostris*). Note the relatively short forelimbs which are of limited use for locomotion on land. (Courtesy of the Natural History Museum of Los Angeles County.)

Fig. 2.4 Skeleton of a living otariid, the Steller sea lion (*Eumetopias jubata*). Note the elongate forelimbs with prominent processes on the limb elements that serve as attachment points for the powerful muscles used for both aquatic and terrestrial locomotion. (Courtesy of the Natural History Museum of Los Angeles County.)

groups, the Arctocephalinae (fur seals) and the Otariinae (sea lions). The walrus family includes mostly extinct species consisting of three groups called the Dusignathinae, the Odobeninae and a paraphyletic collection of basal walruses referred to as the Imagotariinae. The imagotarines are described as archaic non-tusked walruses retaining the piscivorous nature of the phocids and otariids. The Dusignathinae were a group of four-tusked walrus species specializing in squirt/suction feeding behaviour. These two groups are distinguished from the Odobeninae which developed tusks only in the upper jaw and eventually gave rise to the modern walrus, *Odobenus rosmarus*.

2.2.1.1 Pinniped origins

Local coastal upwelling increases primary productivity, and consequently the availability of food for higher trophic levels. These regions of cold water

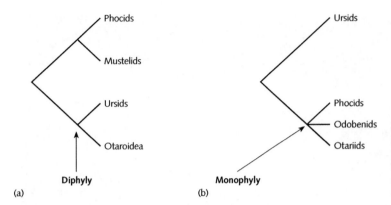

Phocids

Mustelids

Ursids

Otaroidea

Diphyly

(a)

Ursids

Phocids

Odobenids

Otariids

Monophyly

(b)

Fig. 2.5 The prevailing alternative hypotheses for the origin of the pinnipeds: (a) a diphyletic origin involving the mustelids and ursids, and (b) a monophyletic origin involving only the ursids.

with high productivity may have attracted colonization by early pinniped populations which adapted to the aquatic environment from terrestrial ancestors (Costa 1993). The number of terrestrial ancestors giving rise to the pinnipeds has been a matter of long-standing debate. Two schools of thought are apparent from the literature (Fig. 2.5). One, citing primarily biogeographical and morphological evidence, supports a diphyletic origin, attributing the sea lions, fur seals and walruses to an ursid (bear-like) ancestor and the true seals to a mustelid (weasels, skunks, otters and ferrets) ancestor (Flower 1869; Mivart 1885; McLaren 1960; Ray 1976; Tedford 1976; King 1983a; Barnes 1989; Wozencraft 1989; Nojima 1990). The other school, using biomolecular, karyological and recent morphological evidence, supports a monophyletic origin for all three pinniped families stemming from ursids (Sarich 1969a, 1969b; Wyss 1987, 1988; Berta 1991; Vrana *et al.* 1994; Árnason *et al.* 1995), mustelids (Árnason & Widegren 1986; Bininda-Emonds & Russell 1996) or an unresolved arctoid ancestor (Lento *et al.* 1995). Recent molecular studies (Vrana *et al.* 1994; Árnason *et al.* 1995; Lento *et al.* 1995; Ledje & Árnason 1996a, 1996b) and morphological studies (Wyss & Flynn 1993; Berta & Wyss 1994; Bininda-Emonds & Russell 1996) have led to the general acceptance of pinniped monophyly.

Historically, the proponents of the diphyletic hypothesis drew heavily on the available fossil record for the pinnipeds. The otarioids appeared to have originated in the eastern North Pacific (because this is where the oldest fossils were found) about 22 (23.5?) mya, while the phocid fossil record showed their first appearance in the North Atlantic in the late Oligocene (Repenning *et al.* 1979; Koretsky & Sanders 1997). Morphological studies supporting diphyly considered overall similarities among the pinnipeds to be convergent adaptations to the aquatic environment and therefore dismissed large numbers of potentially useful characters. Incorrect assignment of character polarity appears to be the main criticism of morphological studies supporting diphyly (Berta & Wyss 1994).

The true seals (phocids) are the most aquatically adapted family (Repenning *et al.* 1979; Berta & Deméré 1986; Kohno 1993). Costa (1993) suggests, based on diving physiology and breeding patterns, that the eared seal (fur seals and sea lions) physique and breeding pattern evolved first. Changes in body size and foraging strategies later lead to the derivation of the phocid (true seals) body form and breeding behaviour. There is no mention of the role of walruses in Costa's evolutionary description. The use of physiology to study evolution and evolutionary mechanisms has received renewed interest (e.g. Hochachka 1997/98).

There have been several suggestions based on palaeontological, morphological and molecular data, that the carnivores, and the pinnipeds within them, experienced rapid adaptive radiations immediately following the advent of the first carnivore- and pinniped-like body plans/lifestyles (Ray 1976; Wayne *et al.* 1989, 1991; Janczewski *et al.* 1992; Árnason *et al.* 1995; Lento *et al.* 1995; Bininda-Emonds & Russell 1996; Ledje & Árnason 1996a, 1996b). These studies indicate that the arctoid families diverged rapidly sometime between 37 and 40 mya. A rapid radiation in the distant geological past makes this node of the phylogenetic tree very hard to resolve.

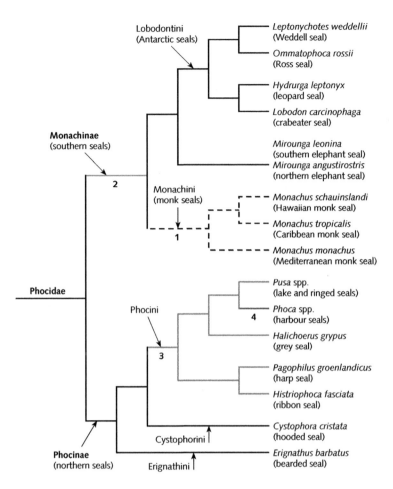

Fig. 2.6 A cladogram depicting the balance of current understanding of the extant phocid relationships based on a compilation of molecular and non-molecular data. Textured lineages indicate involvement in contentious points of evolution: 1 (dashed lines), reflects the uncertainty of relationships among the monk seals; 2, indicates the section of the tree involved in uncertainty among the Monachinae; 3, indicates the section of the tree involved in uncertainty among the Phocini; and 4, refers to suggestions of uncertainty among species level systematics. (Adapted from Bininda-Emonds & Russell 1996.)

2.2.1.2 Phocid evolution

The phocid seals have enjoyed significantly more taxonomic and systematic attention as a group than the other two pinniped families. The oldest fossil phocid is represented by isolated femora of late Oligocene age found in South Carolina (Koretsky & Sanders 1997). A recent monograph (Bininda-Emonds & Russell 1996) provides a clear and concise framework for understanding phocid systematics and evolution. It will be used here as a guideline for our current understanding. Additional evidence from recent biomolecular studies focuses on each of the contentious points.

The phylogenetic tree in Fig. 2.6 represents the classic understanding of extant phocid phylogeny. The tree in Fig. 2.7, which includes all extinct phocid genera and is adapted from Berta and Wyss

(1994), is presented to provide an evolutionary context for the extant phocids. Figure 2.8 presents Bininda-Emonds and Russell's (1996) results which includes support for some previously proposed relationships as well as some novel relationships among the Phocidae.

Analyses of phocid fossils (de Muizon 1982; Wyss 1988; Berta & Wyss 1994) indicate that the hypothesis of phocid monophyly is very stable. Below this level, the tribal distinctions have also remained robust, but the subfamily groupings above the tribes and the generic distinctions below the tribes present some of the current questions concerning phocid systematics (Fig. 2.6). These include the interrelationships of the genus *Monachus* (monk seals) and, consequently, the systematics of the Monachinae (the southern seals), the resolution among the tribe Phocini, and species level phylogeny.

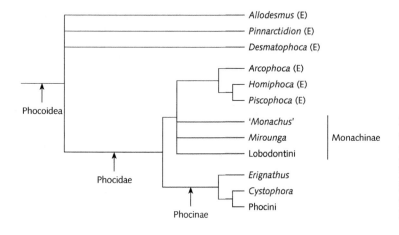

Fig. 2.7 A cladogram including all extinct phocid genera, which provides an evolutionary context for the extant phocids. This cladogram is based on 143 morphological characters. The letter 'E' following taxa names means this taxon is extinct. (Adapted from Berta & Wyss 1994.)

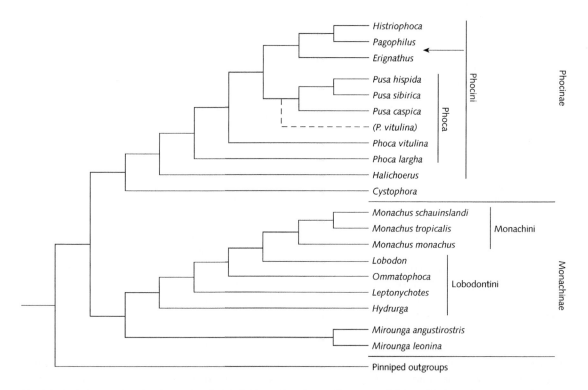

Fig. 2.8 A cladogram showing the results of Bininda-Emonds and Russell's survey of 196 morphological characters. The dashed line leading to *Pusa vitulina* indicates the alternative placement of this species, which represents the only difference between two equally parsimonious trees in their analyses. An arrow indicates the novel placement of the genus, *Erignathus*. Other groups discussed in the text are indicated by the divisions on the right. Pinniped and carnivore outgroups are also indicated on the right. (Adapted from Bininda-Emonds & Russell 1996.)

Tribe Monachini

The monk seals (genus *Monachus*) are generally considered to be the most basal of the extant phocid species, retaining some characters even more primitive than many fossil phocids (Repenning *et al.* 1979; King 1983a; Wyss 1988). The two living species are extremely rare with the Caribbean monk seal (*M. tropicalis*) presumed extinct (Kenyon 1977). Some early morphological studies proposed a sister relationship between *M. schauinslandi* and *M. tropicalis* to the exclusion of *M. monachus* (Davies 1958; Repenning & Ray 1972; King 1983a). More recently, Wyss (1988) has suggested that *M. schauinslandi* should be considered the most basal of the three based on morphology of the ear region. However, the results of Bininda-Emonds and Russell's survey (1996) uphold the original suggestion of a *M. schauinslandi*–*M. tropicalis* sister group with *M. monachus* as an earlier divergence (Fig. 2.8). There have been no comparative molecular studies of the genus *Monachus*.

Since its original description, the status of the subfamily Monachinae has oscillated between including only the species of *Monachus* or including *Monachus* with the southern hemisphere seals (Lobodontini) and the elephant seals, *Mirounga*. Membership of the Monachinae may also be confounded by apparent relationships of the extant monachines to archaic seal species as described by Berta and Wyss (1994), who indicated that the archaic lobodontines *Acrophoca*, *Homiphoca* and *Piscophoca* are more closely related to *Monachus* and *Mirounga* than to the extant Lobodontini. These authors noted that this arrangement was weakly supported in their study. The results of Bininda-Emonds and Russell place the genus *Monachus* as a terminal divergence within the subfamily and instead place *Mirounga* as a basal lineage (Fig. 2.8). This placement renders, at least, the Lobodontini paraphyletic, which corroborates the work of Berta and Wyss (1994).

Tribe Phocini

The tribe Phocini consists of the northern hemisphere seals of the genera *Halichoerus*, *Histriophoca*, *Pagophilus*, *Phoca* and *Pusa*. The monophyly of this group is supported by both morphological and karyological evidence (Burns & Fay 1970; Árnason 1974a, 1977; McLaren 1975). The grey seal *Halichoerus* is generally considered to be the first diverging of these genera (Scheffer 1958; King 1972, 1983a; Bonner 1981). Several studies, both molecular and morphological, have proposed a closer relationship between *Halichoerus*, *Phoca* and *Pusa* to the exclusion of *Histriophoca* and *Pagophilus*, which share numerous morphological similarities.

The analysis of morphological characters (Fig. 2.8) supports the alliance of *Histriophoca* with *Pagophilus* but places *Erignathus* as basal to these, disrupting the monophyly of the genus *Phoca*. Their tree also indicates an alliance of *Pusa* with the *Histriophoca–Pagophilus–Erignathus* clade. *P. vitulina* is either included as basal to *Pusa* (dashed line in Fig. 2.8) or as a basal divergence to the entire *Pusa–Histriophoca–Pagophilus–Erignathus* clade. *P. largha* appears as basal to the whole of this group and, in turn, *Halichoerus* as basal to the group plus *P. largha*, leaving *Cystophora* as the most basal genus of the entire Phocinae.

Evolution at the species level

With any widely distributed species, there often is contention regarding species level systematics. An exemplary case is that of *Phoca vitulina*, the harbour seal. Populations from the Atlantic and Pacific are readily distinguishable, but the subdivisions of the Pacific group have remained unresolved. This uncertainty impacts primarily on *P. largha*, the larga seal, which is distinguishable from other Pacific subspecies of *P. vitulina* on the basis of geographical, ecological, morphological and behavioural evidence, but remains of uncertain taxonomic status. Studies alternately suggest *P. largha* is a subspecies of *P. vitulina* or is a separate species itself. Bininda-Emonds and Russell (1996) considered *P. largha* as a separate species in order to determine its status with respect to *P. vitulina*. A recent survey of the worldwide patterns of mitochondrial control region in *P. vitulina* shows that harbour seals are in fact genetically differentiated and regionally philopatric (Stanley *et al.* 1996) (see Chapter 11).

2.2.1.3 Otariid evolution

The extant Otariidae classically comprises two groups, the fur seals (Arctocephalinae) and the sea lions

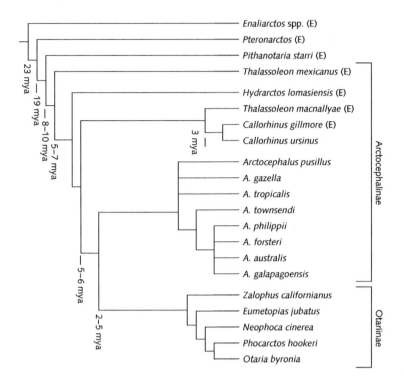

Fig. 2.9 A cladogram of the relationships among extant and fossil otariids based on 41 osteological and soft anatomical characters. This tree has been adapted from Berta and Demére (1986) and Berta and Wyss (1994) to reflect the growing suspicion of paraphyly for the Arctocephalinae and its related fossils (see below). Dates of divergences on this tree are approximated from geological dates reported for the fossil taxa included. An obvious exception to this is the lineage leading to *Hydrarctos lomasiensis*, a single Peruvian fossil dated to 1.5 mya (de Muizon 1985). It is the only southern hemisphere fossil that has been attributed as belonging to the Otariids, and analyses of morphological characters place this taxon in close association with the extinct genus *Thalassoleon*. The letter 'E' following taxa names means the taxon is extinct. The polytomies amongst the extant Arctocephalinae reflect an approximate consensus of most studies that report any information on these relationships within the context of studies of higher taxonomic focus. (Adapted from Berta & Demére 1986.)

(Otariinae). There are nine species or subspecies of fur seals from near-equatorial regions and the southern hemisphere (*Arctocephalus* spp.) and a single monotypic genus of fur seal in the northern hemisphere (*Callorhinus ursinus*). The six living sea lion species are distributed roughly equally between the North Pacific and the southern hemisphere. Morphological resolution of relationships among the arctocephaline fur seals is very low (Berta & Demére 1986). Based on fossil dates of their nearest extinct ancestors, all 16 species and subspecies of living otariids have evolved within the last 2–3 mya. The relationships depicted in Fig. 2.9 reflect a consensus of studies on arctocephaline relationships

(Davies 1958; Repenning *et al.* 1971, 1979; Berta & Demére 1986).

The traditional systematic view groups the fur seals together in one clade and the six sea lions together in a separate clade. The distinguishing morphological features most commonly used to characterize division of the groups are the presence of underhair in the fur seals (absent in sea lions) and a significantly larger body size among sea lions. Within each subfamily, the Otariinae are distinguished on the basis of largely non-overlapping geographical distributions. Davies (1958) suggests close alliances between *Eumetopias* and *Otaria* (the Steller and South American sea lions) and between

Zalophus (the California sea lion) and *Neophoca* (the Australian sea lion) based on biogeographical hypotheses. Preliminary surveys of fossil material from the Steller sea lion and the California sea lion show some evidence for a close relationship between these two species (N. Kohno, personal communication).

Two studies of the overall phylogenetic relationships and species distinctions among the otariids have been undertaken, both using morphological data. In the most recent study, Berta and Deméré (1986) used 41 osteological, soft anatomical and behavioural characters from extant specimens and fossil material in a cladistic analysis that supports the traditional monophyletic groupings within the Otariidae. However, following subsequent analyses that have suggested paraphyly, at least among the Arctocephalinae, recent reanalyses of these and additional data have been undertaken to address the question of monophyly of both groups (A. Berta & P. Adam, in preparation). In a study published some years earlier, Repenning *et al.* (1971) examined arctocephaline fur seal species and based species distinctions on several morphometric features, vocalizations and physiological traits. Previous studies provide molecular evidence of apparent paraphyly among the eared seals (Lento 1995; Lento *et al.* 1997). A recent study including mitochondrial DNA sequences from all extant otariids (G.M. Lento, in preparation) does not uniquely support the traditional grouping of fur seals and sea lions, but instead shows evidence for paraphyly among the Arctocephalinae and the Otariinae (Fig. 2.9).

Evolution at the species level

Recently there have been a number of molecular studies focusing on a single otariid species or comparisons between a few otariids species (Maldonado *et al.* 1995; Lento *et al.* 1997; Bernardi *et al.* 1998). Among these, the genus *Zalophus* presents an interesting case. The genus consists of three subspecies, the California sea lion (*Z. californianus californianus*), the Galápagos sea lion (*Z. c. wollebaeki*) and the recently extinct Japanese sea lion (*Z. c. japonicus*). A survey of the mtDNA control region in the California sea lion reveals significant genetic differentiation between populations in the Pacific Ocean and those in the Gulf of California on the eastern

side of the Baja Peninsula. Itoo's (1985) morphological evaluation suggests that the Japanese sea lion may be significantly different from both the California sea lion and the Galápagos sea lion, enough to elevate it to full species status, *Zalophus japonicus*.

2.2.1.4 Odobenid evolution

Though well-known among the childhood pantheon of charismatic megafauna, the modern walrus, *Odobenus rosmarus*, is actually represented by two extant subspecies: the Pacific walrus (*Odobenus rosmarus divergens*) and the Atlantic walrus (*O. r. rosmarus*). But much less well known is that the modern walrus is, as one author described it, 'a relict species, a lone survivor of a formerly diverse group of odobenid pinnipeds' (Deméré 1994). The modern walruses are distinguished from the rest of the pinnipeds by the development of squirt/suction feeding behaviour during which they squirt strong jets of water into the muddy bottom to uncover clams and then efficiently suck the meat out of the shell or alternately squirt water into the shell to open it (Oliver *et al.* 1983; Nelson & Johnson 1987; Kastelein *et al.* 1994). The later diverging walruses of the subfamily Odobeninae are distinguished by the development of tusks, which in the extant walrus are used to pull themselves up onto ice floes, to enlarge breathing holes, defend their young, and for dominance display among males.

The placement of the walrus as a sister species to the phocids or to the otariids is still unresolved (Fig. 2.10). All proponents of the diphyletic origin for the pinnipeds suggest an odobenid/otariid relationship, whereas most of those supporting pinniped monophyly put forward a phocid/odobenid relationship (Berta 1991; Cozzuol 1992; Berta & Wyss 1994; Vrana *et al.* 1994). However, some studies supporting pinniped monophyly provide evidence for the *Odobenus*–otariid clade (Árnason *et al.* 1995; Lento *et al.* 1995; Bininda-Emonds & Russell 1996).

Recent discoveries of walrus fossils from the North Pacific rim (reviewed in Kohno *et al.* 1995) suggest that an otarioid ancestor entered the North Pacific Ocean around 22–27 mya. Walruses diverged from this lineage around 14 mya. One walrus group remained in the North Pacific, becoming extinct

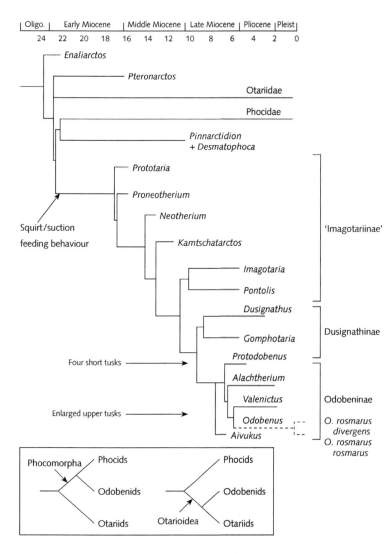

Fig. 2.10 A phylogenetic tree of odobenid evolution. The thick lines indicate the Odobenidae lineage and the thin lines show the position of the Odobenidae relative to the other pinniped families. This work, based on the fossil record and morphology of described fossils, supports a sister relationship between the Odobenids and Phocids evolving shortly after the appearance of the Otariids, though the author notes the extremely short period of time during which all three families may have diverged (21–24 mya). Arrows indicate the points at which key innovations in walrus development occurred. Dashed lines indicate the only remaining extant genus, *Odobenus*, consisting of two subspecies. Inset: The sister relationships among the pinniped subfamilies—the phocomorpha (left) and the otarioidea (right). (Adapted from Kohno, in press.)

around 4 mya. The other dispersed southward, entered the Caribbean via the Central American Seaway, and dispersed northward again until eventually moving back into the Pacific via the Arctic basin about 0.6 mya, accounting for the present-day distribution of the Atlantic and Pacific subspecies.

Kohno *et al.* (1995) also suggest that the tusked walruses originated in the eastern North Pacific, replacing the non-tusked walruses, the Imagotariines and the Dusignathinae, during the late Miocene and never disappeared from this region. The modern genus *Odobenus*, they suggest, might have originated in the North Pacific and spread to the

North Atlantic via either the Central American Seaway or the Arctic Ocean.

Evolution at the species level

The Pacific subspecies of the modern walrus resides in the Bering and Chukchi Seas. The less abundant Atlantic subspecies has a fragmented semicircumpolar distribution stretching from the Canadian Archipelago eastward to Greenland and the Soviet Union as far as the Leptev Sea. In 1994, Cronin *et al.* showed fixed genetic differences between the Atlantic and Pacific populations, indicating not only very restricted gene flow but also a low level of genetic divergence between the subspecies. They calculate the divergence of the two subspecies to be between 500 000 and 785 000 years ago. This timing supports the early evolution of the walrus in the Pacific followed by the appearance of the walrus in the Atlantic by the early Pleistocene. More recently, Andersen *et al.* (1998) propose that walruses in the northwest Greenland area are evolutionarily distinct from walruses in the other three areas. One individual sampled in eastern Greenland exhibited a Pacific haplotype offering evidence for a connection between the Pacific walrus and walruses in eastern Greenland (for further discussion see Chapter 11).

2.2.1.5 Pinniped evolution: a synthesis of thought

The diphyletic hypothesis for the origin of the pinnipeds proposes convergence as an explanation for similarity. Proponents of diphyly argue for convergence to present forms from separate (but presumably similar) terrestrial arctoid ancestors, citing parallel adaptations to a common aquatic environment. Convergence and parallelism have also been proposed to explain relationships between the pinniped families and their fossil relatives (Mitchell 1975).

A paraphyletic group includes those descendants of a common ancestor that retain ancestral characters, whereas others with derived character states are excluded. The biological implication of paraphyly is either: (i) that some lineages have experienced an acceleration in evolution relative to others; (ii) that the evolutionary rate of some lineages has slowed down; (iii) that some lineages have experienced a sweep of character reversals to ancestral forms, termed 'retrogressive' evolution by Howell (1929); or (iv) that selection may have pruned some changes while allowing others to proceed. Distinguishing between these alternatives can be difficult. Heterochrony (differential rates of evolution) can be suggested as a mechanism for alternatives (i) and (ii), although how heterochrony applies to speciation is still controversial. Bininda-Emonds and Russell's (1996) placement of *Monachus* as a derived group among the Monachinae offers a compelling case for exploring heterochrony. Berta and Wyss (1994) suggest that retrogressive evolution may explain the surplus of character reversals (versus convergences) apparent across most nodes of their phylogenetic tree of pinnipeds. However, the extreme number of reversals at some nodes is intellectually troubling.

For each of the areas of problematic phylogenetic resolution discussed above, there has been a suggestion that a rapid radiation occurred. The elusive origin of the pinnipeds is couched in a rapid expansion of arctoid carnivore families between 37 and 40 mya. A radiation at the same geological time has been proposed for the feloid carnivores as well (Wayne *et al.* 1991). A rapid diversification of pinniped subfamilies has been proposed based on interpretations of the fossil record (N. Kohno, personal communication). As a result, the sister group of the odobenids remains unconfirmed. As well, the differentiation of species within the phocids and otariids is suggested to have spanned only the last 2–5 million years.

The fossil record

Many authors have noted the paucity of the fossil record for pinnipeds (Berta & Wyss 1994; Miyazaki *et al.* 1994; Vrana *et al.* 1994; Bininda-Emonds & Russell 1996). New fossil evidence has already had an impact on prevailing biogeographical theories for the pinnipeds. A study of recently discovered walrus fossils proposes a dispersal exactly opposite to the present dogma of Pacific walrus dispersal (Kohno *et al.* 1995). Like the pre-1995 fossil record of the walruses, the fossil record of the eared seals is notably depauperate (Miyazaki *et al.* 1994). Almost all described fossil otariids have been found in the northern hemisphere. A single fur seal fossil (*'Arctocephalus' Hydrarctos lomasiensis*) is reported from Pliocene formations of Peru (de Muizon 1978). The only other reports of southern hemisphere otariid fossils are of modern Australian sea lions, *Neophoca*, dating to Pleistocene formations of New Zealand (King 1983b). Additional fossils from the southern hemisphere could make a profound difference in the current understanding of pinniped phylogeny and biogeography. Several fossil phocids have recently been discovered in New Zealand, an area which it was previously accepted had never been colonized by phocids. These fossils are thought to predate the occurrence of otariids in New Zealand (J. McKee, unpublished data; E. Fordyce, personal communication). Several authors have acknowledged that the overwhelming majority of historical effort to recover fossil material has been centred in the northern hemisphere. One may note that the pinnipeds are not alone in this situation. A southern hemisphere radiation of the modern avian orders during the early Cretaceous has been proposed (Cooper & Penny 1997).

2.3 ORDER CETACEA

Although Aristotle noted that whales are air-breathing, warm-blooded animals that nurse their young, it was commonly asserted until quite recently that whales were fish and not mammals. The renowned taxonomist Linnaeus himself considered cetaceans to be fish in earlier versions of his *Systema Natura* until he incorporated the findings made by the British naturalist John Ray in 1693. Such confusion about the true phylogenetic affinities of whales and dolphins were commonplace in the non-scholarly realms until recently. Today, the cetaceans (whales, dolphins and porpoises) are readily recognized as marine mammals. Their relationship to the ungulates (hooved terrestrial mammals including horses, cows, pigs, camels, elephants, and many others) is generally accepted, though their closest relation among the living ungulates remains an active topic of research. The order Cetacea consists of three suborders: the Archaeoceti (an extinct group of archaic whales), the Odontoceti (toothed whales) and the Mysticeti (baleen whales).

2.3.1 Cetacean origins

The assumption that the order Cetacea is monophyletic is supported by an overwhelming amount of morphological (Van Valen 1968; Barnes & Mitchell 1978; Fordyce 1980; Heyning 1997), cytological (Árnason 1969, 1974b, 1987; Duffield-Kulu 1972; Árnason *et al.* 1984) and molecular sequence (Milinkovitch *et al.* 1993, 1994; Gatesy 1998) evidence. The karyotypes of cetaceans are amazingly conservative when compared to other groups of mammals. The typical chromosome count for most cetaceans is 2N = 44. The exceptions (2N = 42) are sperm whales (Physeteridae and Kogiidae), beaked whales (Ziphiidae) and right whales (Balaenidae). For right whales and beaked whales, the lower counts are a result of the fusion of different chromosome pairs respective to each clade (Duffield-Kulu 1972; Árnason 1987). The derivation of the sperm whale karyotype from that of the typical cetacean is currently unresolved (Árnason 1987). The chromosome banding pattern among cetaceans is also astonishingly conservative.

2.3.2 Relationships of cetaceans to other ungulates

In modern cetaceans, the body shape is fish like with fin-like flippers and flukes. The hindlimbs are but bone vestiges tucked within the body wall, and their nostrils are situated high upon their heads (see Chapter 1). Cetaceans differ so strikingly from their terrestrial kin that it is difficult to discern intuitively which, among the other orders of mammals, are their closest relatives. Leigh Van Valen (1966) was the first to propose a relationship between the cetaceans and the Mesonychidae (variously raised up to the rank of order Mesonychia or Acreodi, e.g. McKenna 1975), within the extinct paraphyletic order Condylartha, which includes all primitive ungulates. He noted many similarities between whales and at least some mesonychids in such features as the construction of the cheekteeth, the loss of the entepicondylar foramen on the humerus, and the venous drainage of the skull. An analysis of the postcranial skeleton of the mesonychid *Pachyaena* (O'Leary & Rose 1995) suggests that this animal may have had a body form similar to tapirs, capybaras or suids (pigs), many of which are excellent swimmers!

There is now convincing fossil evidence that mesonychians are closely related to the ancestor of whales, either as the sister taxon to the whale ancestor, or that whales arose from within a paraphyletic Mesonychidae. Besides cetaceans and mesonychians, the other extant descendants of the Condylarths are the other orders of ungulates (supraorder Ungulata, see McKenna & Bell 1997) which diverged in the Palaeocene. The close relationship of cetaceans to living ungulates has been supported by immunological data (Boyden & Gemeroy 1950), DNA–DNA hybridization data (Milinkovitch 1992), mtDNA sequence data (e.g. Honeycutt & Adkins 1993; Milinkovitch *et al.* 1993; Randi *et al.* 1996) and

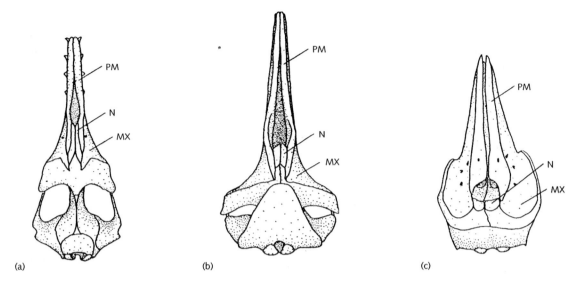

Fig. 2.11 Dorsal views of the skulls of (a) an archaeocete, (b) a mysticete, and (c) an odontocete. The skulls of archaeocetes represent the primitive condition of lacking any telescoping of the bones. In mysticetes, the occiptal bone moves forward to cover much of the parietals and frontal bones over the braincase. In odontocetes, the maxilla and premaxilla telescope backwards over the frontals and parietals. MX, maxilla; N, nares (nasal bones); PM, premaxilla.

nuclear DNA sequence data (e.g. Gatesy 1997, 1998).

While the first fossil specimens of ancient whales were being discovered, it was still very unclear which other group of living mammals the cetaceans are most closely related to. The famous British mammalogist William Henry Flower (1883a) was the first to suggest that cetaceans were related to artiodactyls (even-toed ungulates). Studies using molecular data have not yielded a consistent phylogenetic relationship of the cetacea to other ungulates. Over the decades, a plethora of studies has resulted in a bewildering array of proposed phylogenies (see Gatesy 1998 for a review). Unfortunately, the vast majority of older studies using molecular data are extremely poor in taxon sampling. Several studies which have incorporated a wider sample of taxa have been focused on the higher order relationships across all mammals rather than on the specific task of determining the closest ungulate relative of the cetaceans. These larger studies have resulted in an unresolved polytomy of artiodactyls/perissodactyls/cetaceans/paenungulata (Miyamoto & Goodman 1986; Honeycutt & Adkins 1993; Springer & Kirsch 1993).

Some molecular studies indicate that the cetacea may have actually evolved from within the order Artiodactyla, thus implying that artiodactyls are a paraphyletic assemblage. One study suggested a sister taxon relationship of dolphins to camels or hippos based on mitochondrial cytochrome *b* DNA sequence data (Irwin *et al.* 1991; Irwin & Árnason 1994). Another scenario suggests the sister taxa relationship of cetaceans, ruminants, and hippos to the exclusion of camels (tylopods) and pigs (suids) based on short interspersed elements (SINES), also known as retroposons (Shimamura *et al.* 1997; Nikaido *et al.* 1999) and combined molecular datasets (Gatesy 1998). This proposed phylogeny is not supported using a dataset including both morphological and molecular data (O'Leary 1999; O'Leary & Geisler 1999). Recent finds of early archaeocete hindlimbs reveal that they uniquely share the character of a 'double-pulley' astragulus with artiodactyls (Gingerich *et al.* 2001; Thewissen *et al.* 2001). However, some soft tissue structures, such as an air sinus system around the inner ear, are shared with perissodactyls and rooting the entire ungulate tree has been problematic for all datasets (Heyning 1999).

2.3.3 Suborder Archaeoceti

Archaeocetes represent a basal grade taxon that includes all cetaceans that lack the derived telescoped pattern of the bones of the skull of either mysticetes or odontocetes (Fig. 2.11). This taxon consists of five nominal families: the Pakicetidae, Ambulocetidae, Protocetidae, Remingtonocetidae and Basilosaurocetidae. Archaeocetes are further characterized by an elongate snout, a narrow braincase, a skull with large temporal fossa and well-defined sagittal and lambdoidal crests, a broad supraorbital process of the frontal bone, bony nares situated some distance posterior to the tip of the snout, and a slender zygomatic process of squamosal with a prominent postglenoid process (Kellogg 1936). The oldest archaeocetes are about 50 mya (late Early Eocene) and the entire archaeocete grade was 'extinct' by the end of the Eocene, although their descendant lineages of toothed whales and baleen whales have survived into the present. Archaeocete systematics and biology have recently been reviewed (Thewissen 1998b).

Most Pakicetids have been unearthed from terrestrial, freshwater or nearshore deposits of the eastern Tethys, a massive epicontinental sea which was between the Eurasian and African–Indian continents (Gingerich & Russell 1981; Thewissen 1998a; Thewissen & Hussain 1998). In 1981, the braincase and associated lower jaw fragment of the oldest and most archaic whale, *Pakicetus inachus*, was described (Gingerich & Russell 1981). The Ambulocetidae occupied tidal and estuary habitats (Williams 1998). The enigmatic *Ambulocetus natans* differs from all other known archaeocetes in that its orbits are elevated above the overall profile of the skull, and its hind feet are elongate. These hindlimbs were probably important for aquatic locomotion (Thewissen *et al.* 1996a). Many of these fossils from the eastern Tethys Sea, some with known hindlimbs, are significant in that they serve as morphological transitions between land-dwelling mammals and fully aquatic cetaceans.

The Protocetidae are defined by details of the orbits and teeth, a sacrum composed of less than four fused vertebrae, and changes in tooth structure (Thewissen 1998b). The fine serrations along the cutting edge of the cheekteeth of two protocetids,

Eocetus (Barnes & Mitchell 1978; Uhen 1999) and *Georgiacetus* (Hulbert 1998), have been interpreted as incipient denticles which are found in more advanced archaeocetes suggesting that these archaeocetes represent transitional forms between the Protocetidae and the more derived Basilosauridae, which possess well-developed denticles. In the eastern Tethys, protocetids are found in nearshore waters, whereas protocetids from the western Tethys died in shallow offshore regions and their North American kin have been found in shallow nearshore deposits (Williams 1998).

Remingtonocetidae are known from rock units of middle Eocene age from Pakistan and India. These archaeocetes are found in sediments indicating a tidal coastal environment or nearshore shallow marine deposits (Williams 1998). Two genera of remingtonocetids both have elongate rostra and mandibular symphyses, toothrows terminating in front of the eye and reduced tooth size (Kumar & Sahni 1986; Bajpai & Thewissen 1998).

Though they are interpreted to be near the crown of the archaeocete evolution, the Basilosauridae were among the first archaeocete fossils to be discovered. Based on series of vertebrae, Harlan named the first archaeocete *Basilosaurus* (Latin for 'king of the lizards') *cetoides* thinking these were the remains of a massive reptile. Subsequently, an associated skull was unearthed and the famous anatomist Sir Richard Owen correctly identified it as a mammal, not a reptile. Basilosaurids are typically divided into two subfamilies: the Dorudontinae with unspecialized vertebrae and the Basilosaurinae with extremely elongate vertebral bodies (Barnes & Mitchell 1978).

The durodontines are typified by the well-known genera *Zygorhiza* and *Durodon*. This subfamily truly represents a grade level as there are no synapomorphies to unite them. Barnes and Mitchell (1978) suggested that durodontines were ancestral to both mysticetes and odontocetes because they possessed the complex denticles found in archaic members of the modern suborders and lacked the derived character state of elongate vertebrae unique to basilosaurids. However, the loss of the third molar and the toothrow extending onto the zygomatic arch are derived character states not found in the most primitive odontocetes and mysticetes. This suggests that the durodontines represent the sister taxa to the modern Cetacea clade (Uhen 1998, 1999).

2.3.3.1 Archaeocete trends

Two of the most dramatic changes of early cetaceans was the shift from quadruped locomotion to the axial undulation of swimming and the ability to drink sea water. These changes can be reconstructed by close examination of the vertebral column, limb structure, the osteological correlates of tail flukes, and isotopic signatures within the bones. One striking conclusion is how quickly these transitions occurred. At the dusk of the early Eocene, *Pakicetus* and its contemporaries were quadruped animals that drank fresh water. A few million years later in the middle Eocene, *Rodhocetus* and its kin were swimming with the aid of tail flukes and drinking sea water (Thewissen *et al.* 1996b; Thewissen & Fish 1997). By the late Eocene, basilosaurines possessed only small hindlimbs that barely protruded from the body wall and had only three well-formed digits (Gingerich *et al.* 1990). Based on the small size of the hindlimbs and the fact that forelimbs had limited movement at the elbow, it is unlikely that basilosaurids could move on land. Hence, one can then speculate that by the late Eocene, cetaceans had already become fully aquatic and had already evolved the ability to give birth to their young in the water, therefore severing their last ties to land.

One hypothesis is that archaeocetes first evolved in fluvial or estuarine environments of the eastern Tethys and subsequently dispersed as more morphologically and physiologically advanced forms conquered the oceans. All of the most primitive and chronologically oldest fossil archaeocetes are found along the shores of the eastern Tethys, whereas the more morphologically derived and fully marine protocetids and basilosaurids of the middle and late Eocene are found in rocks from Asia, North Africa, North America and New Zealand (Thewissen 1998a).

2.3.4 Suborder Odontoceti

Since ancient times, it has been recognized that some cetaceans possess teeth (odontocetes), whereas others have baleen for filtering out prey (mys-

ticetes). The retention of teeth, *per se*, does not completely define the Odontoceti because the presence of teeth is an ancestral character. Nonetheless, there are numerous other synapomorphies that unite this clade (Heyning 1997). These synapomorphies include, in advanced odontocetes, that the maxilla (upper jaw) has telescoped back over the frontal bone of the skull so that the frontal bone is completely obscured in a dorsal view of the skull (Fig. 2.11) (Miller 1923). All extant and all but the most primitive fossil odontocetes have asymmetrical skulls, and all modern species have asymmetrical facial soft anatomy (Mead 1975b; Heyning 1989). This asymmetry results from a hypertrophy (enlargement) of the facial soft anatomy on the right side. The degree of asymmetry in the cranium appears to be correlated with the elevation of the cranium's vertex (Heyning & Mead 1990). All odontocetes have a complex series of diverticula off their nasal passages distal to the bony nares (see Lawrence & Schevill 1956; Mead 1975b; Heyning 1989; Cranford *et al.* 1996; for reviews). All extant odontocetes possess a hypertrophied fatty melon in front of the nasal passages which is distinctly different from the diminutive melon-like structure observed in mysticetes (Heyning & Mead 1990; Heyning 1997). Odontocetes are unique among all tetrapods (marine and terrestrial) in that the distal narial passages coalesce to form a single nostril or blowhole. The asymmetry, large melon and the single blowhole appear to be correlated with the ability of odontocetes to echolocate.

2.3.4.1 Archaic odontocetes

Archaic odontocetes are divided into seven families: the Agorophoriidae, Squalodelphinidae, Eoplatanistidae, Acrodelphidae, Squalodontidae, Eurhinodelphinidae and, the most recently described, Waipatiidae. The relationship of these archaic odontocetes to modern odontocetes is poorly known. The grade level family Agorophoriidae includes those species that express the incipient stages of odontocete-type skull telescoping (Miller 1923), with the maxillae covering most of the frontal bone, yet the parietal bones remain prominently exposed. Agorophiids have other very primitive odontocete characters in the formation of the parietal bones,

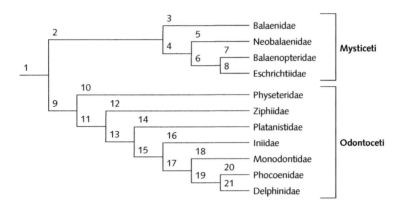

Fig. 2.12 Relationships among the families of living cetaceans based on analyses of morphological and molecular datasets (Heyning 1997; Gatesy 1998; Messenger & McGuire 1998). The morphological characters for the nodes are modified from Heyning (1997): (1) Arterial blood supply to the brain via the intravertebral rete; tail terminating into flukes; no external hindlimbs; front limbs modified into flippers lacking movement except at the shoulder joint; sebaceous glands absent; sudoriferous glands absent; blubber present; rete mirabile present in the thoracic and basicranial region; rostrum elongate with mesorostral gutter; anterior palatine foramen greatly reduced; posterior, pterygoid and peribullar air sinuses present; supraspinatus fossa reduced; acromion and coracoid processes parallel and anteriorly directed; melon present; nasal plugs present; nasal plug muscles present; enlarged mandibular foramen; sigmoid process present on tympanic bulla; throat grooves present. (2) Reduced mandibular foramen; plate-like infraorbital process of maxillae present; baleen present; groove on mandibular symphysis present; mandibles not in contact at symphysis; tympanic membrane modified into glove finger. (3) Highly arched skull; rostrum narrow; baleen long and narrow, with fine fringe; supraorbital process of the frontal sweeps downward at a steep angle from the narial region; dorsal fin absent; coronoid process on the mandible extremely reduced; occipital shield attenuated and extends anterior to the supraorbital process. (4) Loss of one digit in manus; throat grooves present. (5) Ribs expanded and overlapping. (6) Complex rostral sutures. (7) Deep temporal fossa; numerous throat grooves. (8) Postanal gland. (9) Olfactory nerve lost; maxillae overlap frontal in supraorbital region; maxilla does not contribute to wall of orbit; single blowhole; hypertrophied maxolabialis muscles; hypertrophied melon; facial asymmetry; anterior process of periotic does not articulate with braincase; epicranial diverticula present off the nasal passages, including proximal sacs; middle sinus present; nasal bones not roofing the bony nares; arytenoid cartilage and epiglottis elongate; premaxillary foramen present. (10) Mastoid process greatly enlarged; fusion of jugal and lacrimal bones; supracranial basin present; loss of one nasal bone; spermaceti organ present (the spermaceti organ may be homologous with an extremely enlarged right posterior bursa found in other odontocetes); proximal sac modified into frontal sac; distal sac present; mandibular symphysis long. (11) Nasal passages confluent; blowhole ligament present; premaxillary sacs present; proximal sac modified into inferior vestibule/nasofrontal sac–posterior nasal sac complex. (12) Mastoid process greatly enlarged; pterygoid hamuli enlarged; apical mandibular teeth enlarged; vertex region of the skull that includes elevated premaxillae; one pair of anteriorly converging throat grooves; temporal fossa roofed over by expansion of the maxillae; loss of 50 base pair region near the 5′ end of mtDNA control region. (13) Throat grooves present; moderate reduction of the posterior process of the tympanic bulla; cranial hiatus present; preorbital and postorbital lobes present off the pterygoid air sinus. (14) Mandibular symphysis long; pneumatic maxillary crests present; pterygoids enlarged to cover palatines. (15) Vestibular sac present; extreme reduction of the posterior process of the tympanic bulla. (16) Right side of the vestibular sac hypertrophied; premaxillae displaced laterally, not in contact with the nasals. (17) Zygomatic process of squamosal moderately reduced. (18) Pterygoid on palate widely separated; facial plane flat or convex. (19) Extreme reduction of zygomatic process of squamosal. (20) Pterygoid on palate widely separated; premaxillary bosses present; dorsal intrusion of air sinus system along the frontal bone; floor of vestibular sac rigid and deeply folded; posterior edge of premaxillae reduced, not forming lateral edge of bony nares; teeth spatulate. (21) Posterior nasal sac absent; posterior end of left premaxilla reduced, not in contact with left nasal; bridle pigmentation pattern present on head; 160 base pair deletion in a satellite DNA repeat.

the bony nares and the cheekteeth (Whitmore & Sanders 1976). Several agorophorid genera—*Agorophius*, *Xenorophus* and *Archaeodelphis*—are known from the late Oligocene (Miller 1923). These taxa probably represent a paraphyletic grade of basal odontocetes and the family Agorophiidae requires revision.

Squalodontidae, or the shark-toothed dolphins, are differentiated from agorophiids by the more derived condition of skull telescoping (Whitmore & Sanders 1976). In squalodonts, the maxillae extend posteriorly over the parietals to reach the exoccipitals in adult specimens. This aptly named group is typified by relatively larger shark-like, sectorial cheekteeth with stoutly built skulls to accommodate large jaw muscles. Squalodonts were a rather common and widely distributed group that flourished during the late Oligocene through the Miocene. In 1840, J.P.S. Grateloup described a fossilized rostral fragment that he referred to as a large dinosaur related to *Iguanodon*. Thus, squalodonts began their taxonomic existence misidentified as a dinosaur, the same fate as archaeocetes. Squalodonts were clearly high-level carnivores in their time. Kellogg (1923), in his review of the group, mentioned that this family occurred in the North Atlantic (eastern North America and Europe), the South Atlantic (Patagonia) and the South Pacific (New Zealand), but not in North Pacific waters. Although, less abundant, squalodonts have also been recorded from the North Pacific (Barnes 1976).

Eurhinodelphids are relatively small, extremely long-snouted cetaceans of Miocene times (Kellogg 1928). The upper jaw often exceeded in length the lower jaw by a considerable amount. In this feature, eurinodelphids resemble the extinct ichthyosaurs of the genus *Eurhinosaurus* or the billfish of today.

The family Waipatiidae was recently described based on the beautifully preserved skull of *Waipatia maerewhenua* found in New Zealand (Fordyce 1994). In a cladistic analysis, this species fell into a clade with the Squalodontidae, Platanistidae and Squalodelphinidae. Waipatiids possess a moderately long rostrum and moderate polydonty (a number of teeth that exceeds the basic mammalian number). The maxillae are telescoped over the frontals but the parietals are partially exposed on the vertex. This group includes primitive characters such as heterodont dentition with double-rooted cheekteeth.

2.3.4.2 Modern odontocetes

Two recent phylogenetic analyses, one incorporating only morphological data (Heyning 1997) and another incorporating combined morphological and molecular data (Messenger & McGuire 1998), both provide a similar cladogram resulting in a classification of five superfamilies to include all extant families of odontocetes. These superfamilies are the Physeteroidea, Ziphioidea, Platanistoidea, Inioidea and Delphinoidea. This is the classification presented here (Fig. 2.12). An alternative cladogram is provided by Fordyce (1994) which differs in that it groups the sperm whales (Physeteroidae) with the beaked whales (Ziphiidae) based on morphological data, primarily the anatomy of the ear region.

Superfamily Physeteroidea
The Physeteroidea (Physeteridae and Kogiidae) are defined by several anatomical features (Heyning

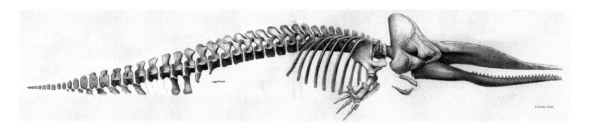

Fig. 2.13 Skeleton of a modern odontocete, the sperm whale (*Physeter macrocephalus*). (From van Beneden & Gervais 1880.)

1989, 1997) including two hallmark features: the presence of the supracranial basin and the spermaceti organ. All extant and fossil sperm whales have a supracranial basin that is formed by the raised facial border at the back of the skull that does not include a distinct and well-formed vertex formed from the the nasal bones and premaxillae (Fig. 2.13). From a dorsal view, this feature imparts an amphitheatre-like appearance to skulls of physeteroids. This structural arrangement is found only in physeteroids and is obviously highly derived. The spermaceti organ is found in all extant physeteroids and kogiids. It is not homologous to the melon (Heyning 1989) but is instead homologous to an extremely enlarged right posterior bursa found in other odontocetes based on structural and positional evidence (Cranford *et al.* 1996). Another notable feature, the elongate mandibular symphysis, is found in all species of the physeteroids and may be ancestral for the family. The reduced mandibular symphysis found in the kogiids may be a reversal related to the overall fore-shortening of the skull in this clade (Heyning 1997).

Family Physeteridae (sperm whales) Sperm whales are moderate to large odontocetes that are found worldwide. The single extant species, *Physeter macrocephalus*, is the largest of the odontocetes, fossil or living. The skull from an adult male can attain a length of 5 m. The next largest physterid is the Miocene species, *Ontocetus oxymycterus*, with a skull estimated to have been 4 m in length (Kellogg 1925). *Physeter* is a deep diving species that is rarely found in shallow waters. It has been recorded diving to a depth of at least 2000 m (Watkins *et al.* 1993), and can stay submerged for at least 90 min, but average dive times are 20–30 min. However, remains of the Miocene sperm whale *Aulophyseter* are not uncommon in the Sharktooth Hill Bonebed which was then a shallow inland sea. Fossil physeteroids are reviewed by Kellogg (1925).

A series of papers using mtDNA data suggested that sperm whales are more closely related to baleen whales than to other toothed whales (e.g. Milinkovitch *et al.* 1993, 1994, 1995). While receiving much publicity, this finding was incongruent with the morphological data (Heyning 1997), nuclear DNA data (Gatesy 1998) and analyses of combined datasets (Messenger & McGuire 1998). Cerchio and Tucker (1998) subsequently proposed that the phylogeny from Milinkovitch *et al.* (1994) was unstable due to a few regions of ambiguous sequence alignment.

Family Kogiidae (dwarf and pygmy sperm whales) The extant members of this clade are the two strikingly similar species of *Kogia*. These are robustly built small to medium-sized (2.7–4 m) cetaceans with distinctly short heads. This clade is defined by the following synapomorphies: (i) both nasals are lacking, and (ii) the presence of a distinct sagittal septum within the supracranial basin. Some classifications (e.g. Heyning 1997) retain dwarf sperm whales as a subfamily within the Physeteridae. However, both the long-term temporal separation and the degree of molecular divergence have been used as evidence for family level status for the Kogiidae (Rice 1998).

Superfamily Ziphioidea

Family Ziphiidae (beaked whales) The extant Ziphiidae include at least 20 species within five or six genera. There are numerous named fossil genera and species, often represented only by the secondarily dense, ossified rostra of adult males. Due to their rarity in museum collections, wide distribution with concurrent geographical variation, and often extreme sexual dimorphism, the phylogenetic relationships within this family are imperfectly understood. To date there has been no rigorous cladistic review of this family.

The Ziphiidae are defined primarily by the following synapomorphies (Fig. 2.12): an enlarged pterygoid hamulus on the bottom of the skull, an elevated vertex that includes elevated premaxillae, an enlarged pair of apical mandibular teeth, and a pair of throat grooves that converge anteriorly. The mitochondrial control region DNA sequence of this family has lost a 50 base pair region near the 5′-end (Dalebout *et al.* 1998). Most ziphiids exhibit an extreme reduction in dentition. Only the extant species, *Tasmacetus sheperdi*, Shepard's beaked whale, retains a full compliment of functional teeth within each toothrow. Vestigial teeth are sometimes found in other ziphiids, notably *Hyperoodon*, the bottlenose whale, and *Mesoplodon grayi*, Gray's beaked whale (Boschma 1950, 1951).

The genus *Mesoplodon* is the most species diverse genus of modern cetaceans with 13 named species. Moore (1968) removed the species *M. pacificus*, known from only two incomplete skulls, and placed it into the monotypic genus *Indopacetus*. Mead (1975a) suggested that the fossil genera *Choneziphius*, *Eboroziphius* and *Pelycorhamphus* were related to the modern species *Ziphius cavirostris* as all these genera share the derived feature of a prenarial basin. Mead also noted similarities between the fossil genera *Belemnoziphius* and *Proroziphius* to the modern genera *Mesoplodon* and *Hyperoodon*. Preliminary phylogenetic analyses based on the mtDNA control region suggest that some beaked whale genera and species may require revision (Dalebout *et al.* 1998).

Superfamily Platanistoidea (the 'river' dolphins)

The status of the superfamily Platanistoidea is controversial (Messenger 1994). In classic usage, this superfamily includes all the extant 'river' dolphins. The four modern genera of this taxon have been variously classified into one to four monotypic families. Virtually all phylogenetic analyses to date have indicated that *Platanista* is not closely related to the other extant river dolphins (de Muizon 1985; Heyning 1989, 1997; Messenger & McGuire 1998).

Family Platanistidae The modern members of this taxon include the river dolphins of the Indian subcontinent of the genus *Platanista*. These riverine cetaceans have an extremely long snout with teeth of the greatest length at the end of the snout. In this regard, *Platanista* resembles the crocodilian gavials that inhabit the same waters. *Platanista* is also called the blind river dolphin as it lacks a functional eye lens and therefore cannot discern images, only degrees of light and dark. As this dolphin inhabits waters that are consistently turbid, sight has limited value (see Chapter 1). Two synapomorphies unite this family (Heyning 1989, 1997).

Superfamily Inioidea

Family Iniidae The Iniidae, as defined here, are united by three distinct morphological synapomorphies (Fig. 2.12). In the classification used here,

there are three extant species of river dolphins, *Inia geoffrensis*, *Pontoporia blainvillei* and *Lipotes vexillifer* in the family Iniidae. Other workers consider there to be three families: the Iniidae, Pontoporiidae and Lipotidae (e.g. Rice 1998), although such taxonomic ranking does not preclude a sister taxa relationship of these families. The family Iniidae is currently placed in the superfamily, called Inioidea. In this classification scheme, the Inioidea is united with the Delphinoidea by five synapomorphies which provide good evidence that the two superfamilies share a common ancestry.

The late Miocene or Pliocene *Ischyrorhynchus* and *Saurodelphis* from fluvial deposits of South America are considered to be related to *Inia* (Cozzuol 1985). The only fossil taxon related to *Lipotes* is *Prolipotes*, based on a fragment of mandible recovered from freshwater Neogene deposits of China (Zhou 1984). The extinct long-snouted genus *Parapontoporia* is considered to be intermediate between *Pontoporia* and *Lipotes* by Barnes (1985a). Conversely, de Muizon (1988b) considered *Parapontoporia* to be more closely related to *Lipotes*. *Parapontoporia* is found in late Miocene to Pliocene marine deposits along the eastern Pacific.

Superfamily Delphinoidea

There is general consensus that the Delphinoidea are monophyletic. Four morphological synapomorphies define the Delphinoidea (Fig. 2.12). The superfamily is the most diverse group of cetaceans, and is relatively conservative in general morphology, making a clear phylogenetic analysis among its members difficult. The conservative pattern of facial anatomy did not allow Mead (1975a) to make substantial statements regarding the systematics of this group based on his data. Previous workers have listed the three currently recognized extant families, Monodontidae, Phocoenidae and Delphinidae. Barnes (1978, 1984) has separated out two families of fossil delphinoids, the Kentriodontidae and Albireonidae. Two characters, the reduction of the zygomatic process and the presence of a bridled pigmentation pattern, suggest that the Delphinidae and the Phocoenidae are sister taxa (Heyning 1997).

Family Monodontidae (beluga and narwhal) The two genera of modern monodontids, the beluga

Delphinapterus leucas, and the narwhal, *Monodon monoceros*, are strictly Arctic in distribution (see Chapter 1). Both are moderately sized delphinoids about 3–4 m in length that lack dorsal fins. The fossil monodontid *Denebola* lived in temperate to subtropical waters of the eastern Pacific (Barnes 1984; de Muizon 1988b). Barnes (1984) retained *Orcaella* in the Monodontidae; however, all subsequent analyses incorporating both morphological and molecular data have shown *Orcaella* to be a delphinid (Heyning 1989; Arnold & Heinsohn 1996). Monodontids were defined cladistically as delphinoids with facial planes that are flat or convex in profile.

Family Phocoenidae (porpoises) The Phocoenidae, or porpoises, include six extant species whose geographical range is almost worldwide (see Chapter 1). The vernacular name porpoise may be a bastardization of the name *porc poisson* for 'pig fish'. Porpoises are small (1.5–2.5 m) animals which lack the distinctive long 'beak' (rostrum) typical of most small delphinids. The oldest fossils are of the genera *Salumiphocaena* and *Australlithax* from the late Miocene of the eastern Pacific (Barnes 1985b; de Muizon 1988a). The Phocoenidae are united by numerous synapomorphies (Fig. 2.12) (Barnes 1985b). However, one character mentioned by Barnes, the flattened hard palate with the pterygoids separated broadly by the palatines, is also found in the species of the delphinid genera *Cephalorhynchus* (Flower 1883b) and *Orcaella*, as well as in both species of monodontids.

Porpoises have been separated into two subfamilies, the Phocaeninae and Phocaenoidinae, by Barnes (1985b). The species *Phocoena dioptrica* was transferred to its own genus, *Australophocaena*, and aligned with *Phocoenoides dalli* within the Phocaenoidinae based on cranial features. However, based on analyses of a portion of the mitochondrial genome, Rosel *et al.* (1995) found no support for these subfamily groupings with *A. dioptrica* clustering within the genus *Phocoena*.

Family Delphinidae (oceanic dolphins) The Delphinidae, or oceanic dolphins, are the most species-rich family of modern cetaceans and their definitive taxonomy is far from stable. There are at least 33 species, with several species complexes, such as *Tursiops*, awaiting revision that are likely to increase the named diversity. The family is worldwide in distribution and includes some populations of *Orcaella brevirostris* and *Sotalia fluviatilis* living in freshwater rivers. Delphinids range in size from about 1.5 m for the diminutive members of the genus *Cephalorhynchus* to almost 9 m for the killer whale, *Orcinus orca*. Delphinids have been found in rocks as old as the late Middle Miocene (Barnes 1976). The living species have been reviewed numerous times (Flower 1883b; True 1889; Fraser & Purves 1960; Kasuya 1973; Mead 1975b). These studies have resulted in a variety of proposed subfamilies. However, all are based on superficial phenetic similarities and modern cladistic analyses based on morphology have not yet been done. There is no comprehensive review of fossil delphinids although Barnes (1990) did review fossils assigned to the genus *Tursiops*.

The Delphinidae are defined by two morphological synapomorphies (Heyning 1997), and Árnason *et al.* (1984) presented evidence of a 160 base pair deletion in a satellite DNA repeat that is unique to delphinids. Recent analyses of mtDNA cyt *b* sequence data has provided results contrary to long-standing taxonomic dogma. For example, the genus *Lagenorhynchus* appears to be polyphyletic (Fig. 2.14) (Cipriano 1997; LeDuc *et al.* 1999). These results have focused welcome scrutiny onto this genus for which there are no morphological synapomorphies.

2.3.5 Suborder Mysticeti

The mysticetes, with their edentulous mouths lined with filtering baleen, are one of the distinct clades among the mammals. The evolutionary transition from capturing single prey to filtering numerous prey items out of mouthfuls of sea water has ramifications, not only in the morphology, but also on the ecology and behaviour of these, the largest of all animals. Mysticetes evolved from cetaceans that possessed teeth as this is the primitive condition among mammals. Certain cranial synapomorphies predate the loss of teeth in the mysticete clade and, therefore, the structurally most primitive mysticetes retain the ancestral condition of possessing teeth.

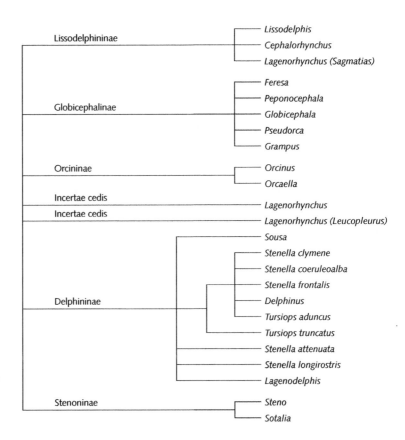

Fig. 2.14 Relationships among the genera of living Delphinidae based on analyses of cytochrome *b* (LeDuc *et al.* 1999). To date there are no comparable cladistic analyses of the morphological data. Note that *Sagmatias* and *Leucopleurus* are new generic classifications proposed by LeDuc *et al.* (1999).

The oldest known mysticete is the toothed species *Llanocetus denticrenatus* from the late Eocene of Seymour Island, Antarctica (Mitchell 1989). The next oldest specimens are those of the Oligocene cetotheres whose wide, edentulous palates strongly imply the presence of baleen. A large number of relatively well-preserved tooth-bearing fossil mysticetes have now been described (Barnes *et al.* 1994). These extinct mysticete taxa may or may not have had incipient baleen. As baleen is made of the protein keratin, it typically decomposes with the other soft tissues leaving no fossil trace with few noteworthy exceptions. These fossil discoveries now represent a moderately good morphological series from the archaeocetes to modern mysticetes.

2.3.5.1 Archaic mysticetes

Three families of extinct mysticetes are recognized. They are the Llanocetidae, Aetiocetidae and Ceto-

theriidae. The family Llanocetidae is based on one species, *Llanocetus denticrenatus*, from a study of a section of mandible containing two widely spaced teeth, and an associated endocast of a cranium from Antarctica described by Mitchell (1989). Judging from the massive size of the mandible, *Llanocetus* was a large cetacean. The complete skull, now under study, is about 2 m long with an estimated total length of perhaps 10 m (Fordyce 1989).

The Aetiocetidae are relatively small-toothed mysticetes known only from the shorelines of the North Pacific (Barnes *et al.* 1994). *Aetiocetus cotylaveus* was first classified within the extinct paraphyletic suborder Archaeoceti (Emlong 1966) because it possessed teeth. It was subsequently reclassified as a mysticete (Van Valen 1968) based on the derived pattern of skull bone telescoping. *Chonecetus* and some species of the genus *Aetiocetus* retain the primitive eutherian tooth count. The species *A. polydentatus* with its expanded tooth count (Barnes

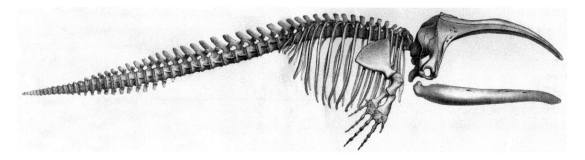

Fig. 2.15 Skeleton of a modern mysticete, the bowhead whale (*Balaena mysticetus*). (From van Beneden & Gervais 1880.)

et al. 1994) exhibits an incipient stage of the derived feature of polydonty.

The Cetotheriidae represent a phylogenetically heterogenous assemblage that is truly a 'wastebasket' taxon with over 60 named species within 30 or so genera of unknown affinities (Fordyce & Barnes 1994). The rostrum is typically broad and flat, not dissimilar to that found in the primitive aetiocetids and modern balaenopterids. The oldest cetotheres are *Cetotheriopsis lintianus* from Austria and the relatively complete skull of *Mauicetus lophocephalus* from New Zealand, both from the late Oligocene. Cetotheres are basically a taxon that accommodates all toothless mysticetes that lack any of the derived characters found in any of the extant families. They undoubtedly constitute a paraphyletic assemblage or perhaps even a polyphyletic group. Among the families of cetaceans, the cetotheres are arguably the most in need of revision.

2.3.5.2 Modern mysticetes

The modern mysticetes are divided into four families: the Balaenidae, the monotypic Neobalaenidae, the monotypic Eschrichtiidae and the Balaenopteridae. The systematics of these families are much less contentious than those for the modern odontocetes, though a few taxa remain elusive in their placement within and among other modern mysticetes.

Family Balaenidae (right whales)
Balaenids are heavy-bodied mysticetes with large heads and cavernous mouths to accommodate the extremely large filtering surface formed by the extra-

ordinary long baleen plates. Balaenids lack throat grooves. This clade is also the only extant group of mysticetes that retains the primitive tetrapod condition of having five digits on the forelimb. There are two species of modern balaenids, the bowhead (Fig. 2.15) and the right whale. These are often separated into two genera *Balaena* (the bowheads) and *Eubalaena* (the right whales). Although Rice (1998) suggests that morphological differences between the two genera are slight and do not warrant generic distinction, a study based on mtDNA control region sequences shows clear reciprocal monophyly between these two genera (Rosenbaum *et al.* 2000). Further, the genetic study indicates reciprocal monophyly and fixed nucleotide differences between *E. glacialis* in the North Atlantic, *E. galcialis* in the North Pacific and *E. australis* (see Chapter 11). The oldest fossil of an extant mysticete family is the balaenid *Morenocetus parvus* from the earliest Miocene of Argentina (McLeod *et al.* 1993).

Family Neobalaenidae (pygmy right whales)
This family is represented solely by the poorly known southern hemisphere pygmy right whale, *Caperea marginata*. Some workers have considered *Caperea* to be a balaenid. However, there is ever-growing consensus, based on morphology and molecular features, that *Caperea* is not an 'aberrant' right whale, but is instead more probably the sister group to the balaenopterid/eschrichtid clade (e.g. Árnason & Best 1991; Adegoke *et al.* 1993; Árnason & Gullberg 1994). The dorsal fin is small, yet distinctive. The rostrum is only somewhat arched, inter-

mediate to the conditions seen between gray whales (Eschrichtiidae) and right whales (Balaenidae). The ribs are unique among cetaceans, living or fossil, in that they are broad and overlap each other in profile (see Beddard 1901, Plate VII). The supraorbital shield is narrow and extends far forward of the orbit, anterior to the position seen in balaenids. The mandible of *Caperea* is rather stout for a baleen whale with a reduced coronoid process.

Family Eschrichtiidae (gray whales)

The enigmatic gray whale, *Eschrichtius robustus*, is the sole member of this family. Gray whales are characterized by their overall mottled grey colour and a very low and small dorsal fin followed by a series of dorsal ridges or knuckles. The two to five throat grooves are well delineated and are confined to the head region. The rostrum is attenuated and moderately arched. The yellowish to white baleen is relatively short and moderately wide. As found in balaenids and neobalaenids, the coronoid process on the mandible is extremely reduced. Gray whales have lost a digit in their hand.

The only fossil gray whale is a late Pleistocene individual of superb preservation (Barnes & McLeod 1984). However, this animal is undistinguishable from the modern species and its young geological age does not help resolve the relationship of gray whales to other baleen whales. Andrews (1914), Miller (1923) and Kellogg (1928) all suggested that gray whales are related to cetotheres based on characters that would now be considered primitive. Barnes and McLeod (1984) noted that this apparent similarity was no grounds to assume a close relationship.

Family Balaenopteridae (rorqual whales)

The Balaenopteridae, also known as the rorquals, are recognized by their numerous throat grooves. Balaenopterids are the 'greyhounds of the sea'. Their bodies are the sleekest among living mysticetes. The dorsal fin is always present and tends to be inversely proportional to body size. The distinctive throat grooves are numerous, ranging from 14–22 grooves in the humpback whale (*Megaptera novaengliae*) to 56–100 grooves in the fin whale (*Balaenoptera physalus*). The rostrum is extremely broad and flat. These throat grooves also extend posteriorly well beyond the gular region, all the way to the umbilicus in many species. The baleen is relatively short and wide and forms a continuous U-shape on the upper palate with the right and left racks uniting in the front by short tuft-like platelets of baleen.

Rorqual whales have a very complex interdigitating pattern of bony sutures between the rostral bones and those of the braincase proper. In transverse cross-section, the frontal bone dips steeply downward from the narial region and then flattens out forming a horizontal plane over the expanse of the supraorbital region. Only four digits are present in the flippers. The humpback is unique among balaenopterids in that it has extremely elongate flippers and has lost both the acromion and coronoid processes on the scapula.

2.3.5.3 Higher level mysticete systematics

There are three major cranial character suites that have been used to ascertain the relative degree of 'primitiveness' of fossil baleen-bearing whales. These are the position and shape of the occipital shield, the complexity of interdigitation between the bones of the rostrum (nasals, premaxillae and maxillae) and the braincase proper, and the slope angle of the supraorbital process of the frontal bone. Ancestrally, the occipital shield does not extend very far anteriorly, providing dorsal midline exposure of the parietals and frontals. The most primitive character state of this feature is seen on the various toothed mysticetes. In the most advanced character state, the occipital is in contact with the nasals and premaxillae on the vertex. This condition is found in modern balaenopterids, neobalaenids, but also in some cetotheres. The complex interdigitation of the bony sutures is clearly derived and is found in balaenopterids. Incipient interdigitation of the rostrum and braincase is seen in cetotheres and eschrichtiids. Modern balaenopterids possess a supraorbital process of the frontal bone that is flat and horizontal until it reaches the braincase and then abruptly turns dorsally to the skull vertex. The result is a large region over the supraorbital process for the greatly enlarged temporalis muscle. The cetothere *Cetotherium* has distinct crests along the temporal ridge along the contact with the frontals,

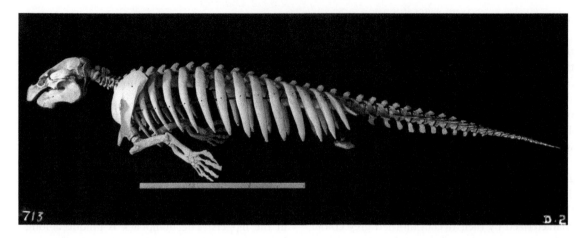

Fig. 2.16 Figure of a modern sirenian, the manatee (*Trichechus manatus*). (Courtesy of the Natural History Museum of Los Angeles County.)

which Kellogg (1928) suggests to be a condition that foreshadows the state found in modern balaenopterids.

Although differing somewhat in detail, most morphologically based phylogenies of baleen whales (Barnes & McLeod 1984; McLeod *et al.* 1993; Geisler & Luo 1996), as well as molecular-based studies (Árnason & Best 1991; Árnason *et al.* 1992; Árnason & Gullberg 1994; Milinkovitch *et al.* 1995), suggest that the balaenids were the first clade to diverge, followed by the Neobalaenidae, and then the Eschrichtiidae and Balaenopteridae as sister taxa.

Several studies using molecular sequence data (Árnason & Gullberg 1994; Milinkovitch *et al.* 1995) have implied that the ancestry of the gray whale (*Eschrichtiidae*) is located near or within the genus *Balaenoptera* (Balaenopteridae). This is a provocative hypothesis, yet there is little morphological evidence that truly conflicts with this assertion of relationship. Having an individual species within a clade evolve a radically different morphology due to different selection pressures is not unknown among organisms, with humans serving as a perfect example.

The evidence from the molecular data that the balaenids were the first clade to diverge among extant mysticetes may have some support in the morphological data. The remaining families (Neo-

balaenidae, Eschrichtiidae and Balaenopteridae) all share a few synapomorphies, such as the loss of one digit in the flipper and the more complicated pattern of rostral sutures, at least within the Eschrichtiidae and Balaenopteridae. A 'total evidence' study combining both morphological and sequence data from a limited number of mysticetes also suggests that balaenids diverged first followed by the eschrichtiid/balaenopterid clade (Messenger & McGuire 1998). Although providing no proof, the oldest fossils among the extant mysticetes are those of balaenids, which is in concordance with the above hypothesis.

2.4 ORDER SIRENIA

The sirenians, manatees, dugongs and their extinct relatives, are a fully aquatic herbivorous group of mammals. The living species are restricted to tropical and subtropical waters; however, fossil species appear to have inhabited temperate waters, and the range of the recently extinct Steller's sea cow extended into Arctic waters.

Sirenians are part of the ungulate clade called Tethytheria, which also includes the elephants and their extinct relatives, the extinct marine desmostylians, and in some classifications, the hyraxes. Most living groups have a unique form of tooth replace-

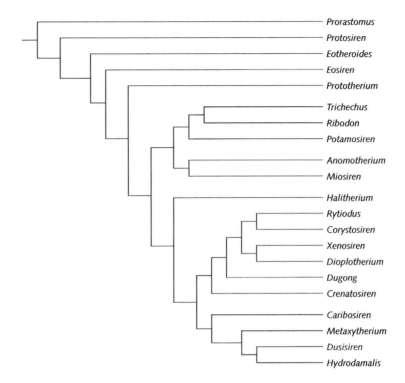

Fig. 2.17 Relationship among the living and fossil genera of sirenians based on morphological data (Domning 1994b).

ment in which new teeth originate at the back of the toothrow, then slowly move forward, and finally the worn down tooth drops out the front.

The oldest and most primitive sirenians are found in Eocene rocks; *Prorastomus* from the West Indies and *Protosiren* from Pakistan, North Africa and Europe. These two genera have been classified in the paraphyletic families Promastomidae and Protosirenidae, respectively. Sirenians of these two families had four limbs and were amphibious (Domning & Gingerich 1994; Domning 2000), but as with the cetaceans, the more modern species have but vestiges of the pelvic girdle (Fig. 2.16) and are propelled entirely by tail flukes. There are two living groups of sirenians: the manatees (family Trichechidae) and the dugong (Dugongidae). The fossil record is rich with taxa, which is not the case for the living species (Fig. 2.17).

There are but two species of dugonids: the dugong, *Dugong dugon*, of the Indo-Pacific and the recently extinct Steller's sea cow, *Hydrodamalis gigas*. This species was hunted to extinction in its last refugia of the Commander Islands in 1768, but

it occurred recently (19 000 years ago) as far south as Monterey (Jones 1967). Extinct relatives are widespread in shallow water deposits of the North Pacific (Domning 1994a). The oldest dugonids are found in the Caribbean and western Atlantic and then appeared to have spread to the Old World and into the Pacific prior to the formation of the Isthmus of Panama. Dugonids typically have a pair of short tusks in their upper jaw.

There are three closely related species of living manatees, the West Indian manatee, *Trichechus manatus*, the African manatee, *T. senegalensis*, and the exclusively freshwater Amazon manatee, *T. inunguis*. The fossil record of manatees is more obscure than that for dugonids. Some fossils of the Miocene age have been found along the Atlantic coasts of South America, North America and West Africa—similar to their modern distribution. There is some evidence that manatees have always been more coastal and riverine than their sea-going dugonids breatheren (Domning 1999). Manatees appear to have an almost unlimited ability to produce new teeth as the anterior teeth wear down.

REFERENCES

Adegoke, J.A., Arnason, U. & Widegren, B. (1993) Sequence organization and evolution, in all extant whalebone whales, of a DNA satellite with terminal chromosome localization. *Chromosoma* 102, 382–388.

Andersen, L.W., Born, E.W., Gjertz, I. *et al.* (1998) Population structure and gene flow of the Atlantic walrus (*Odobenus rosmarus rosmarus*) in the eastern Atlantic Arctic based on mitochondrial DNA and microsatellite variation. *Molecular Ecology* 7, 1323–1336.

Andrews, R.C. (1914) The California gray whale (*Rhachianectes glaucus* Cope): its history, habits, external anatomy, osteology and relationship. *Memoirs of the American Museum of Natural History, New Series 1. Part V. Monographs of the Pacific Cetacea*, pp. 227–287.

Árnason, Ú. (1969) The karyotype of the fin whale. *Hereditas* 62, 273–284.

Árnason, Ú. (1974a) Comparative chromosome studies in Pinnipedia. *Hereditas* 76, 179–226.

Árnason, Ú. (1974b) Comparative chromosome studies in Cetacea. *Hereditas* 77, 1–36.

Árnason, Ú. (1977) The relationship between the four principal pinniped karyotypes. *Hereditas* 87, 227–242.

Árnason, Ú. (1987) The evidence for a common ancestry of toothed and baleen whales based on studies of chromosomes and highly repetitive DNA. *La Kromosomo* II, 1479–1488.

Árnason, Ú. & Best, P.B. (1991) Phylogenetic relationships within the Mysticeti (whalebone whales) based upon studies of highly repetitive DNA in all extant species. *Hereditas* 114, 263–269.

Árnason, Ú. & Gullberg, A. (1994) Relationship of baleen whales established by cytochrome *b* gene sequence comparison. *Nature* 367, 726–728.

Árnason, Ú. & Widegren, B. (1986) Pinniped phylogeny enlightened by molecular hybridizations using highly repetitive DNA. *Molecular Biology and Evolution* 3, 356–365.

Árnason, Ú., Hoglund, M. & Widegren, B. (1984) Conservation of highly repetitive DNA in cetaceans. *Chromosoma* 89, 238–242.

Árnason, Ú., Grétarsdóttir, S. & Widegren, B. (1992) Mysticete (baleen whale) relationships based upon the sequence of the common cetacean DNA satellite. *Molecular Biology and Evolution* 9, 1018–1028.

Árnason, Ú., Bodin, K., Gullberg, A., Ledje, C. & Mouchaty, S. (1995) A molecular view of pinniped relationships with particular emphasis on the true seals. *Journal of Molecular Evolution* 40, 78–85.

Arnold, P. & Heinsohn, G. (1996) Phylogenetic status of the Irrawaddy dolphin *Orcaella brevirostris* (Owen, in Gray): a cladistic analysis. *Memoirs of the Queensland Museum* 39, 141–204.

Bajpai, S. & Thewissen, J.G.M. (1998) Middle Eocene cetaceans from the Harudi and Subathu Formations of India. In: *The Emergence of Whales: Evolutionary Patterns in the Origin of Cetacea* (J.G.M. Thewissen, ed.), pp. 213–233. Plenum Press, New York.

Barnes, L.G. (1976) Outline of eastern North Pacific cetacean assemblages. *Systematic Zoology* 25, 321–343. (Published in 1977.)

Barnes, L.G. (1978) A review of *Lophocetus* and *Liolithax* and their relationship to the delphinoid family Kentriodontidae (Cetacea: Odontoceti). *Science Bulletin, Natural History Museum of Los Angeles County* 28, 1–35.

Barnes, L.G. (1984) Fossil odontocetes (Mammalia: Cetacea) from the Almejas Formation, Isla Cedros, Mexico. *Paleobios* 42, 42 pp.

Barnes, L.G. (1985a) Fossil pontoporiid dolphins (Mammalia: Cetacea) from the Pacific coast of North America. *Contributions in Science, Natural History Museum of Los Angeles County* 363, 34 pp.

Barnes, L.G. (1985b) Evolution, taxonomy and antitropical distributions of the porpoises (Phocoenidae, Mammalia). *Marine Mammal Science* 1, 149–165.

Barnes, L.G. (1989) A new Enaliarctine pinniped from the Astoria Formation, Oregon, and a classification of the otariidae (Mammalia: Carnivora). *Contributions in Science, Natural History Museum of Los Angeles County* 403, 1–26.

Barnes, L.G. (1990) The fossil record and evolutionary relationships of the genus *Tursiops*. In: *The Bottlenose Dolphin* (R.R. Reeves & S. Leatherwood, eds), pp. 3–26. Academic Press, San Diego, CA.

Barnes, L.G. & McLeod, S.A. (1984) The fossil record and phyletic relationships of gray whales. In: *The Gray Whale Eschrichtius robustus* (M.L. Jones, S.L. Swartz & S. Leatherwood, eds), pp. 3–32. Academic Press, Orlando, Fl.

Barnes, L.G. & Mitchell, E.D. (1978) Cetacea. In: *Evolution of African Mammals* (V.J. Maglio & H.B.S. Cooke, eds), pp. 582–602. Harvard University Press, Cambridge, MA.

Barnes, L.G., Kimura, M., Furusawa, H. & Sawamura (1994) Classification and distribution of Oligocene Aetiocetidae (Mammalia; Cetacea; Mysticeti) from the western North America and Japan. *The Island Arc* 3, 392–431. (Published in 1995.)

Beddard, F.E. (1901) Contribution towards a knowledge of the osteology of the pygmy right whale (*Neobalaena marginata*). *Transactions of the Zoological Society of London* 16, 87–115.

Bernardi, G., Fain, S.R., Galloreynoso, J.P., Figueroacarranza, A.L. & LeBoeuf, B.J. (1998) Genetic variability in Guadalupe fur seals. *Journal of Heredity* 89, 301–305.

Berta, A. (1991) New *Enaliarctos** (Pinnipediamorpha) from the Oligocene and Miocene of Oregon and the role of 'Enaliarctids' in pinniped phylogeny. *Smithsonian Contributions to Paleobiology* No. 69. Smithsonian Institute Press, Washington, DC.

Berta, A. & Deméré, T.A. (1986) *Callorhinus gilmorei* n. sp., (Carnivora: Otariidae) from the San Diego Formation (Blancan) and its implications for otariid phylogeny.

Transactions of the San Diego Society of Natural Histology **21**, 111–126.

Berta, A. & Morgan, G.S. (1985) A new sea otter (Carnivora: Mustelidae) from the late Miocene and early Pliocene (Hemphillian) of North America. *Journal of Paleontology* **59**, 809–819.

Berta, A. & Wyss, A.R. (1994) Pinniped phylogeny. In: *Contributions in Marine Mammal Paleontology Honoring Frank C Whitmore Jr* (A. Berta & T. Deméré, eds), pp. 33–56. Proceedings of the San Diego Society of Natural History, Vol. 29.

Bininda-Emonds, O.P.R. & Russell, A.P. (1996) A morphological perspective on the phylogenetic relationships of the extant phocid seals (Mammalia: Carnivora: Phocidae). *Bonner Zoologische Monographien*, No. 41. Zoologisches Forschungsinstitut und Museum Alexander Koenig, Bonn.

Bonner, W.N. (1981) Grey seal—*Halichoerus grypus*. In: *Handbook of Marine Mammals* (S.H. Ridgeway & R.J. Harrison, eds), pp. 111–144. Academic Press, London.

Boschma, H. (1950) Maxillary teeth in specimens of *Hyperoodon rostratus* (Müller) and *Mesoplodon grayi* von Haast stranded on the Dutch Coast. *Koninklijke Nederlandse Akademie Van Wetenschappen* **53**, 775–786.

Boschma, H. (1951) Rows of small teeth in ziphoid whales. *Zoologische Mededelingen* **31** (14), 139–148.

Boyden, A. & Gemeroy, D. (1950) The relative position of the Cetacea among the orders of Mammalia as indicated by precipitin tests. *Zoologica* **35**, 145–151.

Burns, J.J. & Fay, F.H. (1970) Comparative morphology of the skull of the ribbon seal, *Histriophoca fasciata*, with remarks on systematics of Phocidae. *Journal of Zoology, London* **161**, 363–394.

Cerchio, S. & Tucker, P. (1998) Influence of alignment on the mtDNA phylogeny of Cetacea: questionable support for a Mysticeti? Physeteriodea clade. *Systematic Biology* **47**, 336–344.

Cipriano, F. (1997) Antitropical distributions and speciation in dolphins of the genus *Lagenorhynchus*: a preliminary analysis. In: *Molecular Genetics of Marine Mammals* (A.E. Dizon, S.J. Chivers & W.F. Perrin, eds), pp. 305–316. Special Publication No. 3. Society for Marine Mammalogy, Lawrence, KA.

Cooper, A. & Penny, D. (1997) Mass survival of birds across the Cretaceous–Tertiary boundary: molecular evidence. *Science* **275**, 1109–1113.

Costa, D.P. (1993) The relationship between reproductive and foraging energetics and the evolution of the pinnipedia. *Symposium of the Zoological Society of London* **66**, 293–314.

Cozzuol, M.A. (1985) The Odontoceti of the 'Mesopotamiense' of the Parana River ravines, systematic review. *Investigations on Cetacea* **17**, 39–51.

Cozzuol, M.A. (1992) The oldest seal in the southern hemisphere: implications to phocid phylogeny and dispersal. *Journal of Vertebrate Paleontology* **12**, 25A–26A.

Cranford, T.W., Amundin, M. & Norris, K.S. (1996) Functional morphology and homology in the odontocete nasal complex: implications for sound generation. *Journal of Morphology* **228**, 223–285.

Cronin, M.A., Hills, S., Born, E.W. & Patton, J.C. (1994) Mitochondrial DNA variation in Atlantic and Pacific walruses. *Canadian Journal of Zoology* **72**, 1035–1043.

Dalebout, M.L., van Helden, A., van Waerebeek, K. & Baker, C.S. (1998) Molecular genetic identification of southern hemisphere beaked whales (Cetacea: Ziphiidae). *Molecular Ecology* **7**, 687–694.

Davies, J.L. (1958) The pinnipedia: an essay in zoogeography. *Geographical Review* **48**, 474–493.

de Muizon, C. (1978) *Arctocephalus* (*Hydrarctos*) *lomasiensis*, subgen. Nov. et nov. sp., un nouvel Otariidae du Mio-Pliocene de Sacaco (Perou). *Institut Francais d'Etudes Andines Bulletin* **7** (3/4), 168–188.

de Muizon, C. (1982) Phocid phylogeny and dispersal. *Annals of the South African Museum* **89**, 175–213.

de Muizon, C. (1985) Nouvelles donnees sur le diphyletisme des dauphins de riviere (Odontoceti, Cetacea, Mammalia). *Comptes Rendus l'Academie Des Sciences, Series 2* **301**, 359–362.

de Muizon, C. (1988a) Le polyphyletisme des Acrodelphinidae, Odontocetes longirostres du Miocene Europeen. *Bulletin d'Museum Nationale d'Histoire Naturelle, Paris, Series 4* **10C**, 31–88.

de Muizon, C. (1988b) Les relations phylogenetiques des Delphinida (Cetacea, Mammalia). *Annales de Paleontologie* **74**, 159–227.

Deméré, T.A. (1994) The family Odobenidae: a phylogenetic analysis of fossil and living taxa. *Proceedings of the San Diego Society of Natural History* **29**, 99–123.

Domning, D.P. (1994a) Summary of taxa and distribution of Sirenia in the North Pacific Ocean. *Island Arc* **3**, 506–512. (Published in 1995.)

Domning, D.P. (1994b) A phylogenetic analysis of the Sirenia. *Contributions in Marine Mammal Paleontology Honoring Frank C. Whitmore Jr* (A. Berta & T. Deméré, eds), pp. 177–189. Proceedings of the San Diego Society of Natural History Vol. 29.

Domning, D.P. (1999) Fossils explained 24: Sirenians (Seacows). *Geology Today* **March–April**, 75–79.

Domning, D.P. (2000) The readaptation of Eocene Sirenians to life in water. *Historical Biology* **14**, 115–119.

Domning, D.P. & Gingerich, P.D. (1994) *Protosiren smithae*, new species (Mammalia, Sirenia), from the late middle Eocene of Wadi Hitan, Egypt. *Contributionns from the Museum of Paleontology, the University of Michigan* **29** (3), 69–87.

Donoghue, M.J., Doyle, J.A., Gauthier, J., Kluge, A.G. & Rowe, T. (1989) The importance of fossils in phylogenetic reconstruction. *Annual Review of Ecological Systematics* **20**, 431–460.

Duffield-Kulu, D. (1972) Evolution and cytogenetics. In: *Mammals of the Sea, Biology and Medicine* (S.H. Ridgway, ed), pp. 503–527. Charles C. Thomas, Springfield, IL.

Emlong, D. (1966) A new archaic cetacean from the Oligocene of northwest Oregon. *Museum of Natural History, University of Oregon Bulletin* **3**, 51 pp.

Flower, W.H. (1869) On the value of the characters of the base of the cranium in the classification of the order Carnivora, and on the systematic position of *Bassaris* and other disputed forms. *Proceedings of the Zoological Society of London* **1869**, 4–37.

Flower, W.H. (1883a) On whales, past and present, and their probable origin. *Nature* **28**, 226–230.

Flower, W.H. (1883b) On the characters and divisions of the family Delphinidae. *Proceedings of the Zoological Society of London* **1883**, 466–513.

Fordyce, R.E. (1980) Whale evolution and the Oligocene Southern Ocean environment. *Palaeogeography, Palaeoclimatology, Palaeoecology* **31**, 319–336.

Fordyce, R.E. (1989) Origins and evolution of Antarctic marine mammals. In: *Origins and Evolution of Antarctic Biota* (J.A. Crame, ed.), pp. 269–281. Geological Society Special Publication No. 47.

Fordyce, R.E. (1994) *Waipatia maerewhenua*, new genus and new species (Waipatiidae, new Family), an archaic Late Oligocene dolphin (Cetacea: Odontoceti: Platanistoidea) from New Zealand. In: *Contributions in Marine Mammal Paleontology Honoring Frank C. Whitmore Jr* (A. Berta & T. Deméré, eds), pp. 147–176. Proceedings of the San Diego Society of Natural History.

Fordyce, R.E. & Barnes, L.G. (1994) The evolutionary history of whales and dolphins. *Annual Review of Earth and Planetary Sciences* **22**, 419–455.

Fraser, F.C. & Purves, P.E. (1960) Hearing in cetaceans: evolution of the accessory air sacs and structure of the outer and middle ear in Recent cetaceans. *Bulletin of the British Museum (Natural History), Zoology* **7**, 1–140.

Gatesy, J. (1997) More DNA support for a cetacea/ Hippopotimidae clade: the blood-clotting protein gene γ-fibrinogen. *Molecular Biology and Evolution* **14**, 537–543.

Gatesy, J. (1998) Molecular evidence for the phylogenetic affinities of Cetacea. In: *The Emergence of Whales: Evolutionary Patterns in the Origin of Cetacea* (J.G.M. Thewissen, ed.), pp. 63–111. Plenum Press, New York.

Gauthier, J., Kluge, A.G. & Rowe, T. (1988) Amniote phylogeny and the importance of fossils. *Cladistics* **4**, 105–209.

Geisler, J.H. & Luo, Z. (1996) The petrosal and inner ear of *Herpetocetus* sp. (Mammalia: Cetacea) and their implications for the phylogeny and hearing of archaic mysticetes. *Journal of Paleontology* **70**, 1045–1066.

Gingerich, P.D. & Russell, D.E. (1981) *Pakicetus inachus*, a new archaeocete (Mammalia: Cetacea) from the early–Middle Eocene Kuldana Formation of Kohot (Pakistan). *Contributions from the Museum of Paleontology, University of Michigan* **25** (11), 235–246.

Gingerich, P.D., Smith, B.H. & Simons, E.L. (1990) Hind limbs of Eocene *Basilosaurus*: evidence for feet in whales. *Science* **249**, 154–157.

Gingerich, P.D., ul Haq, M., Zalmout, I.S., Khan, I.H. & Malkani, M.S. (2001) Origins of whales from early artiodactyls: hands and feet of Eocene Protocetidae from Pakistan. *Science* **293**, 2239–2242.

Heyning, J.E. (1989) Comparative facial anatomy of beaked whales (Ziphiidae) and a systematic revision among the families of extant Odontoceti. *Contributions in Science, Natural History Museum of Los Angeles County* No. 405, 64 pp.

Heyning, J.E. (1997) Sperm whale phylogeny revisted: analysis of the morphological evidence. *Marine Mammal Science* **13**, 596–613.

Heyning, J.E. (1999) Whale origins—response. *Science* **283**, 1642–1643.

Heyning, J.E. & Mead, J.G. (1990) Evolution of the nasal anatomy of cetaceans. In: *Sensory Abilities of Cetaceans: Laboratory and Field Evidence* (J. Thomas & R. Kastelein, eds), pp. 67–79. Plenum Press, London.

Heyning, J.E. & Thacker, C. (1999) Phylogenies, temporal data, and negative evidence. *Science* **285**, 1179A.

Hochachka, P.W. (1997–1998) Is evolutionary physiology useful to mechanistic physiology? The diving response in pinnipeds as a test case. *Zoology (Jena)* **100**, 328–335.

Honeycutt, R.L. & Adkins, R.M. (1993) Higher level systematics of eutherian mammals: an assessment of molecular characters and phylogenetic hypotheses. *Annual Review of Ecology and Systematics* **24**, 279–305.

Howell, A.B. (1929) Contribution to the comparative anatomy of the eared and earless seals (genera Zalophus and Phoca). *Proceedings of the United States Museum* **73**, 1–142.

Huelsenbeck, J.P. (1991) When are fossils better than extant taxa in phylogenetic analysis? *Systematic Zoology* **40**, 458–469.

Huelsenbeck, J.P., Bull, J.J. & Cunningham, C.W. (1996) Combining data in phylogenetic analysis. *Trends in Ecology and Evolution* **11**, 152–158.

Hulbert, R.C. Jr (1998) Postcranial osteology of the North American Middle Eocene Protocetid *Georgiacetus*. In: *The Emergence of Whales: Evolutionary Patterns in the Origin of Cetacea* (J.G.M. Thewissen, ed.), pp. 235–267. Plenum Press, New York.

Irwin, D.M., Kocher, T.D. & Wilson, A.C. (1991) Evolution of the cytochrome *b* gene of mammals. *Journal of Moecular Evolution* **32**, 128–144.

Irwin, D.M. & Árnason, Ú. (1994) Cytochrome b gene of marine mammals: phylogeny and evolution. *Journal of Mammal Evolution* **2**, 37–55.

Itoo, T. (1985) New cranial materials of the Japanese sea lion, *Zalophus californianus japonicus* (Peters, 1866). *Journal of the Mammalian Society of Japan* **10**, 135–148.

Janczewski, D.N., Yuhki, N., Gilbert, D.A., Jefferson, G.T. & O'Brien, S.J. (1992) Molecular phylogenetic inference from saber-toothed cat fossils of Rancho La Brea. *Proceedings of the National Acadamy of Sciences* **89**, 9769–9773.

Jones, R.E. (1967) A *Hydrodamalis* skull fragment from Monterey Bay, California. *Journal of Mammalogy* **48**, 143–144.

Kastelein, R.A., Muller, M. & Terlouw, A. (1994) Oral suction of a Pacific walrus (*Odobenus rosmarus divergens*) in air and under water. *Zeitschrift Fur Saugetierkunde* 59, 105–115.

Kasuya, T. (1973) Systematic considerations of recent toothed whales based on the morphology of the typano-periotic bone. *Scientific Reports of the Whales Research Institute* 25, 1–103.

Kellogg, R. (1923) Descriptions of two squalodonts recently discovered in the Calvert Cliffs, Maryland; and some notes on the shark-toothed cetaceans. *Proceedings of the US National Museum* 62, 1–69.

Kellogg, R. (1925) Two fossil physeterid whales from California. In: *Additions to the Tertiary History of the Pelagic Mammals on the Pacific Coast of North America*, pp. 1–34. Carnegie Institution of Washington, Washington, DC.

Kellogg, R. (1928) The history of whales—their adaptation to life in the water. *Quarterly Review of Biology* 3 (1), 29–76; concluded 3 (2), 174–208.

Kellogg, R. (1936) A review of the Archaeoceti. *Carnegie Institution of Washington Publication* No. 482.

Kenyon, K.W. (1977) Caribbean monk seal extinct. *Journal of Mammalogy* 58, 97–98.

King, J.E. (1972) Observations on phocid skulls. In: *Functional Anatomy of Marine* Mammals (R.J. Harrison, ed.), Vol. 1, pp. 81–115. Academic Press, London.

King, J.E. (1983a) *Seals of the World*, 2nd edn. Cornell University Press, Ithaca, NY.

King, J.E. (1983b) The Ohope skull—a new species of Pleistocene sea lion from New Zealand. *New Zealand Journal of Marine and Freshwater Research* 17, 105–120.

Kluge, A.G. (1989) A concern for evidence and a phylogenetic hypothesis of relationships among *Epicrates* (Boidae, Serpentes). *Systematic Zoology* 38, 7–25.

Kluge, A.G. & Wolf, A.J. (1993) Cladistics: what's in a word. *Cladistics* 9, 183–199.

Kohno, N. (1993) Approaches to reconstruct Japanese sea lions (4), phylogeny of the Otariidae, 1: Phylogenetic relationships of the extinct otariids to the Recent sea lions. *Aquabiology* 15, 415–420.

Kohno, N. (in press) Evolution of feeding innovations in the walruses. *Palaeobiology*, in press.

Kohno, N., Tomida, Y., Hasegawa, Y. & Furusawa, H. (1995) Pliocene tusked odobenids (Mammalia: Carnivora) in the Western Pacific, and their paleobiogeography. *Bulletin of the National Science Museum, Tokyo Series C* 21, 111–131.

Koretsky, I.A. & Sanders, A.E. (1997) Pinniped bones from the late Oligocene of South Carolina: the oldest true seal (Carnivora, Phocidae). *Journal of Vertebrate Paleontology* 17 (Suppl. 3), 58.

Kumar, K. & Sahni, A. (1986) *Remingtonocetus harudiensis*, new combination, a middle Eocene archaeocete (Mammalia: Cetacea) from western Kutch, India. *Journal of Vertebrate Paleontology* 6, 326–349.

Kurten, B. (1964) The evolution of the polar bear, *Ursus maritimus* Phipps. *Acta Zoologica Fennica* 108, 1–26.

Lambert, W.D. (1997) The osteology and paleoecology of the giant otter *Enhydritherium terraenovae*. *Journal of Vertebrate Paleontology* 17, 738–749.

Lawrence, B. & Schevill, W.E. (1956) Functional anatomy of the delphinid nose. *Bulletin of the Museum of Comparative Zoology* 114, 103–151.

Ledje, C. & Árnason, Ú. (1996a) Phylogenetic relationships within caniform carnivores based on analyses of the mitochondrial 12S rRNA gene. *Journal of Molecular Evolution* 43, 641–649.

Ledje, C. & Árnason, Ú. (1996b) Phylogenetic analyses of complete cytochrome b genes of the order Carnivora with particular emphasis on the Caniformia. *Journal of Molecular Evolution* 42, 135–144.

LeDuc, R.G., Perrin, W.F. & Dizon, A.E. (1999) Phylogenetic relationships among the delphinid cetaceans based on full cytochrome *b* sequences. *Marine Mammal Science* 15 (3), 619–648.

Lento, G.M. (1995) *Molecular Systematic and Population Genetic Studies of Pinnipeds: a Phylogeny of Our Fin-Footed Friends and Their Surreptitious 'Species' Status* PhD Thesis. Victoria University, Wellington, New Zealand.

Lento, G.M., Haddon, M., Chambers, G.K. & Baker, C.S. (1997) Genetic variation, population structure, and species identity of southern hemisphere fur seals, *Arctocephalus* spp. *Journal of Heredity* 88, 202–208.

Lento, G.M., Hickson, R.E., Chambers, G.K. & Penny, D. (1995) Use of spectral analysis to test hypotheses on the origin of pinnipeds. *Molecular Biology and Evolution* 12, 28–52.

McKenna, M. (1975) Toward a phylogenetic classification of the Mammalia. In: *Phylogeny of the Primates* (W.P. Luckett & F.S. Szalay, eds), pp. 21–46. Plenum Press, New York.

McKenna, M.C. & Bell, S.K. (1997) *Classification of Mammals Above the Species*. Columbia University Press, New York.

McLaren, I.A. (1960) Are the pinnipedia biphyletic? *Systematic Zoology* 9, 18–28.

McLaren, I.A. (1975) A speculative overview of phocid evolution. *Rapports et Proces-Verbaux des Réunions, Conseil International pur l'Exploration de la Mer* 169, 43–48.

McLeod, S.A., Whitmore, F.C. & Barnes, L.G. (1993) Evolutionary relationships and classification. In: *The Bowhead Whale* (J.J. Burns, J.J. Montague & C.J. Cowles, eds), pp. 45–70. The Society for Marine Mammalogy Special Publication No. 2. Society for Marine Mammalogy, Lawrence, KA.

Maldonado, J.E., Davila, F.O., Stewart, B.S., Geffen, E. & Wayne, R.K. (1995) Intraspecific genetic differentiation in California sea lions (*Zalphous californianus*) from southern California and the Gulf of California. *Marine Mammal Science* 11, 46–58.

Mead, J.G. (1975a) A fossil beaked whale (Cetacea: Ziphiidae) from the Miocene of Kenya. *Journal of Paleontology* 49, 745–751.

Mead, J.G. (1975b) Anatomy of the external nasal passages and facial complex in the Delphinidae (Mammalia: Cetacea). *Smithsonian Contributions to Zoology* **207**, 72 pp.

Messenger, S.L. (1994) Phylogenetic relationships of Platanistoid river dolphins: assessing the significance of fossil taxa. In: *Contributions in Marine Mammal Paleontology Honoring Frank C Whitmore Jr* (A. Berta & T. Deméré, eds), pp. 125–134. Proceedings of the San Diego Society of Natural History Vol. 29.

Messenger, S.L. & McGuire, J.A. (1998) Morphology, molecules, and the phylogenetics of cetaceans. *Systematic Biology* **47**, 90–124.

Milinkovitch, M.C. (1992) DNA–DNA hybridizations support ungulate ancestry of Cetacea. *Journal of Evolutionary Biology* **5**, 149–160.

Milinkovitch, M.C., Guillermo, O. & Meyer, A. (1993) Revised phylogeny of whales suggested by mitochondrial ribosomal DNA sequences. *Nature* **361**, 346–348.

Milinkovitch, M.C., Meyer, A. & Powell, J.R. (1994) Phylogeny of all major groups of cetaceans based on DNA sequence from three mitochondrial genes. *Molecular Biology and Evolution* **11**, 939–948.

Milinkovitch, M.C., Ortí, G. & Meyer, A. (1995) Novel phylogeny of whales revisited but not revised. *Molecular Biology and Evolution* **12** (3), 518–520.

Miller Jr, G.S. (1923) The telescoping of the cetacean skull. *Smithsonian Miscellaneous Collections* **76**, 1–71.

Mitchell, E.D. (1975) Parallelism and convergence in the evolution of the otariidae and phocidae. *Rapports Proces-Verbaux des Reunions, Conseil International pur l'Exploration de la Mer* **169**, 12–26.

Mitchell, E.D. (1989) A new cetacean from the Late Eocene La Meseta Formation, Seymour Island, Antarctic Peninsula. *Canadian Journal of Fisheries and Aquatic Sciences* **46**, 2219–2235.

Mivart, St.G. (1885) Notes on the pinnipedia. *Proceedings of the Zoological Society of London* **1885**, 484–500.

Miyamoto, M.M. & Goodman, M. (1986) Biomolecular systematics of eutherian mammals: phylogenetic patterns and classification. *Systematic Zoology* **35**, 230–240.

Miyazaki, S., Horikawa, H., Kohno, N. *et al.* (1994) Summary of the fossil record of pinnipeds of Japan, and comparisons with that from the eastern North Pacific. *Island Arc* **3**, 361–372.

Moore, J.C. (1968) Relationships among the living genera of beaked whales with classifications, diagnoses and keys. *Fieldana: Zoology* **53** (4), 209–298.

Moritz, C. & Hillis, D.M. (1996) Molecular systematics: context and controversies. In: *Molecular Systematics*, 2nd edn (D.M. Hillis, C. Moritz & B.K. Marble, eds), pp. 1–13. Sinauer Associates, Inc., Sunderland, MA.

Nelson, C.H. & Johnson, K.R. (1987) Whales and walruses as tillers of the sea floor. *Scientific American* **256**, 112–117.

Nikaido, M., Rooney, A.P. & Okada, N. (1999) Phylogenetic relationships among interspersed elements: hippopotamuses are the closest living relatives to whales. *Proceedings of the National Academy of Sciences* **96**, 10261–10266.

Nojima, T. (1990) A morphological consideration of the relationships of pinnipeds to other carnivorans based on the bony tentorium and bony falx. *Marine Mammal Science* **6**, 54–74.

Novacek, M.J. (1992) Fossils as critical data for phylogeny. In: *Phylogeny and Extinction* (M.J. Novacek & Q.D. Wheeler, eds), pp. 46–88. Columbia University Press, New York.

O'Leary, M.A. (1999) Parsimony analysis of total evidence from extinct and extant taxa and the cetacean–artiodactyl question. *Cladistics* **15**, 315–330.

O'Leary, M.A. & Geisler, J.H. (1999) The position of Cetacea within Mammalia: phylogenetic analysis of morphological data from extinct and extant taxa. *Systematic Biology* **48**, 455–490.

O'Leary, M.A. & Rose, K.D. (1995) Postcranial skeleton of the early Eocene Mesonychid *Pachyaena* (Mammalia, Mesonychia). *Journal of Vertebrate Paleontology* **15**, 401–430.

Oliver, J.S., Slattery, P.N., O'Connor, E.F. & Lowry, L.F. (1983) Walrus, *Odobenus rosmarus*, feeding in the Bering Sea: a benthic perspective. *Fishery Bulletin* **81**, 501–512.

Pamilo, P. & Nei, M. (1988) Relationships between gene trees and species trees. *Molecular Biology and Evolution* **5**, 568–583.

Randi, E., Lucchini, V. & Diong, C.H. (1996) Evolutionary genetics of the Suiformes using mtDNA. *Journal of Mammalian Evolution* **3**, 163–194.

Ray, C.E. (1976) Geography of phocid evolution. *Systematic Zoology* **25**, 391–406. (Published in 1977.)

Repenning, C.A. & Ray, C.E. (1972) On the origin of the Hawaiian monk seal. *Proceedings of the Biological Society of Washington* **89**, 667–688.

Repenning, C.A., Peterson, R.S. & Hubbs, C.L. (1971) Contributions to the systematics of the southern fur seals, with particular reference to the Juan Fernandez and Guadalupe species. In: *Antarctic Pinnipedia* (W.H. Burt, ed.), pp. 1–34. Antarctic Research Series Vol. 18. American Geophysical Union, Washington DC.

Repenning, C.A., Ray, C.E. & Grigorescu, D. (1979) Pinniped biogeography. In: *Historical Biogeography, Plate Tectonics, and the Changing Environment* (J. Gray & J. Boucot, eds), pp. 357–369. State University Press, Corvalis, OR.

Rice, D.W. (1998) *Marine Mammals of The World: Systematics and Distribution*. Special Publication No. 4. Society for Marine Mammalogy, Lawrence, KA.

Rosel, P.E., Hatgood, M.G. & Perrin, W.F. (1995) Phylogenetic relationships among the true porpoises (Cetacea: Phocoenidae). *Molecular Phylogenetics and Evolution* **4**, 463–474.

Rosenbaum, H.C., Brownell, J.R., Brown, M.W. *et al.* (2000) World-wide genetic differentiation of *Eubalaena*: questioning the number of right whale species. *Molecular Ecology* **9**, 1793–1802.

Sarich, V.M. (1969a) Pinniped origins and the rate of evolution of carnivore albumins. *Systematic Zoology* **18**, 286–295.

Sarich, V.M. (1969b) Pinniped phylogeny. *Systematic Zoology* **18**, 416–422.

Scheffer, V.B. (1958) *Seal, Sea Lions, and Walruses: a Review of the Pinnipedia.* Stanford University Press, Stanford.

Shimamura, M., Yasue, H., Ohshima, K. *et al.* (1997) Molecular evidence from retroposons that whales form a clade within even toed ungulates. *Nature* **388**, 666–670.

Smith, A.B. & Littlewood, D.T.J. (1994) Paleontological data and molecular phylogenetic analysis. *Paleobiology* **20**, 259–273.

Springer, M.S. & Kirsch, J.A.W. (1993) A molecular perspective on the phylogeny of placental mammals based on mitochondrial 12s rDNA sequences, with a special reference to the problem of the paenungulata. *Journal of Mammalian Evolution* **1**, 149–166.

Stanley, H.F., Casey, S., Carnahan, J.M. *et al.* (1996) Worldwide patterns of mitochondrial DNA differentiation in the harbor seal (*Phoca vitluna*). *Molecular Biology and Evolution* **13**, 368–382.

Tedford, R.H. (1976) Relationship of pinnipeds to other carnivora (Mammals). *Systematic Zoology* **25**, 363–374. (Published in 1977.)

Thewissen, J.G.M. (1998a) Cetacean origins: evolutionary turmoil during the invasion of the oceans. In: *The Emergence of Whales: Evolutionary Patterns in the Origin of Cetacea* (J.G.M. Thewissen, ed.), pp. 451–464. Plenum Press, New York.

Thewissen, J.G.M., ed. (1998b) *The Emergence of Whales: Evolutionary Patterns in the Origin of Cetacea.* Plenum Press, New York.

Thewissen, J.G.M. & Fish, F.E. (1997) Locomotor evolution in the earliest cetaceans: functional model, modern analogues, and paleontological evidence. *Paleobiology* **23**, 482–490.

Thewissen, J.G.M. & Hussain, S.T. (1998) Systematic review of the Packicetidae, Early and Middle Eocene cetacea (Mammalia) from Pakistan and India. *Bulletin of the Carnegie Museum of Natural History* **34**, 220–238.

Thewissen, J.G.M., Madar, S.I. & Hussain, S.T. (1996a) *Amblocetus natans*, an Eocene cetacean (Mammalia) from Pakistan. *Courrier Forschungsinstitut Senckenberg* **191**, 1–86.

Thewissen, J.G.M., Roe, L.J., O'Nell, J.R. *et al.* (1996b) Evolution of cetacean osmoregulation. *Nature* **381**, 379–380.

Thewissen, J.G.M., Williams, E.M., Roes, L.J. & Hussain, S.T. (2001) Skeletons of terrestrial cetaceans and the relationship of whales to artiodactyls. *Nature* **413**, 277–281.

True, F.W. (1889) Contributions to the natural history of the cetaceans, a review of the family Delphinidae. *Bulletin of the United States National Museum* **36**, 191 pp.

Uhen, M.D. (1998) Middle to Late Eocene Basilosaurines and Dorudontines. In: *The Emergence of Whales: Evolutionary Patterns in the Origin of Cetacea* (J.G.M. Thewissen, ed.), pp. 29–62. Plenum Press, New York.

Uhen, M.D. (1999) New species of protocetid archaeocete whale, *Eocetus wardii*, (Mammalia: Cetacea) from the Middle Eocene of North Carolina. *Journal of Paleontology* **73**, 512–528.

van Beneden, P.J. & Gervais, P. (1880) *Ostéographie des Cétacés Vivant et Fossil Comprenant la Description et l'Iconographie Squelette et du Systeme Dentaire de ces Animaux Ainsi que des Documents Relatifs a leur Histoire Naturelle.* Arthus Bertrand, Paris.

Van Valen, L. (1966) Deltatheria, a new order of Mammals. *Bulletin of the American Museum of Natural History* **132** (1), 1–126.

Van Valen, L. (1968) Monophyly or diphyly in the origin of whales. *Evolution* **22**, 37–41.

Vrana, P.B., Milinkovitich, M.C., Powell, J.R. & Wheeler, W.C. (1994) Higher level relationships of the arctoid carnivora based on sequence data and 'total evidence'. *Molecular Phylogenetics and Evolution* **3**, 47–58.

Watkins, W.A., Daher, M.A., Fristrup, K.M. & Howald, T.J. (1993) Sperm whales tagged with transponders and tracked underwater by sonar. *Marine Mammal Science* **9** (1), 55–67.

Wayne, R.K., Van Valkenburgh, B., Kat, P.W. *et al.* (1989) Genetic and morphological divergence among sympatric canids. *Journal of Heredity* **80**, 447–454.

Wayne, R.K., Van Valkenburgh, B. & O'Brien, S.J. (1991) Molecular distance and divergence time in carnivores and primates. *Molecular Biology and Evolution* **8**, 297–319.

Whitmore, F.C. & Sanders, A.E. (1976) Review of Oligocene Cetacea. *Systematic Zoology* **25**, 304–320. (Published in 1977.)

Wiley, E.O. (1981) *Phylogenetics: the Theory and Practice of Phylogenetic Systematics.* John Wiley & Sons, New York.

Wiley, E.O., Siegel-Causey, D., Brooks, D.R. & Funk, V.A. (1991) *The Complete Cladist.* University of Kansas Museum of Natural History, Special Publication No. 19.

Wilkinson, M. (1995) Coping with abundant missing entries in phylogenetic inference using parsimony. *Systematic Biology* **44**, 501–514.

Williams, E.M. (1998) Synopsis of the earliest cetaceans: Pakicetidae, Ambulocetidae, Remingtonocetidae, and Protocetidae. In: *The Emergence of Whales: Evolutionary Patterns in the Origin of Cetacea* (J.G.M. Thewissen, ed.), pp. 1–28. Plenum Press, New York.

Wozencraft, C. (1989) The phylogeny of the recent carnivora. In: *Carnivore Behavior, Ecology, and Evolution* (J. L. Gittleman, ed.), pp. 495–535. Cornell University Press, Ithaca, NY.

Wyss, A.R. (1987) The walrus auditory region and the monophyly of pinnipeds. *American Museum Novitates* **2871**, 1–13.

Wyss, A.R. (1988) Evidence from flipper structure for a single origin of pinnipeds. *Nature* **334**, 427–428.

Wyss, A.R. (1989) Flippers and pinniped phylogeny: has the problem of convergence been overrated? *Marine Mammal Science* **5**, 343–360.

Wyss, A.R. & Flynn, J.J. (1993) A phylogenetic analysis and definition of the Carnivora. In: *Mammal Phylogeny* (F.S. Szalay, M.J. Novacek & M.C. McKenna, eds), pp. 32–52. Springer-Verlag, New York, NY.

Zhou, K. (1984) Classification and phylogeny of the superorder Platanistoidea, with notes on evidence of the monophyly of the Cetacea. *Scientific Reports of the Whales Research Institute* **34**, 93–108.

CHAPTER 3

Anatomy and Physiology: the Challenge of Aquatic Living

Terrie M. Williams and Graham A. J. Worthy

3.1 INTRODUCTION

Sixty million years ago marine mammals made the transition from terrestrial specialists to intermediate forms capable of moving both in air and water, and from these intermediate forms to aquatic specialists (Repenning 1976; Berta *et al.* 1989; Thewissen *et al.* 1994). Marked anatomical and physiological modifications were necessary during these evolutionary stages to meet the physical demands of living in water instead of air. Water is 800 times denser and 60 times more viscous than air (Dejours 1987) making locomotor movements comparatively more difficult. On land, gravity is the primary physical force to be overcome by the body and limbs during running. In water, buoyancy and drag are the major physical forces challenging the swimmer. Furthermore, changes in hydrostatic pressure as the marine mammal dives will influence locomotion. Differences in the thermal characteristics of air and water also have a profound impact on the physiology and anatomy of mammals. The thermal conductivity of water is 24 times that of air at the same temperature. As a result, aquatic living mammals are exposed to exceptionally high levels of heat transfer; simply keeping warm was undoubtedly a major evolutionary hurdle as mammals moved from land to sea. The high salinity of the marine environment also necessitated modification of the terrestrial kidney to allow marine mammals to maintain osmotic homeostasis (internal water and electrolyte balance) in the absence of fresh water. In sum, the unique physical characteristics of water make the oceans a challenge for mammalian systems that originated on land.

Remarkably, not just one but several major lineages, the pinnipeds, cetaceans and sirenians, made the transition from land to sea. Polar bears and sea otters are also specialized for a marine lifestyle. In this chapter we examine some of the major anatomical and physiological solutions used by these mammalian groups to meet the challenge of aquatic living. We focus on key features that were influenced by the physical characteristics of water. These include: (i) swimming locomotion, (ii) diving, (iii) thermoregulation, and (iv) osmoregulation. Each of these represents a hurdle that required both morphological and physiological modifications for the mammal to change from an efficient terrestrial predator to a marine predator.

3.2 LOCOMOTION: DESIGNING THE MAMMALIAN BODY FOR AQUATIC PERFORMANCE

One of the most obvious differences between marine and terrestrial mammals is the shape of the body and appendages. Lanky limbs and small plantar (foot) surfaces characteristic of elite terrestrial athletes such as the cheetah have been replaced with a markedly reduced appendicular skeleton, streamlined bodies and enlarged propulsive appendages (Fig. 3.1). An example of the evolutionary transition in general body form can be traced in the cetacean lineage as illustrated in Fig. 3.2. The closest relatives of whales, mesonychid condylarths, were terrestrial quadrupeds that shared features of modern ungulates (see Chapter 2 for details on evolution). With increased aquatic specialization we find a gradual reduction in the length of the fore- and hindlimbs and an increase in surface area of the appendages. *Ambulocetus natans* provides fossil evidence of these changes and literally means 'walking whale'. It is likely that *Ambulocetus*

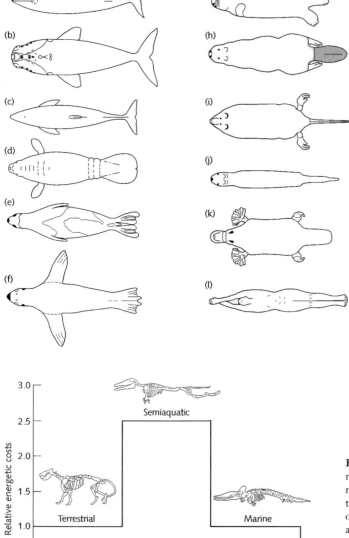

(a)

(b)

(c)

(d)

(e)

(f)

(g)

(h)

(i)

(j)

(k)

(l)

Fig. 3.1 General body shape of mammalian swimmers. Marine mammals are shown in (a)–(g) and display more body streamlining than terrestrial and semiaquatic mammals (h)–(l). (a) Minke whale; (b) right whale; (c) harbour porpoise; (d) Florida manatee; (e) harp seal; (f) California sea lion; (g) sea otter; (h) beaver; (i) muskrat; (j) mink; (k) platypus; (l) human. (From Fish 1993a, with permission.)

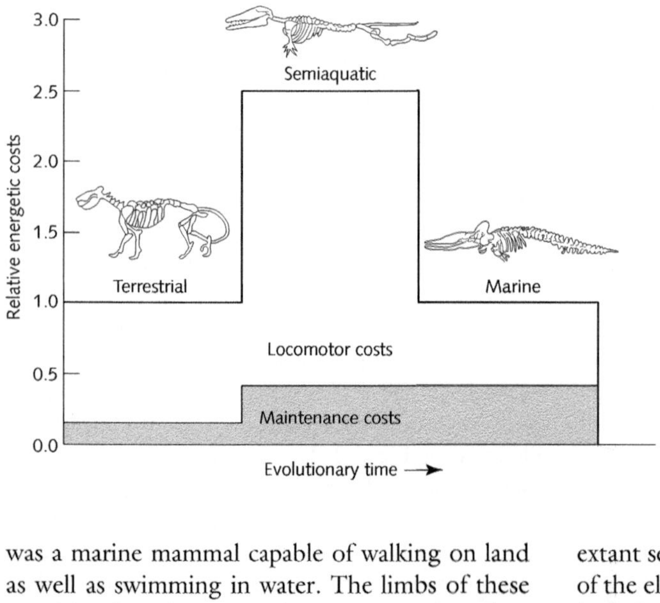

Relative energetic costs

3.0
2.5
2.0
1.5
1.0
0.5
0.0

Semiaquatic

Terrestrial

Marine

Locomotor costs

Maintenance costs

Evolutionary time ⟶

Fig. 3.2 Changes in locomotor costs and basal maintenance costs with the evolution of marine mammals. The evolutionary pathway assumes that ancestral marine mammals included an obligate terrestrial form that was followed by a semiaquatic form (i.e. *Ambulocetus natans*), and finally an obligate marine form. Note the similarity in total energetic costs for terrestrial and marine specialists despite the change in relative contribution of locomotor and maintenance costs. (From Williams 1999, with permission.)

was a marine mammal capable of walking on land as well as swimming in water. The limbs of these transitional marine mammals were more robust than they are in extant species. The shape of the vertebral column indicates that *Ambulocetus* used dorsoventral undulations similar to the swimming movements of extant sea otters (Thewissen *et al.* 1994). Positions of the elbow and femur suggest a sprawling gait on land similar in form to the otariids. In comparison, obligate marine mammals as exemplified by modern cetaceans show vestigial hindlimbs that are not externally visible and forelimbs and elbow joints rendered functionally obsolete in stiffened pectoral fins. The skeletons of ancestral pinnipeds such as *Potamotherium* (Repenning 1976) and *Enaliarctos mealsi* (Berta *et al.* 1989) indicate both otter-like and seal-like locomotor patterns for archaic pinnipeds.

3.2.1 Hydrodynamics and body shape

The simple explanation for these dramatic morphological changes is the mechanical and energetic advantage afforded by body streamlining. By smoothing body contours, the magnitude of drag forces is reduced during swimming. Four types of drag are encountered by swimming marine mammals: (i) frictional drag, (ii) pressure drag, (iii) induced drag, and (iv) wave drag. The first two are associated with the physical characteristics of water surrounding the body of the swimmer. Consequently, the size and shape of the animal will affect the magnitude of frictional and pressure drag. When swimming submerged, these two types of drag predominate (Fish

1993a). The third type of drag, induced drag, is associated with water flow around the flippers, fins and flukes of marine mammals. Many of these body features act as hydrofoils to create lift and thrust during swimming. A consequence of this design, however, is induced drag which results from the pressure difference between the two surfaces of the hydrofoil and the formation of vortices at its tips (Webb 1975).

As the swimmer moves near or on the water surface they experience an additional type of resistance, wave drag. Energy that could have been directed towards moving the animal forward is wasted in producing waves. Total body drag is 4–5 times higher for a body moving on or near the water surface than for the same body submerged due to wave drag (Hertel 1966). This has been demonstrated for humans and harbour seals (Williams & Kooyman 1985) and sea otters (Williams 1989) by towing the subjects on the water surface and submerged (Box 3.1). Wave drag is reduced considerably by submerging and becomes negligible

---

BOX 3.1 BODY STREAMLINING AND DRAG REDUCTION

Swimmers are subject to physical forces, termed drag, that resist forward movement in water and are described by the equation:

$$\text{Drag} = 1/2pV^2ACd$$

where p = density of the fluid, V = velocity of the swimmer, A = area of the body (surface area or frontal area, and Cd = drag coefficient, a term that takes into account the flow characteristics of the fluid around the body.

It is obvious from this equation that velocity has a comparatively large impact on body drag. As the animal attempts to move faster through the water, drag forces on the body increase exponentially. This, in turn, affects locomotor effort and the energetic cost to the swimmer.

The sea otter provides a good example of the effects of swimming velocity on drag forces. In Fig. 1, drag forces for surface (closed circles) and submerged (open circles) sea otters moving through the water at different velocities are compared. Note the non-linear increase in drag force with velocity of the animal. The dashed line denotes the drag forces routinely encountered by wild sea otters based on their preferred surface and submerged swimming speeds. These values represent form drag

(drag force from the body moving through the water) only and do not account for active drag (drag associated with swimming movements) (Williams 1989).

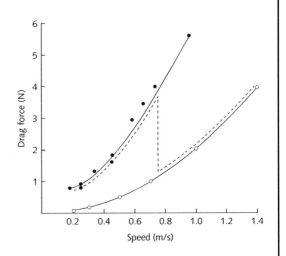

Fig. 1 The effects of swimming speed on drag forces. (Data from Williams 1989.)

once an animal has moved three body diameters down into the water column. The interrelationships between position in the water column, wave drag and subsequent effort have been used effectively by human swimmers. To increase speed during competition in the 1996 Atlanta Olympics, human athletes relied on prolonged periods of submergence and an undulatory dolphin kick following each flip turn off of the pool wall. Sea otters (*Enhydra lutris*) also use this trick of submergence and undulatory swimming to increase speed. Surface swimming by this mammal is relatively slow and usually does not exceed 0.8 m/s. When sea otters want to move fast, they switch to a submerged undulatory style of swimming and can reach speeds of 1.4 m/s.

Another means of minimizing total body drag is to streamline the body and appendages. The morphological modifications required for streamlining are consistent among very different mammalian lineages. Otariids, phocids, odontocetes, and even mysticete whales, have remarkably similar body shapes consisting of rounded leading edges that taper progressively towards the tail (Fig. 3.1). The optimum body shape minimizes drag for a maximum body volume and is described by the fineness ratio (FR) where:

FR = Length of the body/Maximum body diameter

The optimum range of FR for fast swimming vertebrates is 3–7 with an ideal value of 4.5 (Webb 1975). A survey conducted by Fish (1993a) demonstrates that many marine mammals have well streamlined body dimensions within the theoretical optimum FR range. Odontocetes, otariids, phocids, sirenians and the sea otter have body shapes with FRs that range from 3.3 to 8.0. The FRs of mysticetes range from 4.8 to 8.1 for Balaenopteridae and 3.3–8.0 for Balaenidae. Exceptions include the northern right whale dolphin (*Lissodelphis borealis*) and semiaquatic mustelids (mink and river otters) whose body shapes approach snake-like proportions and FRs of 9.0–11.0.

3.2.2 Locomotor movements and the cost of swimming

Body streamlining is not enough to guarantee athletic prowess when moving through water. The mechanism that propels the animal forward must be efficient and powerful enough to counter the effects of drag. Different lineages of marine mammals have solved the problem of aquatic propulsion in a variety of ways. Common features to all are specialized, enlarged propulsive surfaces that oscillate to create thrust. This oscillatory mode of swimming differs considerably in mechanical efficiency, thrust production and energetic cost from terrestrial or semiaquatic mammals that rely on paddling limbs for propulsion (Fish 1996). During paddling, as used by surface-swimming sea otters and polar bears (*Ursus maritimus*), the stroke cycle consists of alternating power and recovery phases. The power phase enables the animal to move forward. The recovery phase is used primarily to reposition the appendage for the following stroke and as such represents a portion of the cycle that does not contribute to the forward movement of the swimmer.

Stroke efficiency is improved in highly derived marine mammals by oscillating the appendages and modifying their shape to serve as hydrofoils (Fish 1993a). Rather than paddling, pinnipeds, cetaceans and sirenians use hydrodynamic lift-based momentum exchange to move through the water (Webb 1984; Fish 1993a). The fore-flippers of otariids, the flukes and pectoral fins of cetaceans, and the paired hind-flippers of phocid seals act as hydrofoils to generate thrust. However, the mechanics of each mode of swimming are very different. Otariids, such as the California sea lion (*Zalophus californianus*), use fore-flipper propulsion (Feldkamp 1987). Walruses (*Odobenus rosmarus*) and phocid seals use alternate lateral sweeps of the hind-flippers in which the posterior half of the body may flex. Cetaceans use dorsoventral movements of the posterior third of the body and fluke to produce thrust. An important feature of the swimming modes of both pinnipeds and cetaceans is the absence of a prolonged recovery phase during the stroke cycle. Thus, thrust can be produced throughout the stroke cycle and mechanical efficiency is increased.

Because marine mammals use so many different modes of swimming, we might expect that cruising speeds would vary according to style. However, this is not what is observed for marine mammals at sea. The range of cruising speeds for marine mammals varies little over a wide range of body sizes and

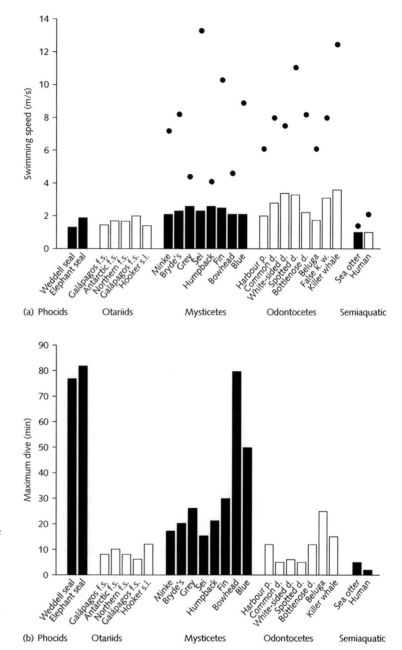

Fig. 3.3 (a) Swimming speeds and (b) maximum dive duration of marine mammals. Phocid seals, otariids, mysticetes, odontocetes and semiaquatic mammals are compared. Bars represent cruising speed while circles represent sprint speeds for the animals in (a). The mass of the animal increases within each taxonomic group.

styles of swimming (Fig. 3.3a). Average swimming speeds for phocid seals using lateral undulation, otariids using fore-flipper propulsion, and mysticetes and odontocetes using dorsoventral undulation range from 1.3 to 3.6 m/s. In comparison with terrestrial mammals, this represents a narrow range of locomotor speeds when the range of body mass (30 kg fur seals to 145 t blue whales) is taken into account.

Sprint speeds of marine mammals are considerably higher and may reach over 10 m/s in odontocetes and mysticetes. Once again body size does not

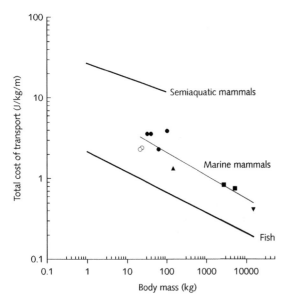

Fig. 3.4 Total cost of transport (COT_{TOT}) in relation to body mass for different classes of swimmers. Individual marine mammals are compared to regressions for semiaquatic mammals and salmonid fish. Marine mammal specialists include phocid seals (closed circles), California sea lions (open circles), bottlenose dolphins (upward triangle), killer whales (squares) and gray whales (downward triangle). The line through the data points for marine mammals is the least squares regression and is described by $COT_{TOT} = 7.87$ Mass$^{-0.29}$. The lower solid line represents the extrapolated regression for salmonid fish where $COT_{TOT} = 2.15$ Mass$^{-0.25}$. The upper solid line shows the regression for swimming semiaquatic mammals where $COT_{TOT} = 26.81$ Mass$^{-0.18}$. (From Williams 1999, with permission.)

necessarily correspond to the fastest swimmers. Sprint speeds for the 120 kg spotted dolphin (*Stenella attenuata*) reach 11.1 m/s and rival the speeds of 5 t killer whales and 68 t fin whales. The slowest sprint speed recorded for a cetacean is 5.1 m/s for the narwhal (*Monodon monoceros*), a comparatively sluggish marine mammal that moves slowly within the Arctic pack ice (Dietz & Heide-Jorgensen 1995). Not surprisingly, the swimming speeds of marine mammals exceed even the best efforts of human athletes. For the sake of comparison, we note that the average human swimmer moves at a cruising speed of 1.0 m/s. The gold medal performance for sprint swimming during the 1996 Olympics in Atlanta was just over 2.0 m/s.

With increased specialization in body morphology for aquatic movements there is a corresponding change in the energetic cost of locomotion (see Chapter 9). The total cost of transport (COT_{TOT}) is defined as the amount of fuel it takes to transport one unit of body mass over a unit distance (Schmidt-Nielsen 1972). COT_{TOT} for swimming mammals falls into two distinct groups based on where the animal swims in the water column and the degree of specialization of the propulsor (Fish 1993a, 1996; Williams 1999). Semiaquatic mammals (minks, muskrats, humans and sea otters swimming on the water surface) have elevated transport costs that are 2–5 times higher than those observed for marine mammals (Fig. 3.4). The lower swimming costs of marine mammals are described by a different regression. Interestingly, this single regression explains the cost of dorsoventral undulation in cetaceans (Fish 1993b, 1998), fore-flipper propulsion in otariids (Feldkamp 1987) and lateral undulation of paired hind-flippers in phocid seals (Fish *et al.* 1988). This may seem unusual in view of the very different mechanics associated with each mode of swimming. However, similar results have been reported for other groups of exercising animals. Transport costs do not vary greatly with the style of swimming in fish (Schmidt-Nielsen 1972, 1984; Bennett 1985) or with bipedal or quadrupedal running in terrestrial mammals (Taylor & Rowntree 1973; Fedak & Seeherman 1979).

Surprisingly, the relationship describing transport costs in swimming marine mammals does not differ from that reported for running terrestrial mammals (Williams 1999). We find that the COT_{TOT} of swimming harbour seals are similar to those of a running goat of equal size; COT_{TOT} for a swimming dolphin approaches that of a running horse. Transport costs of swimming killer whales are similar to those of running elephants. From these relationships it is not hard to imagine the changes in locomotor energetics that may have occurred as ancestral marine mammals made the transition from land to sea (Fig. 3.2). The energetic trend would have been from the comparatively low costs of the terrestrial specialist to the comparatively high costs of the semiaquatic transitional mammal. As body morphology changed to accommodate streamlining and improve propulsive efficiency, energetic costs were reduced back to the original low cost level of the specialist.

Both hydrodynamics and energetics of swimming indicate that the ability to remain submerged provides a distinct advantage for marine mammals. Yet, this poses an interesting physiological challenge for a mammal. How does an air-breathing animal support metabolic processes while under water? The following section will examine this question for transitional mammals that move between land and water, and for marine mammal specialists that spend over 90% of their lives submerged.

3.3 DIVING PHYSIOLOGY: EVOLUTION OF THE UNDERWATER ATHLETE

An important consequence of the aquatic lifestyle of marine mammals is a marked change in how oxygen is delivered, stored and utilized by the body, especially when compared to terrestrial mammals. In general, the components of the pathway for oxygen in marine mammals are similar to those of terrestrial mammals and reflect the evolutionary connection between the groups. However, the function of the individual components differs between terrestrial and marine mammals due to differences in access to air. Anatomical structures originally intended for use on land must now accommodate the special needs of the diving mammal. Unlike the fixed open system of terrestrial mammals, the oxygen pathway for marine-living mammals performs multiple roles (Fig. 3.5). During activities on land or on the water surface, the pathway is open and can operate as in terrestrial mammals. Oxygen flows from ambient air to the lungs, diffuses across the alveoli into capillaries, and is transported through the cardiovascular system to the skeletal muscles where it is ultimately used in oxidative phosphorylation within the mitochondria (Weibel *et al.* 1987). Dispersed within the tissues of each major anatomical component are oxygen and energy stores that can act as buffers along the oxygen pathway. As discussed below, these oxygen stores play a unique role in supporting prolonged periods of submergence by mammals specialized for a marine lifestyle.

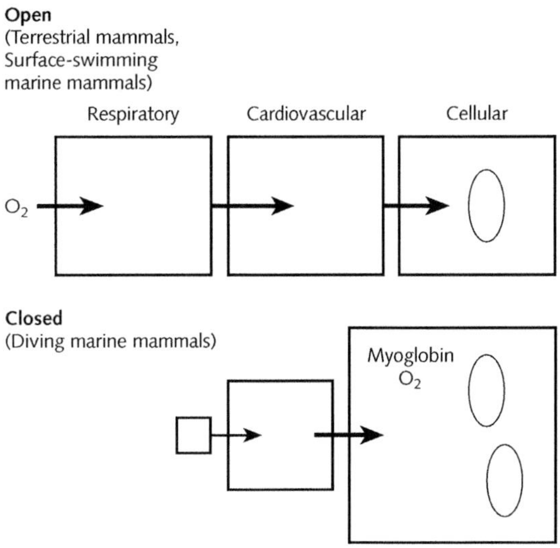

Fig. 3.5 The pathway for oxygen in mammals. Each box represents a major component of the path from ambient air to utilization in the skeletal muscles. The size of the boxes illustrates the relative importance of each component during open and closed states. Note the increased number of mitochondria and myoglobin stores characteristic of the skeletal muscles of marine mammals.

These same pathway components operate as a closed system when the marine mammal performs a dive. Ambient oxygen is no longer available, and the lungs often collapse with increased hydrostatic pressure at depth (Ridgway & Howard 1979; Skrovan *et al.* 1999). Bradycardia (decreased heart rate) associated with the dive response results in a reduction in cardiac output, changes in distribution of the blood, and marked changes in the transport of oxygen to skeletal muscles (see Butler & Jones 1997 for a review). Thus, during the course of a dive, the oxygen pathway of marine mammals is initially open for oxygen loading, closes during submergence when exercise takes place (and oxygen demands are highest), and subsequently reopens after surfacing.

3.3.1 Adapting the lungs for diving

The lungs of marine mammals show several morphological adaptations that support the transition to an aquatic lifestyle. However, the volume of the mammalian lung is dictated more by body size than by preference for a terrestrial or aquatic lifestyle.

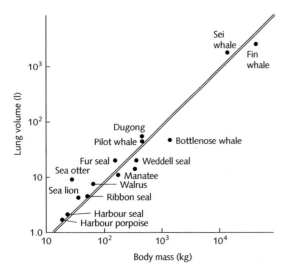

Fig. 3.6 Logarithmic plot of lung volume in relation to body mass for marine mammals. The solid line denotes the regression for all mammals from bats to whales. Note the exceptionally large lungs of the sea otter for its body size. (Redrawn from Kooyman 1973.)

For example, the volume of marine mammal lungs increases with body mass in a manner similar to that of terrestrial mammals (Fig. 3.6). Exceptions include sperm whales (*Physeter catadon*), bottlenose whales (*Hyperoodon* spp.) and fin whales (*Balaenoptera physalus*) whose maximum lung capacities are approximately half that expected for terrestrial mammals (Slijper 1976). Sea otters, dugongs (*Dugong dugon*) and several species of otariid show larger than expected lung volumes.

The size of the lung relative to body mass undoubtedly impacts buoyancy regulation and oxygen storage capacity. The sea otter in particular has an exceptionally large lung volume for its body size. Both lung mass and lung volume are nearly twice that measured for other marine mammals of similar body mass (Kooyman 1973). Such a large lung provides a convenient anatomical float for these animals as they feed on the water surface while using their abdomen as a table.

Although the large lung of the sea otter may also act as an oxygen depot during short dives, it is unlikely that the lungs of deep-diving marine mammals serve a similar role. Morphological evidence suggests that the lungs of many marine mammals collapse at depth, consequently altering buoyancy

and the lung's usefulness as an oxygen store during submergence. Experiments comparing the mechanical properties of isolated lungs from dogs and sea lions demonstrated a progressive collapse in the marine mammal lung as hydrostatic pressure increased during simulated dives. The lungs of the dog trapped air rather than collapsed with changes in pressure (Denison *et al.* 1971). Pressure chamber tests on Weddell seals (*Leptonychotes weddelli*) and northern elephant seals (*Mirounga angustirostris*) also demonstrated tracheal collapse to less than half of its original dimension with exposure to increased hydrostatic pressure (Kooyman *et al.* 1970). Similar tests using lungs from bottlenose dolphins (*Tursiops truncatus*) indicate that the bronchi and trachea as well as the alveoli of the cetacean lung are collapsible. Only the bony nares, comprising a volume of 50 ml, are rigid (Ridgway *et al.* 1969). Changes in the oxygen and carbon dioxide content of expired air of a bottlenose dolphin trained to dive, swim and station at depth indicate that alveolar collapse is complete by a depth of 100 m (Ridgway *et al.* 1969). The lungs of larger whales including the pilot whale (*Globicephala melaena*) (Olsen *et al.* 1969), fin whale and sei whale (*Balaenoptera borealis*) (Scholander 1940) also show evidence of lung collapse in response to increased pressure. As discussed below (see 'Oxygen stores'), the unique collapsible lung of marine mammals enables the animals to avoid a deleterious buildup of nitrogen (N_2) and associated nitrogen narcosis, and decompression sickness (the 'bends') during a dive. Prolonged gliding behaviour during ascent and descent have also been attributed to changes in buoyancy that occur with lung collapse in several species of deep-diving marine mammals (Williams *et al.* 2000).

Mechanical differences in the lungs of mammals correspond to variations in the anatomical structure of the airways. A gradation in the architecture of small and large airways occurs in parallel with the degree of aquatic specialization in mammals. The semiaquatic river otter (*Lutra canadensis*) and shallow-diving sea otter have circular trachea with partially calcified rings. Calcification of the tracheal rings results in structural rigidity that may prohibit deep diving by these mammals. In comparison, deep-diving marine mammals such as the harp seal (*Pagaphilus groenlandicus*), Weddell seal, harbour seal (*Phoca vitulina*) and walrus show decreased calcification of the trachea. This adaptation allows the tracheal rings to bend without breaking during compression at depth (Tarasoff & Kooyman 1973). Among marine mammals the phocid seals have the least modified terminal airway structure compared to terrestrial mammals. In seals a non-cartilaginous portion of bronchiole connects to a respiratory bronchiole and finally the alveoli. Large smooth muscles surround the cartilage-free segments. The otariids and whales represent the other extreme and have cartilaginous reinforced airways leading directly into the alveoli. A series of sphincter muscles are also present in the smaller airways of the cetacean lung (Kooyman 1973). Walrus and sea otter lungs are intermediate to these extremes and show some terminal airways without cartilage and others in which the cartilage extends directly to the alveoli. Whether by muscle or cartilage it is apparent that the airways of marine mammals, especially deep-diving species, are reinforced to ensure patency during lung compression at depth. These adaptations allow a progressive collapse of the lung structures as hydrostatic pressure increases during descent, with initial collapse by the alveoli and subsequently the small and large airways. The pattern then works in reverse as hydrostatic pressure decreases on ascent, and the lungs are able to reinflate in a progressive manner.

3.3.2 Oxygen stores

Despite routine closure of the oxygen pathway, marine mammals can maintain high levels of activity while submerged and preferentially rely on aerobic metabolism to support these activities (Kooyman 1989; Butler & Jones 1997). Maintenance and tight regulation of aerobic metabolism may be more critical in aquatic-adapted species than in terrestrial mammals due to the deleterious effects of anaerobic end-products, particularly on oxygen-sensitive tissues (i.e. brain and heart) during submergence. Unlike terrestrial mammals, sea otters, polar bears, pinnipeds and cetaceans must support the energetic demands of exercise while holding their breath. The ability to balance the conflicting demands for oxygen conservation and utilization during submergence ultimately dictates the diving limitations of the animal (Castellini *et al.* 1985; Hochachka 1986).

Exceptionally large stores of oxygen in the lungs, blood and muscles act as an on-board 'scuba tank' and facilitate aerobic activity by marine mammals during periods of submergence. Total oxygen stores of marine mammals often exceed 2–3 times those of terrestrial species such as dogs and humans. The distribution of these stores varies among the many taxa of marine divers (Fig. 3.7). In dolphins and

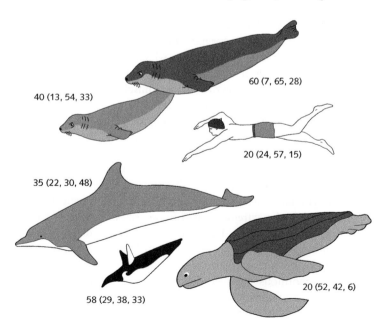

Fig. 3.7 Comparison of oxygen stores for major taxa of diving vertebrates. The numbers outside the parentheses represent the total oxygen store in ml O_2/kg. The numbers in parentheses are the percentage of the total oxygen store located in the lungs, blood and muscles, respectively. (From Kooyman 1989, with permission.)

40 (13, 54, 33)

60 (7, 65, 28)

20 (24, 57, 15)

35 (22, 30, 48)

58 (29, 38, 33)

20 (52, 42, 6)

humans, 22–24% of the total oxygen store is located in the lungs while the remaining 72–78% is sequestered in the blood and skeletal muscles. Only 13% of the total store occurs in the lungs of otariids. This compares to 7% in the lungs of phocid seals. Because nitrogen is stored with oxygen in the lungs, reliance on lung reserves would place the diver at risk of high blood nitrogen tensions at depth. To avoid this, elite divers such as Weddell seals, elephants seals and deep-diving whales have collapsible lungs that move air into the upper airways where it is not in contact with blood. These marine mammals preferentially use the skeletal muscles and blood as the primary oxygen storage sites. Over 87% of the total oxygen reserve of deep divers is distributed between these two tissues (Kooyman 1989).

Myoglobin serves as the primary oxygen carrier in the skeletal muscles of mammals and is exceptionally high in concentration in marine-adapted species. For example, the myoglobin contents of the locomotor muscles of terrestrial mammals often remain below 1.0 g myoglobin/100 g wet muscle regardless of whether the animal is an elite sprinter or endurance athlete (Castellini 1981; Williams *et al.* 1997). In comparison, marine mammals show myoglobin contents that are 3–7 times higher. The skeletal muscles of sea otters have a myoglobin content of 3.1 g myoglobin/100 g wet muscle (Lenfant *et al.* 1970). Among pinnipeds, the myoglobin content of the locomotor muscles correlates with maximum dive duration. Otariids, which are comparatively short divers, maintain myoglobin contents near 3.0 g myglobin/100 g wet muscle. Phocid seals, the elite divers among pinnipeds, have myoglobin contents that average 4–5 g myoglobin/100 g wet muscle (Kooyman 1989). Cetaceans also rely on large oxygen reserves in the muscles to support aerobic metabolism during diving. However, the relationship between dive duration and myoglobin content is less clear for cetaceans and is complicated by the extreme range of body sizes for this group (Noren 1997). Large muscle oxygen reserves in cetaceans are a function of both high myoglobin content and large muscle mass, particularly in the enormous mysticete whales. Among odontocetes, myoglobin concentration in the longissimus dorsi (the primary swimming muscle)

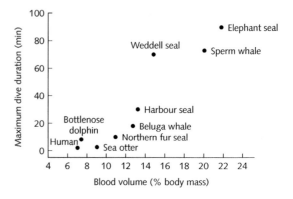

Fig. 3.8 Blood volume of marine mammals as a percentage of body mass in relation to maximum dive duration. (Data from Kooyman 1989.)

ranges from approximately 2.0 g myglobin/100 g wet muscle in the northern right whale dolphin to 8.0 g myglobin/100 g wet muscle in the narwhal (Noren 1997). Myoglobin contents of mysticetes range from 0.9 g myglobin/100 g wet muscle in the sei whale (Tawara 1950) to 3.5 g myglobin/ 100 g wet muscle in the bowhead whale (*Balaena mysticetus*) (Noren 1997; Noren & Williams 2000).

Oxygen storage capacity in the blood is enhanced in marine mammals by increases in: (i) blood volume, (ii) the number of circulating red blood cells, and (iii) haemoglobin concentration. As illustrated in Fig. 3.8 there is a positive correlation between blood volume and maximum dive duration in mammals. This was demonstrated in a study by Ridgway and Johnston (1966) that compared the oxygen storage characteristics of blood from three species of small cetacean. The species selected varied in diving and swimming capabilities. The investigators reported blood volumes of 143 ml blood/kg body mass for the highly active, deep-diving Dall's porpoise (*Phocoenoides dalli*). Values were 108 ml blood/kg body mass for the intermediately active Pacific white-sided dolphin (*Lagenorhynchus obliquidens*) and 71 ml blood/kg body mass for the more sedentary, coastal-dwelling bottlenose dolphin. Haemoglobin concentration, an important component of the oxygen-carrying capacity of blood, also reflected the aquatic behaviour of each species. The highest haemoglobin concentration was found in the Dall's porpoise and the lowest in bottlenose dolphins. In general, haemoglobin concentrations for mammals are 14–17 g/ 100 ml blood for shallow to moderate divers such as humans, sea otters, northern fur seals (*Callorhinus ursinus*) and bottlenose dolphins, and 21–25 g/ 100 ml blood for deeper divers such as harbour seals, elephant seals, Weddell seals, Dall's porpoise and beluga whales (*Delphinopterus leucas*) (Kooyman 1989).

The elite divers among marine mammals, phocid seals, also alter blood oxygen by changing the number of circulating red blood cells during the course of a dive. The most detailed work concerning these changes has been conducted on freely diving Weddell seals in the Antarctic. Haematocrit (the volume of red blood cells per volume of blood) in these superb divers rises as the dive progresses, and declines back to resting levels during the postdive recovery period. The magnitude of these changes in haematocrit depends on the length of the dive. The longer the dive the greater the increase in haematocrit. Because the spleen of the Weddell seal is large (3–4 times larger than terrestrial mammals for its body mass), it can serve as an enormous reservoir of oxygenated red blood cells. During a dive, the spleen of the Weddell seal contracts and injects red blood cells into the circulation (Hurford *et al.* 1996). This results in the characteristic elevation in haematocrit and can induce a 60% increase in haemoglobin concentration within the first 10 min of a dive (Qvist *et al.* 1986). The benefit to the seal is an infusion of circulating oxygen within the red blood cells for the working tissues. Although highly developed in Weddell seals, this physiological mechanism is not unprecedented among mammals and has also been observed in exercising racehorses and dogs. Seals are simply able to take advantage of this physiological mechanism to support aerobic diving. Preliminary evidence suggests that splenic contraction occurs during breath-holding in other species of phocid seals, including the northern elephant seal. It remains to be seen whether the sequestering of red blood cells during rest and their mobilization during submergence provides an advantage for aerobic diving in other semiaquatic or marine mammals.

3.3.3 Physiological responses to submergence

During submergence mammals undergo a suite of physiological changes known as the dive response. Key elements of the response are: (i) breath-holding, which is termed apnoea, (ii) bradycardia, a pronounced reduction in heart rate, and (iii) peripheral vasoconstriction characterized by the selective redistribution of blood to oxygen-sensitive tissues. The level of response is variable and depends on such factors as the degree of aquatic specialization, species, dive duration, behaviour and type of dive. Very different physiological responses occur for voluntary and involuntary dives, with the most extreme diving response displayed during forced submersion.

The physiological response to submergence is a general mammalian phenomenon, although the degree of response differs in terrestrial, semiaquatic and marine mammals. Despite the ubiquitous nature of these physiological events among mammals, they should not be considered part of an invariant reflex. The response is far more complex than that. An unfortunate choice of terminology in earlier studies labelled the physiological changes with submersion as the 'diving reflex'. Later studies involving free-ranging marine mammals, as well as sea lions (Ridgway *et al.* 1975) and bottlenose dolphins (Elsner *et al.* 1966) trained to dive on command, demonstrated a level of conscious control over the intensity of bradycardia developed during submergence. As a result, the term 'diving response' rather than 'diving reflex' is considered more accurate in describing the many physiological changes that occur with submersion (Elsner 1999).

Obviously, the key to successful diving is the ability to breath-hold. In this regard, marine mammals are unrivalled. This is due in part to their large size and to their ability to store oxygen in the lungs, blood and muscles. As found for swimming speed, the duration of breath-hold does not necessarily correlate to body size in marine mammals (Fig. 3.3b). Phocid seals, especially the elephant seals and Weddell seal, tend to show longer dive durations than otariids, odontocetes and even many larger species of mysticetes. Within each of these taxonomic groups, however, body size appears to have an effect on breath-hold ability, and maximum dive

Taxonomic group	n	Mass range (kg)	Regression
Phocids	16	80–4000	Max. duration = 3.39 Mb$^{0.42}$
Otariids	13	30–270	Max. duration = 6.22 Mb$^{0.10}$
Odontocetes	22	60–51 700	Max. duration = 0.51 Mb$^{0.51}$
Mysticetes	9	12 700–145 000	Max. duration = 0.04 Mb$^{0.61}$

Table 3.1 Allometric relationships for maximum dive duration in relation to body mass for pinnipeds and cetaceans. Dive duration is in minutes and body mass (Mb) is in kilograms. (Data from Schreer & Kovacs 1997.)

duration increases predictably with body mass for each group (Table 3.1) (Schreer & Kovacs 1997). Maximum dive durations for phocid seals range from 11 min in the 250 kg crabeater seal (*Lobodon carcinophagus*) (Bengtson & Stewart 1992) to 120 min in a 600 kg female southern elephant seal (*Mirounga leonina*) (Hindell *et al.* 1991). Ranges for other marine mammal groups are 6–16 min for otariids, 2–138 min for odontocetes, and 15–50 min for mysticetes. Size ranges for these groups are shown in Table 3.1. Maximum dive duration for the 1.9 t walrus is 13 min (Wiig *et al.* 1993), and for the 1.6 t manatee (*Trichechus* spp.) is 16 min (Irving 1939). The longest dive duration recorded for a mammal is 138 min for the 51.7 t sperm whale (Watkins *et al.* 1985).

One of the hallmarks of the dive response is bradycardia, a marked reduction in heart rate. Nearly all marine mammals measured to date show a rapid and profound decrease in heart rate upon submergence. Bradycardia is maintained throughout the dive and followed by tachycardia, a rapid increase in heart rate, as the animal surfaces. Often an 'anticipatory tachycardia' occurs during the ascent from a dive as the animal prepares the cardiovascular system for oxygen loading (Fig. 3.9). Likewise, changes in heart rate may also occur during the predive period in anticipation of the upcoming dive. For an elite diver such as the elephant seal, typical heart rates for 150–250 kg animals range from 103 to 112 beats/min on the water surface to 20–50 beats/min during 10–17 min dives (Andrews *et al.* 1997). This compares with 197 kg bottlenose dolphins on trained dives ranging from 1 min to 4 min in which predive heart rates averaged 101–111 beats/min and decreased to 30–37 beats/min during submergence (Williams *et al.* 1999a). A 3.7 t male killer whale freely swimming and diving in a

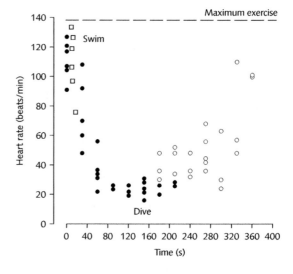

Fig. 3.9 Heart rate in relation to dive duration for bottlenose dolphins freely diving to 210 m. Each point represents the average heart rate for 10 s intervals during the dive. Solid circles are values for heart rate during the descent; open circles are for the ascent phase. Average heart rates for dolphins swimming on the water surface are shown by the squares. The upper dashed line illustrates the maximum heart rates for bottlenose dolphins pushing on a load cell. (From Williams *et al.* 1999a, with permission.)

net pen demonstrated a range of heart rates from approximately 60 beats/min when on the water surface to 30 beats/min when submerged for longer than 15 s (Spencer *et al.* 1967). Cardiovascular responses of an otariid, the California sea lion, are comparatively slower. Predive surface heart rates of 150–250 beats/min gradually slowed to 20–50 beats/min during 1–3 min trained dives by 25–35 kg sea lions (Ponganis *et al.* 1997). The Amazonian manatee (*Trichechus inunguis*) is an exception to the typical pattern of surface tachycardia and marked bradycardia during submergence in marine mammals.

The average heart rate of these comparatively sedentary herbivores is approximately 50 beats/min when breathing on the water surface and decreases slowly to just 30–40 beats/min on voluntary dives (Gallivan & Best 1986). Although the level of bradycardia in freely diving manatees seems attenuated, this sirenian is capable of marked bradycardia equivalent in magnitude to other marine mammals. When frightened during a dive, manatees can decrease their heart rate to as low as 5–6 beats/min.

The length of a dive influences the level of bradycardia developed by marine mammals. Free-ranging Weddell seals (Kooyman & Campbell 1972; Hill *et al.* 1987), grey seals (*Halichoerus grypus*) (Thompson & Fedak 1993), elephant seals (Andrews *et al.* 1997; Hindell & Lea 1998) and bottlenose dolphins (Williams *et al.* 1999a) demonstrate an inverse relationship between heart rate and the length of the dive. The longer the dive, the more intense the bradycardia, and hence, the lower the heart rate during submergence. One of the more impressive demonstrations of this response is in the grey seal. When resting on the water surface the heart rate of this 200 kg phocid seal averages 119 beats/min. Heart rate drops immediately upon submergence and can remain at only 4 beats/min during dives exceeding 15 min (Thompson & Fedak 1993). This intense bradycardia appears to be part of the normal physiological repertoire of the foraging grey seal. Many species of marine mammal, as illustrated by the manatee, also demonstrate exceptionally low heart rates if forcibly submerged or subjected to a behavioural disturbance during the dive (Kooyman 1989; Butler & Jones 1997). These extreme heart rates can be 2–10 times lower than measured during natural dives.

The decrease in heart rate on submergence is accompanied by a selective redistribution of blood in the diving mammal. As observed for heart rate, the level of vascular response depends on the type of dive. In a detailed study of Weddell seals, injected microspheres were used to determine relative blood flow to tissues during involuntary submersions. Significant declines in blood flow from resting values were recorded for all tissues except the brain during the simulated dive (Zapol *et al.* 1979). Evidence from renal and hepatic tests indicates that blood flow to the kidneys and liver is more variable during voluntary diving by Weddell seals (Davis *et al.* 1983). Normal renal and hepatic function, and hence blood flow to these organs, occurs during natural aerobic dives. During prolonged dives that exceed 23 min, kidney function is reduced but hepatic blood flow may be maintained in Weddell seals.

It seems obvious that circulation to the skeletal muscles involved in powering swimming movements would be beneficial for supporting aerobic activity during the dive. Measured indirectly by changes in temperature (Ponganis *et al.* 1993) and partial pressures of nitrogen (Ridgway & Howard 1979), circulation to the skeletal muscles does appear to remain open in freely diving marine mammals. Extreme vasoconstriction can occur during involuntary or exceptionally long dives, and markedly reduces blood flow to the skeletal muscles. A consequence of this vascular shutdown is a sharp rise in blood lactate during the postdive recovery period once blood flow to the muscles is restored (Scholander 1940).

3.3.4 Exercising under water

In view of the marked physiological responses to diving, how does the marine mammal exercise while submerged? In terrestrial mammals, heart rate typically increases as a function of exercise intensity (Brooks *et al.* 1996). This does not always occur for marine mammals, particularly during submergence. Recent studies using miniaturized heart rate monitors and time depth recorders on free-ranging marine mammals have demonstrated that the duration of submergence rather than the level of exercise *per se* dictates many of the cardiovascular and respiratory responses observed during diving. For example, the changes in heart rate for bottlenose dolphins freely diving to 210 m (Fig. 3.9) are similar to those of sedentary dolphins quietly resting on the bottom of a 2 m deep oceanarium pool (Elsner *et al.* 1966). Many diving pinnipeds, including elephant seals (Andrews *et al.* 1997), grey seals (Thompson & Fedak 1993) and California sea lions (Ponganis *et al.* 1997), maintain steady, reduced heart rates during submergence irrespective of locomotor speed. Based on these studies, the physiological adjustments associated with the dive response over-rides those typically associated with an exercise response. This over-ride feature appears to be most developed in highly adapted marine mammals. Other diving vertebrates, including many species of sea birds (Butler & Jones 1997) and the hippopotamus (Elsner 1966), show increased variability in heart rate when exercise is superimposed on breath-holding during submergence.

These findings do not imply that the level of effort has no effect on oxygen utilization or energetic cost to the diver. Rather, the timing of oxygen loading for marine mammals differs from that of terrestrial mammals. Except under extreme levels of effort, oxygen loading occurs simultaneously with exercise in terrestrial mammals. In diving and swimming marine mammals there is a requisite temporal delay in oxygen loading relative to when exercise takes place. As a result, the physiological effects of exercise during diving are manifested primarily during the postdive recovery period. Exceptionally high physiological rates may occur in the period immediately after a dive. For example, the postdive heart rate and respiratory rate of bottlenose dolphins reach the highest levels recorded for any activity (Williams *et al.* 1993). Several studies on pinnipeds (Thompson & Fedak 1993; Andrews *et al.* 1997) have suggested that comparatively high physiological rates during these postdive periods benefit the marine mammal by reducing the requisite recovery time and shortening the interdive surface interval.

A recent study that simultaneously examined the postdive oxygen consumption and locomotor behaviour of Weddell seals has demonstrated the effect of exercise on diving costs. Seals were fitted with a miniaturized, submersible camera and released into an isolated ice hole. The animals freely dived and foraged beneath the Antarctic sea ice while flipper stroking was videotaped from the backward-facing camera (Davis *et al.* 1999). Analysis of the tapes revealed several modes of swimming by the seals including constant stroking, prolonged gliding on descent and burst-and-glide locomotion on ascent. The range of swimming speeds averaged 1.5–2.0 m/s regardless of the mode of swimming. However, the energetic costs of each were very different. The more strokes used by the seal, and consequently the greater the number of muscle contractions, the higher the postdive oxygen consumption (Williams

et al. 2000). Clearly, exercise, even if performed while submerged requires energy for the working muscles and an oxygen payback on surfacing.

Several biochemical and morphological adaptations allow the skeletal muscles of marine mammals to continue working despite the closure of the respiratory pathway (Fig. 3.5). Oxygen storage in myoglobin has already been discussed. Other adaptations include elevated mitochondrial volume density and enhanced aerobic enzyme capacities in the muscles that power swimming movements. The volume density of interfibrillar mitochondria in the locomotor muscles of pinnipeds (Steller sea lions, northern fur seals and harbour seals) is 1.7–2.2 times greater than predicted for terrestrial mammals of similar size (Kanatous *et al.* 1999). Whether an adaptation to hypoxia (Kanatous *et al.* 1999) or endurance exercise (Hochachka 1998), such elevated skeletal muscle mitochondrial densities place these marine mammals among the elite mammalian athletes.

3.4 THERMOREGULATION: THE CHALLENGE OF WARM BODIES IN COLD WATER

One of the most difficult challenges faced by mammals making the transition from land to sea was thermoregulation. In retaining the same high body temperatures of terrestrial mammals, these animals often faced large thermal gradients for heat loss, particularly when living in polar regions (but see also Box 3.2). The physical properties of water, including high heat capacity and thermal conductivity, result in heat transfer rates that reach 24 times that of air at similar temperatures (Dejours 1987). Transitional marine mammals, as well as extant pinnipeds, whose life history includes terrestrial and aquatic periods are further challenged by the conflicting responses necessary for coping with the disparate thermal properties of air and water.

Marine mammals deal with the potentially high rate of heat loss during immersion in two ways, physiologically and morphologically. They can elevate heat production by increasing metabolic rates to compensate for high heat losses. Alternatively, they can increase insulation to help retain body heat

BOX 3.2 TOO HOT OR TOO COLD?

We frequently think of thermoregulation in marine mammals as a problem in keeping warm while immersed. However, under certain conditions marine mammals must also be able to dissipate excess heat. This requires the use of 'thermal windows' to circumvent the insulating layer of fur or blubber. Sparsely haired appendages and enlarged peripheral areas such as the dorsal fins and flukes of cetaceans (Pabst *et al.* 1995) facilitate the transfer of excess heat during periods of high heat production or elevated environmental temperatures (Fig. 1).

These poorly insulated areas are serviced by a specialized countercurrent arrangement of blood vessels (Parry 1949; Scholander & Schevill 1955; Hampton & Whittow 1976). Blood flow to the skin surface is increased during periods of high heat production to provide maximum cooling. The increases in heat flow associated with these circulatory changes have been examined in a wide variety of marine mammals including bottlenose dolphins (Williams *et al.* 1999b), harbour porpoises (Kanwisher & Sundnes 1965) and Hawaiian spinner dolphins (Hampton & Whittow 1976). The use of the dorsal fin and flukes as radiators seems especially important in some odontocetes for the regulation of temperature-sensitive organs such as the intra-abdominal testes (Rommel *et al.* 1994; Pabst *et al.* 1995).

Many pinnipeds, particularly those in polar regions, routinely experience high solar radiation and have high densities of arteriovenous anastomoses (AVAs) in their skin. These specialized blood vessels are used to bypass the insulating blubber layer in times of heat stress and carry excess heat to the skin surface for dissipation (Molyneux & Bryden 1978; Bryden 1979). AVAs are also found in the flippers of otariids and may be used for removing excess heat while swimming (Bryden & Molyneux 1978). In addition, sweat glands on the flippers of otariids aid in heat transfer. On hot days

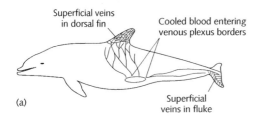

(a)

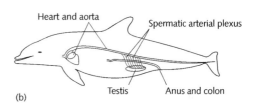

(b)

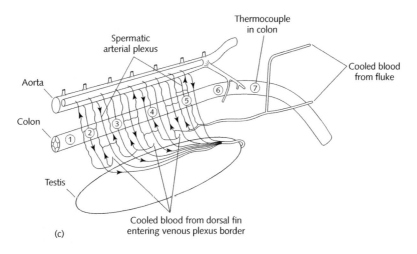

(c)

Fig. 1 Countercurrent exchangers in the extremities and the testes of the bottlenose dolphin. (a) Blood in the superficial veins of the dorsal fin and flukes is cooled by exposure to surrounding water. (b, c) This cooled blood is then passed to a countercurrent exchanger in association with the testes. In this way the peripheral structures form a radiator that help to cool the internal organs. Numbers 1–7 represent measurement sites for assessing variation in the temperature of blood leaving and entering the testicular region. (From Pabst *et al.* 1995, with permission.)

(*continued on p. 88*)

BOX 3.2 (*cont'd*)

sea lions and fur seals can often be seen fanning their flippers and increasing evaporative heat loss at these sites (Blix *et al.* 1979). It should be noted that sweat glands are not found in the skin of cetaceans.

Evaporative cooling resulting from both sweating and respiratory losses accounts for less than 20% of heat production in California sea lions studied under experimental conditions (Matsuura & Whittow 1974; South *et al.* 1976). Instead of relying solely on physiological mechanisms, these animals may use behavioural thermoregulation such as simply entering the water as the primary mechanism for dealing with

high environmental temperatures (see Odell 1974 for a review). Likewise, many tropical or temperate pinniped species are overinsulated for life on land and rely on behavioural mechanisms to prevent overheating. During extended periods of time ashore, as occurs during breeding and molting, these marine mammals may avoid the sun or move into water to cool down (Gentry 1972; Limberger *et al.* 1986; Frances & Boness 1991). Similarly, some species of phocid seals will spend the warmer periods of the day in the water to avoid excessive heat absorption (Watts 1992).

---

when immersed. The former response is an excellent short-term solution, but is energetically expensive if maintained for long periods (see Chapter 9). The latter is the most efficient long-term solution for maintaining a high stable core temperature while living in water.

3.4.1 Increasing metabolic rate

Heat production in marine mammals may be increased by a variety of mechanisms including activity, the processing of food, and shivering and non-shivering thermogenesis (see Chapter 9). The sea otter provides an excellent example of a marine mammal that takes advantage of all of these mechanisms to help maintain its body temperature. This small marine mammal has a highly variable core body temperature (Costa 1982; Costa & Kooyman 1982; Davis *et al.* 1988) that rises during periods of activity and slowly falls during rest periods. To compensate for high heat losses in water, sea otters rely on a basal metabolic rate that is 2.4 times the expected value of a terrestrial mammal. In addition, these mammals increase their level of activity as water temperature decreases, and utilize heat produced from the digestion of food to maintain high core temperatures (Costa & Kooyman 1982, 1984).

The relative importance of these various mechanisms for supplementing heat production in other species of marine mammal is not known. For decades it had been widely accepted that the basal metabolic rates of marine mammals are considerably greater than those of terrestrial mammals of

similar size (Scholander 1940; Scholander *et al.* 1942; Irving & Hart 1957; Hart & Irving 1959; Kanwisher & Sundnes 1965, 1966; Ridgway & Patton 1971; Ridgway 1972; Snyder 1983). These high rates were thought to result from the perceived need to cope with thermal stresses associated with exposure to the cold aquatic environment. However, recent studies suggest that the basal metabolic rates of marine mammals may depend on the individual species and life history patterns (see Fig. 9.1, Chapter 9). Some species appear to have metabolic rates near those predicted for terrestrial mammals of similar size (Oritsland & Ronald 1975; Parsons 1977; Gallivan & Ronald 1979; Gaskin 1982; Lavigne *et al.* 1982; Yasui & Gaskin 1986; Worthy 1987; Kasting *et al.* 1989). This is consistent with measurements by Worthy *et al.* (1987) on juvenile porpoises, but contrasts with Kanwisher and Sundnes (1966) who suggest that porpoises need to function at six times predicted metabolic levels to survive. Other species such as the sea otter have exceptionally high basal metabolic rates. Still others including manatees and dugongs have relatively low metabolic rates that average 25–30% of values predicted for similarly sized terrestrial mammals (Scholander & Irving 1941; Gallivan & Best 1980, 1986; Gallivan *et al.* 1983; Irvine 1983; Miculka & Worthy 1995). Most probably, the herbivorous feeding habits, sedentary lifestyle and tropical distribution of these sirenians contribute to the low rates of heat production observed.

Activity and the processing of food may supplement heat production in many of these species, but will depend on the mobility of the animal as well as composition of the diet. Parry (1949) inferred from his studies on the insulation of harbour porpoise, that small cetaceans are obliged to remain active to maintain core body temperature. The Atlantic bottlenose dolphin and the Hawaiian spinner dolphin (*Stenella longirostris*) depend on the energy produced by activity and the digestion of food, as well as marked control over peripheral blood flow to maintain thermal balance (Hampton & Whittow 1976).

3.4.2 Increasing insulation

The alternative solution to maintaining core temperature during immersion is to insulate the body. For mammals, the type of insulation varies with the degree of aquatic specialization, and undoubtedly changed during the transition from terrestrial to aquatic living. Semiaquatic mammals such as beavers, muskrats and otters that move between the thermally disparate media of air and water, maintain dense pelage for insulation in water. In some of these species, fur insulation alone is inadequate for retaining body heat during prolonged periods of immersion. Small semiaquatic mammals including the water rat and mink show a progressive decrease in core body temperature when immersed, and must shuttle between cooling periods in water and warming periods on land (Williams 1986). In contrast, sea otters spend most of their lives in water and rely exclusively on fur for insulation (Fig. 3.10a). This is made possible by an exceptional, waterproof fur coat that is the densest of any mammal measured to date with over 150 000 hairs/cm² in some anatomical sites (Williams *et al.* 1992). The insulating value of this unique fur is similar to that of blubber but is packaged in a thinner layer. However, relying on dense pelage for insulation while in water comes at an energetic cost to the sea otter. The maintenance of this specialized fur requires that sea otters spend at least 12% of their day grooming to maintain its water repellency and insulating value.

The insulation of choice for obligate marine mammals is blubber. Otariids, particularly the fur seals, combine an external fur layer and an underlying blubber layer to keep warm. Most marine mammals, including phocid seals, sirenians and cetaceans, rely

(a)

(b)

Fig. 3.10 Two types of insulation in marine mammals. (a) Sea otters rely on dense pelage for remaining warm in water while cetaceans (b) and seals utilize a thick layer of blubber. The blubber layer also provides body streamlining and an energy store for marine mammals.

solely on a thick, internal blubber layer to conserve heat (Worthy & Lavigne 1987; Worthy 1991; Watts 1992; Williams *et al.* 1992; Miculka & Worthy 1995). Although blubber acts as the primary insulator for most marine mammals, it also serves several other functions including as an aid for streamlining and buoyancy, and as a major energy store (Fig. 3.10b). Depending on the animal's immediate needs, these roles may conflict and result in regional variations in blubber depth. For example, Ryg *et al.* (1988) reported that some anatomical sites of phocid seals appear to be overinsulated while other sites are underinsulated. Similar results have been found for elephant seals (Gales & Burton 1987) and a variety of cetaceans including beluga whales and narwhals (Doidge 1990), as well as bottlenose dolphins, a species that shows a high degree of seasonal variability in blubber thickness for different body regions.

A large database established for bottlenose dolphins documents seasonal changes in blubber depth (G.A.J. Worthy *et al.*, unpublished data). These changes correspond to seasonal changes in water temperature. However, the magnitude of change in blubber depth seems relatively minor in comparison to the marked seasonal temperature changes experienced by some dolphins. For example, mean overall blubber depths in Florida bottlenose dolphins range from 12.8 mm when the water temperatures approach 33°C to 17.6 mm in the winter when water temperatures decline to 20°C.

These relatively small changes in thickness may be all that is necessary for retaining body heat due to variability in the quality of insulation provided by blubber. The blubber layer is the most obvious store of body fat for marine mammals. Furthermore, the quantity and type of fat dictates the thermal characteristics of blubber (Parry 1949; Kanwisher & Sundnes 1966; Ryg *et al.* 1993). In some species, blubber may account for virtually all of the body's fat stores (Worthy *et al.* 1992). Worthy and Edwards (1990) reported that harbour porpoise blubber is comprised of 81.6 ± 3.6% lipid, while the blubber of spotted dolphins is only 54.9 ± 2.8% lipid. These differences in lipid content in addition to a thicker blubber layer results in an insulating layer in harbour porpoises that is four times more effective in retaining body heat than the blubber of spotted dolphins (Worthy & Edwards 1990). Comparative studies on other cetaceans indicate that the blubber of Pacific white-sided dolphins and common dolphins (*Delphinis delphis*) is a more effective insulator than that of bottlenose dolphins. The differences have been attributed to both blubber thickness and marked differences in lipid content (Worthy 1991).

The importance of insulation to the survival of marine mammals becomes especially apparent during environmental or anthropogenic events that overwhelm the thermoregulatory capabilities of the animal. This was noted during the 1989 *Exxon Valdez* oil spill in Alaska. The disruption of fur insulation following contamination with crude oil was a factor that contributed to the high mortality of sea otters

(Williams & Davis 1995). Unusual water temperatures encountered by wild marine mammals can also be detrimental if insulation is inadequate. This occurred in the winter of 1989–90 when exceptionally cold temperatures froze the coastal waterways of Texas. The thin blubber layer of bottlenose dolphins trapped in frozen bays resulted in high levels of heat loss and led to the deaths of 26 animals.

3.5 OSMOREGULATION: WATER BALANCE WHILE LIVING IN THE OCEAN

A delicate balance between water intake and excretion is required for mammals to maintain the appropriate concentration and volume of internal fluids that bathe the cells. This process, termed osmoregulation, is responsible for maintaining the concentration of water and electrolytes comprising the animal's internal environment. The primary organ responsible for osmoregulation is the kidney. Feeding and water ingestion by an animal alters osmoregulatory balance, which is re-established by the removal of excess fluids and electrolytes through urine, faeces and evaporation.

The high salinity of sea water and the absence of fresh drinking water present major physiological challenges to the osmoregulatory system of marine mammals, and undoubtedly influenced the transition from land to sea in some lineages. As observed for the other organ systems examined in this chapter, both physiological and morphological modifications accompanied the transition from reliance on fresh water to living in sea water.

The size and structure of the kidneys of marine mammals reveal the morphological solution to the problem of water balance when living in highly saline environments. In general, the kidneys of marine mammals are larger than found in terrestrial mammals of similar body mass (Beuchat 1996). The ratio of kidney to body mass ranges from 0.44% in the fin whale to 1.1% in the bottlenose dolphin and white-sided dolphin. This compares with the relatively small kidney to body mass ratio of terrestrial mammals which ranges from 0.3% in elephants to 0.4% in humans, deer and zebras (Slijper 1979). Another difference between marine and terrestrial

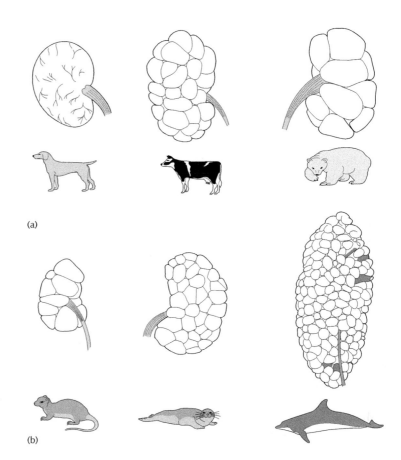

(a)

(b)

Fig. 3.11 External structure of the mammalian kidney. (a) Terrestrial mammals (dog, cow, bear) are compared to (b) semiaquatic (river otter) and marine (seal, dolphin) mammals. Note the increased number of lobes (reniculi) for the kidneys of marine-living mammals. (Redrawn from Slijper 1979.)

mammals is the number of lobes, termed reniculi, that comprise each kidney. Rather than a single, smooth lobe as found for humans and horses, the kidneys of many species of marine mammal are highly subdivided with each reniculus often serving as a complete miniature kidney with a cortex, medulla, papilla and calyx (Fig. 3.11). The number of reniculi is larger in cetaceans than observed for cattle. Thus, we find over 450 reniculi in the kidney of the bottlenose dolphin, and more than 3000 reniculi in mysticete whales. Elephants, bears, West Indian manatees and otters have 6–8 reniculi in each kidney, while cattle may have 25–30 reniculi. The dugong is an exception. Although this mammal lives in a marine environment, it has a smooth kidney. The number of reniculi in the kidneys of marine mammals corresponds roughly to the salinity of the diet, and, therefore, increased capacity for urine excretion. Whales and seals that feed on saline-enriched crustaceans tend to have a more

lobulated kidney than river dolphins or other freshwater animals with a low salt content diet (Slijper 1979).

Not all marine mammals live in a highly saline environment. Amazon River dolphins (*Inia geoffrensis*) and manatees occupy freshwater rivers, and some populations of phocid seals can be found in freshwater lakes. Polar-living phocids such as Weddell seals may chew on ice or snow, and captive seals will drink from a hose or trough (Ridgway 1972). Obviously, fresh water is available for these animals. However, is it absolutely necessary? Do marine mammals need to drink fresh water to maintain internal water and electrolyte balance? The answer to both of these questions is no.

Animals can utilize three basic sources of water for osmoregulation. The most obvious source is free water that the animal drinks. Less obvious, but of great importance to many marine mammals, is preformed and metabolic water in food. Preformed water is a direct component of food. Because most fish and invertebrates consist of 60–80% water, these prey items supply a considerable amount of free water to an animal without actually drinking. Metabolic water is derived from the metabolism of fat, protein or carbohydrate during the digestion of food. Mammals can derive 1.07 g of water from each gram of fat that is broken down, and 0.4 g of water from each gram of protein. The fattier the fish, the more water that is available to the animal. Interestingly, the same process takes place when the animal breaks down internal fat and protein stores, as occurs when marine mammals fast. Marine mammals can derive both energy and water from the catabolism of its own fat reserves during fasting. The catabolic process is so effective that no adverse effects were reported for a sea lion deprived of fresh water and salt water for 45 days (Pilson 1970). Concomitant with fasting by marine mammals is a decrease in urinary output to conserve water. Fasting elephant seal pups will reduce urine output by 84% after 10 weeks of fasting (Adams & Costa 1993) complementing other water-sparing mechanisms (Huntley et al. 1984).

Much of the anatomy and physiology of marine mammals is designed for reducing water loss. The major routes of water loss from the body are through evaporative processes and excretion in urine and faeces. Pinnipeds possess few sweat glands and cetaceans have none. Thus, water loss through surface evaporation is relatively minor in these mammals. Evaporation of water from the respiratory tract is also low due to the presence of countercurrent exchangers that help retain moisture (Huntley et al. 1984). These countercurrent mechanisms are similar in function to those of desert species. Water loss associated with urinary output will also vary according to the degree of protein catabolism. Protein breakdown results in the formation of urea that must be removed in the urine. As a result, increased protein catabolism results in increased water losses for marine mammals. The net effect is a loss of water.

Mariposia (seawater drinking) may be beneficial to animals on a high protein diet, since sea water can provide urinary osmotic space for urea (Hui 1981; Costa 1982). Sea otters have one of the highest reported rates of seawater consumption for any marine mammal, averaging 62 ± 27 ml/kg/day,

with a range of 0–124 ml/kg/day (Costa 1982). It is unlikely that this rate of saltwater ingestion is incidental to swallowing prey since sea otters consume their prey while floating on their backs. Rather, Costa (1982) suggests that sea otters actively consume sea water to aid in the high rates of urea production associated with the animal's high protein diet.

Seawater ingestion has been reported for a number of other marine mammals. Captive northern fur seals consume 1.8 ml sea water/kg body mass/day, and harbour seals will consume 4.8 ml/kg/day incidental to feeding (Depocas et al. 1971). The rate of seawater ingestion is considerably higher for cetaceans. Common dolphins will drink 12–13 ml/kg/day of sea water when not feeding, and take in approximately 73 ml sea water/kg body mass/day across the skin surface (Hui 1981). Similar rates of seawater influx have been reported for harbour porpoises and feeding Atlantic bottlenose dolphins (D.P. Costa & G.A.J. Worthy, unpublished data). The bottlenose dolphins had a water flux of 42. 1–71.3 ml/kg/day, 31% of which was from preformed and metabolic water, and 69% of which was from drinking water or water crossing the skin surface. Other species of seals, sea lions and porpoises either actively ingest sea water or are at least capable of it (Pilson 1970; Ridgway 1972; Gentry 1981).

Water balance is especially interesting in the manatee, a freshwater marine mammal. Little is known about the ability of wild manatees to osmoregulate and maintain water balance, but their anatomy suggests an enhanced capacity to concentrate their urine (Maluf 1989). Captive manatees held in saltwater conditions without access to fresh water and fed a diet of sea grass showed significant increases in plasma osmolality and plasma concentrations of sodium and chloride within 9 days (Ortiz et al. 1998). The manatees eventually refused to eat sea grasses containing high salt concentrations. These data suggest that wild manatees may require regular access to fresh, or perhaps brackish, water to meet water balance needs. In captive situations, this need is met by drinking fresh water or by eating food that is high in free water (e.g. lettuce at approximately 94% water). Manatees living in fresh water and consuming lettuce show the highest rate of water intake (145 ± 12 ml/kg/day) compared to manatees in salt water on a diet of lettuce (45 ± 3 ml/kg/day) or manatees exposed to salt water on a diet of sea grass (21 ± 3 ml/kg/day).

3.6 CONCLUSIONS

It is apparent from this chapter that the physiology of mammals underwent considerable modifications that accompanied marked anatomical changes during evolution. Many of the physiological responses can be attributed to the unique physical characteristics of the marine environment in comparison to life on land. Each modification in body form and physiological function undoubtedly came with an energetic consequence. How these modifications affect metabolic rate and contribute to the total energy budget of marine mammals is discussed in Chapter 9.

REFERENCES

Adams, S.H. & Costa, D.P. (1993) Water conservation and protein metabolism in northern elephant seal pups during the postweaning fast. *Journal of Comparative Physiology B* **163**, 367–373.

Andrews, R.D., Jones, D.R., Williams, J.D. *et al.* (1997) Heart rates of northern elephant seals diving at sea and resting on the beach. *Journal of Experimental Biology* **200**, 2083–2095.

Bengtson, J.L. & Stewart, B.S. (1992) Diving and haulout behavior of crabeater seals in the Weddell sea, Antarctica during March 1986. *Polar Biology* **12**, 635–644.

Bennett, A.F. (1985) Energetics and locomotion. In: *Functional Vertebrate Morphology* (M. Hildebrand, D.M. Bramble, K.F. Liem & D.B. Wake, eds), pp. 173–184. Harvard University Press, Cambridge, MA.

Berta, A., Ray, C.E. & Wyss, A.R. (1989) Skeleton of the oldest known pinniped, *Enaliarctos mealsi. Science* **244**, 60–62.

Beuchat, C.A. (1996) Structure and concentrating ability of the mammalian kidney: correlations with habitat. *American Journal of Physiology* **271**, R157–R179.

Blix, A.S., Gray, H.J. & Ronald, K. (1979) Some aspects of temperature regulation in newborn harp seal pups. *American Journal of Physiology* **236**, R188–R197.

Brooks, G.A., Fahey, T.D. & White, T.P. (1996) *Exercise Physiology: Human Bioenergetics and its Application*, 2nd edn. Mayfield Publishing Co., Mountainview, CA.

Bryden, M.M. (1979) Arteriovenous anastomoses in the skin of seals III. The harp seal *Pagophilus groenlandicus* and the hooded seal *Cystophora cristata* (Pinnipedia: Phocidae). *Aquatic Mammals* **6**, 67–75.

Bryden, M.M. & Molyneux, G.S. (1978) Arteriovenous anastomoses in the skin of seals II. The California sea lion *Zalophus californianus* and the northern fur seal *Callorhinus ursinus* (Pinnipedia: Otariidae). *Anatomical Record* **191**, 253–260.

Butler, P.J. & Jones, D.R. (1997) Physiology of diving of birds and mammals. *Physiological Reviews* **77** (3), 837–899.

Castellini, M.A. (1981) *Biochemical adaptations for diving in marine mammals.* PhD Thesis, University of California, San Diego.

Castellini, M.A., Murphy, B.J., Fedak, M. *et al.* (1985) Balancing the conflicting metabolic demands of diving and exercise in seals. *Journal of Applied Physiology* **58** (2), 392–399.

Costa, D.P. (1982) Energy, nitrogen, and electrolyte flux and sea water drinking in the sea otter, *Enhydra lutris. Physiological Zoology* **55**, 35–44.

Costa, D.P. & Kooyman, G.L. (1982) Oxygen consumption, thermoregulation, and the effects of fur oiling and washing on the sea otter, *Enhydra lutris. Canadian Journal of Zoology* **60**, 2761–2767.

Costa, D.P. & Kooyman, G.L. (1984) Contribution of specific dynamic action to heat balance and thermoregulation in the sea otter, *Enhydra lutris. Physiogical Zoology* **57**, 199–203.

Davis, R.W., Castellini, M.A., Kooyman, G.L. *et al.* (1983) Renal glomerular filtration rate and hepatic blood flow during voluntary diving in Weddell seals. *American Journal of Physiology* **245**, R743–R748.

Davis, R.W., Williams, T.M., Thomas, J.A., Kastelein, R.A. & Cornell, L.H. (1988) The effects of oil contamination and cleaning on sea otters (*Enhydra lutris*). II. Metabolism, thermoregulation, and behavior. *Canadian Journal of Zoology* **66**, 2782–2790.

Davis, R.W., Fuiman, L.A., Williams, T.M. *et al.* (1999) Hunting behavior of a marine mammal beneath the Antarctic fast-ice. *Science* **283**, 993–996.

Dejours, P. (1987) Water and air physical characteristics and their physiological consequences. In: *Comparative Physiology: Life in Water and on Land* (P. Dejours, L. Bolis, C.R. Taylor & E.R. Weibel, eds), pp. 3–11. Fidia Research Series. Springer-Verlag, New York.

Denison, D.M., Warrell, D.A. & West, J.B. (1971) Airway structure and alveolar emptying in the lungs of sea lions and dogs. *Respiration Physiology* **13**, 253–260.

Depocas, F., Hart, J.S. & Fisher, H.D. (1971) Sea water drinking and water flux in starved and in fed harbor seals, *Phoca vitulina. Canadian Journal of Physiology and Pharmacology* **49**, 53–62.

Dietz, R. & Heide-Jorgensen, M.P. (1995) Movements and swimming speed of narwhals, *Monodon monoceros*, equipped with satellite transmitters in Melville Bay, northwest Greenland. *Canadian Journal of Zoology* **73**, 2106–2119.

Doidge, D.W. (1990) Integumentary heat loss and blubber distribution in the beluga, *Delphinapterus leucas*, with comparisons to the narwhal, *Monodon monoceros*. In: *Advances in Research on the Beluga Whale, Delphinapterus Leucas* (T.G. Smith, D.J. St Aubin & J.R. Geraci, eds), pp. 129–140. Canadian Bulletin of Fisheries and Aquatic Science No. 224. Ottawa, Canada.

Elsner, R. (1966) Diving bradycardia in the unrestrained hippopotamus. *Nature* **212**, 408.

Elsner, R. (1999) Living in water, solutions to physiological problems. In: *Biology of Marine Mammals* (J.E. Reynolds III & S.A. Rommel, eds), pp. 73–116. Smithsonian Institution Press, Washington, DC.

Elsner, R., Kennedy, D.W. & Burgess, K. (1966) Diving bradycardia in the trained dolphin. *Nature* **212**, 407–408.

Fedak, M.A. & Seeherman, H.J. (1979) Reappraisal of energetics if locomotion shows identical costs in bipeds and quadrupeds including ostrich and horse. *Nature* **282**, 713–716.

Feldkamp, S.D. (1987) Foreflipper propulsion in the California sea lion, *Zalophus californianus*. *Journal of Zoology (London)* **212**, 43–57.

Fish, F.E. (1993a) Influence of hydrodynamic design and propulsive mode on mammalian swimming energetics. *Australian Journal of Zoology* **42**, 79–101.

Fish, F.E. (1993b) Power output and propulsive efficiency of swimming bottlenose dolphins (*Tursiops truncatus*). *Journal of Experimental Biology* **185**, 179–193.

Fish, F.E. (1996) Transitions from drag-based to lift-based propulsion in mammalian swimming. *American Zoologist* **36**, 628–641.

Fish, F.E. (1998) Comparative kinematics and hydrodynamics of odontocete cetaceans: morphological and ecological correlates with swimming performance. *Journal of Experimental Biology* **201**, 2867–2877.

Fish, F.E., Innes, S. & Ronald, K. (1988) Kinematics and estimated thrust production of swimming harp and ringed seals. *Journal of Experimental Biology* **137**, 157–173.

Frances, J.M. & Boness, D.J. (1991) The effect of thermoregulatory behaviour on the mating system of the Juan Fernandez fur seal, *Arctocephalus philippii*. *Behaviour* **119**, 104–126.

Gales, N.J. & Burton, H.R. (1987) Ultrasonic measurement of blubber thickness of the southern elephant seal, *Mirounga leonina* (Linn.). *Australian Journal of Zoology* **35**, 207–217.

Gallivan, G.J. & Best, R.C. (1980) Metabolism and respiration of the Amazonian manatee (*Trichechus inunguis*). *Physiological Zoology* **56**, 245–253.

Gallivan, G.J. & Best, R.C. (1986) The influence of feeding and fasting on the metabolic rate of the Amazonian manatee (*Trichechus inunguis*). *Physiological Zoology* **59**, 552–557.

Gallivan, G.J., Best, R.C. & Kanwisher, J.W. (1983) Temperature regulation in the Amazonian manatee *Trichechus inunguis*. *Physiological Zoology* **56**, 255–262.

Gallivan, G.J. & Ronald, K. (1979) Temperature regulation in freely diving harp seals (*Phoca groenlandica*). *Canadian Journal of Zoology* **57**, 2256–2263.

Gaskin, D.E. (1982) *The Ecology of Whales and Dolphins*. Heinemann, London.

Gentry, R.L. (1972) Thermoregulatory behavior of eared seals. *Behaviour* **46**, 73–93.

Gentry, R.L. (1981) Seawater drinking in eared seals. *Comparative Biochemistry and Physiology* **68**, 81–86.

Hampton, I.F.G. & Whittow, G.C. (1976) Body temperature and heat transfer in the Hawaiian spinner dolphin, *Stenella longirostris*. *Comparative Biochemistry and Physiology* **55A**, 195–197.

Hart, J.S. & Irving, L. (1959) The energetics of harbour seals in air and in water with special consideration of seasonal changes. *Canadian Journal of Zoology* **37**, 447–457.

Hertel, H. (1966) *Structure, Form, Movement*. Reinhold, New York.

Hill, R.D., Schneider, R.C., Liggins, G.C. *et al.* (1987) Heart rate and body temperature during free diving of Weddell seals. *American Journal of Physiology* **253**, R344–R351.

Hindell, M.A. & Lea, M.-A. (1998) Heart rate, swimming speed, and estimated oxygen consumption of a free-ranging southern elephant seal. *Physiological Zoology* **71** (1), 74–84.

Hindell, M.A., Slip, D.J. & Burton, H.R. (1991) The diving behavior of adult male and female southern elephant seals, *Mirounga leonina* (Pinnipedia: Phocidae). *Australian Journal of Zoology* **39**, 595–619.

Hochachka, P.W. (1986) Balancing the conflicting metabolic demands of exercise and diving. *Federation Proceedings* **45**, 2948–2952.

Hochachka, P.W. (1998) Mechanism and the evolution of hypoxia-tolerance in humans. *Journal of Experimental Biology* **201**, 1243–1254.

Hui, C.A. (1981) Seawater consumption and water flux in the common dolphin, *Delphinus Delphis*. *Physiological Zoology* **54**, 430–440.

Huntley, A.C., Costa, D.P. & Rubin, R.D. (1984) The contribution of nasal countercurrent heat exchange to water balance in the northern elephant seal, *Mirounga angustirostris*. *Journal of Experimental Biology* **113**, 447–454.

Hurford, W.E., Hochachka, P.W., Schneider, R.C. *et al.* (1996) Splenic contraction, catecholamine release, and blood Volume distribution during diving in the Weddell seal. *Journal of Applied Physiology* **80** (1), 298–306.

Irvine, A.B. (1983) Manatee metabolism and its influence on distribution in Florida. *Biological Conservation* **25**, 315–334.

Irving, L. (1939) Respiration in diving mammals. *Physiological Reviews* **19**, 112–134.

Irving, L. & Hart, J.S. (1957) The metabolism and insulation of seals as bare skinned mammals in cold water. *Canadian Journal of Zoology* **35**, 497–511.

Kanatous, S.B., DiMichele, L.V., Cowan, D.F. & Davis, R.W. (1999) High aerobic capacities in the skeletal muscles of pinnipeds: adaptations to diving hypoxia. *Journal of Applied Physiology* **86** (4), 1247–1256.

Kanwisher, J. & Sundnes, G. (1965) Physiology of a small cetacean. *Hvalradwts Skrifter* **48**, 45–53.

Kanwisher, J. & Sundnes, G. (1966) Thermal regulation of whales and dolphins. In: *Whales, Dolphins and Porpoises* (K. Norris, ed.), pp. 397–407. University of California Press, Berkeley, CA.

Kasting, N.W., Adderley, S.A.L., Safford, T. & Hewlett, K.G. (1989) Thermoregulation in beluga (*Delphinapterus leucas*) and killer (*Orcinus orca*) whales. *Physiological Zoology* **62**, 687–701.

Kooyman, G.L. (1973) Respiratory adaptations of marine mammals. *American Zoologist* **13**, 457–468.

Kooyman, G.L. (1989) *Diverse Divers.* Springer-Verlag, Berlin.

Kooyman, G.L. & Campbell, W.B. (1972) Heart rates in freely diving Weddell seals, *Leptonychotes weddelli*. *Comparative Biochemistry and Physiology* **43A**, 31–36.

Kooyman, G.L., Hammond, D.D. & Schroeder, J.P. (1970) Bronchograms and tracheograms of seals under pressure. *Science* **169**, 82–84.

Lavigne, D.M., Barchard, W., Innes, S. & Oritsland, N.A. (1982) Pinniped bioenergetics. *FAO Fisheries Series* **5**, 191–235.

Lenfant, C., Johansen, K. & Torrance, J.D. (1970) Gas transport and oxygen storage capacity in some pinnipeds and the sea otter. *Respiration Physiology* **9**, 277–286.

Limberger, D., Trillmich, F., Biebach, H. & Stevenson, R.D. (1986) Temperature regulation and microhabitat choice by free-ranging Galapagos fur seal pups (*Arctocephalus galapagoensis*). *Oecologia* **69**, 53–59.

Maluf, N.S.R. (1989) Renal anatomy of the manatee, *Trichechus manatus*, Linneaus. *American Journal of Anatomy* **184**, 269–286.

Matsuura, D.T. & Whittow, G.C. (1974) Evaporative heat loss in the California sea lion and harbor seal. *Comparative Biochemistry and Physiology* **48A**, 9–20.

Miculka, T.A. & Worthy, G.A.J. (1995) Metabolic capabilities and the limits to thermoneutrality in juvenile and adult West Indian manatees (*Trichechus manatus*) (abstract). In: *Eleventh Biennial Conference on the Biology of Marine Mammals, Orlando, Florida, 14–18 December 1995*, p. 77.

Molyneux, G.S. & Bryden, M.M. (1978) Arteriovenous anastomoses in the skin of seals I. The Weddell seal *Leptonychotes weddelli* and the elephant seal *Mirounga leonina* (Pinnipedia: Phocidae). *Anatomical Record* **191**, 239–252.

Noren, S. (1997) *Oxygen stores and acid buffering capacities of cetacean skeletal muscle: a hierarchy in adaptations for maximum dive durations.* MS Thesis, University of California, Santa Cruz.

Noren, S.R. & Williams T.M. (2000) Body size and skeletal muscle myoglobin of cetaceans: adaptations for maximizing dive duration. *Comparative Biochemistry and Physiology A* **126**, 181–191.

Odell, D.K. (1974) Behavioral thermoregulation in the California sea lion. *Behavioral Biology* **10**, 231–237.

Olsen, C.R., Hale, F.C. & Elsner, R. (1969) Mechanics of ventilation in the pilot whale. *Respiration Physiology* **7**, 137–149.

Oritsland, N.A. & Ronald, K. (1975) Energetics of the freely diving harp seal (*Pagophilus groenlandicus*). *Rapports et Proces-Verbaux des Reunions, Conseil International pur L'Exploration de la Mer* **169**, 451–455.

Ortiz, R.M., Worthy, G.A.J. & MacKenzie, D.S. (1998) Osmoregulation in wild and captive West Indian manatees (*Trichechus manatus*). *Physiological Zoology* **71** (4), 449–457.

Pabst, D.A., Rommel, S.A., McLellan, W.A. *et al.* (1995) Thermoregulation of the intra-abdominal testes of the bottlenose dolphin (*Tursiops truncatus*) during exercise. *Journal of Experimental Biology* **198**, 221–226.

Parry, D.A. (1949) The structure of whale blubber and a discussion of its thermal properties. *Quarterly Journal of Microscopic Science* **90**, 13–26.

Parsons, J.L. (1977) *Metabolic studies on ringed seals (Phoca hispida).* MSc Thesis, University of Guelph, Guelph, Canada.

Pilson, M.E.Q. (1970) Water balance in California sea lions. *Physiological Zoology* **43**, 257–269.

Ponganis, P.J., Kooyman, G.L., Castellini, M.A. *et al.* (1993) Muscle temperature and swim velocity profiles during diving in a Weddell seal, *Leptonychotes weddellii*. *Journal of Experimental Biology* **183**, 341–348.

Ponganis, P.J., Kooyman, G.L., Winter, L.M. & Starke, L.N. (1997) Heart rate and plasma lactate responses during submerged swimming and trained diving in California sea lions, *Zalophus californianus*. *Journal of Comparative Physiology B* **167**, 9–16.

Qvist, J., Hill, R.D., Schneider, R.C. *et al.* (1986) Hemoglobin concentrations and blood gas tensions of free-diving Weddell seals. *Journal of Applied Physiology* **61**, 1560–1569.

Repenning, C.A. (1976) Adaptive evolution of sea lions and walruses. *Systematic Zoology* **25**, 375–390.

Ridgway, S.H. (1972) Homeostasis in the aquatic environment. In: *Mammals of the Sea*, pp. 590–747. C.C. Thomas, Springfield, IL.

Ridgway, S.H. & Howard, R. (1979) Dolphin lung collapse and intramuscular circulation during free diving: evidence from nitrogen washout. *Science* **206**, 1182–1183.

Ridgway, S.H. & Johnston, D.G. (1966) Blood oxygen and ecology of porpoises of three genera. *Science* **151**, 456–458.

Ridgway, S.H. & Patton, G.S. (1971) Dolphin thyroid: some anatomical and physiological findings. *Zeitschrift fuer Vergleichende Physiologie* **71**, 129–141.

Ridgway, S.H., Scronce, B.L. & Kanwisher, J. (1969) Respiration and deep diving in the bottlenose porpoise. *Science* **166**, 1651–1654.

Ridgway, S.H., Carder, D.A. & Clark, W. (1975) Conditioned bradycardia in the sea lion, *Zalophus californianus*. *Nature* **256**, 370–338.

Rommel, S.A., Pabst, D.A., McLellan, W.A. *et al.* (1994) Temperature regulation of the testes of the bottlenose dolphin (*Tursiops truncatus*): evidence from colonic temperatures. *Journal of Comparative Physiology B* **164**, 130–134.

Ryg, M., Smith, T.G. & Oritsland, N.A. (1988) Thermal significance of the topographical distribution of blubber in ringed seals (*Phoca hispida*). *Canadian Journal of Fisheries and Aquatic Science* **45**, 985–992.

Ryg, M., Lyderson, C., Knutsen, L.O. *et al.* (1993) Scaling of insulation in seals and whales. *Journal of Zoology* **230**, 193–206.

Schmidt-Nielsen, K. (1972) Locomotion: energy cost of swimming, flying, and running. *Science* **177**, 222–228.

Schmidt-Nielsen, K. (1984) *Scaling: Why Is Animal Size So Important?*. Cambridge University Press, New York.

Scholander, P.F. (1940) Experimental investigations on the respiratory function in diving mammals and birds. *Hvalradwts Skrifter* **22**, 1–31.

Scholander, P.F. & Irving, L. (1941) Experimental investigations on the respiration and diving of the Florida manatee. *Journal of Cellular and Comparative Physiology* **17**, 169–191.

Scholander, P.F. & Schevill, W.E. (1955) Counter-current vascular heat exchange in the fins of whales. *Journal of Applied Physiology* **8**, 279–282.

Scholander, P.F., Irving, L. & Grinnell, S.W. (1942) On the temperature and metabolism of the seal during diving. *Journal of Cellular and Comparative Physiology* **19**, 67–78.

Schreer, J.F. & Kovacs, K.M. (1997) Allometry of diving capacity in air-breathing vertebrates. *Canadian Journal of Zoology* **75**, 339–358.

Skrovan, R.C., Williams, T.M., Berry, P.S., Moore, P.W. & Davis, R.W. (1999) The diving physiology of bottlenose dolphins (*Tursiops truncatus*) II. Biomechanics and changes in buoyancy with depth. *Journal of Experimental Biology* **202**, 2749–2761.

Slijper, E.J. (1976) *Whales and Dolphins*. University of Michigan Press, Ann Arbor, MI.

Slijper, E.J. (1979) *Whales*. Cornell University Press, New York.

Snyder, G.K. (1983) Respiratory adaptations in diving mammals. *Respiration Physiology* **54**, 269–294.

South, F.E., Luecke, R.H., Zatzman, M.L. & Shanklin, M.D. (1976) Air temperature and direct partitional calorimetry of the California sea lion (*Zalophus californianus*). *Comparative Biochemistry and Physiology* **54**, 27–30.

Spencer, M.P., Gornall, T.A. & Poulter, T.C. (1967) Respiratory and cardiac activity of killer whales. *Journal of Applied Physiology* **22**, 974–981.

Tarasoff, F.J. & Kooyman, G.L. (1973) Observations on the anatomy of the respiratory system of the river otter, sea otter, and harp seal. II. The trachea and bronchial tree. *Canadian Journal of Zoology* **51**, 171–177.

Tawara, T. (1950) On the respiratory pigments of whale (studies on whale blood II.). *Scientific Reports of the Whales Research Institute* **3**, 96–101.

Taylor, C.R. & Rowntree, V.J. (1973) Running on two or on four legs: which consumes more energy? *Science* **1779**, 186–187.

Thewissen, J.G.M., Hussain, S.T. & Arif, M. (1994) Fossil evidence for the origin of aquatic locomotion in archaeocete whales. *Nature* **263**, 210–212.

Thompson, D. & Fedak, M.A. (1993) Cardiac responses of grey seals during diving at sea. *Journal of Experimental Biology* **174**, 139–164.

Watkins, W.A., Moore, K.E. & Tyack, P. (1985) Investigations of sperm whale acoustic behaviors in the southeast Caribbean. *Cetology* **49**, 1–15.

Watts, P. (1992) Thermal constraints on hauling out by harbour seals (*Phoca vitulina*). *Canadian Journal of Zoology* **70**, 553–560.

Webb, P.W. (1975) Hydrodynamics and energetics of fish propulsion. *Bulletin of the Fisheries Research Board of Canada* **190**, 1–159.

Webb, P.W. (1984) Body form, locomotion and foraging in aquatic vertebrates. *American Zoologist* **28**, 709–725.

Weibel, E.R., Taylor, C.R., Hoppeler, H. *et al.* (1987) Adaptive variation in the mammalian respiratory system in relation to energetic demand: I. Introduction to problem and strategy. *Respiration Physiology* **69**, 1–6.

Wiig, O., Gjertz, I., Griffiths, D. *et al.* (1993) Diving patterns of an Atlantic walrus, *Odobenus rosmarus rosmarus* near Svalbard. *Polar Biology* **13**, 71–72.

Williams, T.D., Allen, D.D., Groff, J.M. & Glass, R.L. (1992) An analysis of California sea otter (*Enhydra lutris*) pelage and integument. *Marine Mammal Science* **8** (1), 1–18.

Williams, T.M. (1986) Thermoregulation in the North American mink during rest and activity in the aquatic environment. *Physiological Zoology* **59**, 293–305.

Williams, T.M. (1989) Swimming by sea otters: adaptations for low energetic cost locomotion. *Journal of Comparative Physiology A* **164**, 815–824.

Williams, T.M. (1999) The evolution of cost efficient swimming in marine mammals: limits to energetic optimization. *Philosophical Transactions of the Royal Society of London B* **354**, 193–201.

Williams, T.M. & Davis, R.W. (1995) *Emergency Care and Rehabilitation of Oiled Sea Otters. A Guide for Oil Spills Involving Fur-Bearing Marine Mammals.* University of Alaska Press, Fairbanks, AK.

Williams, T.M. & Kooyman, G.L. (1985) Swimming performance and hydrodynamic characteristics of harbor seals. *Phoca Vitulina Physiological Zoology* **58** (5), 576–589.

Williams, T.M., Friedl, W.A. & Haun, J.E. (1993) The physiology of bottlenose dolphins (*Tursiops truncatus*): heart rate, metabolic rate and plasma lactate concentration during exercise. *Journal of Experimental Biology* **179**, 31–46.

Williams, T.M., Dobson, G.P., Mathieu-Costello, O. *et al.* (1997) Skeletal muscle histology and biochemistry of an elite sprinter, the African cheetah. *Journal of Comparative Physiology B* **167**, 527–535.

Williams, T.M., Haun, J.E. & Friedl, W.A. (1999a) The diving physiology of bottlenose dolphins, (*Tursiops truncatus*) I. Balancing the demands of exercise for energy conservation at depth. *Journal of Experimental Biology* **202**, 2739–2748.

Williams, T.M., Noren, D., Berry, P. *et al.* (1999b) The diving physiology of bottlenose dolphins, (*Tursiops truncatus*) II.

Thermoregulation at depth. *Journal of Experimental Biology* **202**, 2763–2769.

Williams, T.M., Davis, R.W., Fuiman, L.A. *et al.* (2000) Sink or swim strategies for cost efficient diving by marine mammals. *Science* **288**, 133–136.

Worthy, G.A.J. (1987) Metabolism and growth of young harp and grey seals. *Canadian Journal of Zoology* **65**, 1377–1382.

Worthy, G.A.J. (1991) Thermoregulatory implications of the interspecific variation in blubber composition of odontocete cetaceans (abstract). In: *Ninth Biennial Conference on the Biology of Marine Mammals, 5–12 December 1991*, p. 74. Chicago, IL.

Worthy, G.A.J. & Edwards, E.F. (1990) Morphometric and biochemical factors affecting heat loss in a small temperate cetacean (*Phocoena phocoena*) and a small tropical cetacean (*Stenella attenuata*). *Physiological Zoology* **63**, 432–442.

Worthy, G.A.J. & Lavigne, D.M. (1987) Mass loss, metabolic rate and energy utilization by harp and grey seal pups during the postweaning fast. *Physiological Zoology* **60**, 352–364.

Worthy, G.A.J., Innes, S., Braune, B.M. & Stewart, R.E.A. (1987) Rapid acclimation of cetaceans to an open-system respirometer. In: *Approaches to Marine Mammal Energetics* (A.C. Huntley, D.P. Costa, G.A.J. Worthy & M.A. Castellini, eds), pp. 115–126. Society for Marine Mammalogy Special Publication No. 1. Allen Press, Lawrence, KS.

Worthy, G.A.J., Morris, P.A., Costa, D.P. & LeBoeuf, B.J. (1992) Moult energetics of the northern elephant seal (*Mirounga angustirostris*). *Journal of Zoology (London)* **227**, 257–265.

Yasui, W.Y. & Gaskin, D.E. (1986) Energy budget of a small cetacean, the harbor porpoise, *Phocoena phocoena* (L.). *Ophelia* **25**, 183–197.

Zapol, W.M., Liggins, G.C., Schneider, R.C. *et al.* (1979) Regional blood flow during simulated diving in the conscious Weddell seal. *Journal of Applied Physiology* **47**, 968–973.

CHAPTER 4

Neural Morphology

Ilya I. Glezer

4.1 INTRODUCTION

Comparative studies of the central and peripheral nervous systems have a long history leading back to the classic Greco-Roman sources of medical science. It seems almost improbable, but more than a thousand years ago the first stones of the contemporary understanding of anatomy and physiology of the nervous system were laid down by the ancient geniuses of Hippocrates, Galen and Empedocles. However, it was not until the nineteenth and twentieth centuries that the morphophysiological studies of the central and peripheral nervous systems received a solid technological and experimental basis with the introduction of electrophysiology, light and electron microscopy, biochemistry, clinical and experimental embryology and pathology.

From the very beginning, investigators were struck by the extreme size and convolutional complexity of the neocortex in two groups of mammals: primates (including *Homo sapiens*) and cetaceans (Beauregard 1883; Kükenthal & Ziehen 1889; Elliot Smith 1910; Langworthy 1931, 1932; Le Gros Clark 1932; Ariëns Kappers *et al.* 1936; Filimonoff 1949; Breathnach 1960; Ebner 1969; Pilleri & Kraus 1969; Pilleri & Chen 1982; Ebbesson 1984; Morgane *et al.* 1986a, 1986b). Why and how these two groups of mammals had evolved such complicated brains still remains an unsolved puzzle. Many extremely successful biological groups have existed for millions of years with much less qualitatively and quantitatively developed brains (e.g. dinosaurs, reptiles, amphibians, fishes, birds, etc.). Presumably the ecological niche occupied by primates and cetaceans required a relatively high capacity for information processing. Some general evolutionary trends during the

radiation of mammals are instructive, though any implication of a wholly linear process would be far too simplistic.

Limiting my discussion to the Palaeocene epoch (i.e. the beginning of the extreme mammalian radiations, and following the extinctions of the major dinosaurian groups), the following trends in mammalian brain evolution can be noted:

1 A general trend towards **allometric** (see Box 4.1 for a glossary of terms, given in bold at the first mention in the text) increases in absolute and, in particular, relative mass of the brain with respect to body size. This process, called encephalization, is primarily due to the general increase in the number of cells and the corresponding growth of neuronal processes (axons and dendrites) and synaptic contacts. In turn, this overall amplification of interneuronal connections leads to the increase of non-neuronal elements, such as glial cells and vessels.

2 Among the brain regions, the most intensive growth was found in the **telencephalon** in comparison with the brainstem, termed telencephalization.

3 In the telencephalon the most intensive growth was in the **cerebral cortex** in comparison with the **basal ganglia**, termed corticalization.

4 In the cerebral cortex there was relatively greater development and growth of the **neocortical** areas in comparison with archicortical (**hippocampus**) and palaeocortical (**rhinencephalon**) formations (termed neocorticalization). It should be noted that this process is quite variable among mammalian orders and species, and only the most general characteristics are provided here. As a rule, information projected through modular pathways to the so-called primary neocortical region are specialized for the discrete analysis of specific modalities (visual, auditory, etc.) with **somatotopic** representation of particular

BOX 4.1 GLOSSARY

Afferent: an axon or group of axons (tract) bringing information to given structures.

Allometric: pertaining to the quantitative relationships between the part of the structure or system and the whole developing structure or system.

Basal ganglia: a collection of associated nuclei including the amygdala, globus pallidus, caudate nucleus, putamen, subthalamic nucleus and substantia nigra.

Cerebral aqueduct: a narrow tube in the mesencephalon connecting the third and fourth ventricles of the brain.

Cerebral cortex: a layered plate of grey matter covering the hemispheres of the telencephalon.

Cytoarchitectonic maps: schematic representations of the distribution of structural regions (areas) on the surface of the cerebral cortex.

Efferent: an axon or group of axons (tract) conveying information from given structures.

Gyrification: extent and distribution of the cortical gyri (folds of cortical plate).

Hippocampus: gyrus (fold) of the cortical plate representing the archicortex.

Lissencephalic: smooth cortical surface void of cortical folds (gyri).

Lumbosacral: pertaining to lumbar (lower) and sacral (associated with the pelvis in most vertebrates) regions of vertebrate column or spinal cord.

Mesencephalon (syn. midbrain): part of the brainstem connecting the metencephalon (hindbrain) and diencephalon (intermediate brain).

Neocortical: pertaining to the neocortex, the most phylogenetically recent part of the cerebral cortex.

Occipital: pertaining to the most posterior part of the skull or brain.

Ontogenesis: the process of the individual development of an organism, starting from conception to the adult stage and ending with death.

Paedomorphic: pertaining to the ontogenetic shift of the early stages of development to later stages or even to the adult stage.

Proisocortical: having primitive features of the cortex (in cell distribution, layering and cytoarchitectonic mapping).

Pyramidal tract: major efferent tract originating from the neocortex and synapsing on nuclei of the brainstem and spinal cord.

Rhinencephalon: complex of cortical and subcortical structures related to olfactory function.

Somatic: pertaining to the innervation of skin, joints and striate skeletal muscles.

Somatotopic: topographic organization of the sensory and motor systems where each point (cell group) of the functional map in the cerebral cortex corresponds to each point of the peripheral receptor.

Sulcation: extent and distribution of the furrows (sulci) between cortical folds (gyri) of the cerebral cortex.

Telencephalon (endbrain): part of the brain consisting of the basal ganglia and cerebral cortex.

Vestibular: pertaining to regulation of balance, perception of space relationships, and character of movements (rotation, acceleration, deceleration, gravity).

Visceral: pertaining to the innervation of the inner organs and smooth musculature.

---

functions. From these primary areas, transcortical pathways transferred this information to secondary, tertiary and other areas for more complex analysis and synthesis. The final product of this functional mapping transferred eventually into the memory blocks and **efferent** areas. In most terrestrial species, pinnipeds and sirenians, efferent and **afferent** areas of the neocortex are morphologically specialized in terms of cellular organizations and the laminar distribution of cell populations. In cetaceans, functionally different areas are extremely homogenous in terms of cellular organization and anatomy.

5 Intensive neocortical growth resulted in a complicated process of **gyrification** and **sulcation** of the cerebral hemispheres. This in turn, was a consequence of many complimentary factors, some of them mechanical (packaging of the largest possible brain surface into the smallest volume of skull), and others functional, such as mapping the functional brain areas according to informational inputs and outputs. Overall, in all mammalian groups (with some important exceptions, such as sirenians) the evolution of the brain resulted in the conversion of **lissencephalic** brain surfaces into gyrified and sulcated surfaces. In general, within a given mammalian order, the species with larger body mass have more a gyrified and sulcated brain than smaller species. For all mammalian species, the earlier stages of embryonic and fetal development are characterized by lissencephalic brains which later in **ontogenesis** change into more gyrified ones.

Although there are three major groups of marine mammals: cetaceans, sirenians and pinnipeds, this chapter will concentrate mainly on the central and peripheral nervous systems of cetaceans, since there are incomparably more data for these species. All marine mammals are secondarily aquatic since their ancestors originated from land mammals. The differences between the three groups are: (i) the identity of their terrestrial ancestors, (ii) how far their morphology and physiology changed due to this descent, and (iii) what particular ecological niche the extant species occupy (see Chapters 1 and 2).

Pinnipedia, while being excellently adapted to the marine habitat, are actually semiaquatic mammals reproducing on land and spending at least half of their lives onshore. Their ancestors adapted to the marine environment in the Oligocene, which is much later than for the sirenians and cetaceans. Morphologically and physiologically, their central and peripheral nervous systems do not differ significantly from those of terrestrial carnivores. The Sirenia (manatees and dugongs) are totally aquatic mammals. They descended from early ungulates probably adapting to the water in the early Eocene, about 50–60 million years ago (see Chapter 2). However, their adaptations were relatively narrow in terms of their habitat (only the shallow waters of the shelf) and in terms of locomotion and feeding behaviours (see Chapter 1). In ecological terms, only cetaceans (whales and dolphins) are fully adapted to the life in the deep seas.

4.2 BRAIN SIZE

Table 4.1 shows correlations between body weight and brain weight in cetaceans in comparison with terrestrial mammals. In general, the evolution of odontocete cetaceans was characterized by the disproportionate growth of the brain mass leading to high encephalization indices in extant species. The relative size and shape of marine and terrestrial mammalian brains are illustrated in Fig. 4.1.

The proportions of the different regions of the brain are also distinct from terrestrial species. For example, in both Mysticeti and Odontoceti the weight of the cerebral hemispheres constitutes approximately 68–75%, the cerebellum 18–23% and the brainstem excluding the cerebellum 4–5% (Pilleri & Kraus 1969; Ridgway & Brownson 1984; Schwerdtfeger *et al.* 1984). In humans the mean

weight (volume) of the cerebral hemispheres is approximately 83–88%, the cerebellum 10–11% and the brainstem 1.8–1.9% (Blinkov & Glezer 1968).

For mysticetes, the relative size of the brain is small, even though its absolute size is impressively large. On the other hand, both the relative and absolute size of the odontocete brain is comparable to that of anthropoid primates (Table 4.1). Moreover, in some odontocetes, especially from the family Delphinidae, the relative size of the brain is 2–4 times larger than in most advanced anthropoids (e.g. the chimpanzee, *Pan troglodytes*); even so, the highest allometric index in cetaceans is significantly lower than in *Homo sapiens*. Parametric data are of limited use in this context, however, as the relationships are often exponential and depend on the velocity of brain and body growth in ontogenesis, which varies in different species (Blinkov & Glezer 1968). Relative to body size, odontocetes clearly have the largest brains among the cetaceans (Table 4.1); and among the odontocetes, the bottlenose dolphins (*Tursiops truncatus*) and common dolphins (*Delphinus delphis*) have the highest relative index of cephalization. The lowest odontocete cephalization indices are found in some of the river dolphins (*Plantanista*, *Pontoporia*) and larger species, such as the sperm whale and pilot whale.

In seals and other pinnipeds, the relative weight of the brain is comparable to that of some cetaceans, primates, carnivores and ungulates (Table 4.1), and the relative size of the cerebral cortex represents 90% of the total size of telencephalon (Filimonoff 1949). Manatees and dugongs are characterized by the smallest encephalization indexes, comparable to those of some baleen whales, and of semiaquatic animals such as the hippopotamus and beaver (Table 4.1). However, in the sirenians, the relative volume of the telencephalon is within the range of cetaceans (71%), and the cerebral cortex comprises 90% of the telencephalon (Kamya *et al.* 1979; Reep & O'Shea 1990).

4.3 SHAPE AND OTHER GROSS ANATOMICAL FEATURES OF THE BRAIN

The shape of the cetacean brain, in comparison with the shapes of brains in advanced terrestrial

Table 4.1 Brain and body weight in Cetacea, Pinnipedia and Primates. (Data from Filimonoff 1949; Pilleri 1962, 1966a, 1966b, 1969; Blinkov & Glezer 1968; Pilleri & Gihr 1968, 1969a, 1969b; Pilleri *et al.* 1968; Pilleri & Busnel 1969; Pilleri & Kraus 1969; Kamya *et al.* 1979; Ridgway & Brownson 1984; Schwerdtfeger *et al.* 1984; Reep & O'Shea 1990)

	Brain (E) weight (g)	Body (P) weight (kg)	Ratio E : P	Allometric index (E²/P)	Regression coefficient (LogP vs logE)
Order Cetacea					
Suborder Mysticeti					
Balaenoptera physalus	7111	30 435	1 : 4279	1.66	0.1498
Balaenoptera musculus	3636	50 904	1 : 14000	0.26	
Megaptera novaeangliae	6439	39 311	1 : 6105	1.06	
Suborder Odontoceti					
Physeter catadon	7849	37 206	1 : 4740	1.66	0.1115
Orcinus orca	5620	6000	1 : 1067	5.26	
Globicephala melas	2500	3000	1 : 1200	2.08	
Delphinapterus leucas	2263	500	1 : 220	10.24	
Tursiops truncatus	1560	216	1 : 138	11.2	0.3174
Delphinus delphis	776	51.3	1 : 66	11.7	
Phocoena phocoena	458	27.8	1 : 61	7.55	0.5691
Inia geoffrensis	610	62.4	1 : 102	5.9	0.3191
Pontoporia blainvillei	220	34.9	1 : 159	1.38	
Order Carnivora					
Suborder Pinnipedia					
Erignathus barbatus	460	281	1 : 610	0.75	
Odobenus rosmarus	1126	667	1 : 592	1.9	
Phoca richardi	442	107.3	1 : 243	1.8	
Phoca hispida	253	40	1 : 158	1.6	
Phoca vitulina	271	30	1 : 110	2.24	
Order Sirenia					
Dugong dugon	220.6	187.5	1 : 851	0.26	
Trichechus manatus	302.2	450	1 : 1489	0.20	
Order Ungulata, Artiodactyla					
Hippopotamus amphibius	480.0	1867.5	1 : 3891	0.12	
Order Rodentia					
Castor canadensis	41	11.7	1 : 285	0.14	
Order Primates					
Erythrocebus patas	115	10	1 : 89	1.3	
Pan troglodytes	440	32	1 : 118	3.7	
Gorilla gorilla	406	207	1 : 509	0.8	
Homo sapiens	1400	70	1 : 50	28	0.3174

species (Primates, Carnivora, Ungulata), is characterized by the prevalence of the transverse and frontal diameters over the sagittal (longitudinal) diameters. In other words, the brain of cetaceans (both Mysticeti and Odontoceti) is shortened in the anteroposterior axis, and relatively increased in height and width (Figs 4.1–4.3). The specific proportions of the cetacean brain reflect the overall evolutionary changes in body form, as it adapted to the aquatic environment and the selective pressures of hydrodynamic locomotion. The overall rounded shape of the brain hemispheres in cetaceans is a consequence of the absence of the pointed **occipital** and frontal poles (Figs 4.1–4.3). Only the temporal

Terrestrial mammals

Aquatic, semiaquatic mammals

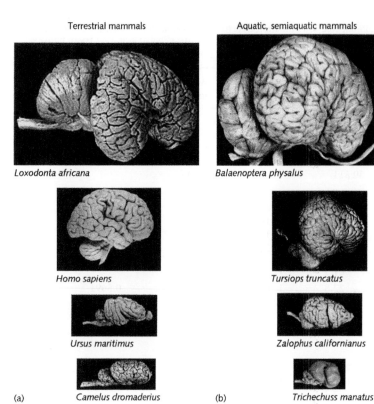

Loxodonta africana

Balaenoptera physalus

Homo sapiens

Tursiops truncatus

Ursus maritimus

Zalophus californianus

(a)   *Camelus dromaderius*   (b)   *Trichechuss manatus*

Fig. 4.1 A lateral view of the brains of: (a) some terrestrial mammals (from top to bottom: African elephant, man, bear and camel), and (b) some aquatic and semiaquatic mammals (from top to bottom: fin whale, bottlenose dolphin, sea lion and manatee).

poles can be identified. Peculiar relationships exist between the axes of the hemispheres and the brainstem. The axis of the **mesencephalon** is almost perpendicular to the axis of the pons, and the anteroposterior axis of the hemisphere is almost perpendicular to the axis of the mesencephalon (Fig. 4.3). As a result, the telencephalon of the cetacean brain is bent ventrally. Again, these peculiar relationships between the axes of the brainstem and cerebral hemispheres relate to the evolutionary hydrodynamic and respiratory changes in cetacean skull morphology. The 'migration' of the external nares (nostrils) to the top of the head, with simultaneous dorsal dislocation of the whole facial skeleton, often referred to as the 'telescoping' of the skull (see Chapter 2), have resulted in profound changes in the overall shape of the cetacean brain and the distribution of the cranial nerves. This evolutionary change in the external shape of the brain also brought about deep modifications in the shape of the ventricles and the **cerebral aqueduct**. For example, the occipital horns of the lateral ventricles are almost non-expressed, while the body of the

lateral ventricle has an unusually narrow and high arc (Fig. 4.3). The inferior (temporal) horn is present but is relatively shorter and narrower than in the brains of terrestrial animals of comparable size (humans, for example). The changes described above are also reflected in the shape of the brainstem regions (Fig. 4.3).

The general shape and proportions of the brain in pinnipeds are the same as in advanced terrestrial carnivores (Figs 4.1 and 4.2). Unlike in cetaceans, the cerebrum of pinnipeds has clearly defined occipital and frontal lobes, and gyrification and sulcation of the cortex have a typical arch-like character. Due to the adaptation of vision in these mammals both for aerial and aquatic habitats, their superior colliculi in the brainstem are much larger than in cetaceans (compare Fig. 4.3). The brain of manatees and dugongs are lissenecephalic with a few well-expressed fissures (Figs 4.1 and 4.2). Although the absolute weight (200–300 g) of the manatee's brain is similar to that of many terrestrial mammals, it is unusual for a brain this size to be void of sulci and gyri (Reep & O'Shea 1990).

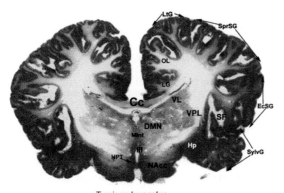

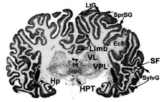

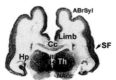

Tursiops truncatus

Zalophus californianus

Trichechuss manatus

Fig. 4.2 Frontal sections at the level of the diencephalon through the brain of three groups of marine mammals (from top to bottom: Cetacea, Pinnipedia and Sirenia). III, third ventricle; ABrSyl, ascending branch of the sylvian fissure; Cc, corpus callosum; DMN, dorsomedial nucleus of thalamus; EcSG, ectosylvian gyrus; Hp, hippocampus; HPT, hypothalamus; LG, limbic gyrus; LtG, lateral gyrus; Mint, massa intermedia; NAcc, nucleus accumbens; OL, oval lobule; SF, sylvian fissure; SprSG, suprasylvian gyrus; SylvG, sylvian gyrus; Th, thalamus; VL, ventrolateral nucleus of thalamus; VPL, ventroposterolateral nucleus of thalamus.

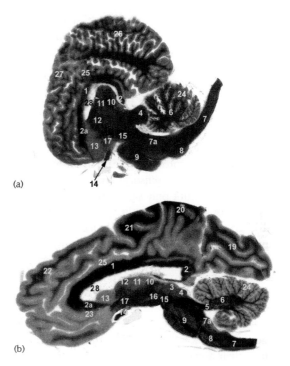

(a)

(b)

Fig. 4.3 Midsagittal sections of: (a) cetacean brain (*Tursiops truncatus*), and (b) primate brain (*Homo sapiens*), shown in the anatomical positions in the skull. 1, corpus callosum; 2, splenium of corpus callosum; 2a, rostrum of corpus callosum; 3, superior colliculus; 4, inferior colliculus; 5, superior cerebellar peduncle; 6, dentate nucleus of cerebellum; 7, medulla; 7a, rhomboid fossa; 8, inferior olive; 9, pons; 10, pulvinar; 11, dorsal thalamus; 12, anterior nucleus of thalamus; 13, caudate nucleus (head); 14, optic nerve; 15, cerebral peduncle; 16, substantia nigra; 17, hypothalamus; 18, mesencephalic cistern; 19, cuneus; 20, precuneus; 21, paracentral lobule; 22, superior frontal gyrus; 23, subcallosal gyrus; 24, cerebellum; 25, limbic gyrus; 26, oval lobule; 27, lateral gyrus; 28, lateral ventricle.

4.4 REGIONAL GROSS ANATOMY AND HISTOLOGY OF THE NERVOUS SYSTEM

4.4.1 The peripheral nervous system and spinal cord

The evolutionary transformation of the terrestrial mammalian body into a hydrodynamically efficient body form is the main factor forming the features of the spinal cord. The disappearance of the hind extremities, modification of the anterior extremities into flippers, development of the powerful tail, and the disappearance of body hairs, are all reflected in the morphology of the spinal cord in cetaceans. The number of spinal segments and paired roots originating from them varies from 40 to 44 in different species of Odontoceti and Mysticeti. In the region of the 2nd or 3rd **lumbosacral** segments, ventral and dorsal roots start to organize into four nervous trunks running caudally along each side of the body. Two of them are located dorsally from transverse vertebral processes, and two are situated ventrally. These trunks innervate all the musculature of the tail. In the region of the upper and lower fins of the tail, these trunks produce an extremely branched terminal nerve plexus.

According to the histological data and the experimental lesion/degeneration data, the motor structures of the spinal cord in cetaceans are more fully developed than the sensory structures. Thus, ventral horns are not only significantly larger than dorsal horns, but also contain very differentiated and large motoneurons (almost twice the size of human motoneurons). However, in comparison with terrestrial mammals, the descending (efferent) tracts of the spinal cord in cetaceans are rather small. Thus, only one-sixth of the **pyramidal tract** of the medulla reaches the spinal cord structures. The weak representation of the pyramidal tract in cetaceans undoubtedly relates to the transformation of their anterior extremities into flippers, and the disappearance of individual fingers. It is well established that in terrestrial mammals the pyramidal tract controls coordinated and complex movements of the distal muscles of the upper extremity.

The ascending tracts of the spinal chord are present in cetaceans, but are much smaller than in terrestrial mammals, and one type (spinocerebellar tracts) are not found in cetaceans at all. The information from the spinal cord to the cerebellum may be transmitted by the dorsal columnar tracts in these species. This fact may explain a relative enlargement of the cerebellum in cetaceans. This part of the brain in these species became the major processor of sensory modalities from the skin and musculoskeletal systems. The reduction of afferent and efferent central connections of the spinal cord is not correlated with a loss in the richness of skin innervation, nor with a reduction in the variety of sensory receptors. However, dolphin skin sensitivity is highest in regions innervated primarily by the cranial nerves (around the eyes, blowhole, genitals and snout), while sensitivity is much lower on the rest of the body where the innervation is from the spinal cord. Evidently, evolutionary transformations of the hindlimbs into flippers also caused the relative shortening of the spinal cord in pinnipeds.

4.4.2 Brainstem

4.4.2.1 Medulla oblongata

In cetaceans, as in terrestrial mammals, the medulla oblongata (myelencephalon) and, partially, pons (see below) are similar to the spinal cord, since these regions also provide a segmental innervation of the cranial part of the cetacean body. The structures in these regions of the brainstem are mainly represented by the **somatic** and **visceral** nuclei of the V–XII pairs of the cranial nerves (see below) and related circuits. The multiple ascending and descending tracts crossing the medulla and the differentiation of its nuclei markedly complicate the organization of this brain region in comparison with the more symmetrical and 'primitive' spinal cord. The medulla (Fig. 4.3) consists of longitudinal columns of neurons localized in the following order (from most ventral and medial to most lateral and dorsal): somatic efferent, visceral efferent, visceral afferent and somatic afferent functional columns. Somatic efferent columns innervate the muscles of the eyes and tongue. Visceral efferent columns are especially important in aquatic mammals, since they innervate the muscles of the jaws, face, larynx and pharynx.

There is a very sharp and dorsally convex cervical flexure between the spinal cord and medulla in cetaceans. As in the spinal cord, the ventral part of medulla is larger than its dorsal part. The components of the dorsal column and pyramidal tracts are significantly reduced compared with terrestrial mammals. The most prominent nuclei in the lower medulla of cetaceans are those in the inferior olive complex (Fig. 4.3). Whereas in terrestrial mammals these nuclei are located laterally, in cetaceans due to the drastic reduction of pyramidal tracts, inferior olives are situated close to the median plane. In the inferior olivary complex, the medial accessory nucleus is extremely large in both Mysticeti and Odontoceti, whereas the principal and dorsal accessory olives are relatively small. Interpolating from terrestrial mammals to cetaceans, some authors consider the extreme development of the medial accessory olives in cetaceans to be due to the intensive development of afferents from the caudate nucleus and spinal cord with following projections to nuclei in the enlarged cerebellar hemispheres (see below).

4.4.2.2 Pons

The development of the pons in the evolution of the mammalian brain is directly and indirectly related to the establishment of inputs from the upper and lower regions of the brain to the cerebellum (Fig. 4.3). The disappearance of the lateral line of fishes and amphibia and its nuclei led to the development of vestibuloacoustic nuclear complexes in the dorsal pons, which took over the balance and auditory modalities in mammals. In aquatic mammals, especially in cetaceans, the nuclei and tracts of this part of the pons are extremely well represented. Due to the anteroposterior contraction and ventral rotation of the cerebrum, the basis of the pons in cetaceans is unusually wide and almost entirely dislocated to the level of the midbrain. The overall large size of the pons in aquatic mammals reflects its 'switchboard' role between the extremely expanded cerebral and cerebellar cortices.

4.4.2.3 Cerebellum

The function of the cerebellum (Fig. 4.3) is dedicated primarily to dealing with the position of the body in space and time. It also integrates sensorimotor commands, providing the basis for constant motor functions. In mammals there are three major sources of inputs to the cerebellum: (i) **vestibular** nuclei and receptors (to the archicerebellum); (ii) spinal cord (to the palaeocerebellum); and (iii) cerebral cortex: auditory, visual and general sensory (to the neocerebellum). The first two are feedback systems, permitting the correction of directions, strength and other parameters of movement during their performance. The third system is 'feed-forward', permitting the correction of movements before they are performed. That is, helping to plan a future motor activity.

As mentioned above, the cerebellum in cetaceans is significantly larger than in terrestrial mammals, relative to the cerebrum (Fig. 4.1). There is not much difference in the structure of the cerebellum of Odontoceti and Mysticeti. Structural differences between the cetacean and terrestrial mammalian cerebellum may be interpreted as consequences of the evolutionary adaptations to the aquatic environment with respect to sensorimotor systems. For example, the weak development of the anterior cerebellar lobe in cetaceans coincides with the transformation of the anterior extremities into simplified flippers.

Histological analysis shows that the general lamination of the cerebellar cortex is the same as in other mammals, although the cellular structure differs in some odontocete species. The general organization and connections of the cerebellum in cetaceans are influenced by peculiarities of the organization of their motor system. Among these are the reduced representation of the motor system in the cerebral cortex and thalamus, and the weak development of the vestibular system and pyramidal tracts. On the other hand, the development of the basal ganglia is extreme in cetaceans (see below).

4.4.2.4 Midbrain

In vertebrates the midbrain (mesencephalon) represents the most anterior segmented neural tube which can still be divided into dorsal (tectum and tegmentum) and ventral regions (basis). The dorsal region is enlarged in brain evolution due to the intensive optic inputs, both from the retina and the concentration of the motor and parasympathetic nuclei regulating eye movements. The tectum of the midbrain is an important processing centre receiving information from the vestibular and auditory systems, as well as crude skin sense.

As in the case of the pons, the midbrain of cetaceans is rather 'contracted' along the anteroposterior axis. There are also several unique macroscopic features of the cetacean midbrain. For example, in the tectum of the midbrain of cetaceans, the posterior (inferior) colliculi (Fig. 4.3) are extremely hypertrophied both in Mysticeti and Odontoceti. The tracts related to the inferior colliculi are also extremely enlarged. Recent studies of histological and immunocytochemical features of the inferior colliculi revealed a modular structure in odontocete cetaceans, which probably relates to their echolocation abilities since, among other mammals, similar features have only been found in bats (Zook *et al.* 1988; Glezer *et al.* 1998). Thus, it was found that in the dolphin the inferior colliculus displays a vertical lamination of its central nucleus, in addition to

Table 4.2 Percentage of total for individual cranial nerve fibres.

	I	II	III	IV	V	VI	VII	VIII	IX	X	XI	XII
Tursiops	0	29.6	1.78	0.59	29.64	0.395	9.881	19.76	3.953	1.482	1.482	1.383
Inia	0	4.00	0.39	0.0002	40.3	0.232	5.425	31	3.487	6.2	6.2	2.583
Balaenoptera	?	37.8	1.89	0.18	33.24	0.988	3.998	13.84	1.662	3.145	2.695	0.539
Homo	0.83	75.5	1.81	0.26	10.57	0.498	0.981	3.77	0.755	2.589	1.887	0.604

the typical horizontal lamination found in terrestrial mammals. It is possible that these vertical domains, which we have defined as 'tectosomes', may represent frequency bands for processing information from echo signals. The superior colliculi (Fig. 4.3) in most species of cetaceans are very well developed, but significantly smaller than the inferior colliculi. The size of the superior colliculi is in direct proportion to the thickness of the optic nerve. Thus, the South American and Asian river dolphins, which have extremely reduced optic nerve and eyes, also have very small superior colliculi.

4.4.2.5 Cranial nerves

Olfactory nerve (I)

In mysticetes, the olfactory tract is present at the adult stage, but the olfactory bulb is found only in fetuses, while in odontocetes both exist only in the fetal stage of development. The reduction or total absence of the peripheral structures of the olfactory system in cetaceans is a dramatic testimony to the evolutionary adaptive changes which occurred in the terrestrial ancestors of cetaceans. Modification of cranial structures ('telescoping', see Chapter 2) led to the repositioning of the nasal bones (nares) at the top of the head, and at the same time peripheral olfactory receptors were cut off from the central structures. Thus, in cetaceans, especially in odontocetes, respiratory adaptive pressures took precedence, as separation of the respiratory tract from the gastrointestinal tract was necessary for survival. However, the evolutionary degeneration of the peripheral olfactory structures did not cause the reduction of central olfactory structures. All so-called tertiary structures of the rhinencephalon are present and well developed in cetaceans. Central

olfactory structures may be involved in reproduction, the control of secretions and/or the circulation of the nasal sac vocalization system in odontoctes. All species of pinnipeds possess well-developed main and accessory olfactory systems, while these are rudimentary in manatees and dugongs.

Optic nerve (II)

The optic chiasma and optic tract (Fig. 4.3) are well developed in most cetaceans, with the exception of some species of river dolphins with markedly reduced eyesight (see Chapter 1). However, even though cetaceans have well-developed eyes, the number of axons in the optic nerve is as few as one-tenth the number in humans (Table 4.2). In most cetacean species the left and right optic nerves are fully crossed at the chiasma. However, in some odontocetes (e.g. 14% in *Phocoena phocoena*), some proportion of the nerve fibres in the optic tract do not cross. Recent studies have also demonstrated an interchange of fibres (decussation) of the optic nerves in cetaceans (Tarpley & Ridgway 1994). An interesting, but as yet unexplained observation, is a significant reduction in the thickness of the optic tract and a corresponding decrease in the number of axons after chiasma in both Mysticeti and Odontoceti (e.g. a 25% decrease in thickness and 19% decrease in the number of fibres for *Tursiops truncatus*).

4.4.2.6 Accessory structures and cranial nerves related to movement and accommodation of the eye

Anatomical examinations of accessory structures in marine mammal eyes demonstrated modifications which can be attributed to the primary demand of underwater vision. There is a large disparity between the Odontoceti and Mysticeti in the development of the oculomotor, trochlear and abducent nerves (III, IV and VI) and their nuclei. Whereas quantitatively these nerves in mysticetes are similar to the corresponding nerves in the human brain, in odontocetes all three nerves are significantly smaller than in humans (Table 4.2). The peculiarity of the development of these nerves relates to the unique construction of the extrinsic and intrinsic muscles of the eye in cetaceans. It was shown that in most cetacean species, 'recti' muscles are inserted into eyelids and 'oblique' and 'retractor' muscles into the sclera (the outer coat of the vertebrate eye). Thus, it appears that only the oblique and retractor muscles are involved in eyeball movements in cetaceans.

Trigeminal nerve (V)

The spinal, main and mesencephalic nuclei of the trigeminal nerve are very well developed both in the Mysticeti and Odontoceti. This correlates extremely well with the representation of the 'ventral posteromedial nucleus' in the thalamus. Since in terrestrial mammals the ventral posteromedial nucleus receives stimuli from the taste system, it is suggestive that the trigeminal nerve in cetaceans may play a role in innervating chemoreceptors in the tongue and oral cavity, analogous to taste receptors. The extreme development of the trigeminal nerve in cetaceans is well correlated with the general anatomy of their body, where the head occupies more than one-third of its length. The adaptation of the terrestrial mammalian body into a streamlined aquatic one could not be achieved without serious morphophysiological sacrifices. The transformation of the anterior extremities into flippers, the extreme reduction or disappearance of the olfactory peripheral receptors, and a limited role for the visual system in the aquatic environment, left evolving cetaceans at a serious disadvantage for channels of sensory information. It is therefore logical that this evolutionary loss should be compensated, and one of these compensations was the hypertrophy of the trigeminal system and transformation of the cetacean head into a giant multifunctional sensor. This idea is well supported by the presence of numerous sensory tubercles and sinus hairs around the lips, innervated by the sensory part of the trigeminal nerve in

Mysticeti. In Odontoceti the tactile sensors receive additional input from the 'melon' or spermacetic organ (in *Physeter*) which functions in echolocation activity (see Chapter 6). It was found that the melon is a motile organ richly innervated by the trigeminal nerve. The presence of large muscles homologous to masseteric muscles concentrating around the mouth in cetaceans explains the hypertrophy of the motor root and nucleus of the trigeminal nerve.

Facial and intermediate nerves (VII)

The complex of facial and intermediate nerves (VII), although well represented in cetaceans, are still less prominent than the visual, trigeminal and vestibulocochlear nerves (Table 2.2). The motor part of the VII nerve complex is represented by two-thirds of the fibres, and the remaining one-third is the 'nervus intermedius' which corresponds to the special visceral sensory part of the complex. It was found that the motor part of this nerve innervates the maxillonasolabial group of muscles which controls the movements of the external nares (blowhole) in cetaceans. The motor nucleus of this nerve is differentiated into several cellular groups. The rich innervation of this muscle group by the motor root of the VIIth nerve is understandable, since the blowhole serves not only as a sphincter for the respiratory system but also participates in sound production (see Chapter 6). In addition, this motor root also innervates the muscles that participate in changing the shape of the melon. Two-way communication between the trigeminal nuclei and the motor nucleus of the facial nerve is expected, providing sensory innervation for the melon and blowhole areas. The function and area of peripheral distribution of the intermediate nerve in cetaceans are not well understood. However, since in terrestrial mammals the intermediate nerve represents a sensory taste modality, it has been suggested that in cetaceans this nerve is related to chemoreception. The presence of sensory receptors has been demonstrated in the mucosa of the cetacean tongue (Yablokov *et al.* 1972).

Vestibulocochlear nerve (VIII)

As expected, the vestibulocochlear (auditory) nerve in cetaceans is one of the thickest cranial nerves. In

Odontoceti, the cochlear root of the nerve takes approximately 60% of the axons and the remaining 40% go to the vestibular root. The vestibular component of the VIIIth nerve, and especially its central nuclei, are prominently reduced in both odontocetes and mysticetes. This coincides with the significant diminishment of the semicircular canals compared to terrestrial mammals. Among the three canals, the largest is the lateral. This feature might reflect the importance of scanning movements of the head of some cetaceans in the horizontal plane.

The brainstem nuclei of the auditory system and their corresponding input and output tracts in cetaceans are grossly enlarged comparative to these structures in terrestrial mammals, especially primates. For example, in *Delphinus delphis*, the surface of the ventral cochlear nucleus, superior olive, ventral nucleus of the lateral lemniscus and inferior colliculus (Fig. 4.3) are, respectively, 15, 156, 272 and 12 times larger than in humans (De Graaf 1967; Ridgway & Brownson 1984). However, some of the auditory nuclei of the cetacean brainstem are either reduced in size or absent (e.g. the dorsal cochlear nucleus and the medial nucleus of the superior olive are not found in cetaceans).

Glossopharyngeal, vagus and accessory nerves (IX, X, XI)

In all aquatic mammals the complex of glossopharyngeal, vagus and accessory nerves (IX, X, XI), which are related to the major visceral functions, are very similar quantitatively and qualitatively in cetaceans to those in terrestrial mammals. The brainstem nuclei of these nerves are very well developed and in the same positions in the medulla as in land mammals.

Hypoglossal nerve (XII)

The hypoglossal nerve (XII) is uniquely organized in cetaceans. It appears on the ventral surface of the medulla, dorsally to the inferior olive, whereas in all terrestrial mammals it appears ventrally between the pyramids and inferior olives, and the hypoglossal nucleus is displaced. Interestingly, this unique organization of the hypoglossal nucleus and nerve is similar to that found in amphibians, reptiles and birds, and may be one of the archetypal mammalian features of cetaceans.

4.4.2.7 Diencephalon

In all vertebrates the diencephalon is divided into four main components: the epithalamus, subthalamus, hypothalamus and dorsal thalamus (Figs 4.2 and 4.3). The latter is subdivided into the palaeothalamus and neothalamus. The following functions are assigned to the subdivisions of the diencephalon. The epithalamus participates mostly in visceral and pineal control over some olfactory reflexes, and sleep is controlled through the secretion of melatonin. The hypothalamus is involved in the autonomic control of endocrine glands, emotional expressions, the regulation of vagal tone and cortical arousal. The subthalamus is an important part of the extrapyramidal motor system (an indirect loop involved in inhibition of the motor activity of the cerebral cortex). The dorsal thalamus has multifaceted functions since it contains major specific and non-specific relay nuclei projecting sensory and motor information to particular areas of the cerebral cortex in a highly organized (somatotopic) fashion. In this manner, thalamic inputs to the cerebral cortex (called 'radiations') define the distribution of the primary, secondary, etc., areas of the functional map of the cerebral cortex. In turn, the cortical areas project back processed information to the homologous nuclei of the thalamus, as well as to the basal ganglia (see below), brainstem nuclei and spinal cord. Practically all somatic and visceral information from internal and external sensory and motor systems must be processed through one or more subdivisions of the thalamus for normal physiological function (visual, auditory, general sensations, metabolism, emotional motivations, etc.).

The topography of the diencephalon in cetaceans is defined by the ventral rotation of the brain, and the extreme expressions of the brainstem flexures. Although division of the diencephalon is the same as in terrestrial mammals, the orientations of these regions and their relationships are unique in cetaceans (Kruger 1959). The major nuclear mass of the diencephalon, the thalamus, is rotated counterclockwise along its transverse axis in such a way that its longitudinal axis is almost perpendicular to the longitudinal axis of the hemispheres and brainstem. Thus, nuclei which in terrestrial species are posterior, became dorsal in cetaceans, and anterior structures became ventral, and so on. In general, the diencephalic structures in cetaceans have very large transverse expansion and dorsoventral diameter. The special feature of the diencephalon in odontocete cetaceans is an enormous 'massa intermedia' occupying almost the entire wall of the thalami, while in some Mysticeti (e.g. *Balaenoptera borealis*) the massa intermedia is absent. In the epithalamus the pineal gland is rudimentary in both odontocetes and mysticetes. The anterior group of nuclei is very prominent in odontocetes, which correlates with the extreme enlargement of the limbic (marginal) lobes in the cetacean cerebral hemispheres. The largest nuclei of the dorsal thalamus in cetaceans are found in the lateral and ventral tiers.

4.4.2.8 Telencephalon (cerebral hemispheres)

In general, in all extant mammalian species the telencephalon contains two major subdivisions with distinct functional characteristics: the basal ganglia and the cerebral cortex, organized into two hemispheres surrounding the two largest cavities of the brain (the lateral ventricles; Fig. 4.3). The two cerebral hemispheres are connected with the two major commissures: the anterior and corpus callosum (Fig. 4.2), whereas a posterior commissure is part of the brainstem.

The end brain (telencephalon) first appears in early vertebrates (cyclostomes). In amphibia, the vesicle is divided into two hemispheres with the two corresponding cavities—the lateral ventricles. The epithelial roof of the ventricles is invaded by nerve cells from the subcortical mass of cells, and is thereby transformed into the cerebral cortex of reptiles and birds. The subcortical mass in turn is differentiated into the basal ganglia (striatum and pallidum). These ganglia reach the highest degree of development in birds. However, the cortical plate (pallium) reaches its highest differentiation and largest absolute and relative size only in the higher mammals, especially in primates and cetaceans.

Commissures of brain

The largest commissure of the mammalian brain, the corpus callosum (Figs 4.2 and 4.3), is drastically reduced in the cetacean brain in terms of its thickness. For example, the estimated number of myelinated fibres in the corpus callosum of *Tursiops truncatus* is approximately 138 million, whereas in humans it is about 445 million (I.I. Glezer, unpublished data). However, this reduction is not universal across all species of cetaceans, and is inversely correlated to the overall size of the brain. Paradoxically, the manatee with its small, lissencephalic brain has a relatively large corpus callosum, comparable to humans.

The anterior commissure in the cetacean brain is present but significantly smaller than in terrestrial mammals, while the posterior commissure, habenular commissure and commissures between the superior and inferior colliculi are larger in cetacean brains.

Basal ganglia

The basal ganglia (striatum) of the mammalian telencephalon are subdivided phylogenetically, functionally and morphologically into three parts: the archistriatum (amygdalar nucleus), the palaeostriatum (globus pallidus) and the neostriatum (putamen and caudate nuclei). The archistriatum relates to the limbic lobe of the brain, and due to its connections to the hypothalamus, cingulate cortex, septum and midbrain reticular formation, participates in emotional states related to the extrapyramidal motor system. On the other hand, the palaeo- and neostriatum are involved in the fine regulation of involuntary and voluntary movements due to extensive connections with the cerebral cortex, subthalamus and substantia nigra.

Basal ganglia in cetaceans include the typical structures of the caudate nucleus, putamen, globus pallidus and amygdala. However, due to the rotation of the whole cerebrum, the basal ganglia are displaced into extreme ventral positions, encroaching partly on the cortical surface of the so-called olfactory lobe. Overall, the size of the basal ganglia in cetaceans (relative to the size of the cerebral hemispheres) are smaller than in primates. The amygdoloid complex of nuclei is very well represented in cetaceans, and similar to terrestrial mammals. Evidently the loss of the peripheral olfactory structures is not reflected in the size of the amygdala in cetaceans.

Table 4.3 Absolute and relative sizes of the cerebral cortex and its regions in marine and terrestrial mammals. (Adapted from Filimonoff 1949; Blinkov & Glezer 1968.)

Species	Total surface (mm²)	Neocortex (mm²)	%	Palaeocortex (mm²)	%	Archicortex (mm²)	%	Intermediate zones (mm²)	%
Tursiops truncatus	46 427	45 696	97.8	438	0.9	386	0.8	206	0.4
Zalophus californianus	15 135	14 284	94.4	266	1.8	337	2.2	249	1.6
Canis vulgaris	6523	5480	84.2	447	6.8	416	6.3	179	2.7
Homo sapiens	83 591	80 201	95.9	480	0.6	1863	2.2	1045	1.3
Erinaceus europaeus	254	82	32.4	76	29.8	51	20.2	45	17.6

Cerebral cortex

The largest part of the telencephalon is represented by the cerebral cortex (or 'pallium'), which is subdivided into the following subsections: (i) the archipallium (synonyms: medial pallium, archicortex, allocortex), which is represented by the hippocampal formation; (ii) the palaeopallium (synonyms: lateral pallium, palaeocortex, allocortex), which is represented by the olfactory cortex, septum and pyriform lobe; and (iii) the neopallium (synonyms: neocortex, isocortex), which is represented by the occipital, parietal, temporal, frontal and limbic cortices. These three subsections constitute a continuous cortical plate, organized into layers and cellular columns (modules). While in the archicortex and palaeocortex there are one to three cortical layers, in the neocortex there are five to eight layers. The evolution of the pallium in higher mammalian species is characterized by great lateral expansion of the neocortex, which is differentiated into multiple morphofunctional areas (maps), and the extreme relative diminishment of both archi and palaeocortices. These occupy only 2–3% of the cortex in primates and cetaceans (Table 4.3).

In many ways the whole structure of the cetacean brain is defined by the large size and extreme gyrification of the cerebral cortex. It is characterized by extremely deep primary fissures with many internal branches, as though the cortical plate in cetaceans was multiply folded on itself during ontogenesis.

However, the general plan of the surface structure of the cetacean cerebral cortex is typically mammalian. The extremely complicated gyrification and sulcation of the cerebral cortex in cetaceans are in paradoxical contrast to its relatively simple histological organization. Many studies on the cytoarchitecture and Golgi organization of the cetacean neocortex have shown features that are similar to those in the more 'primitive' archicortex and palaeocortex of terrestrial mammals (Morgane & Jacobs 1972; Kesarev et al. 1977a, 1977b; Morgane et al. 1985, 1986a, 1986b, 1990; Glezer et al. 1988; Glezer & Morgane 1990; Morgane & Glezer 1990; Hof et al. 1999, 2000). These features include a very narrow cortical plate and homogeneous cytoarchitecture. Further, there is extensive morphological development of layers I (in thickness) and II (in cellularity) compared to other cortical layers, and the absence or very weak development of layer IV in all regions of the neocortex (Fig. 4.4) (Morgane et al. 1985, 1986a, 1986b, 1990; Glezer et al. 1988, 1990, 1993a, 1993b). In our recent study (Glezer et al. 1998) we showed that these features are present in both the visual and auditory cortices of odontocete cetaceans. Attempts to build **cytoarchitectonic maps** analog-ous to the Brodmann and von Economo maps in terrestrial mammals and man were not very successful. In fact, in blind experiments, the neocortex of cetaceans can be easily misdiagnosed as the neocortex of some

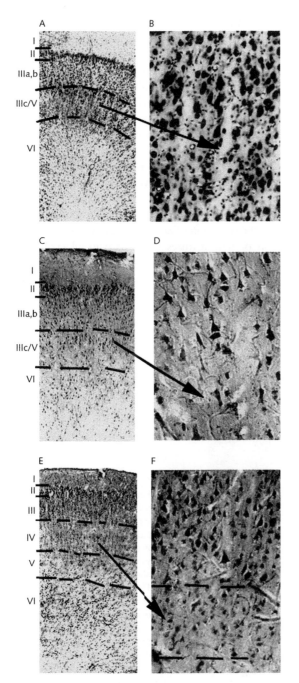

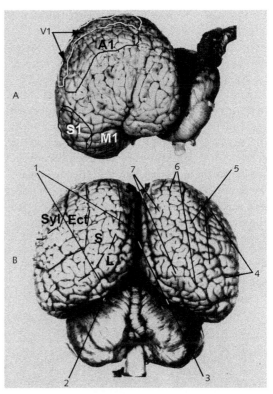

Fig. 4.5 (A) Lateral and (B) superior views of the brain of the bottlenose dolphin (*Tursiops truncatus*). (a) The major primary areas in the cortex of cetacean brain, showing the functional mapping of the primary visual area (V1), primary auditory area (A1), primary motor area (M1) and primary sensory area (S1). (b) The overall parallels with corresponding cortical structures in terrestrial mammals permits the recognition on the laterosuperior surface of the cetacean brain of: the sylvian (Syl), ectosylvian (Ect), suprasylvian (5, S) and lateral gyri (7, L). The prominent sulci and fissures are the lateral (6), suprasylvian (4), interhemispheric (3), transverse cleft (2) and entolateral (1). (Adapted from Morgane *et al.* 1990.)

Fig. 4.4 Cytoarchitectonics of the primary auditory cortex in marine mammals. (A, B) *Delphinapterus leucas*: magnification × 32 (A) and × 320 (B); (C, D) *Balaenoptera physalus*: magnification × 32 (C) and × 320 (D). These demonstrate the very narrow cortical plate, homogeneous cytoarchiteture, extensive morphological development of layers I (in thickness) and II (in cellularity), absense of layer IV, and overall agranularity of the neocortex (see text). (E, F) *Arctocephalus pusillus*: magnification × 32 (E) and × 320 (F). The characteristics of the primary auditory cortex in the seal are typical of terrestrial mammals, as is the well-developed layer IV.

lissencephalic species, such as insectivores and rodents.

Functional mapping of the cetacean neocortex showed very atypical localizations of the primary and secondary visual and auditory areas (Fig. 4.5). Thus, V1 ('visual 1') is located on the surface of the lateral gyrus, in the parietal lobe, i.e. dislocated anteriorly in comparison with terrestrial mammals. A1 ('auditory 1') is found on the lower part of the suprasylvian gyrus, also quite far from the typical location in the temporal lobe of other mammals. The motor area is found in an extreme rostral position, which does not give any space for the frontal lobe areas. In general, however, functional mapping of the cetacean brain has yet to receive a detailed investigation, which will most probably be based on contemporary non-invasive imaging techniques such as MRI (magnetic resonance imaging) and PET (positron emission tomography) scans.

The archetypal cytoarchitectonics of the neocortex in cetaceans has been interpreted differently by various authors. Some considered its 'primitive' features as a sign of evolutionary de-differentiation, while others see them as a sign of the retention of the archetypal features of the so-called 'initial mammalian brain'. However, it is more probable that during the complicated evolutionary transition from a terrestrial ecological niche to an aquatic one, both the retention of ancestral traits and the acquisition of new features took place in all evolving

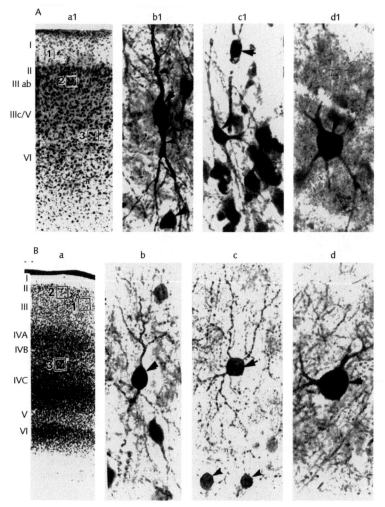

Fig. 4.6 Cytoarchitectonics and immunochemistry of the primary visual cortex (V1) in *Tursiops truncatus* and *Macaca fascicularis*. (A) *Tursiops truncatus*. a1, cytoarchitectonic preparation (Nissl); b1, c1, d1, immunocytochemical reaction to calretinin showing GABAergic neurons. The numbered squares (1–3) on a1 correspond to the locations of neurons b1, c1 and d1 (1–3, respectively). The illustrations show the prevalence of CR- and CB-immunoreactive over PV-immunoreactive neurons. (B) *Macaca fascicularis*. a, cytoarchitectonic preparation (Nissl); b, c, d, immunocytochemical reaction to calretinin showing GABAergic neurons (as above). The numbered squares correspond as for (a). The illustrations show a greater diversity of neuron types.

cetacean biological systems, including the brain. According to the palaeontological record and molecular data, the most probable archetypal groups for the origin of cetaceans were early ungulates (Miyamoto & Goodman 1986; Gingerich *et al.* 1990; Millinkovitch *et al.* 1993). One possible 'preadaptive' feature of ungulate ancestors is the early (precocial) physical maturity of their calves after birth (which probably evolved as an antipredator strategy). Since cetaceans give birth in the water, the sensory and motor systems of the newborn whale will have to be ready for immediate adaptation to the environment to prevent drowning. One possible evolutionary mechanism for a rapid ontogenic development of neural capacity is the process proposed by Severtsov (1939), referred to as paedomorphosis. In this process, characteristics that are initially present in ancestral juvenile forms are shifted to the adult stage (see Gould 1977; Schoch 1986; Roth *et al.* 1993; Schad 1993; Marcus *et al.* 1995; Godfrey & Sutherland 1996). Although there is no direct evidence for this, it would be consistent with other features of the cetacean brain, in particular the **proisocortical** characteristics of neocortex in cetaceans (Morgane *et al.* 1986a, 1986b, 1990; Glezer *et al.* 1993a, 1993b). In different classes of vertebrates the 'price' for fast **paedomorphic** evolutionary changes is the relative simplification of the particular structure or system, especially the skull and central nervous system (Roth *et al.* 1993; Schad 1993). It is possible that the extreme multiplication of relatively simplified cortical modules (Glezer *et al.* 1988) and, as a consequence, the development of the large cortical surface in cetaceans, are compensatory evolutionary features helping to overcome the paedomorphic ontogenetic shift. The relative prevalence of the calzetinin (CR)- and calbindin (CB)-immunoreactive neurons over parvalbumin (PV)-immunoreactive neurons in major cetacean sensory systems (Fig. 4.6) may serve as an additional indication of paedomorphic evolutionary traits of these species. This is supported by the fact that in ontogenesis of mammalian (rat, cat, monkey) central nervous systems, the most precocious are CR-immunoreactive neurons, followed by CB-immunoreactive neurons, while PV-immunoreactive neurons typically appear later (Enderlin *et al.* 1987; Solbach & Celio 1991; Anstrom *et al.* 1997).

Therefore a putative paedomorphic shift in cetacean brain development may have resulted in the embryo-like deficit of PV-immunoreactive neurons in the adult cetacean brain and its enrichment with the CR- and CB-immunoreactive neurons. However, the enrichment of auditory subcortical nuclei in *Tursiops truncatus* in CB- and CR-immunoreactive neurons has another intriguing aspect. In bats, CB- and CR-immunoreactive neurons in these same nuclei participate in the preservation of temporal information during echolocation (Zettel *et al.* 1991).

REFERENCES

Anstrom, K.K., McHaffie, J.G. & Stein, B.E. (1997) Distribution of paravalbumin and calbindin D28k in the superior colliculus of newborn and adult cats and monkeys. *Society of Neoroscience Abstracts* **23**, 1541.

Ariëns Kappers, C.U., Huber, G.C. & Crosby, E.C. (1936) The telencephalon of mammals. In: *The Comparative Anatomy of the Nervous System of the Vertebrates including Man*, Vol. 2, pp. 1428–1436. MacMillan, New York.

Beauregard, H. (1883) Recherches sur l'encéphale des Balanides. *Journal d'Anatomie Physiologique* **19**, 481–516.

Blinkov, S.M. & Glezer, I.I. (1968) *The Human Brain in Figures and Tables*, pp. 342–344. Plenum Press, New York.

Breathnach, A.S. (1960) The cetacean central nervous system. *Biological Reviews* **35**, 187–230.

De Graaf, A.S. (1967) *Anatomical Aspects of the Cetacean Brainstem*, pp. 169. Van Gorcum & Co. Assen, the Netherlands.

Ebbesson, S.O.E. (1984) Evolution and ontogeny of neural circuits. *Behavioral and Brain Sciences* **7**, 321–366.

Ebner, F. (1969) A comparison of primitive forebrain organization in metatherian and eutherian mammals. *Annals of the New York Academy of Science* **167**, 241–257.

Elliot Smith, G. (1910) Some problem relating to the evolution of the brain. *Lancet* **I**, 147–153, 221–227.

Enderlin, S., Norman, A.W. & Celio, M.R. (1987) Ontogeny of the calcium-binding proteins calbindin D-28k in the rat nervous system. *Anatomy and Embryology* **177**, 15–28.

Filimonoff, I.N. (1949) Cytoarchtectonics. General concepts. Classification of cytoarchitectonic formations. In: *Cytoarchitecture of the Cerebral Cortex of Man* (S.A. Sarkisov, I.N. Filimonoff & N.A. Preobrajenskaya, eds), pp. 11–32. Medgiz, Moscow.

Gingerich, P.D., Smith, B.H. & Simons, E.L. (1990) Hind limbs of Eocene *Basilosaurus*. Evidence of feet in Whale. *Science* **249**, 154–156.

Glezer, I.I. & Morgane, P.J. (1990) Ultrastructure of synapses and Golgi analysis of neurons in the neocortex of the lateral gyrus (visual cortex) of the dolphin (*Stenella coeruleoalba*) and the pilot whale (*Globicephala melaena*). *Brain Research Bulletin* **24**, 401–427.

Glezer, I.I., Jacobs, M. & Morgane, P.J. (1988) Implications of the 'initial brain' concept for brain evolution in cetaceans. *Behavioral and Brain Sciences* **11**, 75–116.

Glezer, I.I., Morgane, P.J. & Leranth, C. (1990) Immunocytochemistry of neurotransmitters in visual neocortex of several toothed whales, light and electron microscopic study. In: *Sensory Abilities of Cetaceans. Laboratory and Field Evidence* (J.A. Thomas & R.A. Kastelein, eds), pp. 39–67. Plenum Press, New York.

Glezer, I.I., Hof, P.R., Leranth, C. & Morgane, P.J. (1993a) Calcium-binding protein-containing neuronal populations in mammalian visual cortex: a comparative study in whales, insectivorous, bats, rodents, and primates. *Cerebral Cortex* **3**, 249–272.

Glezer, I.I., Hof, P.R., Leranth, C. & Morgane, P.J. (1993b) Morphological and histochemical features of odontocete visual neocortex: immunocytochemical analysis of pyramidal and non-pyramidal populations of neurons. In: *Marine Mammals Sensory Systems* (J.A. Thomas, R.A. Kastelein & A.Ya. Supin, eds), pp. 1–38. Plenum Press, New York.

Glezer, I.I., Hof, P.R. & Morgane, P.J. (1998) Comparative analysis of calcium-binding protein-immunoreactive neuronal populations in the auditory and visual system of the bottlenose dolphin (*Tursiops truncatus*) and the macaque monkey (*Macaca fascicularis*). *Journal of Chemical Neuroanatomy* **15**, 203–237.

Godfrey, L.R. & Sutherland, M.R. (1996) Paradox of peromorphic paedomorphosis: heterochrony and human evolution. *American Journal of Anthropology* **99**, 17–42.

Gould, S.J. (1977) *Ontogeny and Phylogeny*. Harvard University Press, Cambridge, MA.

Hof, P.R., Glezer, I.I., Conde, F. *et al.* (1999) Cellular distribution of the calcium-binding proteins parvalbumin, calbindin and calretinin in the neocortex of mammals: phylogenetic and developmental patterns. *Journal of Chemical Neuroanatomy* **16**, 77–116.

Hof, P.R., Glezer, I.I., Nimchinsky, E.A. & Erwin, J.W. (2000) Neurochemical and cellular specializations in the mammalian neocortex reflect phylogenetic relationships: evidence from primates, cetaceans and artiodactyls. *Brain, Behaviour and Evolution* **55**, 300–310.

Kamya, T., Uchida, S. & Kataoka, T. (1979) Organ weights of *Dugong dugong*. *Science Report of the Whales Research Institute* **31**, 129–132.

Kesarev, V.S., Malofeyeva, L.I. & Trykova, O.V. (1977a) Structural organization of the cerebral neocortex in cetaceans (in Russian). *Archiv Anatomii, Histologi i Embryologii* **73**, 23–30.

Kesarev, V.S., Malofeyeva, L.I. & Trykova, O.V. (1977b) Ecological specificity of cetacean neocortex. *Journal für Hirnforschung* **18**, 477–460.

Kruger, L. (1959) The thalamus of the dolphin (*Tursiops truncatus*) in comparison with other mammals. *Journal of Comparative Neurology* **111**, 133–194.

Kükenthal, W. & Ziehen, T. (1889) Das zentralnervensystem der cetaceen. In: *Makroskopische und Mikroskopische Anatomie des Zentralnervensystems*, pp. 153–178. Gustav Fischer, Jena.

Langworthy, O.R. (1932) A description of the central nervous system of the porpoise (*Tursiops truncatus*). *Journal of Comparative Neurology* **54**, 437–488.

Le Gros Clark, W. (1932) The structure and connections of the thalamus. *Brain* **55**, 406–413.

Marcus, J.A., Deacon, T.W., Reep, R.L. & Marshall, C.D. (1995) Patterns of interspecific variation in cortical neurons number. *Society of Neuroscience Abstracts* **21**, 154.

Millinkovitch, M.C., Orti, G. & Meyer, A. (1993) Revised phylogeny of whales suggested by mitochondrial ribosomal DNA sequences. *Nature* **361**, 346–348.

Miyamoto, M. & Goodman, M. (1986) Biomolecular systematics of eutherian mammals: phylogenetic patters and classification. *Systematic Zoology* **35**, 230–240.

Morgane, P.J. & Jacobs, M.S. (1972) The comparative anatomy of the cetacean nervous system. In: *Functional Anatomy of Marine Mammals* (R.J. Harrison, ed.), pp. 117–244. Academic Press, London.

Morgane, P.J. & Glezer, I.I. (1990) Sensory neocortex in dolphin brain. In: *Sensory Abilities of Cetaceans. Laboratory and Field Evidence* (J.A. Thomas & R.A. Kastelein, eds), pp. 101–137. Plenum Press, New York.

Morgane, P.J., Jacobs, M.S. & Galaburda, A.M. (1985) Conservative features of neocortical evolution. *Brain, Behaviour and Evolution* **26**, 176–184.

Morgane, P.J., Jacobs, M.S. & Galaburda, A.M. (1986a) Evolutionary morphology of the dolphin brain. In: *Dolphin Cognition and Behaviour: a Comparative Approach* (R.J. Schusterman, F.G. Woods & J.A. Thomas, eds), pp. 5–29. Lawrence Erlbaum, Hillsdale, NY.

Morgane, P.J., Jacobs, M.S. & Galaburda, A.M. (1986b) Evolutionary aspects of cortical organization in the dolphin brain. In: *Research on Dolphins* (M. Bryden & R.J. Harrison, eds), pp. 71–98. Oxford University Press, Oxford.

Morgane, P.J., Glezer, I.I. & Jacobs, M.S. (1990) Comparative and evolutionary anatomy of visual cortex of dolphin. In: *Cerebral Cortex. Comparative Structure and Evolution of Cerebral Cortex* (A. Peters & E.G. Jones, eds), Vol. 8B, Part II, pp. 215–258. Plenum Press, New York.

Pilleri, G. (1962) Die zentralnervose Rangordnung der Cetacea (Mammalia). *Acta Anatomica* **53**, 449–508.

Pilleri, G. (1966a) Morphologie des Gehirnes des Buckelwals, *Megaptera novaeangliae* Borowski (Cetacea, Mysticeti, Balaenopteridae). *Experientia* **22**, 849–851.

Pilleri, G. (1966b) Morphologie des Gehirnes des Seiwals, *Balaenoptera borealis* Lesson (Cetacea, Mysticeti, Balaenopteridae). *Journal für Hirnforschung* **8**, 437–491.

Pilleri, G. (1969) Das Zentralnervensystem der Zahn und Bartenwale. *Revue Swiss de Zoologie* **76**, 995–1041.

Pilleri, G. & Busnel, R.G. (1969) On the brain of the Amazon dolphin *Inia geoffrensis* de Blainville 1817 (Cetacea, Susuidae). *Acta Anatomica* **73**, 92–97.

Pilleri, G. & Chen, P. (1982) The brain of Chinese finless porpoise. In: *Investigations on Cetacea* (G. Pilleri, ed.), Vol. 13, pp. 27–78. Der Bund, Bern.

Pilleri, G. & Gihr, M. (1968) The structure of the cerebral cortex of the Ganges dolphin Susu (*Plantagenista gangetica* Lebeck 1801). *Experientia* 24, 932–934.

Pilleri, G. & Gihr, M. (1969a) Relatives Hirngewicht der Cetacea. *Revue Suisse de Zoologie* 76, 767–779.

Pilleri, G. & Gihr, M. (1969b) Das Zentralnervensystem der Zahn- und Bartenwale. *Revue Suisse de Zoologie* 76, 995–1037.

Pilleri, G. & Kraus, D.A. (1969) Zum Aufbau des Cortex beim Cetaceen. *Revue Suisse de Zoologie* 76, 760–767.

Pilleri, G., Kraus, C. & Gihr, M. (1968) Considerations sur le cerveau et le comportement du *Delphinus delphis*. *Zeitschrift für Mikroskopische und Anatomische Forschung* 79, 373–388.

Reep, R.L. & O'Shea, T.J. (1990) Regional brain morphometry and lissencephaly in the Sirenia. *Brain, Behaviour and Evolution* 35, 185–194.

Ridgway, S.H. (1990) The central nervous system of the bottlenose dolphin. In: *The Bottlenose Dolphin*, pp. 69–97. Academic Press, New York.

Ridgway, S.H. & Brownson, R.H. (1984) Relative brain sizes and cortical surface areas in odontocetes. *Acta Zoologica Fennica* 172, 149–152.

Roth, G., Nashikawa, K.C., Naujoks-Manteuffel, C., Schmidt, A. & Wake, D.B. (1993) Paedomorphosis and simplification in the nervous system of salamander. *Brain, Behaviour and Evolution* 42, 137–170.

Schad, W. (1993) Heterochronical patterns of evolution in the transitional stages of vertebrate classes. *Acta Biotheoretica* 41, 383–389.

Schoch, R.M. (1986) *Phylogeny Reconstruction in Paleontology*. Van Nostrand Reinhold, New York.

Schwerdtfeger, W.K., Oelschläger, H.A. & Stephan, H. (1984) Quantitative neuroanatomy of the brain of the La Plata dolphin *Pontoporia blainvillei*. *Anatomy and Embryology (Berlin)* 170, 11–19.

Severtsov, A. (1939) *Morphological Principles of Evolution*. Nauka, Moscow.

Solbach, S. & Celio, M.R. (1991) Ontogeny of calcium-binding proteins parvalbumin in the rat nervous system. *Anatomy and Embryology* 184, 103–124.

Tarpley, R.J. & Ridgway, S.H. (1994) Corpus callosum size in delphinid cetaceans. *Brain, Behaviour and Evolution* 44, 156–165.

Yablokov, A.V., Belkovitch, V.M. & Borisov, V.I. (1972) *Whales and Dolphins*. Nauka, Moscow.

Zettel, M.L., Carr, C.E. & O'Neill, W.E. (1991) Calbindin-like immunoreactivity in the central auditory system of the mustached bat, *Pteronotus parnelli*. *Journal of Comparative Neurology* 313, 1–16.

Zook, J.M., Jacobs, M.S., Glezer, I.I. & Morgane, P.G. (1988) Some comparative aspects of auditory brainstem cytoarchitecture in echolocating mammals, speculations on the morphological basis of time-domain signal processing. In: *Animal Sonar Processes and Performance* (P.E. Nachtigal & P.W.B. Moore, eds), pp. 311–316. Plenum Press, New York.

CHAPTER 5

Sensory Systems

Guido Dehnhardt

5.1 INTRODUCTION

Behavioural ecology, an evolutionary approach to animal behaviour, presumes that decisions animals make in their natural environment are based on detailed knowledge and result in optimal (in terms of fitness) behavioural responses to environmental variation (Roitblat 1987; Kamil 1988; Krebs & Kacelnik 1991). Since the acquisition of knowledge must be based on subtle sensory information, *sensory biology* has a key position in our understanding of the complex world of marine mammals.

Different sensory modalities are treated separately in this chapter. However, the complex problems marine mammals are faced with, for example navigation (see Chapters 6 and 7), can not be understood without an organismic or multimodality approach. Although animals can perform certain tasks when experimentally restricted to only one sensory system, under natural conditions most behavioural skills are based on multimodal convergence. This means that sensory information from different modalities either complement each other or have to be compared simultaneously so that the resulting behaviour is a function of all contributing modalities in combination with the animal's centrally stored information. This chapter takes this into consideration by treating the classic sensory modalities (mechanoreception, vision and chemoreception) in a balanced way.

5.2 MECHANORECEPTION

5.2.1 The sense of touch

With the exception of vibrissal follicles (Fig. 5.1), the receptor units of the mammalian sense of touch are not merged in sensory organs but are distributed over the entire body surface. As for other sensory modalities, different types of tactile sensation are mediated by different types of receptors. Human skin receptors respond to vibratory stimuli in the range of 0.3–3.0 Hz (Merkel cells), 3.0–40 Hz (Meissner corpuscles) and 10–500 Hz (Pacinian corpuscles), mediating sensations like 'pressure', 'flutter' and 'vibration', respectively (Bolanowski *et al.* 1988). Electrophysiological studies have shown that in the bottlenose dolphin (Ridgway & Carder 1990) and the northern fur seal (Ladygina *et al.* 1985) the entire body surface is sensitive to different kinds of mechanical stimulation. Based on absolute sensitivity (dolphin) and the extent of cortical representation (fur seal), these studies revealed that in both species, areas of the head are most significant for the reception of tactile information. Whereas in dolphins, skin sensitivity is suggested to be involved in drag reduction (Ridgway & Carder 1990), interest in pinnipeds and manatees focused on the significance of the facial vibrissae or sinus hairs ('whiskers', see Fig. 5.1).

The occurrence and distribution of sensory hairs in marine mammals is diverse (Fig. 5.1) (Ling 1977). Baleen whales, for instance, possess about 100 very thin (*c.* 0.3 mm in diameter) and immobile vibrissae distributed over the lower and upper jaw, whereas most odontocetes lose these hairs postnatally and only an array of two (*Phocoena phocoena*) to 10 (*Delphinus delphis*) follicles are situated on each side of the upper jaw (Fig. 5.1). River dolphins are an exception in that both of their jaws are more or less covered by immobile thin bristles. However, it has yet to be shown whether these bristles are true sinus hairs.

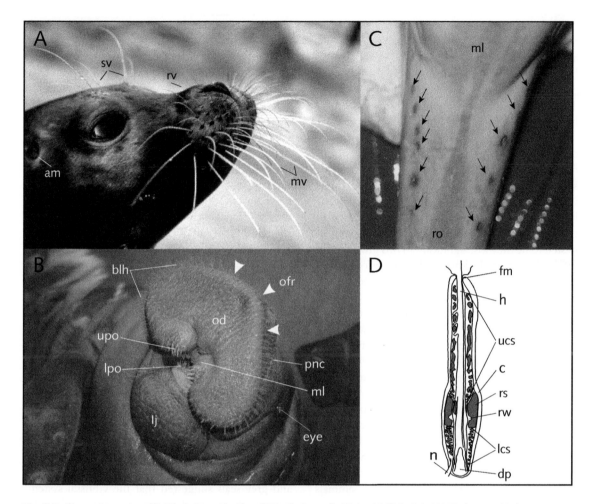

Fig. 5.1 Outer appearance of facial vibrissae (A–C) and morphology of a vibrissal follicle (D). (A) Harbour seal (*Phoca vitulina*): am, auditory meatus; mv, mystacial vibrissae (fully protruded, about 44 on each side of the muzzle); rv, rhinal vibrissae (one on each side of the back of the muzzle); sv, supraorbital vibrissae (five above each eye). (B) Muzzle of an Antillean manatee breaking through the water surface: blh, bristle-like hairs of the oral disc (od); lj, lower jaw; lpo, perioral bristles of the lower jaw; ml, muscular lips; ofr and white arrows, orofacial ridge (the boundary between the oral disc and the supradisc region); pnc, postnasal crease; upo, perioral bristles of the upper jaw. (Adapted from Bachteler & Dehnhardt 1999.) (C) Dorsal view of the head of a dolphin (*Sotalia fluviatilis guianensis*) where each arrow marks the mouth of a vibrissal follicle on the upper jaw of the dolphin. ml, melon; ro, rostrum. (D) Schematic view of a longitudinal section of a vibrissal follicle of the ringed seal: c, capsule; dp, dermal papilla; fm, mouth of follicle at the skin surface; h, vibrissal hair shaft; lcs, lower cavernous sinus; n, deep vibrissal nerve penetrating the capsule; rs, ring sinus; rw, ring wulst; ucs, upper cavernous sinus. Inside the capsule the hair is anchored to the dermal papilla and has a second fulcrum at the mouth of the follicle. A deflection of the outer hair shaft is proposed to be transformed into a contrarotating bowing action of the hair shaft inside the follicle, resulting in a compression of mechanoreceptors on the leading edge of the bending shaft, while on all other sides around the shaft stretch should be the effective stimulus. (Adapted from Dehnhardt *et al.* 1998a.)

Because of their external appearance, the vibrissae of species belonging to the order Sirenia are of particular interest. These obligate herbivore marine mammals possess sinus hairs all over the body (Ling 1977; R. L. Reep, personal communication) and show a distinct morphological differentiation of their facial vibrissae (Reep *et al.* 1998). The entire outer muzzle—divided into the supradisc and the oral disc (Fig. 5.1)—is densely covered with quite flexible hairs labelled 'bristle-like hairs'. Additionally, sirenians possess a second type of facial sinus hair, so called perioral bristles, located in six pads on the upper lip, the oral cavity and the lower jaw. These hairs are extraordinarily rigid bristles and are highly moveable.

The facial vibrissae of pinnipeds (Fig. 5.1) can be divided into three groups: the mobile mystacial vibrissae on the muzzle, and the more or less immobile supraorbital vibrissae (above each eye) and rhinal vibrissae (only in phocids, vertically situated on the back of the muzzle). According to the thickness of the hair shaft, vibrissae of pinnipeds have been suggested to be the most significant in this group (Ling 1977). However, thickness and structure of the hair does not reflect significance but has to be considered as a mechanical adaptation to the signals the hair can receive and transmit to the receptors within the hair follicle. In this respect the vibrissae of most phocid species are interesting. While vibrissal hair shafts of the species of Otariidae and Odobenidae, as well as those of phocids like the bearded seal (*Erignathus barbatus*) and the monk seals (*Monachus* spp.), are oval in diameter and smooth in outline, those of all other phocid species are extremely flattened and have waved surfaces (Watkins & Wartzok 1985; Hyvärinen 1989; Dehnhardt & Kaminski 1995).

In general, vibrissal follicles (Fig. 5.1) of mammals are intricately structured and densely innervated sensory organs, though the functional significance of most follicle structures is not known yet. Although their basic morphology seems to be similar across species, interspecific differences in the structure of the sinus system as well as the degree and pattern of vibrissal innervation may reflect differences in sensitivity and function (Rice *et al.* 1986). There is some information on the structure of vibrissal follicles in the different orders of marine mammals (Ling 1977), but sufficient descriptions of the morphology of vibrissal follicles are only available for pinnipeds (*Zalophus californianus*: Stephens *et al.* 1973; and *Phoca hispida*: Hyvärinen & Katajisto 1984; Hyvärinen 1989). In contrast to the bipartite blood sinus of most terrestrial mammals, composed of a ring sinus close to the apical end of the follicle and a cavernous sinus located below it, follicles of the pinniped species studied so far possess an additional cavernous sinus situated above the ring sinus. This upper cavernous sinus takes about 60% of the total length of the follicle (which can measure up to 2 cm) and is not supplied with sensory elements. Therefore, the ring sinus area, where most mechanosensitive receptors are located, is inserted much deeper in the capsule than is normal for terrestrial mammals. At the level where the deep vibrissal (a branch off the infraorbital branch of the trigeminal nerve) nerve is passing through the capsule, the number of nerve fibres (1000–1600) of *Phoca hispida* exceeds that calculated for well-endowed terrestrial species like the rat and cat (Rice *et al.* 1986) by a factor of 10. This difference in innervation density remains unaltered at the receptor level: up to 20 000 Merkel cells—the predominant receptor found in vibrissal follicles—compared with 500–2000 in the follicle of a terrestrial mammal (Dehnhardt *et al.* 1999). The same relationship has been found for other sensory elements like free nerve endings, lanceolate and lamellated nerve endings. This difference in innervation already indicates that the vibrissal system is of special importance for aquatic mammals.

The functional significance of vibrissae remained obscure for a long time. Whereas the name 'vibrissae' emphasizes the reception of vibrational information, results from single unit recordings at the infraorbital branch of the trigeminal nerve of harbour seals and grey seals suggested that the vibrissal system is primarily designed for the perception of tactile information which the animals receive through physical contact of the vibrissae to objects in the environment (Dykes 1975). Dykes postulated active touch achievements such as the recognition of surface structures as well as the discrimination of macrospatial stimuli like the shape and size of objects.

Active touch achievements are a function of the so-called haptic sense, integrating cutaneous mechanosensation as well as kinaesthetic information (Gibson 1962; Loomis & Lederman 1986), thus providing a good example of behavioural skills based on a synergism of different sensory modalities. Sensory elements of the kinaesthetic modality are proprioceptors like joint receptors and muscle spindles that provide information about the position and movement of parts of the body.

Together with the highly mobile muscular lips, the different types of facial vibrissae of manatees and dugongs form a haptic system unique among marine mammals. Manatees use their vibrissal–muscular complex, consisting of the rigid perioral bristles and the muscular lips, in a prehensile manner (Marshall *et al.* 1998). This suggests a complex interaction of mechanosensory and kinaesthetic information during exploration and object manipulation. These prehensile properties of the vibrissal–muscular complex are primarily used to grasp and bring food into the oral cavity when feeding on submerged or floating vegetation. Bachteler and Dehnhardt (1999) demonstrated that, similar to the employment of mystacial vibrissae in pinnipeds, manatees use the bristle-like hairs of the oral disc for the discrimination of textured surfaces. This is consistent with the results of Marshall *et al.* (1998) suggesting that the bristle-like hairs of the oral disc are involved in the tactile exploration of objects. It is conceivable that the two different types of facial hairs of sirenians serve different functions in active touch processes. However, it is likely that in many cases the two hair types are employed synergistically in a two-stage haptic process, with the bristle-like hairs providing first sensory information before an object is grasped and manipulated for further evaluation by means of the perioral bristles (Bachteler & Dehnhardt 1999).

In accordance with the neurophysiological results that Dykes (1975) obtained for seals, psychophysical experiments have shown that walruses (Kastelein & van Gaalen 1988), California sea lions (Dehnhardt 1990, 1994; Dehnhardt & Dücker 1996) and harbour seals (Dehnhardt & Kaminski 1995; Dehnhardt *et al.* 1997, 1998a) are capable of identifying the size, shape and surface structure of an object by active touch with their mystacial vibrissae. In the California sea lion, tactile shape recognition is as fast and reliable as by vision. Vibrissal size discrimination capabilities in harbour seals are comparable to those of the prehensile hands of some primate species (Carlson *et al.* 1989) and comes close to the visual resolving power of pinnipeds. From the model described in Fig. 5.2 it becomes clear that haptic size discrimination in these pinnipeds relies on vibrissal mechanosensory information, whereas its accuracy is

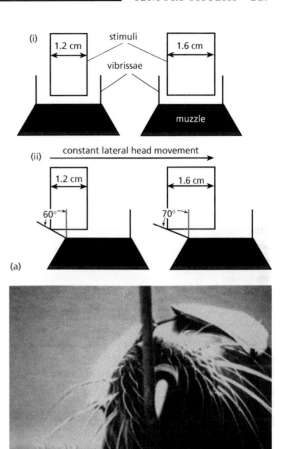

(a)

(b)

Fig. 5.2 (a) Model for the perception of size differences based on the tactual behaviour of seals and sea lions during the comparison of circular discs with diameters of 1.2 cm and 1.6 cm. The vibrissae are considered to be rigid levers. (i) Centered positioning of the head in relation to the disc. (ii) Identical lateral head movements cause greater deflections of the outer vibrissal shafts in the disc with the larger diameter than is the case for the smaller disc. These different deflections of the outer vibrissal hair shafts result in mechanical stimulations of differing intensitiy inside the follicles which lead to different responses of the mechanoreceptors located in the area of the ring sinus. In contrast, humans measure the angles of the finger joints when required to make size discriminations with the thumb and index finger (kinaesthetic discrimination) (John *et al.* 1989), while cutaneous mechanosensory information (the amount of compression of the fingerpads) is held constant at objects to be compared. (b) A California sea lion touching a disc. Information about the localization of a disc was received by the long posterior vibrissae, while after the localization of a disc, the sea lion identified it using only the shorter vibrissae located anteriorly.

codetermined by the kinaesthetic control of exploratory movements.

Vibrissae provide pinnipeds with tactile information primarily in the aquatic environment, where the animals are in danger of hypothermia when exposed to low water temperatures (see Chapters 3 and 9). The animals avoid heat dissipation by a reduction of the skin blood flow and an effective insulation (blubber). Additionally—unlike most terrestrial mammals—they allow their outermost tissue layers to cool down close to ambient temperature (Kvadsheim *et al.* 1997). Since studies on tactile sensitivity in humans demonstrated that a decrease in skin temperature leads to severe deterioration of tactile sensitivity (Green *et al.* 1979; Gescheider *et al.* 1997), temperature regulation in pinnipeds raises the question of whether and how they maintain vibrissal tactile sensitivity in their thermally hostile environment. Dehnhardt *et al.* (1998a) showed that the ability of harbour seals to discriminate textured surfaces remained essentially unaltered under different thermal conditions. Even at water temperatures of about 1°C the seals achieved 'Weber fractions' of 0.09 (indicating 9% stimulus difference). Infrared thermography revealed that, contrary to the dictate of thermal economy, the mystacial and supraorbital vibrissal pads of harbour seals are areas of excessive heat loss. Thermally clearly defined against the rest of the head, the high temperatures measured at the surfaces of vibrissal pads vividly demonstrate that in these sensory areas no vasoconstriction occurs during cold acclimation, indicating a separate vibrissal blood circulation (Mauck *et al.* 2000). Selective heating of vibrissal pads is suggested to be a function of the upper cavernous blood sinus (Fig. 5.1), which is free of receptors and thus may primarily serve as a thermic insulator for the receptor area below it. The fact that seals do not allow cooling of their vibrissal follicles indicates that without this adaptation this mechanosensory system would be subject to similar deficiencies known from the human hand in response to cooling. Heat loss from these comparatively small skin regions is the energetic price seals have to pay for keeping their vibrissal system working. This indicates that a permanent access to tactile information is of biological importance for these marine mammals.

In which context are active touch capabilities of importance for seals in the wild? Beside their social function described by Miller (1975), vibrissae may be an indispensable sensory system in prey detection. Observations in the wild suggest that southern sea lions (*Otaria flavescens*) and walruses use their vibrissae when foraging at the sea bottom (Lindt 1956; Fay 1982). On the basis of otoliths found in faecal samples of harbour seals, Härkönen (1987) demonstrated that benthic prey, especially flatfish, are of supreme importance for the animals studied. The finding that blind seals are well-nourished throughout the year (Newby *et al.* 1970) also suggests the potential significance of haptic vibrissal information for the detection and identification of benthic prey.

5.2.2 Hydrodynamic reception

Several experiments have failed to find evidence of a sonar system in pinnipeds (Wartzok *et al.* 1984) (see Chapter 6), so the question remains whether vibrissae could provide sensory information for the detection of pelagic fish. In the aquatic environment, one utilizable source of sensory information consists of water disturbances, inevitably caused by any moving organisms. Consequently, hydrodynamic sensory systems, like the lateral line of fish, have evolved many times in aquatic animals (Bleckmann 1994). Although it has been well demonstrated that the vibrissae of harbour seals respond to vibrations mediated by a rod directly contacting the hair (Dykes 1975; Renouf 1979; Mills & Renouf 1986) their function as a hydrodynamic receptor system has only been demonstrated recently for the first time with a technique commonly used to study the fish lateral line (Dehnhardt *et al.* 1998b). Water movements (10–100 Hz) were generated with a constant volume oscillating sphere positioned 5–50 cm in front of the vibrissae of a harbour seal. In terms of particle displacement the seal responded to water movements of < 1 μm. The shape of the tuning curve obtained for the harbour seal (Fig. 5.3) is similar to those determined for other aquatic animals equipped with a hydrodynamic sense (Bleckmann 1994), and characterize the vibrissae as a hydrodynamic receptor system with a spectral sensitivity well tuned to the frequency range of fish-generated water movements.

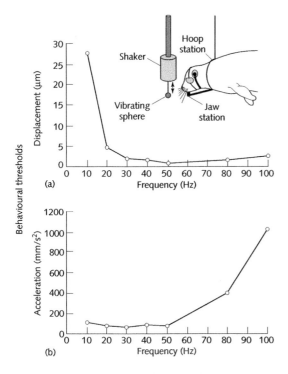

Behavioural thresholds

(a)

(b)

Fig. 5.3 Hydrodynamic reception. Inset: At the beginning of a trial the harbour seal placed its head in a hoop opposite to a sphere. Additionally the seal was trained to press the tip of its lower jaw to a little knob mounted on the hoop and to correct the position whenever it lost contact with the jaw station. In this way a defined distance between the tip of the vibrissae and the vibrating sphere was guaranteed. Knowing the exact distance between the vibrissae and the sphere, the effective stimulus amplitude was calculated. Optical and acoustic cues were excluded by providing the seal with removable eye caps and headphones (pink noise masking). The seal indicated the detection of a water movement by leaving the hoop station during the period of stimulus presentation (go/no-go paradigm). Tuning curves: (a) behavioural displacement and (b) acceleration thresholds of a harbour seal to sinusoidal water movements. In the frequency range 10–50 Hz the seal essentially responded proportionally to the acceleration of water movements, while at higher frequencies it responded proportionally to the water displacements. A similar change in operation mode is known from the fish lateral line.

Although hydrodynamic information is assumed to be important only in the vicinity of the receiving animal, the wakes of fishes persist for several minutes (Hanke *et al.* 2000), thus representing trackable hydrodynamic trails of considerable length. Dehnhardt *et al.* (2001) used a miniature submarine for

the generation of hydrodynamic trails and showed that a blindfolded harbour seal can use its whiskers to detect and track trails as long as 40 m. These results demonstrate for the first time that hydrodynamic information can be used for long-distance object location, thus establishing a new system for spatial orientation in the aquatic environment that might help explain successful feeding of pinnipeds in dark and murky waters.

5.2.3 Audition

The return of terrestrial mammals to an aquatic environment required adaptations of their auditory systems not only with regard to the new aquatic lifestyle, for example reducing hydrodynamic resistance, but also to new physical conditions for underwater signal transmission. Furthermore, differences in the hearing abilities of marine mammals may have evolved due to the selection pressure of species-specific ecological and social demands.

Most marine mammals dive to considerable depth, where they are exposed to high hydrostatic pressure. Therefore, closing mechanisms in the area of the ear canal (auditory meatus, see Fig. 5.4 for a comparison with the human ear), as, for example, the auricular muscles in pinnipeds, are developed so that the ear is protected from penetrating water. Fast swimming on the other hand requires a streamlined body so that a selection pressure on the reduction of 'disturbing body apendages' can be expected. Consequently the external ear or pinna is reduced in marine mammals (Fig. 5.1) and is only present in eared seals and otters. Because external ear tissue is acoustically transparent under water, and thus is not important for underwater sound reception, there should have been no trade-off between hydrodynamic adaptation and sensory function.

Depending on the degree of aquatic adaptation of each species, ranging from the amphibious otariids to pelagic whales (Ketten 1991), the sound-receiving systems of marine mammals have adapted to the physical demands of water. Current ideas of how the problem of sound transmission to the inner ear is solved in different orders of marine mammals will be described primarily on the basis of 'sound localization'.

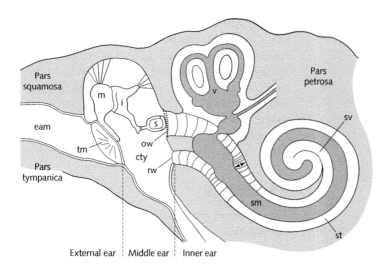

Fig. 5.4 Fundamental structures of the human outer, middle and inner ear. In humans, the bones surrounding the auditory structures are fused into the temporal bone. cty, cavum tympani; eam, external auditory meatus; i, incus; m, malleus; ow, oval window; rw, round window; s, stapes; sm, scala media; st, scala tympani; sv, scala vestibuli; tm, tympanic membrane; v, vestibule. (Adapted from Silbernagel & Despopulus 1983.)

In the horizontal plane, directional information can be processed by the auditory system on the basis of interaural time and intensity differences (binaural effects). Because of an animal's head dimensions, a sound will reach the more distant ear later, or it will be shadowed by the head and will thus be of lower intensity at the more distant ear. Since sound travels about 4.5 times faster in water than in air, it is obvious that information processing based on time or sound level differences must work more precisely under water than in air. For the bottlenose dolphin it has been shown that both kinds of auditory information processing are superior in this odontocete species compared to any other mammal yet tested (Moore *et al.* 1995).

The few studies dealing with directional sensitivity in odontocetes show that the bottlenose dolphin can discriminate two pure tones (10–100 kHz) presented from underwater speakers separated by an angle of 2° to 3°. For echolocation clicks at 64 kHz even smaller minimum audible angles have been determined (0.7° to 0.9°) (Renaud & Popper 1975). It is surprising that this excellent sound localization ability was found to be as good in the vertical plane as in the horizontal plane, because binaural effects are not present in the vertical plane. The mechanism underlying this sound localization ability is still unclear (Au 1993).

A submerged terrestrial mammal experiences a loss of directional sensitivity, because there is only a neglectable impedance mismatch between water and cranial tissues. Consequently, underwater sound waves enter nearly all parts of the head and proceed via bone conduction to the organ of corti from all directions at nearly the same time. In adaptation to these conditions, auditory organs in odontocetes are largely isolated from the skull, so that bone conduction via the skull is avoided. This raises the question of how a sound wave is transmitted to the inner ear. Electrophysiological studies obtained minimum absolute hearing thresholds when sounds below 20–30 kHz were presented near the external meatus of a dolphin (Bullock *et al.* 1968; Popov & Supin 1990). This indicates that the extremely narrow ear canal, which at least in adult animals is additionally filled with cellular debris, still works as an auditory pathway. However, for echolocation signals (> 20–30 kHz), Norris (1964, 1968) suggested bone or tissue conduction through the mandibular fat channels of the lower jaw and the pan bone as a new, modified pathway in toothed whales (cf. Chapter 6). The impedance of the mandibular fat is close to that of water and thus may guide the sound wave over the pan bone to the petrotympanic bullae. This hypothesis is supported by electrophysiological and behavioural evidence. For example, directional sensitivity in the bottlenose dolphin is optimal when tones arrive from 15° off-midline (Renaud & Popper 1975). This is consistent with the finding that ultrasonic stimuli (> 20 kHz) presented to the lower jaw result in significantly greater evoked potentials in the central auditory system (Bullock *et al.* 1968) as well as greater cochlear potentials (McCormick *et al.* 1970). Furthermore, if during behavioural tests the lower jaw is covered by a neoprene hood impervious to acoustical signals, echolocation performance of a bottlenose dolphin is significantly impaired (Brill 1988). From all these results, Ketten (1991) concluded that in toothed whales two parallel sound-receiving systems exist, one for ultrasonic signals (jaw hearing) and one for lower frequency sounds (ear canal).

Compared to the ear of odontocetes, the acoustic isolation of the ear of baleen whales (Ketten 1991), manatees (Ketten *et al.* 1992) and pinnipeds (Ramprashad *et al.* 1972; Repenning 1972) is less complete. Although the path of sound reception in mysticetes is still unknown, the highly derived tympanic membrane that connects to the external auditory canal suggests that the conventional pathway of terrestrial mammals is functional in baleen whales (Ketten 1992).

Electrophysiological and anatomical evidence indicates that in manatees a novel auditory pathway developed. These fully aquatic mammals possess a lipid-filled zygomatic process, which is connected to the squamosal–periotic complex (Domning & Hayek 1986; Ketten *et al.* 1992). This structure is suggested to be functionally equivalent to the mandibular fat channel of odontocetes, thus serving as a modified pathway to the inner ear (Ketten *et al.* 1992). This is consistent with the finding that evoked potentials in the central auditory system are greater when sounds are presented above the area of the zygomatic process (Bullock *et al.* 1980; Klishin *et al.* 1990).

Pinnipeds do not appear to have a novel pathway for sound transmission. Because in pinnipeds the periotic bones and tympanic bulla are not isolated from the skull as found in odontocetes and the cochlear round window is up to three times the size of the oval window (this ratio is about 1 : 1 in most terrestrial carnivores, see Fig. 5.4), Repenning (1972) suggests that underwater bone conduction seems to be the most important sound-transmitting mechanism. However, though bone conduction via the skull should limit directional sensitivity, good sound-localization abilities have been determined in the harbour seal (Møhl 1964, 1968; Terhune 1974; Richardson *et al.* 1995), while in the California sea

lion directional sensitivity is poorer and more variable (Table 5.1) (Gentry 1967; Bullock *et al.* 1971; Moore 1975; Moore & Au 1975).

Behavioural studies showed that the ear of pinnipeds is also functional in air, but underwater sensitivity is always better compared to aerial sensitivity (Watkins & Wartzok 1985). The aerial audiograms obtained in some studies (*Phoca vitulina*: Møhl 1968; *Zalophus californianus*: Schusterman *et al.* 1972; *Callorhinus ursinus*: Moore & Schusterman 1987) demonstrated a decrease in auditory sensitivity at 4 kHz (Fig. 5.5). In addition to the general difference between underwater and aerial audiograms, this particular decrease in sensitivity has been interpreted as indicative of two different auditory pathways to the inner ear in the two media (Watkins & Wartzok 1985; Moore & Schusterman 1987). As mentioned above, bone conduction is proposed for underwater sound reception, whereas aerial hearing is assumed to rely on the conventional mammalian pathway (Nachtigall 1986).

If the conventional auditory pathway is functional in baleen whales, the large distance between the two ears makes sound localization less problematic (Gourevitch 1980). Field studies provide evidence that baleen whales can localize the sources of a sound. It has been shown that they precisely swim towards sounds projected by underwater speakers and swim away from sounds of killer whales (Frankel *et al.* 1995; Richardson *et al.* 1995).

Habitats of marine mammals range from rivers and estuaries to shallow coastal and deep open ocean waters. As has been shown for bats (Neuweiler 1984) and rodents (Heffner *et al.* 1994), properties and achievements of the auditory system reflect the habitat of a species. Starting from this assumption, Ketten and Wartzok (1990) proposed that audition in marine mammals has not only evolved in adaptation to the physical demands of water, but that the auditory capacity of marine mammals should correlate with their specific marine environments and lifestyles, and consequently with corresponding cochlea characteristics.

This hypothesis has been tested for cetaceans. Based on different cochlear measurements (e.g. the number of turns, basilar membrane width and thickness, outer osseous spiral lamina length, basal/apical width, axial height) two different cochlea

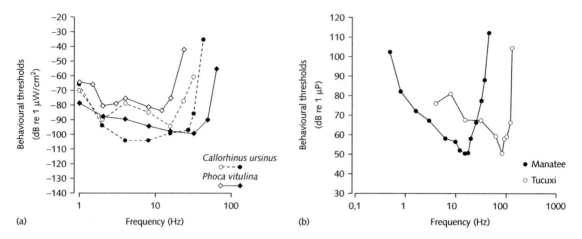

Fig. 5.5 (a) Underwater (filled symbols) versus aerial (open symbols) audiograms for a phocid and an otariid pinniped species, considered as amphibious mammals, and (b) underwater audiograms for a tucuxi and a manatee, both considered as fully aquatic. The comparison of aerial and underwater hearing thresholds is made on the basis of intensities (dB re: 1 μW/cm²), whereas underwater thresholds in (b) are calculated as sound pressure level (SPL). (Data for *Callorhinus* from Moore & Schusterman 1987; for *Phoca* from Møhl 1968; for *Sotalia* from Sauerland & Dehnhardt 1998; and for the manatee from Gerstein *et al.* 1999.)

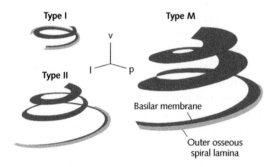

Fig. 5.6 Basilar membrane and outer osseous spiral lamina of odontocetes (type I and II) and mysticetes (type M). Spirals of type I cochleas are characterized by a small ratio of axial height to basal turn diameter (< 0.4), an axial height < 0.1% of the body length, less than two spiral turns, and an outer osseous spiral lamina measuring > 60% of the length of the basilar membrane. Type II cochleas are characterized by an axial height to basal turn diameter ratio of > 0.5, an axial height > 2% of the body length, typically 2.5 spiral turns, and an outer osseous spiral lamina measuring about 25% of the length of the basilar membrane. (Adapted from Ketten 1991.)

types have been discribed for echolocating odontocetes (type I and II) and one cochlea type for the potentially infrasonic mysticetes (type M) (Fig. 5.6) (Ketten & Wartzok 1990; Ketten 1991, 1992).

Cochlea type I is found in species like the Amazon river dolphin and the coastal harbour porpoise, which produce extraordinary high-frequency echolocation signals. The type II cochlea is found in species living offshore or in open ocean waters, like the bottlenose dolphin, where low-frequency echolocation signals are more suitable for a long-range detection of objects.

Up to now, there has not been sufficient behavioural evidence supporting the idea of dividing the odontocetes into two different, frequency-specific acoustic groups. For example the upper hearing limit of the Amazon river dolphin (*c.* 100 kHz) is much lower than expected from the study by Ketten and Wartzok (1990). However, extensive comparative psychophysical data are only available for the absolute hearing sensitivity of toothed whales (Table 5.1). Absolute thresholds are probably not the appropriate hearing function for the demonstration of environmental effects. A more promising approach would be to compare discriminative capacities of species, e.g. for frequency, but more comparative data are required concerning this function of hearing.

As a direct measure of the frequency range of a species, basilar membrane dimensions are most

Table 5.1 Comparison of different characteristics of hearing in odontocetes, pinnipeds and some terrestrial vertebrates, as determined by psychophysical and evoked potential (source in *italics*) techniques.

Species	Test range (kHz)	Absolute sensitivity		Differential sensitivity (frequency discrimination)	Directional sensitivity (MAA)	Source
		Range of best hearing (kHz)	High-frequency cut-off (kHz)			
Odontocetes						
Beluga	1–123	11–105	120			White et al. 1978
	16–110	50–80	100			*Popov & Supin 1990*
Bottlenosed dolphin	0.075–150	15–110	150			Johnson 1967
	20–100			0.28% at 5 kHz		Jacobs 1972
					2.1° at 20 kHz	Renaud & Popper 1975
					0.9° at 64 kHz	
	1–70					Herman & Arbeit 1972
	1–105			0.09% at 27 kHz		Thompson & Herman 1975
	5–140	32–100	120	0.21% at 4 kHz		*Popov & Supin 1990*
	1–105	50–102	105			Bullock et al. 1968
Harbour porpoise	1–150	3–70	150			Andersen 1970
					3° at 2 kHz	Dudok van Heel 1962
					7.9° at 6 kHz	
False killer whale	10–140	125–130	140	0.1% at 125–130 kHz		Popov et al. 1986
	2–115	17–74	115			Thomas et al. 1988
Killer whale	0.5–32	15–30	120			Hall & Johnson 1972
						(Hf cut-off after Bain et al. 1993)
Striped dolphin	10–105	50–80	c. 105			Bullock et al. 1968
Chinese river dolphin	1–200	16–32	c. 100			Ding Wang et al. 1992
Risso's dolphin	1.6–110	4–64	100			Nachtigall et al. 1995
Amazon river dolphin	1–105	12–64	100			Jacobs & Hall 1972
	8–120	15–35, 60–100	105			*Popov & Supin 1990*
Tucuxi (coastal)	4–135	64–105	125			Sauerland & Dehnhardt 1998
Tucuxi (riverine)	5–160	30–100				*Popov & Supin 1990*

Pinnipeds (underwater spp.)

Harbour seal	1–180 (1–57)	4–45	64	1% at 32 kHz	3.1° at 2 kHz	Møhl 1964, 1968
					4.1° for clicks	Terhune 1974
Ringed seal	1–90	2.8–45	64	3% at 8 kHz		Terhune & Ronald 1975
Harp seal	1–100	1–40	64			Terhune & Ronald 1972
Monk seal	2–46	16–24	40			Thomas et al. 1990
Grey seal	2–120	8–40				Ridgway & Joyce 1975
California sea lion	0.25–64	1–28	32		10° at 6 kHz	Schusterman et al. 1972
						Gentry 1967
					6–9° for clicks	Moore 1975
					3.5° at 1 kHz	Moore & Au 1975
				1.65% at 24 kHz		Schusterman & Moore 1978
Northern fur seal	1–42	2–32	32			Moore & Schusterman 1987
	0.5–40	5–17		1–2% at 1–30 kHz		Babushina et al. 1991

Pinnipeds (above water spp.)

Harbour seal	1–22.5	2–2.8, 4–16	22.5		4.8° at 0.5 kHz	Møhl 1964, 1968
					1 1/2° for clicks	Terhune 1974
Harp seal	1–32	1.4–11.3	–			Terhune & Ronald 1971
California sea lion	1–32	2–8	32		4° at 1 kHz	Moore & Schusterman 1987
					9° for clicks (1 kHz)	Moore & Au 1975
						Moore 1975
Northern fur seal	0.5–32	2, 8–16	16			Moore & Schusterman 1987
	0.1–25	2–16		1.6–1.8% at 3–5 kHz		Babushina et al. 1991

Other mammals (in air)

Cat	0.065–60	2–16	60	0.75 % at 20 kHz	8° at 0.5 kHz	Neff & Hind 1955
						Elliott et al. 1960
						Casseday & Neff 1973
Dog	0.04–46	1–16	46			Heffner 1983
Human	0.064–16	1–4				Sivian & White 1933
	0.25–10				0.9° at 0.75 kHz	Mills 1958
	0.1–4			0.6% at 0.5 kHz		Sinnott & Aslin 1985

Hf, high frequency; MAA, minimum audible angle.

important (Keidel & Neff 1974). The better support of the basilar membrane by the outer spiral lamina in type I odontocetes, and its narrower apical end compared to type II species, indicates a specialization for high-frequency hearing. In mysticetes, the basilar membrane is much wider and thinner than in odontocetes, suggesting low-frequency specialization (Ketten 1992). This is consistent with the width of the apical end of the basilar membrane, which is about five times as wide in baleen as it is in toothed whales (Fig. 5.6).

Although all evidence concerning the shape of baleen whale audiograms is indirect, behavioural reactions of free-ranging whales to underwater playbacks from conspecifics or other natural sounds lend support to the hypothesis of low-frequency specialization in these large cetaceans (Watkins 1981; Richardson & Greene 1993; Frankel et al. 1995) (see Chapter 6). Gray whales, for example, react to tones from 0.8 kHz to 1.8 kHz (95–142 dB re 1 μPa) with the best sensitivities at 1–1.5 kHz (Dahlheim & Ljungblad 1990). Some of the large whales also show behavioural responses to higher frequencies ranging from 3 kHz (Lien et al. 1990; Maybaum 1993) up to 28 kHz (Watkins 1986).

Audiograms of odontocetes generally show that these cetaceans can hear in a wider range of frequencies than any other mammal tested so far (Fig. 5.5). In the bottlenose dolphin best hearing (arbitrarily defined as 10–20 dB from maximum sensitivity) ranges from 12 kHz to 75 kHz with minimum thresholds of 39–55 dB re 1 μPa (Table 5.1). This frequency spectrum covers relatively low-frequency communication signals (e.g. whistles) and high-frequency echolocation signals. However, high-frequency hearing abilities related to echolocation signals are exceptionally good up to 80–150 kHz (Table 5.1). As to be expected for echolocating animals, frequency discrimination abilities of the bottlenose dolphin, measured as just noticeable differences (JNDs, which equal absolute difference thresholds), and the harbour porpoise (auditory evoked potentials), are particularly good. Calculated as Weber fractions (the ratio of the lowest frequency difference detected by the dolphin to the starting frequency of the sound), the bottlenose dolphin can distinguish two sounds differing in fre-

quency by 0.2–0.8% in a wide range of frequencies (2 kHz up to 130 kHz). Frequency sensitivity is related to the number of hair cells in the organ of Corti. In *Tursiops truncatus* the mean number of inner hair cells is 3451 and the number of outer hair cells is 13 933 (Wever et al. 1971). In the harp seal the corresponding numbers are 3654 and 14 318, and in the ringed seal 3232 and 13 947 (Ramprashad et al. 1972). However, although the ratio of inner hair cells to outer hair cells is similar in *Tursiops* and seals, frequency discrimination in pinnipeds is less precise and restricted to a lower range of frequencies (Table 5.1).

Based on cochlea and middle ear anatomy, Ketten et al. (1992) proposed poor sensitivity and poor localization abilities in a relatively narrow frequency range for the manatee (*Trichechus manatus*). However, Gerstein et al. (1999) point out that, at least, absolute hearing thresholds of West Indian manatees are lower than suggested (Fig. 5.5). Their psychophysical tests revealed that the West Indian manatees' hearing range extends from 0.15 kHz to 46 kHz with best sensitivity at 6–20 kHz. Bullock et al. (1982) obtained auditory evoked potentials from the same species and found best sensitivity at 1 kHz, a cutoff at 4–8 kHz, and improved sensitivity at 36 kHz. In the Amazonian manatee responses to rather low (< 200 Hz, Bullock et al. 1980) as well as high frequencies (50 kHz, Klishin et al. 1990; Popov & Supin 1990) have been determined, but the range of best frequencies was narrower (Popov & Supin 1990).

It is important for marine mammals to detect and localize sounds and sound sources in the masking background of their noisy environment (Fay 1992; Richardson et al. 1995). Odontocetes and pinnipeds are generally better at detecting signals in noise at higher frequencies than terrestrial mammals (best at 0.1–10 kHz vs human and cat at 0.1–2 kHz) (Richardson et al. 1995). However, adaptations of the auditory system are apparently not sufficient on their own. Toothed whales developed behavioural strategies, such as changing the conventional straight-line echolocation (Penner et al. 1986) or shifting the frequency of echolocation clicks into a frequency range not masked by ambient noise (Richardson et al. 1995). There is also evidence that belugas (Johnson 1991) and harbour seals

(Turnbull & Terhune 1993) reduce the effect of masking by emitting a series of vocalizations.

5.3 VISION

Like most other forms of energy, the electromagnetic radiation making up visible light experiences significant changes in quality and intensity when entering the marine environment. Depending on wavelength, light is attenuated by absorption and scattering so that with depth it becomes more and more monochromatic and its spectrum is shifted to shorter wavelengths (Jerlov 1968). However, depending on factors such as suspended particles, there should be sufficient light intensities even at great depth to make vision an important source of information for aquatic animals (Duntley 1963). Because several species of pinnipeds and cetaceans can detect and discriminate objects under water as well as in air, the eyes of marine mamals have to meet twofold requirements in order to effectively operate in the two media. Accordingly, adaptations

to these requirements have been found in diverse species of marine mammals in terms of spectral sensitivity of visual pigments, retinal organization, refraction characteristics of the dioptric apparatus and visual acuity under various conditions.

It has been repeatedly proposed that the spectral sensitivity of the visual pigments of fishes is adapted to their particular photic environment (e.g. Wald *et al.* 1957). The maximum absorption of pigments should correlate with the spectral distribution of light at the depth where the respective species is found. Correspondingly, blue-sensitive pigments can be extracted from deep-sea fish (wavelength of maximum absorbance λ_{max}: 477–487 nm), while marine fish living near the surface have pigments more sensitive to green light (Denton & Warren 1956; Munz 1964; Beatty 1969; Lythgoe 1972). This so-called sensitivity hypothesis has also been suggested for cetaceans (McFarland 1971) and pinnipeds (Lavigne & Ronald 1975a).

Table 5.2 shows the maximum absorption values λ_{max} for extracted visual pigments of marine mammals, as well as visual sensitivities determined in

Table 5.2 Comparison of different characteristics of vision in marine mammals as determined by biochemical, morphological (*italics*) and psychophysical techniques.

Species	Wavelength of maximum absorbance	Visual acuity and *colour vision* (?)
PINNIPEDIA		
Phocidae		
Harp seal (*Pagophilus groenlandicus*)	496.6 (blue–green) (Lavigne & Ronald 1975a)	
Harbour seal (*Phoca vitulina*)	495.7 (blue–green) (Lavigne & Ronald 1975a)	8'3" in air (Schusterman 1972) 8'3" under water (Schusterman & Balliet 1970)
Grey seal (*Halichoerus grypus*)	496.6 (blue–green) (Lavigne & Ronald 1975a)	
Weddell seal (*Leptonychotes weddelli*)	495–496 (blue–green) (Lythgoe & Dartnall 1970)	
Southern elephant seal (*Mirounga leonina*)	485–486 (blue) (Lythgoe & Dartnall 1970)	
Spotted seal (*Phoca largha*)	530 (green) (Wartzok 1979)	
Otariidae		
California sea lion (*Zalophus californianus*)	497, 501 (blue–green) (Crescitelli 1958)	Dichromatic colour vision (blue–green) (Griebel & Schmidt 1992) 4'8"–7'8" / 4'6"–7'0" (Schusterman 1972)
Steller's sea lion (*Eumetopias jubatus*)		7'1" in air (Schusterman 1972)

(*continued*)

Table 5.2 (*cont'd*)

Species	Wavelength of maximum absorbance	Visual acuity and *colour vision (?)*
South American fur seal (*Arctocephalus australis*)		Dichromatic colour vision (blue–green), 5′44″ in air (minimum visibile), 7′2″ in air (minimum separabile) (Busch & Dücker 1987)
South African fur seal (*Arctocephalus pusillus*)		Dichromatic colour vision (blue–green), 5′45″ in air (minimum visibile), 6′59″ in air (minimum separabile) (Busch & Dücker 1987)
Northern fur seal (*Callorhinus ursinus*)	499.9 (blue–green), (Lavigne & Ronald 1975a)	
SIRENIA		
Manatee (*Trichechus manatus*)	505 nm (greenish) (Piggins *et al.* 1983)	Most probably dichromatic colour vision (Griebel & Schmidt 1996)
CETACEA		
Mysticeti		
Humpback whale (*Megaptera novaeangliae*)	492 (blue to blue–green) (Dartnall 1962)	
Gray whale (*Eschrichtius robustus*)	497 (blue–green) (McFarland 1971)	*11′ in water (Mass & Supin 1997)*
Odontoceti		
Beaked whale (*Berardius bairdii*)	481 (blue) (McFarland 1971)	
Risso's dolphin (*Grampus griseus*)		18′7″ in air (Nachtigall 1989)
Pacific white-sided dolphin (*Lagenorhynchus obliquidens*)	485.5 (blue) (McFarland 1971)	6′ under water, bright light (Spong & White 1971) *11′2″ under water (Murayama & Somiya 1998)*
Killer whale (*Orcinus orca*)		5′5″ under water (White *et al.* 1971)
Bottlenose dolphin (*Tursiops truncatus*)	486 (blue) (McFarland 1971)	12′6″ in air, 9′0″ under water (Herman *et al.* 1975) 18′ in air (Pepper & Simons 1973) *9′ under water, 12′ in air (Mass & Supin 1995)*
False killer whale (*Pseudorca crassidens*)	486.5 (blue) (McFarland 1971)	*9′3″ under water (Murayama & Somiya 1998)*
Piebald (Commerson's) dolphin (*Cephalorhynchus commersonii*)	485.4 (blue) (McFarland 1971)	
Saddleback dolphin (*Delphinus delphis*)	488.9 (blue) (McFarland 1971)	*8′ under water (Dral 1983)*
Dusky dolphin (*Lagenorhynchus australis*)	485.9 (blue) (McFarland 1971)	
Narrow-snouted dolphin (*Stenella attenuata*)	485.7 (blue) (McFarland 1971)	
Tucuxi (*Sotalia fluviatilis*)		*25′ under water, 33′ in air (Mass & Supin 1999)*
Amazon river dolphin (*Inia geoffrensis*)		*40′ under water, 53′ in air (Mass & Supin 1989)*
Dall's porpoise (*Phocoenoides dalli*)		*11′5″ under water (Murayama et al. 1995)*
Harbour porpoise (*Phocoena phocoena*)		*11′ under water, 14′7″ in air (Mass & Supin 1986)*
Hawaiian spinner dolphin (*Stenella* spp.)	485.4 (blue) (McFarland 1971)	

Species names according to King (1983) and Watson (1981).

behavioural experiments. It becomes clear that pigments of species living in coastal waters and diving only to shallow or medium depths have absorption maxima in the blue–green part of the light spectrum, as reported for most coastal fishes (Munz 1964). The Weddell seal's λ_{max} of 495–496 nm seems to contradict the sensitivity hypothesis, because this species dives to depths of 300–600 m (Kooyman 1989). However, the spectral region of greatest intensity in polar waters is shifted from the blue (475 nm) of tropical ocean waters to rather green (500–560 nm), which may explain the apparent contradiction. Similarly, the absorption maxima of delphinid visual pigments, which cluster around 486 nm, have been interpreted as an adaptation to the open ocean and diving habits of the species under examination (McFarland 1971), while the absorption maxima of 481 nm in the beaked whale and 486 nm in the southern elephant seal are indicative of deep-sea visual pigments in these deep-diving species (Lythgoe & Dartnall 1970; McFarland 1971).

However, there seem to be some differences between the results of maximum absorption analyses of extracted visual pigments and thresholds for spectral sensitivity determined mainly by psychophysical methods. Wartzok (1979) found rod-dominated spectral sensitivity curves for the harbour seal and the spotted seal (or largha seal, *Phoca largha*) (King 1983) with best sensitivities near 530 nm (green) for both species. Similarly, Lavigne and Ronald (1972) reported maximum photopic sensitivity in the harp seal near 550 nm, while scotopic sensitivity peaks in the region of 500–525 nm. Nevertheless, these authors interpreted their results as indicative of an adaptation to green coastal waters as well.

There has been considerable effort to clarify which photoreceptors are present in the retinas of marine mammals. The eye of most marine mammal species resembles the eye of a nocturnal mammal in its large size (Walls 1942; Jamieson & Fisher 1972), its dilatable pupil which maximizes light collection, and a choroid (located between the retina and outer coating of the eye) provided with a light-reflecting layer of neatly orientated collagen fibres, the so-called tapetum lucidum (Walls 1942). Thus, a retina dominated by rod-like receptors usually found in nocturnal and crepuscular animals could be expected to be adaptive for marine mammals as well (Jamieson & Fisher 1971, 1972). Photoreceptors are among those parts of the retina which show early postmortem autolysis. Probably because of this, some histological studies, mostly examining pinniped retinas using light microscopic techniques, failed to demonstrate the presence of cones (Walls 1942; Landau & Dawson 1970; Nagy & Ronald 1970). However, in cetacean retinas, some early authors described cone-like elements (see Dral 1977), which was corroborated by Perez *et al.* (1972). While there is limited anatomical information favouring the presence of two different types of receptors (rods and cones) in seal retinas (Jamieson & Fisher 1971), this assumption is supported by physiological evidence based on critical flicker frequency in the harp seal (Bernholz & Matthews 1975), and as concluded from visual acuity functions in the California sea lion (Lavigne & Ronald 1975b). Van Esch and De Wolf (1979) provided physiological evidence for cone function in *Tursiops*.

Most non-primate mammals possess short wavelength-sensitive (S) and middle to long wavelength-sensitive (M/L) cones. However, a complete absence of S-cones has been reported for some nocturnal primates (Wikler & Rakic 1990; Jacobs *et al.* 1993, 1996), rodents (Szél *et al.* 1994, 1996; Calderone & Jacobs 1995) and carnivores (Jacobs & Deegan 1992). The absence of S-cones, therefore, may be associated with nocturnality. Recently, the absence of functional S-cones has been reported for a number of whales, dolphins and pinnipeds (Fasick *et al.* 1998; Levenson *et al.* 1998; Peichl & Moutairo 1998; Peichl *et al.* 2001). The lack of S-cones has been discussed as being adaptive with respect to the 'redder' coastal waters (Loew & McFarland 1990), considered to be the main habitat of some of these species (Peichl & Moutairo 1998). The S-cone loss in marine species from two distant mammalian orders argues for convergent evolution and an adaptive advantage of that trait in the marine visual environment. However, as the spectral composition of light in clear ocean waters is increasingly blue-shifted with depth, an S-cone loss would seem particularly disadvantageous for species inhabiting the open ocean. This paradox might be explained by an evolutionary consideration: if the

S-opsin deletion is a phylogenetically old event, its adaptive advantage may be related to an early coastal phase in marine mammal evolution rather that to present lifestyles (Peichl *et al.* 2001).

Anatomical evidence for at least two receptor types may suggest the capability of functional colour vision but is certainly not conclusive, particularly not for trichromatic colour vision of the primate type. It has been suggested that colour vision emerged from the development of more than one visual pigment, not to provide colour discrimination but to facilitate an animal's object detection ability by both sensitivity and contrast methods (Easter 1975). Using psychophysical methods, colour discrimination has been demonstrated in the spotted seal (Wartzok & McCormick 1978), while Madsen and Herman (1980) failed to demonstrate colour discrimination in a bottlenose dolphin. Busch and Dücker (1987) considered the ability of their fur seals (*Arctocephalus pusillus* and *A. australis*) to discriminate blue and green—but not red and yellow—from various shades of grey to be an adaptation to their aquatic habitat near the coast. Recently, colour discrimination has also been demonstrated in the California sea lion (Griebel & Schmid 1992) and in the manatee (*Trichechus manatus*) (Griebel & Schmid 1996). However, the provision of marine mammals with different photoreceptors, their ability to discriminate colours, and the adaptive value for their aquatic or amphibious lifestyle is still a matter of debate.

Anatomical examinations of marine mammal eyes demonstrated modifications which can be attributed to the primary demand of underwater vision. Due to the fact that the cornea and encased fluids in the eye have roughly the same refractive index as sea water, it is suggested they become optically inefficient in this medium and the refractive power of a submerged marine mammal eye is restricted to the lens. As in fishes, in most pinnipeds and cetaceans a large, almost spherical, lens with a corresponding high refractive power has evolved (Walls 1942; Jamieson & Fisher 1972). Because a spherical lens may compensate for the lack of strong corneal refractive power, these animals may achieve emmetropia (normal sightedness) under water. Regaining its refractive power in air, the cornea should produce severe myopia (near sightedness) as

well as astigmatism. Corneal astigmatism is a distortion of the retinal image due to unequal refraction of the strongly curved cornea as part of the dioptric (refractive) system of the eye. However, rather than being without function under water or impairing vision in air, Kröger and Kirschfeld (1992) provided evidence for the eye of the harbour porpoise (*Phocoena phocoena*) that due to a different curvature of the anterior and posterior corneal surfaces, the cornea obtains a negative refractive power that counteracts myopia in air. As the mid-nasal region of the cornea in the California sea lion (Dawson *et al.* 1987) and the southern sea lion (G. Dehnhardt *et al.*, unpublished data) corresponds to a flat circular area (6–8 mm in diameter), these pinnipeds may achieve a similar effect. These results suggest that for a better understanding of vision in marine mammals detailed studies on the dioptric system are needed in a variety of species (Sivak *et al.* 1989). In adaptation to their murky river habitats, where vision is restricted to the detection of light and dark for orientation to the water surface, the eyes of the river dolphins *Platanista gangetica* and *P. indi* have no lens.

There is some evidence that pinnipeds and some odontocete species have good visual acuity both under water and in air. Herman *et al.* (1975) supposed that the double-split pupil of the bottlenose dolphin (Fig. 5.7) (Rivamonte 1976) observed at high illumination, is responsible for equivalent aerial and underwater vision under daylight conditions. When dilated, the pupil is a horizontally orientated crescent, but under bright illumination it closes completely in its central portion leaving a small, irregularly shaped aperture at its nasal and temporal extreme. In accordance with this pupil reaction to bright illumination, studies on the density of ganglion cells in total retina preparations of the eyes of marine mammals discovered two areas centralis (area of high cell density and thus high visual resolution) in several small odontocetes as well as in the gray whale (Fig. 5.7) (Dral 1977, 1983; Mass & Supin 1990, 1999), while the northern fur seal and the Amazon river dolphin (Mass & Supin 1990) possess only one area centralis. A double-split pupil and two areas centralis in the nasal and temporal part of the retina correspond to the behavioural evidence that dolphins primarily

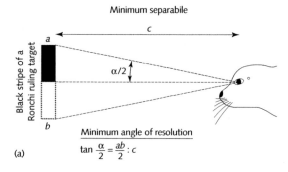

(a)

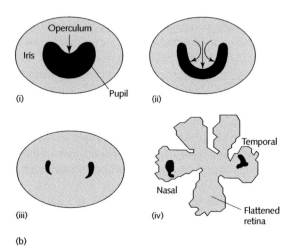

(b)

Fig. 5.7 (a) Determination of visual acuity. For the determination of the minimum separabile, the subject is required to discriminate stimuli made up of equally spaced alternating black and white stripes (Ronchi rulings). Achieving a mean angle of resolution (MAR) of 8′3″ means that an animal is able to discriminate stripes 0.25 cm wide at a viewing distance of 100 cm. The same calculation is applicable to the determination of the minimum visible, defined as the angle of resolution under which a dot-shaped object can be detected. (b) The pupil of a bottlenose dolphin at different stages of increasing illumination (from i to iii). At high ambient light levels the pupil of the eye of most pinniped species constricts to a narrow vertical aperture (inverted drop form). The restricted pupils of marine mammals result in so-called stenopaic vision, which functions in an analogous fashion to pinhole cameras, where a small aperture provides good depth-of-field. (Adapted from Rivamonte 1976.) (iv) Flattened retina of a tucuxi (*Sotalia fluviatilis fluviatilis*) showing, in accordance with the double slit pupil (iii), two areas (in black) of high cell density. (Adapted from Mass & Supin 1999.)

use peripheral parts of the retina when they focus an eye on something (Dral 1975).

Visual acuity can be defined as the eye's spatial resolving power or perception of fine detail at various distances (Fig. 5.7). However, early behavioural studies (Feinstein & Rice 1966; Jamieson & Fisher 1970; White *et al.* 1971) might have measured the animals' abilities of brightness discrimination instead of performing acuity measurements. For example, a brightness cue is provided in a task that requires the test animal to discriminate between two black discs of varying size presented on white backgrounds. Schusterman and Balliet (1970) prevented a harbour seal and a Steller sea lion from using brightness cues by presenting for the first time gratings of alternating black and white stripes against grey stimuli of equal brightness (Table 5.2). Schusterman (1972) reported California sea lions, harbour seals and Steller sea lions to be capable of resolving gratings subtending visual angles of from 5′ to 9′ of arc. However, while both underwater and aerial visual acuity was found to depend on luminance, deterioration of visual resolution was much stronger in air than under water. All marine cetaceans tested so far by psychophysical methods fall within the 5′ to 9′ range of underwater daylight acuity reported for various pinniped species. Although these angles of visual resolution are considerably inferior to those of diurnal primates (man, 20″; chimpanzee, 28″; rhesus monkey, 34″), marine mammals seem to compare quite favourably with several species of land mammals reputed to have rather sharp aerial vision, such as the elephant (10′20″), antelope (13′54″), red deer (11′8″) and cat (5′3″) (Rahmann 1967).

5.4 CHEMORECEPTION

In contrast to what is known about chemoreception in fish (Hara 1992), olfaction (in mammals: smelling chemicals dispersed in air) and gustation (tasting dissolved chemicals) in marine mammals have not as yet drawn much attention.

5.4.1 Olfaction

Early anatomical descriptions of central and/or

peripheral chemosensory structures suggest that in comparison with most terrestrial mammals the olfactory system is somewhat reduced in pinnipeds, even more reduced in baleen whales, and absent in odontocetes (see reviews by Lowell & Flanigan 1980; Watkins & Wartzok 1985; Nachtigall 1986).

In baleen whales the olfactory tract is present in adults, but the olfactory bulb was found only in fetuses. In toothed whales both the olfactory tracts and bulbs exist only in the fetal stage of development. However, small fibrotic vestiges of the olfactory tracts were sometimes found in odontocetes. It should be stressed that evolutionary degeneration of the peripheral olfactory structures did not cause the reduction of the central olfactory structures. All so-called tertiary structures of the rhinencephalon are present and well developed in both Mysticeti and Odontoceti (see Chapter 4).

All species of pinnipeds posses well-developed main and accessory olfactory systems (Mackay-Sim et al. 1985), including a vomeronasal organ (also known as the 'organ of Jacobson', which often functions to detect olfactory sensations from food in the oral cavity). Although the morphological basis for a sense of smell is still a matter of some debate it is quite realistic that in pinnipeds olfaction is functional and plays, for example, a role in mother–pup recognition. Whereas the morphology of peripheral chemosensory structures still requires elucidation, behavioural experiments would be best suited to elucidate the presence and function of olfaction in marine mammals.

In manatees and dugongs the main and accessory olfactory systems are rudimentary and a vomeronasal organ is not present (Mackay-Sim et al. 1985).

5.4.2 Gustation

In mammals, the sense of taste receives information about dissolved substances via taste buds in the oral cavity (Pfaffmann et al. 1971). In comparison with terrestrial mammals, taste buds on the tongues of small odontocetes and pinnipeds are modified in structure and their number is reduced, suggesting a limited sense of taste. However, the bottlenose dolphin, the common dolphin and the harbour porpoise have been shown by behavioural and electrophysiological (galvanic skin response) methods

to be able to distinguish several chemicals from sea water (Kuznetsov 1990). With regard to the four primary tastes humans perceive as sour, bitter, salty and sweet, the Steller sea lion should be able to sense all of them but sweet.

Using psychophysical techniques, Nachtigall and Hall (1984) and Friedl et al. (1990) determined detection thresholds for the four primary tastes in the bottlenose dolphin and the California sea lion. In comparison with humans, the threshold of the dolphin for 'sour' (citric acid) was about seven times higher (0.3 ppt, parts per thousand), while the threshold of the sea lion (0.05 ppt) was similar to that of humans. Thresholds for 'bitter' (quinine sulphat or quinine monohydrochloride dihydrate) were approximately twice as high in the dolphin (human: 1.4×10^{-4} ppt, dolphin: 2.4×10^{-4} ppt) and were extraordinarily high in the sea lion (0.4 ppt). Compared with humans, thresholds determined for 'salty' (sodium chloride) were higher by a factor of c. 10 in the dolphin (1.6 ppt) and 20 times higher in the sea lion (3.6 ppt). In accordance with the results obtained for the Steller sea lion (Kuznetsov 1990), the California sea lion did not respond to 'sweet' (sucrose), while the threshold of the dolphin (1.6 ppt) was about 10 times as high as that of humans.

Although the studies reported so far show that marine mammals taste chemicals dissolved in water, the functional significance of this sensory ability under natural conditions remains unclear. Beside its potential role in communication, studies on marine fishes and invertebrates suggest that chemoreception may also be of significance in orientation (Atema 1988).

Spatial salinity variation in oceanic basins is one potential clue for spatial orientation in the marine habitat. Vertically stratified density layers (see model in Fig. 5.8), as well as river plume fronts, typically indicated by extremely steep salinity gradients, are well suited to provide locational information, especially for orientation within the foraging range of marine animals (see Chapter 7). Vertically stratified density layers are suggested to be involved in long-distance orientation of fishes (Westerberg 1984). In addition to their possible function in orientation, fronts may indicate areas of high biological productivity and hence good foraging areas. Sims

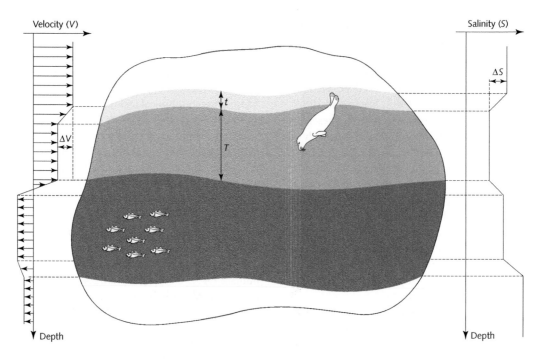

Fig. 5.8 Adaptation of Westerberg's (1984) 'pancake model for salmon homing' to potential orientation mechanisms in pinnipeds. Oceanic basins are by no means homogenous water masses, but are vertically stratified in layers colloquially called 'pancakes' (T, 1–10 m thick and up to 100 km or even more in horizontal scale). In addition to a constant current direction and velocity, each pancake is characterized by a relative homogeneity of temperature and salinity. However, these parameters differ from pancake to pancake, which are separated by thin shear layers (t, a few centimetres thick) signalling the change in, for example, current velocity (V) and salinity (S). Based on these oceanographic data it is conceivable that during a vertical dive the high sensitivity for salinity differences permits a seal to sample several pancake layers, to choose the one that tastes right (see Atema 1988, p. 4), and to travel horizontally in the selected pancake until it terminates and new sampling is required. The current shear in the thin layers (t) could be measured by the seal's vibrissal system, thus providing additional information about the edge of a pancake as well as the relative motion between two pancakes. For fishes it is suggested that, in conjunction with a coarse compass function (e.g. sun or magnetic compass), these environmental informations could be sufficient for orientation at the mesoscale, for example in an animal's foraging range. As indicated by the sensory abilities shown, at least for pinnipeds, this model also provides interesting and testable hypotheses for orientation in marine mammals.

and Quayle (1998) have shown that basking sharks (*Cetorhinus maximus*) orientate and forage along fronts, a possibility also suggested for sperm whales (Griffin 1999). To make use of spatial salinity variations, marine mammals must possess a high differential sensitivity for this kind of environmental information. This has recently been shown for two harbour seals (Sticken & Dehnhardt 2000). Requiring the seals to discriminate seawater solutions differing in salinity, difference thresholds were obtained for five different levels of salinity (15–35‰) and were compared to human thresholds. While differential sensitivity in humans was best at low salinity levels and had already decreased significantly at 25‰, sensitivity in the seals increased with increasing salinity up to a level of 30‰. Furthermore, the lowest thresholds achieved by humans were as high as 18% salinity difference, which is in good accordance with results obtained for NaCl discrimination in humans (Pfaffmann *et al.* 1971). In contrast, the seals were able to detect minimum salinity differences of 4% (*c.* 1.0‰ salinity difference at 30‰), the lowest threshold ever determined in a mammal. These data show that rather than being functionally reduced, the sense of taste in pinnipeds may be highly specialized, and well tuned to the natural occurrence of chemical information in marine habitats.

5.5 CONCLUSIONS

This chapter has primarily described the classic sensory modalities with direct morphological and physiological evidence for their function. However, it should be noted that marine mammals may use sensory information not accessible to humans, as is indicated by the use of ultra- and infrasonic sound as well as hydrodynamic information. For the significance of other information channels only indirect evidence exists. For example, correlations between coastal locations of cetaceans, live strandings and negative magnetic anomalies suggest that these animals possess a magnetic sensory system (Klinowska 1985; Kirschvink et al. 1986), but experimental evidence is still missing.

As most general reviews (e.g. Fobes & Smock 1981; Schusterman 1981; Watkins & Wartzok 1985; Nachtigall 1986; Wartzok & Ketten 1999) reveal, research on sensory systems of marine mammals has concentrated on audition and vision. We have just started to discern other information channels animals may use for solving problems they are faced with in their natural environments. In this respect, a great deal of inspiration can be drawn from a **comparative approach**. Marine mammals share their aquatic environments with numerous orders of other vertebrate species all of which can be expected to be well adapted to their environment. Sharing the same environment may result in a convergent evolution of utilizing the same information channels, as suggested by hydrodynamic reception in fish and seals. Consequently, a broad approach to the sensory interaction of marine mammals with their natural environments should be encouraged.

REFERENCES

Andersen, S. (1970) Directional hearing in the harbour porpoise *Phocoena phocoena*. In: *Investigations on Cetacea* (G. Pilleri, ed.), Vol. 2, pp. 260–263. The Brain Anatomy Institut, Bern.

Atema, J. (1988) Distribution of chemical stimuli. In: *Sensory Biology of Aquatic Animals* (J. Atema, R.R. Fay, A.N. Popper & W.N. Tavolga, eds), pp. 29–56. Springer-Verlag, New York.

Au, W.W.L. (1993) *The Sonar of Dolphins*. Springer-Verlag, New York.

Babushina, Y.e.S., Zaslavskii, G.L. & Yurkevich, L.I. (1991) Air and underwater hearing characteristics of the northern fur seal: audiograms, frequency and differential thresholds. *Biophysics* 36 (5), 909–913.

Bachteler, D. & Dehnhardt, G. (1999) Active touch performance in the Antillean manatee: evidence for a functional differentiation of facial tactile hairs. *Zoology* 102, 61–69.

Bain, D.E., Kriete, B. & Dahlheim, M.E. (1993) Hearing abilities of killer whales (*Orcinus orca*). *Journal of the Acoustical Society of America* 94 (3), 1829.

Beatty, D.D. (1969) Visual pigments of three species of cartilaginous fishes. *Nature* 222, 285.

Bernholz, C.D. & Matthews, M.L. (1975) Critical flicker frequency in a harp seal: evidence for duplex retinal organisation. *Vision Research* 15, 733–736.

Bleckmann, H. (1994) Reception of hydrodynamic stimuli in aquatic and semiaquatic animals. *Progress in Zoology* 41, 115 pp.

Bolanowski Jr, S.J., Gescheider, G.A., Verrillo, R.T. & Checkosky, C.M. (1988) Four channels mediate the mechanical aspect of touch. *Journal of the Acoustical Society of America* 84, 1680–1694.

Brill, R.L. (1988) The jaw-hearing dolphin: preliminary behavioural and acoustical evidence. In: *Animal Sonar* (P.E. Nachtigall & P.W.D. Moore, eds), pp. 281–287. Plenum Press, New York.

Bullock, T.H., Grinnell, A.D., Ikezono, E. *et al.* (1968) Electrophysiological studies of central auditory mechanisms in cetaceans. *Zeitschrift für Vergleichende Physiologie* 59, 117–156.

Bullock, T.H., Ridgway, S.H. & Suga, N. (1971) Acoustically evoked potentials in midbrain auditory structures in sea lions (Pinnipedia). *Zeitschrift für Vergleichende Physiologie* 74, 372–387.

Bullock, T.H., Domning, D.P. & Best, R.C. (1980) Evoked brain potentials demonstrate hearing in a manatee (*Trichechus manatus*). *Journal of Mammalogy* 61 (1), 130–133.

Bullock, T.H., O'Shea, T.J. & McClune, M.C. (1982) Auditory evoked potentials in the West Indian manatee (*Trichechus manatus*). *Journal of Comparative Physiology A* 148 (4), 547–554.

Busch, H. & Dücker, G. (1987) Das visuelle Leistungsvermögen der Seebären (*Arctocephalus pusillus* und. *Arctocephalus Australis*). *Zoologischer Anzeiger* 219, 197–224.

Calderone, J.B. & Jacobs, G.H. (1995) Photopigments of two types of hamster. *Investigative Ophthalmology and Visual Science* 36, S276.

Carlson, S., Tanila, H., Linnankoski, I., Pertovaara, A. & Kehr, A. (1989) Comparison of tactile discrimination ability of visually deprived and normal monkeys. *Acta Physiologica Scandinavica* 135, 405–410.

Casseday, J.H. & Neff, W.D. (1973) Localization of pure tones. *Journal of the Acoustical Society of America* 54, 365–372.

Crescitelli, F. (1958) Natural history of visual pigments. *Annals of the New York Academy of Sciences* 74, 230–255.

Dahlheim, M.E. & Ljungblad, D.K. (1990) Preliminary hearing study on gray whales (*Eschrichtius robustus*) in the field. In: *Sensory Abilities of Cetaceans* (J. Thomas & R. Kastelein, eds), pp. 335–346. Plenum Press, New York.

Dartnall, H.J.A. (1962) The photobiology of visual processes. In: *The Eye* (H. Davson, ed.), pp. 321–533. Academic Press, New York.

Dawson, W.W., Schroeder, J.P. & Sharpe, S.N. (1987) Corneal surface properties of two marine mammal species. *Marine Mammal Science* 3, 186–197.

Dehnhardt, G. (1990) Preliminary results from psychophysical studies on the tactile sensitivity in marine mammals. In: *Sensory Abilities of Cetaceans* (J.A. Thomas & R.A. Kastelein, eds), pp. 435–446, Plenum Press, New York.

Dehnhardt, G. (1994) Tactile size discrimination by a California sea lion (*Zalophus californianus*) using its mystacial vibrissae. *Journal of Comparative Physiology* 175, 791–800.

Dehnhardt, G. & Dücker, G. (1996) Tactual discrimination of size and shape by a California sea lion (*Zalophus californianus*). *Animal Learning and Behaviour* 24, 366–374.

Dehnhardt, G. & Kaminski, A. (1995) Sensitivity of the mystacial vibrissae of harbour seals (*Phoca vitulina*) for size differences of actively touched objects. *Journal of Experimental Biology* 198, 2317–2323.

Dehnhardt, G., Sinder, M. & Sachser, N. (1997) Tactual discrimination of size by means of mystacial vibrissae in harbour seals: in air versus underwater. *Zeitschrift für Säugetierkunde* 62 (Suppl. 2), 40–43.

Dehnhardt, G., Mauck, B. & Hyvärinen, H. (1998a) Ambient temperature does not affect the tactile sensitivity of mystacial vibrissae in harbour seals. *Journal of Experimental Biology* 201, 3023–3029.

Dehnhardt, G., Mauck, B. & Bleckmann, H. (1998b) Seal whiskers detect water movements. *Nature* 394, 235–236.

Dehnhardt, G., Hyvärinen, H. & Palviainen, A. (1999) Structure and innervation of the vibrissal follicle–sinus complex in the Australian waterrat (*Hydromys chrysogaster*). *Journal of Comparative Neurology* 411, 550–562.

Dehnhardt, G., Mauck, B., Hanke, W. & Bleckmann, H. (2001) Hydrodynamic trail-following in harbor seals (*Phoca vitulina*). *Science* 293, 102–104.

Denton, E.J. & Warren, F.J. (1956) Visual pigments of deep sea fish. *Nature* 178, 1059.

Ding Wang, Wang, K., Xiao, Y. & Sheng, G. (1992) Auditory sensitivity of a Chinese river dolphin, *Lipotes vexillifer*. In: *Marine Mammal Sensory Systems* (J. Thomas, R.A. Kastelein & A.Ya. Supin, eds), pp. 213–212. Plenum Press, New York.

Domning, D.P. & Hayek, L.C. (1986) Interspecific and intraspecific morphological variation in manatees (*Sirenia: Trichechus*). *Marine Mammal Science* 2 (2), 87–144.

Dral, A.D.G. (1975) Some quantitative aspects of the retina of *Tursiops truncatus*. *Aquatic Mammals* 2, 28–31.

Dral, A.D.G. (1977) On the retinal anatomy of Cetacea (mainly *Tursiops truncatus*). In: *Functional Anatomy of Marine Mammals* (R.J. Harrison, ed.), Vol. 3, pp. 80–134. Academic Press, London.

Dral, A.D.G. (1983) The retinal ganglion cells of *Delphinus delphis* and their distribution. *Aquatic Mammals* 10, 57–68.

Dudock van Heel, W.H. (1962) Sound and cetacea. *Netherlands Journal of Sea Research* 1 (4), 407–507.

Duntley, S.Q. (1963) Light in the sea. *Journal of the Optical Society of America A: Optics and Image Science* 53, 214–233.

Dykes, R.W. (1975) Afferent fibres from mystacial vibrissae of cats and seals. *Journal of Neurophysiology* 38, 650–662.

Easter, S.S. (1975) Retinal specialization for aquatic vision: theory and facts. In: *Vision In Fishes. New Approaches in Research* (M.A. Ali, ed.), pp. 609–618. Plenum Press, New York.

Elliot, D., Stein, L. & Harrison, M. (1960) Determination of absolute-intensity thresholds and frequency difference thresholds in cats. *Journal of the Acoustical Society of America* 32, 380–384.

Fasick, J.I., Cronin, T.W., Hunt, D.M. & Robinson, P.R. (1998) The visual pigments of the bottlenose dolphin (*Tursiops truncatus*). *Visual Neuroscience* 15, 643–652.

Fay, F.H. (1982) *Ecology and Biology of the Pacific Walrus, Odobenus rosmarus divergens Illiger*. Fish and Wildlife Service Publication No. 74. Fish and Wildlife Service, Washington, DC.

Fay, R.R. (1992) Structure and function in sound discrimination among vertebrates. In: *The Evolutionary Biology of Hearing* (D.B. Webster, R.R. Fay & A.N. Popper, eds), pp. 229–263. Springer-Verlag, Berlin.

Feinstein, S.H. & Rice, C.E. (1966) Discrimination of area differences by harbour seal. *Psychonomic Science* 4, 379–380.

Fobes, J.L. & Smock, C.C. (1981) Sensory capacity of marine mammals. *Psychological Bulletin* 89, 288–307.

Frankel, A.S., Mobley, J.R. & Herman, L.M. (1995) Estimation of auditory response threshold in humpback whales using biologically meaningful sounds. In: *Sensory Systems of Aquatic Mammals* (R.A. Kastelein, J.A. Thomas & P.E. Nachtigall, eds), pp. 55–70. De Spil Publishers, Woerden.

Friedl, W.A., Nachtigall, P.E., Moore, P.W.B. & Chun, N.K.W. (1990) Taste reception in the bottlenose dolphin (*Tursiops truncatus gilli*) and the California sea lion (*Zalophus californianus*). In: *Sensory Abilities of Cetaceans* (J.A. Thomas & R.A. Kastelein, eds), pp. 447–451, Plenum Press, New York.

Gentry, R.L. (1967) Underwater auditory localization in the California sea lion (*Zalophus californianus*). *Journal of Auditory Research* 7, 187–193.

Gerstein, E.R., Gerstein, L., Forsythe, S.E. & Blue, J.E. (1999) The underwater audiogram of the West Indian manatee (*Trichechus manatus*). *Journal of the Acoustical Society of America* 105, 3575–3583.

Gescheider, G.A., Thorpe, J.M., Goodarz, J. & Bolanowski Jr, S.J. (1997) The effects of skin temperature on the detection and discrimination of tactile stimulation. *Somatosensory and Motor Research* 14, 181–188.

Gibson, J.J. (1962) Observations on active touch. *Psychological Review* **69**, 477–491.

Gourevitch, G. (1980) Directional hearing in terrestrial mammals. In: *Comparative Studies of Hearing in Vertebrates* (A.N. Popper & R.R. Fay, eds), pp. 375–373. Springer-Verlag, New York.

Green, B.G., Lederman, S.J. & Stevens, J.C. (1979) The effect of skin temperature on the perception of roughness. *Sensory Processes* **3**, 327–333.

Griebel, U. & Schmid, A. (1992) Color vision in the California sea lion (*Zalophus californianus*). *Vision Research* **32**, 477–482.

Griebel, U. & Schmid, A. (1996) Color vision in the manatee (*Trichechus manatus*). *Vision Research* **36**, 2747–2757.

Griffin, R.B. (1999) Sperm whale distribution and community ecology associated with a warm-core ring off Georges bank. *Marine Mammal Science* **15**, 33–51.

Hall, J.D. & Johnson, C.S. (1972) Auditory thresholds of a killer whale *Orcinus orca* Linnaeus. *Journal of the Acoustical Society of America* **51** (2), 515–517.

Hanke, W., Brücker, C. & Bleckmann, H. (2000) The ageing of the low-frequency water disturbances caused by swimming goldfish and its possible relevance to prey detection. *Journal of Experimental Biology* **203**, 1193–1200.

Hara, T.J. (1992) *Fish Chemoreception*. Chapman & Hall, London.

Härkönen, T.J. (1987) Seasonal and regional variations in the feeding habits of the harbour seal, *Phoca vitulina*, in the Skagerrak and Kattegat. *Journal of Zoology, London* **213**, 535–543.

Heffner, H.E. (1983) Hearing in large and small dogs: absolute thresholds and size of the tympanic membrane. *Behavioural Neuroscience* **97**, 310–318.

Heffner, R.S., Heffner, H.E., Contos, C. & Kearns, D. (1994) Hearing in prairie dogs: transitions between surface and subterrenean rodents. *Hearing Research* **73**, 185–189.

Herman, L.M. & Arbeit, W.R. (1972) Frequency difference limens in the bottlenose dolphin: 1–70 kHz. *Journal of Auditory Research* **12** (2), 109–120.

Herman, L.M., Peacock, M.F., Yunker, M.P. & Madsen, C.J. (1975) Bottlenose dolphin: double-split pupil yields equivalent aerial and underwater diurnal acuity. *Science* **189**, 650–652.

Hyvärinen, H. (1989) Diving in darkness: whiskers as sense organs of the ringed seal (*Phoca hispida*). *Journal of Zoology, London* **218**, 663–678.

Hyvärinen, H. & Katajisto, H. (1984) Functional structure of the vibrissae of the ringed seal (*Phoca hispida*, Schr.). *Acta Zoologica Fennica* **171**, 27–30.

Jacobs, D.W. (1972) Auditory frequency dicrimination in the Atlantic bottlenose dolphin, *Tursiops truncatus* Montague: a preliminary report. *Journal of the Acoustical Society of America* **52** (2), 696–698.

Jacobs, G.H. & Deegan, J.F. (1992) Cone photopigments in nocturnal and diurnal procyonids. *Journal of Comparative Physiology A* **171**, 351–358.

Jacobs, G.H., Deegan, J.F., Neitz, J., Crognale, M.A. & Neitz, M. (1993) Photopigments and color vision in the nocturnal monkey, *Aotus*. *Vision Research* **33**, 1773–1783.

Jacobs, D.W. & Hall, J.D. (1972) Auditory thresholds of a fresh water dolphin, *Inia geoffrensis* Lainville. *Journal of the Acoustical Society of America* **51** (2), 530–533.

Jacobs, G.H., Neitz, M. & Neitz, J. (1996) Mutations in S-cone pigment genes and the absence of colour vision in two species of nocturnal primate. *Proceedings of the Royal Society London Series B: Biological Sciences* **263**, 705–710.

Jamieson, G. & Fisher, H.D. (1970) Visual discrimination in the harbour seal, *Phoca vitulina*, above and below water. *Vision Research* **10**, 1175–1180.

Jamieson, G.S. & Fisher, H.D. (1971) The retina of the harbour seal, *Phoca vitulina*. *Canadian Journal of Zoology* **49**, 19–23.

Jamieson, G. & Fisher, H.D. (1972) The pinniped eye: a review. In: *The Functional Anatomy of Marine Mammals* (R. Harrison, ed.), Vol. 1, pp. 245–262. Academic Press, London.

Jerlov, N.G. (1968) *Optical Oceanography*. Elsevier, Amsterdam.

John, K.T., Goodwin, A.W. & Darian-Smith, I. (1989) Tactile discrimination of thickness. *Experimental Brain Research* **78**, 62–68.

Johnson, C.S. (1967) Sound detection thresholds in marine mammals. In: *Marine Bio-Acoustics* (W.N. Tavolga, ed.), Vol. , pp. 247–260. Pergamon Press, Oxford.

Johnson, C.S. (1991) Hearing thresholds for periodic 60-kHz tone pulses in the beluga whale. *Journal of the Acoustical Society of America* **89** (6), 2996–3001.

Kamil, A.C. (1988) Behavioural ecology and sensory biology. In: *Sensory Biology of Aquatic Animals* (J. Atema, R.R. Fay, A.N. Popper & W.N. Tavolga, eds), pp. 189–201. Springer-Verlag, New York.

Kastelein, R.A. & van Gaalen, M.A. (1988) The sensitivity of the vibrissae of a Pacific walrus (*Odobenus rosmarus divergens*). *Aquatic Mammals* **14**, 123–133.

Keidel, W.D. & Neff, W.D., eds (1974) *Handbook of Sensory Physiology*, Vol. V/1. Springer-Verlag, Berlin.

Ketten, D.R. (1991) The marine mammal ear: spezializations for aquatic audition and echolocation. In: *The Evolutionary Biology of Hearing* (D.B. Webster, R.R. Fay & A.N. Popper, eds), pp. 717–747. Springer-Verlag, Berlin.

Ketten, D.R. (1992) The cetacean ear: form, frequency, and evolution. In: *Marine Mammal Sensory Systems* (J. Thomas, R.A. Kastelein & A.Ya. Supin, eds), pp. 53–75. Plenum Press, New York.

Ketten, D.R. & Wartzok, D. (1990) Three-dimensional reconstructions of the dolphins ear. In: *Sensory Abilities of Cetaceans* (J. Thomas & R. Kastelein, eds), pp. 81–105. Plenum Press, New York.

Ketten, D.R., Odell, D.K. & Domning, D.P. (1992) Structure, function, and adaption of manatee ear. In: *Marine Mammal Sensory Systems* (eds Thomas, J.R.A. Kastelein, A.Ya. Supin), pp. 77–95. Plenum Press, New York.

King, J.E. (1983) *Seals of the World*. Cornell University Press, Ithaca, NY.

Kirschvink, J.L., Dizon, A.E. & Westphal, J.A. (1986) Evidence from strandings for geomagnetic sensitivity in cetaceans. *Journal of Experimental Biology* **120**, 1–24.

Klinowska, M. (1985) Cetacean live stranding dates relate to geomagnetic disturbances. *Aquatic Mammals* **11** (3), 109–119.

Klishin, V.O., Pezo Diaz, R., Popov, V.V., Supin, A.Ya. (1990) Some characteristics of hearing of the Brazilian manatee, *Trichechus inunguis*. *Aquatic Mammals* **16** (3), 139–144.

Kooyman, G. (1989) *Diverse Divers; Physiology and Behaviour*. Springer-Verlag, Berlin.

Krebs, J.R. & Kacelnik, A. (1991) Decision-making. In: *Behavioural Ecology, an Evolutionary Approach* (J.R. Krebs & N.B. Davis, eds), pp. 105–136. Blackwell Science, Oxford.

Kröger, R.H.H. & Kirschfeld, K. (1992) The cornea as an optical element in the cetacean eye. In: *Marine Mammal Sensory Systems* (J. Thomas, ed.), pp. 97–105. Plenum Press, New York.

Kuznetzov, V.B. (1990) Chemical sense of dolphins: quasi-olfaction. In: *Sensory Abilities of Cetaceans* (J.A. Thomas & R.A. Kastelein, eds), pp. 481–503, Plenum Press, New York.

Kvadsheim, P.H., Gotaas, A.R.L., Folkow, L.P. & Blix, A.S. (1997) An experimental validation of heat loss models for marine mammals. *Journal of Theoretical Biology* **184**, 15–23.

Ladygina, T.F., Popov, V.V. & Supin, A.Y. (1985) Somatotopic projections in the cerebral cortex of the fur seal (*Callorhinus ursinus*). *Neurophysiology* **17**, 344–351 (in Russian).

Landau, D. & Dawson, W.W. (1970) The histology of retinas from the order Pinnipedia. *Vision Research* **10**, 691–702.

Lavigne, D.M. & Ronald, K. (1972) The harp seal, *Pagophilus groenlandicus* (Erxleben 1777). XXIII. Spectral sensitivity. *Canadian Journal of Zoology* **50**, 1197–1206.

Lavigne, D.M. & Ronald, K. (1975a) Pinniped visual pigments. *Comparative Biochemistry and Physiology* **52**, 325–329.

Lavigne, D.M. & Ronald, K. (1975b) Evidence of duplicity in the retina of the California sea lion (*Zalophus californianus*). *Comparative Biochemistry and Physiology* **50**, 65–70.

Levenson, D.H., Crognale, M.A., Ponganis, P.J., Deegan, J.F. & Jacobs, G.H. (1998) Cone photopigment of harbour seals: implications for color vision. *Investigative Ophthalmology and Visual Science* **39**, S978.

Lien, J., Todd, S. & Guigne, J. (1990) Inferences about perception in large cetaceans, especially humpback whales, from incidental catches in fixed fishing gear, enhancement of nets by 'alarm' devices, and the acoustics of fishing gear. In: *Sensory Abilities of Cetaceans* (J. Thomas & R. Kastelein, eds), pp. 347–362. Plenum Press, New York.

Lindt, C.C. (1956) Underwater behaviour of the southern sea lion, *Otaria jubata*. *Journal of Mammalogy* **37**, 287–288.

Ling, J.K. (1977) Vibrissae of marine mammals. In: *Functional Anatomy of Marine Mammals* (R.J. Harrison, ed.), Vol. 3, pp. 387–415. Academic Press, London.

Loew, E.R. & McFarland, W.N. (1990) The underwater visual environment. In: *The Visual System of Fish* (R.H. Douglas & M.B.A. Djamgoz, eds), pp. 1–43. Chapman & Hall, London.

Loomis, J.M. & Lederman, S.J. (1986) Tactual perception. In: *Handbook of Perception and Human Performance* (K.R. Boff, L. Kaufman & J.R. Thomas, eds), Vol. 2, pp. 1–41. Wiley, New York.

Lowell, W.R. & Flanigan Jr, W.F. (1980) Marine mammal chemoreception. *Mammal Review* **10**, 53–59.

Lythgoe, J.N. (1972) The adaptation of visual pigments to the photic environment. In: *The Handbook of Sensory Physiology, VII. I. The Photochemistry of Vision* (H.J.A. Dartnall, ed.), pp. 566–603. Springer, Hamburg.

Lythgoe, J.N. & Dartnall, H.J.A. (1970) A 'deep sea rhodopsin' in a marine mammal. *Nature* **227**, 995–996.

McCormick, J.G., Weaver, E.G., Palin, G. & Ridgway, S.H. (1970) Sound conduction in the dolphin ear. *Journal of the Acoustical Society of America* **48**, 1418–1428.

McFarland, W.N. (1971) Cetacean visual pigments. *Vision Research* **11**, 1065–1076.

Mackay-Sim, A., Duvall, D. & Graves, B.M. (1985) The West Indian manatee, *Trichechus manatus*, lacks a vomeronasal organ. *Brain, Behavior and Evolution* **27**, 186–194.

Madsen, C.J. & Herman, L.M. (1980) Social and ecological correlates of cetacean vision and visual appearance. In: *Cetacean Behaviour: Mechanisms and Functions* (L.M. Herman, ed.), pp. 101–147. John Wiley & Sons, New York.

Marshall, C.D., Huth, G.D., Edmonds, V.M., Halin, D.L. & Reep, R.L. (1998) Prehensile use of perioral bristles during feeding and associated behaviours of the Florida manatee (*Trichechus manatus latirostris*). *Marine Mammal Science* **14**, 274–289.

Mass, A.M. & Supin, A.Ya. (1986) Topographic distribution of size and density of ganglion cells in the retina of a porpoise, *Phocoena phocoena*. *Aquatic Mammals* **12**, 95–102.

Mass, A.M. & Supin, A.Ya. (1989) Distribution of ganglion cells in the retina of an Amazon river dolphin *Inia geoffrensis*. *Aquatic Mammals* **15**, 49–56.

Mass, A.M. & Supin, A.Ya. (1990) Best vision zones in the retinae of some cetaceans. In: *Sensory Abilities of Cetaceans* (J. Thomas & R. Kastelein, eds), pp. 505–517. Plenum Press, New York.

Mass, A.M. & Supin, A.Ya. (1995) Retinal cell topography of the retina in the bottlenose dolphin *Tursiops truncatus*. *Brain, Behaviour and Evolution* **45**, 257–265.

Mass, A.M. & Supin, A.Ya. (1997) Ocular anatomy, retinal ganglion cell distribution, and visual resolution in the grey whale *Eschrichtius gibbosus*. *Aquatic Mammals* **23**, 17–28.

Mass, A.M. & Supin, A.Ya. (1999) Retinal topography and visual acuity in the riverine tucuxi (*Sotalia fluviatilis*). *Marine Mammal Science* **15**, 351–365.

Mauck, B., Eysel, U. & Dehnhardt, G. (2000) Selective heating of vibrissal follicles in seals (*Phoca vitulina*) and dolphins (*Sotalia fluviatilis guianensis*). *Journal of Experimental Biology* **203**, 2125–2131.

Maybaum, H.L. (1993) Responses of humpback whales to sonar sounds. *Journal of the Acoustical Society of America* **94** (3), 1848–1849.

Miller, E.H. (1975) A comparative study of facial expressions of two species of pinnipeds. *Behaviour* **53**, 268–284.

Mills, A.M. (1958) On the minimum audible angle. *Journal of the Acoustical Society of America* **30**, 237–246.

Mills, F. & Renouf, D. (1986) Determination of the vibration sensitivity of harbour seal (*Phoca vitulina*) vibrissae. *Journal of Experimental Marine Biology and Ecology* **100**, 3–9.

Møhl, B. (1964) Preliminary studies on hearing in seals. *Videnskabelige Meddelelser Fra Dansk Naturhistorisk Forening I Kjøbenhaven* **127**, 283–294.

Møhl, B. (1968) Auditory sensitivity of the common seal in air and water. *Journal of Auditory Research* **8** (1), 27–38.

Moore, P.W.D. (1975) Underwater localization of click and pulsed pure-tone signals by the California sea lion (*Zalophus californianus*). *Journal of the Acoustical Society of America* **57** (2), 406–410.

Moore, P.W.D. & Au, W.W.L. (1975) Underwater localization of pulsed pure tones by the California sea lion (*Zalophus californianus*). *Journal of the Acoustical Society of America* **58** (3), 721–727.

Moore, P.W.B., Pawloski, D.A. & Dankiewicz, L. (1995) Interaural time and intensity difference thresholds in the bottlenose dolphin (*Tursiops truncatus*). In: *Sensory Systems of Aquatic Mammals* (R.A. Kastelein, J.A. Thomas & P.E. Nachtigall, eds), pp. 11–23. De Spil Publishers, Woerden.

Moore, P.W.B. & Schusterman, R.J. (1987) Audiometric assessment of northern fur seals, *Callorhinus ursinus*. *Marine Mammal Science* **3**, 31–53.

Munz, F.W. (1964) The visual pigments of epipelagic and rocky-shore fishes. *Vision Research* **8**, 983–996.

Murayama, T. & Somiya, H. (1998) Distribution of ganglion cells and object localizing ability in the retina of three cetaceans. *Fisheries Science* **64**, 27–30.

Murayama, T., Somiya, H., Aoki, I. & Ishii, T. (1995) Retinal cell size and distribution predict visual capabilities of Dall's porpoise. *Marine Mammal Science* **11**, 136–149.

Nachtigall, P.E. (1986) Vision, audition, and chemoreception in dolphins and other marine mammals. In: *Dolphin Cognition and Behaviour: a Comparative Approach* (R.J. Schusterman, J.A. Thomas & F.G. Woods, eds), pp. 79–113. Erlbaum, Hillsdale, NJ.

Nachtigall, P.E. (1989) Risso's dolphins (*Grampus griseus*) vision. In: *Eighth Bienniel Conference on the Biology of Marine Mammals, Monterey, CA*. Society for Marine Mammalogy, Lawrence, KA.

Nachtigall, P.E. & Hall, R.W. (1984) Taste reception in the bottlenose dolphin. *Acta Zoologica Fennica* **172**, 147–148.

Nachtigall, P.E., Au, W.W.L., Pawloski, J.L. & Moore, P.W.B. (1995) Risso's dolphin (*Grampus griseus*) hearing thresholds in Kaneohe Bay, Hawaii. In: *Sensory Systems of Aquatic Mammals* (R.A. Kastelein, J.A. Thomas & P.E. Nachtigall, eds), pp. 49–53. De Spil Publishers, Woerden.

Nagy, A.R. & Ronald, K. (1970) The harp seal, *Pagophilus groenlandicus* (Erxleben 1777). VI. Structure of the retina. *Canadian Journal of Zoology* **48**, 367–370.

Neuweiler, G. (1984) Foraging, echolocation and audition in bats. *Naturwissenschaften* **71**, 446–455.

Newby, T.C., Hart, F.M. & Arnold, R.A. (1970) Weight and blindness of harbour seals. *Journal of Mammalogy* **51**, 152.

Norris, K.S. (1964) Some problems of echolocation in cetaceans. In: *Marine Bio-Acoustics* (W.N. Tavolga, ed.), pp. 317–336. Pergamon Press, Oxford.

Norris, K.S. (1968) The evolution of acoustic mechanisms in odontocete cetaceans. In: *Evolution and Environment. Peabody Museum Centenary Celebration Volume* (E.T. Drake, ed.), pp. 225–241. Yale University Press, New Haven, CT.

Peichl, L. & Moutairo, K. (1998) Absence of short-wavelength sensitive cones in the retinae of seals (Carnivora) and African giant rats (Rodentia). *European Journal of Neuroscience* **10**, 2586–2594.

Peichl, L., Behrmann, G. & Kröger, R.H.H. (2001) For whales and seals the ocean is not blue: a visual pigment loss in marine mammals. *European Journal of Neuroscience* **13**, 1520–1528.

Penner, R.H., Turl, C.W. & Au, W.W.L. (1986) Target detection by the beluga using a surface-reflected path. *Journal of the Acoustical Society of America* **80** (6), 1842–1843.

Pepper, R.L. & Simmons Jr, J.V. (1973) In-air visual acuity of the bottlenose dolphin. *Experimental Neurology* **41**, 271–276.

Perez, J.M., Dawson, W.W. & Landau, D. (1972) Retinal anatomy of the bottlenose dolphin (*Tursiops truncatus*). *Cetology* **11**, 1–11.

Pfaffmann, C., Bartoshuk, L.M. & McBurney, D. (1971) Taste psychophysics. In: *Handbook of Sensory Physiology.* (L.M. Beidler, ed.), Vol. 4, pp. 73–99. Springer-Verlag, Berlin.

Piggins, D., Muntz, W.R.A. & Best, R.C. (1983) Physical and morphological aspects of the eye of the manatee *Trichechus inunguis* Natterer 1883: (Sirenia: Mammalia). *Marine Behaviour and Physiology* **9**, 111–130.

Popov, V.V., Ladygina, T.F. & Supin, A.Ya. (1986) Evoked potentials of the auditory cortex of the porpoise, *Phocoena phocoena*. *Journal of Comparative Physiology A* **158**, 705–711.

Popov, V. & Supin, A. (1990) Electrophysiological studies of hearing in some cetaceans and a manatee. In: *Sensory Systems of Aquatic Mammals* (R.A. Kastelein, J.A. Thomas & P.E. Nachtigall, eds), pp. 405–415. De Spil Publishers, Woerden.

Rahmann, H. (1967) Die Sehschärfe bei Wirbeltieren. *Naturwissenschaftliche Rundschau* **20**, 8–14.

Ramprashad, F., Corey, S. & Ronald, K. (1972) Anatomy of the seal's ear (*Pagophilus groenlandicus*). In: *Functional Anatomy of Marine Mammals* (R.J. Harrison, ed.), pp. 264–306. Academic Press, London.

Reep, R.L., Marshall, C.D., Stoll, M.L. & Whitaker, D.M. (1998) Distribution and innervation of facial bristles and hairs in the Florida manatee (*Trichechus manatus latirostris*). *Marine Mammal Science* **14**, 257–273.

Renaud, D.J. & Popper, A.N. (1975) Sound localization by the bottlenose porpoise *Tursiops truncatus*. *Journal of Experimental Biology* **63** (3), 569–585.

Renouf, D. (1979) Preliminary measurements of the sensitivity of the vibrissae of harbour seals (*Phoca vitulina*) to low frequency vibrations. *Journal of Zoology, London* **188**, 443–450.

Repenning, C.A. (1972) Underwater hearing in seals: functional morphology. In: *Functional Anatomy of Marine Mammals* (R.J. Harrison, ed.), pp. 307–331. Academic Press, London.

Rice, F.L., Mance, A. & Munger, B.L. (1986) A comparative light microscopic analysis of the innervation of the mystacial pad. I. Vibrissal follicles. *Journal of Comparative Neurology* **252**, 154–174.

Richardson, W.J. & Greene, C.R. (1993) Variability in behavioural reaction thresholds of bowhead whales to man-made underwater sounds. *Journal of the Acoustical Society of America* **94** (3), 1848.

Richardson, W.J., Greene, C.R., Malme, C.I. & Thomson, D.H. (1995) *Marine Mammals and Noise*. Academic Press, London.

Ridgway, S.H. & Carder, D.A. (1990) Tactile sensitivity, somatosensory response, skin vibrations, and the skin surface ridges of the bottlenose dolphin, *Tursiops truncatus*. In: *Sensory Abilities of Cetaceans* (J. Thomas & R. Kastelein, eds), pp. 163–179. Plenum Press, New York.

Ridgway, S.H. & Joyce, P.L. (1975) Studies on seal brain by radiotelemetry. *Rapports et Proces-Verbaux des Réunions, Conseil International pur l' Exploration de la Mer* **169**, 81–91.

Rivamonte, L.A. (1976) Eye model to account for comparable aerial and underwater acuities of the bottlenose dolphin. *Netherlands Journal of Sea Research* **10** (Suppl. 4), 491–498.

Roitblat, H.L. (1987) *Introduction to Comparative Cognition*. Freeman, New York.

Sauerland, M. & Dehnhardt, G. (1998) Underwater audiogram of a tucuxi (*Sotalia fluviatilis guianensis*). *Journal of the Acoustical Society of America* **103**, 1199–1204.

Schusterman, R.J. (1972) Visual acuity in pinnipeds. In: *Behaviour of Marine Animals* (H.E. Winn & B.L. Olla, eds), Vol. 2, pp. 469–491. Plenum Publishing Corporation, New York.

Schusterman, R.J. (1981) Behavioural capabilities of seals and sea lions: review of their hearing, visual, learning and diving skills. *Psychological Record* **31**, 125–143.

Schusterman, R.J. & Balliet, R.F. (1970) Visual acuity of the harbour seal and the Steller sea lion under water. *Nature* **226**, 563–564.

Schusterman, R.J., Balliet, R.F. & Nixon, J. (1972) Underwater audiogram of the California sea lion by the conditioned vocalization technique. *Journal of Experimental Analysis of Behaviour* **17**, 339–350.

Silbernagel, S. & Despopulus, A. (1983) *Taschenatlas der Physiologie*. Thieme-Verlag, Stuttgart.

Sims, D.W. & Quayle, V.A. (1998) Selective foraging behaviour of basking sharks on zooplankton in a small-scale front. *Nature* **393**, 460–464.

Sinnott, J.M. & Aslin, R.N. (1985) Frequency and intensity discrimination in human infants and adults. *Journal of the Acoustical Society of America* **78**, 1986–1992.

Sivak, J.G., Howland, H.C., West, J. & Weerheim, J. (1989) The eye of the hooded seal, *Cystophora christata*, in air and water. *Journal of Comparative Physiology A* **165**, 771–777.

Sivian, L.J. & White, S.D. (1933) On minimum audible fields. *Journal of the Acoustical Society of America* **4**, 288–231.

Spong, P. & White, D. (1971) Visual acuity and discrimination learning in the dolphin (*Lagenorhynchus obliquidens*). *Experimental Neurology* **31**, 431–436.

Stephens, R.J., Beebe, I.J. & Poulter, T.C. (1973) Innervation of the vibrissae of the California sea lion, *Zalophus californianus*. *Anatomical Record* **176**, 421–442.

Sticken, J. & Dehnhardt, G. (2000) Salinity discrimination in harbour seals: a sensory basis for spatial orientation in the maine environment? *Naturwissenschaften* **87**, 499–502.

Szél, Á., Csorba, G., Caffé, A.R., Szél, G., Röhlich, P. & Van Veen, T. (1994) Different patterns of retinal cone topography in two genera of rodents, *Mus* and *Apodemus*. *Cell and Tissue Research* **276**, 143–150.

Szél, Á., Röhlich, P., Caffé, A.R. & Van Veen, T. (1996) Distribution of cone photoreceptors in the mammalian retina. *Microscopy Research and Technique* **35**, 445–462.

Terhune, J.M. (1974) Directional hearing of a harbour seal in air and water. *Journal of the Acoustical Society of America* **56** (6), 1862–1865.

Terhune, J.M. & Ronald, K. (1971) The harp seal, *Phagophilus groenlandicus* (Erxleben, 1777). X. The air audiogram. *Canadian Journal of Zoology* **49** (3), 385–390.

Terhune, J.M. & Ronald, K. (1972) The harp seal, *Phagophilus groenlandicus* (Erxleben, 1777). III. The underwater audiogram. *Canadian Journal of Zoology* **50** (5), 565–569.

Terhune, J.M. & Ronald, K. (1975) Underwater hearing sensitivity of two ringed seals (*Pusa hispida*). *Canadian Journal of Zoology* **53** (3), 227–231.

Thomas, J.A., Chun, N., Au, W. & Pugh, K. (1988) Underwater audiogram of a false killer whale (*Pseudoorca crassidens*). *Journal of the Acoustical Society of America* **84** (3), 936–940.

Thomas, J.A., Moore, P.D.W., Withrow, R. & Stoermer, M. (1990) Underwater audiogram of a Hawaiin monk seal (*Monachus schauinslandi*). *Journal of the Acoustical Society of America* **87** (1), 417–420.

Thompson, R.K.R. & Herman, L.M. (1975) Underwater frequency discrimination in the bottlenose dolphin (1–140 kHz) and the human (1–8 kHz). *Journal of the Acoustical Society of America* **57** (4), 943–948.

Turnbull, S.D. & Terhune, J.M. (1993) Repetition enhances hearing detection thresholds in a harbour seal

(*Phoca vitulina*). *Canadian Journal of Zoology* **71** (5), 926–932.

Van Esch, A. & De Wolf, J. (1979) Evidence for cone function in the dolphin retina—a preliminary report. *Aquatic Mammals* 7, 35–37.

Wald, G., Brown, P.K. & Brown, P.S. (1957) Visual pigments and depth of habitat of marine fishes. *Nature* **180**, 969–971.

Walls, G.L. (1942) *The Vertebrate Eye and its Adaptive Radiation*. Cranbrook Institute of Science Bulletin No. 19. Reprinted 1963 by Hafner Press, New York.

Wartzok, D. (1979) Phocid spectral sensitivity curves. In: *Third Biennial Conference on the Biology of Marine Mammals, Seattle*, p. 62. Society for Marine Mammalogy, Lawrence, KA.

Wartzok, D. & Ketton, D.R. (1999) Marine mammal sensory systems. In: *Biology of Marine Mammals* (J.E. Reynolds III & S.A. Rommel, eds), pp. 117–175. Smithsonian Institute Press, Washington, DC.

Wartzok, D. & McCormick, M.G. (1978) Color discrimination by a Bering Sea spotted seal, *Phoca largha*. *Vision Research* 18, 781–784.

Wartzok, D., Schusterman, R.J. & Gailey-Phipps, J. (1984) Seal echolocation? *Nature* 308, 753.

Watkins, W.A. (1981) Reactions of three species of whales *Balaenoptera physalus, Megaptera novaeangliae* and *Balaenoptera edeni* to implanted radio tags. *Deep-Sea Research* 28A (6), 589–599.

Watkins, W.A. (1986) Whale reactions to human activities in Cape Cod waters. *Marine Mammal Science* 2 (4), 251–262.

Watkins, W.A. & Wartzok, D. (1985) Sensory biophysics of marine mammals. *Marine Mammal Science*. 1, 219–260.

Watson, L. (1981) *Whales of the World—a Complete Guide to the World's Living Whales, Dolphins and Porpoises*. Hutchinson & Co, London.

Westerberg, H. (1984) The orientation of fish and the vertical stratification at fine- and micro-structure scales. In: *Mechanisms of Migration in Fishes* (J.D. McCleave, G.P. Arnold, J.J. Dodson, & W.H. Neill, eds), pp. 179–203. Plenum Press, New York.

Wever, E.G., McCormick, J.G., Palin, J. & Ridgway, S.H. (1971) The cochlea of the dolphin *Tursiops truncatus*: hair cells and ganglion cells. *Proceedings of the National Academy of Science USA* 68, 2908–2912.

White, D.O.N., Cameron, N., Spong, P. & Bradford, J. (1971) Visual acuity of the killer whale (*Orcinus orca*). *Experimental Neurology* 32, 230–236.

White, M.J., Norris, J., Ljungblad, D., Baron, K. & di Sciara, G. (1978) Auditory thresholds of two beluga whales (*Delphinapterus leucas*). In: *HSWRI Technical Report*, pp. 78–109. Report from Hubbs/Sea World Research Institute, San Diego, CA for the US Naval Ocean Systems Center, San Diego.

Wikler, K.C. & Rakic, P. (1990) Distribution of photoreceptor subtypes in the retina of diurnal and nocturnal primates. *Journal of Neuroscience* 10, 3390–3401.

CHAPTER 6

Vocal Anatomy, Acoustic Communication and Echolocation

Peter L. Tyack and Edward H. Miller

6.1 INTRODUCTION

In this chapter, we take the functional perspective of behavioural ecology to explore how marine mammals communicate, orientate, and echolocate in the sea. Most of the problems faced by marine animals are identical to those facing terrestrial animals: finding and capturing prey, avoiding predators, orienting in geographical space, finding and selecting a mate, and taking care of offspring. Yet the marine environment differs from the terrestrial environment in ways that have modified the costs and benefits of different sensory modalities that might be used to solve these problems. In particular, the marine environment favours the acoustic channel for rapid transmission of information over ranges of greater than a few tens of metres, and this has led to the evolution of auditory and vocal specializations in many marine mammals. These specializations are reflected in the morphology, functional anatomy and physiology of marine mammals (see Chapter 5).

Different communication problems have different requirements for how far signals should be transmitted, how long signals should last, how rapidly signals should travel, etc. For example, an alarm call intended for conspecific recipients would ideally travel rapidly only to them and would not be detectable by the predator, which is often more distant. To reduce the risk of interception by predators, alarm calls should be quiet, brief and difficult to locate (Klump & Shalter 1984). By contrast, reproductive advertisement displays of males are likelier to be detected by females if the displays are lengthy or repeated and have a large broadcast range; selection would also favour displays that can easily be located by females.

Animals live in different environments and move over different scales; these differences alter the costs and benefits of different sensory modalities for solving a specific communication or sensory problem. Many marine mammal species are social and highly mobile, so potentially large ranges for communication may be socially and ecologically important to them. Table 6.1 summarizes important differences between air and water for several sensory modalities. Small marine organisms use electrical and chemical signals to communicate over ranges of a few metres, but their limited ranges make them unsuitable for rapid communication over long ranges. The maximal ranges of propagation for long-distance optical or acoustic signals differ greatly in air versus water. Sunlight illuminates optical displays in open terrestrial environments during the daytime, often allowing them to be detected at ranges of a kilometre or more. Light scarcely attenuates as it passes through kilometres of clear air, but marine animals can seldom see an object more than a few tens of metres away. In contrast, sound attenuates less and travels faster in water than in air. Maximal ranges for detecting vocalizations of terrestrial animals are typically 1–10 km, but underwater microphones, called hydrophones, can detect low-frequency vocalizations of marine mammals at ranges of up to hundreds of kilometres (Tyack & Clark 2000). The limited range of vision and excellent propagation of sound under water not only favour acoustics for long-range communication, but also have selected for the evolution of echolocation in some marine mammals (Section 6.4).

When terrestrial mammals invaded the sea, they encountered new constraints and opportunities for sensing signals. Some marine mammals (sirenians, cetaceans, phocid seals and the walrus, *Odobenus*

Table 6.1 Comparison of the range, speed and locatability of different sensory modalities in air and under water. This table summarizes different physical attributes of acoustic, light, chemical and electrical cues in these two environments, along with features that are relevant for antidetection or location strategies.

Sensory modality	Mode of sensing	Mode of production	Range		Speed (m/s)		Locatability	Antidetection and location strategies
			Air	Water	Air	Water		
Acoustic	Hearing	Vocalization	1 km?	10–100 km	340	1500	Moderate	Faint hard-to-localize signal
Light	Vision	Luminescence	1–10 km	1–100 m	3×10^8	2.25×10^8	Good	Disruptive coloration; counter-illumination
Chemical	Olfaction	Pheromone	1 km?	10–100 m	Slow (wind)	Slow (current)	Moderate	Highly specific chemical signal and sense
Electrical	Electrical sense	Electric organ discharge	NA	Few metres	NA	Fast	Moderate	Jamming?

NA, not applicable.

rosmarus) evolved specializations for using sound to communicate under water and to explore the marine environment; others, including the otariid pinnipeds, sea otter (*Enhydra lutris*) and polar bear (*Ursus maritimus*), still communicate mainly in air. All sirenians and cetaceans live their entire lives in the sea, and cetaceans show the most elaborate and extreme specializations for acoustic communication under water (and rarely produce in-air vocalizations; Watkins 1967a). For example, the number of fibres in the auditory versus optic nerve is 2–3 times as high in cetaceans as in land mammals, suggesting an increased investment in audition compared to vision (Ketten 1997, 2000) (Chapter 4). Other than sirenians, the sea otter and cetaceans, all marine mammals spend critical parts of their lives on land or ice and some (mainly phocid seals) need to communicate both in air and under water. Kastak and Schusterman (1998) compared in-air and underwater hearing of three pinniped species. They argued that the California sea lion (*Zalophus californianus*) is adapted to hear best in air; the common seal (*Phoca vitulina*) can hear equally well in air and under water; and the auditory system of the northern elephant seal (*Mirounga angustirostris*) is adapted for underwater sensitivity at the expense of aerial hearing. These differences appear to reflect the relative importance of hearing in the two different environments for the different species.

6.2 SCOPE OF ACOUSTIC COMMUNICATION

6.2.1 Communicatively significant sounds: a continuum from incidental to highly specialized

> *When we speak of communicating, we commonly imply the use of specialized signals. These are not strictly necessary; we could define communication as any sharing of information from any source (Smith 1977, p. 13).*

Communication is usually defined in terms of a signal that provides information from a signaller to a receiver. Communication signals originally evolved in animals from useful but inadvertent sources of information, sources such as the sounds of respiration, locomotion or ingestion (Kenyon 1969; Watkins & Schevill 1976). If the sender benefits from transmitting information with a signal, evolution may select for more specialized signals or ritualized displays. A display that is 'ritualized' is defined as evolutionarily 'specialized in form . . . or frequency as an adaptation expressly to permit or facilitate communication' (Moynihan 1970, p. 86); ritualization can also take the form of a signal— even a simple signal—being used only in highly specific circumstances (Smith 1986, 1991a; Bain

1992). For example, when pinnipeds dive quickly upon detecting a potential predator the sounds they produce at the water surface commonly elicit reactions in conspecific animals, such as sudden diving or swimming away. When alarmed, some species (e.g. the harp seal, *Pagophilus groenlandicus*) loudly slap the water surface as they dive. Rapid submergence presumably was the original source of ritualized slapping of the surface by fore- or hind-flippers in pinnipeds, as in some aquatic displays of adult male common seals (Hewer 1974; Hanggi & Schusterman 1992, 1994). Male sea otters patrol the territory while swimming on the back with 'unusually vigorous kicking and splashing actions' (Calkins & Lent 1975, p. 529).

Information provided to receivers from sources other than displays is extremely important in communication over short distances. Bradbury and Vehrencamp (1998) recognized inadvertent **cues** as an important kind of non-display information. Examples of acoustic cues include:
1 Sounds produced by sea otters striking a food item on the belly with a stone while floating on the back.
2 Movement-related hydrodynamic or surface sounds (Watkins 1981; Watkins & Wartzok 1985).
3 Sounds produced by otariids rapidly walking on gravel (displacing it) or on a rock surface (especially if the animal is wet).

Sounds of locomotion may not be evolutionarily specialized as displays, but nevertheless can provide valuable information to receivers, for example in helping them to appraise body mass or to assess the success of feeding animals. Physical contact between fighting male marine mammals can generate loud percussive sounds by bodies or body parts (such as flippers, Miller 1991) making sudden forceful contact (e.g. humpback whales, *Megaptera novaeangliae*; Tyack & Whitehead 1983). Other males observing such interactions may be able to obtain information from them that is useful in their own future interactions with the fighting males.

There is a continuum in animal communication from unritualized cues to displays that are highly specialized for their signalling function. Distinctions between displays, unritualized actions and cues are conceptually useful but they may blur during the evolution of ritualized displays or countermeasures against interception (Andrew 1972; Beer 1982).

For example, seismic vibrations are generated incidentally by locomotion of northern elephant seals on land and are perceptible to recipients up to 20 m distant (Shipley *et al.* 1992). In some instances vibrations appear to be produced deliberately. Males and females occasionally 'raise their forequarters and slap the sand with their head or chest without vocalizing' (Shipley *et al.* 1992, p. 559). These movements also generate sounds and may represent a ritualized display.

Some of the most obvious displays of cetaceans to human observers on a boat are the so-called 'aerial' displays of breaching, lobtailing and flipper slapping. We use vision to see these displays in the air, but the primary signal a cetacean sends to a conspecific whale with these 'aerial' displays is more likely to be the sound of the body hitting the water surface. Some of these sounds can be quite loud; for example, fluke and flipper slaps of humpback whales range in source level from 183 to 192 dB re 1 μPa at 1 m (Thompson *et al.* 1986), compared with a range of 144–174 dB for the species' song (Levenson 1972). These 'aerial' displays may be heard kilometres away, but animals at close range may sense both the optical and acoustic components.

Respiration and breathing have provided the raw material for the evolution of displays involving bubbles. 'Bubble blowing' in various forms occurs in displaying common seals, harp seals, bearded seals (*Erignathus barbatus*) and probably other pinniped species (Fig. 6.1a) (Ray *et al.* 1969; Hewer 1974; Møhl *et al.* 1975; Hanggi & Schusterman 1992, 1994). Cetaceans also produce bubble displays. Bottlenose dolphins (*Tursiops truncatus*) occasionally blow streams of bubbles that are highly synchronized with the production of whistle vocalizations (Caldwell & Caldwell 1972a; McCowan 1995; Herzing 1996). The bubble streams are a highly visible marker identifying who is vocalizing, but it is not known whether other dolphins use the information for this purpose or as an additional signal used to modulate interpretation of the whistle. During the breeding season, humpback whales form groups in which males compete for access to a female (Tyack & Whitehead 1983). Males in these competitive groups produce streams of bubbles up to 30 m long (Fig. 6.1b). When marine mammals exhale forcefully, they also create

(a)

(b)

Fig. 6.1 Underwater exhalation of air is a common acoustic and optical display in marine mammals. (a) Adult male harp seal (*Pagophilus groenlandicus*) releasing air bubbles while vocalizing in threat. (From Merdsoy *et al.* 1976). (b) Adult male humpback whale (*Megaptera novaeangliae*) producing a stream of bubbles while challenging another male for access to a female in a competitive group (Tyack & Whitehead 1983). (© D. Glockner-Ferrari, Center for Whale Studies)

a non-vocal sound as the bubbles rise to the surface. Presumably air bubbles thus constitute a combined optical and acoustical display. Most terrestrial mammals exhale when they vocalize, but many marine mammals can vocalize without releasing bubbles and so must have anatomical specializations allowing the air passing the sound source to be held in some internal reservoir.

6.2.2 Vocal and non-vocal sounds

The terrestrial ancestors of marine mammals had vocal sound-production mechanisms typical of

terrestrial mammals—air was passed under pressure from the lungs past laryngeal vocal folds that vibrated to produce sound energy (Kelemen 1963; Schneider 1964; Dorst 1973). It is possible to distinguish between 'vocal' sounds produced by specializations of the respiratory tract, and 'non-vocal' sounds produced by moving or striking body parts against one another or against some external structure. The term 'vocalization' is sometimes restricted to glottally produced sounds in the larynx, but we retain a broad interpretation of 'vocal sounds' in marine mammals to refer to sounds produced by air passing through specialized sound-producing structures in the body, whether these are vocal folds or some other organ. Some sounds we treat as vocal have simple mechanisms of production outside the larynx (e.g. snorts). For example, a male Forster's fur seal (*Arctocephalus forsteri*) arriving at a colony site in the spring may snort once or several times to announce its presence and willingness to compete for a territory (Winn & Schneider 1977). The pulse sequences produced under water by rutting male walruses using a non-laryngeal respiratory mechanism, and laryngeally produced sound sequences in other pinniped species can express complex ritualized temporal patterning even though the sounds themselves may be very simple (Fig. 6.2) (Bartholomew & Collias 1962; Le Boeuf & Petrinovich 1974; Ray & Watkins 1975; Sandegren 1976; Schusterman 1977; Shipley *et al.* 1981; Stirling *et al.* 1983; Miller 1991; Sanvito & Galimberti 2000a). Cetologists debate about whether the vocalizations of cetaceans are produced as air passes vocal folds in the larynx or as air flows past hard tissue in the nasal passages. The distinction of sounds as vocal or non-vocal based upon laryngeal origin is not useful for cetaceans given our current state of ignorance about the details of sound-production mechanisms.

6.3 SOUND PRODUCTION

6.3.1 Laryngeal anatomy and mechanisms of sound production

The larynx of terrestrial mammals functions both for respiration and sound production; in marine mammals the larynx also exhibits adaptations for

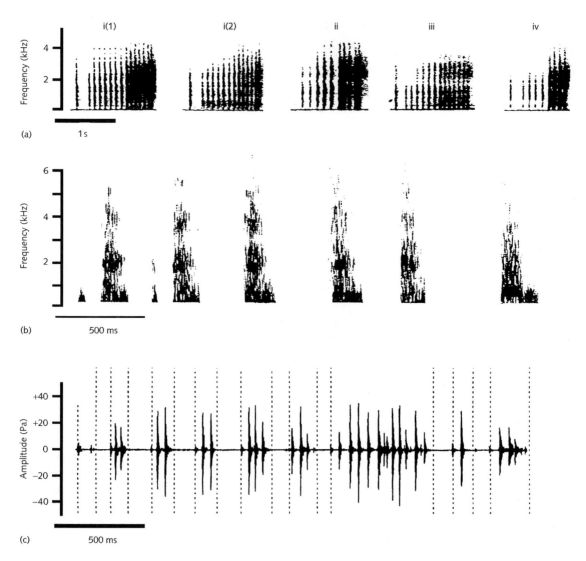

Fig. 6.2 Ritualization of acoustic displays can be expressed as stereotyped temporal patterning of simple sounds. (a) Spectrograms of pulse trains in underwater displays of rutting male Atlantic walruses (*Odobenus rosmarus rosmarus*). Four different individuals (i–iv) are represented; spectrograms i(1) and i(2) are from one male in successive years. (From Stirling *et al.* 1987.) (b) Spectrogram of chuffing vocalization of adult male polar bear (*Ursus maritimus*; the low-frequency sounds before the first two chuffs represent 'plop' component). (Adapted from Peters 1978.) (c) Waveform of drumming vocalization of a rutting male southern elephant seal (*Mirounga leonina*); the vertical dashed lines demarcate groupings of pulses in this airborne utterance. (From Sanvito & Galimberti 2000b.)

diving. The following review pertains mainly to pinnipeds and cetaceans because they have been studied the most and have diverse anatomical characteristics related to the production of vocalizations. For observations on the larynx and hyoid bone of sirenians and the polar bear see Kaiser (1974), Negus (1949) and Schneider (1963, 1964).

Cartilages and muscles of the mammalian pharynx, larynx and hyoid are well described, particularly for domestic species. The laryngeal skeleton comprises four main cartilages: thyroid, cricoid and paired arytenoids; smaller sesamoid cartilages and inter-arytenoid cartilages may also be present (Fig. 6.3). The musculus vocalis is present in pinnipeds and

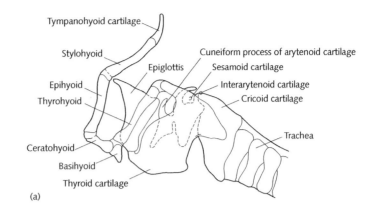

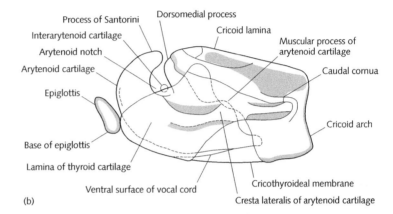

Fig. 6.3 (a) Hyoid, laryngeal cartilages and anterior end of the trachea in a left lateral view of the dog (*Canis familiaris*). (From Evans 1993.) (b) The laryngeal skeleton of a California sea lion (*Zalophus californianus*) in left lateral view. (Adapted from Odend'hal 1966.)

other Carnivora and plays an important role in the production and control of vocalizations (Schneider 1962, 1963, 1964; Odend'hal 1966; Piérard 1969; King 1983). The vocal folds are attached between the thyroid cartilage and the vocal process of the arytenoid cartilage on the left and right sides. The configuration of the laryngeal cavity, the space between the vocal folds, and the tension on the vocal folds are affected by the positions and shapes of the main cartilages and state of contraction of the laryngeal muscles. In pinnipeds, the size and shape of the laryngeal cartilages and muscles vary between species (Murie 1874; Chiasson 1955; Piérard 1965; Schneider 1963, 1964; Piérard 1969; King 1972, 1983), suggesting important species differences in the mechanisms of vocalization. There may also be differences within a species; in the northern fur seal (*Callorhinus ursinus*), for example, sesamoid cartilages are present only in males older than 5 years of age (Piérard 1969).

The larynx of baleen whales is similar to that of terrestrial mammals, as can be seen in Fig. 6.4, which compares the larynx of a terrestrial ungulate (the horse) with mysticete and odontocete cetaceans. Over two centuries ago, Hunter (1787) noticed an unusual laryngeal sac, called the diverticulum, on the lower side of the mysticete trachea, connected to the respiratory tract by an opening on the lower side of the thyroid cartilage (marked 'D' in Fig. 6.4b). Aroyan *et al.* (2000) modelled sound production in the blue whale (*Balaenoptera musculus*), and suggested that this laryngeal sac and the nasal passages may act as a resonator. Odontocetes have a larynx that differs from terrestrial mammals in that the arytenoid and epiglottal cartilages are elongated to form a beak-like structure that is held in the nasal duct by a sphincterlike palatopharyngeal muscle (Fig. 6.4c, d). This separates the respiratory tract from the mouth and oesophagus, reducing the risk of choking and

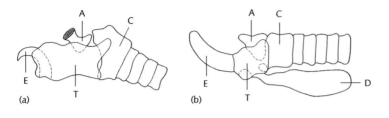

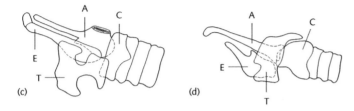

Fig. 6.4 Laryngeal anatomy in: (a) a terrestrial ungulate (horse), (b) a mysticete, and in two odontocete cetaceans, (c) a narwhal and (d) a pilot whale. A, arytenoid cartilage; C, cricoid cartilage; D, diverticulum; E, epiglottis; T, thyroid cartilage. (From Slijper 1979.)

allowing these animals to breathe and swallow at the same time.

There has been considerable debate about whether the larynx or nasal passages are the source of sounds such as clicks or whistles in odontocetes. The odontocete larynx is complex with a powerful musculature, and since the larynx is the source of vocalizations in most terrestrial mammals, many researchers have suggested that odontocete vocalizations are produced by the larynx (Lawrence &

Schevill 1956; Purves 1966; Purves & Pilleri 1973, 1983). Reidenberg and Laitman (1988) reported vocal folds in the odontocete larynx, and argued that this supported the laryngeal-source hypothesis for odontocetes. Odontocetes also have a complex and specialized upper respiratory tract (Fig. 6.5). Since Norris *et al.* (1961) first proposed it, a growing group of biologists have suggested that the nasal passages may be the source of most odontocete vocalizations (Cranford 2000). In contrast to

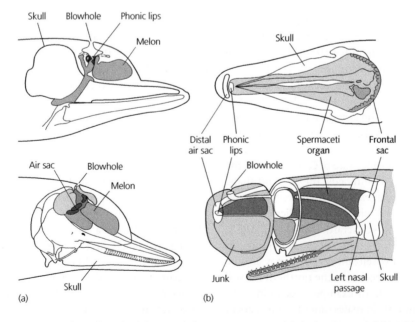

Fig. 6.5 Functional anatomy of sound production in two odontocete cetaceans: (a) the bottlenose dolphin, *Tursiops truncatus*, and (b) the sperm whale, *Physeter macrocephalus*. (Adapted from Au 1993; Cranford 2000.)

the primarily anatomical evidence for sound production in the cetacean larynx, there is strong physiological evidence for sound production in the nasal passages of odontocetes. This is reviewed in Section 6.3.2.3.

6.3.2 Anatomical mechanisms and specializations for sound production

6.3.2.1 Pinnipeds

The mechanism of vocalization in pinnipeds is essentially unstudied. Poulter (1965) localized the bark of California sea lions and suggested that the bark originates in the larynx. Brauer *et al.* (1966) reported on click vocalizations of California sea lions breathing heliox (a helium–oxygen mixture). Because of the low density of heliox, resonance effects due to gas-filled cavities are detectable by an upward shift in the frequency spectrum of vocalizations (Nowicki 1987). Sea lion clicks made in heliox displayed a significant upward shift in the frequency spectrum, suggesting resonance in air. California sea lions can bark successively under water without emitting bubbles and can emit two kinds of calls under water simultaneously, suggesting the possibility of alternative or multiple sound-generation mechanisms that do not require air to be expelled (Brauer *et al.* 1966).

Motor patterns that accompany the production of otariid roars and barks illustrate the complexity and interspecific diversity in mechanisms of sound production (Miller & Phillips 2001). For example, when breeding male South American sea lions (*Otaria flavescens*) bark they depress the lower jaw and simultaneously extend the tongue and flatten it enough to permit oral exhalation (and a bark); the nostrils remain completely or nearly closed. By contrast, the Australian sea lion (*Neophoca cinerea*) is reported to produce barks 'by vibrating the posterior part of the tongue against the soft palate at a rate of about 3 per second' (Marlow 1975, p. 186).

Walruses produce a great diversity of airborne vocalizations at all times of year (Miller 1985; Kastelein *et al.* 1995), and have diverse and elaborate displays below and at the water surface, especially during the rut (Fig. 6.2a) (Miller 1975, 1985, 1991; Ray & Watkins 1975; Fay 1982; Miller & Boness 1983; Stirling *et al.* 1983; Fay *et al.* 1984;

Kastelein *et al.* 1995; Verboom & Kastelein 1995). The walrus may have more ways of producing sounds than any other pinniped species, including whistles produced by blowing air through the lips and gong-like sounds produced by pharyngeal sacs (Fig. 6.6a). Walruses have highly mobile tongues for feeding; supralaryngeal filtering by varied tongue shapes and positions may explain why even simple brief calls of the species are so diverse (Miller 1985; Kastelein *et al.* 1997).

Some phocid species have tracheal mechanisms for sound generation that were probably enabled by two respiratory adaptations for diving. One was the adaptive development of compressible airways in the form of flexible cartilage and membranes (Kooyman 1981). The other was increased tracheal width for rapid inspiration and expiration at the surface (Kooyman & Andersen 1969; Bryden & Felts 1974; Kooyman 1981; Ray 1981). Vocal folds are present in the Weddell seal (*Leptonychotes weddellii*) but may not generate vocalizations under water; rather, air movements between the larynx and trachea may generate sounds by vibration of the anterior tracheal membranes (Piérard 1969). Sufficient residual air is present for this kind of sound production even during dives to > 300 m or > 3030 kPa pressure (Kooyman *et al.* 1970). A similar mechanism of sound production may occur in the Ross seal (*Ommatophoca rossii*) (Bryden & Felts 1974; Ray 1981). The spectacular long (sometimes > 1 min) continuous underwater song of male bearded seals is followed, after an interval, by a terminal 'moan' when air is released (see Fig. 6.14a) (Ray *et al.* 1969). The song is probably produced by vibration of the expansive dorsal tracheal membrane that extends posteriorly to the lungs (Fig. 6.6b) (Burns 1981a). Adult male ribbon seals (*Histriophoca fasciata*) have a large air sac on the right side, which is connected via a valvular slit to the trachea near the tracheo bronchial junction (Fig. 6.6c) (Sleptsov 1940; Abe *et al.* 1977; Burns 1981b). This sac is absent or more weakly developed in females, but it is unknown how this sac may contribute to sound production.

Intriguing observations on pulsation of the throat or chest in some vocalizing phocids suggest that involvement in sound production by the trachea and other parts of the respiratory tract may be widespread (Boness & James 1979; Gailey-Phipps 1984; Miller & Job 1992; Terhune *et al.* 1994a).

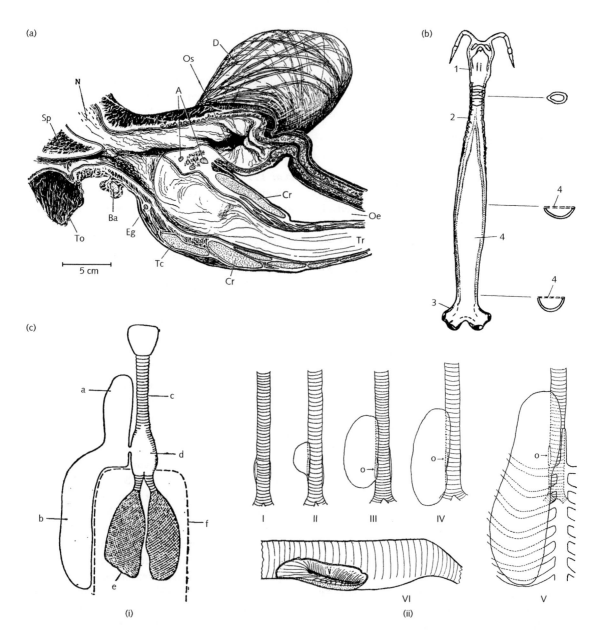

Fig. 6.6 Anatomical specializations for sound production in pinnipeds (the role of the air sac in sound production of the ribbon seal is likely but not established). (a) A sagittal section of the pharynx in the Pacific walrus (*Odobenus rosmarus divergens*; adult male), in left lateral view; note the partly distended pharyngeal pouch on the right side. A, arytenoid cartilage (cut); Ba, basihyoid (cut); Cr, cricoid cartilage (cut); D, diverticulum; Eg, epiglottis (cut); N, nasopharynx; Oe, oesophagus; Os, ostium diverticuli; Sp, soft palate; Tc, thyroid cartilage (cut); To, tongue; Tr, trachea. (From Fay 1960.) (b) The trachea (plus larynx, etc.) of the bearded seal (*Erignathus barbatus*) in dorsal aspect, showing the membranous section of the trachea between the larynx and lungs. 1, larynx; 2, trachea; 3, left bronchus; 4, membranous connective tissue. (From Sokolov *et al.* 1970.) (c) The tracheal air sac in the male ribbon seal (*Histriophoca fasciata*; ventral views): (i) in schematic outline, and (ii) of different sizes (specimens increase in size from I to V, except specimen III which was slightly larger than IV). The opening between the air sac and the trachea in a large male is shown in VI (right lateral aspect; membranous sac removed to reveal valve, 'v'). a, cervical portion of air sac; b, thoracic portion of air sac; c, trachea; d, expanded section of posterior trachea; e, lungs; f, rib cage; o, opening to air sac. (From Sleptsov 1940; Abe *et al.* 1977.)

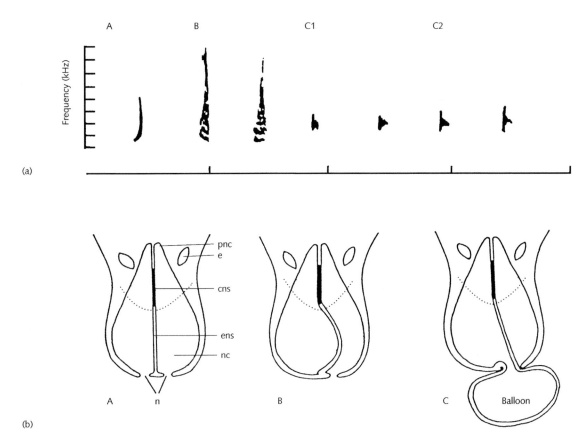

Fig. 6.7 Several kinds of mechanical non-vocal sounds are produced by the nasal hood and air-filled balloon (membranous nasal septum) of adult male hooded seals (*Cystophora cristata*). (a) Spectrograms of vocalizations produced: A, when fully inflated the hood is slightly deflated ('bloop'); B, when fully inflated the hood and septum are rapidly whipped downward ('whoosh'; two examples given); C, at the top of the upsweep of the hood and septum ('ping'; C1 and C2 represent separate recordings with the left spectrogram of each pair recorded in air and the other recorded simultaneously under water). (From Ballard & Kovacs 1995.) (b) Possible mechanism of balloon extrusion (schematic diagram, dorsal aspect): beginning from a relaxed position (A), one nostril is closed and the internal air pressure forces the anterior elastic nasal septum to bulge outwards (B), until it is extruded through one nostril as a large balloon (C). cns, cartilaginous part of nasal septum; e, eye; ens, elastic portion of nasal septum; n, nostril; nc, anterior part of nasal cavity; pnc, posterior part of nasal cavity. (Adapted from Berland 1966.)

Male hooded seals (*Cystophora cristata*) have a specialized inflatable nasal hood and septum that are used for a combined visual and acoustic display. Part of the nasal septum is highly elastic and can be extruded through one nostril as a large air-filled bladder, fiery red in colour (Fig. 6.7b) (Berland 1958, 1966; Popov 1961; Mohr 1963; Reeves & Ling 1981; Kovacs & Lavigne 1986; Lavigne & Kovacs 1988; Reeves *et al.* 1992). Distinctive 'bloop', 'ping' and 'whoosh' noises are produced during inflation or deflation of the hood and septum (Fig. 6.7a) (Terhune & Ronald 1973; Ballard & Kovacs 1995).

6.3.2.2 *Polar bears and sirenians*

The airborne sounds of the polar bear appear to be produced by laryngeal mechanisms typical of terrestrial mammals. The rapidly repeated chuff vocalization of the polar bear 'is produced with the mouth slightly open, and is characterized by visible chest and abdominal contractions and vertical and lateral motions of the upper lips overlying the cheek teeth . . . it appears that air is shuttled back and forth past the larynx with the emission of each burst of sound'; lip vibration may contribute to the sound

(Wemmer *et al.* 1976, p. 426). Sirenians vocalize under water and do not release air when vocalizing (Hartman 1971), but nothing is known about the functional anatomy of sirenian vocalization. Caribbean manatees (*Trichechus manatus*), Amazon manatees (*T. inunguis*) and dugongs (*Dugong dugon*) all contract the area behind the nostrils while vocalizing (R. S. de Sousa Lima, personal communication).

6.3.2.3 Cetaceans

Mechanisms of sound production in cetaceans are poorly understood. There has been considerable debate about which anatomical structures are the source of particular vocal sounds such as the clicks and whistles of odontocetes. Most cetaceans possess a well-developed larynx with vocal folds that might be capable of vibrating to produce sound when air passes over them (Reidenberg & Laitmann 1988). Aroyan *et al.* (2000) move beyond the debate about which structures excite vibrations, and they present a preliminary model of sound production in the blue whale. They suggest that when a blue whale vocalizes, much of the air stored in the lungs must flow through a valve that excites large pressure fluctuations in the nasal passages. They suggest that the relatively rigid nasal passages and the laryngeal sac act as a resonator for this sound. They note that the anatomical sites most commonly suggested as the oscillating valve for sound production are the vocal folds and the arytenoid cartilage, but either could function for this model. This work provides a good start for modelling sound production in mysticetes.

Over the past several decades, cetacean biologists have argued that odontocete cetaceans produce sound when air flows past the nasal passages in the skull. Evans and Maderson (1973) proposed that dolphin clicks were produced by a friction-based mechanism as air moved nasal plugs against the bony nares, with sound produced by the contact of hard tissues, as with stridulation in invertebrates. Cranford (1992, 1999, 2000) and Cranford *et al.* (1996) pointed out that the odontocete nasal passage can be blocked by a pair of internal nasal lips that he calls the 'phonic lips'. Cranford (2000) argued against the friction mechanism and suggested that odontocete cetaceans produce sound when air flows past these 'phonic' lips in a process similar to that by which terrestrial mammals produce glottal pulses with the larynx. Cranford (2000) used a high-speed endoscope to show that each pulse produced by a bottlenose dolphin is coincident with a movement that 'begins with the lips parting, followed by an explosion of air and fluid erupting from the gap between the lips, and concludes with closure of the lips' (p. 133).

Physiological experiments with dolphins support the hypothesis that the nasal passages are a source of sound production in odontocetes. Dormer (1979) made X-ray motion pictures of dolphins as they vocalized, and showed that air moved from the bony nares into the upper nasal passages. The X-ray images showed movement of the nasal plugs in the bony nares during vocalization, but did not show any movement of the larynx. Dormer (1979) associated movement of the right nasal plug with click production and movement of the left nasal plug with whistle production. MacKay and Liaw (1981) confirmed Dormer's results that the right nasal plug moves when clicks are produced. Ridgway *et al.* (1980) and Amundin and Andersen (1983) measured air pressure above and below the bony nares and measured electrical activity of the surrounding muscles. Before vocalization, air pressure built up below the nares. During vocalization, muscle activity occurred near the nares, but not near the larynx, and air pressure changed across the nares, suggesting that air flows across the nares and into the upper nasal sacs during vocalization. Aroyan *et al.* (1992, 2000) and Aroyan (1996) used structural acoustic models to calculate that the measured beam pattern from clicks of the common dolphin (*Delphinus delphis*) are consistent with a sound source about 1 cm below the right phonic lips. No such physiological studies have been made on mysticetes, which may use the larynx and the associated sac in sound production as proposed by Aroyan *et al.* (2000).

Identifying the source of sound is only part of the sound production story. Mechanisms for sound production must also match the acoustic impedance to the medium of air or sea water, and they may function to direct some sounds in a beam. The echolocation clicks of dolphins are known to be highly directional, pointing ahead of the dolphin in a beam about 6–10° in width (Au 1993). The dolphin has a fatty body called the melon in its forehead (Fig. 6.5a). This melon contains unusual fats with a sound velocity similar to that of sea water. The melon couples acoustic energy from the nasal area to sea water by matching the acoustic impedance of these different media. There is a gradient in sound velocity within the melon that causes some refraction of high-frequency sound, but Au (1993) states that this refraction is not enough on its own to account for the directionality of the dolphin click. The beam pattern of dolphin clicks stems from a complex interaction of reflection from the skull and air sacs, coupled with refraction in the soft tissues. Aroyan (1996) argues that the skull plays a predominant role in forming the beam of dolphin echolocation clicks.

We have a more detailed model of sound production for sperm whales (*Physeter macrocephalus*) than for other cetacean species. Sperm whales have a large organ called the spermaceti organ, which lies dorsal and anterior to the skull. Below the spermaceti organ is the 'junk', which is composed of a series of fatty structures separated by columns of dense connective tissue. Norris and Harvey (1972) argue that the spermaceti organ generates the clicks that are the dominant vocalization of sperm whales. Sperm whale clicks comprise a burst of pulses with equally spaced interpulse intervals (IPIs) (Backus & Schevill 1966; Goold & Jones 1995). Norris and Harvey (1972) suggest that the invariance of IPIs may result from reverberation within the spermaceti organ. They suggest that the spermaceti organ has a reflector of sound at the posterior end (frontal sac) and a partial reflector of sound at the anterior end (distal sac) (Fig. 6.5b). They further propose that the source of the sound energy in the click comes from a strong valve (i.e. the phonic lips of Cranford 1999) in the right nasal passage at the anterior end of the spermaceti organ (Fig. 6.5b). Norris and Harvey (1972) propose that the phonic lips act as a check valve, and that as pressure builds up behind these cornified lips in the right nasal passage, they open and close rapidly, producing an initial pulse of sound energy. Some of the energy from the first pulse within the click is hypothesized to be transmitted directly into the water. The remaining pulses are hypothesized to result as the initial sound reflects between the posterior and anterior reflectors within the spermaceti organ. Cranford (1999) proposes that some of the sound energy reflecting off the frontal sac becomes directed into the junk, and may be projected into the ocean medium, following a path that bypasses the reflective distal sac. Some of the sound energy reflecting back off the anterior reflector will reflect forward from the posterior reflector, leading to multiple pulses within a click. If this hypothesis is correct, then the IPI could represent an accurate indicator of the length of the spermaceti organ. Gordon (1991) measured the length of sperm whales in the wild and found the expected correlation between IPI and estimated size of the spermaceti organ.

The recent work of Cranford (1999), Cranford *et al.* (1996) and Aroyan *et al.* (1992) uses new techniques involving non-invasive three-dimensional imaging of animals, inferring the acoustic parameters from these images, and modelling the acoustic properties of the anatomy using structural mechanical models. These new techniques offer promise that this field will move from the decades of controversy regarding the source of sound production to a full analysis of how marine mammals produce sounds with particular time-frequency and directional qualities. This research is also shedding light on the evolution of sound production in odontocetes. In spite of the differences in form of sound-producing structures in dolphins and sperm whales (Fig. 6.5), Cranford *et al.* (1996) and Cranford (1999) suggest that the phonic lips of dolphins and sperm whales are homologous and that the junk in sperm whales may be homologous with the melon of dolphins.

6.3.2.4 Source–filter models of sound production

The controversy over the source of sound production in cetaceans appears to have inhibited the development of more sophisticated models that include not only the source but also the effects of filtering. Supralaryngeal filters are extremely important in shaping the acoustic properties of vocalizations of terrestrial mammals. Section 6.3.2.1 described how pinnipeds may change the shape of their oral and nasal cavities as they vocalize. As discussed above, the sperm whale has organs in

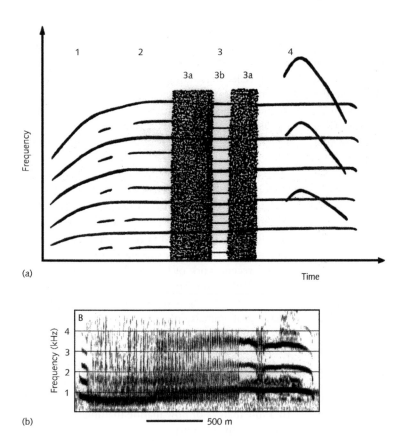

(a)

(b)

500 m

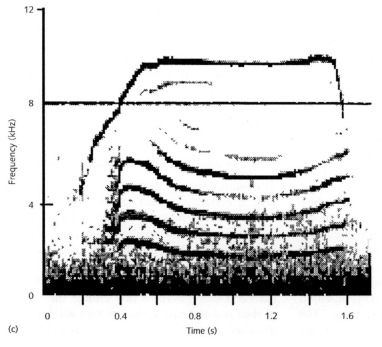

(c)

Fig. 6.8 Many communicatively important features of vocalizations result from non-linear processes in the vocal tract. (a) Schematic narrow-band spectrogram illustrating periodic phonation (1, 3b) and three phenomena resulting from non-linear processes: subharmonics (2), chaos (3a) and biphonation (4). (From Wilden *et al.* 1998.) (b) Spectrogram of the vocalization of an adult female South American fur seal (*Arctocephalus australis*) to her pup. A sharp transition from periodic to chaotic phonation occurs early in the call. (From E.H. Miller, unpublished data; see also Miller & Murray 1995.) (c) Spectrogram of a stereotyped, pulsed call N1 from a killer whale (*Orcinus orca*) showing a tonal high-frequency component harmonically unrelated to the pulsed call. (From data courtesy of P. Miller.)

the head that appear to be specialized to modify sounds emanating from the sound source. Odontocete cetaceans have a complex upper respiratory tract, with many air sacs, but their role in sound production has rarely been discussed for marine mammals (Miller & Job 1992). In humans, many of the effects of acoustic filtering are predictable from the shapes, dimensions and elasticity of the resonating cavities (Rubin & Vatikiotis-Bateson 1998). One important result of filtering is the production of **formants**—parts of the frequency spectrum that are reinforced by resonant properties of the vocal tract (Cherry 1978; Pierce 1983; Lieberman 1984). Supralaryngeal filters change continuously in size and shape during many vocalizations, imparting distinctive temporal patterns to formants. Because supralaryngeal filtering is effected mainly by a few key cavities, resulting vocalizations are informative about facial expression, tongue position and shape, body size, and other behaviourally and anatomically telling features of the pharynx of vocalizing animals (Andrew 1963, 1976; Rossing 1990; Shipley *et al.* 1991; Fitch 1997; Terhune *et al.* 1994a). Many cetacean vocalizations also appear to have formant-like features, but a more detailed model of vocal production in cetaceans is required in order to interpret such features in terms of vocal tract filtering.

Many features of vocalizations are not explainable just in terms of linear analysis of the sound source and resonant filters. Additional explanations in terms of non-linear phenomena are necessary. Three important phenomena that result from non-linear dynamics during mammalian vocalization are subharmonics, biphonation and deterministic chaos (Fee *et al.* 1998; Wilden *et al.* 1998). Figure 6.8 gives a schematic illustration of how these phenomena appear on spectrograms, along with examples of such phenomena in vocalizations of pinnipeds and cetaceans (for sirenians, see Anderson & Barclay 1995). In the South American fur seal (*Arctocephalus australis*) there is 'residual' harmonic structure following the abrupt transition from periodic phonation (Fig. 6.8b), which provides evidence of chaotic phonation rather than turbulent noise (Wilden *et al.* 1998). Many stereotyped pulsed calls of killer whales (*Orcinus orca*) also have a tonal high-frequency component (Fig. 6.8c) (Hoelzel & Osborne 1986) similar to the biphonation phenomenon indicated in Fig. 6.7a (see also Tyack 1991).

6.4 ECHOLOCATION

The term 'echolocation' was coined by Donald Griffin (1958) to describe the ability of flying bats to locate obstacles and prey by listening for echoes returning from high-frequency clicks that they emitted. A similar form of echolocation has been well documented for dolphins in captivity. Most dolphins produce high-frequency clicks produced in a narrow beam in front of the head. After producing a click, a dolphin usually waits and listens for echoes that may backscatter from targets within the click's beam. The large and successful body of experimental research on dolphin echolocation has been well summarized by Au (1993). When biologists think of echolocation, they think of an animal listening for echoes from its own high-frequency directional clicks, backscattered from targets ahead of it. Yet this is only one narrow example of the broad set of sonars that human engineers have developed to explore the environment (Tyack 1997). For example, sonars need not be high in frequency. Low-frequency sound can be propagated over greater distances than high frequencies in the sea, and man-made sonars designed to detect large targets at ranges of more than a few kilometres typically use much lower frequencies than the frequencies of dolphin or bat sonars. Some low-frequency animal sounds may be used in a similar manner, not just for communication, but also to explore or orientate in the ocean environment. In the rest of this section, we move from examples of marine mammal sonars with solid experimental data to more theoretical analyses of how marine mammals may use sound to explore their environment.

6.4.1 Dolphin echolocation

The best studied echolocation system in marine mammals is that of the bottlenose dolphin (Au 1993). Bottlenose dolphins produce clicks that can have very high peak-to-peak sound pressure levels (up to > 220 dB re 1 μPa at 1 m) but over very short periods (several microseconds), with a total click duration of only 50–80 μs (Au 1993). The clicks often have a relatively broad bandwidth (30–40 kHz) centred on high frequencies (often > 100 kHz).

BOX 6.1 UNDERWATER ACOUSTICS

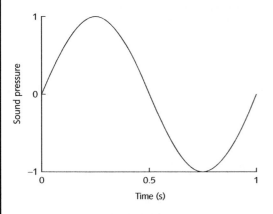

Frequency (Hz, kHz), wavelength and bandwidth:
A sound that we perceive as a pure tone has a sinusoidal
pattern of pressure fluctuations. One full cycle of a
sinusoidal sound is shown above. The *frequency* of these
pressure fluctuations is measured in cycles per second.
If the time it takes the sound to make a full cycle is
t seconds, then the frequency of the sound is $f = 1/t$.
The modern name for the unit of frequency is the *Hertz*,
and just as 1000 metres are called a kilometre, 1000
Hertz are called a *kiloHertz*, abbreviated kHz. The
sound illustrated above took 1 s for a full cycle, so it has
a frequency of 1 Hz. The *wavelength* of this tonal sound
is the distance from one measurement of the maximal
pressure to the next maximum. Sound passes through a
homogeneous medium with a constant speed, c. The
speed of sound in water is approximately 1500 m/s,
roughly five times the value in air, 340 m/s. The speed
of sound c relates the frequency f to the wavelength λ by
the following formula: $c = \lambda f$. Not all sounds are tonal.
Sounds that have energy in a range of frequencies, say in
the frequency range between 200 Hz and 300 Hz,
would be described as having a *bandwidth* of 100 Hz.
Sound intensity, sound pressure: Sound *intensity* is
the amount of energy per unit time (power) flowing
through a unit of area. The intensity of a sound (I)
equals the acoustic pressure (P) squared divided by a
proportionality factor which is specific for each medium.
This factor is called the specific acoustic resistance of the

medium, and equals the density of the medium, ρ, times
the speed of sound, c.

$$I = \frac{P^2}{\rho c} \qquad \text{(eqn 1)}$$

**Decibel (dB); reference pressure = 1 microPascal
(μPa)**: The primary definition of the decibel is as a
ratio of intensities. The decibel always compares a
pressure or intensity to a reference unit. Both intensities
and pressures are referred to a unit of pressure. For
underwater sound, the standard reference pressure is 1
μPa; for sound in air, the standard reference is 20 μPa,
which is about the faintest sound a human can hear. The
microPascal is a unit of pressure: 1 μPa = 10^{-6} Pa = 10^{-6}
Newtons/m². The standard underwater reference I_{ref} is
the intensity of a sound having a pressure level of 1 μPa
(Urick 1983). If I and I_{ref} are two intensities, their
difference in dB is calculated as follows:

$$\text{Intensity difference in dB} = 10 \left[\log \frac{I}{I_{ref}} \right] \qquad \text{(eqn 2)}$$

For the intensity levels and pressure levels to be
comparable in dB, the difference in sound pressure
between a measured and reference pressure is defined in
decibels as follows:

$$\text{Pressure difference in dB} = 20 \left[\log \frac{P}{P_{ref}} \right] \qquad \text{(eqn 3)}$$

Most acoustic measurements actually measure the
pressure fluctuations induced by a sound wave.
Calibrated hydrophones relate the output voltage from a
sound to the reference pressure of 1 μPa. Use of this with
eqn 3 allows one to report the equivalent sound level for
intensity or pressure. The difference in multiplier for
intensity (10 log) and pressure (20 log) maintains the
appropriate proportionality of intensity and pressure for
sounds in the same medium. If $I \propto P^2$ by eqn 1, then log
$I \propto 2 \log P$, or 10 log $I \propto 20 \log P$. As an example, take a
sound measured to be 10 times the pressure reference.
This would be 20 dB re 1 μPa by eqn 3. Since intensity is
proportional to pressure squared, the intensity of this
sound would be 10^2 or 100 times the intensity of the
reference. This would still be 20 dB re the reference
intensity, by the definition of intensity in eqn 2.

Use of these high frequencies allows dolphins to
produce a directional click and to detect small
objects. Sound energy will reflect efficiently from a
rigid object with a circumference greater than or
equal to the wavelength of the sound. Box 6.1
explains that the wavelength λ of a signal equals the

speed of sound divided by the frequency. Since the
speed of sound is about 1500 m/s in sea water, the
wavelength of a 100 kHz sound = 1500/100 000
= 0.015 m or 1.5 cm. Thus dolphins producing
clicks with energy above 100 kHz should be able to
efficiently detect targets with a circumference of the

order of 1 cm or larger. These kinds of calculations have been tested in careful experimental work with captive dolphins. Suggestive data on dolphin echolocation were collected in the 1950s, but the critical breakthrough came when Norris *et al.* (1961) blindfolded dolphins to rule out the possibility that they were using vision in detection tasks. Dolphins tested in an experimental apparatus for echolocation have demonstrated sophisticated echolocation abilities; e.g. an echolocating dolphin can detect a 2.5 cm metal target about 72 m away (Murchison 1980).

The brain of bottlenose dolphins is capable of very rapid auditory processing—their ears integrate high-frequency energy over an interval of about 0.25 ms (Nachtigall *et al.* 2000). This specialized hearing is an adaptation for receiving echolocation clicks. A bottlenose dolphin echolocates by producing a click and then waiting to detect the returning echo. If the dolphin detects a target, it will usually click again after a short delay for processing the echo information. Thus, if a dolphin does not detect an obstacle or a target nearby, it will usually produce clicks with a slow repetition rate. As it closes in on a target, the dolphin will often produce a train of clicks with an accelerating tempo and a decreasing interclick interval (ICI). Work with captive dolphins has shown that the ICI equals the round trip travel time of the click and echo (RT) plus a lag time (LT), which may represent the time the animal requires to process the sonar data. Bottlenose dolphins echolocating on targets 20–120 m away may have lag times of 19–45 ms (Au 1993).

$$ICI = RT + LT \qquad \text{(eqn 6.4)}$$

Knowledge of the echolocation pattern just described has allowed investigators in the field to deduce when animals may be finding prey. Miller *et al.* (1995) studied patterns of clicks in wild narwhals (*Monodon monoceros*). They noted slow series of clicks with equal ICIs interspersed with accelerating rapid series, and suggested that these patterns reflect (respectively) orientation/detection and pursuit phases of foraging mediated by echolocation. However, the assumptions behind eqn 6.4 are not always justified. Beluga whales (*Delphinapterus leucas*) emit series of clicks with ICIs less than the round trip travel time to the sonar target (Turl & Penner 1989). Beluga whales do not have to process each

pulse independently, waiting for the echo to return before emitting the next pulse, but rather appear to use a different kind of sonar processing in which they can process whole series of pulses together. Some other odontocetes also appear to have echolocation signals that differ from bottlenose dolphins, and they also may differ in their sonar signal processing. Pulsed signals of phocoenid porpoises and *Cephalorhynchus* dolphins are 5–10 times longer (150–600 µs) and roughly half the bandwidth (10–20 kHz) of bottlenose dolphin clicks, and their sound pressure levels range between 150 and 170 dB re 1 µPa at 1 m—several orders of magnitude weaker than the loudest bottlenose dolphin clicks (for *Phocoena phocoena*: Møhl & Andersen 1973; Kamminga & Wiersma 1981; Amundin 1991; for *Phocoenoides dalli*: Hatakeyama & Soeda 1990; for *Cephalorhynchus commersonii*: Kamminga & Wiersma 1982; Evans *et al.* 1988; for *C. hectori*: Dawson & Thorpe 1990). Bottlenose dolphins may produce echolocation clicks with spectral peaks well below 100 kHz, but porpoise and *Cephalorhynchus* clicks tend to have peaks above 100 kHz. Ketten (1994, 2000) describes porpoises and *Cephalorhynchus* as having inner ears that are specialized for high-frequency audition (> 100 kHz). Variability in the echolocation signals and hearing specializations of odontocetes suggests caution in assuming that all species process sonar sounds the same way as bottlenose dolphins do.

6.4.2 Do sperm whales echolocate?

Sperm whales produce click sounds that many biologists think are used for echolocation. However, unlike dolphins, sperm whales have not been kept in captivity for long enough to allow experimental demonstration of echolocation. We know that when sperm whales dive and forage, they tend to produce long series of click sounds with relatively invariant ICIs of 0.5–2.0 s (Whitehead & Weilgart 1990, 1991). As sperm whales start a dive, they often begin these clicks as they reach depths of 150–300 m (Papastavrou *et al.* 1989). Recordings of particularly loud sperm whale clicks often include reverberation from the sea floor (Fig. 6.9), so it is likely that sperm whales hear echoes if they approach the bottom during a dive. Sperm whales are large

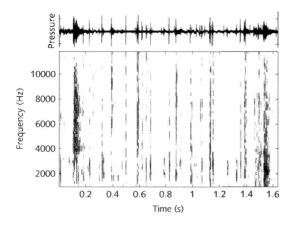

Fig. 6.9 Spectrogram of a slow click from a sperm whale,
Physeter macrocephalus, in waters near the island of Dominica
in the Caribbean. The direct arrival of the click is visible
on the left of the spectrogram, and an echo of the click
reflecting off the sea floor is visible on the right of the
spectrogram at 1.5 s. The shorter clicks in the middle
of the spectrogram are regular clicks from sperm whales.
(From data courtesy of William Watkins.)

(up to 45 t) and dive rapidly (1–2 m/s) (Papastavrou
et al. 1989; Watkins *et al.* 1993). If sperm whales
could not detect proximity to the bottom, there
would be a risk of collision. One way to test whether sperm whales use clicks to detect the sea floor
assumes that sperm whales follow eqn 6.4 and
produce a click after hearing the bottom echo
from the previous click. As a sperm whale dives, the
distance between the whale and the sea floor
decreases; this means that the round trip travel time
to the bottom should decrease. If the repetition
rate of these clicks is correlated with the round trip
travel time, then the repetition rate should increase
as the whale dives, a prediction for which Gordon
et al. (1992) and Gordon and Tyack (2002) found
supportive evidence.

Sperm whales may also detect their squid prey
in the deep dark ocean using echolocation (Backus
& Schevill 1966; Gordon 1987; Whitehead &
Weilgart 1990). Sperm whales feed on squid at
depths of 400 m or more during dives that typically
last 40–50 min (Papastavrou *et al.* 1989). During
these dives, sperm whales produce clicks with a pattern that Gordon (1987), Gordon *et al.* (1992) and
Jaquet *et al.* (1999) interpret as consistent with
detecting prey. On the other hand, Watkins (1980)

argues that sperm whale clicks are not well suited
to echolocation of prey. For example, the clicks
of sperm whales are lower in frequency, longer in
duration and have been thought to be much less
directional than the high-frequency clicks of dolphins. However, Møhl *et al.* (2000) report suggestive evidence that sperm whale clicks may be much
more directional than has been thought, with properties better suited to echolocation. Clearly, more
careful work is needed to determine whether and
how sperm whales use echo information to detect
prey or locate targets.

6.4.3 Can odontocetes stun or injure prey with loud sounds?

An original theory of how dolphins may use sound
that has appealed to the popular press is the 'acoustic
stunning hypothesis' of Norris and Møhl (1983).
These investigators reviewed evidence that odontocetes can produce sounds with very high sound
pressure levels such as the 220+ dB bottlenose
dolphin clicks discussed above. However, these
high-frequency clicks have a very rapid rise time and
only achieve such high source levels for several
microseconds, so the total energy delivered by the
signal is comparable to the less intense, longer
sounds produced by many cetaceans (Au 1993).
Sperm whales and dolphins also produce intense
lower frequency clicks, with durations of tens of
milliseconds. There are few reliable calibrated estimates of the source level of these sounds, so it is difficult to analyse their possible acoustic effects. Norris
and Møhl (1983) review observations that fish being
preyed upon by dolphins may appear disorientated
or incapacitated. They propose that this debilitation
is caused by exposure to the intense, pulsed sounds
of odontocetes. However, there are other explanations for the fish behaviour, such as low levels of
oxygen in dense fish schools (Würsig 1986). Testing
the acoustic stunning hypothesis requires exposing
fish to sounds of odontocetes where the received
level at the fish is measured with a calibrated hydrophone, coupled with subsequent behavioural and
anatomical testing of the fish subjects. Until this kind
of carefully controlled experiment is conducted,
'acoustic stunning' must remain in the 'interesting
but untested' category of theories.

6.4.4 Low-frequency echolocation

The only echolocation system that has been demonstrated in cetaceans involves the use of high-frequency clicks by small odontocetes. These animals clearly have evolved a highly specialized system for echolocation. However, sound may be used to explore the environment even among cetaceans that are not specialized for high-frequency echolocation. Norris (1967, 1969), Payne and Webb (1971) and Thompson *et al.* (1979) have all suggested that whales might be able to sense echoes of low-frequency vocalizations from distant bathymetric features to orientate or navigate. Migrating bowhead whales (*Balaena mysticetus*) appear to use echoes from their calls to detect ice obstacles (Clark 1989; George *et al.* 1989). Ellison *et al.* (1987) used acoustic models to show that deep-keeled ice may produce strong echoes from the low-frequency calls of migrating bowhead whales, and they suggest that bowhead whales may use these echoes to sense and avoid deep ice.

6.4.5 The chequered history of research on seal echolocation

Pinnipeds are readily available for experiments in a captive setting, and the excitement about dolphin echolocation stimulated studies of echolocation in pinnipeds. The first papers published on this topic (Poulter 1963a, 1963b, 1966) presented evidence for echolocation in the California sea lion. In response, Schusterman (1967) and Schevill (1968) published papers arguing that Poulter's conclusions were not justified by his data. Undeterred, Poulter & Del Carlo (1971) published a paper confidently entitled 'Echo ranging signals: sonar of the Steller sea lion, *Eumetopias jubata*'. An independent group followed up on this work a decade later, publishing a paper entitled 'Evidence that seals may use echolocation' (Renouf & Davis 1982). Wartzok *et al.* (1984) published a rebuttal questioning whether this evidence was convincing enough to force a reassessment of earlier studies that found no experimental evidence for echolocation in seals (Evans & Haugen 1963; Schusterman 1967; Oliver 1978; Scronce & Ridgway 1980). Publication of a paper on a topic such as echolocation in marine mammals will usually

stimulate studies that attempt to replicate the finding. In the case of dolphin echolocation, the findings were confirmed by more careful experimental designs, but in the case of echolocation by pinnipeds, most studies have failed to replicate the results or found flaws in the design of the experiments.

6.5 COMMUNICATION

6.5.1 Introduction

The etymological root of 'communicate' means 'to share', and communication can be defined as 'any sharing of information . . . between individual animals' (Smith 1977, p. 11). Social behaviour and social structure establish the selective pressures that shape and maintain characteristics of signals and signalling behaviour (Evans & Bastian 1969; Green & Marler 1979; McKinney 1992). Communication is almost always framed as a process in which a sender transmits information to a receiver via a signal (Owings & Morton 1998). In studying any communicative exchange, it is important to consider the costs and benefits of the information transfer to both sender and receiver. Some authors limit the term 'true communication' to exchanges of information that benefit both the signaller and the receiver (Smith 1977). When communication benefits the signaller at a cost to the receiver, it has been termed 'manipulation' (Dawkins & Krebs 1978; Krebs & Davies 1993). Signals that benefit the signaller through true communication or manipulation may become evolutionarily specialized as displays. When communication benefits a receiver at a cost to the signaller, this may be called 'eavesdropping' (Bradbury & Vehrencamp 1998) or 'interception' (Myrberg 1981). Unintentional cues that may inform a receiver at a cost to the signaller are unlikely to evolve into ritualized displays; rather selection would favour decreasing the signalling value of the cue, making it more cryptic. As was discussed in Section 6.2.1, signals form a continuum from highly ritualized displays to subtle incidental cues. Much information used by communicating animals comes from the context of the interaction, from cues such as scents or markings used in individual recognition, or from simple unritualized behaviours such as intention movements of advance or retreat. A biologist analysing a communicative interaction cannot just focus on the most obvious display, but must attend to all of these sources of information.

Wilson (1975, p. 111) argues that communication has occurred when a signal changes the probabilities of subsequent behaviour in a receiver. This is a behavioural interpretation of the formal definition of communication used in information theory, where a signal is described as providing a receiver with information that reduces uncertainty (although it may be misleading information in the case of manipulation) and allows the receiver to make decisions. Wilson's (1975) definition cannot provide a practical criterion for assessing whether communication occurs in a particular interaction between animals. Communication in animals is not limited to situations where one signal alters the probabilities of the subsequent response from one receiver. A female songbird may listen to thousands of songs from many males before choosing a mate. A young songbird listening to a song may alter an auditory template that will only be expressed in the following year. Even when an animal shows no obvious response to a communicative exchange, students of animal cognition may be able to design tests to indicate that an animal has learned from it. It makes little sense to argue that communication occurred only after the test was given. We favour adopting a more cognitive perspective on these issues, in which the information from the signal detected by the receiver is viewed as communication, whether or not it has elicited immediate behavioural responses.

6.5.2 Structure, function, adaptation and phylogeny in acoustic signals

'All scientists use models, whether they realize it or not' (James & McCulloch 1985, p. 44).

Textbook descriptions of research on animal communication may describe a simple sequential process in which a display is described, the social context in which the display is produced and responses of receivers are analysed, and finally the function is inferred. In practice, however, communicative sounds are seldom differentiated by quantitative analysis of acoustic measurements alone, but are often described using a combination of qualitative information on acoustic structure and social uses. For example, if a fieldworker hears calls that sound similar when her study animal sees a predator, then she might call these 'alarm' calls. This becomes a problem if researchers are not explicit about the interaction between descriptive and functional analyses. For this reason, it is usually recommended for ethologists to use one set of terms for descriptive analyses of displays and different terms to refer to the assumed function (Martin & Bateson 1993).

Two main questions trouble researchers as they describe repertoires: 'How many kinds of calls are there?' and 'What do the calls mean?' There is no single answer to these questions. The first question is based on the erroneous assumption that one can always define a unique set of calls based upon acoustic structure. This is not always the case; the number of call types recognizable on structural grounds depends upon how fine or coarse a level of description is used (e.g. how many variables are used). In turn, this is set by the nature of the investigation; in studies of phylogeny or adaptation, 'one needs behavioural events that strike an optimum between being diverse enough to reveal change and conservative enough to expose relationships' (Barlow 1992, p. 368). Some workers have used multivariate methods such as cluster analysis to help determine repertoire size. However, cluster analyses depend upon which acoustic variables and which clustering algorithm are selected, so the analytical results (including the hierarchical nature of most clustering applications) may reflect the subjective choices of researchers (Sneath & Sokal 1973; Miller 1979, 1988; de Queiroz & Good 1997; Janik 1999).

The second question is complicated because there may not be a one-to-one correlation between call type and function. For example, when a territorial male songbird sings a song, the song may have different functions depending upon the potential audience: females, other territorial males or young males. The songs of some songbirds appear to be interchangeable; in this case, different songs may have the same functional role (Smith 1996). If single call types can have multiple functions, and if multiple call types can have the same function in the same species (Altmann 1967; Smith 1977,

1997; Green & Marler 1979), then it is clearly necessary to separate the structural description of calls from the functional interpretation.

6.5.2.1 From calls to repertoires: acoustic structure across scales

Descriptions and physical analyses of acoustic signals are integral parts of comprehensive ethograms. Many invaluable 'first accounts' of vocalizations placed communicative sounds in a few convenient categories, described in terms of both acoustic structure and social context, and presented as representative spectrograms (Bartholomew & Collias 1962; Schevill & Watkins 1965; Schevill *et al.* 1966;

Peterson & Bartholomew 1969; Ray *et al.* 1969; Stirling & Warneke 1971; Lisitsina 1973; Stirling 1973). Early workers did not have the advantage of digital computers and a range of quantitative techniques for analysis and interpretation (Watkins 1967b; Marler 1969; Beecher 1988; Beeman 1998; Stoddard 1998). These quantitative techniques of acoustic analysis enable a more explicit acoustic analysis separated from the functional context that may colour the interpretations of the human observer.

Traditionally, the starting point for structural description of a species' acoustic repertoire has been the identification of signalling acts at a low level, such as the vocal classes shown in Fig. 6.10.

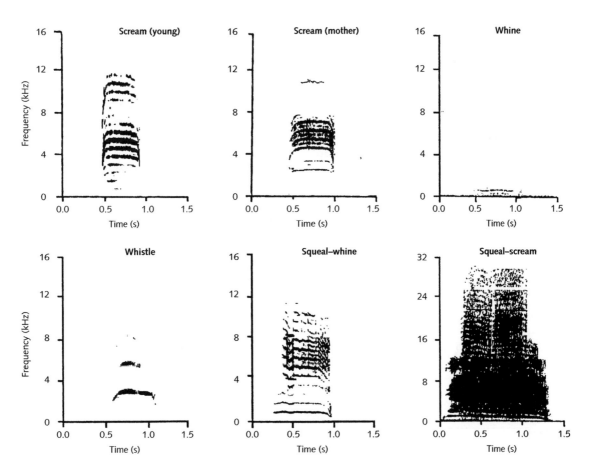

Fig. 6.10 Part of the vocal repertoire of the sea otter (*Enhydra lutris*), shown as spectrograms. The 'scream' call (upper left two spectrograms) could be considered as a single call type; the 'whine' often merges into 'squeal–whine', so these may illustrate variation in signal form within a single broadly defined call type. (From McShane *et al.* 1995.)

The vocalizations of some animals appear to fall into easily distinguishable general categories, such as the pulses and tonal whistles of dolphins. However, the decision of what comprises the smallest unit for this low-level repertoire of signal units (Smith 1986, p. 316) is seldom discussed in detail, and this decision can strongly influence later results. For example, does one count an individual dolphin pulse or train of pulses as the basic unit? There are likely to be hundreds more pulses than trains, so counts of the calls would be significantly affected by this choice. Some dolphin whistles are composed of subunits that may be separated by gaps of silence. Caldwell *et al.* (1990) called these subunits 'loops'. When individual dolphins are isolated, they often produce an individually distinctive whistle with a stereotyped initial loop, a variable number of repetitions of a central loop, followed by a terminal loop. Others studying dolphin whistles may split whistles whenever there is a moment of silence (McCowan & Reiss 1995). This seemingly minor change can lead to very different conclusions about whistles. McCowan and Reiss (1995) analysed as whistles what Caldwell *et al.* (1990) would have called loops, and concluded that there were few individually distinctive whistles in a group of captive dolphins.

The reason for this problem regarding lumping or splitting the basic unit of vocalization is that many species have repertoires of patterned combinations of signal units (Smith 1986, p. 324). This has traditionally been recognized for bird song, where the basic unit of vocalization, the song, is made up of a sequence of individual sounds (often termed syllables). Examples of patterned sequences in marine mammals include the underwater songs of Weddell seals (Green & Burton 1988; Morrice *et al.* 1994) and humpback whales (Payne & McVay 1971). Patterned displays may also combine optical and acoustic components, such as air-bladder displays of male hooded seals (Fig. 6.7) or the jaw-clap display of bottlenose dolphins (Overstrom 1983). These patterned combinations of signal units are governed by organizational rules (Smith 1986, 1991b) such as those reviewed by Hailman *et al.* (1985). Patterned combinations need not be elaborate; in Hawaiian monk seals (*Monachus schauinslandi*) a common airborne threat vocalization is a series of bubble sounds followed by a guttural expiration (Miller & Job 1992). These patterned combinations of low-level signal units need not be separated by silence, but little research has analysed the issue of whether individual utterances of animals are made up of 'chunks' that may be recombined in different patterns.

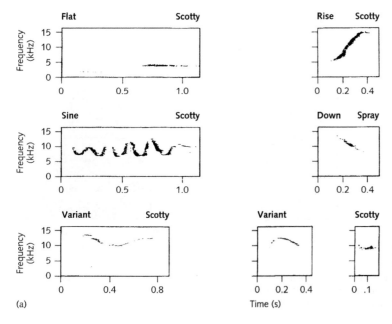

Fig. 6.11 (a) Spectrograms of whistles from bottlenose dolphins (*Tursiops truncatus*) arranged in acoustically distinct categories. (Adapted from Tyack 1986.) (*continued*)

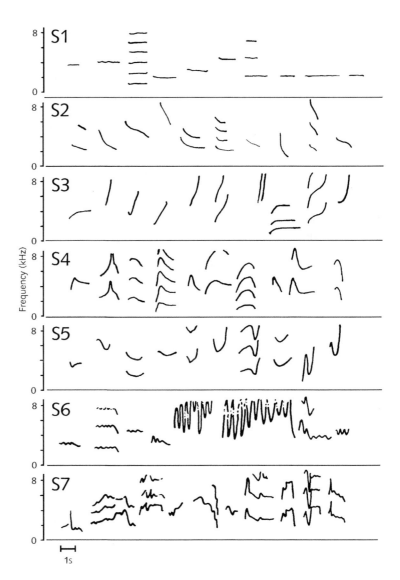

Fig. 6.11 (*cont'd*) (b) Contours from pilot whales (*Globicephala melas*) arranged as a graded series. (Adapted from Weilgart & Whitehead 1990.)

Most analyses of animal sounds tend to focus on categorizing vocalizations into discrete categories. Another way to approach repertoire description is to describe variations of signal form across signal units (Smith 1986, p. 317). Following this approach, variation could be described in acoustic features such as presence, number or distribution of harmonic structures; frequency or amplitude modulation; presence of broadband noise, biphonation or sub-harmonics; and rate of vocalization. Such acoustic qualities can vary substantially across repetitions of a single display type. Viewed in this way, the 'squeal' and 'squeal–whine' of the sea otter may be variations on a single class of signal, as the former often merges into the latter (Fig. 6.10) (McShane *et al.* 1995). The dolphin whistles illustrated in Fig. 6.11a separate whistles with downward, flat or upward frequency modulation as discrete categories. However, as Taruski (1979), Weilgart and Whitehead (1990) and others have noted, odontocete whistles can also be viewed as a continuum of changing frequency modulation (Fig. 6.11b). When whistles can be attributed to an individual, each individual may produce series of whistles that are very similar in some acoustic features (contour of the loops) and variable in others (such as number of loops, duration, etc.). In this context, the stereotyped features may carry information useful for individual recognition, while the other features may carry other information. Variation in signal form greatly increases the information available in a communication system; the importance of variation in form has been well stated by Hailman and Ficken (1996, p. 141): 'apparent types of vocalizations do carry information, but the details, and in some cases perhaps the most important information, lies in the variations on a type'. Brownlee and Norris (1994, p. 180) make this point for dolphin whistles: 'Variations in how whistles are emitted can carry graded . . . information about emotional state, level of alertness, hierarchy, the presence of food or danger . . . the modulation or temporal patterning of the whistles allows transmission of a variety of context-specific information. . . . Such a signal system can be modulated in intensity, frequency, and frequency pattern through time to produce a complex system of great potential information-carrying capacity . . .'.

Not only can an individual animal produce a sequence of signals, but two or more animals may have repertoires of formalized interactions (Smith 1986, p. 325). Examples are vocal interchanges between mothers and their offspring, boundary displays between territorial male otariids, precopulatory and copulatory behaviour between males and females, and so on (Gentry 1975; Lisitsina 1981; Miller 1991). Formalized interactions can involve a call and response of several signals between two animals over a few seconds, but they can involve more than two parties, and can involve more signals over a longer time. For example, Brownlee and Norris (1994, p. 182) suggest that spinner dolphins (*Stenella longirostris*) monitor whistle rates across an entire school over periods of a few minutes to an hour or more as they swim in a zig-zag path in order to time the behavioural transition from inshore resting to travelling offshore to feed: 'We interpret the fluctuating occurrence of contagious whistles during zig-zag swimming as partially social facilitation, with dolphins testing each other's alertness until the school is primed for its transit out to sea.'

6.5.2.2 Social functions and functional analysis

How we view the 'meaning' or 'function' of a display depends upon which model of animal communication we adopt. Researchers studying animal communication may employ widely differing models with different assumptions and descriptive/analytical approaches (for example, compare the linguistic model used by Evans & Bastian (1969) to more behavioural ecological models such as those used by Miller (1991) and Tyack (1999) for marine mammals). Our language biases and constrains our interpretations of what communication is for, how it functions and how it is structured (Golani 1992). 'Alarm' calls, 'dialect', and 'call type' are examples of widely (and casually!) used language in animal communication research that bias research programmes from conceptualization and data collection through to data analysis and interpretation.

Calls used by mothers and pups of otariids can serve as an example of how interpretation can be embedded in the name of a call. After fasting while nursing their pup for a week or two following birth, otariid females begin a cycle of alternately feeding offshore and nursing the pup on land. When females return from their marine feeding trips, they and their pups need to find one another, and do so initially through loud vocalizations; females also tend to return to where they and their pups habitually nurse, and pups tend to return there in anticipation of their mothers' return. Vocalizations used by females and pups in this circumstance are commonly referred to as 'pup attraction calls' and 'female attraction calls', respectively (Fig. 6.8b). Confusingly, 'pup attraction calls' can mean 'calls used by pups to attract females', and 'female attraction calls' can mean 'calls used by females to attract pups'. Another problem with these terms is that they define the calls in terms of context rather than structure, and this assumes that there is a one-to-one mapping of this particular context onto a specific call with one specific function (Lisitsina 1988; Insley 1996; Fernández-Juricic *et al.* 1999; Phillips & Stirling 2000, 2001). Interestingly, the vocalizations in question are very similar in structure to a call type used by territorial males variously termed roar, full threat call, etc. (E.H. Miller & A.V. Phillips, unpublished

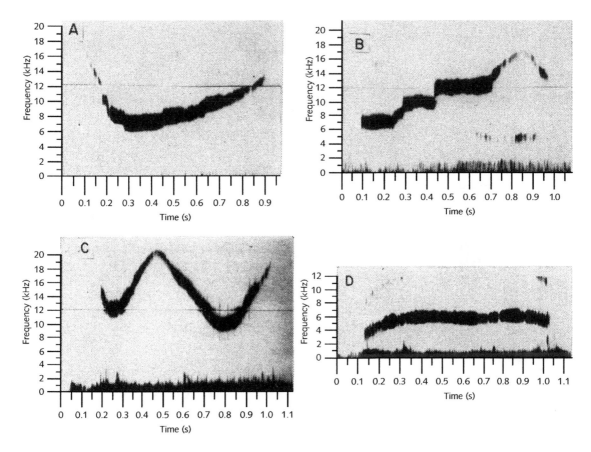

Fig. 6.12 Contours and spectrograms of signature whistles recorded from different captive common dolphins (*Delphinus delphis*). (From Caldwell & Caldwell 1968.)

data). Why should the calls be so similar in structure? The answer may be that the calls of females, pups and males are the same in terms of the kinds of behavioural information made available to receivers. That information could simply be 'I am seeking interaction or am willing to interact', with information about location (e.g. in the case of territorial males, or of females and pups reuniting), and from non-behavioural messages in the vocalization also used by the receiver in selecting an appropriate response. For example, sex, age or body size may be indicated by frequency and spectral features; individuality may be reflected in a unique combination of acoustic features, and so on (St Clair Hill *et al.* 2001).

Another example of problems with mixing functional and descriptive terms involves dolphin whistles. Early descriptions of dolphin whistles often used

recorded whistles when animals were harmed by the human observers—these whistles were called 'distress whistles' (e.g. Lilly 1963; Busnel & Dziedzic 1968). Caldwell and Caldwell (1965, 1968) noted that in their recordings of captive dolphins, each individual dolphin tended to produce an individually distinctive whistle, which they termed a 'signature whistle' (Fig. 6.12). More recent observations suggest that wild and captive dolphins tend to produce these 'signature whistles' when they are isolated (Janik & Slater 1998) or involuntarily restrained by humans (Sayigh *et al.* 1990), so many of the so-called 'distress whistles' are likely to have been the individually distinctive 'signature whistle' of those individuals (Caldwell *et al.* 1990). Many ethological studies of communication assume that their goal should be to define a repertoire of species-specific

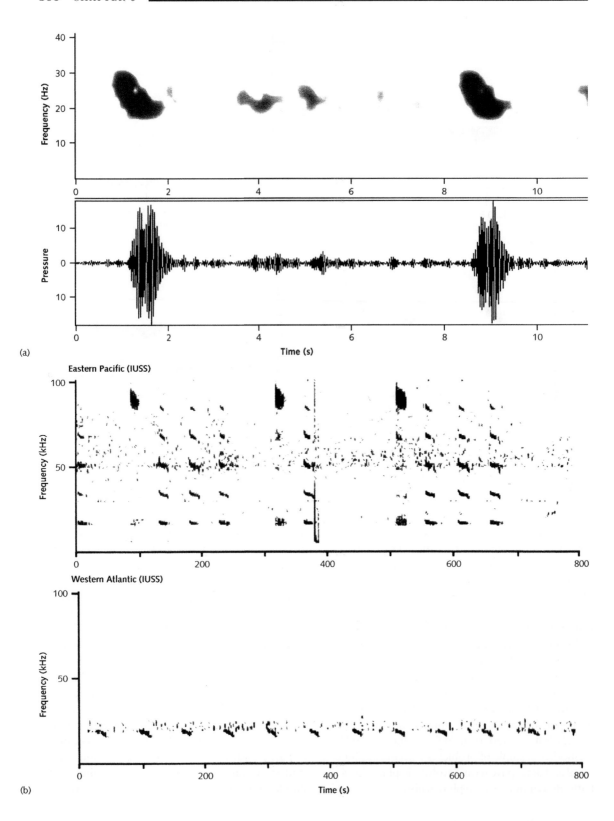

(a)

Eastern Pacific (IUSS)

Western Atlantic (IUSS)

(b)

calls that are shared within an age–sex class. However, recordings from isolated dolphins suggest a different pattern where individuals have highly distinctive whistles. Rather than being defined using acoustic information independent from knowledge of who made the call, signature whistles are defined operationally as a highly stereotyped and individually distinctive whistle typically made by some dolphins when isolated. As with the seal calls, dolphins may use similar whistles in a variety of social contexts (mother–young, male pair, etc.) to make similar information available to a partner, often maintaining contact and expressing willingness to interact.

Many studies of communication in marine mammals provide general correlations of vocal production and broad behavioural states, such as feeding, travelling, socializing and resting. Bengtson and Fitzgerald (1985) found strong correlations between rates of vocalization and general behavioural categories of West Indian manatees. Hoelzel and Osborne (1986) and Ford (1989) found that both rates and types of vocalizations in killer whales varied as a function of behavioural context. The patterns of click production are very different when sperm whales are socializing near the surface or feeding while diving (Whitehead & Weilgart 1991). These coarse analyses can help identify whether a particular call is more probably involved in behavioural states such as feeding versus travelling, but more fine-grained analyses are required to tease apart the pattern of signal and response by which short-range communication can mediate short-term social interactions. Fine-grained analyses require techniques for identifying which individual animal makes each display (e.g. Burgess *et al.* 1998; Miller & Tyack 1998) and requires following the behaviour of an individual signaller and potential receivers over time. Fine-grained analyses of marine mammal displays have been conducted for communication in air between pinnipeds (e.g. Trillmich 1981) and for agonistic interactions of dolphins (Overstrom 1983; Samuels & Gifford 1997).

Details of adaptive 'design features' may vary interspecifically because of differences in environment, effective range or intended receivers. There is an important difference in the functional analysis of signals evolved for communication over long versus short ranges. Inferring the function of vocalizations is relatively simple for long-distance signals but can be very difficult for those used over short distances. The reasons are simple: long-range signals are adapted in form and pattern of delivery to withstand degradation and attenuation, and little information outside the signal itself is available to recipients. Thus long-distance signals are typically loud, stereotyped, spectrally simple, long, repetitive and temporally patterned (e.g. repeated rhythmically or with strong sequential ordering). All of these features describe the sequences of low-frequency calls produced by finback and blue whales during the breeding season (Fig. 6.13) (Cummings & Thompson 1971; Edds 1982; Watkins *et al.* 1987; Tyack & Clark 2000). We do not know the effective range of these calls for whales, but humans can easily detect these calls at ranges of hundreds of kilometres (Tyack & Clark 2000). Long-range signals may be adapted to reach different classes of recipients over different distances (Terhune & Ronald 1986). For example, the underwater songs of male Weddell seals and the airborne roars of territorial male otariids communicate over shorter distances to females than to competing males, because the females to whom a sound may be directed tend to be closer than the nearest adult males.

Most research on acoustic communication focuses on loud specialized signals that have evolved for long-range communication. Our knowledge and understanding of short-range sounds are severely limited but cues and soft vocalizations that evolved for short transmission distances are likely to be important in group-living or gregarious species. Short-range

Fig. 6.13 (*opposite*) Spectrograms of low-frequency calls of finback (*Balaenoptera physalus*) and blue (*Balaenoptera musculus*) whales. (a) Waveform and spectrogram illustrating two low-frequency (nominally 20 Hz) pulses of finback whales recorded in the North Atlantic. The direct arrivals of pulses are visible at 1–2 s and 9–10 s, and echoes from reverberation of the first pulse are visible between 4 s and 5 s. (b) Low-frequency calls from blue whales recorded from different ocean basins using arrays of US navy hydrophones mounted on the sea floor in the eastern North Pacific and western North Atlantic. There are systematic differences in the calls from the different oceans. The Pacific call has a 90 Hz unit that is absent from the Atlantic calls, and the lower frequency components of the Pacific and Atlantic calls have different rhythmic patterns. (From Tyack & Clark 2000.)

signals are altered little by the physical environment, and so can be acoustically more complex and can encode much information through subtle variation in acoustic structure. Short-range communication is often multimodal; non-acoustic information about the signaller is often directly available to recipients nearby. Recipients can use this information outside the signal itself when appraising a signal's meaning —information such as a signaller's distance and orientation, the presence or proximity of other animals, knowledge of the individual identity of the signaller and past experience with the signaller, etc. The adaptive functions of short-range sounds can be difficult to assess because their proximate effects are often subtle and varied; they are often accompanied by optical, chemical or tactile information; they include many sounds produced in the course of other activities; and their significance to receivers is greatly affected by information outside the sounds themselves (Smith 1977, 1997; Cheney & Seyfarth 1990; Hauser 1996; Bradbury & Vehrencamp 1998). Some short range signals are produced almost continuously; variation in these tonic signals may communicate about a calling animal's state (Schleidt 1973). The bark vocalization of territorial male otariids and chuff vocalization of polar bears are examples of tonic signalling (Wemmer et al. 1976; Peters 1984; Miller 1991).

Long-range signals are often selected to be loud, stereotyped and simple, but paradoxically many long-distance signals are shaped by sexual selection, which often promotes signal diversification and complexity (Andersson 1994; Bradbury & Vehrencamp 1998). Thus males must often be under selection to produce effective long-range signals that are poorly designed for long-range transmission! One way to solve this paradox is to make a signal contrast with the ambient environment. In many species of seal and several species of whale, males are known to repeat songs during the breeding season. This regular repetition of songs may make them easier to detect in random noise. If the receiver detects a pattern of expected repetition, it can more reliably detect songs in noise. Harp seals produce vocalizations that contrast with ambient noise not only by repetition, but also by increasing amplitude, frequency and pulse rate over the course of each call (Watkins & Schevill 1979; Terhune et al. 1987;

Terhune 1994). Signal complexity can be increased by changing successive repetitions in some predictable way. The songs of Weddell seals and humpback whales follow syntactical rules for ordering different types of elements (Payne & McVay 1971; Guinee et al. 1983; Morrice et al. 1994). Gradual changes in acoustic structure within long songs, termed 'drift' by Andrew (1969) and Lemon (1975) and 'progressive change' by Payne et al. (1983), enable directions and rates of change to be readily tracked by recipients (Fig. 6.14) (Marler & Tenaza 1977; Payne et al. 1983; Payne & Payne 1985).

6.5.2.3 Evolution of acoustic signals: from variation to phylogeny

Genetically based phenotypic variation within populations is a prerequisite for population differentiation and ultimately for speciation (see Chapter 11). Therefore, the description and measurement of natural variations provide important information for evolutionary studies. The vocalizations of most non-human terrestrial mammals have acoustic features that appear to be influenced primarily by genetic factors. Some marine mammalogists have advocated using vocalizations as an indicator of population structure (Payne & Guinee 1983). However, many marine mammals can learn to imitate acoustic models in their environment (Janik & Slater 1997), so one must be careful in selecting features from vocalizations for evolutionary studies. Few measures of acoustic variation exist for marine mammal vocalizations, but a general trend seems to be toward greater variation in temporal than in frequency variables (Miller 1986, 1991).

Important components of variation within populations include sex, individuality, age, size and social context. Differences in vocalizations between the sexes of marine mammals are pronounced for the sperm whale, in which adult males have a head and spermaceti organ much larger than adult females (Nishiwaki et al. 1963). An unusual kind of click, called a 'slow click' by Weilgart and Whitehead (1988), is reportedly only produced by adult males. Female Amazonian manatees are reported to have higher frequency calls than males (R.S. Sousa Lima, personal communication). If these differences between sounds of adult males and females stem

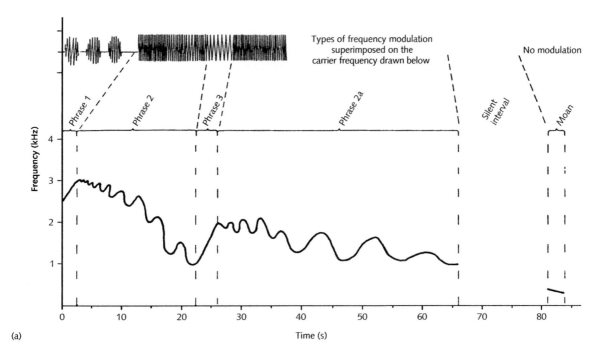

(a)

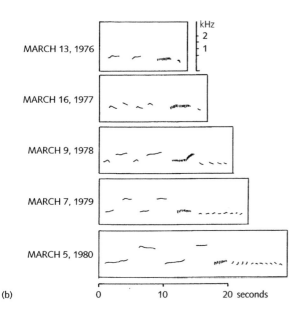

(b)

Fig. 6.14 Song repetition, syntax and gradual changes ('drift') within songs of the bearded seal (*Erignathus barbatus*) and humpback whale (*Megaptera novaeangliae*). These features of song are typical of reproductive advertisement displays in marine mammals, and may be adaptations to assist receivers to accurately receive a complex long-distance signal. (a) The song of the bearded seal exhibits syntactical organization in the initial bursts of frequency modulation (FM; phrase 1) of the carrier frequency, continuous but varying FM (phrase 2), slower but regular FM (phrase 3), etc. The song also undergoes progressive changes (sometimes with slow reversals) internally over its course, e.g. in carrier frequency (depicted in the schematic representation of a spectrogram, below). (From Ray *et al.* 1969.) (b) Each row represents a phrase from one theme of a humpback whale's song. This phrase is repeated a variable number of times before the whale switches to a different theme. Each row illustrates a typical phrase from each of 5 years in the songs of humpback whales in Hawaiian waters, showing progressive changes in the song over periods of years. (From Payne *et al.* 1983.)

from differences in the sound-producing organs, then they are likely to be good candidates for phylogenetic analyses. Long-distance underwater displays used to attract mates are restricted to males in most species studied to date, for example humpback whales (Glockner 1983; Baker *et al.* 1991;

Palsbøll *et al.* 1992), bearded and Weddell seals and walruses (Section 6.3.2.1). Female Weddell seals do not produce all of the vocalizations produced by males (Thomas & Kuechle 1982), but where both sexes produce the same class of sounds, almost no sexual differences have been reported.

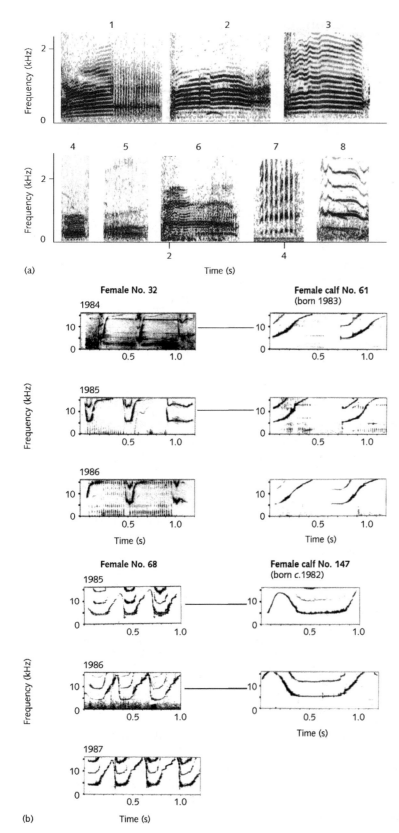

Fig. 6.15 Individuality is nearly universal in acoustic structure. (a) Spectrograms of South American sea lion (*Otaria flavescens*): primary call from three mothers (1–3); and grunts from two females (4, 5), a primary call from a yearling (6), and primary calls from two pups (7, 8). (From Fernández-Juricic *et al.* 1999.) (b) Individually distinctive whistles from two adult female bottlenose dolphins (*Tursiops truncatus*) and their daughters recorded in several different years. (From Sayigh *et al.* 1990.)

Numerous studies report individual differences in the vocalizations of marine mammals, some of which are known to persist over years (Fig. 6.15) (Sandegren 1976; Shipley *et al.* 1981; Trillmich 1981; Stirling *et al.* 1987; Caldwell *et al.* 1990; Sayigh *et al.* 1990; Fernández-Juricic *et al.* 1999; Insley 2000; Phillips & Stirling 2000; Sanvito & Galimberti 2000b; St Clair Hill *et al.* 2001). There are slight differences in the vocal tracts of different individuals, and these differences lead to individually distinctive characteristics that are called 'voice cues' among terrestrial animals. However, these involuntarily distinctive vocal characteristics may be less distinctive in underwater than in airborne vocalizations (Tyack 1999). As mammals dive, gases in the vocal tract halve in volume for every doubling of pressure. These depth-induced changes in the volume and shape of the vocal tract are likely to outweigh the subtle developmental differences that lead to differences in voice. For example, the whistles of a beluga whale recorded at different depths show strong differences in their frequency spectra (Ridgway 1997). If there is a functional need for diving animals to have individually distinctive calls, they may be unable to use voice cues and may need to create distinctive calls by learning to modify acoustic features under voluntary control, such as the frequency modulation of whistles. Janik and Slater (1997) and Tyack (1999) argue that this may have provided an important selection pressure for vocal learning in marine mammals.

Evidence from closely related species of birds suggests that animals evolve systems for parent–offspring recognition if the ecological setting involves a sufficient risk of providing care to the wrong offspring or of withholding care from the correct one (Loesche *et al.* 1991). Job *et al.* (1995), McCulloch and Boness (2000), Phillips and Stirling (2000) and Insley (1992, 2000, 2001) have studied vocal individuality in pinnipeds on the basis of predictions from this model. Insley (1992) predicted that the calls of female northern fur seals would be more individually different from one another than those of female northern elephant seals, because in the former species females and pups separate repeatedly and must reunite largely on the basis of vocalizations. His prediction was met.

Effects of size and age on acoustic structure are largely caused by variation in the anatomy of sound production in mammals, and account for a substantial proportion of natural variation. Little attention has been paid to the effects of development on the acoustic structure of vocalizations in bears and pinnipeds (Rasa 1971; Shipley *et al.* 1986). In spite of limited study, marine mammals stand out among non-human mammals in the extent to which vocal learning may play a role in vocal development (Janik & Slater 1997; Tyack & Sayigh 1997). Male Weddell seals lengthen some underwater vocalizations in response to conspecific vocalizations (Terhune *et al.* 1994b). The strongest evidence of modification of pinniped vocalization resulting from exposure to auditory models stems from Hoover, a common seal raised by humans, who started to imitate speech sounds with a Maine accent as he reached sexual maturity (Ralls *et al.* 1985). There is similar evidence that odontocete cetaceans can imitate novel man-made sounds (beluga whale: Eaton 1979; bottlenose dolphin: Evans 1967; Caldwell & Caldwell 1972b; Herman 1980; Richards *et al.* 1984).

Population differentiation in signals can result from genetic divergence or cultural divergence, or both (Lynch 1996). The term 'dialect' generally refers to the latter. Killer whales in the Puget Sound area have repertoires of stereotyped calls that are distinctive for each matrilineal group (Ford 1989); these have been called dialects, although the groups may be sympatric. These killer whale calls are generally thought to be learned, with enough copying error so that the longer groups have been separated, the less similarity there is between call repertoires (Ford 1991). Based on observed geographical differences alone, it is impossible to infer whether these group-distinctive call repertoires are rooted in genetic or cultural change, but acoustic divergence in the face of high gene flow or independent evidence of vocal learning make the latter more likely. Because geographical differentiation can result from either or both processes, one cannot infer population or stock discreteness solely from geographical differences in vocalizations; additional information on dispersal or development, or experimental studies are also required (James 1991).

Studies on geographical variation in marine mammal vocalizations have focused on the songs of humpback whales and a few species of pinnipeds. The underwater songs of rutting male Weddell seals, bearded seals, harp seals, leopard seals (*Hydrurga leptonyx*) and walruses vary geographically, as do some long-distance vocal displays of elephant seals (Le Boeuf & Peterson 1969; Le Boeuf & Petrinovich 1975; Ray & Watkins 1975; Stirling *et al.* 1983, 1987; Thomas & Stirling 1983; Thomas *et al.* 1988; Morrice *et al.* 1994; Terhune 1994; Rogers *et al.* 1995; Thomas & Golladay 1995; Pahl *et al.* 1997; Perry & Terhune 1999; Sanvito & Galimberti 2000b). The best documented and most striking geographical patterns are in Weddell seals, which show distinctly different call types in different parts of their range and show some evidence of geographical variation even on a small spatial scale.

There is strong evidence for geographical and temporal variation in the songs of humpback whales (Helweg *et al.* 1998). Guinee *et al.* (1983) tracked changes over 1–2 years in the songs of a small number of individual whales, and found that the song of each whale was more similar to other whales recorded at the same time than to its own song from a different year. This verifies that individual whales modify their songs to match the current version. The convergence of songs from different whales sung during the same period, coupled with progressive and pervasive changes in songs over time, provide evidence that geographical variation in humpback songs reflects cultural change (Payne *et al.* 1983).

Acoustic characters measured from vocalizations are candidates for characters that will show informative patterns of ancestry and descent, but few systematic or phylogenetic studies of vocalizations in related species have been carried out for marine mammals. These studies are complicated by the clear evidence for vocal learning in odontocetes, mysticetes and pinnipeds. On the other hand, most other mammals (especially non-human terrestrial mammals) appear to inherit pattern generators for stereotyped vocal output, and will develop the species-typical repertoire with little or no auditory input from conspecific individuals (Janik & Slater 1997). A species-typical vocalization that develops independently of exposure must represent a highly coordinated set of inherited traits involving central pattern generators, and neural control of vocal and respiratory musculature and of the sound-production apparatus. Such a complex and coordinated system

may evolve relatively slowly, especially for acoustic features closely tied to the sound-production apparatus. These kinds of features are therefore likely to be informative about higher level phylogenetic relationships. Other features such as the sequential patterning of sound units may evolve more rapidly and thus be more useful for phylogenetic studies at the population or species level. Stirling (1971) described vocalizations of several species of southern hemisphere fur seals and Miller and Phillips (2001) addressed evolutionary conservatism of vocalizations in *Arctocephalus* plus the northern fur seal. These studies suggest that evolutionary conservatism is likely to be found in general purpose signals that are shared by different classes of senders and receivers; these will often be short-range signals (Moynihan 1973; Fernández-Juricic *et al.* 1999). In contrast, sexually selected displays, like long-distance broadcast displays of whales and rutting male pinnipeds, are likeliest to evolve most quickly. Improving knowledge of phylogenetic relationships will permit explicit tests of evolutionary rates and trends in acoustic communication of marine mammals (Brooks & McLennan 1991; Ford 1991; Irwin 1996; Prum 1997). To do so will require more quantitative acoustic analyses of the similarity of calls than are currently typically performed, and more comprehensive descriptions of repertoire structure and signalling behaviour.

6.6 CONCLUSIONS

6.6.1 Gaps in our knowledge

We know a lot about the diversity of sounds produced by marine mammals and have a detailed understanding of how captive dolphins echolocate on artificial targets, but lack knowledge in many areas:

1 Functional anatomy of sound-producing and sound-filtering structures; mechanisms of sound production.

2 Acoustic perception (including categorization of natural sounds).

3 Use of echolocation by odontocetes in the wild (what are the targets for which this ability evolved?).

4 Use of low-frequency sounds to orientate in and explore the environment.

5 Vocal development and vocal learning.

6 Functional analyses of vocalizations and vocal behaviour; the behavioural significance of variation in acoustic displays.

7 Social recognition through vocalizations.

All these areas present opportunities for future research. Recent advances in electronics and signal processing enable quantitative analysis of vocal repertoires and enable techniques for field identification of who is vocalizing, and for relating vocal behaviour to social interaction.

As researchers develop larger databases of animal sounds with associated information, there is an increasing need for recordings and data to be made available for other research purposes (Kroodsma et al. 1996; Bradbury et al. 1999). In the pioneering studies on marine mammal bioacoustics, sounds were recorded on magnetic tapes that are deteriorating. Since many marine mammal populations have vocal repertoires that may change over time, these early recordings can be extremely valuable. It can also be very expensive to go to sea, and much research depends upon serendipitous circumstances that can be difficult to duplicate. Many recordings obtained in this work have aesthetic and scientific value well beyond the use made by any one investigator. The effort to develop techniques for categorizing marine mammal sounds would also benefit enormously from some standardized sets of recorded sounds, so the different techniques could be evaluated with the same set of stimuli. Bradbury et al. (1999) and Kroodsma et al. (1996) argue for the development of systematic collections of archived marine mammal sounds that are readily accessible to scientists and to the general public.

6.6.2 The importance of scale in marine mammal communication and echolocation

The oceans have few boundaries, on scales from molecules or micrometres to oceanic basins or megametres. Even simple ocean processes such as diffusion may be important over temporal scales from milliseconds to millenia. Thus it is a commonplace of oceanography that one must be explicit about the spatial and temporal scales over which one is analysing data. When students think of animal communication, they often imagine one animal sending one signal to a recipient a few metres away that responds within seconds. Yet the adaptive significance of many activities, such as feeding and reproduction, accumulate over the lifetime of an animal. There are many cases where the short-term cost of an interaction, say male–male competition for dominance in a breeding system, looms large on short temporal scales, and is only outweighed by the long-term benefit of increased access to females. When one analyses the function(s) of animal calls, it is critical to think carefully about which temporal scales are relevant, both for costs and benefits. We have pointed out examples from dolphin whistles and humpback song where the simple act of categorizing calls depends upon choice of temporal scales, a choice that is seldom discussed explicitly. These choices may have profound influences upon our results, and we should think carefully about our reasons for choosing one particular scale for an analysis. We also need to think about the spatial scales that are appropriate for solving particular problems. For example, an echolocating dolphin uses high-frequency clicks to detect an obstacle or prey at a range of 1–100 m. Most baleen whales migrate thousands of kilometres on an annual basis. Unlike dolphin echolocation, which functions well for detecting small targets within 100 m, baleen whales face problems of orienting over hundreds of kilometres, problems that would be better addressed with low-frequency signals.

6.6.3 Importance of bioacoustic research to conservation

Many marine mammal populations were heavily exploited over the past centuries (see Chapter 14). Some, such as the Steller sea cow (*Hydrodamalis gigas*) were driven extinct, and others, such as the northern right whale (*Eubalaena glacialis*), have poor prospects of surviving another century (Caswell et al. 1999). For decades it has been recognized that there is an urgent need to monitor the populations of marine mammals. Where pinnipeds congregate on a few breeding beaches, it may be easy to monitor their populations, but many marine mammals are dispersed in the oceans. All marine mammals must surface to breathe, so many census techniques rely upon sighting animals from ships or airplanes. Yet most marine mammals at sea are only visible at the surface for 1–10% of the time, so it is difficult to count them efficiently or accurately. Other techniques that rely upon sighting animals can track individuals by photographing natural markings and can use mark-recapture techniques to estimate population size. This approach can be useful for small localized populations, but can suffer from sighting biases.

Acoustic monitoring of the vocalizations of marine mammals has proven to be a useful tool for estimating the distribution and abundance of some species. For example, Clark *et al.* (1996) spent more than a decade tracking vocalizations of bowhead whales migrating past Point Barrow, Alaska, and these acoustic detections yielded much higher population estimates than visual detections from the same area. Ships may combine a visual survey with using an array of hydrophones to detect and locate vocalizing whales (Norris *et al.* 1999). These acoustic data can in some circumstances be used exactly as visual sightings, with much higher detection rates for animals such as sperm whales (Leaper *et al.* 1992).

An appreciation for the amazing ways in which marine mammals have adapted to make use of the acoustic properties of the ocean can help us to appreciate that man-made noise in the sea may present a conservation problem for marine mammals. Over the past century, the acoustic environment of the sea has changed with the advent of motorized shipping, underwater explosives and the development of sound sources such as sonars, air guns, etc. (Richardson *et al.* 1995). Any or all of these sources of noise might interfere with how marine mammals use sound in the sea (Richardson *et al.* 1995), so noise pollution might have serious short- and long-term impacts on activity budgets, communication, echolocation, feeding and social structure in marine mammals.

The study of how marine mammals use sound is not just fascinating in its own right as basic research, but it also is important for protecting these animals and encouraging the recovery of endangered populations.

REFERENCES

Abe, H., Hasegawa, Y. & Wada, K. (1977) A note on the air-sac of ribbon seal. *Scientific Reports of the Whales Research Institute* **29**, 129–135.

Altmann, S.A. (1967) The structure of primate social communication. In: *Social Communication among Primates* (S.A. Altmann, ed.), pp. 325–362. University of Chicago Press, Chicago, IL.

Amundin, M. (1991) *Sound production in odontocetes with emphasis on the harbour porpoise, Phocoena phocoena.* PhD Thesis, University of Stockholm, Stockholm, Sweden.

Amundin, M. & Andersen, S.H. (1983) Bony nares air pressure and nasal plug muscle activity during click production in the harbour porpoise, *Phocoena phocoena*, and the bottlenosed dolphin, *Tursiops truncatus. Journal of Experimental Biology* **105**, 275–282.

Anderson, P.K. & Barclay, R.M.R. (1995) Acoustic signals of solitary dugongs: physical characteristics and behavioral correlates. *Journal of Mammalogy* **76**, 1226–1237.

Andersson, M. (1994) *Sexual Selection.* Princeton University Press, Princeton, NJ.

Andrew, R.J. (1963) Trends apparent in the evolution of vocalization in the Old World monkeys and apes. *Symposia of the Zoological Society of London* **10**, 89–101.

Andrew, R.J. (1969) The effects of testosterone on avian vocalizations. In: *Bird Vocalizations: Their Relation to Current Problems in Biology and Psychology* (R.A. Hinde, ed.), pp. 97–130. Cambridge University Press, Cambridge, UK.

Andrew, R.J. (1972) The information potentially available in mammal displays. In: *Non-Verbal Communication* (R.A. Hinde, ed.), pp. 179–206. Cambridge University Press, Cambridge, UK.

Andrew, R.J. (1976) Use of formants in the grunts of baboons and other nonhuman primates. *Annals of the New York Academy of Sciences* **280**, 673–693.

Aroyan, J.L. (1996) *Three-dimensional numerical simulation of biosonar signal emission and reception in the common dolphin.* PhD Thesis, University of California at Santa Cruz, Santa Cruz, CA.

Aroyan, J.L., Cranford, T.W., Kent, J. & Norris, K.S. (1992) Computer modeling of acoustic beam formation in *Delphinus delphis. Journal of the Acoustical Society of America* **92**, 2539–2545.

Aroyan, J.L., McDonald, M.A., Webb, S.C. *et al.* (2000) Acoustic models of sound production and propagation. In: *Hearing in Whales and Dolphins* (A.N. Popper, R.H. Fay & W.W.L. Au, eds), pp. 409–469. Springer-Verlag, New York.

Au, W.W.L. (1993) *The Sonar of Dolphins.* Springer-Verlag, New York.

Au, W.W.L. (2001) Echolocation. In: *Encyclopedia of Marine Mammals* (W.F. Perrin, B. Würsig & H.G.M. Thewissen, eds), pp. 358–367. Academic Press, San Diego.

Backus, R. & Schevill, W.E. (1966) *Physeter* clicks. In: *Whales, Dolphins, and Porpoises* (K.S. Norris, ed.), pp. 510–528. University of California Press, Berkeley, CA.

Bain, D.E. (1992) Multi-scale communication by vertebrates. In: *Marine Mammal Sensory Systems* (J.A. Thomas, K.A. Kastelein & A.Ya. Supin, eds), pp. 601–629. Plenum Press, New York.

Baker, C.S., Lambertsen, R.H., Weinrich, M.T. *et al.* (1991) Molecular genetic identification of the sex of humpback whales (*Megaptera novaeangliae*). In: *Genetic Ecology of Whales and Dolphins* (A.R. Hoelzel, ed.), pp. 105–111. Reports of the International Whaling Commission Special Issue No. 13. International Whaling Commission, Cambridge, UK.

Ballard, K.A. & Kovacs, K.M. (1995) The acoustic repertoire of hooded seals (*Cystophora cristata*). *Canadian Journal of Zoology* 73, 1362–1374.

Barlow, G.W. (1992) Is the mobility gradient suitable for general application? *Behavioral and Brain Sciences* 15, 267–268.

Bartholomew, G.A. & Collias, N.E. (1962) The role of vocalization in the social behaviour of the northern elephant seal. *Animal Behaviour* 10, 7–14.

Beecher, M.D. (1988) Spectrographic analysis of animal vocalizations: implications of the 'uncertainty principle'. *Bioacoustics* 1, 187–208.

Beeman, K. (1998) Digital signal analysis, editing, and synthesis. In: *Animal Acoustic Communication: Sound Analysis and Research Methods* (S.L. Hopp, M.J. Owren & C.S. Evans, eds), pp. 59–103. Springer-Verlag, Berlin.

Beer, C.G. (1982) Conceptual issues in the study of communication. In: *Acoustic Communication in Birds, Vol. 2: Song Learning and its Consequences* (D.E. Kroodsma & E.H. Miller, eds), pp. 279–310. Academic Press, New York.

Bengtson, J.L. & Fitzgerald, S.M. (1985) Potential role of vocalizations in West Indian manatees. *Journal of Mammalogy* 66, 816–819.

Berland, B. (1958) The hood of the hooded seal, *Cystophora cristata* Erxl. *Nature* 182, 408–409.

Berland, B. (1966) The hood and its extrusible balloon in the hooded seal—*Cystophora cristata* Erxl. *Norsk Polarinstitutt, Årbok, Oslo* 1965, 95–102.

Boness, D.J. & James, H. (1979) Reproductive behaviour of the grey seal (*Halichoerus grypus*) on Sable Island, Nova Scotia. *Journal of Zoology* 188, 477–500.

Bradbury, J.W. & Vehrencamp, S.L. (1998) *Principles of Animal Communication*. Sinauer Associates Inc., Sunderland, MA.

Bradbury, J., Budney, G.F., Stemple, D.W. & Kroodsma, D.E. (1999) Organizing and archiving private collections of tape recordings. *Animal Behaviour* 57, 1343–1344.

Brauer, R.W., Jennings, R.A. & Poulter, T.C. (1966) The effect of substituting helium and oxygen for air on the vocalization of the California sea lion, *Zalophus californianus*. In: *Proceedings of the Third Annual Conference on Biological Sonar and Diving Mammals* (T.C. Poulter,

ed.), pp. 68–72. Stanford Research Institute, Fremont, California.

Brooks, D.R. & McLennan, D.A. (1991) *Phylogeny, Ecology, and Behavior. A Research Program in Comparative Biology*. University of Chicago Press, Chicago, IL.

Brownlee, S.M. & Norris, K.S. (1994) The acoustic domain. *The Hawaiian Spinner Dolphin* (K.S. Norris, B. Würsig, R.S. Wells & M. Würsig, eds), pp. 161–185. University of California Press, Berkeley, CA.

Bryden, M.M. & Felts, W.J.L. (1974) Quantitative anatomical observations on the skeletal and muscular systems of four species of Antarctic seals. *Journal of Anatomy* 118, 589–600.

Burgess, W.C., Tyack, P.L., LeBoeuf, B.J. & Costa, D.P. (1998) A programmable acoustic recording tag and first results from free-ranging northern elephant seals. *Deep-Sea Research* 45, 1327–1351.

Burns, J.J. (1981a) Bearded seal *Erignathus barbatus* Erxleben, 1777. In: *Handbook of Marine Mammals, Vol 2: Seals* (S.H. Ridgway & R.J. Harrison, eds), pp. 145–170. Academic Press, London.

Burns, J.J. (1981b) Ribbon seal *Phoca fasciata* Zimmermann, 1783. In: *Handbook of Marine Mammals, Vol 2: Seals* (S.H. Ridgway & R.J. Harrison, eds), pp. 89–109. Academic Press, London.

Busnel, R.-G. & Dziedzic, A. (1968) Etude des signaux acoustiques associé à des situations détresse chez certain cétacés odontocètes. *Annales de l'Institut Océanographique, Monaco* 46, 109–144.

Caldwell, D.K. & Caldwell, M.C. (1972a) *The World of the Bottlenose Dolphin*. Lippincott, Philadelphia.

Caldwell, M.C. & Caldwell, D.K. (1965) Individualized whistle contours in bottlenosed dolphins (*Tursiops truncatus*). *Nature* 207, 434–435.

Caldwell, M.C. & Caldwell, D.K. (1968) Vocalizations of naïve captive dolphins in small groups. *Science* 159, 1121–1123.

Caldwell, M.C. & Caldwell, D.K. (1972b) Vocal mimicry in the whistle mode by an Atlantic bottlenosed dolphin. *Cetology* 9, 1–8.

Caldwell, M.C., Caldwell, D.K. & Tyack, P.L. (1990) A review of the signature whistle hypothesis for the Atlantic bottlenose dolphin, *Tursiops truncatus*. In: *The Bottlenose Dolphin: Recent Progress in Research* (S. Leatherwood & R. Reeves, eds), pp. 199–234. Academic Press, San Diego.

Calkins, D. & Lent, P.C. (1975) Territoriality and mating behavior in Prince William Sound sea otters. *Journal of Mammalogy* 56, 528–529.

Caswell, H., Fujiwara, M. & Breault, S. (1999) Declining survival probability threatens the North Atlantic right whale. *Proceedings of the National Academy of Science, USA* 96, 3308–3313.

Cheney, D.L. & Seyfarth, R.M. (1990) *How Monkeys See the World: Inside the Mind of Another Species*. University of Chicago Press, Chicago, IL.

Cherry, C. (1978) *On Human Communication: a Review, a Survey, and a Criticism*. MIT Press, Cambridge, MA.

Chiasson, R.B. (1955) *The morphology of the Alaskan fur seal.* PhD Thesis, Stanford University, Stanford, CA.

Clark, C.W. (1989) The use of bowhead whale call tracks based on call characteristics as an independent means of determining tracking parameters. *Reports of the International Whaling Commission* 39, 111–113.

Clark, C.W., Charif, R., Mitchell, S. & Colby, J. (1996) Distribution and behavior of the bowhead whale, *Balaena mysticetus*, based on analysis of acoustic data collected during the 1993 spring migration off Point Barrow, Alaska. *Reports of the International Whaling Commission* 46, 541–552.

Cranford, T.W. (1992) *Functional morphology of the odontocete forehead: implications for sound generation.* PhD Thesis, University of California at Santa Cruz, Santa Cruz, CA.

Cranford, T.W. (1999) The sperm whale's nose: sexual selection on a grand scale? *Marine Mammal Science* 15, 1133–1157.

Cranford, T.W. (2000) In search of impulse sound sources in odontocetes. In: *Hearing in Whales and Dolphins* (A.N. Popper, R.H. Fay & W.W.L. Au, eds), pp. 109–155. Springer Verlag, New York.

Cranford, T.W., Amundin, M. & Norris, K.S. (1996) Functional morphology and homology in the odontocete nasal complex: implications for sound generation. *Journal of Morphology* 228, 223–285.

Cummings, W. & Thompson, P.O. (1971) Underwater sounds from the blue whale, *Balaenoptera musculus. Journal of the Acoustical Society of America* 50, 1193–1198.

Dawkins, R. & Krebs, J.R. (1978) Animal signals: information or manipulation. In: *Behavioural Ecology: an Evolutionary Approach* (J.R. Krebs & N.B. Davies, eds), pp. 282–309. Sinauer Associates, Sunderland, MA.

Dawson, S. & Thorpe, C.W. (1990) A quantitative analysis of the sounds of Hector's dolphin. *Ethology* 86, 131–145.

de Queiroz, K. & Good, D.A. (1997) Phenetic clustering in biology: a critique. *Quarterly Review of Biology* 72, 3–30.

Dormer, K.J. (1979) Mechanism of sound production and air recycling in delphinids: cineradiographic evidence. *Journal of the Acoustical Society of America* 65, 229–239.

Dorst, J. (1973) Appareil respiratoire. In: *Traité de Zoologie, Vol 16: Mammifères, Splanchnologie* (J. Anthony, L. Arvy, J. Dorst, M. Gabe, R. Weill & P. Zeitoun, eds), pp. 484–600. Masson, Paris.

Eaton, R.L. (1979) A beluga whale imitates human speech. *Carnivore* 2, 22–23.

Edds, P.L. (1982) Vocalizations of the blue whale, *Balaenoptera musculus*, in the St Lawrence River. *Journal of Mammalogy* 63, 345–347.

Ellison, W.T., Clark, C.W. & Bishop, G.C. (1987) Potential use of surface reverberation by bowhead whales, *Balaena mysticetus*, in under-ice navigation. *Reports of the International Whaling Commission* 37, 329–332.

Evans, H.E. (1993) *Miller's Anatomy of the Dog.* W.B. Saunders, Philadelphia, PA.

Evans, W.E. (1967) Vocalization among marine mammals. In: *Marine Bioacoustics* (W.N. Tavolga, ed.), Vol. 2, pp. 159–186. Pergamon Press, Oxford.

Evans, W.E. & Bastian, J. (1969) Marine mammal communication: social and ecological factors. In: *The Biology of Marine Mammals* (H.T. Andersen, ed.), pp. 425–475. Academic Press, New York.

Evans, W.E. & Haugen, R. (1963) An experimental study of echolocation ability of a California sea lion, *Zalophus californianus* (Lesson). *Bulletin of the Southern California Academy of Sciences* 62, 165–175.

Evans, W.E. & Maderson, P.F.A. (1973) Mechanisms of sound production in delphinid cetaceans: a review and some anatomical considerations. *American Zoologist* 13, 1205–1213.

Evans, W.E., Awbrey, F.T. & Hackbarth, H. (1988) High frequency pulse produced by free ranging Commerson's dolphin *Cephalorhynchus commersonii* compared with those of phocoenids. *Reports of the International Whaling Commission Special Issue* 9, 173–181.

Fay, F.H. (1960) Structure and function of the pharyngeal pouches of the walrus (*Odobenus rosmarus* L.). *Mammalia* 24, 361–371.

Fay, F.H. (1982) *Ecology and biology of the Pacific Walrus, Odobenus rosmarus divergens Illiger.* North American Fauna No. 74. US Department of the Interior, Fish and Wildlife Service, Washington, DC.

Fay, F.H., Ray, G.C. & Kibal'chich, A.A. (1984) Timing and location of mating and associated behavior of the Pacific walrus, *Odobenus rosmarus* Illiger. In: *Soviet–American Cooperative Research on Marine Mammals, Vol. 1: Pinnipeds* (F.H. Fay & G.A. Fedoseev, eds), pp. 89–99. NOAA Technical Report NMFS No. 12. US Department of Commerce, National Oceanic and Atmospheric Administration, National Marine Fisheries Service, Washington, DC.

Fee, M.S., Shraiman, B., Pesaran, B. & Mitra, P.P. (1998) The role of nonlinear dynamics of the syrinx in the vocalizations of a songbird. *Nature* 395, 67–71.

Fernández-Juricic, E., Campagna, C., Enriquez, V. & Ortiz, C.L. (1999) Vocal communication and individual variation in breeding South American sea lions. *Behaviour* 136, 495–517.

Fitch, W.T. (1997) Vocal tract length and formant frequency dispersion correlate with body size in rhesus macaques. *Journal of the Acoustical Society of America* 102, 1213–1222.

Ford, J.K.B. (1989) Acoustic behavior of resident killer whales (*Orcinus orca*) off Vancouver Island, British Columbia. *Canadian Journal of Zoology* 67, 727–745.

Ford, J.K.B. (1991) Vocal traditions among resident killer whales (*Orcinus orca*) in coastal waters of British Columbia. *Canadian Journal of Zoology* 69, 1454–1483.

Gailey-Phipps, J.J. (1984) *Acoustic communication and behavior of the spotted seal (Phoca largha).* PhD Thesis, Johns Hopkins University, Baltimore, MD.

Gentry, R.L. (1975) Comparative social behavior of eared seals. *Rapports et Procès-Verbaux Des Réunions, Conseil International Pour l'Exploration de la Mer* 169, 188–194.

George, J.C., Clark, C., Carroll, G.M. & Ellison, W.T. (1989) Observations on the ice-breaking and ice navigation behavior of migrating bowhead whales (*Balaena mysticetus*) near Point Barrow, Alaska, spring 1985. *Arctic* **42**, 24–30.

Glockner, D.A. (1983) Determining the sex of humpback whales (*Megaptera novaeangliae*) in their natural environment. In: *Communication and Behavior of Whales* (R. Payne, ed.), pp. 447–464. Westview Press, Boulder, CO.

Golani, I. (1992) A mobility gradient in the organization of vertebrate movement: the perception of movement through symbolic language. *Behavioral and Brain Sciences* **15**, 249–266.

Goold, J.C. & Jones, S.E. (1995) Time and frequency domain characteristics of sperm whale clicks. *Journal of the Acoustical Society of America* **98**, 1279–1291.

Gordon, J.C.D. (1987) *Behaviour and ecology of sperm whales off Sri Lanka*. PhD Thesis, University of Cambridge, Cambridge, UK.

Gordon, J.C.D. (1991) Evaluation of a method for determining the length of sperm whales (*Physeter catodon*) from their vocalizations. *Journal of Zoology* **224**, 301–314.

Gordon, J.C.D., Leaper, R., Hartley, F.G. & Chappell, O. (1992) Effects of whale watching vessels on the surface and underwater acoustic behaviour of sperm whales off Kaikoura New Zealand. *New Zealand Department of Conservation, Science and Research Series* **52**, 64.

Gordon, J. & Tyack, P.L. (2002) Sounds and cetaceans. In: *Marine Mammals: Biology and Conservation* (P.G.H. Evans & J.A. Raga, eds), pp. 139–196. Kluwer Academic/ Plenum Press, New York.

Green, K. & Burton, H.R. (1988) Do Weddell seals sing? *Polar Biology* **8**, 165–166.

Green, S. & Marler, P. (1979) The analysis of animal communication. In: *Handbook of Behavioral Neurobiology, Vol 3: Social Behavior and Communication* (P. Marler & J.G. Vandenberghe, eds), pp. 73–158. Plenum Press, New York.

Griffin, D.R. (1958) *Listening in the Dark*. Yale University Press, New Haven. (Reprint edition, 1974. Dover Publications, New York.)

Guinee, L., Chu, K. & Dorsey, E.M. (1983) Changes over time in the songs of known individual humpback whales (*Megaptera novaeangliae*). In: *Communication and Behavior of Whales* (R. Payne, ed.), pp. 59–80. Westview Press, Boulder, CO.

Hailman, J.P. & Ficken, M.S. (1996) Comparative analysis of vocal repertoires, with reference to chickadees. In: *Ecology and Evolution of Acoustic Communication in Birds* (D.E. Kroodsma & E.H. Miller, eds), pp. 136–159. Comstock Publishing Associates, Ithaca, NY.

Hailman, J.P., Ficken, M.S. & Ficken, R.W. (1985) The 'chick-a-dee' calls of *Parus atricapillus*: a recombinant system of animal communication compared with written English. *Semiotica* **56**, 191–224.

Hanggi, E.B. & Schusterman, R.J. (1992) Underwater acoustical displays by male harbour seals (*Phoca vitulina*): initial results. In: *Marine Mammal Sensory Systems* (J.A. Thomas, R.A. Kastelein & A.Y. Supin, eds), pp. 449–457. Plenum Press, New York.

Hanggi, E.B. & Schusterman, R.J. (1994) Underwater acoustic displays and individual variation in male harbour seals, *Phoca vitulina*. *Animal Behaviour* **48**, 1275–1283.

Hartman, D.S. (1971) *Behavior and ecology of the Florida manatee, Trichechus manatus latirostris (Harlan), at Crystal River, Citrus County*. PhD Thesis, Cornell University, Ithaca, NY.

Hatakeyama, Y. & Soeda, H. (1990) Studies on echolocation of porpoises taken in salmon gillnet fisheries. In: *Sensory Abilities of Cetaceans* (J.A. Thomas & R. Kastelein, eds), pp. 269–281. Plenum Press, New York.

Hauser, M.D. (1996) *The Evolution of Communication*. MIT Press, Cambridge, MA.

Helweg, D.A., Cato, D.H., Jenkins, P.F., Garrigue, C. & McCauley, R.D. (1998) Geographic variation in South Pacific humpback whale songs. *Behaviour* **135**, 1–27.

Herman, L.M. (1980) Cognitive characteristics of dolphins. In: *Cetacean Behavior: Mechanisms and Functions* (L.M. Herman, ed.), pp. 363–429. Wiley-Interscience, New York.

Herzing, D.L. (1996) Vocalizations and associated underwater behavior of free-ranging Atlantic spotted dolphins, *Stenella frontalis* and bottlenose dolphins, *Tursiops truncatus*. *Aquatic Mammals* **22**, 61–79.

Hewer, H.R. (1974) *British Seals*. William Collins & Sons, Glasgow, UK.

Hoelzel, A.R. & Osborne, R.W. (1986) Killer whale call characteristics: implications for cooperative foraging. In: *Behavioral Ecology of Killer Whales* (B.C. Kirkevold & J.S. Lockhard, eds), pp. 373–403. Alan R. Liss, New York.

Hunter, J. (1787) Observations on the structure and oeconomy of whales. *Philosophical Transactions* **77**, 371–450.

Insley, S.J. (1992) Mother–offspring separation and acoustic stereotypy: a comparison of call morphology in two species of pinnipeds. *Behaviour* **120**, 103–122.

Insley, S.J. (2000) Long-term vocal recognition in the northern fur seal. *Nature* **406**, 404–405.

Insley, S.J. (2001) Mother–offspring vocal recognition in northern fur seals is mutual not asymmetrical. *Animal Behaviour* **61**, 129–137.

Irwin, R.E. (1996) The phylogenetic content of avian courtship display and song evolution. In: *Phylogenies and the Comparative Method in Animal Behavior* (E.P. Martins, ed.), pp. 234–252. Oxford University Press, New York.

James, F.C. (1991) Complementary descriptive and experimental studies of clinal variation in birds. *American Zoologist* **31**, 694–706.

James, F.C. & McCulloch, C.E. (1985) Data analysis and the design of experiments in ornithology. In: *Current Ornithology* (R.F. Johnston, ed.), pp. 1–63. Plenum Publishing, New York.

Janik, V.M. (1999) Pitfalls in the categorization of behavior: a comparison of dolphin whistle categorization methods. *Animal Behavior* **57**, 133–143.

Janik, V.M. & Slater, P.J.B. (1997) Vocal learning in mammals. *Advances in the Study of Behavior* **26**, 59–99.

Janik, V.M. & Slater, P.J.B. (1998) Context-specific use suggests that bottlenose dolphin signature whistles are cohesion calls. *Animal Behaviour* **56**, 829–838.

Jaquet, N., Dawson, S., Slooten, E. & Douglas, L. (1999) Sperm whale vocal behavior: why do they click? In: *Abstracts, Thirteenth Biennial Conference on the Biology of Marine Mammals*, pp. 89–90. Society for Marine Mammalogy, Lawrence, KA.

Job, D.A., Boness, D.J. & Francis, J.M. (1995) Individual variation in nursing vocalizations of Hawaiian monk seal pups, *Monachus schauinslandi* (Phocidae, Pinnipedia), and lack of maternal recognition. *Canadian Journal of Zoology* **73**, 975–983.

Kaiser, H.E. (1974) *Morphology of the Sirenia. A Macroscopic and X-ray Atlas of the Osteology of Recent Species*. S. Karger, Basel.

Kamminga, C. & Wiersma, H. (1981) Investigations on cetacean sonar II. Acoustical similarities and differences in odontocete sonar signals. *Aquatic Mammals* **8**, 41–62.

Kamminga, C. & Wiersma, H. (1982) Investigations on cetacean sonar V. The true nature of the sonar sound of *Cephalorhynchus commersonii*. *Aquatic Mammals* **9**, 95–104.

Kastak, D. & Schusterman, R.J. (1998) Low-frequency amphibious hearing in pinnipeds: methods, measurements, noise, and ecology. *Journal of the Acoustical Society of America* **103**, 2216–2228.

Kastelein, R.A., Dubbeldam, J.L. & de Bakker, M.A.G. (1997) The anatomy of the walrus head (*Odobenus rosmarus*). Part 5: the tongue and its function in walrus ecology. *Aquatic Mammals* **23**, 29–47.

Kastelein, R.A., Postma, J. & Verboom, W.C. (1995) Airborne vocalizations of Pacific walrus pups (*Odobenus rosmarus divergens*). In: *Sensory Systems of Aquatic Mammals* (R.A. Kastelein, J.A. Thomas & P.E. Nachtigall, eds), pp. 265–285. De Spil, Woerden, the Netherlands.

Kelemen, G. (1963) Comparative anatomy and performance of the vocal organ in vertebrates. In: *Acoustic Behaviour of Animals* (R.-G. Busnel, ed.), pp. 489–521. Elsevier, Amsterdam.

Kenyon, K.W. (1969) *The Sea Otter in the Eastern Pacific Ocean*. North American Fauna No. 68. US Department of the Interior, Fish and Wildlife Service, Washington, DC.

Ketten, D.R. (1994) Functional analyses of whale ears: adaptations for underwater hearing. *IEEE Proceedings in Underwater Acoustics* **1**, 264–270.

Ketten, D.R. (1997) Structure and function in whale ears. *Bioacoustics* **8**, 103–135.

Ketten, D.R. (2000) Cetacean ears. In: *Hearing in Whales and Dolphins* (A.N. Popper, R.H. Fay & W.W.L. Au, eds), pp. 43–108. Springer Verlag, New York.

King, J.E. (1972) On the laryngeal skeletons of the leopard seal, *Hydrurga leptonyx*, and the Ross seal, *Ommatophoca rossi*. *Mammalia* **36**, 146–156.

King, J.E. (1983) *Seals of the World*. Cornell University Press, Ithaca, NY.

Klump, G.M. & Shalter, M.D. (1984) Acoustic behaviour of birds and mammals in the predator context. I Factors affecting the structure of alarm signals; II The functional significance and evolution of alarm signals. *Zeitschrift für Tierpsychologie* **66**, 189–226.

Kooyman, G.L. (1981) Leopard seal *Hydrurga leptonyx* Blainville, 1820. In: *Handbook of Marine Mammals, Vol 2: Seals* (S.H. Ridgway & R.J. Harrison, eds), pp. 261–274. Academic Press, London.

Kooyman, G.L. & Andersen, H.T. (1969) Deep diving. In: *The Biology of Marine Mammals* (H.T. Andersen, ed.), pp. 65–94. Academic Press, New York.

Kooyman, G.L., Hammond, D.D. & Schroeder, J.P. (1970) Bronchograms and tracheograms of seals under pressure. *Science* **169**, 82–84.

Kovacs, K.M. & Lavigne, D.M. (1986) *Cystophora cristata*. *Mammalian Species* **258**, 1–9.

Krebs, J.R. & Davies, N.B. (1993) *An Introduction to Behavioural Ecology*. Blackwell Scientific Publications, Oxford.

Kroodsma, D.E., Budney, G.F., Grotke, R.W. *et al.* (1996) Natural sound archives: guidance for recordists and a request for cooperation. In: *Ecology and Evolution of Acoustic Communication in Birds* (D.E. Kroodsma & E.H. Miller, eds), pp. 474–486. Comstock Publishing Associates, Ithaca, NY.

Lavigne, D.M. & Kovacs, K.M. (1988) *Harps and Hoods: Ice-breeding Seals of the Northwest Atlantic*. University of Waterloo Press, Waterloo, Ontario.

Lawrence, B. & Schevill, W.E. (1956) The functional anatomy of the delphinid nose. *Bulletin of the Museum of Comparative Zoology, Harvard* **114**, 103–151.

Le Boeuf, B.J. & Peterson, R.S. (1969) Dialects in elephant seals. *Science* **166**, 1655–1656.

Le Boeuf, B.J. & Petrinovich, L.F. (1974) Elephant seals: interspecific comparisons of vocal and reproductive behavior. *Mammalia* **38**, 16–32.

Le Boeuf, B.J. & Petrinovich, L.F. (1975) Elephant seal dialects: are they reliable? *Rapports et Proces-Verbaux des Réunions, Conseil International pur l'Exploration de la Mer* **169**, 213–218.

Leaper, R., Chappell, O. & Gordon, J.C.D. (1992) The development of practical techniques for surveying sperm whale populations acoustically. *Reports of the International Whaling Commission* **42**, 549–560.

Lemon, R.E. (1975) How birds develop song dialects. *Condor* **77**, 385–406.

Levenson, C. (1972) *Characteristics of Sounds Produced by Humpback Whales (Megaptera novaeangliae)*. NAVOCEANO Technical Note No. 7700-6-72. Naval Oceanographic Office, Washington, DC.

Lieberman, P. (1984) *The Biology and Evolution of Language*. Harvard University Press, Cambridge, MA.

Lilly, J.C. (1963) Distress call of the bottlenose dolphin: stimuli and evoked behavioral responses. *Science* **139**, 116–118.

Lisitsina, T.Y. (1973) The behavior and acoustical signals of the northern fur seal *Callorhinus ursinus* on the rookery. *Zoologicheskii Zhurnal* **52**, 1220–1228 (in Russian).

Lisitsina, T.Y. (1981) Structure of the breeding grounds and the social behavior of the eared seals (Otariidae). In: *Ecology, Population Structure and Infraspecific Communication Processes in Mammals* (N.P. Naumov & V.E. Sokolov, eds), pp. 99–150. Nauka, Moscow.

Lisitsina, T.Y. (1988) Situation-dependent changes of sound signals of females and the young of *Callorhinus ursinus* (Pinnipedia, Otariidae). *Zoologicheskii Zhurnal* 67, 274–286 (in Russian).

Loesche, P., Stoddard, P.K., Higgins, B.J. & Beecher, M.D. (1991) Signature versus perceptual adaptations for individual vocal recognition in swallows. *Behaviour* 118, 15–25.

Lynch, A. (1996) The population memetics of birdsong. In: *Ecology and Evolution of Acoustic Communication in Birds* (D.E. Kroodsma & E.H. Miller, eds), pp. 181–197. Comstock Publishing Associates, Ithaca, NY.

McCowan, B. (1995) A new quantitative technique for categorizing whistles using simulated signals and whistles from captive bottlenose dolphins (Delphinidae, *Tursiops truncatus*). *Ethology* 100, 177–193.

McCowan, B. & Reiss, D. (1995) Quantitative comparison of whistle repertoires from captive adult bottlenose dolphins (Delphinidae, *Tursiops truncatus*): a reevaluation of the signature whistle hypothesis. *Ethology* 100, 193–209.

MacKay, R.S. & Liaw, C. (1981) Dolphin vocalization mechanisms. *Science* 212, 676–678.

McKinney, F. (1992) Courtship, pair formation, and signal systems. In: *Ecology and Management of Breeding Waterfowl* (B.D.J. Batt, A.D. Afton, M.G. Anderson, C.D. Ankney, D.H. Johnson, J.A. Kadlec & G.L. Krapu, eds), pp. 214–250. University of Minnesota Press, Minneapolis, MN.

McShane, L.J., Estes, J.A., Riedman, M.L. & Staedler, M.M. (1995) Repertoire, structure, and individual variation of vocalizations in the sea otter. *Journal of Mammalogy* 76, 414–427.

Marler, P. (1969) Tonal quality of bird vocalizations. In: *Bird Vocalizations: their Relations to Current Problems in Biology and Psychology* (R.A. Hinde, ed.), pp. 5–18. Cambridge University Press, Cambridge.

Marler, P. & Tenaza, R. (1977) Signaling behavior of apes with special reference to vocalization. In: *How Animals Communicate* (T.A. Sebeok, ed.), pp. 965–1033. University of Indiana, Bloomington, IN.

Marlow, B.J. (1975) The comparative behaviour of the Australasian sea lions *Neophoca cinerea* and *Phocarctos hookeri* (Pinnipedia: Otariidae). *Mammalia* 39, 159–230.

Martin, P. & Bateson, P. (1993) *Measuring Behaviour. An Introductory Guide*, 2nd edn. Cambridge University Press, Cambridge, UK.

McCulloch, S. & Boness, D.J. (2000) Mother–pup vocal recognition in the grey seal (*Halichoerus grypus*) of Sable Island, Nova Scotia, Canada. *Journal of Zoology* 251, 449–455.

Merdsoy, B.R., Curtsinger, W.R. & Renouf, D. (1976) *Preliminary Underwater Observations of the Breeding Behavior of the Harp Seal (Pagophilus groenlandicus)*. NTIS,

ACMRR/MM/SC118. FAO Advisory Committee on Marine Resources Research, Bergen, Norway.

Miller, E.H. (1975) Walrus ethology. I. The social role of tusks and applications of multidimensional scaling. *Canadian Journal of Zoology* 53, 590–613.

Miller, E.H. (1979) An approach to the analysis of graded vocalizations of birds. *Behavioral and Neural Biology* 27, 25–38.

Miller, E.H. (1985) Airborne acoustic communication in the walrus *Odobenus rosmarus*. *National Geographic Research* 1, 124–145.

Miller, E.H. (1986) Components of variation in nuptial calls of the least sandpiper (*Calidris minutilla*; Aves, Scolopacidae). *Systematic Zoology* 35, 400–413.

Miller, E.H. (1988) Description of bird behavior for comparative purposes. In: *Current Ornithology* (R.F. Johnston, ed.), Vol. 5, pp. 347–394. Plenum Press, New York.

Miller, E.H. (1991) Communication in pinnipeds, with special reference to non-acoustic communication. In: *The Behaviour of Pinnipeds* (D. Renouf, ed.), pp. 128–235. Chapman & Hall, London.

Miller, E.H. & Boness, D.J. (1983) Summer behavior of Atlantic walruses *Odobenus rosmarus rosmarus* (L.) at Coats Island, N.W.T. (Canada). *Zeitschrift für Säugetierkunde* 48, 298–313.

Miller, E.H. & Job, D.A. (1992) Airborne acoustic communication in the Hawaiian monk seal, *Monachus schauinslandi*. In: *Marine Mammal Sensory Systems* (J.A. Thomas, R.A. Kastelein & A.Y. Supin, eds), pp. 485–531. Plenum Press, New York.

Miller, E.H. & Murray, A.V. (1995) Structure, complexity, and organization of vocalizations in harp seal (*Phoca groenlandica*) pups. In: *Sensory Systems of Aquatic Mammals* (R.A. Kastelein, J.A. Thomas & P.E. Nachtigall, eds), pp. 237–264. De Spil, Woerden, the Netherlands.

Miller, P.J. & Tyack, P.L. (1998) A small towed beamforming array to identify resident killer whales (*Orcinus orca*) concurrent with focal behavioral observations. *Deep-Sea Research* 45, 1389–1405.

Miller, P.J. & Phillips, A.V. (2001) *Sexual selection and vocal evolution in fur seals and sea lions (Carnivora Otariidae)*. Unpublished manuscript.

Miller, L.A., Pristed, J., Møhl, B. & Surlykke, A. (1995) The click-sounds of narwhals (*Monodon monoceros*) in Inglefield Bay, Northwest Greenland. *Marine Mammal Science* 11 (4), 491–502.

Møhl, B. & Andersen, S. (1973) Echolocation: high frequency component in the click of the harbour porpoise (*Phocoena phocoena* L.). *Journal of the Acoustical Society of America* 54, 1368–1372.

Møhl, B., Terhune, J.M. & Ronald, K. (1975) Underwater calls of the harp seal, *Pagophilus groenlandicus*. *Rapports et Procès-Verbaux des Réunions, Conseil International pour l'Exploration de la Mer* 169, 533–543.

Møhl, B., Wahlberg, M., Madsen, P.T., Miller, L.A., Surlykke, A. (2000) Sperm whale clicks: directionality and source level revisited. *Journal of the Acoustical Society of America* 107, 638–648.

Mohr, E. (1963) Beiträge zur Naturgeschichte der Klapp-mütze, *Cystophora cristata* Erxl. 1777. *Zeitschrift für Säugetierkunde* **28**, 65–84.

Morrice, M.G., Burton, H.R. & Green, K. (1994) Micro-geographic variation and songs in the underwater vocalisa-tion repertoire of the Weddell seal (*Leptonychotes weddellii*) from the Vestfold Hills, Antarctica. *Polar Biology* **14**, 441–446.

Moynihan, M. (1970) Control, suppression, decay, disap-pearance and replacement of displays. *Journal of Theoretical Biology* **29**, 85–112.

Moynihan, M. (1973) The evolution of behavior and the role of behavior in evolution. *Breviora (Museum of Comparative Zoology)* **415**, 1–29.

Murchison, A.E. (1980) Detection range and range resolu-tion of echolocating bottlenose porpoise (*Tursiops trun-catus*). In: *Animal Sonar Systems* (R.-G. Busnel & J.F. Fish, eds), pp. 43–70. Plenum, New York.

Murie, J. (1874) Researches upon the anatomy of the Pinnipedia. Part III. Descriptive anatomy of the sea-lion (*Otaria jubata*). *Transactions of the Zoological Society of London* **8**, 501–582.

Myrberg Jr, A.A. (1981) Sound communication and inter-ception in fishes. In: *Hearing and Sound Communication in Fishes* (W.N. Tavolga, A.N. Popper & R.R. Fay, eds), pp. 395–425. Springer-Verlag, New York.

Nachtigall, P.E., Lemonds, D.W. & Roitblat, H.L. (2000) Psychoacoustic studies of dolphin and whale hearing. In: *Hearing in Whales and Dolphins* (A.N. Popper, R.H. Fay & W.W.L. Au, eds), pp. 330–363. Springer-Verlag, New York.

Negus, V.E. (1949) *The Comparative Anatomy and Physio-logy of the Larynx.* Heinemann, London.

Nishiwaki, M., Ohsumi, S. & Maeda, Y. (1963) Change of form in the sperm whale accompanied with growth. *Scientific Reports of the Whales Research Institute Tokyo* **17**, 1–4.

Norris, K.S. (1967) Some observations on the migration and orientation of marine mammals. In: *Animal Orientation and Navigation* (R.M. Storm, ed.), pp. 101–125. Oregon State University Press, Corvallis, OR.

Norris, K.S. (1969) The echolocation of marine mammals. In: *The Biology of Marine Mammals* (H.T. Andersen, ed.), pp. 391–423. Academic Press, New York.

Norris, K.S. & Harvey, G.W. (1972) A theory for the func-tion of the spermaceti organ of the sperm whale. In: *Animal Orientation and Navigation* (S.R. Galler, K. Schmidt-Koenig, G.J. Jacobs, R.E. Belleville, eds), p. 262. NASA Special Publication, Washington DC.

Norris, K.S. & Møhl, B. (1983) Can odontocetes debilitate prey with sound? *American Naturalist* **122**, 85–104.

Norris, K.S., Prescott, J.H., Asa-Dorian, P.V. & Perkins, P. (1961) An experimental demonstration of echolocation behavior in the porpoise, *Tursiops truncatus* (Montagu). *Biological Bulletin* **120**, 163–176.

Norris, T.F., McDonald, M. & Barlow, J. (1999) Acoustic detections of singing humpback whales (*Megaptera novaeangliae*) in the eastern North Pacific during their northbound migration. *Journal of the Acoustical Society of America* **106**, 506–514.

Nowicki, S. (1987) Vocal tract resonances in oscine bird sound production: evidence from birdsongs in a helium atmosphere. *Nature* **325**, 53–55.

Odend'hal, S. (1966) The anatomy of the larynx of the California sea lion (*Zalophus californianus*). In: *Proceedings of the Third Annual Conference on Biological Sonar and Diving Mammals* (T.C. Poulter, ed.), pp. 55–67. Stanford Research Institute, Fremont, CA.

Oliver, G.W. (1978) Navigation in mazes by a gray seal, *Halichoerus grypus* (Fabricius). *Behaviour* **67**, 97–114.

Overstrom, N.A. (1983) Association between burst-pulse sounds and aggressive behavior in captive Atlantic bottle-nosed dolphins (*Tursiops truncatus*). *Zoo Biology* **2**, 93–103.

Owings, D.H. & Morton, E.S. (1998) *Animal Vocal Com-munication: a New Approach.* Cambridge University Press, Cambridge, UK.

Pahl, B.C., Terhune, J.M. & Burton, H.R. (1997) Reper-toire and geographic variation in underwater vocalisations of Weddell seals (*Leptonychotes weddellii*, Pinnipedia: Phocidae) at the Vestfold Hills, Antarctica. *Australian Journal of Zoology* **45**, 171–187.

Palsbøll, P.J., Vader, A., Bakke, I. & El-Gewely, M.R. (1992) Determination of gender in cetaceans by the poly-merase chain reaction. *Canadian Journal of Zoology* **70**, 2166–2170.

Papastavrou, V., Smith, S.C. & Whitehead, H. (1989) Diving behaviour of the sperm whale, *Physeter macrocephalus*, off the Galapagos Islands. *Canadian Journal of Zoology* **67**, 839–846.

Payne, K. & Payne, R. (1985) Large scale changes over 19 years in the songs of humpback whales in Bermuda. *Zeitschrift für Tierpsychologie* **68**, 89–114.

Payne, K.B., Tyack, P. & Payne, R.S. (1983) Progressive changes in the songs of humpback whales. In: *Commun-ication and Behavior of Whales* (R. Payne, ed.), pp. 9–59. Westview Press, Boulder, CO.

Payne, R.S. & Guinee, L.N. (1983) Humpback whale (*Megaptera novaeangliae*) songs as an indicator of 'stocks'. In: *Communication and Behavior of Whales* (R. Payne, ed.), pp. 333–358. Westview Press, Boulder, CO.

Payne, R.S. & McVay, S. (1971) Songs of humpback whales. *Science* **173**, 585–597.

Payne, R.S. & Webb, D. (1971) Orientation by means of long range acoustic signalling in baleen whales. *Annals of the New York Academy of Sciences* **188**, 110–141.

Perry, E.A. & Terhune, J.M. (1999) Variation of harp seal (*Pagophilus groenlandicus*) underwater vocalizations among three breeding locations. *Journal of Zoology, London* **249**, 181–186.

Peters, G. (1978) Einige Beobachtungen zur Lautge-bung der Bären—Bioakustische Untersuchungen im Zoologischen garten. *Zeitschrift Des Kölner Zoo* **21**, 45–51.

Peters, G. (1984) On the structure of friendly close range vocalizations in terrestrial carnivores (Mammalia: Carnivora: Fissipedia). *Zeitschrift für Säugetierkunde* **49**, 157–182.

Peterson, R.S. & Bartholomew, G.A. (1969) Airborne vocal communication in the California sea lion *Zalophus californianus*. *Animal Behaviour* **17**, 17–24.

Phillips, A.V. & Stirling, I. (2000) Vocal individuality in mother and pup South American fur seals, *Arctocephalus australis*. *Marine Mammal Science* **16**, 592–616.

Phillips, A.V. & Stirling, I. (2001) Vocal repertoire of South American fur seals, *Arctocephalus australis*: structure, function and context. *Canadian Journal of Zoology* **79**, 420–437.

Piérard, J. (1965) La cavité sous-cricoïdienne (antrum sub-cricoideum) de l'otarie de Steller (*Eumetopias jubata* Schreber). *Mammalia* **29**, 429–431.

Piérard, J. (1969) Le larynx du phoque de Weddell (*Leptonychotes weddelli*, Lesson, 1826). *Canadian Journal of Zoology* **47**, 77–87.

Pierce, J.R. (1983) *The Science of Musical Sound*. Scientific American Books, New York.

Popov, L.A. (1961) Materials to the general morphology of *Cystophora cristata*, Erxl. of Greenland Sea. *Trudy Soveshchaniya Po Ekologii I Promyslu Morskikh Mlekopitayushchikh* (E.N. Pavlovskiy & S.E. Kleinenberg, eds), pp. 180–191. Izdatel'stvo Akademii Nauk SSR, Moscow.

Poulter, T.C. (1963a) Sonar signals of the sea lion. *Science* **139**, 753–755.

Poulter, T.C. (1963b) The sonar of the sea lion. *IEEE Transactions in Ultrasonic Engineering* **10**, 109–111.

Poulter, T.C. (1965) Location of the point of origin of the vocalization of the California sea lion, *Zalophus californianus*. In: *Proceedings of the Second Annual Conference on Biological Sonar and Diving Mammals* (T.C. Poulter, ed.), pp. 41–48. Stanford Research Institute, Fremont, CA.

Poulter, T.C. (1966) The use of active sonar by the California sea lion, *Zalophus californianus*. *Journal of Auditory Research* **6**, 165–173.

Poulter, T.C. & Del Carlo, D.G. (1971) Echo ranging signals: sonar of the Steller sea lion, *Eumetopias jubata*. *Journal of Auditory Research* **11**, 43–52.

Prum, R.O. (1997) Phylogenetic tests of alternative intersexual selection mechanisms: trait macroevolution in a polygynous clade (Aves: Pipridae). *American Naturalist* **149**, 668–692.

Purves, P.E. (1966) Anatomical and experimental observations on the cetacean sonar system. In: *Animal Sonar Systems* (R.-G. Busnel, ed.), pp. 197–269. Laboratoire de Physiologie Acoustique, Jouy-en-Josas, France.

Purves, P.E. & Pilleri, G. (1973) Observations on the ear, nose, and throat and eye of *Platanista indi*. *Investigations on Cetacea* **5**, 13–57.

Purves, P.E. & Pilleri, G. (1983) *Echolocation in Whales and Dolphins*. Academic Press, London.

Ralls, K., Fiorelli, P. & Gish, S. (1985) Vocalizations and vocal mimicry in captive harbour seals, *Phoca vitulina*. *Canadian Journal of Zoology* **63**, 1050–1056.

Rasa, O.A.E. (1971) Social interaction and object manipulation in weaned pups of the northern elephant seal *Mirounga angustirostris*. *Zeitschrift für Tierpsychologie* **29**, 82–102.

Ray, G.C. (1981) Ross seal *Ommatophoca rossi* Gray, 1844. In: *Handbook of Marine Mammals, Vol 2: Seals* (S.H. Ridgway & R.J. Harrison, eds), pp. 237–260. Academic Press, London.

Ray, G.C. & Watkins, W.A. (1975) Social function of underwater sounds in the walrus *Odobenus rosmarus*. *Rapports et Procès-Verbaux des Réunions, Conseil International pour l'Exploration de la Mer* **169**, 524–526.

Ray, G.C., Watkins, W.A. & Burns, J.J. (1969) The underwater song of *Erignathus* (bearded seal). *Zoologica (New York)* **54**, 79–83.

Reeves, R.R. & Ling, J.K. (1981) Hooded seal *Cystophora cristata* Erxleben, 1777. In: *Handbook of Marine Mammals, Vol 2: Seals* (S.H. Ridgway & R.J. Harrison, eds), pp. 171–194. Academic Press, London.

Reeves, R.R., Stewart, B.S. & Leatherwood, S. (1992) *The Sierra Club Handbook of Seals and Sirenians*. Sierra Club Books San Francisco, CA.

Reidenberg, J.S. & Laitmann, J.T. (1988) Existence of vocal folds in the larynx of odontoceti (toothed whales). *Anatomical Record* **221**, 886–891.

Renouf, D. & Davis, M.B. (1982) Evidence that seals may use echolocation. *Nature* **300**, 635–637.

Richards, D.G., Wolz, J.P. & Herman, L.M. (1984) Vocal mimicry of computer-generated sounds and vocal labeling of objects by a bottlenosed dolphin, *Tursiops truncatus*. *Journal of Comparative Psychology* **98**, 10–28.

Richardson, W.J., Greene Jr, C.R., Malme, C.I. & Thomson, D.H. (1995) *Marine Mammals and Noise*. Academic Press, New York.

Ridgway, S. (1997) First audiogram for marine mammals in the open ocean and at depth: hearing and whistling by two white whales down to 30 atmospheres. *Journal of the Acoustical Society of America* **101**, 3136 (abstract).

Ridgway, S.H., Carder, D.A., Green, R.F. *et al.* (1980) Electromyographic and pressure events in the nasolaryngeal system of dolphins during sound production. In: *Animal Sonar Systems* (R.G. Busnel & J.F. Fish, eds), pp. 239–250. Plenum Publishing, New York.

Rogers, T., Cato, D.H. & Bryden, M.M. (1995) Underwater vocal repertoire of the leopard seal (*Hydrurga leptonyx*) in Prydz Bay, Antarctica. In: *Sensory Systems of Aquatic Mammals* (R.A. Kastelein, J.A. Thomas & P.E. Nachtigall, eds), pp. 223–236. De Spil, Woerden, the Netherlands.

Rossing, T.D. (1990) *The Science of Sound*, 2nd edn. Addison-Wesley, Reading, MA.

Rubin, P. & Vatikiotis-Bateson, E. (1998) Measuring and modeling speech: production. In: *Animal Acoustic Communication: Sound Analysis and Research Methods* (S.L. Hopp, M.J. Owren & C.S. Evans, eds), pp. 251–290. Springer-Verlag, Berlin.

Samuels, A. & Gifford, T. (1997) A quantitative assessment of dominance relations among bottlenose dolphins. *Marine Mammal Science* **13**, 70–99.

Sandegren, F.E. (1976) Agonistic behavior in the male northern elephant seal. *Behaviour* **57**, 136–158.

Sanvito, S. & Galimberti, F. (2000a) Bioacoustics of southern elephant seals. I. Acoustic structure of male aggressive vocalizations. *Bioacoustics* **10**, 259–282.

Sanvito, S. & Galimberti, F. (2000b) Bioacoustics of southern elephant seals. II. Individual and geographic variation in male aggressive vocalizations. *Bioacoustics* **10**, 287–307.

Sayigh, L.S., Tyack, P.L., Wells, R.S. & Scott, M.D. (1990) Signature whistles of free-ranging bottlenose dolphins, *Tursiops truncatus*: stability and mother–offspring comparisons. *Behavioral Ecology and Sociobiology* **26**, 247–260.

Schevill, W.E. (1968) Sea lion echo ranging? *Journal of the Acoustical Society of America* **43**, 1458–1459.

Schevill, W.E. & Watkins, W.A. (1965) Underwater calls of *Leptonychotes* (Weddell seal). *Zoologica* **50**, 45–46.

Schevill, W.E., Watkins, W.A. & Ray, C. (1966) Analysis of underwater *Odobenus* calls with remarks on the development and function of the pharyngeal pouches. *Zoologica (New York)* **51**, 103–106.

Schleidt, W.M. (1973) Tonic communication: continual effects of discrete signs in animal communication systems. *Journal of Theoretical Biology* **42**, 359–386.

Schneider, R. (1962) Vergleichende Untersuchungen am Kehlkopf der Robben (Mammalia, Carnivora, Pinnipedia). *Jahrbuch Morphologisches* **103**, 177–262.

Schneider, R. (1963) Morphologische Anpassungserscheinungen am Kehlkopf einiger aquatiler Säugetiere. *Zeitschrift für Säugetierkunde* **28**, 257–267.

Schneider, R. (1964) Der Larynx der Säugetiere. *Handbuch der Zoologie* Band **8**, Lieferung **35**, 7(5), 1–128.

Schusterman, R.J. (1967) Perception and determinants of underwater vocalization in the California sea lion. In: *Animal Sonar Systems* (R.-G. Busnel, ed.), pp. 535–617, Laboratoire de Physiologie Acoustique, Jouy-en-Josas, France.

Schusterman, R.J. (1977) Temporal patterning in sea lion barking (*Zalophus californianus*). *Behavioral Biology* **20**, 404–408.

Scronce, B.L. & Ridgway, S.H. (1980) Grey seal, *Halichoerus*: echolocation not demonstrated. In: *Animal Sonar Systems* (R.-G. Busnel & J. Fish, eds), pp. 991–993, Plenum Press, London.

Shipley, C., Hines, M. & Buchwald, J.S. (1981) Individual differences in threat calls of northern elephant seal bulls. *Animal Behaviour* **29**, 12–19.

Shipley, C., Hines, M. & Buchwald, J.S. (1986) Vocalizations of northern elephant seal bulls: development of adult call characteristics during puberty. *Journal of Mammalogy* **67**, 526–536.

Shipley, C., Carterette, E.C. & Buchwald, J.S. (1991) The effects of articulation on the acoustical structure of feline vocalizations. *Journal of the Acoustical Society of America* **89**, 902–909.

Shipley, C., Stewart, B.S. & Bass, J. (1992) Seismic communication in northern elephant seals. In: *Marine Mammal Sensory Systems* (J.A. Thomas, R.A. Kastelein & A.Y. Supin, eds), pp. 553–562. Plenum Press, New York.

Sleptsov, M.M. (1940) On adaptations of pinnipeds to swimming. *Zoologicheskii Zhurnal* **19**, 379–386 (in Russian).

Slijper, E.J. (1979) *Whales*, 2nd edn. Cornell University Press, Ithaca, NY.

Smith, W.J. (1977) *The Behavior of Communicating: an Ethological Approach*. Harvard University Press, Cambridge, MA.

Smith, W.J. (1986) Signaling behavior: contributions of different repertoires. In: *Dolphin Cognition and Behavior: a Comparative Approach* (R.J. Schusterman, J.A. Thomas & F.G. Wood, eds), pp. 315–330. Lawrence Erlbaum Associates, Hillsdale, NJ.

Smith, W.J. (1991a) Animal communication and the study of cognition. In: *Cognitive Ethology: the Minds of Other Animals* (C.A. Ristau, ed.), pp. 209–230. Lawrence Erlbaum Associates, Hillsdale, NJ.

Smith, W.J. (1991b) Singing is based on two markedly different kinds of signaling. *Journal of Theoretical Biology* **152**, 241–253.

Smith, W.J. (1996) Using interactive playback to study how songs and singing contribute to communication about behavior. In: *Ecology and Evolution of Acoustic Communication in Birds* (D.E. Kroodsma & E.H. Miller, eds), pp. 377–397. Comstock Publishing Associates, Ithaca, NY.

Smith, W.J. (1997) The behavior of communicating, after twenty years. In: *Perspectives in Ethology, Vol. 12: Communication* (D.H. Owings, M.D. Beecher & N.S. Thompson, eds), pp. 7–53. Plenum Press, New York.

Sneath, P.H.A. & Sokal, R.R. (1973) *Numerical Taxonomy: the Principles and Practice of Numerical Classification*. W.H. Freeman, San Francisco, CA.

Sokolov, A.S., Kosygin, G.M. & Shustov, A.P. (1970) Structure of lungs and trachea of Bering Sea pinnipeds. In: *Pinnipeds of the North Pacific* (English translation of *Vsesoyuznyi Nauchno-Issledovatel'skii Institut Morskogo Rybnogo Khozyaistva I Okeanografii (VINRO), Vol. 68 (1968)*) (V.A. Arsen'ev & K.I. Panin, eds), pp. 250–262. Israel Program for Scientific Translations, Jerusalem, Israel.

St Clair Hill, M., Ferguson, J.W.H., Bester, M.N. & Kerley, G.I.H. (2001) Preliminary comparison of calls of the hybridizing fur seals *Arctocephalus tropicalis* and *A. gazella*. *African Zoology* **36**, 45–53.

Stirling, I. (1971) Studies on the behaviour of the South Australian fur seal, *Arctocephalus forsteri* (Lesson) II. Adult females and pups. *Australian Journal of Zoology* **19**, 267–273.

Stirling, I. (1973) Vocalization in the ringed seal (*Phoca hispida*). *Journal of the Fisheries Research Board of Canada* **30**, 1592–1594.

Stirling, I. & Warneke, R.M. (1971) Implications of a comparison of the airborne vocalizations and some aspects of the behaviour of the two Australian fur seals, *Arctocephalus* spp., on the evolution and present taxonomy of the genus. *Australian Journal of Zoology* **19**, 227–241.

Stirling, I., Calvert, W. & Cleator, H. (1983) Underwater vocalizations as a tool for studying the distribution and relative abundance of wintering pinnipeds in the high arctic. *Arctic* **36**, 262–274.

Stirling, I., Calvert, W. & Spencer, C. (1987) Evidence of stereotyped underwater vocalizations of male Atlantic walruses (*Odobenus rosmarus rosmarus*). *Canadian Journal of Zoology* **65**, 2311–2321.

Stoddard, P.K. (1998) Application of filters in bioacoustics. In: *Animal Acoustic Communication: Sound Analysis and Research Methods* (S.L. Hopp, M.J. Owren & C.S. Evans, eds), pp. 105–127. Springer-Verlag, Berlin.

Taruski, A.G. (1979) The whistle repertoire of the North Atlantic pilot whale (*Globicephala melaena*) and its relationship to behavior and environment. In: *Behavior of Marine Animals, Vol 3: Cetaceans* (H.E. Winn & B.L. Olla, eds), pp. 345–368. Plenum Press, New York.

Terhune, J.M. (1994) Geographical variation of harp seal underwater vocalizations. *Canadian Journal of Zoology* **72**, 892–897.

Terhune, J.M. & Ronald, K. (1973) Some hooded seal (*Cystophora cristata*) sounds in March. *Canadian Journal of Zoology* **51**, 319–321.

Terhune, J.M. & Ronald, K. (1986) Distant and near-range functions of harp seal underwater calls. *Canadian Journal of Zoology* **64**, 1065–1070.

Terhune, J.M., MacGowan, G., Underhill, L. & Ronald, K. (1987) Repetitive rates of harp seal underwater vocalizations. *Canadian Journal of Zoology* **65**, 2119–2120.

Terhune, J.M., Burton, H. & Green, K. (1994a) Weddell seal in-air call sequences made with closed mouths. *Polar Biology* **14**, 117–122.

Terhune, J.M., Grandmaitre, N.C., Burton, H.R. & Green, K. (1994b) Weddell seals lengthen many underwater calls in response to conspecific vocalizations. *Bioacoustics* **5**, 223–226.

Thomas, J.A. & Golladay, C.L. (1995) Geographic variation in leopard seal (*Hydrurga leptonyx*) underwater vocalizations. In: *Sensory Systems of Aquatic Mammals* (R.A. Kastelein, J.A. Thomas & P.E. Nachtigall, eds), pp. 201–221. De Spil, Woerden, the Netherlands.

Thomas, J.A. & Kuechle, V.B. (1982) Quantitative analysis of Weddell seal (*Leptonychotes weddelli*) underwater vocalizations at McMurdo Sound, Antarctica. *Journal of the Acoustic Society of America* **72**, 1730–1738.

Thomas, J.A. & Stirling, I. (1983) Geographic variation in the underwater vocalizations of Weddell seals (*Leptonychotes weddelli*) from Palmer Peninsula and McMurdo Sound, Antarctica. *Canadian Journal of Zoology* **61**, 2203–2212.

Thomas, J.A., Puddicombe, R.A., George, M. & Lewis, D. (1988) Variations in underwater vocalizations of Weddell seals (*Leptonychotes weddelli*) at the Vestfold Hills as a measure of breeding population discreteness. *Hydrobiologia* **165**, 279–284.

Thompson, P.O., Cummings, W.C. & Ha, S.J. (1986) Sounds, source levels, and associated behavior of humpback whales, southeast Alaska. *Journal of the Acoustic Society of America* **80**, 735–740.

Thompson, T.J., Winn, H.E. & Perkins, P.J. (1979) Mysticete sounds. In: *Behavior of Marine Animals, Vol. 3: Cetaceans* (H.E. Winn & B.L. Olla, eds), pp. 403–431. Plenum Press, New York.

Trillmich, F. (1981) Mutual mother–pup recognition in Galapagos fur seals and sea lions: cues used and functional significance. *Behaviour* **78**, 21–42.

Turl, C.W. & Penner, R.H. (1989) Differences in echolocation click patterns of the beluga (*Delphinapterus leucas*) and the bottlenosed dolphin (*Tursiops truncatus*). *Journal of the Acoustical Society of America* **86** (2), 497–502.

Tyack, P. (1986) Whistle repertoires of two bottlenosed dolphins, *Tursiops truncatus*: mimicry of signature whistles? *Behavioral Ecology and Sociobiology* **18**, 251–257.

Tyack, P. (1991) Use of a telemetry device to identify which dolphin produces a sound. In: *Dolphin Societies: Discoveries and Puzzles* (K. Pryor & K.S. Norris, eds.), pp. 319–344. University of California Press, Berkeley, CA.

Tyack, P.L. (1997) Studying how cetaceans use sound to explore their environment. *Perspectives in Ethology, Vol. 12. Communication* (D.H. Owings, M.D. Beecher & N.S. Thompson, eds), pp. 251–297. Plenum Press, New York.

Tyack, P.L. (1999) *Functional Aspects of Cetacean Communication*. In: *Cetacean Societies: Field Studies of Whales and Dolphins* (J. Mann, R. Connor, P.L. Tyack & H. Whitehead, eds), pp. 270–307, University of Chicago Press, Chicago, IL.

Tyack, P.L. & Clark, C.W. (2000) Communication and acoustic behavior of dolphins and whales. In: *Hearing in Whales and Dolphins* (A.N. Popper, R.H. Fay & W.W.L. Au, eds), pp. 156–224. Springer-Verlag, New York.

Tyack, P.L. & Sayigh, L.S. (1997) Vocal learning in cetaceans. In: *Social Influences on Vocal Development* (C.T. Snowdon & M. Hausberger, eds), pp. 208–233. Cambridge University Press, Cambridge, UK.

Tyack, P.L. & Whitehead, H. (1983) Male competition in large groups of wintering humpback whales. *Behaviour* **83**, 132–154.

Urick, R.J. (1983) *Principles of Underwater Sound*, 3rd edn. McGraw-Hill, New York.

Verboom, W.C. & Kastelein, R.A. (1995) Rutting whistles of a male Pacific walrus (*Odobenus rosmarus divergens*). In: *Sensory Systems of Aquatic Mammals* (R.A. Kastelein, J.A. Thomas & P.E. Nachtigall, eds), pp. 287–298. De Spil, Woerden, the Netherlands.

Wartzok, D., Schusterman, R.J. & Gailey-Phipps, J. (1984) Seal echolocation? *Nature* **308**, 753.

Watkins, W.A. (1967a) Air-borne sounds of the humpback whale, *Megaptera novaeangliae*. *Journal of Mammalogy* **48**, 573–578.

Watkins, W.A. (1967b) The harmonic interval: fact or artifact in spectral analysis of pulse trains. In: *Marine Bio-Acoustics* (W.N. Tavolga, ed.), Vol. 2, pp. 15–42. Pergamon Press, London.

Watkins, W.A. (1980) Acoustics and the behavior of sperm whales. In: *Animal Sonar Systems* (R.-G. Busnel & J.F. Fish, eds), pp. 283–290. Plenum Press, New York.

Watkins, W.A. (1981) Activities and underwater sounds of fin whales. *Scientific Reports of the Whales Research Institute* **33**, 83–117.

Watkins, W.A. & Schevill, W.E. (1976) Right whale feeding and baleen rattle. *Journal of Mammalogy* **57**, 58–66.

Watkins, W.A. & Schevill, W.E. (1979) Distinctive characteristics of underwater calls of the harp seal, *Phoca groenlandica*, during the breeding season. *Journal of the Acoustical Society of America* **66**, 983–988.

Watkins, W.A. & Wartzok, D. (1985) Sensory biophysics of marine mammals. *Marine Mammal Science* **1**, 219–260.

Watkins, W.A., Tyack, P., Moore, K.E. & Bird, J.E. (1987) The 20-Hz signals of finback whales (*Balaenoptera physalus*). *Journal of the Acoustical Society of America* **82**, 1901–1912.

Watkins, W.A., Daher, M.A., Fristrup, K.M., Howald, T.J. & di Sciara, G.N. (1993) Sperm whales tagged with transponders and tracked underwater by sonar. *Marine Mammal Science* **9**, 55–67.

Weilgart, L. & Whitehead, H. (1988) Distinctive vocalizations from mature male sperm whales (*Physeter macrocephalus*). *Canadian Journal of Zoology* **66**, 1931–1937.

Weilgart, L.S. & Whitehead, H. (1990) Vocalizations of the North Atlantic pilot whale (*Globicephala melas*) as related to behavioral contexts. *Behavioral Ecology and Sociobiology* **26**, 399–402.

Wemmer, C., Von Ebers, M. & Scow, K. (1976) An analysis of the chuffing vocalization in the polar bear (*Ursus maritimus*). *Journal of Zoology, London* **180**, 425–439.

Whitehead, H. & Weilgart, L. (1990) Click rates from sperm whales. *Journal of the Acoustical Society of America* **87**, 1798–1806.

Whitehead, H. & Weilgart, L. (1991) Patterns of visually observable behavior and vocalizations in groups of female sperm whales. *Behaviour* **118**, 275–296.

Wilden, I., Herzel, H., Peters, G. & Tembrock, G. (1998) Subharmonics, biphonation, and deterministic chaos in mammal vocalization. *Bioacoustics* **9**, 171–196.

Wilson, E.O. (1975) *Sociobiology*. Harvard University Press, Cambridge, MA.

Winn, H.E. & Schneider, J. (1977) Communication in sirenians [*sic*], sea otters, and pinnipeds. In: *How Animals Communicate* (T.A. Sebeok, ed.), pp. 809–840. Indiana University Press, Bloomington, IN.

Würsig, B. (1986) Delphinid foraging strategies. In: *Dolphin Cognition and Behavior: a Comparative Approach* (R.J. Schusterman, J.A. Thomas & F.G. Wood, eds), pp. 347–359. Lawrence Erlbaum Associates, Hillsdale, NJ.

CHAPTER 7

Patterns of Movement

Peter T. Stevick, Bernie J. McConnell and Philip S. Hammond

7.1 INTRODUCTION

Resources in the marine environment tend to be patchily distributed. Patchiness is simply a form of variability where the resources are clumped rather than distributed randomly or systematically, and is a feature of many, if not all, natural environments. But environmental variability is a function of scale, both temporal and spatial. Patches of good or poor habitat (for a given activity) may extend over large areas of ocean and last for months, or they may be confined to particular parts of estuaries or bays and change daily or weekly. The scale at which habitat suitable for one or more activities varies is a powerful driving force behind a species' movement patterns.

Another important feature of the environment influencing patterns of movement is its relative predictability. The variability of the oceans is itself dynamic; patches of good or poor habitat change over time and space. While increased productivity is predictably associated with continental shelf breaks, the location of increased productivity in the open oceans associated with mesoscale features (e.g. frontal systems) is often unpredictable and has to be actively sought. Movement can thus be viewed as a species' adaptation to a patchy, unpredictable environment over a range of temporal and spatial scales.

Thus an understanding of movement is integrally connected with an understanding of the reasons why animals move, and the major section of this chapter is devoted to the forces driving the patterns of movement we observe in marine mammals. These driving forces include the need to execute activities in certain types of habitat at specific times, and the need to optimize the use of resources in that habitat

to maximize fitness. They also include the need to move away from areas that are potentially detrimental to survival such as to avoid predation or disease, or to avoid physically harsh conditions.

Animals move only as much as they must to maximize their access to resources. However, resource requirements will vary from one activity to another. For species that occupy one broad habitat type that is suitable for all activities, the benefits of well-developed movement abilities may be slight. But for some species, habitat that is ideal for breeding may be poor for feeding, or vice versa. For example, species that have evolved to separate breeding from feeding in time (i.e. some of the baleen whales and phocid seals) have also evolved the ability to navigate over appropriate distances and, in some cases, to withstand the physical and energetic demands of movement over great distances. This long-range, directed travel is an important pattern of movement in marine mammals and one section of this chapter is specifically on such migration.

Although attention is often focused on long-distance travel in the description and analysis of marine mammal movements, it is important to recognize that the majority of marine mammal species do not make such large-scale movements. Movements on smaller scales are often related to feeding or breeding activity which are treated in more detail in Chapters 8 and 10. Similarly, we do not discuss diving, which is covered in Chapters 3 and 9.

This chapter begins with an overview of the methods used for the study of marine mammal movements. The concluding sections consider the role of exploration by animals, describe the cues potentially used by marine mammals for finding their way in the marine environment, and potential human influence on movement patterns.

We have not attempted to catalogue the vast and rapidly expanding literature on marine mammal movement. Instead, we have tried to cover key aspects and to select examples that best illustrate them. Inevitably, a number of well-studied species (humpback whale, gray whale, grey seal and elephant seal) contribute disproportionately to the examples.

7.2 SOURCES OF INFORMATION

We generally only get glimpses into the lives of marine mammals. They spend much of their time beneath the surface of the sea and many inhabit pelagic or polar regions where direct observation over extended periods is difficult or impossible. Developments in methods in recent years, particularly in telemetry, have greatly increased our knowledge of the movements of individual animals. The use of natural markings for recognizing individual animals has also led to significant advances. But the number of species for which these methods of study have been possible is limited. Movements can often only be inferred indirectly from observations of distribution patterns of groups of animals.

7.2.1 Movements inferred from distribution patterns

From the earliest days, whalers and sealers, like all hunters, have recorded the location and date of captures to take advantage of repeatable patterns of occurrence and thereby maximize their hunting efficiency. Analysis of whaling records, in particular, underpinned much of our early knowledge of the movement patterns of cetaceans (e.g. Harrison Matthews 1938; Mackintosh 1942; Tomilin 1957).

Information on marine mammal (mostly cetacean) movements has often been inferred from simple observation of animals (alive or dead) in certain places at certain times. For example, pygmy right whale stranding records from South Africa, Australia and New Zealand have been used to infer seasonal movements (Baker 1985). Mitchell (1991) reviewed winter stranding and sighting records of minke whales in the North Atlantic to investigate seasonal migration.

The use of survey ships (and aircraft) to estimate the abundance of cetaceans has grown substantially from the late 1970s (International Whaling Commission 1982) to the present day and can yield information about movements. For example, surveys for minke whales in the eastern North Atlantic in 1989 and 1995 showed quite different distributions over a large scale (Øien 1991; Schweder et al. 1997), implying substantial variation in movements to feeding areas from year to year. Details of sighting survey methodology can be found in Buckland et al. (1993a).

A useful source of information comes from data collected by observers on ships that are at sea for a purpose other than studying cetaceans, so-called platforms of opportunity. These data can be extensive and can result in good temporal and spatial coverage, thus providing information about movements on large scales (e.g. Northridge et al. 1995).

7.2.1.1 Passive acoustics

Many species of marine mammals have recognizable vocalizations, and recordings of these can be used as a means of locating or tracking the distribution and movements of animals (Whitehead & Gordon 1986; Clapham & Mattila 1990; Moore et al. 1998). While normally this method only provides information on species distribution, individual animals may also be tracked acoustically. A single blue whale was tracked over 3000 km for 43 days across the mid Atlantic by determining the position of its vocalizations (Clark 1995).

7.2.1.2 Population characteristics

Differences in population characteristics between individuals from different areas have been used to infer the level of movement or isolation between these areas, though the spatial resolution of these methods is generally low. Characteristics used include parasite infestations (Balbuena et al. 1995; Araki et al. 1997), morphological features and irregularities (Veinger 1980; Lockyer 1995), genetic markers (Árnason 1995; Palsbøll et al. 1995), contaminant burdens (Aguilar 1987), vocalizations (Winn et al. 1981; Weilgart & Whitehead 1997) (see Chapter 6), coloration (Yochem et al. 1990;

Allen *et al.* 1994), and diet assessed through stomach contents (Sekiguchi *et al.* 1996), stable isotope ratios (Best & Schell 1996; Hobson & Schell 1998) (see Chapter 8) and fatty acid analysis (Iverson *et al.* 1997).

7.2.2 Movements of individuals

Old harpoons found embedded in animals during whaling operations have given some clues to their prior movements (Ingebrigtsen 1929; Tomilin 1957; Martin 1982), and were the inspiration for the most extensively used artificial tag for cetaceans; the Discovery tag (Brown 1978). This uniquely numbered tag was fired into the whale, penetrating the skin and blubber, and was recovered on flensing. Between 1932 and 1984, over 23 000 whales of at least 11 species were marked with Discovery tags (Brown 1977; Lockyer 1979; Kato *et al.* 1993).

Discovery marking provided information on only two locations for any individual animal, the second at time of death. More information is obtained from multiple re-sightings of live individuals. Tags for this purpose have included a modification of the Discovery tag with a colour-coded streamer, spaghetti tags similar to those used for marking fish, and fin and flipper tags (Brown 1978; Hobbs 1988; Testa & Rothery 1992; Jeffries *et al.* 1993). Branding has been used successfully to mark pinnipeds and cetaceans (Hobbs 1988; Pomeroy *et al.* 1994). Pinnipeds can be marked temporarily by dye marking or gluing marks to the pelage that are shed at the following moult (Hall *et al.* 2000). Branding and attaching tags normally requires capture and handling of the animals and is therefore not appropriate for the larger cetaceans.

7.2.2.1 *Identification of individuals by natural markings*

In most species, individual animals are distinguishable from one another by variations in patterns of natural markings. This characteristic can be used as a natural tag. The earliest reported use of this technique in the study of marine mammals was by Moore (1956), who used sketches of body scarring to identify individual manatees. Schevill and Backus

(1960) followed the movements of a single humpback whale over 10 days in 1958, recognizing it in the field by dorsal fin shape, body markings and fluke pattern. In the 1970s, interest in non-lethal methods to study marine mammals accelerated and the use of natural markings (especially for studies of cetaceans) developed rapidly, with pioneering work on orcas, humpback and right whales (see Hammond *et al.* 1990). Studies using individual identification have substantially expanded our understanding of movement patterns. For example, in humpback whales, this technique has demonstrated a transit of over 8300 km, the longest mammalian migration on record (Stone *et al.* 1990), site fidelity to feeding grounds (Katona & Beard 1990; Clapham *et al.* 1993), and changes in feeding distribution in response to prey shifts (Weinrich *et al.* 1997).

The markings used for identification vary widely from species to species (Fig. 7.1). The entire body is mottled in blue and gray whales and in several species of seal (e.g. female grey, harbour and ringed seals). There is distinctive pigmentation on the ventral flukes of humpback whales, the chevron patterns of fin whales, and the saddle markings of the orca. Serrations and notches in the flukes are the primary means of identification in sperm whales. In the right whales, patches of roughened skin on the head (callosities) provide attachment sites for cyamids, which contrast with the skin pigment providing a means of identification. Nicks, notches and other irregularities in dorsal fins are critical to the identification of some less variably pigmented species (especially dolphins), as are acquired scars in some species (e.g. bottlenose dolphins, Risso's dolphins, sei whales, pilot whales and manatees).

As the number of identified individuals grows, the task of manually comparing each new individual with all previously identified individuals becomes increasingly labour-intensive. As a result, there is considerable interest in the use of computer technology for comparison of images (for a variety of approaches see Hiby & Lovell 1990; Mizroch *et al.* 1990; Whitehead 1990; Huele & deHaes 1998) (Fig. 7.2). Pattern recognition on a three-dimensional object which may be photographed from different angles and under different lighting conditions is a complex task, however, and working systems are available for only a few species.

Fig. 7.1 Natural markings are widely used to recognize individual animals for studies of movement patterns. (a) Humpback whales have distinctively pigmented areas on the ventral side of the flukes, often complemented by acquired scars. The scars on the upper right of this individual were caused by the teeth of an orca. The serrations in the trailing edge of the fluke are also useful in identification. (b) In bottlenose dolphins the shape of the dorsal fin, along with any nicks or notches in its edge, combined with body scarring, are used for distinguishing individuals. Skin lesions can also be useful over shorter time periods. (c) In right whales, areas of roughened skin on the head, called callosities, provide means of identification. (d) In some pinnipeds, such as the grey seal, the pelage is strongly patterned and stable over time. Grey seal markings are sexually dimorphic and only females can reliably be identified. (Courtesy of K. Grellier, R. Harris and P. Stevick, Sea Mammal Research Unit and Aberdeen University.)

7.2.2.2 Genetic tagging

Following recognition of the potential for using DNA extracted from tissue samples for individual identification (Amos & Hoelzel 1990), recent advances in molecular genetics have now made this a reality. Identification from nuclear microsatellite loci has been used successfully to determine individual movements of humpback whales within feeding and breeding grounds and between winter and summer habitats (Palsbøll *et al.* 1997). Sex can be determined from the same tissue sample, so gender differences in movements can also be investigated in species where sex discrimination in the field is difficult. Sampling is invasive but the skin samples required are very small and little short-term and no

(a)

(b)

Fig. 7.2 Computer-assisted matching of grey seals after Hiby and Lovell (1990). A three-dimensional 'cage' is manipulated onto the photograph of the seal's head (a) so that a standard section of the fur can be identified (b), extracted, placed in standard orientation (inset) and stored digitally for later comparison.

long-term reactions have been reported from biopsy collection (Brown *et al*. 1991; Weinrich *et al*. 1991; Gauthier & Sears 1999) (see Chapter 11).

7.2.2.3 Telemetry

Over the last 10 years, telemetry devices have provided exciting new insights into the movements and behaviour of individual marine mammals at sea. Such studies are technologically and analytically demanding (White & Garrott 1990). The devices must be small, light and robust, with minimal hydrodynamic drag and the hardware, software and transmission medium must be designed to minimize energy use and thus battery size. Attachment is usually through gluing to the fur in pinnipeds (Fig. 7.3) (Fedak *et al*. 1983); tusks have also been used as attachment sites in walruses (Wiig *et al*. 1996). Tags have been tethered to manatees by means of a strap around the peduncle (Deutsch *et al*. 1998) and applied to polar bears using neck collars (Messier *et al*. 1992). For cetaceans, attachments include pins through the dorsal fin or dorsal ridge (Martin *et al*. 1993; Read & Westgate 1997), suction cups (Stone *et al*. 1994; Hooker & Baird 1999) and subdermal devices (Watkins 1981; Mate *et al*. 1997).

Location may be determined either on-board the device itself, or externally. External location fixes use either an acoustic or radio signal from the device. Such signals may also contain individual

(a)

(b)

Fig. 7.3 Telemetry devices attached to (a) a beluga and (b) a grey seal. (Courtesy of SMRU.)

animal codes and behavioural data from on-board sensors, such as pressure, temperature, swim speed and heart rate (Thompson & Fedak 1993). Geolocation data recorders monitor the time of sunrise and sunset and these are later used to compute location (Delong 1992), the accuracy of which may be improved by temperature data (Boyd & Arnbom 1991; Hindell *et al.* 1991).

Radio signals are attenuated under water and thus tracking is limited to those times when the animal is at the surface. Location may be obtained using handheld directional aerials or automatic receiving stations within a limited line of sight range (usually < 30 km) (Thompson *et al.* 1991; Nicholas *et al.* 1992). This restriction is overcome in the Argos satellite system (Argos 1989) which provides a global coverage capable of tracking many individuals concurrently. The Argos system consists of UHF receivers on-board polar-orbiting satellites which pick up signals (uplinks) from transmitters within their field of view. Data may be encoded within an uplink, which is relayed via ground stations and a computer network to the user. Computer visualization techniques can allow the tracks and behaviour of animals to be replayed within the ecological context of underwater terrain maps and remotely sensed data (Fedak *et al.* 1996).

7.3 FACTORS DETERMINING PATTERNS OF MOVEMENT

The movement patterns we observe have evolved in response to the needs of animals to survive and reproduce in the environment in which they live. Thus we examine movements in relation to the factors which lead animals to move. We often know little about the activities of the animals being observed, and many activities will overlap in time and space. Simple explanations for movements of individuals may often, therefore, be difficult to find.

7.3.1 Foraging

Obtaining an adequate supply food is one of the most basic determinants of the survival and success of an individual (see Chapter 8). Foraging movements are driven by the spatial and temporal distribution of food and its predictability.

Marine mammals are often associated with physical forms such as banks, canyons or the edge of the continental shelf which may concentrate prey (Kenney & Winn 1986; McConnell & Fedak 1996; Hooker *et al.* 1999; Campagna *et al.* 2000). They may also be associated with transient oceanographic

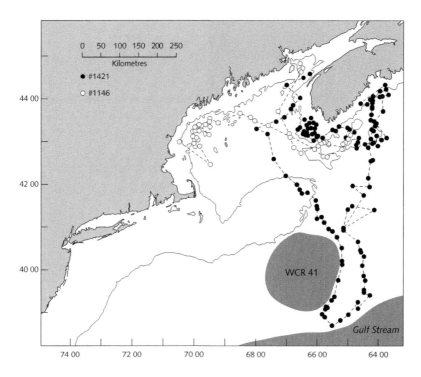

Fig. 7.4 Tracks of northern right whales monitored by satellite telemetry in the western North Atlantic. The path of whale No. 1421 follows the cold water intrusion along the leading edge of a warm core ring. Such areas may serve to concentrate copepods, which are the primary prey for right whales in the region. This track illustrates the ability of these animals to locate and orientate to such transitory oceanographic phenomenon in deep water, far from what is traditionally thought of as the habitat for this species. (From Mate *et al.* 1997, courtesy of The Wildlife Society.)

features such as mesoscale frontal systems or larger scale features such as the Antarctic polar front (Hindell *et al.* 1991). In spite of the very different foraging habits of the two species, both sperm whales and right whales have been shown to be associated with the boundaries of warm core rings, eddies separated from warm water currents (Mate *et al.* 1997; Griffin 1999). Figure 7.4 shows the movement of a right whale along the leading edge of a warm core ring. The cold water entrainment along the edge of these rings is associated with higher levels of productivity than surrounding regions. Such associations indicate the ability to recognize and follow these transient features.

That some non-transient, physical features may predictably lead to concentrations of prey, means that return to specific sites where prey has been abundant in the past may be a useful foraging

strategy. Individually identified humpback and fin whales in the North Atlantic are consistently re-sighted in successive years close to the feeding area in which they were first identified (Agler *et al.* 1990; Clapham *et al.* 1993). In Great Britain, grey seal movements consist mainly of repeated short trips to localized foraging areas (McConnell *et al.* 1999) (Fig. 7.5), while polar bears off Greenland regularly return to or remain in specific fjords (Born *et al.* 1997).

While foraging, the movements of marine mammals are determined by the size and density of prey patches, and the area over which patches may be predictably located (Mayo & Marx 1990; Whitehead 1996; Jaquet & Whitehead 1999). When in patches of high prey density, animals move so as to increase their chances of staying within the patch. They turn more frequently and at greater

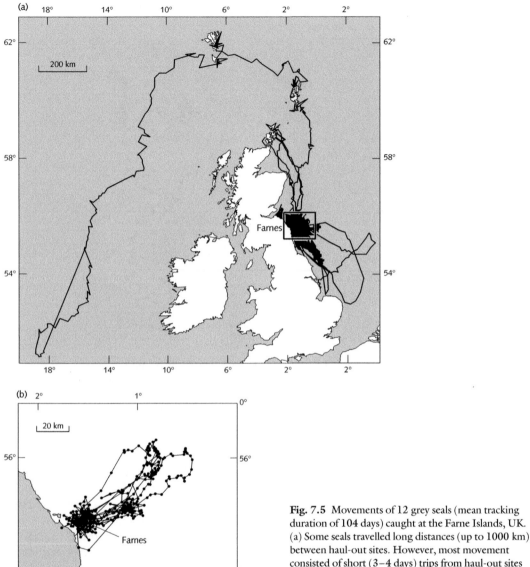

Fig. 7.5 Movements of 12 grey seals (mean tracking duration of 104 days) caught at the Farne Islands, UK. (a) Some seals travelled long distances (up to 1000 km) between haul-out sites. However, most movement consisted of short (3–4 days) trips from haul-out sites to local foraging sites and an example, an adult male, is shown in (b). (Adapted from McConnell *et al.* 1999.)

angles. Upon leaving a high-density patch they move so as to maximize their chances of encountering another patch. This may entail travelling in a relatively straight path or a broad curve. Figure 7.6 illustrates the movements of right and sperm whales during foraging. Note the similarity in patterns of movement and the relationship between movement and foraging success, even at enormously different spatial scales; compare these with the tracks of

elephant seals in the open ocean shown in Fig. 7.7. Although there is no information on foraging success through these tracks, the seals may also be searching for suitable patches of prey (primarily squid).

Within a species, preference for specific prey may also influence the movement patterns of individuals or groups. As an extreme example, sympatric groups of orca have been shown to display differences in seasonal occurrence, habitat use patterns and

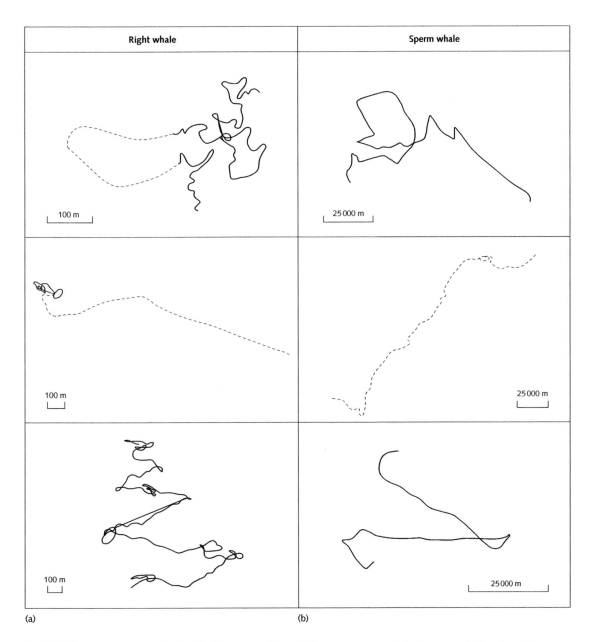

Right whale	Sperm whale

100 m

25 000 m

100 m

25 000 m

100 m

25 000 m

(a)   (b)

Fig. 7.6 Movement patterns related to feeding success. The solid lines represent tracks during successful foraging. Movements in these areas are characterized by frequent changes in direction. The dotted lines represent tracks in areas of lower food density and unsuccessful foraging. In these areas direction changes are fewer and less dramatic. Note the similarity in pattern at highly different spatial scales. (a) Northern right whales. (Adapted from Mayo & Marx 1990.) (b) Sperm whales in the tropical Pacific. (Adapted from Jaquet & Whitehead 1999.)

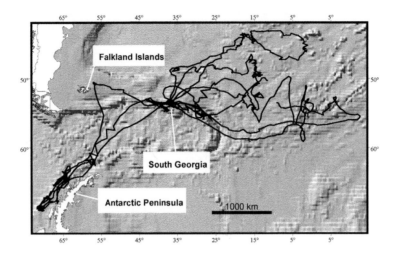

Fig. 7.7 The movements of 12 southern elephant seals tracked from South Georgia in the South Atlantic for an average of 119 days (McConnell & Fedak 1996). Initially the nine females moved rapidly and in a directed manner from South Georgia. Then, after about 15 days they slowed down. Five females travelled into the deep open waters of the Southern Ocean, up to 3000 km from South Georgia. However, four travelled 1500 km to the continental shelf of the Antarctic peninsula. One of the latter group then proceeded to the Falkland Islands where it spent periods in localized areas on the continental shelf with little or no travel before returning to South Georgia. This seal returned to the Antarctic peninsula the following year. While the females in these studies travelled far, the two adult males in the later study remained on or near the continental shelf around South Georgia. (Adapted from McConnell & Fedak 1996.)

propensity for long-distance movement related to whether they specialize on fish or marine mammal prey (Dalheim & Heyning 1999).

7.3.1.1 Influence of size, sex and reproductive demands

A number of studies have shown that foraging movements are influenced by body size, sex and the demands of reproduction. Foraging range is positively correlated with body size in harbour seals and is significantly greater in males (average 26 km) than females (15 km) (Thompson *et al.* 1998b). Female otariids may nurse their pups for many months and they are constrained to be central place foragers during their prolonged lactation periods (see Chapter 8). Boyd *et al.* (1998) showed that nursing female Antarctic fur seals foraged within 350 km of the breeding site at South Georgia, whereas two of three tagged males travelled towards Signy Island about 900 km south. Thompson *et al.* (1998a), however, have shown that nursing southern sea lions do not always return directly to their pups after foraging trips, but may haul out elsewhere first.

There are also strong sexual differences in migratory movements to foraging areas in sperm whales, belugas and elephant seals, as described in Section 7.4.4.

7.3.1.2 Temporal variation in foraging movements

Foraging movements may vary temporally in response to availability of prey, which may vary in cycles ranging from tidal to seasonal. These changes may be of regular occurrence or irregular and unpredictable in nature.

Tidal and diurnal movements

Prey may be more active, concentrated or accessible during certain times of day or at certain points in the tidal cycle. Tucuxi (a river dolphin) move from open-water areas of ocean or deep river channels to bays or lakes in early morning, returning in late afternoon (Da Silva & Best 1994). The reverse is true of spinner dolphins in several regions (Norris 1991; Perrin & Gilpatric 1994) and Hector's dolphin in New Zealand (Stone *et al.* 1995) which move inshore in the morning and offshore in the afternoon. Such diurnal movements are likely to be the result of vertical migration of prey (Gannier 1999). As prey move to the surface in the evening they become available to mammalian predators. As they return to depths beyond the diving capabilities of the animal, or simply to regions where foraging becomes more energetically expensive, individuals may choose to move to areas where they can rest, haul out, socialize or reduce predation risk.

While less common, foraging movements associated with the tide are also reported. Humpback dolphins move inshore on the rising tide to feed, occasionally moving up into mangrove channels, and then move offshore with the falling tide (Ross *et al.* 1994). Tidal changes may simply reflect the use of habitat which is available only at high water, or it may reflect food concentrations along tidal fronts. Tidal patterns of movement in seals may be related to the tidal availability of haul-out sites. However the pattern of daytime foraging in harbour seals (Thompson *et al.* 1991) and grey seals (Sjöberg *et al.* 1999) may be influenced by the shoaling behaviour of clupeoid fish.

Seasonal movements

Cycles of productivity follow predictable seasonal trends. Thus we might expect marine mammals to follow these seasonally available resources. Latitudinal shifts in distribution are common. Harbour seals, for example, move from the coast of Nova Scotia and Maine in the summer to Massachusetts and New Hampshire in winter (Rosenfeld *et al.* 1988; Payne & Selzer 1989). Seasonal movements between shallow inshore water and deeper offshore water are also reported, for example in the Hector's dolphin (Dawson 1991). Along the British coast, offshore movement of common dolphins during autumn is correlated with regularly occurring seasonal changes in oceanographic systems (Goold 1998).

However, a resource that is predictable for one or two seasons, or longer, may not always be so. In response, areas of high marine mammal density may shift in an unpredictable manner. Changes in the distribution of humpback, fin, right and sei whales off the northeastern US coast were associated with a relative abundance of copepods, herring and sand eels in the region (Payne *et al.* 1986, 1990; Weinrich

et al. 1997), while areas utilized by harbour seals in the Moray Firth, Scotland were related to changes in the relative abundance of sand eels and sprats (Tollit *et al.* 1997).

Exceptional oceanographic conditions in the Barents Sea caused the capelin stock to collapse in 1987. With no alternative prey available, starving harp seals then moved south down the coast of Norway (Nilssen *et al.* 1998). This invasion may have triggered the transfer of phocine distemper virus to the vulnerable European harbour seal population (Goodhart 1988) which resulted in the deaths of 18 000 harbour seals in 1988.

7.3.2 Breeding

Movement patterns related to breeding vary considerably within marine mammals, depending upon the social system of the species (see Chapters 10 and 12). Many marine mammals form aggregations during the breeding season, often travelling long distances to do so. This has obvious advantages for locating mates. There are also advantages in numbers in terms of protection for very young pups or calves, and there is a strong seasonality in the reproductive cycle in most marine mammals so that giving birth occurs shortly before mating. Hence many species of seal breed colonially and there are specific breeding areas for some baleen whales.

While cetaceans and sirenians give birth in suitable areas of water, breeding areas for other marine mammals must be on land or ice. Beyond these obvious constraints, the factors which lead animals to use specific breeding habitats are often poorly understood. For example, it has been widely proposed that large whales must travel to tropical or subtropical waters for calving to reduce the thermal stress on newborn calves, although there is little evidence to support this. Several species of small cetaceans, in which sustaining body temperatures should be more difficult than in larger whales (see Chapter 9), regularly calve in polar or subpolar waters. Winter sightings of orca with small calves well within the Antarctic ice, and humpback whales with newborn calves at high latitudes, suggest that calves can survive even in extreme temperature conditions (Williamson 1961; Gill & Thiele 1997).

There is considerable variability in the degree to which feeding and breeding areas are separated from one another. For pinnipeds, the necessity to give birth and raise a pup on land or ice may require them to travel to sites far removed from areas where they forage during the rest of the year (e.g. elephant seals), and many phocid seals do not feed during the pupping season. Such movements are equivalent to the migrations of baleen whales which calve (and do not feed) in low-latitude waters and forage in high-latitude feeding areas. On the other hand, some pinniped breeding colonies or rookeries may be adjacent to foraging areas. Indeed, for otariids, in which the pups are ashore for extended periods and the nursing females must continue to feed, such proximity is essential. This strategy is equivalent to that characterized by pelagic dolphins, in which reproduction may occur year-round and therefore does not take place in specific areas removed from foraging areas. Even in baleen whales, feeding and breeding habitats may not be mutually exclusive. Coastal populations of sei whales feed inshore throughout the year and demonstrate little periodicity in breeding (Best 1996).

7.3.2.1 Fidelity to breeding areas

Fidelity to breeding areas is common among animals and is thought to confer considerable reproductive advantages, though the pattern and extent of fidelity varies depending on the reproductive system of the species and the sex of the individual (Greenwood 1980). Most species of cetaceans do not have specific breeding areas or, if they do, we do not know where they are. Humpback, gray and southern right whales, however, aggregate for breeding and calving and most individuals seem to return to the same breeding area year after year (Chittleborough 1965; Swartz 1986; Perry et al. 1990) (see Chapter 11 for a discussion of the genetic implications of this fidelity). But such fidelity is not absolute; a small number of humpback whales have also been identified in different breeding grounds (Perry et al. 1990; Salden et al. 1999). The lack of evidence from the historic whaling records for humpback whales near the Hawaiian Islands in winter even suggests recent colonization of a new breeding area (Herman 1979).

Site fidelity on a finer scale is seen in many pinnipeds. Individual Weddell seals (Croxall & Hiby 1983; Testa 1987), northern and southern elephant seals (Hindell & Little 1988; Le Boeuf & Reiter 1988) and Antarctic fur seals (Lunn & Boyd 1991) have been shown to return to the same breeding site within a colony. Some of the best evidence for pinniped breeding site fidelity comes from studies of grey seals. Boness and James (1979) found that 68% of female grey seals pupped within 50 m of their previous pupping site on Sable Island, Canada and Pomeroy et al. (1994) found that 75% of females pupped within 83 m of previous pupping sites on the island of North Rona in the British Isles.

There is little to suggest fidelity to sites within a breeding area for cetaceans, however. In humpback whales, a progression through the breeding area during a season has been suggested (Baker & Herman 1981; Mattila & Clapham 1989), but movement of individuals in the opposite direction has also been shown (Cerchio et al. 1998; Mate et al. 1998). In gray whales, there is considerable movement between lagoons, particularly by males. Mothers with newborn calves often remain in a limited area of a single lagoon but movement between lagoons by mothers with calves late in the season is common (Swartz 1986).

7.3.3 Moult

Phocid seals undergo an annual moult. During this period the growth of new hair requires increased supplies of blood to the skin. If animals were to remain in the water, this could result in excessive heat loss (Boily 1995) and seals generally spend more time ashore when moulting. The drive to haul out during moult is exemplified by the elephant seal; both northern and southern species return to moult on land as much as 1000 km away from foraging areas. In species where foraging areas are close to terrestrial moulting sites (for example the grey seal and harbour seal), movements related to moulting may be no different from those related to hauling out during the rest of the year (see below).

Cetaceans generally shed skin continuously, and movements associated with moulting are uncommon. Beluga whales, however, migrate into the high Arctic during summer where they congregate at large, freshwater river mouths. They undergo a major skin moult at this time, though the specific features of this habitat which contribute to the moult are unknown (Smith & Martin 1994).

7.3.4 Pinnipeds—the need to haul out

In addition to time spent involved in reproduction and moulting, pinnipeds spend a significant amount of time hauling out on land or ice. For example, McConnell *et al.* (1999) found that grey seals outside the moult/breeding period spent more than 40% of their time close to or on a haul-out site, 50 km or more from the main foraging areas. The patterns of movement that emerge are species-specific and are likely to be influenced by sex, size and condition (Thompson *et al.* 1998b).

The benefits of this behaviour that outweigh the costs of travel and reduced foraging time are not clear. Possible reasons include thermal regulation, predator avoidance, rest (reduced metabolic rate), social interaction and parasite reduction (see Watts 1996). Tide, time of day, weather, availability of ice and disturbance may modify an animal's behavioural response to these factors (Schneider & Payne 1983; Pauli & Terhune 1987; Bengtson & Stewart 1992; Grellier *et al.* 1996; Born *et al.* 1999). Brasseur *et al.* (1996) showed that captive harbour seals temporarily deprived of available haul-out space, compensated by hauling out for longer periods than usual. This suggests some physiological basis to the need to haul out.

There is a high degree of variability in haul-out patterns among pinniped species. Harbour seals in the Moray Firth, Scotland, forage within a 60 km radius of haul-out sites, with trips generally lasting 1–3 days (Thompson *et al.* 1996, 1998b). Grey seal movements off Britain and in the Baltic Sea consist mainly of short (2–5 day) trips from a haul-out site to foraging areas, usually less than 50 km away (Fig. 7.5) (Sjöberg *et al.* 1995; McConnell *et al.* 1999). Hooded seals spend much longer at sea, travelling as far as 1000 km from their moult and breeding sites on the sea ice off eastern Greenland to open ocean foraging areas in the Norwegian Sea and Northeast Atlantic on trips with a mean duration of 47 days (Folkow *et al.* 1996).

Elephant seals do not haul out at all (or at least very little) during their foraging phases. They spend many consecutive months travelling thousands of kilometres through the open ocean (Fig. 7.7) (Stewart & Delong 1995; McConnell & Fedak 1996). Although elephant seals do not appear to haul out on land or ice during their foraging phases, the physiological needs satisfied by hauling out in other species may be partially satisfied by occasional extended surface intervals (Le Boeuf *et al.* 1988; Boyd & Arnbom 1991; Hindell *et al.* 1991).

7.3.5 Avoidance

Movement is not always toward a region or resource, it may be driven by the desire or need to leave a region where environmental conditions are hostile.

7.3.5.1 *Predators and disease*

Little is known about the influence of predation or disease on movement patterns, but there is circumstantial evidence for a selective advantage for long-distance movement in avoiding disease. It has been suggested that the long period of fasting during migration and in the breeding lagoons leads to a lower internal parasite load in gray whales than in other mysticetes (Wolman 1985). This is supported by the heavy parasite burden found in the apparently non-migratory population of humpback whales in the northern Indian Ocean compared to conspecifics which follow typical migration patterns (Mikhalev 1997).

Burns *et al.* (1999) suggested that juvenile Weddell seals disperse from natal areas in the Ross Sea along coastal fast ice to avoid killer whale predation as well as allowing foraging in the shallower waters. Predation by killer whales on young calves has also been proposed as a major determinant of baleen whale migration (Corkeron & Connor 1999).

7.3.5.2 *Thermoregulation*

Although marine mammals are well adapted to cope with thermal stress (see Chapter 9), they are best adapted to specific temperature regimes, and may be forced to vacate otherwise suitable habitat if temperatures fall outside the ranges to which they are adapted (see also Section 7.3.3). For example, manatees range over several hundred kilometres through marine, brackish and fresh water during the course of a year but retreat to warm freshwater springs or thermal industrial outfalls during periods of low temperature. The use of these warm-water sites for energy conservation may be crucial to survival, at least in exceedingly cold years, and is independent of the presence of food or other resources at these sites (Garrott et al. 1994).

7.3.5.3 Ice

The productivity of polar waters, especially at the ice edge, leads to large concentrations of marine mammals in summer months. For cetaceans, which cannot haul out, this carries the danger of being trapped when the ice extends with approaching winter unless they move away. Individuals that linger near the ice late in the season, or wander too far into leads may find themselves trapped in closing leads or between the ice and land, unable to breathe. Such ice entrapments have been observed in many species of both mysticetes and odontocetes, and are routine enough in the beluga whale to be utilized by hunters in Greenland (Beamish 1979; Siegstad & Heide-Jørgensen 1994).

Ice cover is not an absolute impediment, however. Some species have an impressive ability to move through areas of apparently unbroken ice, allowing them to exploit resources that would otherwise be inaccessible. Several species including minke whales, Arnoux's beaked whales and orca, along with crabeater seals and leopard seals have been observed in small open-water areas well into the Antarctic ice pack, in some cases over 400 km from the ice edge (Gill & Thiele 1997; Thiele & Gill 1999). Bowhead whales are known to travel extensively under ice, surfacing beneath the ice and cracking it with the head, making raised areas in which to breathe (George et al. 1989). Other species have been observed to utilize small access holes, some so covered in ice as to be effectively invisible from the surface (Hobson & Martin 1996). Beluga whales have been tracked for extended periods under apparently solid ice (A.R. Martin, personal communication). These observations suggest either that areas of open water are a sufficiently regular

occurrence that animals swimming beneath the ice can count on finding a breathing hole before they deplete their oxygen reserves, or that they have an ability to locate open water at considerable distance.

7.3.5.4 Water levels

Species inhabiting shallow, coastal or riverine environments may utilize areas that are only accessible during part of the season. For example, Amazonian tucuxi spend much of the rainy season in lakes. They must move into the main river channels before the onset of the dry season, however, or risk being isolated in water bodies which may become stagnant, shallow, depleted of food or even dry up entirely before the water level rises again (Da Silva & Best 1994).

7.4 MIGRATION

The large-scale migration of many of the baleen whales and some seals is the most dramatic pattern of marine mammal movement. Gray whales are a regular sight along much of the west coast of North America as they travel from the Arctic to the tropics, while harp seals travel in their tens of thousands during the spring from their southern breeding grounds to their northern Arctic feeding grounds.

The term migration has been used almost synonymously with movement, describing movement patterns as diverse as passage through a feeding area and anomalous long-distance wandering. Migration is used here to describe large-scale directed movements characterized by a regular, generally seasonal, return, as exemplified by humpback and gray whales and harp and elephant seals.

Migratory travel may be through areas of few or poor resources, but fasting during migration is apparently restricted to the mysticetes. Even in this group, while humpback whales have seldom been observed feeding during migration, they will when food is available (Best et al. 1995), and gray whales rarely feed on the southbound migration, but do so more regularly when returning northbound (Jones et al. 1984). Physiological and anatomical constraints and condition factors (primarily size) affect an animal's ability to endure long-distance movements over barren ground. Baleen whales, especially, are able to travel great distances without feeding because of their ability to store large quantities of energy as blubber (Lockyer 1981) (see Chapter 9). Similarly, the large blubber store in elephant seals enables them to travel many thousands of kilometres without the requirement to feed frequently or regularly.

Migration is energetically expensive and may be dangerous. Why do animals undertake such demanding journeys? The separation of food resources from appropriate breeding habitats is clearly critical in most cases. However, the specific factors leading to the choice of such segregated locations are rarely known, and it is often unclear why other regions in closer proximity are not equally appropriate. Some of the factors outlined above such as predation pressure and thermoregulation may, therefore, be critical to understanding migration patterns. It has been suggested that current migrations reflect historic patterns of prey distribution, the utility of which has been lost due to continental drift and changes in global climate (Lipps & Mitchell 1976). However, the undertaking of energetically expensive long-distance migration would rapidly be selected against unless it conveyed some current advantage.

7.4.1 Migratory routes

Precise migration routes are rarely known, though telemetry is providing new insights into this. Cetacean migration between winter and summer grounds has normally been supposed to occur between areas which are closest in longitude, leading to nearly north–south migrations (Mackintosh 1942; Tomilin 1957). This has largely been supported by Discovery tagging and individual identification, even in cases where other regions of similar habitat were closer (Chittleborough 1965; Baker et al. 1986; Buckland & Duff 1989; Kaufman et al. 1990). Migration can, however, encompass a substantial longitudinal component in humpback whales (Dawbin 1964; Darling et al. 1996; Stevick et al. 1999) and minke whales (Clark 1995). Telemetry has shown remarkably direct movement between migratory end-points, indicating very little deviation from the direct route used during transits (Brillinger & Stewart 1998; Mate et al. 1998).

Humpback whales appear to follow the migration route that they learned during their first year throughout their lives (Fig. 7.8). Feeding area philopatry in humpback whales is very strong; up to 90% annual return has been reported for the Gulf of Maine (Clapham et al. 1993). The origin of this postweaning site fidelity appears to be through maternal association (Clapham et al. 1993; Weinrich 1998). This implies that the feeding destination and presumably the route to it are learned by the calf in early life.

Unlike birds (Drury & Nisbet 1964; Aidley 1981) and fish (Arnold 1981), there is little evidence that whales utilize current systems to aid in migration. Townsend (1935) and Dawbin (1966) were unable to establish relationships between movements of whales and current systems. North Atlantic humpback whales fail to take advantage of the Gulf Stream in their northbound transit, favouring instead a more direct course for the feeding aggregations off Canada, Greenland, Iceland and Norway, even though the Gulf Stream could add substantially to travel speed (Clapham & Mattila 1990). However, the routes followed by minke whales through the West Indies in winter follow the North Atlantic gyre, westward with the Equatorial Current, then north with the Gulf Stream (Clark 1995).

Even for species which do not normally associate in groups, remaining near conspecifics during migration may be advantageous. Sound may be used to maintain contact, even over very large distances, potentially maintaining the cohesion of social groups during migration (Payne & Webb 1971). As some marine mammals have been shown to be especially sensitive to chemicals found in urine and faeces, they may be able to detect and follow the trail of conspecifics, similarly helping to maintain the cohesion of groups during transits (Pryor 1990).

There is evidence of segregation by age and sex in migration routes. Female gray whales accompanying young calves travel closer to the shore and follow the contours of the coastline more closely than do other individuals, and immature animals are also found closer to the shore than adults. It has been proposed that this pattern is linked to the avoidance of orca by the more vulnerable young (Poole 1984).

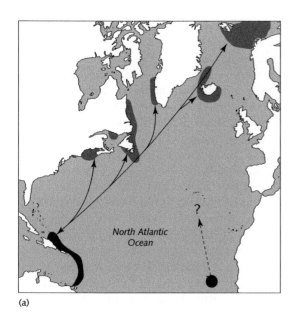

(a)

Fig. 7.8 Humpback whale migrations: (a) North Atlantic. (b) North Pacific. Feeding areas are represented by light stipple and breeding areas by dark shading. Broken arrows indicate undocumented or poorly documented migrations. In most regions, actual travel routes are not known, and arrows merely connect migratory end-points.

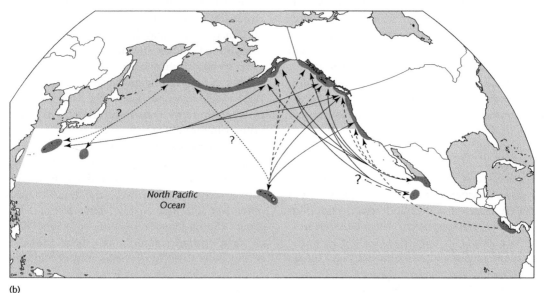

(b)

7.4.2 Distance

Gray whales travel up to 9000 km in the eastern Pacific as they migrate between feeding areas in high latitudes and breeding areas in lower latitudes (Swartz 1986). There are several documented cases of 8000 km journeys in humpback whales (Stone *et al.* 1990; Darling *et al.* 1996; Stevick *et al.* 1999). Male sperm whales travelling from 60° latitude to the equator cover about 6500 km. Elephant seals can travel up to 3000 km from breeding and moulting sites on land to remote feeding/foraging areas (McConnell & Fedak 1996; Le Boeuf *et al.* 2000). Some harp seals travel 4000 km from feeding grounds at as much as 70° north to breeding grounds at 40° north (Sergeant 1991).

Within a species, migratory distances can vary greatly between individuals and populations. Humpback whale migrations may be as little as 1200 km between California and Mexico in the North Pacific or as great as 8300 km between South America and the Antarctic (Perry *et al.* 1990; Stone *et al.* 1990). Non-migratory groups of baleen whales may exist in tropical, high productivity, upwelling systems (Papastavrou & Warebeek 1997). For example, in the northern Indian Ocean and the Arabian Sea a group of humpback whales is thought to spend the entire year near the equator (Mikhalev 1997). Offshore populations of Bryde's whales move hundreds of kilometres seasonally, while coastal populations are locally resident, and move little throughout the year (Best 1996). In some regions, bottlenose dolphins are seasonally migratory (Kenney 1990; Barco *et al.* 1999). In other regions they range widely and irregularly over hundreds of kilometres (Defran *et al.* 1999). And in other areas, they demonstrate a high degree of site fidelity year-round (Würsig & Harris 1990; Wells 1991; Wilson *et al.* 1997).

Individual variability in migration patterns is also evident. The observation of an individual humpback whale which made two round trips between Japan and British Columbia, while the typical migration route is approximately north–south, illustrates the extent of individual variation (Darling *et al.* 1996).

7.4.3 Speed

Travel speeds are constrained by hydrodynamics and the physiology of the animal. In long-distance travel, animals might be expected to move at the speed which allows them to cover the maximum distance at the minimum metabolic cost. This minimum cost of transport (mct) speed has been calculated for only a few species because of experimental difficulties (Williams *et al.* 1992; Yazdi *et al.* 1999). Speeds of long-distance travel reported for marine mammals are presented in Table 7.1. The range varies considerably within species and is strongly influenced by the method used for determining speed. Interestingly, however, telemetry results can show a remarkably uniform speed within a species in animals undertaking long-distance travel (McConnell *et al.* 1992; B.J. Mate, personal communication). McConnell *et al.* (1992) observed that the long-distance travel speed of grey seals accorded with mct estimates derived from captive studies. The correlation is less clear with cetaceans. Studies

on bottlenose dolphins suggest optimal swimming speeds substantially higher than the daily mean speeds reported from telemetry for a dolphin tracked over 4200 km (Wells *et al.* 1999; Yazdi *et al.* 1999). Indeed few of the speeds reported for the animal fell within the estimated mct range. Thus it is not clear if the observed consistency in long-distance travel speed is related to optimizing transport cost.

Cetaceans are able to use wave energy, including the bow waves of vessels, to substantially reduce the energy required for swimming (Bose & Lien 1990; Williams *et al.* 1992). At low swimming speeds near the surface, large whales may be able to utilize wave forces for as much as a third of the energy needed for propulsion (Bose & Lien 1990).

Animals do not necessarily maintain a constant speed throughout migration. The travel speed of gray whales is higher during the early part of the southward migration. Average rates are approximately 7.1 km/h between the Arctic and central California, slowing to 2.2 km/h for the transit from there to the breeding lagoons in Mexico (Swartz 1986). Minke whales travel more than twice as fast when south of the Antarctic convergence than when north of it (Kasamatsu *et al.* 1995). In both cases, the speed slows in areas where activity associated with breeding presumably increases.

Most evidence suggests that long-distance movements show little daily pattern in travel speed. However, tagged fin whales have been observed to travel more slowly during the night during long-distance transits, even in the Arctic summer when there is little darkness (Watkins *et al.* 1981, 1996), although this may be related to diurnal movements of prey. Gray whales migrate more slowly during the day than during the evening, but only during the later part of the southern migration (Buckland *et al.* 1993b; Perryman *et al.* 1999). This may be explained by the higher incidence of social behaviour during daylight hours leading to a slower rate of progress during the day, contributing to the overall slower speed discussed above.

7.4.4 Intraspecific differences in migratory behaviour

In some species, there are differences in migration patterns related to sex, age or reproductive status.

Table 7.1 Rates of travel reported for selected species of marine mammals. The data were collected using a number of different techniques and during different activities. Also, some are reported as means over longer time intervals, others as a range of observed speeds. Note that these rates are normally calculated from time elapsed between two locations and do not represent measures of speed through the water.

Species	Speed (km/h)	Source
Mysticetes		
Humpback whale	2.3–6.4	Dawbin 1966; Gabriele *et al.* 1996; Mate *et al.* 1998
Gray whale	2.7–7.9	Jones *et al.* 1984; Swartz *et al.* 1987
Fin whale	0.7–12.6	Watkins *et al.* 1981; Watkins *et al.* 1996
Right whale	0.8–4.6	Mate *et al.* 1997
Minke whale	1.5–3.9	Kasamatsu *et al.* 1995
Bowhead whale	1.3–6	Reeves & Leatherwood 1985
Odontocetes		
Sperm whale	2.5–5.2	Papastavrou *et al.* 1989; Jaquet & Whitehead 1999
Beluga	1.1–6.0	Martin *et al.* 1993; Smith & Martin 1994
Pan-tropical spotted dolphin	2.3–3.9	Perrin & Hohn 1994
Bottlenose dolphin	1.5–4.5	Wells *et al.* 1999
Harbour porpoise	0.6–2.3	Read & Westgate 1997; Westgate *et al.* 1998
Otariids		
South American sea lion	3.6	Thompson *et al.* 1998a
Galápagos sea lions	4.7–5.7	Ponganis *et al.* 1990
Hooker's sea lions	3.6–5.4	Ponganis *et al.* 1990
Northern fur seal (pups)	1.5–1.7	Ragen *et al.* 1995
Phocids		
Southern elephant seal		
adults	3.1–4.2	McConnell & Fedak 1996
pups	3.0–4.0	B.J. McConnell, unpublished data
Weddell seal	0.7	Burns *et al.* 1999
Grey seal	3.1–4.2	McConnell *et al.* 1992, 1999; Sjöberg *et al.* 1995
Hooded seal	3.5	Folkow *et al.* 1996
Sirenians		
Florida manatee	0.9–2.1	Reid *et al.* 1991; Deutsch *et al.* 1998

These may be a result of different energy requirements of animals of different sizes, particularly because of sexual dimorphism.

In some cases, only one sex migrates. In sperm whales, adult males regularly travel to polar waters, penetrating as far as the ice edge in summer to feed, and returning to tropical regions in winter, while the females and young animals travel extensively, but remain in tropical and temperate waters (Rice 1989). The migration of northern right whales to calving grounds off the southeastern US coast consists almost exclusively of females and immature whales, with males remaining in the north throughout the winter (Winn *et al.* 1986). In autumn, male walruses travel from Svalbard to Franz Josef Land for breeding, while females are resident in Franz Joseph Land throughout the year (Wiig *et al.* 1996).

In other cases, males and females migrate to different locations. Male beluga whales in the Canadian high Arctic travel in summer to a particular deepwater trench in Viscount Melville Sound, whereas females and juveniles migrate to the Amundsen Gulf (A.R. Martin, personal communication). This difference is believed to be because the greater size and, therefore breath-hold capacity, of adult males allows them to exploit the abundant food reserves at depths that are out of range for the smaller females and immature animals.

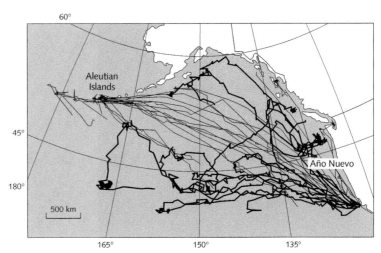

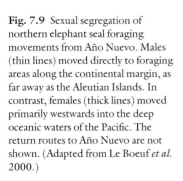

Fig. 7.9 Sexual segregation of northern elephant seal foraging movements from Año Nuevo. Males (thin lines) moved directly to foraging areas along the continental margin, as far away as the Aleutian Islands. In contrast, females (thick lines) moved primarily westwards into the deep oceanic waters of the Pacific. The return routes to Año Nuevo are not shown. (Adapted from Le Boeuf *et al.* 2000.)

Similarly, sexual segregation in migratory movements of elephant seals may be related to their extreme sexual dimorphism. Males may be up to 10 times heavier than females. As a consequence, they have greater daily energy requirements (Boyd *et al.* 1994), and may require richer feeding grounds than the females. Male southern elephant seals migrate to foraging areas over the continental shelf, while females migrate to deep-water foraging areas off the shelf edge or near the Antarctic polar front (Hindell *et al.* 1991; Campagna *et al.* 1995, 2000). Likewise, in northern elephant seals from California, females travel in a westerly direction to the deep waters of the eastern Pacific Ocean, whereas males migrate northwest as far as the Aleutian Islands, to the shallower waters of the continental shelf (Fig. 7.9) (Le Boeuf *et al.* 1993; Stewart & Delong 1995). Such sexual differences in movements are apparently already evident in juvenile elephant seals (Stewart 1997).

Immature animals are not constrained to be in breeding areas and have different energy requirements from adults, and so may have different migratory patterns. Adult and immature minke whales and Dall's porpoise have been shown to follow different migratory routes to separate destinations (Amano & Kuramochi 1992; Hatanaka & Miyashita 1997). Immature baleen whales are found further north in the Antarctic than are adults and closer to coasts in temperate and tropical areas (Mackintosh 1966). Immature harp seals do not migrate with the bulk of the adults; they arrive earlier than adults off

eastern Newfoundland and later in the Gulf of St Lawrence (Sergeant 1991).

An individual may choose to migrate in some years and not in others. Acoustic surveys show many fin, blue and minke whales remaining in cold water throughout the winter, while others move substantial distances toward the tropics where breeding and calving are thought to occur (Clark 1995). It has also been observed that female humpback whales are under-represented during migration or in breeding areas, implying that not all may migrate or complete migration in a given year (Brown *et al.* 1995; Craig & Herman 1997; Smith *et al.* 1999).

The timing of migration is often a function of age, sex and reproductive class. In mysticetes, females tend to migrate earlier than males and adults earlier than immature animals (Fig. 7.10). Newly pregnant females usually arrive first on the feeding grounds, followed by adult males. There does not appear to be a gender difference in timing for immature whales, which all travel in the middle of the migration period, though on average before the adult males in humpbacks and after them in gray whales. Females with newborn calves are generally the last to arrive on the feeding grounds (Lockyer 1981; Swartz 1986; Dawbin 1997). Migration toward the breeding ground occurs in roughly the reverse order. Late lactating females tend to travel first, followed by immature animals and adult males, then resting females, with pregnant females travelling last (Dawbin 1997). A slightly different pattern

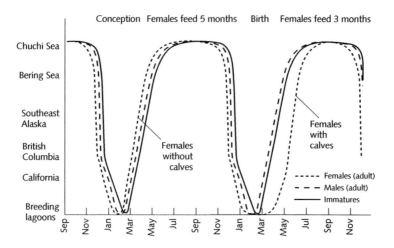

Fig. 7.10 Schematic representation of the annual migration cycle of mysticetes as exemplified by the well-documented migration of the eastern Pacific gray whale. Adult females follow a different migration pattern in the 2 years of the reproductive cycle. They are the first group to migrate north and southbound except after the birth of a calf, when they are the last group to migrate. This leads to a near doubling of the time spent in the breeding lagoons and nearly 2 months less in the feeding grounds in the year that a female is accompanied by a calf. The rate of southbound migration slows as the animals reach central California, about 1500 km from the breeding lagoons. (Data from Rice & Wolman 1971; Poole 1984; Swartz 1986.)

is seen in the gray whale, with pregnant females moving to the breeding ground first, yet leaving last, and therefore staying on the breeding range for considerably longer than other individuals (Wolman 1985).

The timing of colonization at pinniped breeding sites also varies by age and sex classes (see Chapter 10).

7.5 EXPLORATION

Successful foraging requires some knowledge of the variability in prey availability. Thus an animal may benefit from investing some time and energy in exploring the environment and familiarizing itself with the relationships between habitats and resources. This information can be used to maximize efficiency of movement in the future. In many marine mammal species, a young animal accompanies its mother during foraging and thus has the opportunity to learn the locations of foraging areas and productive searching strategies from her and perhaps from other conspecifics. For example, humpback whales have been shown to return preferentially to the specific banks visited with their mother prior to weaning (Weinrich 1998).

In most phocid seals, however, lactation is brief and weaned pups have to fend for themselves, apparently devoid of subsequent parental or peer guidance. These naïve animals must explore and learn how and where to forage before they exhaust the stores of energy provided by their mothers. We assume that initially their spatial map is blank, and thus we would expect that a greater part of their time would be spent exploring. Grey seal (Hickling 1962) and harbour seal (Bonner & Witthames 1974; Thompson et al. 1994) pups can move many hundreds of kilometres in their first months at sea and we speculate that exploration may play an important role in these movements. The dispersal of southern elephant seal pups from Macquarie Island is rapid and directed. The routes travelled by different individuals can often be remarkably similar to one another across distances of 1500 km (B.J. McConnell, unpublished data). But dispersal from natal areas may not always be an active process. Harp seals pups are passively transported by northward ice drift from the White Sea into the productive waters of the Barents Sea (Popov 1970).

If animals do learn from exploration while young, we might expect to see the result of this in consistent patterns of movement later in life. There is some evidence for this, despite the great individual variability in movement patterns. Individual grey seals in British waters often return to the same foraging sites, although different seals may use different areas, and some undertake distant travel between haul-out sites while others do not (McConnell *et al.* 1999). Similarly, some elephant seals apparently prefer to stay near to or on the continental shelf edge while others prefer open ocean meandering (see references in Section 7.4.4). Perhaps most interestingly, individual elephant seals that were tracked over more than one foraging season have shown consistency in their movements from one year to the next (Stewart & Delong 1995; McConnell & Fedak 1996; Jonker & Bester 1998).

When animals are identified in unexpected places— hooded seal off California (Dudley 1992), pan-tropical spotted dolphin off Alaska (Perrin *et al.* 1987) and spinner dolphin off New Zealand (Cawthorn 1992), for example—these individuals may simply be lost, or they may be involved in exploration. Often these animals are young, and while many are found dead, some tagged individuals have returned to more traditional habitats, consistent with exploratory behaviour (Ridoux *et al.* 1998; Stevick & Fernald 1998).

But exploration is unlikely to be restricted to young animals. Unpredictability in resource availability dictates that animals will benefit from continuing to expand the areas with which they are familiar. There is no direct evidence for this, but in telemetric studies animals are occasionally observed to undertake large, seemingly atypical, movements. One grey seal (illustrated in Fig. 7.5) left traditional habitat and travelled hundreds of kilometres over deep water (McConnell *et al.* 1992, 1999). Similarly, one Florida manatee was tracked over 2500 km from Florida to Rhode Island (Deutsch *et al.* 1998). These observations have become sufficiently commonplace to suggest that such movements may be a normal feature of marine mammal life.

7.6 NAVIGATION

The ability to undertake large-scale movements and particularly to return to specific locations, indicates that marine mammals have sophisticated means of navigation. A number of species of pinnipeds have demonstrated excellent homing behaviour after being translocated distances of up to several hundred kilometres (e.g. Boyd & Arnbom 1991; Hindell & Pemberton 1997). There is no direct evidence for how marine mammals find their way in the sea, however, as most studies are based on natural behaviour rather than on experimental manipulation.

Some inferences about navigation can be made from the behaviour of animals, their sensory systems and associations between the characteristics of the physical environment and the timing of movement and location of animals. These associations do not allow cause and effect to be established, however. For example, when a whale is observed to travel along the continental shelf edge, it may be using the bottom topography as a positional cue, or the shelf edge may be simply associated with increased prey availability.

It is reasonable, however, to assume that marine mammals make use of a variety of senses and information sources. Other animals displaying wide-ranging movements (for example birds: Aidley 1981; Able & Able 1996) are known to utilize many features. Cues potentially available to marine mammals include celestial bodies, bottom topography, landmarks, currents, temperature and salinity gradients, odours and tastes, sounds and geomagnetism.

Baker (1978) defines orientation (recognizing and maintaining direction) and navigation (identifying the direction of a given point in space) as the critical abilities necessary for directed movement. We do not discuss these abilities separately and simply group them under the term navigation.

7.6.1 Underwater topography and landmarks

Physical features, such as the nature of the sea bed, coastline or ice cover, could be detected by marine mammals using sight or hearing (see Chapter 5), thus providing information, both general and specific, about their location.

An association with seabed topography appears to be widespread and has been demonstrated in a number of species, with examples including southern elephant seals, fin, humpback, sperm, northern bottlenose, long-finned pilot and minke whales, and striped, Atlantic white-sided and common dolphins

(Selzer & Payne 1988; McConnell & Fedak 1996; Jonker & Bester 1998; Hooker *et al.* 1999). However, in some cases at least, the features in question are below the dive capabilities demonstrated for the species, and most of these observations are made in foraging areas and so may be related to prey availability. During large-scale transits, harbour porpoises were observed to follow the 92 m isobath suggesting that some marine mammals may use bottom contours to navigate (Read & Westgate 1997); however, Walker *et al.* (1992) found no correlation between fin whale distribution during migration and bottom topography. Changes in turbidity and in the type or density of organisms in the water column associated with underwater features may be detected by marine mammals in situations where the geological features themselves may not be (Madsen & Herman 1980).

Cetaceans that live in coastal areas may use prominent land features to aid navigation (Pryor 1990), though evidence for this is limited. Large-scale movements along coasts are rare in cetaceans (with the notable exception of the gray whale). Distant movements of grey seals off Britain are mainly within sight of land (McConnell *et al.* 1992, 1999). Ringed and Weddell seals use visual landmarks to locate breathing holes in the ice (Wartzok *et al.* 1992), but when blindfolded, both these species were able to use acoustic cues to locate breathing holes.

7.6.2 The sun and other celestial bodies

The sun provides clues to location, both through its position in the sky and through photo-period. Dawbin (1966) proposed that whales use changing day length in polar regions as a cue to initiate migration because it changes more rapidly and predictably during the time of early migration than does water temperature or other available cues. However, no correlation between migratory movements and photoperiod has been established for bottlenose dolphins in the western North Atlantic (Barco *et al.* 1999). Diurnal patterns of movement recorded for seals and cetaceans suggest the use of information from the sun, but these movements are just as likely to reflect a response to changes in prey distribution (Gannier 1999), and are discussed above in Section 7.3.1.

7.6.3 Currents and other oceanographic features

Water masses, currents and frontal systems may provide important clues for the navigation and orientation of marine mammals through gradients of temperature, salinity, turbidity and other factors. In many species, seasonal distribution is strongly influenced by water temperature indicating an ability to detect and respond to thermal gradients (Perrin & Gilpatric 1994; Barco *et al.* 1999; Hooker *et al.* 1999). Though cetaceans lack the typical mammalian receptors for taste and smell, cetaceans and pinnipeds have demonstrated the ability to detect chemicals in sea water (Friedl *et al.* 1990; Kuznetzov 1990). This may allow them to detect changes in water masses along travel routes, use salinity gradients as navigational cues and to locate chemical 'landmarks' (see Chapter 5).

7.6.4 Geomagnetism

There has long been interest in the possibility that marine mammals could use the earth's magnetic field to aid navigation because of the potential lack of other stimuli, particularly for pelagic species which inhabit large expanses of deep open water. It has been proposed that areas of low geomagnetic intensity and gradient, which are often orientated in a north–south direction, may be used as a guide by migrants. While this theory remains controversial, a correlation has been demonstrated between the position of fin whales and low geomagnetic gradients, particularly in autumn, but also in winter and spring (Walker *et al.* 1992). This correlation is strongest when excluding whales observed feeding, consistent with the association being related to migration. In addition, associations have been found between sites where geomagnetic field strength is low and the locations of live strandings (see below) and drive fisheries for cetacea (reviewed by Klinowska 1990).

7.6.5 Navigation failure?

Does the navigation system of marine mammals break down? Extralimital observations of marine mammals may represent lost individuals. If so this would provide evidence of failure to navigate. Alternatively, they may represent animals exploring new areas or perhaps responding to rare circumstances such as a shortage of food, as discussed above.

7.6.5.1 Strandings

Are strandings a result of navigation failure? Strandings of cetaceans have fascinated humans since at least the days of Aristotle. Animals that are already dead when they strand are simply the product of mortality at sea. But live strandings, especially of groups of animals (so-called mass strandings), raise questions about why animals that spend their entire lives at sea come to beach themselves ashore.

Repeated strandings in particular locations suggest that there is something in the geology or topography of these sites that contributes to the strandings. Gently sloping sandy beaches, often with an adjacent sand spit or peninsula are common to many mass stranding sites worldwide. Other oceanographic and geomorphological characteristics, including complex topography, nearshore intrusions of deep water, funnel-shaped basins, turbidity, heavy surf and wind-driven onshore currents have been suggested to influence strandings. Some of these conditions may interfere with echolocation or give misleading echolocation signals. In other areas, whale 'traps' may be formed by a combination of complex channels, extensive tidal flats, strong currents and high tidal ranges. Evidence for these features is reviewed by Best (1982), Brabyn and McLean (1992) and Geraci and Lounsbury (1993). However, no single feature appears to be common to all sites.

Klinowska (1986) found a highly significant relationship between live strandings around the coast of Britain and grid squares where magnetic contours crossed the coast at angles between 45° and 90°. Klinowska (1986) made the analogy that a live stranding is equivalent to a car driver travelling in the 'right' direction and encountering an unexpected hazard. Similar investigations in other areas have produced more equivocal results. In New Zealand, for example, no relationship could be found between magnetic fields and locations of either mass or solitary strandings (Brabyn & Frew 1994).

Parasites that may cause neurological dysfunction have been implicated in some strandings (Ridgway & Dailey 1972). But the lack of baseline information on the incidence of parasitic and pathogenic infection in cetaceans makes it difficult to evaluate the importance of these factors. Offshore species strand much more frequently than inshore species (Geraci & St Aubain 1979; Geraci & Lounsbury 1993). This may be because their previous experience of deep water, pelagic environments gives them less knowledge of navigating or orientating in waters close to land.

It seems clear that a number of interacting factors contribute to strandings. The failure of a single sense, or confusion in a single source of information is unlikely to explain the loss of navigation ability, where so many other sources of information are available and an integration of all information is likely to be used. In birds, individuals experimentally deprived of one sense nearly always compensate and navigate successfully using alternative cues (Able & Able 1996; Walcott 1996).

It is also possible that the causes of strandings are unrelated to navigation failure. Some tracking studies of previously stranded animals following successful rehabilitation have shown typical behaviour patterns and a return to typical habitats, indicating no long-term inability to navigate (e.g. Westgate et al. 1998). Helping behaviour has been observed in some social odontocetes, and group cohesion is high (see Chapter 12); it is possible that this could trigger or contribute to a stranding. Other proposals such as suicide or escape during stress are untestable and difficult to credit (see reviews by Best 1982; Geraci & Lounsbury 1993).

7.7 ANTHROPOGENIC IMPACTS

Humans have altered, and continue to alter, the environment in ways that may influence patterns of marine mammal movements (see Chapter 14). During the last century, there have been major changes in the quality of coastal waters and habitats throughout the world (GESAMP 1990). These have been caused by general urban growth along coasts, and more specifically by such activities as fishing, the discharge of industrial waste, oil and gas extraction, transport and recreation. All these activities have the potential to disturb or even destroy habitat that is important for marine mammals and may thus affect their movements.

On a global scale, the increase in human-generated noise in coastal waters and the oceans may make some habitats undesirable and may even affect navigational ability (Richardson *et al.* 1995). Climate change also has the potential to alter the distribution of suitable habitat types and thus the movement patterns of marine mammals (Tynan & Demaster 1997). There is the intriguing possibility of the North-West Passage opening up to allow movement of marine mammals between the Atlantic and the Pacific. This could have far-reaching implications for the dynamics of certain populations.

7.8 CONCLUSIONS

Movement patterns of marine mammals are an adaptation to an unpredictable, variable environment over a range of temporal and spatial scales, with respect to resources required for key activities (feeding, breeding, resting, etc.). Where variability is high movements tend to be greater than where it is low, enabling animals to search for and locate good areas and move rapidly through poor areas, as exemplified by elephant seals foraging over large expanses of the Southern Ocean. Where predictability is high we see regular directed movements to particular places, such as humpback and gray whales migrating to specific warm, sheltered localized areas to calve and mate.

The scale at which habitats are suitable for one or more activity varies, and this will determine the scale of a species' movement patterns. Humpback and gray whales, and elephant and harp seals migrate thousands of kilometres between breeding and feeding areas because the resources required for these two activities are separated widely in space, though the specific features that lead to the migration are still the topic of debate (e.g. Corkeron & Connor 1999). The seasonality of ocean productivity and of the reproductive cycle drives the temporal scale of these movements. At the other extreme, we observe tidal or diurnal movements and foraging movements at fine spatial scales. The bottom line

is that marine mammals move as little or as much as required to take maximum advantage of the resources required for key activities.

The study of marine mammal movements has an exciting future. The development of smaller and cheaper telemetry devices and of novel attachment techniques will allow the movements of a greater range of species to be tracked, while long-term studies will add to our understanding of changes in movement patterns with age. However, the study of movement is not an end in itself, but a means to understand and connect other basic functions such as feeding, breeding, avoiding predation, etc. Movement patterns of species have profound effects on the dynamics, epidemiology, gene flow (see Chapter 11) and, ultimately, the viability of populations. They can also affect the logistics of how we measure marine mammal population characteristics. The more we understand patterns of movement, the better will be our understanding of marine mammal populations.

REFERENCES

Able, K.P. & Able, M.A. (1996) The flexible migratory orientation system of the savannah sparrow (*Passerculus sandwichensis*). *Journal of Experimental Biology* **199** (1), 3–8.

Agler, B.A., Beard, J.A., Bowman, R.S. *et al.* (1990) Fin whale (*Balaenoptera physalus*) photographic identification: methodology and preliminary results from the western North Atlantic. *Report of the International Whaling Commission, Special Issue* **12**, 349–356.

Aguilar, A. (1987) Using organochlorine pollutants to discriminate marine mammal populations: a review and critique of methods. *Marine Mammal Science* **3** (3), 242–262.

Aidley, D.J., ed. (1981) *Animal Migration. Society for Experimental Biology Seminar Series 13.* Cambridge University Press, Cambridge, UK.

Allen, J.M., Rosenbaum, H.C., Katona, S.K., Clapham, P.J. & Mattila, D.K. (1994) Regional and sexual differences in fluke pigmentation of humpback whales (*Megaptera novaeangliae*) from the North Atlantic Ocean. *Canadian Journal of Zoology* **72** (2), 274–279.

Amano, M. & Kuramochi, T. (1992) Segregative migration of Dall's porpoise (*Phocenoides dalli*) in the Sea of Japan and Sea of Okhotsk. *Marine Mammal Science* **8** (2), 143–151.

Amos, B. & Hoelzel, A.R. (1990) DNA fingerprinting cetacean biopsy samples for individual identification. *Report of the International Whaling Commission, Special Issue* **12**, 79–85.

Araki, J., Kuramochi, T., Machida, M., Nagasawa, K. & Uchida, A. (1997) A note on the parasitic fauna of the western North Pacific minke whale (*Balaenoptera acutorostrata*). *Report of the International Whaling Commission* **47**, 565–567.

Argos (1989) *Guide to the Argos System.* CLS Argos, Toulouse, France.

Árnason, A. (1995) Genetic markers and whale stocks in the North Atlantic ocean: a review. In: *Whales, Seals, Fish and Man. Developments in Marine Biology 4* (A.S. Blix, L. Walløe & Ø. Ulltang, eds), pp. 91–103. Elsevier Science, Amsterdam.

Arnold, G.P. (1981) Movement of fish in relation to water currents. In: *Animal Migration. Society for Experimental Biology Seminar Series 13* (D.J. Aidley, ed.), pp. 55–80. Cambridge University Press, Cambridge, UK.

Baker, A.N. (1985) Pygmy right whale *Capera marginata* (Gray, 1846). In: *Handbook of Marine Mammals, Vol. 3: the Sirenians and Baleen Whales* (S.H. Ridgway & R. Harrison, eds), pp. 345–354. Academic Press, London.

Baker, C.S. & Herman, L.M. (1981) Migration and local movement of humpback whales (*Megaptera novaeangliae*) through Hawaiian waters. *Canadian Journal of Zoology* **59** (3), 460–469.

Baker, C.S., Herman, L.M., Perry, A. *et al.* (1986) Migratory movement and population structure of humpback whales (*Megaptera novaeangliae*) in the central and eastern North Pacific. *Marine Ecology Progress Series* **31** (2), 105–119.

Baker, R.R. (1978) *The Evolutionary Ecology of Animal Migration.* Hodder & Stoughton, London.

Balbuena, J.A., Aznar, F.J., Fernández, M. & Raga, J.A. (1995) Parasites as indicators of social structure and stock identity of marine mammals. In: *Whales, Seals, Fish and Man. Developments in Marine Biology 4* (A.S. Blix, L. Walløe & Ø. Ulltang, eds), pp. 133–139. Elsevier Science, Amsterdam.

Barco, S.G., Swingle, W.M., McLellan, W.A., Harris, R.N. & Pabst, P.A. (1999) Local abundance and distribution of bottlenose dolphins (*Tursiops truncatus*) in the nearshore waters of Virginia Beach, Virginia. *Marine Mammal Science* **15** (2), 394–408.

Beamish, P. (1979) Behavior and significance of entrapped baleen whales. In: *Behavior of Marine Animals* (H.E. Winn & B.L. Olla, eds), pp. 291–309. Plenum Press, New York.

Bengtson, J.L. & Stewart, B.S. (1992) Diving and haulout behavior of crab-eater seals in the Weddell Sea, Antarctica, during March 1986. *Polar Biology* **12** (6–7), 635–644.

Best, P.B. (1982) Whales: why do they strand? *African Wildlife* **36** (3), 96–101.

Best, P.B. (1996) Evidence of migration by Bryde's whales from the offshore population in the southeast Atlantic. *Report of the International Whaling Commission* **46**, 315–322.

Best, P.B. & Schell, D.M. (1996) Stable isotopes in southern right whale (*Eubalaena australis*) baleen as indicators of seasonal movements, feeding and growth. *Marine Biology* **124** (4), 483–494.

Best, P.B., Sekiguchi, K. & Findlay, K.P. (1995) A suspended migration of humpback whales *Megaptera novaeangliae* on the west coast of South Africa. *Marine Ecology Progress Series* **118** (1–3), 1–12.

Boily, P. (1995) Theoretical heat-flux in water and habitat selection of phocid seals and beluga whales during the annual molt. *Journal of Theoretical Biology* **172** (3), 235–244.

Boness, D.J. & James, H. (1979) Reproductive behaviour of the grey seal (*Halichoerus grypus*) on Sable Island, Nova Scotia. *Journal of Zoology, London* **188** (4), 477–500.

Bonner, W.N. & Witthames, S.R. (1974) Dispersal of common seals (*Phoca vitulina*) tagged in the Wash, East Anglia. *Journal of Zoology, London* **174** (4), 528–531.

Born, E.W., Wiig, O. & Thomassen, J. (1997) Seasonal and annual movements of radio-collared polar bears (*Ursus maritimus*) in northeast Greenland. *Journal of Marine Systems* **10** (1–4), 67–77.

Born, E.W., Riget, F.F., Dietz, R. & Andriashek, D. (1999) Escape responses of hauled out ringed seals (*Phoca hispida*) to aircraft disturbance. *Polar Biology* **21** (3), 171–178.

Bose, N. & Lien, J. (1990) Energy absorption from waves: a free ride for cetaceans. *Proceedings of the Royal Society of London, Series B* **240** (1299), 591–605.

Boyd, I.L. & Arnbom, T. (1991) Diving behavior in relation to water temperature in the southern elephant seal—foraging implications. *Polar Biology* **11** (4), 259–266.

Boyd, I.L., Arnbom, T.A. & Fedak, M.A. (1994) Biomass and energy consumption of the South Georgia stock of southern elephant seals. In: *Elephant Seals: Population Ecology, Behavior and Physiology* (B.J. Le Boeuf & R.M. Laws, eds), pp. 98–120. University of California Press, Los Angeles.

Boyd, I.L., McCafferty, D.J., Reid, K., Taylor, R. & Walker, T.R. (1998) Dispersal of male and female Antarctic fur seals (*Arctocephalus gazella*). *Canadian Journal of Fisheries and Aquatic Sciences* **55** (4), 845–852.

Brabyn, M. & Frew, R.V.C. (1994) New Zealand herd strandings sites do not relate to geomagnetic topography. *Marine Mammal Science* **10** (2), 195–207.

Brabyn, M.W. & McLean, I.G. (1992) Oceanography and coastal topography of herd-stranding sites for whales in New Zealand. *Journal of Mammalogy* **73** (3), 469–476.

Brasseur, S., Creuwels, J., Vanderwerf, B. & Reijnders, P. (1996) Deprivation indicates necessity for haul-out in harbor seals. *Marine Mammal Science* **12** (4), 619–624.

Brillinger, D.R. & Stewart, B.S. (1998) Elephant seal movements: modelling migration. *Canadian Journal of Statistics* **26** (3), 431–443.

Brown, M.R., Corkeron, P.J., Hale, P.T., Schultz, K.W. & Bryden, M.M. (1995) Evidence for a sex-segregated migration in the humpback whale (*Megaptera novaeangliae*). *Proceedings of the Royal Society of London, Series B* **259** (1355), 229–234.

Brown, M.W., Kraus, S.D. & Gaskin, D.E. (1991) Reaction of North Atlantic right whales (*Eubalaena glacialis*) to skin biopsy sampling for genetic and pollutant analysis. *Report of the International Whaling Commission, Special Issue* **13**, 81–89.

Brown, S.G. (1977) Whale marking: a short review. In: *A Voyage of Discovery* (M. Angel, ed.), pp. 569–581. Pergamon Press, Oxford.

Brown, S.G. (1978) Whale marking techniques. In: *Recognition Marking of Animals in Research* (B. Stonehouse, ed.), pp. 71–80. Macmillan, London.

Buckland, S.T. & Duff, E.I. (1989) Analysis of the southern hemisphere minke whale mark–recovery data. *Report of the International Whaling Commission, Special Issue* **11**, 121–143.

Buckland, S.T., Anderson, D.R., Burnham, K.P. & Laake, J.L. (1993a) *Distance Sampling*. Chapman & Hall. London.

Buckland, S.T., Breiwick, J.M., Cattanach, K.L. & Laake, J.L. (1993b) Estimated population size in the California gray whale. *Marine Mammal Science* **9** (3), 235–249.

Burns, J.M., Castellini, M.A. & Testa, J.W. (1999) Movements and diving behavior of weaned Weddell seal (*Leptonychotes weddellii*) pups. *Polar Biology* **21** (1), 23–36.

Campagna, C., Le Boeuf, B.J., Blackwell, S.B., Crocker, D.E. & Quintana, F. (1995) Diving behavior and foraging location of female southern elephant seals from Patagonia. *Journal of Zoology, London* **236** (1), 55–71.

Campagna, C., Fedak, M.A. & McConnell, B.J. (2000) Post-breeding distribution and diving behaviour of adult male southern elephant seals from Patagonia. *Journal of Mammalogy* **80** (1), 1341–1352.

Cawthorn, M.W. (1992) New Zealand. Progress report on cetacean research. *Report of the International Whaling Commission* **42**, 357–360.

Cerchio, S., Gabriele, C.M., Norris, T.F. & Herman, L.M. (1998) Movements of humpback whales between Kauai and Hawaii: implications for population structure and abundance estimation in the Hawaiian Islands. *Marine Ecology Progress Series* **175** (1–3), 13–22.

Chittleborough, R.G. (1965) Dynamics of two populations of the humpback whale, *Megaptera novaeangliae* (Borowski). *Australian Journal of Marine and Freshwater Research* **16**, 33–128.

Clapham, P.J. & Mattila, D.K. (1990) Humpback whale songs as indicators of migration routes. *Marine Mammal Science* **6** (2), 155–160.

Clapham, P.J., Baraff, L.S., Carlson, C.A. *et al.* (1993) Seasonal occurrence and annual return of humpback whales, *Megaptera novaeangliae*, in the southern Gulf of Maine. *Canadian Journal of Zoology* **71** (2), 440–443.

Clark, C.W. (1995) Application of US Navy underwater hydrophone arrays for scientific research on whales. *Report of the International Whaling Commission* **45**, 210–212.

Corkeron, P.J. & Connor, R.C. (1999) Why do baleen whales migrate? *Marine Mammal Science* **15** (4), 1228–1245.

Craig, A.S. & Herman, L.M. (1997) Sex differences in site fidelity and migration of humpback whales (*Megaptera novaeangliae*) to the Hawaiian Islands. *Canadian Journal of Zoology* **75** (11), 1923–1933.

Croxall, J.P. & Hiby, A.R. (1983) Fecundity, survival and site fidelity in Weddell seals *Leptonychotes weddelli*. *Journal of Applied Ecology* **20** (1), 19–32.

Da Silva, V.M.F. & Best, R.C. (1994) Tucuxi *Sotalia fluviatilis* (Gervais, 1853). In: *Handbook of Marine Mammals, Vol. 5: The First Book of Dolphins* (S.H. Ridgway & R. Harrison, eds), pp. 43–70. Academic Press, London.

Dalheim, M.E. & Heyning, J.E. (1999) Killer Whale *Orcinus orca* (Linnaeus, 1758). In: *Handbook of Marine Mammals, Vol. 6: The Second Book of Dolphins and Porpoises* (S.H. Ridgway & R. Harrison, eds), pp. 281–322. Academic Press, San Diego.

Darling, J.D., Calambokidis, J., Balcomb, K.C. *et al.* (1996) Movement of a humpback whale (*Megaptera novaeangliae*) from Japan to British Columbia and return. *Marine Mammal Science* **12** (2), 281–287.

Dawbin, W.A. (1964) Movements of humpback whales marked in the south west Pacific Ocean 1952–62. *Norsk Hvalfangst-Tidende* **53** (3), 68–78.

Dawbin, W.A. (1966) The seasonal migratory cycle of humpback whales. In: *Whales, Dolphins and Porpoises* (K. Norris, ed.), pp. 145–170. University of California Press, Berkley, CA.

Dawbin, W.H. (1997) Temporal segregation of humpback whales during migration in southern hemisphere waters. *Memoirs of the Queensland Museum* **42** (1), 105–138.

Dawson, S.M. (1991) Incidental catch of Hector's dolphins in inshore gillnets. *Marine Mammal Science* **7** (2), 118–132.

Defran, R.H., Weller, D.W., Kelly, D.L. & Espinosa, M.A. (1999) Range characteristics of Pacific coast bottlenose dolphins (*Tursiops truncatus*) in the southern California Bight. *Marine Mammal Science* **15** (2), 381–393.

Delong, R.L. (1992) Documenting migrations of northern elephant seals using day length. *Marine Mammal Science* **8** (2), 155–159.

Deutsch, C.J., Bonde, R.K. & Reid, J.P. (1998) Radio-tracking manatees from land and space: tag design, implementation, and lessons learned from long-term study. *Marine Technology Society Journal* **32** (1), 18–29.

Drury, W.H. & Nisbet, I.T.C. (1964) Radar studies of orientation of songbird migrants in southeastern New England. *Bird Banding* **35**, 69–199.

Dudley, M. (1992) First Pacific record of a hooded seal, *Cystophora cristata* Erxleben, 1977. *Marine Mammal Science* **8** (2), 164–168.

Fedak, M.A., Anderson, S.S. & Curry, M.G. (1983) Attachment of a radio tag to the fur of seals. *Journal of Zoology, London* **200** (2), 298–300.

Fedak, M.A., Lovell, P. & McConnell, B.J. (1996) MAMVIS: a marine mammal behaviour visualization system. *Journal of Visualization and Computer Animation* **7** (3), 141–147.

Folkow, L.M., Mårtensson, P.-E. & Blix, A.S. (1996) Annual distribution of hooded seals (*Cystophora cristata*) in the Greenland and Norwegian Seas. *Polar Biology* **16** (3), 179–189.

Friedl, W.A., Nachtigall, P.E., Moore, P.W.B. *et al.* (1990) Taste reception in the Pacific bottlenose dolphin (*Tursiops truncatus gilli*) and the California sea lion (*Zalophus californianus*). In: *Sensory Abilities of Cetaceans. Laboratory and Field Evidence* (J.A. Thomas & R.A. Kastelein, eds), pp. 447–454. Plenum Press, New York.

Gabriele, C.M., Straley, J.M., Herman, L.M. & Coleman, R.J. (1996) Fastest documented migration of a North Pacific humpback whale. *Marine Mammal Science* **12** (3), 457–464.

Gannier, A. (1999) Diel variations of the striped dolphin distribution off the French Riviera (northwestern Mediterranean Sea). *Aquatic Mammals* **25** (3), 123–134.

Garrott, R.A., Ackerman, B.B., Cary, J.R. *et al.* (1994) Trends in counts of Florida manatees at winter aggregation sites. *Journal of Wildlife Management* **58** (4), 642–654.

Gauthier, J. & Sears, R. (1999) Behavioral response of four species of balaenopterid whales to biopsy sampling. *Marine Mammal Science* **15** (1), 85–101.

George, J.C., Clark, C., Carroll, G.M. & Ellison, W. (1989) Observations on the ice-breaking and ice navigation behavior of migration bowhead whales (*Balaena mysticetus*) near Point Barrow, Alaska, spring 1985. *Arctic* **42** (1), 24–30.

Geraci, J.R. & Lounsbury, V.J. (1993) *Marine Mammals Ashore.* Texas A & M Seagrant, Galveston.

Geraci, J.R. & St Aubain, D.J. (1979) *Biology of Marine Mammals: Insights through Strandings.* Report No. MMC 77/13. Marine Mammal Commission, Bethesda, MD.

GESAMP (Group of Experts on the Scientific Aspects of Marine Environmental Protection) (1990) *The State of the Marine Environment.* UNEP Regional Seas Reports and Studies, Report No. 115. United Nations Environment Program, New York.

Gill, P.C. & Thiele, D. (1997) A winter sighting of killer whales (*Orcinus orca*) in Antarctic sea ice. *Polar Biology* **17** (5), 401–404.

Goodhart, C.B. (1988) Did virus transfer from harp seals to common seals. *Nature* **336** (6194), 21.

Goold, J.C. (1998) Acoustic assessment of populations of common dolphin off the west Wales coast, with perspectives from satellite infrared imagery. *Journal of the Marine Biological Association of the United Kingdom* **78** (4), 1353–1368.

Greenwood, P.J. (1980) Mating systems, philopatry and dispersal in birds and mammals. *Animal Behaviour* **28** (4), 1140–1162.

Grellier, K., Thompson, P.M. & Corpe, H.M. (1996) The effect of weather conditions on harbour seal (*Phoca vitulina*) haul-out behavior in the Moray Firth, north east Scotland. *Canadian Journal of Zoology* **74** (10), 1806–1811.

Griffin, R.B. (1999) Sperm whale distribution and community ecology associated with a warm-core ring off Georges Bank. *Marine Mammal Science* **15** (1), 33–51.

Hall, A., Moss, S. & McConnell, B. (2000) A new tag for identifying seals. *Marine Mammal Science* **16** (1), 254–257.

Hammond, P.S., Mizroch, S.A. & Donovan, G.P., eds (1990) Individual recognition of cetaceans: use of photo-identification and other techniques to estimate population parameters. *Report of the International Whaling Commission, Special Issue* **12**.

Harrison Matthews, L. (1938) The sperm whale (*Physeter catadon*). *Discovery Reports* **17**, 93–168.

Hatanaka, H. & Miyashita, T. (1997) On the feeding migration of the Okhotsk Sea—west Pacific stock of minke whales, estimates based on length composition data. *Report of the International Whaling Commission* **47**, 557–564.

Herman, L.M. (1979) Humpback whales in Hawaiian waters: a study in historical ecology. *Pacific Science* **33** (1), 1–15.

Hiby, L. & Lovell, P. (1990) Computer aided matching of natural markings: a prototype system for grey seals. *Report of the International Whaling Commission, Special Issue* **12**, 57–62.

Hickling, G. (1962) *Grey Seals and the Farne Islands.* Routledge & Kegan Paul, London.

Hindell, M.A. & Little, G.J. (1988) Longevity, fertility and philopatry of 2 female southern elephant seals (*Mirounga leonina*) at Macquarie Island. *Marine Mammal Science* **4** (2), 168–171.

Hindell, M.A. & Pemberton, D. (1997) Successful use of a translocation program to investigate diving behavior in a male Australian fur seal, *Arctocephalus pusillus doriferus*. *Marine Mammal Science* **13** (2), 219–228.

Hindell, M.A., Burton, H.R. & Slip, D.J. (1991) Foraging areas of southern elephant seals, *Mirounga leonina*, as inferred from water temperature data. *Australian Journal of Marine and Freshwater Research* **42** (2), 115–128.

Hobbs, L. (1988) Tags on whales, dolphins and porpoises. Appendix A. In: *Whales, Dolphins, and Porpoises of the Eastern North Pacific and Adjacent Arctic Waters* (S. Leatherwood, R.R. Reeves, W.F. Perrin & W.E. Evans, eds), pp. 218–230. Dover Publications, New York.

Hobson, K.A. & Schell, D.M. (1998) Stable carbon and nitrogen isotope patterns in baleen from eastern Arctic bowhead whales (*Balaena mysticetus*). *Canadian Journal of Fisheries and Aquatic Sciences* **55** (12), 2601–2607.

Hobson, R.P. & Martin, A.R. (1996) Behaviour and dive times of Arnoux's beaked whales, *Berardius arnuxii*, at narrow leads in fast ice. *Canadian Journal of Zoology* **74** (2), 388–393.

Hooker, S.K. & Baird, R.W. (1999) Deep-diving behaviour of the northern bottlenose whale, *Hyperoodon ampullatus* (Cetacea: Ziphiidae). *Proceedings of the Royal Society of London, Series B* **266** (1420), 671–676.

Hooker, S.K., Whitehead, H. & Gowans, S. (1999) Marine protected area design and the spatial and temporal distribution of cetaceans in a submarine canyon. *Conservation Biology* **13** (3), 592–602.

Huele, R. & deHaes, H.U. (1998) Identification of individual sperm whales by wavelet transform of the trailing edge of the flukes. *Marine Mammal Science* **14** (1), 143–145.

Ingebrigtsen, A. (1929) Whales caught in the North Atlantic and other seas. *Rapports et Proces-Verbaux des Reunions Conseil Permanent International Pour L'exploration de la Mer* **56**, 1–26.

International Whaling Commission (1982) Report of the workshop on the design of sighting surveys. *Report of the International Whaling Commission* **32**, 533–549.

Iverson, S.J., Frost, K.J. & Lowry, F.L. (1997) Fatty acid signatures reveal fine scale structure of foraging distribution of harbor seals and their prey in Prince William Sound Alaska. *Marine Ecology Progress Series* **151** (1–3), 255–271.

Jaquet, N. & Whitehead, H. (1999) Movements, distribution and feeding success of sperm whales in the Pacific Ocean, over scales of days and tens of kilometers. *Aquatic Mammals* **25** (1), 1–13.

Jeffries, S.J., Brown, R.F. & Harvey, J.T. (1993) Techniques for capturing, handling and marking harbour seals. *Aquatic Mammals* **19** (1), 21–25.

Jones, M.L., Swartz, S.L. & Leatherwood, S., eds (1984) *The Gray Whale (Eschrichtius robustus)*. Academic Press, Orlando.

Jonker, F.C. & Bester, M.N. (1998) Seasonal movements and foraging areas of adult southern female elephant seals, *Mirounga leonina*, from Marion Island. *Antarctic Science* **10** (1), 21–30.

Kasamatsu, F., Nishiwaki, S. & Ishikawa, H. (1995) Breeding areas and southbound migrations of southern minke whales *Balaenoptera acutorostrata*. *Marine Ecology Progress Series* **119** (1–3), 1–10.

Kato, H., Tanaka, E. & Sakuramoto, K. (1993) Movement of southern minke whales in the Antarctic feeding grounds from mark-recapture analyses. *Report of the International Whaling Commission* **43**, 335–342.

Katona, S.K. & Beard, J.A. (1990) Population size, migrations and feeding aggregations of the humpback whale (*Megaptera novaeangliae*) in the western North Atlantic Ocean. *Report of the International Whaling Commission, Special Issue* **12**, 295–305.

Kaufman, G.D., Osmond, M.G., Ward, A.J. & Forestell, P.H. (1990) Photographic documentation of the migratory movement of a humpback whale (*Megaptera novaeangliae*) between East Australia and Antarctic area V. *Report of the International Whaling Commission, Special Issue* **12**, 295–305.

Kenney, R.D. (1990) Bottlenose dolphin off the northeastern United States. In: *The Bottlenose Dolphin* (S. Leatherwood & R.R. Reeves, eds), pp. 369–386. Academic Press, San Diego.

Kenney, R.D. & Winn, H.E. (1986) Cetacean high-use habitats of the northeastern continental shelf. *Fishery Bulletin, USA* **84** (2), 345–357.

Klinowska, M. (1986) The cetacean magnetic sense—evidence from strandings. In: *Research on Dolphins* (M.M. Bryden & R. Harrison, eds), pp. 401–432. Oxford University Press, Oxford.

Klinowska, M. (1990) Geomagnetic orientation in cetaceans: behavioral evidence. In: *Sensory Abilities of Cetaceans. Laboratory and Field Evidence* (J.A. Thomas & R.A. Kastelein, eds), pp. 651–663. Plenum Press, New York.

Kuznetzov, V.B. (1990) Chemical sense of dolphins: quasi-olifaction. In: *Sensory Abilities of Cetaceans. Laboratory and Field Evidence* (J.A. Thomas & R.A. Kastelein, eds), pp. 481–503. Plenum Press, New York.

Le Boeuf, B.J. & Reiter, J. (1988) Lifetime reproductive success in northern elephant seals. In: *Reproductive Success* (T.H. Clutton-Brock, ed.), pp. 344–362. University of Chicago Press, Chicago.

Le Boeuf, B.J., Costa, D.P., Huntley, A.C. & Feldkamp, S.D. (1988) Continuous, deep diving in female northern elephant seals, *Mirounga angustirostris*. *Canadian Journal of Zoology* **66** (2), 446–458.

Le Boeuf, B.J., Crocker, D.E., Blackwell, S.B., Morris, P.A. & Thorson, P.H. (1993) Sex differences in diving and foraging behaviour of northern elephant seals. *Symposium of the Zoological Society of London* **66**, 149–178.

Le Boeuf, B.J., Crocker, D.E., Costa, D.P. *et al.* (2000) Foraging ecology of northern elephant seals. *Ecological Monographs* **70**, 353–382.

Lipps, L.H. & Mitchell, E. (1976) Trophic model for the adaptive radiations and extinctions of pelagic marine mammals. *Paleobiology* **2** (2), 147–155.

Lockyer, C. (1979) Response of orcas to tagging. *Carnivore* **2** (3), 19–21.

Lockyer, C. (1981) Growth and energy budgets of large baleen whales from the southern hemisphere. *FAO Fisheries Series 5, Mammals in the Seas* **3**, 379–487.

Lockyer, C. (1995) A review of factors involved in zonation in odontocete teeth, and an investigation of the likely impact of environmental factors and major life events on harbor porpoise tooth structure. *Report of the International Whaling Commission, Special Issue* **16**, 511–529.

Lunn, N.J. & Boyd, I.L. (1991) Pupping-site fidelity of Antarctic fur seals at Bird Island, South Georgia. *Journal of Mammalogy* **72** (1), 202–206.

McConnell, B.J. & Fedak, M.A. (1996) Movements of southern elephant seals. *Canadian Journal of Zoology* **74** (8), 1485–1496.

McConnell, B.J., Chambers, C., Nicholas, K.S. & Fedak, M.A. (1992) Satellite tracking of grey seals (*Halichoerus grypus*). *Journal of Zoology, London* **226** (2), 271–282.

McConnell, B.J., Fedac, M.A., Lovell, P. & Hammond, P.S. (1999) Movements and foraging areas of grey seals in the North Sea. *Journal of Applied Ecology* **36** (4), 573–590.

Mackintosh, N.A. (1942) The southern stocks of whalebone whales. *Discovery Reports* **22**, 197–300.

Mackintosh, N.A. (1966) The distribution of southern blue and fin whales. In: *Whales, Dolphins and Porpoises* (K. Norris, ed.), pp. 125–144. University of California Press, Berkeley.

Madsen, C.J. & Herman, L.M. (1980) Social and ecological correlates of cetacean vision and visual appearance. In: *Cetacean Behavior: Mechanisms and Functions* (L.M. Herman, ed.), pp. 101–147. John Wiley & Sons, New York.

Martin, A.R. (1982) A link between the sperm whales occurring off Iceland and the Azores. *Mammalia* **46** (2), 259–260.

Martin, A.R., Smith, T.G. & Cox, O.P. (1993) Studying the behaviour and movements of high Arctic belugas with satellite telemetry. *Symposium of the Zoological Society of London* **66**, 195–210.

Mate, B.R., Nieukirk, S.L. & Kraus, S.D. (1997) Satellite-monitored movements of the northern right whale. *Journal of Wildlife Management* **61** (4), 1393–1405.

Mate, B.R., Gisiner, R. & Mobley, J. (1998) Local and migratory movements of Hawaiian humpback whales tracked by satellite telemetry. *Canadian Journal of Zoology* **76** (5), 863–868.

Mattila, D.K. & Clapham, P.J. (1989) Humpback whales, *Megaptera novaeangliae*, and other cetaceans on Virgin Bank and in the northern Leeward Islands, 1985 and 1986. *Canadian Journal of Zoology* **67** (9), 2201–2211.

Mayo, C.A. & Marx, M.K. (1990) Surface foraging behaviour of the North Atlantic right whale, *Eubalaena glacialis*, and associated zooplankton characteristics. *Canadian Journal of Zoology* **68** (10), 2214–2220.

Messier, F., Taylor, M.K. & Ramsay, M.A. (1992) Seasonal activity patterns of female polar bears (*Ursus maritimus*) in the Canadian Arctic as revealed by satellite telemetry. *Journal of Zoology, London* **226** (2), 219–229.

Mikhalev, Y.A. (1997) Humpback whales *Megaptera novaeangliae* in the Arabian Sea. *Marine Ecology Progress Series* **149** (1–3), 13–21.

Mitchell, E.D. (1991) Winter records of the minke whale (*Balaenoptera acutorostrata acutorostrata* Lacépéde 1804) in the southern North Atlantic. *Report of the International Whaling Commission* **41**, 455–457.

Mizroch, S.A., Beard, J.A. & Lynde, M. (1990) Computer assisted photo-identification of humpback whales. *Report of the International Whaling Commission, Special Issue* **12**, 63–70.

Moore, J.C. (1956) Observations of manatees in aggregations. *American Museum Noviates* **1811**, 1–24.

Moore, S.E., Stafford, K.M., Dalheim, M.E. *et al.* (1998) Seasonal variation in reception of fin whale calls at five geographic areas in the North Pacific. *Marine Mammal Science* **14** (3), 617–627.

Nicholas, K.S., Fedak, M.A. & Hammond, P.S. (1992) An automatic recording station for detecting and storing radio signals from free ranging animals. In: *Wildlife Telemetry: Remote Monitoring and Tracking of Animals* (I.G. Priede & S.M. Swift, eds), pp. 76–78. Ellis Horwood, Chichester, UK.

Nilssen, K.T., Haug, T., Oritsland, T., Lindblom, L. & Kjellqwist, S.A. (1998) Invasions of harp seals *Phoca groenlandica* Erxleben to coastal waters of Norway in 1995: ecological and demographic implications. *Sarsia* **83** (4), 337–345.

Norris, K.S. (1991) *Dolphin Days. Life and Times of the Spinner Dolphin.* Norton, New York.

Northridge, S.P., Tasker, M.L., Webb, A. & Williams, J.M. (1995) Seasonal distribution and relative abundance of harbor porpoises, white-sided dolphins and minke whales in waters around the British Isles. *ICES Journal of Marine Science* **52** (1), 55–66.

Øien, N. (1991) Abundance of the northeastern Atlantic stock of minke whales based on shipboard surveys conducted in July 1989. *Report of the International Whaling Commission* **41**, 433–437.

Palsbøll, P.J., Clapham, P.J., Mattila, D.K., *et al.* (1995) Distribution of mt DNA haplotypes in North Atlantic humpback whales: the influence of behavior on population structure. *Marine Ecology Progress Series* **116** (1–3), 1–10.

Palsbøll, P.J., Allen, J., Bérubé, M. *et al.* (1997) Genetic tagging of humpback whales. *Nature, London* **388** (6644), 767–769.

Papastavrou, V. & Warebeek, K.V. (1997) A note on the occurrence of humpback whales (*Megaptera novaeangliae*) in tropical and subtropical areas: the upwelling link. *Report of the International Whaling Commission* **47**, 945–947.

Papastavrou, V., Smith, S.C. & Whitehead, H. (1989) Diving behaviour of the sperm whale (*Physeter macrocephalus*) off the Galapagos Islands. *Canadian Journal of Zoology* **67** (4), 839–846.

Pauli, B.D. & Terhune, J.M. (1987) Meterological influences on the harbour seal haul out. *Aquatic Mammals* **13**, 114–118.

Payne, P.M. & Selzer, L.A. (1989) The distribution, abundance and selected prey of the harbor seal, *Phoca vitulina concolor*, in southern New England. *Marine Mammal Science* **5** (2), 173–192.

Payne, P.M., Nicholas, J.R., O'Brien, L. & Powers, K.D. (1986) Distribution of the humpback whale, *Megaptera novaeangliae*, on Georges Bank and in the Gulf of Maine in relation to densities of the sand eel, *Ammodytes americanus*. *Fishery Bulletin, USA* **84** (2), 271–277.

Payne, P.M., Wiley, D.N., Young, S.B. *et al.* (1990) Recent fluctuations in the abundance of baleen whales in the southern Gulf of Maine in relation to changes in selected prey. *Fishery Bulletin, USA* **88** (4), 687–696.

Payne, R.S. & Webb, D. (1971) Orientation by means of long range acoustic signalling in baleen whales. *Annals of the New York Academy of Sciences* **188**, 110–142.

Perrin, W.F. & Gilpatric, J.W. (1994) Spinner dolphin *Stenella longirostris* (Gray, 1828). In: *Handbook of Marine Mammals, Vol 5: The First Book of Dolphins* (S.H. Ridgway & R. Harrison, eds), pp. 99–128. Academic Press, London.

Perrin, W.F. & Hohn, A.A. (1994) Pantropical spotted dolphin *Stenella attenuata*. In: *Handbook of Marine Mammals, Vol 5: The First Book of Dolphins* (S.H. Ridgway & R. Harrison, eds), pp. 71–98. Academic Press, London.

Perrin, W.F., Mitchell, E.D., Mead, J.G. *et al.* (1987) Revision of the spotted dolphins, *Stenella* spp. *Marine Mammal Science* **3** (2), 99–170.

Perry, A., Baker, C.S. & Herman, L.M. (1990) Population characteristics of individually identified humpback whales in the central and eastern North Pacific: a summary and critique. *Report of the International Whaling Commission, Special Issue* **12**, 307–317.

Perryman, W.L., Donahue, M.A., Laake, J.L. & Martin, T.E. (1999) Diel variation in migration rates of eastern Pacific gray whales measured with thermal imaging sensors. *Marine Mammal Science* **15** (2), 426–445.

Pomeroy, P.P., Anderson, S.S., Twiss, S.D. & McConnell, B.J. (1994) Dispersion and site fidelity of breeding female grey seals (*Halichoerus grypus*) on North Rona, Scotland. *Journal of Zoology, London* **233** (3), 429–447.

Ponganis, P.J., Ponganis, E.P., Ponganis, K.V. *et al.* (1990) Swimming velocities in otariids. *Canadian Journal of Zoology* **68** (10), 2105–2112.

Poole, M.M. (1984) Migration corridors of gray whales along the central California coast, 1980–1982. In: *The Gray Whale (Eschrichtius robustus)* (M.L. Jones, S.L. Swartz & S. Leatherwood, eds), pp. 389–407. Academic Press, Orlando.

Popov, L.A. (1970) Soviet tagging of harp and hooded seals in the North Atlantic. *Fiskeridirektoratets Skrifter, Serie Havundersokelser* **16**, 1–9.

Pryor, K. (1990) Concluding comments on vision, tactition, and chemoreception. In: *Sensory Abilities of Cetaceans. Laboratory and Field Evidence* (J.A. Thomas & R.A. Kastelein, eds), pp. 561–569. Plenum Press, New York.

Ragen, T.J., Antonelis, G.A. & Kiyota, M. (1995) Early migration of northern fur seal pups from St Paul Island, Alaska. *Journal of Mammalogy* **76** (4), 1137–1148.

Read, A.J. & Westgate, A.J. (1997) Monitoring the movements of harbour porpoises (*Phocoena phocoena*) with satellite telemetry. *Marine Biology* **130** (2), 315–322.

Reeves, R.R. & Leatherwood, S. (1985) Bowhead whale *Balaena mysticetus* (Linnaeus, 1758). In: *Handbook of Marine Mammals, Vol 3: The Sirenians and Baleen Whales* (S.H. Ridgway & R. Harrison, eds), pp. 305–344. Academic Press, London.

Reid, J.P., Rathbun, G.B. & Wilcox, J.R. (1991) Distribution patterns of individually identifiable West Indian manatees (*Trichechus manatus*) in Florida. *Marine Mammal Science* **7** (2), 180–190.

Rice, D.W. (1989) Sperm whale *Physeter macrocephalus* Linnaeus, 1758. In: *Handbook of Marine Mammals, Vol 4: River Dolphins and the Larger Toothed Whales* (S.H. Ridgway & R. Harrison, eds), pp. 177–233. Academic Press, London.

Rice, D.W. & Wolman, A.A. (1971) The life history and ecology of the gray whale (*Eschrichtius robustus*). *American Society of Mammalogists, Special Publication* **3**, 1–142.

Richardson, W.J., Greene, C.R., Malme, C.I. & Thompson, D.H., eds (1995) *Marine Mammals and Noise*. Academic Press, San Diego.

Ridgway, S.H. & Dailey, M.D. (1972) Cerebral and cerebellar involvement of trematode parasites in dolphins and their possible role in stranding. *Journal of Wildlife Diseases* **8**, 33–43.

Ridoux, V., Hall, A.J., Steingrimsson, G. & Olafsson, G. (1998) An inadvertant homing experiment with a young ringed seal. *Marine Mammal Science* **14** (4), 883–888.

Rosenfeld, M., George, M. & Terhune, J.M. (1988) Evidence of autumnal harbor seal, *Phoca vitulina*, movement from Canada to the United States. *Canadian Field-Naturalist* **102** (3), 527–529.

Ross, G.J.B., Heinsohn, G.E. & Cockroft, V.G. (1994) Humpback dolphins *Sousa chinensis* (Osbeck, 1765), *Sousa plumbea* (G. Cuvier, 1829) and *Sousa teuszii* (Kukenthal,

1892). In: *Handbook of Marine Mammals, Vol 5: The First Book of Dolphins* (S.H. Ridgway & R. Harrison, eds), pp. 23–42. Academic Press, London.

Salden, D.R., Herman, L.M., Yamaguchi, M. & Sato, F. (1999) Multiple visits of individual humpback whales (*Megaptera novaeangliae*) between the Hawaiian and Japanese winter grounds. *Canadian Journal of Zoology* **77** (3), 504–508.

Schevill, W.E. & Backus, R.H. (1960) Daily patrol of a megaptera. *Journal of Mammalogy* **41** (2), 279–281.

Schneider, D.C. & Payne, P.M. (1983) Factors affecting haul-out of harbor seals at a site in southeastern Massachusetts. *Journal of Mammalogy* **64** (3), 518–520.

Schweder, T., Skaug, H.J., Dimakos, H.J., Langaas, M. & Øien, N. (1997) Abundance of northeastern Atlantic minke whales, estimates for 1989 and 1995. *Report of the International Whaling Commission* **47**, 453–479.

Sekiguchi, K., Klages, N.T.W. & Best, P.B. (1996) The diet of strap-toothed whales (*Mesoplodon layardii*). *Journal of Zoology, London* **239** (3), 453–463.

Selzer, L.A. & Payne, P.M. (1988) The distribution of white-sided (*Lagenorhynchus acutus*) and common dolphin (*Delphinus delphis*) versus environmental features of the continental shelf of the northeastern United States. *Marine Mammal Science* **4** (2), 141–153.

Sergeant, D.E. (1991) *Harp Seals, Man and Ice*. Department of Fisheries and Oceans Canada, Ottawa.

Siegstad, H. & Heide-Jørgensen, M.P. (1994) Ice entrapments of narwhals (*Monodon monoceros*) and white whales (*Delphinapterus leucas*) in Greenland. *Meddelelser Om Grønland, Bioscience* **39**, 151–160.

Sjöberg, M., Fedak, M.A. & McConnell, B.J. (1995) Movements and diurnal behavior patterns in a baltic grey seal (*Halichoerus grypus*). *Polar Biology* **15** (8), 593–595.

Sjöberg, M., McConnell, B. & Fedak, M. (1999) Haulout patterns of grey seals *Halichoerus grypus* in the Baltic Sea. *Wildlife Biology* **5** (1), 37–47.

Smith, T.D., Allen, J., Clapham, P.J. *et al.* (1999) An ocean-basin-wide mark-recapture study of the North Atlantic humpback whale (*Megaptera novaeangliae*). *Marine Mammal Science* **15** (1), 1–32.

Smith, T.G. & Martin, A.R. (1994) Distribution and movements of belugas, *Delphinapterus leucas*, in the Canadian high Arctic. *Canadian Journal of Fisheries and Aquatic Sciences* **51** (7), 1653–1663.

Stevick, P.T. & Fernald, T.W. (1998) Increase in extralimital records of harp seals in Maine. *Northeastern Naturalist* **5** (1), 75–82.

Stevick, P.T., Øien, N. & Mattila, D.K. (1999) Migratory destinations of humpback whales from Norwegian and adjacent waters: evidence for stock identity. *Journal of Cetacean Research and Management* **1** (3), 147–152.

Stewart, B.S. (1997) Ontogeny of differential migration and sexual segregation in northern elephant seals. *Journal of Mammalogy* **78** (4), 1101–1116.

Stewart, B.S. & Delong, R.L. (1995) Double migrations of the northern elephant seal, *Mirounga angustirostris*. *Journal of Mammalogy* **76** (1), 196–205.

Stone, G.S., Flórez-González, L. & Katona, S. (1990) Whale migration record. *Nature, London* **346** (6286), 705.

Stone, G.S., Brown, J. & Yoshinaga, A. (1995) Diurnal movement patterns of Hector's dolphin as observed from clifftops. *Marine Mammal Science* **11** (3), 395–402.

Stone, G.S., Goodyear, J., Hutt, A. & Yoshinaga, A. (1994) A new non-invasive tagging method for studying wild dolphins. *Marine Technology Society Journal* **28** (1), 11–16.

Swartz, S.L. (1986) Gray whale migratory, social and breeding behavior. *Report of the International Whaling Commission, Special Issue* **8**, 207–229.

Swartz, S.L., Jones, M.L., Goodyear, J., Withrow, D.E. & Miller, R.V. (1987) Radio-telemetric studies of gray whale migration along the California coast: a preliminary comparison of day and night migration rates. *Report of the International Whaling Commission* **37**, 295–299.

Testa, J.W. (1987) Long term reproductive patterns and sighting bias in Weddell seals (*Leponychotes weddelli*). *Canadian Journal of Zoology* **65** (5), 1091–1099.

Testa, J.W. & Rothery, P. (1992) Effectiveness of various cattle ear tags as markers for Weddell seals. *Marine Mammal Science* **8** (4), 344–353.

Thiele, D. & Gill, P.C. (1999) Cetacean observations during a winter voyage into Antarctic sea ice south of Australia. *Antarctic Science* **11** (1), 48–53.

Thompson, D. & Fedak, M.A. (1993) Cardiac responses of grey seals during diving at sea. *Journal of Experimental Biology* **174**, 139–164.

Thompson, P.M., Pierce, G.J., Hislop, J.R.G., Miller, D. & Diack, J.S.W. (1991) Winter foraging by common seals (*Phoca vitulina*) in relation to food availability in the inner Moray Firth, NE Scotland. *Journal of Animal Ecology* **60** (1), 283–294.

Thompson, P.M., Kovacs, K.M. & McConnell, B.J. (1994) Natal dispersal of harbor seals (*Phoca vitulina*) from breeding sites in Orkney, Scotland. *Journal of Zoology, London* **234** (4), 668–673.

Thompson, D., Duck, C.D., McConnell, B.J. & Garrett, J. (1998a) Foraging behaviour and diet of lactating female southern sea lions (*Otaria flavescens*) in the Falkland Islands. *Journal of Zoology, London* **246** (2), 135–146.

Thompson, P.M., Mackay, A., Tollit, D.J., Enderby, S. & Hammond, P.S. (1998b) The influence of body size and sex on the characteristics of harbour seal foraging trips. *Canadian Journal of Zoology* **76** (6), 1044–1053.

Thompson, P.M., McConnell, B.J., Tollit, D.J., Mackay, A., Hunter, C. & Racey, P.A. (1996) Comparative distribution, movements and diet of harbour and grey seals from the Moray Firth, NE Scotland. *Journal of Applied Ecology* **33** (6), 1572–1584.

Tollit, D.J., Greenstreet, S.P.R. & Thompson, P.M. (1997) Prey selection by harbour seals, *Phoca vitulina*, in relation to variations in prey abundance. *Canadian Journal of Zoology* **75** (9), 1508–1518.

Tomilin, A.G. (1957) *Mammals of the USSR and Adjacent Countries*. National Technical Information Service, Washington, DC.

Townsend, C.H. (1935) The distribution of certain whales as shown by logbook records of American whaleships. *Zoologica* **19** (1), 1–50.

Tynan, C.T. & Demaster, D.P. (1997) Observations and predictions of Arctic climatic change: potential effects on marine mammals. *Arctic* **50** (4), 308–322.

Veinger, G.N. (1980) Intraspecies structural data of sperm whales in the North Pacific. *Report of the International Whaling Commission, Special Issue* **2**, 103–105.

Walcott, C. (1996) Pigeon homing: observations, experiments and confusions. *Journal of Experimental Biology* **199** (1), 21–27.

Walker, M.M., Kirschvink, J.L., Ahmed, G. & Dizon, A.E. (1992) Evidence that fin whales respond to the geomagnetic field during migration. *Journal of Experimental Biology* **171**, 67–78.

Wartzok, D., Elsner, R., Stone, H., Kelly, B.P. & Davis, R.W. (1992) Under-ice movements and the sensory basis of hole finding by ringed and Weddell seals. *Canadian Journal of Zoology* **70** (9), 1712–1722.

Watkins, W.A. (1981) Reaction of 3 species of whales *Balaenoptera physalus*, *Megaptera novaeangliae*, and *Balaenoptera edeni* to implanted radio tags. *Deep-Sea Research Part A, Oceanographic Research Papers* **28** (6), 589–599.

Watkins, W.A., Moore, K.E., Wartzok, D. & Johnson, J.H. (1981) Radio tracking of finback (*Balaenoptera physalus*) and humpback (*Megaptera novaeangliae*) whales in Prince William Sound, Alaska. *Deep-Sea Research Part A, Oceanographic Research Papers* **28** (6), 577–588.

Watkins, W.A., Sigurjónsson, J., Wartzok, D. *et al.* (1996) Fin whale tracked by satellite off Iceland. *Marine Mammal Science* **12** (4), 564–569.

Watts, P. (1996) The diel hauling-out cycle of harbour seals in an open marine environment: correlates and constraints. *Journal of Zoology, London* **240** (1), 175–200.

Weilgart, L. & Whitehead, H. (1997) Group-specific dialects and geographical variation in coda repertoire in South Pacific sperm whales. *Behavioral Ecology and Sociobiology* **40** (5), 277–285.

Weinrich, M. (1998) Early experience in habitat choice by humpback whales (*Megaptera novaeangliae*). *Journal of Mammalogy* **79** (1), 163–170.

Weinrich, M., Lambertsen, R.H., Baker, C.S., Schilling, M.R. & Belt, C.R. (1991) Behavioural responses of humpback whales (*Megaptera novaeangliae*) in the southern Gulf of Maine to biopsy sampling. *Report of the International Whaling Commission, Special Issue* **13**, 91–97.

Weinrich, M., Martin, M., Griffiths, R., Bove, J. & Schilling, M. (1997) A shift in distribution of humpback whales, *Megaptera novaeangliae*, in response to prey in the southern Gulf of Maine. *Fishery Bulletin, USA* **95** (4), 826–836.

Wells, R.S. (1991) The role of long-term study in understanding the social structure of a bottlenose dolphin community. In: *Dolphin Societies: Discoveries and Puzzles* (K. Pryor & K.S. Norris, eds), pp. 199–225. University of California Press, Berkeley, CA.

Wells, R.S., Rhinehart, H.L., Cunningham, P. *et al.* (1999) Long distance offshore movements of bottlenose dolphins. *Marine Mammal Science* **15** (4), 1098–1114.

Westgate, A.J., Read, A.J., Cox, T.M. *et al.* (1998) Monitoring a rehabilitated harbor porpoise using satellite telemetry. *Marine Mammal Science* **14** (3), 599–604.

White, G.C. & Garrott, R.A. (1990) *Analysis of Wildlife Radio Tracking Data.* Academic Press, San Diego.

Whitehead, H. (1990) Computer assisted individual identification of sperm whale flukes. *Report of the International Whaling Commission, Special Issue* **12**, 71–78.

Whitehead, H. (1996) Variation in the feeding success of sperm whales: temporal scale, spatial scale and relationship to migrations. *Journal of Animal Ecology* **65** (4),429–438.

Whitehead, H. & Gordon, J. (1986) Methods of obtaining data for assessing and modelling sperm whale populations which do not depend on catches. *Report of the International Whaling Commission, Special Issue* **8**, 149–166.

Wiig, O., Gjertz, I. & Griffiths, D. (1996) Migration of walruses (*Odobenus rosmarus*) in the Svalbard and Franz Josef Land area. *Journal of Zoology, London* **238** (4), 769–784.

Williams, T.M., Friedl, W.A., Fong, M.L. *et al.* (1992) Travel at low energetic cost by swimming and wave-riding bottlenose dolphins. *Nature, London* **355** (6363), 821–823.

Williamson, G.R. (1961) Winter sighting of a humpback whale suckling its calf on the Grand Bank of Newfoundland. *Norsk Hvalfangst-Tidende* **50** (8), 335–341.

Wilson, B., Thompson, P.M. & Hammond, P.S. (1997) Habitat use by bottlenose dolphins: seasonal distribution and stratified movement patterns in the Moray Firth, Scotland. *Journal of Applied Ecology* **34** (6), 1365–1374.

Winn, H.E., Thompson, T.J., Cummings, W.C. *et al.* (1981) Song of the humpback whale—population comparisons. *Behavioral Ecology and Sociobiology* **8**, 41–46.

Winn, H.E., Price, C.A. & Sorenson, P.W. (1986) The distributional biology of the right whale (*Eubalaena glacialis*). *Report of the International Whaling Commission, Special Issue* **10**, 129–138.

Wolman, A.A. (1985) Gray whale *Eschrichtius robustus* (Lilljeborg, 1861). In: *Handbook of Marine Mammals, Vol 3: The Sirenians and Baleen Whales* (S.H. Ridgway & R. Harrison, eds), pp. 67–90. Academic Press, London.

Würsig, B. & Harris, G. (1990) Site and association fidelity in bottlenose dolphins off Argentina. In: *The Bottlenose Dolphin* (S. Leatherwood & R.R. Reeves, eds), pp. 361–365. Academic Press, San Diego, CA.

Yazdi, P., Kilian, A. & Culik, B.M. (1999) Energy expenditure of swimming bottlenose dolphins (*Tursiops truncatus*). *Marine Biology* **134** (4), 601–607.

Yochem, P.K., Stewart, B.S., Mina, M. *et al.* (1990) Non-metrical analyses of pelage patterns in demographic studies of harbor seals. *Report of the International Whaling Commission, Special Issue* **12**, 87–90.

CHAPTER 8

Feeding Ecology

W. Don Bowen, Andy J. Read and Jim A. Estes

8.1 INTRODUCTION

Marine mammals forage at widely different temporal and spatial scales and in so doing exploit many different kinds of prey. Given their taxonomic and geographical diversity (see Chapter 1), processes acting on both evolutionary and ecological time-scales have undoubtedly influenced the patterns of foraging we see today and placed different constraints on the foraging behaviour of various groups.

Our approach in this chapter is to examine marine mammal foraging from an evolutionary perspective. We assume that characteristics of feeding behaviour are under selection and that variability in feeding success affects survival probability and reproductive performance of individuals; that is, it affects fitness. We also recognize that historical uncertainty limits the extent to which extant patterns and processes can be interpreted as evolutionary adaptations (Gould & Lewontin 1979; Janzen & Martin 1982). An evolutionary perspective is needed to understand behavioural and physiological responses of individuals to environmental change. However, evolutionary changes in populations occur within the context of ecosystems, thus it is important to bridge the gap between evolutionary biology and ecosystem science (Levin 1992).

Foraging success determines an organism's intake of nutrients, while life-history patterns result from the expenditure of these nutrients on fitness-related activities (Boggs 1992). The allocation of limited resources among competing activities links these two such that an understanding of feeding ecology is central to an understanding of the evolution of life history patterns. Despite its importance, the feeding ecology of most marine mammals is poorly known.

With few exceptions (e.g. sea otters), feeding usually occurs at depth, in remote areas of the ocean, and over large spatial scales. All of these factors make direct observation difficult or impossible. As a result, although the main types of prey eaten by many marine mammals are known, the factors that influence foraging behaviour and diet, and the consequences of variation in feeding success on fitness have been investigated only recently and for a handful of species.

Given the difficulties associated with studying marine mammal foraging, it is not surprising that our rapidly growing knowledge relies heavily on recently developed technology and new methods of diet analysis (Boxes 8.1–8.3). These new tools have made it possible to investigate the spatial and temporal scales of foraging by marine mammals in relation to phylogenetic and ecological constraints.

8.2 PHYLOGENY AND FEEDING

Extant groups of marine mammals show diverse morphological, physiological and behavioural adaptations to their aquatic way of life. The major anatomical adaptations of the feeding apparatus in cetaceans clearly have had important consequences for their diets and foraging behaviour. Among baleen whales the shape, arrangement and structure of the baleen plates reflect both the types of prey eaten and the methods used to capture food (Fig. 8.1) (Pivorunas 1979). Baleen whales feed primarily on planktonic crustaceans, such as copepods, euphausiids and amphipods, although in some species (e.g. humpback, *Megaptera novaeangliae*, and minke, *Balaenoptera acutorostrata*, whales: Haug *et al.* 1995) fish are important prey. All baleen whales require

BOX 8.1 INVESTIGATING FORAGING BEHAVIOUR AND DIETS USING FATTY ACID SIGNATURE ANALYSIS

Lipids in marine organisms are characterized by their diversity (> 60 types) and high levels of long-chain and polyunsaturated fatty acids which originate in various unicellular phytoplankton and seaweeds (Ackman 1980). Unlike proteins that are readily broken down during digestion, dietary fatty acids pass into the circulation intact and those of carbon chain length greater that 14 are often deposited in animal tissue with little modification.

Iverson (1988, 1993) suggested that the pattern of tissue fatty acids might be used to investigate diet composition and foraging ecology of marine mammals. The pattern of prey fatty acids can be thought of as a prey signature (see below) that is deposited within the blubber of marine mammals in a predictable way (Iverson *et al.* 1995). Three tissues (i.e. blood, blubber and milk), each capable of providing information about foraging on a different temporal scale, can be used for fatty acid analysis. Fatty acids stored in blubber represent the integration of feeding over periods of weeks to perhaps months depending on the rate and degree of lipid deposition (Kirsch *et al.* 2000). However, not all blubber contains information on dietary fatty acids. For example, the outer layer of the blubber of many

cetaceans (particularly the smaller odontocetes, Koopman *et al.* 1996) appears to be metabolically inactive, such that fatty acid composition is not responsive to changes in diet. Fatty acids in milk provide a short-term view of the diet if the mother is feeding (e.g. a lactating fur seal, Iverson *et al.* 1997a) or a longer-term view of previous feeding if the mother is fasting and thus mobilizing lipids from body stores (e.g. a lactating phocid).

As with any method there are disadvantages associated with fatty acid analysis. One is that not all prey signatures may be unique. Another is that prey signatures may vary significantly in relation to age, reproductive condition and geographical location. Accurate identification of prey components in the diet of a marine mammal requires that a comprehensive library of reference prey signatures be assembled. The method may often provide only qualitative information on diet composition, although quantitative estimation is possible (S.J. Iverson, personal communication). Finally, metabolic alteration of the signature may occur and, thus, a clear understanding of the lipid metabolism of marine mammals is required for the confident use of the method (Iverson 1993).

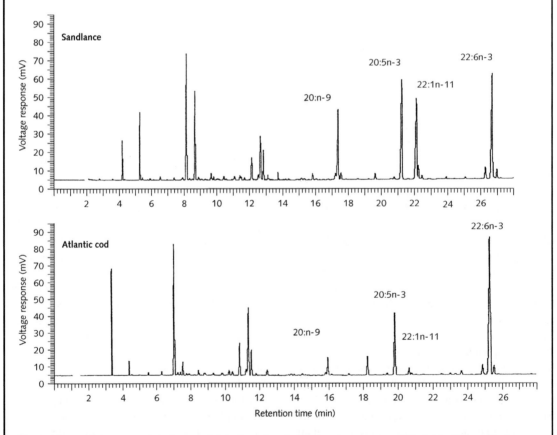

Fig. 1 Fatty acid signatures of sandlance and Atlantic cod, each showing the more than 60 individual fatty acids that make up prey signatures. Four fatty acids are labelled to help illustrate the differences between these two prey species. (Courtesy of S.J. Iverson.)

BOX 8.2 STABLE ISOTOPE RATIOS REVEAL TROPHIC LEVEL OF FORAGING

The carbon isotope ratio, $^{13}C : ^{12}C$ (denoted $d^{13}C$), and the nitrogen isotope ratio, $^{15}N : ^{14}N$ (denoted $d^{15}N$), of various animal tissues can be useful in diet studies because they reflect the foods that have been assimilated by the predator (DeNiro & Epstein 1978, 1981; Minagawa & Wada 1984; Peterson & Fry 1987). The $d^{15}N$ composition of an individual animal is typically about 3–4 parts per thousand greater than that of its diet (Minagawa & Wada 1984; Fry 1988). As this enrichment occurs at each trophic level within a food web, $d^{15}N$ values provide a good indication of the trophic level at which the predator feeds (Wada $et\ al.$ 1987; Fry 1988). On the other hand, $d^{13}C$ values do not show the same kind of predictable trophic enrichment (Fry 1988; Hobson & Welch 1992), but do seem to vary geographically such that the location of feeding can often be deduced (e.g. Schell $et\ al.$ 1989). Naturally occurring stable isotopes of carbon and nitrogen have been used in the study of the foraging ecology of several marine mammals (Ramsay & Hobson 1991; Hobson & Welch 1992; Rau $et\ al.$ 1992; Ostrom $et\ al.$ 1993; Ames $et\ al.$ 1996). Burton and Koch (1999) recently used stable nitrogen and carbon isotope analysis to demonstrate latitudinal (ocean temperature) and onshore/offshore (productivity) differences among species, populations and sexes for several northeast Pacific pinnipeds.

BOX 8.3 NEW TECHNOLOGICAL TOOLS TO STUDY FORAGING BEHAVIOUR

Very high-frequency and satellite-linked telemetry and microprocessor-based data loggers have played an important role in the study of the foraging behaviour of marine mammals (Croxall 1995) (see Chapter 7). Nevertheless, there are many aspects of foraging behaviour that cannot be addressed in this way (e.g. handling time, capture success and how these vary with prey type). Our ability to study these aspects of foraging has been revolutionized by the use of animal-borne video cameras and data-logging systems (Marshall 1998; Davis $et\ al.$ 1999). In addition to video, these systems collect data on water temperature, swim speed, dive characteristics, orientation and sound (e.g. Burgess $et\ al.$ 1998). Underwater and overhead video observations are also providing information on the strategies employed by foraging bottlenose dolphins for the first time (Rossbach & Herzing 1997; Nowacek 1999). These data are providing our first glimpse of how marine mammals search and capture prey and how foraging behaviour changes as a function of prey type.

Fig. 1 Harbour seal with Crittercam underwater video apparatus attached. (Courtesy of W.D. Bowen.)

dense concentrations of prey, which they capture by engulfing (i.e. gulpers) or skimming (i.e. skimmers). Gulpers, such as the blue whales (*Baleanoptera musculus*), posses ventral throat grooves and a specialized tongue that allow tremendous distention of the throat area during feeding (Lambertsen 1983; Orton & Brodie 1987). The engulfed water is expelled from the buccal cavity by the upward and backward motion of the tongue, such that food is moved posteriorly toward the oesophagus. The gray whale (*Eschrichtius robustus*) is unusual in that it filters amphipods and other crustaceans from the benthos by engulfing sediment and straining prey with its short and relatively stiff baleen plates (Nerini 1984). The skimmers (e.g. bowhead whale, *Balaena mysticetus*) have long plates of fine baleen that allow continuous straining of small prey, such as copepods.

Unlike the heterodont condition of most mammals, toothed whales (odontocetes) possess long rows of

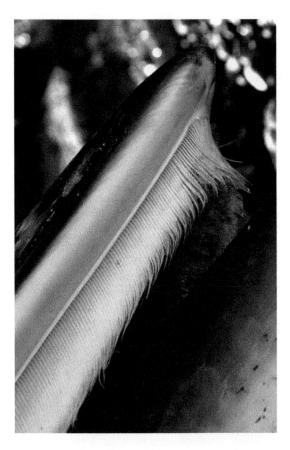

Fig. 8.1 The short stiff baleen of a minke whale adapted to a more piscivorous diet. (Courtesy of A. Read.)

Fig. 8.2 Bottlenose dolphin illustrating the homodont dentition of small odontocetes. (Courtesy of A. Read.)

uniformly shaped teeth (homodonts) designed for grasping and holding prey such as fish or squid (Fig. 8.2). Toothed whales usually consume their prey whole, eliminating the need for specialized teeth to shear or grind food. Odontocete genera, such as *Stenella*, *Delphinus* and *Lagenorhynchus*, are typical of this group with 20–65 pairs of sharp conical teeth in each jaw. Although primarily fish eating (piscivorous), these species also consume significant amounts of squids and crustaceans (Fitch & Brownell 1968; Miyazaki *et al.* 1973; Perrin *et al.* 1973; Recchia & Read 1989; Cockcroft & Ross 1990; McKinnon 1993; Young & Cockcroft 1994; Aarefjord *et al.* 1995).

Species such as the pilot whales (*Globicephala*), which feed primarily on squids (Desportes & Mouritsen 1993; Gannon *et al.* 1998), have fewer (7–12 pairs per tooth row), but larger teeth. This is also true of another squid specialist, the sperm whale (*Physeter macrocephalus*), where 20–25 teeth are found in each lower jaw, but all teeth in the upper jaws are vestigial and rarely erupt through the gums (Rice 1989). Both the number and size of teeth is further reduced in females of some species. This reduction in dentition is most extreme in many beaked whales (Ziphiidae) in which only a single pair of teeth protrudes from the lower jaw, usually only in males. Most beaked whales appear to employ suction to capture prey, rather than grasping items with their teeth. Although relatively little is known of most ziphiid diets, mesopelagic and bathypelagic squid and fish appear to be important food in all species (Gaskin 1982; Mead 1989).

Sirenians have highly modified mouthparts adapted to feeding on plants. The foraging mode of sirenians is reflected by the degree of rostral deflection of their skull (Domning 1980); for example, dugongs (*Dugong dugon*) appear to be obligate bottom feeders as a consequence of their sharply down-turned rostrum, whereas West Indian manatees (*Trichechus manatus*), with their moderately down-turned rostrum, feed equally well on the bottom, at the surface or in the water column (Fig. 8.3). This has been confirmed by Marshall *et al.* (1998) in studies of the use of the perioral bristle fields (modified vibrissae) of the Florida manatee (*T. m. latirostris*) during feeding. They found that the bristle fields on the upper lip were used in a prehensile

Fig. 8.3 Perioral bristle fields (modified vibrissae) of a Florida manatee adapted to feeding on both surface and bottom vegetation. (Courtesy of G. Worthy.)

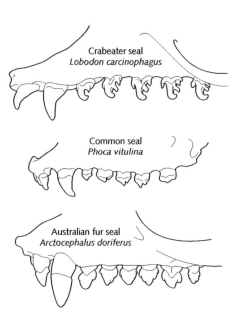

Fig. 8.4 Comparative tooth structure of three pinniped species. (After King 1983.)

manner and that the fields used during feeding on surface vegetation differed from those used during bottom grazing. They concluded that this plasticity in bristle use had evolved to increase the efficiency of manatees as a generalist herbivore.

Perhaps the most extreme variation in this group occurred in the now-extinct hydrodamalines, a lineage of dugongid sireneans that included Steller's sea cow (Domning 1978). The earliest hydrodamalines, known from the Eocene period, were tropical and probably fed exclusively on sea grasses, much as modern-day dugongs. With the onset of Miocene polar cooling, the marine flora in the North Pacific Ocean began to shift from domination by sea grasses to macroalgae. Evolutionary change in the foraging structures of hydrodamalines tracked this environmental shift. Grinding teeth (necessary for proper mastication of the fibrous sea grasses) were gradually replaced by a cornified plate (more efficient for mashing the soft, cellulose-free macroalgae) of epidermal origin.

Modification of the feeding apparatus has been less extensive in pinnipeds. Nevertheless, several Antarctic species show an adaptation functionally similar to the filter-feeding action of baleens. Crabeater seals (*Lobodon carcinophagus*) appear to special-ize on krill, a euphausiid (*Euphausia superba*) that occurs in dense swarms. The importance of krill in the diet of this species is evidenced by the highly modified structure of maxillary dentition (Fig. 8.4) that appears to be an adaptation to some form of filter feeding (Laws 1984). The walruses (Odobenidae) also illustrate a departure from the usual pinniped dentition (see Chapter 1). In contrast with the typical pinniped pattern of feeding on mobile prey (e.g. fishes, cephalopod molluscs and crustaceans) in demersal or pelagic habitats, the odobenids feed largely on sessile or weakly motile benthic invertebrates in soft sediments. Their dental morphology reflects the differing constraints and needs of feeding on these different prey groups. The walruses' molars and premolars have large, flattened crowns adapted for feeding on bivalve molluscs. Extreme elongation of the walruses' canine teeth into 'tusks' is another obvious departure in dental morphology from the typical pinniped (and carnivore) dentition. Although walrus tusks are commonly thought to be adaptations for digging and furrowing in the soft sediments, their evolutionary explanation probably stems more from the fact that benthic feeding must have freed the walruses from the constraints that tusks would have imposed on the pursuit and capture

Fig. 8.5 A sea otter with a rock tool on its abdomen. (Courtesy of Friends of the Sea Otter, Carmel, California.)

benthic invertebrates such as gastropod and bivalve molluscs. In contrast with the carnassial (shearing) form of the molars and premolars in all other otters and nearly all other carnivores, the sea otter has heavy bunodont (flat, crushing) molars and vestigial premolars, thus facilitating consumption of their hard-shelled prey. In addition, sea otters commonly carry rocks from the sea floor to the ocean surface to assist in breaking into heavy-shelled prey (Fig. 8.5). Except for marine-living populations of Cape clawless otters, such tool-using behaviour among mammals is found only in primates (Alcock 1972; Estes 1989).

of mobile prey. Unconstrained by such detriments to foraging efficiency, enlarged canines provide a variety of social advantages to mammals (Repenning 1976; Packer 1983).

The otters, despite their more recent phylogenetic diversification compared with cetaceans, pinnipeds and sirenians, show a variety of interesting foraging patterns. The dozen or so extant otter species display a dichotomous foraging mode, some are mainly piscivores while others feed principally on benthic invertebrates. These groups have fundamentally different sensory–motor capabilities, as seen in the structure and evolution of their brains (Radinsky 1968). The piscivorous otters, which include such species as the Old and New World river otters (*Lutra lutra* and *Lontra canadensis*) and the giant otter (*Pteroneura brasiliensis*), capture highly elusive prey (mostly fishes) in their mouth, thus requiring detailed sensory–motor coordination in the head and facial region. In contrast, the invertebrate feeding otters (including the sea otter; Asian small-clawed otter, *Aonyx cinerea*; and Cape clawless otter, *Aonyx capensis*), use their forepaws to sense and capture prey. Brain structure and function in these groups are organized accordingly. In all mammals, sensory–motor function is laid out along the brain's prefrontal gyrus with the medial and lateral regions of the gyrus corresponding with respective body regions. For the piscivorous otters, the greatest structural complexity occurs in the medial region of the gyrus, whereas invertebrate feeders show the opposite pattern (see Chapters 4 and 5).

Both the behaviour and dentition of sea otters are highly modified for feeding on hard-shelled

8.3 LIFE HISTORY AND FEEDING

Marine mammals share many of the life history characteristics of other large mammals (see Chapter 10). Life history variation and physiology are strongly influenced by body size (e.g. Peters 1983). For example, body size affects both oxygen storage and utilization and thus can be expected to influence dive depth and duration (Boyd & Croxall 1996) (see Chapter 9), which in turn affects the range of foraging options available to different species. Body size also influences the evolution of foraging strategies of marine mammals through its effects on fasting ability and seasonal energy storage (e.g. Brodie 1975). The abilities to store energy and withstand periods of food shortage affect the temporal scales of foraging, which in turn affect reproductive strategies (see Chapters 9 and 10).

Animals can be regarded as capital or income breeders based on their capacity to store energy during feeding for use in reproduction (Sibly & Calow 1986). This capacity is largely a function of body size as mass-specific metabolic expenditure decreases with body mass, whereas fat storage increases with body mass to the power of 1.1 (Prothero 1995). In the case of marine mammals, capital breeders include the larger baleen whales, polar bears, the larger species of phocid seals (Fig. 8.6) and adult male otariids. In these species, the energy and nutrients needed for reproduction are stored prior to the breeding season and then used during the breeding season. Income breeders, by contrast, have few energy and nutrient reserves and must continue to feed during the breeding season. Sea otters, most odontocetes,

(a)

(b)

Fig. 8.6 The grey seal females shown here in (a) early and (b) late lactation are examples of a capital breeder. The female stores all the energy and nutrient requirements for reproduction prior to giving birth and expends stores during lactation. (Courtesy of W.D. Bowen.)

smaller phocids and adult female otariids are generally income breeders.

One example of the influence of body size on foraging strategies is found among baleen whales. The annual migration of baleen whales from low latitudes in winter to high latitudes in summer enables these filter feeders to exploit the high seasonal productivity of high-latitude areas (see Chapter 7). Brodie (1975) argued that these species migrate to the low latitudes during the winter because prey densities sufficient to permit feeding at high latitudes are restricted to the summer. Prey density at low latitudes are seldom sufficient to permit efficient feeding, so individuals fast until prey densities increase at high-latitude feeding areas. According to Brodie (1975), selection for larger body size among baleen whales was a way of adjusting specific metabolic rate and permitting efficient use of lipid stores during extended exclusion from feeding grounds. Thus, intraspecific variation in body size, such as that found in blue and fin whales, where the Antarctic forms are larger than those in the northern hemisphere, may be a response to differences in the duration of the

feeding and fasting periods (Brodie 1975). Larger forms in the Antarctic feed over a much more restricted period of the summer than those in the northern hemisphere, where somewhat lower prey densities occur for longer periods of the year.

Body size and phylogeny both appear to have had strong influences on the evolution of foraging strategies of pinnipeds during the breeding season (Boness & Bowen 1996; Boyd 1998) (see Chapter 10). The harbour seal provides an interesting example of the shared influence of body size and phylogeny on foraging strategies during reproduction. The harbour seal is a relatively small phocid species with adult females and males weighing about 85 kg and 110 kg, respectively (Coltman *et al.* 1998). Female harbour seals give birth to single young that they nurse for an average of 24 days. Mating occurs at sea. Unlike the larger phocid species, females cannot support the entire cost of lactation from body reserves (Bowen *et al.* 1992) and begin to undertake regular foraging trips after fasting for about 1 week (Fig. 8.7) (Boness *et al.* 1994). The frequency and duration of these foraging trips are similar to those found in some lactating otariids. By contrast, most adult male harbour seals devote considerable time to foraging early in the season (Coltman *et al.* 1997) and in so doing most males maintain or increase body mass during this period (Walker & Bowen 1993). During the latter part of the season when oestrous females become increasingly available, males reduce foraging and devote more time to reproductive behaviour while at sea. As a result, males may lose up to 1 kg of mass daily (Walker & Bowen 1993). Given their body size, Coltman *et al.* (1998) show that most adult male harbour seals could fast for only about 19 days or about 60% of the period when oestrous females are available. Thus, most males appear to balance foraging and reproductive behaviour in such a way as to maximize potential encounter rates with oestrous females (Coltman *et al.* 1997).

The sea otter provides another interesting example of the interplay between body size (the smallest of all marine mammals), phylogeny (the most recent expatriate from land to water) and foraging. Sea otters must forage actively and often to maintain body temperature in the frigid coastal waters of the North Pacific Ocean and Bering Sea. An unusually high metabolic rate, acquired at least in part from

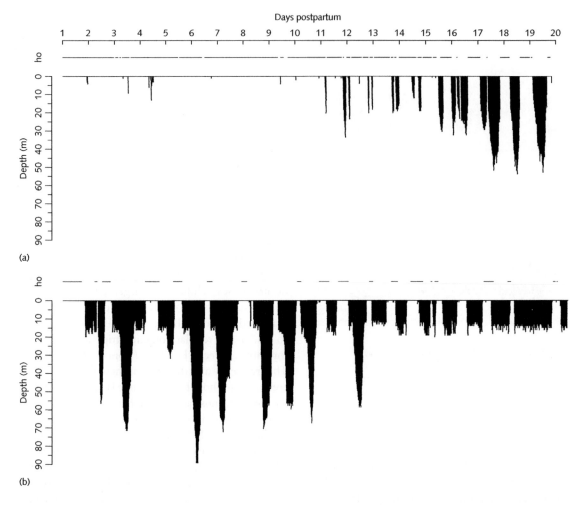

Fig. 8.7 Contrasting diving behaviour of male and female harbour seals over a 20-day period during the breeding season on Sable Island, Canada. Data are compressed such that every 150th sampled depth is plotted: (a) an adult female, (b) an adult male. Periods when the seals haul out on land are represented by the broken line above each dive plot. (Adapted from Boness *et al.* 1994; Coltman *et al.* 1997.)

their mustelid ancestors, fuels the demands of this mismatch between body size and heat loss, in turn requiring the sea otter to consume 20–25% of its body weight daily in prey (Kenyon 1969; Costa 1982). The apparent consequence of these phylogenetic and morphological constraints is a fundamentally different link between foraging and reproduction than that seen in any other marine mammal. Sea otters must feed almost continuously. Even in the days and weeks immediately postpartum, they must interrupt parental care to replenish their energy and nutrient stores. Thus, virtually the entire

cost of lactation must be met using the income from feeding.

8.4 SPATIAL AND TEMPORAL SCALES OF MARINE MAMMAL FORAGING BEHAVIOUR

To understand the ecology of apex predators in large marine ecosystems, we need information on the scales at which they forage. Environmental variation at different temporal and spatial scales affects the survival and reproductive success of individuals and thus defines evolutionary selection pressures (e.g. Whitehead 1996). Few groups of animals exhibit such dramatic variation in body size as marine mammals. The range in size from sea otters to blue whales spans four orders of magnitude in body mass. Thus, it is likely that a wide range of spatial and temporal scales will be relevant to different marine mammals. Nevertheless, in all marine mammals, the environmental variation observed by individuals will be a consequence of their own scales of experience. Life history adaptations such as dispersal, migration and iteroparous reproduction (i.e. species in which individuals reproduce more than once) effectively modify the realized scales of variability experienced by individuals (Levin 1992).

Large-bodied, long-lived species with low reproductive rates must have evolved adaptations that enable them to deal with environmental variability over temporal scales of less than their lifetimes and spatial scales of less than their home ranges (Whitehead 1996). This temporal and spatial flexibility enables large, long-lived species to forage at multiple scales and thereby reduce the effects of local variability in prey abundance. Many species of marine mammals have these life history characteristics, but the sperm whale is a particularly good example. Sperm whales are deep divers that feed on meso- and bathypelagic cephalopods (Clarke 1980). It is only possible for sperm whales to track long-term environmental variability by changing vital rates over several decades or longer. Individual whales must deal with variability at shorter timescales (i.e. months or years). Whitehead (1996) examined the variability in the feeding success of sperm whales, as indicated by defaecation rates, off the Galápagos Islands at temporal scales of 5 h to 4 years and spatial scales of 100–5000 km, i.e. within an individual's potential range. In this population, feeding success is reduced during warm conditions (Smith & Whitehead 1993). However, variability in feeding success at temporal scales of up to several months and spatial scales of 100 km was about 60% of the long-term mean, but at scales of 2–4 years and distances > 500 km was 130% of the long-term mean. Despite its large size, Whitehead (1996) estimated that a female sperm whale can survive only about 3 months on lipid stores. Juveniles

can survive for even shorter periods. Given their life history, it seems unlikely that long periods of food shortage caused by environmental variability result in substantial mortality. Further, the lack of coherence in environmental variability over spatial scales of 500 km or more suggests that an appropriate strategy for sperm whales is to move to better feeding areas when faced with food shortages. Movements on this scale have been observed (Best 1979) and would take only 5–6 days at observed rates of travel (Whitehead 1996) (see Chapter 7). Behavioural plasticity in response to variation in the distribution and abundance of prey allows these large, long-lived species to succeed in an unpredictable and patchy environment on both ecological and evolutionary timescales.

Large body size and the ability to store large amounts of fat in the form of blubber enable some species of marine mammals (e.g. baleen whales, larger phocid species) to feed irregularly and thus to exploit distant foraging locations and patchy resources. For example, northern elephant seals (*Mirounga angustirostris*), one of the largest pinnipeds, exhibit extreme size dimorphism with adult males more than five times the body mass of adult females. They make two long-distance migrations each year between breeding and moulting sites in California and pelagic foraging areas in the North Pacific (Stewart & DeLong 1993, 1995) (see Chapter 7). This annual double migration of 18 000–21 000 km is the longest reported for a mammal (Stewart & DeLong 1995). Both sexes use the California Current as a corridor to areas further north. However, adult females forage in different areas of the North Pacific from males (the much larger sex), perhaps to reduce intraspecific competition for food (see Fig. 7.9). Females remain south of 50°N latitude during both migrations, but tend to forage in areas farther west during the postmoult period. There is preliminary evidence from three individuals that some females use similar migration routes and feeding areas between years. Males migrate farther north than females throughout the year with most travelling to the northern Gulf of Alaska and the eastern Aleutian Islands. These distribution patterns seem to correspond with three water masses in the North Pacific and with the distribution of cephalopods, which are the main prey of elephant seals. Females tend to feed near the Sub-arctic Current, between 40° and 50°N latitude, an area of cooler water to the north of warmer sub-tropical waters. Males aggregate along the offshore boundary of the Alaska Stream. These open-ocean frontal regions are known or thought to be areas of high biological productivity (Stewart & DeLong 1995).

Body size and sex also appear to influence the char-acteristics of foraging trips in less size-dimorphic species. Thompson *et al.* (1998) found a positive relationship between the proportion of time at sea during the non-breeding season and body mass for both male and female harbour seals in the Moray Firth, Scotland. Foraging trips of males (mean = 61.1 h) were about twice the duration of female trips, and males travelled an average of 25.5 km to foraging locations compared to only 13.9 km for females. Larger animals had significantly longer trips and travelled further to foraging areas, consistent with theoretical predictions for a central place for-ager (i.e. an individual returning to the same central location between successive foraging trips) (Orians & Pearson 1979). Thompson *et al.* (1998) speculated that these relationships reflect energy constraints acting on males and females of different body size (also see Coltman *et al.* 1997). Diving ability scales to body size in pinnipeds with an exponent similar to that of metabolic rate (Boyd & Croxall 1996). In the Moray Firth, water depth gradually deepens away from the haul-out areas such that it may be energetically advantageous for larger seals to forage at the deeper, more offshore sites.

Recent work by Iverson *et al.* (1997b) has demonstrated mesoscale partitioning of foraging habitats by harbour seals in Prince William Sound, Alaska. Prince William Sound is a small (90 km by 130 km) complex embayment of the Gulf of Alaska where harbour seals haul out or give birth at over 50 sites. Fatty acid signatures (Box 8.1) from harbour seal blubber biopsies differed within the Sound at a spatial scale of about 40–50 km, but also differed between haul outs separated by only 9–15 km. These differences in fatty acid signatures undoubtedly reflect differences in the composition of diets and indicate a level of habitat partitioning that was unexpected as individual harbour seals could easily forage throughout the

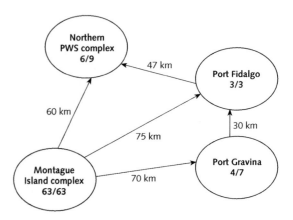

Fig. 8.8 Illustration of the spatial partitioning of the diets of harbour seals in Prince William Sound (PWS), Alaska based on fatty acid signature analysis. All 63 seals from five haul-out sites near the northwestern end of Montague Island in southern Prince William Sound were correctly identified as having distinct fatty acid signatures from haul outs 60–70 km away. Initial estimates, based on small samples, also suggest the partitioning of foraging areas in other areas of the Sound. (Adapted from Iverson *et al.* 1997.)

Sound (Fig. 8.8). Satellite telemetry studies of the harbour seals within the Sound have confirmed the scale of foraging activity indicated by the fatty acid signature analysis (Frost *et al.* 1998). If such restricted foraging ranges are typical of longer-term behaviour, seals may benefit from increased foraging success through knowledge of local prey distribution or reduced risk of exposure to pred-ators such as killer whales.

Reduction in food availability or an increase in foraging costs will change the optimal behaviour of a predator. Boyd *et al.* (1997) examined how foraging effort in lactating Antarctic fur seals, a central place forager during the breeding season, responded to experimentally increased foraging costs. Foraging costs were increased by gluing a small (250 g) cuboidal block of wood on the back of females. They examined the effects at three temporal scales: individual dives, bout of dives and complete foraging trips lasting about 5 days. At the scale of individual dives, treated females made shorter, shallower dives, and compensated for reduced swimming speed by diving at a steeper angle. Females appeared to adjust their behaviour to maximize the time spent at the bottom of the dive. At the scale of bouts, there was no difference between the treated and control females, but at the scale of complete trips, treated females spent more time at sea. Nevertheless, there was no difference in the growth rate of pups between the two groups of females. Thus behaviour adjustment at the scale of individual dives enabled females to compensate for the increased foraging costs (Boyd *et al.* 1997).

It is usually difficult to simultaneously measure the foraging behaviour of a marine predator and the availability of its prey, but temporal variation in the behaviour of a predator may provide an insight into the spatial distribution of prey (Boyd 1996). In species like fur seals, diving is thought to reflect foraging activity because these species remain at the surface while travelling or resting (Boyd 1996). In the case of Antarctic fur seals, lactating females appear to feed mainly on krill, which is patchily distributed. Both of these features make it possible to use the diving behaviour of females to investigate the way fur seals respond to within-season and interannual variation in the patchiness of their prey. Using time–depth recorders (TDRs), some of which were able to measure swimming speed, Boyd (1996) found that the distribution of travel durations between bouts of diving changed over a 5-year period in such a way as to suggest that the spatial distribution of krill swarms varied between years. The temporal behaviour of females suggested that prey patches were clumped at fine scales but that there was a more even distribution of prey at the mesoscale level. During a year of low krill abundance, there were indications that females had difficulty finding enough food as evidenced by reduced pup growth rate. However, the foraging behaviour of females (i.e. longer time spent at each patch) suggested that there was no reduction in the number of patches of krill in that year, but that the patches were of poorer quality. Thus, females were behaving in a way that would maximize their average rate of energy intake, as predicted by optimal foraging theory (e.g. Charnov 1976).

Simultaneous measurements of both predator and prey distributions, although often expensive and challenging, are becoming more common and permit more detailed study of foraging over a variety of spatial scales. Hunt *et al.* (1992) continuously measured the abundance of krill, macaroni penguins and Antarctic fur seals (*Arctocephalus gazella*) in the waters around Bird Island, South Georgia during the breeding season. They found that the highest correlations between fur seals and krill occurred over the scales of 10–100 km (effect of distance from colony statistically removed). They noted, however, that the weak correlations at small spatial scales may have been underestimated because they could not consistently identify predators that were feeding, surveys were not conducted at night, and the measure of krill density included krill that would not have been available to these predators.

Sea otters also provide an example of the complex relationships between prey availability and foraging behaviour. Otter populations were hunted nearly to extinction in the Pacific maritime fur trade (Kenyon 1969). By 1911, when further hunting was prohibited, only a few colonies survived, widely scattered across the Pacific rim. Because of the sea otter's limited dispersal ability, recovery was highly discordant in space and time. Areas near the remnant colonies recovered early whereas more distant locations recovered later. This historical accident, coupled with the fact that sea otter predation greatly alters prey assemblages (Estes & Palmisano 1974), created a natural experiment by which the effect of prey availability on foraging behaviour could be evaluated. Estes *et al.* (1981) utilized this approach by contrasting the diet and foraging behaviour of high- and low-density otter populations in the western Aleutian Islands, Prince William Sound and central California. Although prey species varied among these three regions, several interesting patterns emerged. Dietary diversity was consistently greater in long-established populations. Furthermore, much of the dietary variation within populations was attributable to individuals, which typically feed on a single prey type throughout a foraging bout. When prey switching did occur, it happened most commonly following an unsuccessful dive. Subsequent work using this same approach has shown that time spent foraging by sea otters varies strongly as a function of population status. Otter populations at or near equilibrium density spent > 50% of their time foraging whereas recently established and still growing populations spent < 20% of their time foraging (Estes *et al.* 1982).

8.5 EFFECTS OF ENVIRONMENTAL VARIABILITY AND PREY ABUNDANCE ON FORAGING DECISIONS

Survival and reproduction depend on successful foraging. Thus, we would predict the foraging behaviour of marine mammals to change in response to changes in prey abundance and distribution. Furthermore, based on foraging theory, we might expect these changes to maximize one of three currencies: efficiency (the ratio of energy gain over energy expenditure), net rate of energy intake or average rate of delivery of energy to offspring (Ydenberg et al. 1994; Houston 1995).

In the case of central place foragers (Orians & Pearson 1979) such as lactating Antarctic fur seals, interannual changes in food availability can have dramatic effects on fitness and population dynamics (Costa et al. 1989; Boyd et al. 1994; McCafferty et al. 1998). Antarctic fur seals are well suited for this type of study because during the breeding season they have a specialized diet consisting almost entirely of Antarctic krill (Euphausia superba), and the abundance of krill varies significantly between years. For example, the austral summer of 1990–91 was a period of low krill abundance. During 1990–91, female fur seals spent significantly greater effort both in terms of more time and increased activity while foraging than usual (Boyd et al. 1994). Despite this greater foraging effort (estimated at one-third to one-half greater than normal), mothers were unable to meet the energy demands of lactation resulting in lower pup growth rates and increased pup mortality (Lunn & Boyd 1993; Lunn et al. 1994). Although females made a greater number of dives on longer foraging trips during a year of low krill abundance, there was no difference in the number of dives per hour or per cent of time submerged. This implies that when food was scarce, fur seal females simply spent longer searching for food to meet both their own requirements and those of their offspring (McCafferty et al. 1998).

Not all central place foragers respond to changes in prey abundance in the same way. Costa and Gentry (1986) found that unlike Antarctic fur seals, northern fur seal (Callorhinus ursinus) females kept foraging trip duration constant during a season of reduced prey abundance, but worked harder (i.e. increased their field metabolic rate) to support lactation. This was interpreted as a response, in part, to increased cost of foraging associated with changes in the target prey species (Costa et al. 1989) as different foraging patterns may have different metabolic costs (Costa 1988).

In non-central place foragers, interannual changes in the abundance of important prey can also have energetic consequences to the extent that the net energy gain from consuming different types of prey may differ significantly. Thompson and colleagues have studied the causes and consequence of variation in the diet and foraging behaviour of harbour seals in the Moray Firth, Scotland (Thompson et al. 1991, 1996). Interannual variation in the composition of harbour seal diets was reflected by differences in the winter abundance of clupeid fishes such as Atlantic herring (Clupea harengus) and spratt (Sprattus sprattus). In years of high clupeid abundance, there was an increase in the proportion of these species in the diet and seals foraged further offshore. Thompson et al. (1996) also showed a significant difference in the condition of seals caught in the spring following good and bad clupeid years, suggesting that there are energetic advantages and disadvantages to foraging on certain prey types.

Events in offshore oceanic ecosystems strongly influence diet, body condition and thus demography of coastal-feeding marine mammals. During most years, sea otters in the Aleutian archipelago consume prey associated with nearshore kelp forest ecosystems, such as sea urchins, crabs and kelp forest fishes. The abundance and quality of these prey species regulate local otter population size through their effects on body condition and increased postpartum and postweaning mortality rates in systems at or near carrying capacity (Monson et al. 2000). The sea otter's diet is subsidized on rare occasions by Pacific smooth lumpsuckers (Aptocyclus ventricosus), which typically live far offshore but undergo episodic inshore spawning migrations (Yoshida & Yamaguchi 1985). These inshore migrations occur during winter and early spring, the very time of the year when food scarcity is most severely felt by the otters. During these episodes, vast numbers of lumpsuckers appear in the coastal system, and because of their sluggish behaviour, are easily captured by sea otters. Lumpsuckers become the otter's main source of sustenance in such times (Watt *et al.* 2001). As a consequence, time spent foraging and foraging bout length decline, body condition (as measured by age-specific length : mass ratio) increases, and the normally high late winter/spring mortality rate of young otters declines to a very low value. In effect, this oceanic food subsidy temporarily releases sea otters in the nearshore ecosystem from food limitation.

A similar event occurs in central California during strong El Niño conditions. In this case, vast swarms of pelagic red crabs (*Pleurencodes*) are transported northward from their more typically southern range in the anomalously warm surface water that flows northward along the Pacific coast of North America during El Niño events. Numerous consumers, including sea otters, switch to diets largely or wholly comprised of pelagic crabs during this period (Lyons 1991). The phenomenon is so striking that the normally white guano-covered roost of seabirds turn red from the crabs' pigments. The consequences (benefits?) to sea otters of this prey regime shift are currently unknown.

8.6 PREY SELECTION

How do predators select which prey to eat? Emlen (1973) suggests that selection for efficient foraging acts along three dimensions: (i) increased ability to recognize food requirements and the best time to feed, (ii) increased ability to chose the best foods, and (iii) increased ability to locate, capture and ingest foods. To these, we would add increased ability to digest and assimilate foods. Each of these components of feeding may be affected by the sex, age, experience and genetic make-up of an individual, its reproductive status, intra- and interspecific competition, prey characteristics (size, nutritional quality, behaviour, relative abundance and distribution, and handling time) and the risk of predation.

For the most part, ecologists use indirect methods to study the foraging ecology of marine mammals. Divers have observed foraging gray whales (Darling 1977) and feeding sometimes can be inferred from surface observations of behaviour (e.g. Simila &

Ugarte 1993; Würsig & Clark 1993; Baird & Dill 1995). But, with few exceptions (e.g. sea otters: Estes *et al.* 1981; Riedman & Estes 1990; Kvitek *et al.* 1993; manatees: Hartman 1979; northern right whales (*Eubalaena glacialis*) while skim feeding: Mayo & Marx 1990), direct observation of feeding is not possible. The most common methods used to determine the diet of marine mammals rely on the identification of prey structures that are resistant to digestion and can be collected from stomachs, intestines or faeces (see Boxes 8.1 and 8.2 for new approaches). Sagittal otoliths are the most commonly used prey structure, but bones, scales and lenses also provide a means of prey identification (Fitch & Brownell 1968; reviews by Pierce & Boyle 1991; Pierce *et al.* 1993; Bowen & Siniff 1999). Where cephalopods are eaten, beaks can be used for prey identification (e.g. Clarke 1986). Our reliance on indirect methods to determine what is eaten, coupled with the difficulty in estimating the availability of prey at scales that are relevant to marine mammals, has made this aspect of feeding ecology particularly challenging to study.

Most pinniped species feed on a variety of prey. For example, more than 100 taxa of crustaceans, cephalopods and teleosts have been identified from the stomach contents of 6457 harp seals (*Pagophilus groenlandicus*) in the northwest Atlantic collected over the past 40 years (reviewed by Wallace & Lavigne 1992). Over 40 taxa of fish and invertebrates were found in 682 grey seal stomachs examined from eastern Canada (Benoit & Bowen 1990) and more than 48 taxa were identified from 2841 faecal samples collected from harbour seals in British Columbia (Olesiuk *et al.* 1990). Lowry *et al.* (1990) found 52 prey types in 1476 faecal samples collected from California sea lions. Over 60 taxa, mainly fish, were found in the stomachs of northern fur seals sampled at sea from 1958 to 1974 (Kajimura 1984). Such studies have led to the view that pinnipeds are generalist predators. Given their wide-ranging foraging behaviour and large size there are theoretical reasons for expecting that pinnipeds would consume a wide variety of prey types. As Emlen (1973) noted, if one accepts that such predators are food limited then we would expect them to use a wide variety of prey to maximize their fitness.

Although pinnipeds may consume a wide variety of prey, relatively few species (usually less than five and often only two or three) account for most of the energy ingested in any one season or geographical location. For example, of the 24 taxa eaten by grey seals on the Eastern Scotian Shelf, Canada, two to four species accounted for over 80% of the biomass of prey eaten in each of eight collections ($n = 365$) between June 1991 and January 1993 (Bowen & Harrison 1994). Hammond and Prime (1990) and Hammond *et al.* (1994) also found evidence of seasonal and geographical variation in the number of species that comprise the diet in Scotland. For example, at Donna Nook six species accounted for 80% of the biomass eaten by grey seals in some months, whereas only two species comprised that percentage of the diet in other months (Hammond & Prime 1990). Only five of 52 species accounted for more than 80% of the relative frequency of prey in the diets of California sea lions, although both the species and their relative contribution to the diet differed before, during and after the El Niño of the early 1980s (Lowry *et al.* 1990).

Although pinnipeds appear to be generalist predators for the most part, several species appear to specialize. Both the crabeater seal (*Lobodon carinophagus*) and lactating female Antarctic fur seals feed almost exclusively on krill, mainly *Euphausia superba* (Laws 1984; Doidge & Croxall 1985). Ringed seals (*Phoca hispida*), and to a lesser extent bearded seals (*Erignathus barbatus*), are the main food of the polar bear (Stirling & Archibald 1977; Smith 1980; Hammill & Smith 1991). During periods of hyperphagia in the spring, polar bears often consume only the blubber layer of their prey, thereby exhibiting an even further level of specialization (Fig. 8.9).

Many odontocetes also feed on a wide variety of prey. For example, 20 fish, squid and crustacean species have been recovered from the stomachs of more than 500 harbour porpoises from the Bay of Fundy and Gulf of Maine (Smith & Gaskin 1974; Recchia & Read 1989; Smith & Read 1992; Gannon *et al.* 1997). Nevertheless, only a few species are common in the diet and Atlantic herring (*Clupea harengus*), a lipid- and energy-rich fish, contributed approximately two-thirds of the mass ingested by these porpoises. Bottlenose dolphins

Fig. 8.9 An adult harp seal that had been killed by a polar bear on the ice of southern Labrador during the breeding season. Only the skin and blubber over the thorax of the seal were eaten. (Courtesy of W.D. Bowen.)

are also described as exhibiting a diverse diet; Barros and Odell (1990) identified 43 taxonomically diverse prey species (from 25 families) in 76 stomachs of stranded dolphins from the southeastern United States. Like harbour porpoises, however, the diet of these dolphins is often dominated by a few species. In Sarasota, Florida, for example, pinfish (*Lagodon rhomboides*) comprised nearly 70% of all prey items (Barros & Wells 1998). Until we can make direct observations of prey choice made by wild dolphins and porpoises (Box 8.3), it is difficult to draw conclusions regarding prey selection.

Dietary studies of sea otters provide a strikingly different view of prey selection. Riedman and Estes (1990), in a review of available information on sea otter foraging through the 1980s, listed more than 160 known prey species. This remarkable diversity is contributed in part by habitat variation, both on a geographical scale (from the warm temperate Channel Islands of southern California to boreal/subarctic Alaska and Russia) and between the rocky and soft sediment benthos. Nonetheless, even within a particular region and habitat type, sea otters often consume a diversity of prey. For example, more than 50 prey types are consumed by sea otters in kelp forests of the southern Monterey Bay region in central California. However, a more detailed look has shown that the great majority of this dietary diversity is contributed by variation among individuals (Lyons 1991). Longitudinal records of tagged sea

Fig. 8.10 A group of bottlenose dolphins stranding fish at Hiltonhead, South Carolina. (Courtesy of K. Uriam.)

otters indicate that most individuals specialize on one to three prey types. Records for some individuals span nearly a decade, thus suggesting that these individual patterns and preferences persist for a lifetime. Furthermore, individualized dietary preferences do not appear to be driven by small-scale spatial variation in food availability as otters with nearly coincident home ranges often show no overlap whatsoever in diet. The accumulating data indicate that individualized dietary patterns in sea otters are matrilineally inherited, which is not surprising given the relatively long period of mother–young association (about 6 months) and the apparently great need for learning by otter pups during this period. The more perplexing yet intriguing questions of: 'How do these individual differences arise?' 'What are their associated costs and benefits?' and 'What are their ecological consequences for both otters and the kelp forest ecosystem?' remain unanswered.

As long-lived predators, most species of marine mammals are likely to exhibit individual variation in foraging behaviour. The repertoire of capture techniques used by bottlenose dolphins, for example, includes: intentional beaching, in which dolphins follow stranded prey onto muddy banks (Fig. 8.10) (Hoese 1971); fish whacking, in which fish are struck clear of the water with the dolphin's flukes (Wells *et al.* 1987); and sponge carrying, in which dolphins use sponges on their rostra while foraging along rocky reefs (Smolker *et al.* 1997). Recent work in Sarasota, Florida indicates that individual dolphins favour particular strategies and that these

preferences may be passed from mother to calf (Nowacek 1999). The development and rapid transmission of novel foraging strategies has also been documented for humpback whales in the Gulf of Maine (Weinrich *et al.* 1992).

Baleen whales also specialize during foraging. Hoelzel *et al.* (1989) found that individual minke whales specialized in one of two types of feeding behaviour: exploiting concentrations of fish below flocks of feeding birds or lunge feeding during which the whale actively concentrated prey near the surface. Individual whales were consistent in their use of a particular strategy, with bird-associated feeding representing the use of ephemeral concentrated patches and lunge feeding representing the use of more predictable patches of food.

To begin to understand prey selection in marine mammals, attributes of potential prey types must be measured at scales that are relevant to these aquatic predators. There have been several attempts to do this at different scales. Bailey and Ainley (1982) reported that the interannual decline in California sea lion predation on Pacific hake (*Merluccius productus*) was related to the decline in the abundance of 2–4-year-old hake, the most common ages eaten. Sinclair *et al.* (1994) found that interannual variation in the importance of walleye pollock (*Theragra chalcogramma*) in the diet of northern fur seals (*Callorhinus ursinus*) was positively related to year-class size of pollock. They also found that the species and size composition of prey taken by fur seals was similar to that in mid-water trawl collections, but differed from bottom trawl catches, both conducted at the same locations that the fur seals were sampled. Based on these results, Sinclair *et al.* (1994) concluded that fur seals were size selective, mid-water feeders during the summer and autumn in the eastern Bering Sea. Bowen and Harrison (1994) compared the diet of grey seals in the vicinity of Sable Island with estimates of the abundance of prey species from ground fish, bottom trawl surveys conducted in the same area. Although there were exceptions, generally these comparisons suggested that the more abundant and widespread species were more frequently eaten by grey seals. Lawson *et al.* (1998) reported evidence of positive selection of both capelin (*Mallotus villosus*) and Arctic cod (*Boreogadus saida*) by harp seals at an inshore area off northeastern Newfoundland, although Arctic cod dominated the diet. Capelin was also positively selected offshore, where it dominated the diet; however, the sample size of harp seals examined offshore was too small to reach firm conclusions.

At a finer scale, Tollit *et al.* (1997) estimated harbour seal diets in the Moray Firth between 1987 and 1995 from regular collections of faecal samples. In January 1992 and 1994, and June 1992, they also estimated the biomass of fishes using a combination of acoustic and bottom trawl gear. One to three fish species dominated the diet (> 85%) of harbour seals in each of the months. Based on an odds ratio selection index, Altantic cod (*Gadus mohua*) were strongly selected in January 1992, but not in 1994 when cod were 4.5 time more abundant. By contrast, herring (*Clupea harengus*) were ignored by seals in 1992, but were selected in January 1994, despite a fivefold decrease in estimated abundance. Sprats (*Sprattus sprattus*) were taken in proportion to their abundance when most abundant, but were excluded from the diet in favour of sand eels (*Ammodytes marinus*) in June 1992, when sand eels were taken in proportion to their abundance. Comparisons of the reconstructed fish lengths with those from trawl catches indicated that seals selected larger than average individuals from the local fish populations. Sea otters also selectively consume the larger sizes of sea urchins (Estes & Duggins 1995) and mussels (VanBlaricom 1988).

Different foraging tactics may be used to feed on different types of prey. We might expect these tactics to reflect differences in prey characteristics, such as size, energy content and behaviour (e.g. schooling vs cryptic), such that individuals maximize the rate of net energy intake. During lactation, female northern fur seals alternate between foraging trips of about 7 days and periods of 2 days ashore to suckle their pups. During these foraging trips females use two different modes of feeding, although a mixture of these two modes is also seen (Gentry *et al.* 1986; Goebel *et al.* 1991). Some females generally dive to depths of < 75 m, i.e. 'shallow divers', whereas others generally dive to deeper depths, so-called 'deep divers'. Shallow divers tend to dive less deeply during the night than during the day suggesting that they are feeding on prey that migrates toward the surface at night. Females diving in deep

water beyond the continental shelf primarily showed the shallow-diving pattern. Depth of diving in deep divers, on the other hand, typically showed little change with respect to time of day suggesting that feeding occurred on non-vertically migrating prey. These females generally foraged at or near the continental shelf break. Although the diet of the females studied was not known, these different diving patterns suggest that individual females may target different types of prey. Data from pelagic sampling of fur seals in the Bering Sea (Kajimura 1980) indicate that females feeding beyond the continental shelf consume vertically migrating oceanic squids or deep-sea smelts. By contrast, deep-diving females were more likely to feed on walleye pollock, Pacific herring and capelin, which are largely demersal species. It should be noted, however, that these conclusions are based on measurements from a small number of females, over only a portion of lactation. Additional data may reveal that these patterns form more of a continuum of foraging behaviour mediated by the relative availability of different prey types.

What are the energetic consequences of shallow versus deep diving for lactating fur seal females? Costa and Gentry (1986) used doubly labelled water turnover to measure food intake and metabolic rate of lactating fur seals instrumented with TDRs. Two deep divers consumed less food and expended less energy than the two shallow divers, but all females gained similar body mass during the single foraging trip. Similar conclusions were reached by Goebel *et al.* (1991) from observations of behaviour, i.e. deep divers expended less foraging effort than did shallow divers. These differences in foraging effort between shallow and deep divers might be explained by differences in the energy content of prey, prey size or rate of prey capture. Unfortunately the data necessary to evaluate these possibilities are not available, but the use of animal-borne video systems (Box 8.3) and fatty acids (Box 8.1) combined with energetic measurements could provide answers to these questions.

Evidence of the use of different foraging tactics has also been reported in harbour seals. Tollit *et al.* (1997) interpreted a dramatic change in diet as indicating different foraging tactics by harbour seals to respond to prevailing feeding conditions. Such responses are common in many other predators (Krebs & Davies 1979; Stephens & Krebs 1986; Berger 1998). Tollit *et al.* (1997) based their conclusion not only on what was eaten, but what was not eaten. Clearly, their data showed that the prey choices made by seals were not dependent upon absolute abundance of prey types. They suggest that seals adopted either a 'demersal' or 'pelagic' strategy of feeding. Although the fitness consequences of different foraging tactics are unknown, there is evidence that short-term changes in body condition are associated with changes in diet (Thompson *et al.* 1997). Similar associations between diet and body condition have been shown for sea otters (see above).

Bottlenose dolphins exhibit considerable plasticity in their foraging techniques. In the shallow, tidal creeks of South Carolina and Georgia, bottlenose dolphins chase fish onto mud banks to capture them (Fig. 8.10), before sliding back into the water (Hoese 1971). Along the Gulf coast of Florida, dolphins may stun or kill prey with their flukes, launching fish up to 9 m into the air (Wells *et al.* 1987). In the Bahamas, bottlenose dolphins capture infaunal prey by burying themselves into the sandy bottom, sometimes up to their pectoral fins (Rossbach & Herzing 1997). In offshore waters, bottlenose dolphins feed on mesopelagic fishes and squid, but the methods they use to capture these prey are unknown.

Group hunting behaviour has been observed in a number of taxa. The net benefits of group foraging include more efficient capture of large, dangerous prey, increased capture success or increased encounter rate. Many species of odontocetes appear to forage cooperatively, although quantitative documentation of this behaviour is rare. In his review of foraging behaviour in delphinids, Würsig (1986) noted that cooperation can be beneficial in both detecting and capturing prey. The distinction between detection and capture is important and the degree of cooperation in these two phases of foraging may vary within taxa. For example, Hoelzel (1993) observed resident killer whales in British Columbia working together to detect salmon, but not to necessarily capture them. Off Punta Norte, Argentina, one killer whale in each pod did most of the hunting of southern sea lions (*Otaria flavescens*),

the most common prey, but provisioned or shared food with others in the pod (Hoelzel 1991). In contrast, Simila and Ugarte (1993) described a feeding behaviour, known as the carousel, used by groups of killer whales in Norway to concentrate schools of herring near the surface where they are more easily captured. Cooperation in prey capture is also critical to the foraging success of killer whales when they attack baleen whales (e.g. Silber *et al.* 1990). In such attacks, there may be a considerable risk to the predators; coordination among pod members may reduce their risk of injury. Cooperation in prey capture has also been well documented for many smaller delphinids such as the dusky dolphin (*Lagenorhynchus obscurus*) that feeds on small, schooling prey (Würsig & Würsig 1980). Hatfield *et al.* (1998) reported an intriguing case of group hunting by killer whales on sea otters. In this instance, a group of three whales approached a small number of otters hauled out on a barely emergent shoal. The whale group split, with two of the individuals proceeding to the other side of the shoal. The remaining whale rushed the shoal at high speed, creating a wave of sufficient size to wash the otters into the sea, whereupon at least one of the otters was eaten by the awaiting whales.

Group feeding in humpback whales produces a variety of bubble structures that are thought to perhaps contain prey such as herring, thereby making lunge foraging (Fig. 8.11) more efficient (e.g. Jurasz & Jurasz 1979; Hain *et al.* 1982) (cf. Fig. 6.1). Known as 'bubble feeding', individual

Fig. 8.11 A group of humpback whales lunge feeding in southeast Alaska. (Coutesy of A.R. Hoelzel.)

whales expel air under water to form clouds, nets or curtains of bubbles. However, the function of such bubble structures is a matter of considerable speculation. Recently, Sharpe and Dill (1997) conducted experiments on the effects of bubbles on the behaviour of captive schools of herring. In these experiments, herring showed a strong avoidance of bubbles and could be contained within a circular bubble net similar to that made by foraging humpback whales.

An intriguing example of prey selection occurs in the killer whale. Two forms of killer whales are recognized in several parts of their range. In the eastern North Pacific, these two sympatric forms, called 'transients' and 'residents' can be distinguished by differences in seasonal distribution, social structure, behaviour, morphology, mitochondrial and nuclear DNA, and diet (Bigg 1982; Bigg et al. 1987; Baird & Stacey 1988; Stevens et al. 1989; Hoelzel et al. 1998) (see Chapter 11). In fact, differences in diet are thought to form the basis for these separate forms. Killer whales are known to feed on a variety of prey including cetaceans, pinnipeds, sea otters, dugongs, fishes, seabirds and cephalopods (Jefferson et al. 1991). Resident killer whales are primarily piscivorous, with salmon (*Oncorhynchus* spp.) being important prey (Ford et al. 1994; Nichol & Shackleton 1995; Ford et al. 1998). The transient form feeds mainly on marine mammals, especially harbour seals (Bigg et al. 1987; Ford et al. 1998). Seasonal changes in the distribution of transient killer whales off southern Vancouver Island appeared to be related in late summer to increased predation on young harbour seals (Baird & Dill 1995), which appear to meet the whales' energy requirements at that time.

The use of echolocation varies dramatically between the two forms (Barrett-Lennard et al. 1996). Residents produce long, frequent echolocation trains, but transients produce infrequent and isolated echolocation clicks. This difference in vocal behaviour is believed to reflect the probability of alerting prey; the prey of transients (seals, porpoises and dolphins) are able to detect high-frequency echolocation, but the prey of residents (salmon) are unable to detect these signals. However, it could also be that transients rely more on vision than echolocation for locating marine mammal prey.

8.7 EFFECTS OF MARINE MAMMAL FORAGING ON COMMUNITY STRUCTURE AND ECOSYSTEMS

The relative influence of top-down versus bottom-up control of ecosystem structure and function has been an important question in ecology for decades (e.g. Hariston et al. 1960; Hunter & Price 1992). In the top-down view, consumers control the abundance and diversity of species at lower trophic levels, whereas, in the bottom-up view, consumers are limited by the availability of resources at lower trophic levels (Fig. 8.12). As noted earlier, marine mammals are a diverse collection of taxa, varying in body size and life history. Nevertheless, they are large, long-lived, iteroparous vertebrates that are often abundant. Thus, we might expect some species to exert top-down control on the ecosystems within which they live (reviewed by Bowen 1997).

8.7.1 Some theory and general background

Without autotrophs, higher level consumers could not exist. In this sense, all ecosystems are under bottom-up control. A food web solely under bottom-up control ought to appear as a classic Eltonian pyramid, which is to say that some combination of abundance and biomass should decline monotonically at progressively higher trophic levels for the simple reason that neither materials nor energy are perfectly conserved as they pass from prey to consumer. In contrast, food webs under strong top-down control with three or more trophic levels should display more of a constricted or oscillating trophic structure. For this reason, the potential length of top-down food webs is necessarily shorter than those under bottom-up control. Another feature of systems under top-down control is that the strength of plant–herbivore interactions at the base of the food web should be weak or strong depending on whether the respective number of trophic levels is odd or even (Fretwell 1977, 1987). While the former prediction has not been seriously evaluated, there is growing evidence for the latter from a variety of aquatic and terrestrial systems (Power 1990;

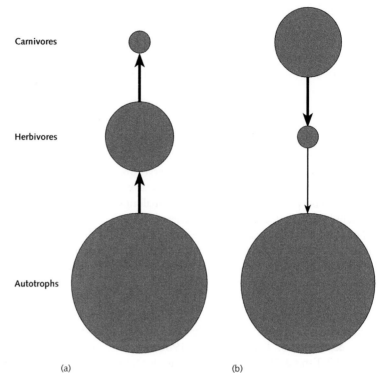

Fig. 8.12 Bottom-up verus top-down food webs. Circle sizes represent crude approximations to biomass or production. Arrows indicate the direction of limiting interactions; bold and light arrows represent strong and weak interactions, respectively. (a) In bottom-up driven systems, production and efficiency of energy transfer drives the abundance of successively higher trophic levels. (b) In top-down systems, intermediate trophic levels are limited by their consumers.

Carnivores

Herbivores

Autotrophs

(a)   (b)

Carpenter & Kitchell 1993; Estes *et al.* 1998; Lindberg *et al.* 1998; Terborgh *et al.* 1999).

8.7.2 Top-down effects

Evidence for the top-down control of biological communities by marine mammals is scant, with a few notable exceptions (see below). As we have indicated earlier, the reasons for this are not difficult to appreciate. Furthermore, in attempting to determine this role of marine mammals in aquatic ecosystems, we are faced with the intersection of no less than three inexact disciplines: marine ecology, marine mammal population biology and fisheries ecology (see Katona & Whitehead 1988). In each of these complex fields, research is expensive, manipulative experiments are rarely possible, interactions occur at quite different spatial and temporal scales making measurement of system properties difficult, and there is an inherent indeterminacy in the behaviour of these complex systems which makes simplifying deterministic explanations problematic. Although these are features that are not unique to the sea,

none the less, they mean that understanding comes slowly and mostly through the correlation of observed events (Bowen 1997).

Our most complete understanding of the top-down control of community structure by a marine mammal comes from studies of the sea otter (Estes & Palmisano 1974; Simenstad *et al.* 1978; Estes & Duggins 1995). As described further below, the influence of sea otter predation on coastal ecosystems is dramatic and diverse, and thus sea otters have been widely cited as one of the better examples of a 'keystone species'; i.e. a species whose numerical abundance is low but whose impact on the ecosystem is high (Power *et al.* 1996). Several features of otters and coastal ecosystems have made this understanding possible. First, and as mentioned previously (Section 8.4), the history of sea otter exploitation and recovery in the fur trade fortuitously created a fragmented spatial array of populations. Understanding the sea otter's role as a predator in coastal ecosystems thus was a simple matter of contrasting areas where otters were abundant with nearby areas where once-abundant populations remained absent. This approach has been supplemented with longitudinal studies at particular areas, using recolonization and subsequent population growth to capture the sea otter's influence on both kelp forest (Estes & Duggins 1995) and soft sediment (Kvitek & Oliver 1992) systems.

Sea otters strongly limit their prey populations in both rocky and soft sediment systems. Our knowledge of the influence of otters in soft sediment systems is limited largely to the reduction in size and abundance of infaunal (Kvitek & Oliver 1988, 1992) or epifaunal (VanBlaricom 1988) bivalves. Similar effects on such prey as sea urchins, abalones and crabs have been documented for numerous rocky reef systems. In some areas, these effects have led to serious conflicts with commercial and recreational shellfisheries (Estes & VanBlaricom 1985). A host of indirect effects from sea otter predation are also known for rocky reef systems. Where sea otters are absent, kelp deforestation often occurs because of the unregulated growth of sea urchin populations (Estes & Palmisano 1974; Duggins 1980; Breen et al. 1982; Estes & Duggins 1995), in turn leading to a host of effects on other invertebrates, fishes, birds and marine mammal species. These various indirect effects in large measure result from the central role of kelp as a habitat and the source of primary production in coastal marine ecosystems (see Estes 1996 for a review).

Over a sufficiently long period of time, such strong top-down control inevitably should have exerted strong selective influences on the key players in this system. This expectation has been investigated for the coevolution of plants and their herbivores by contrasting species characteristics and interactions between northern and southern hemisphere kelp forests. Warm, tropical seas have prevented wholesale biotic interchange between these regions. In contrast with North Pacific kelp forests, which presumably have been subjected to strong top-down effects by sea otters and their ancestors since the late Miocene (roughly 15 million years), no predators of comparable influence are known from southern hemisphere kelp forests (Estes & Steinberg 1988). Thus, northern hemisphere kelp forests are historically three trophic level (odd numbered) systems in which plant–herbivore interactions were predictably weak through evolutionary time.

Southern hemisphere systems, by contrast, were two trophic level (even numbered) systems during this same period of earth history, with the predicted consequence of strong plant–herbivore interactions through evolutionary time. By this view, southern hemisphere plants were expected to have evolved under strong selection for the evolution of antiherbivore defences whereas plants in the northern hemisphere were not. Steinberg's (1989) contrast of phlorotannin (algal secondary metabolites, known to deter herbivores and produced by all brown algal species) concentrations between the northeast Pacific and Australasian kelp forest plants (an order of magnitude greater in Australasian species) provided the first supporting evidence for this macroevolutionary model. Subsequent experimental research has further shown that southern hemisphere herbivores are more resistant to phlorotannins than their northern hemisphere counterparts (Steinberg & van Altna 1992; Steinberg et al. 1995). Collectively, this evidence suggests that sea otter predation decoupled what otherwise would have been a coevolutionary arms race in North Pacific kelp forests, perhaps explaining why the plants in this system are so susceptible to overgrazing.

After nearly a century of recovery from overhunting, sea otter populations in western Alaska have declined during the 1990s. Currently available evidence suggests that increased killer whale predation is the proximate cause (Estes et al. 1998). The ultimate reason for this change is less certain, although probably relates to changes in the oceanic ecosystem. In any case, the addition of killer whales as sea otter predators has caused the coastal food chain to grow from three to four trophic levels, with predictable effects on plant–herbivore interactions (Fig. 8.13). With the otter's recent decline, sea urchin populations have exploded, again resulting in large-scale deforestation of the kelp bed system.

Gray whales (Eschrichtius robustus) may also exert important ecological influences on benthic ecosystems through predation. In the Bering Sea, gray whales forage by engulfing sediment, which contains ampeliscid amphipods. The sediment is filtered through the baleen, which retains the amphipods. This feeding activity has both direct and indirect effects on the benthic ecology of the Bering Sea. Using models of the energetic requirements of gray

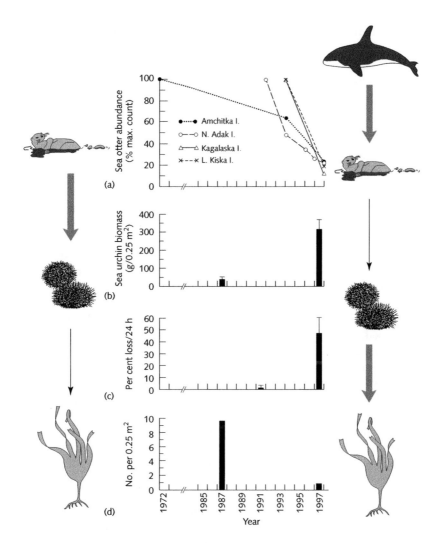

Fig. 8.13 Food web changes in the coastal ecosystems of the western Aleutian archipelago, following entry of killer whales into the system and their reduction of sea otter populations. The trophic structures before and after this event are shown by the drawings on either side of the figure. (a) Declines in the sea otter populations on four islands. (b) Changes in sea urchin biomass at Adak Island during the decline. (c) Changes in the rate of plant tissue loss to herbivory at Adak Island before and after the sea otter population decline. (d) Change in kelp density before and after the sea otter population decline. (From Estes *et al.* 1998, with permission.)

whales and empirical observations of amphipod production, Highsmith and Coyle (1992) estimated that the whales crop from 8% to 18% of amphipod productivity each year. This represents a significant removal of biomass and energy from the benthic community and is near the commonly accepted maximum rate of transfer efficiency (10%) between adjacent trophic levels. In addition to these direct effects, the disruption of the benthos by gray whale feeding provides opportunities for benthic scavengers, including other amphipods (Oliver & Slattery 1985). This periodic disturbance of the benthos may actually increase the carrying capacity of ampeliscid amphipods by resuspending sediments and trapping organic debris in the pits formed by feeding whales.

Walruses may structure the benthic fauna by selectively feeding on older individuals of a few species of bivalve molluscs (Fay & Stocker 1982). Ingestion and defaecation by walruses may result in substantial redistribution of sediment, which may favour colonization of some species, but not others (Fay *et al.* 1977; Oliver *et al.* 1983). Oliver *et al.* (1985) experimentally investigated the effects of walrus feeding on macrobenthic assemblages at several sites in the Bering Sea. Walrus feeding appeared to affect community structure in three ways: by providing food for scavengers such as sea stars (*Asterias amurensis*) and brittle stars (*Amphiodia craterodmeta*), by providing habitats under discarded bivalve shells, and by reducing the abundance of macroinvertebrates in feeding pits compared to surrounding sediments. However, the effect of walrus feeding behaviour on macroinvertebrate assemblages over periods of greater than a few months and at large spatial scales remains poorly understood.

Herbivory can have strong impacts on the structure of terrestrial plant communities. Sirenians are major consumers of the seagrasses in tropical coastal ecosystems. Preen (1995) studied the impact of dugong foraging on a seagrass community at the southern limit of their range in eastern Australia. Dugongs are the main consumers of seagrasses. They generally feed in large groups and may graze at the same location for extended periods. Compared to non-grazed areas, grazing resulted in a reduced density of seagrass shoots by 65–95%, above-ground biomass by 73–96% and below-ground biomass by 31–71% at three sites. In exclosure experiments, intensive grazing resulted in a change in the species composition of the seagrass complex, favouring the more nutritious, rapidly growing species preferred by dugongs. Preen (1995) suggested that dugong foraging be termed cultivation grazing because it can improve the quality of the dugong's diet by preventing the expansion of the dominant, but less nutritious, slow-growing species of seagrass. Additional studies over longer time periods will be needed to confirm this suggestion, but it does appear the dugong grazing, and similar foraging by manatees, may well have strong impacts on the structure of seagrass communities.

At a much larger scale, the top-down control of marine mammals on ecosystem structure has been inferred from the changes that followed intense commercial exploitation of marine mammals in the Southern Ocean. This exploitation reduced some populations of seals and whales to very low levels in the nineteenth and early twentieth centuries (Laws 1977) and amounted to nothing less than an enormous uncontrolled 'experiment' (Laws 1985). The abundance of baleen whales and the sperm whale probably declined by more than 50% between 1904 and 1973. Because the largest species were taken first, the cetacean biomass declined from an estimated 45 million tonnes to only 9 million tonnes over this same period (Laws 1985). Both the Antarctic fur seal (*Arctocephalus gazella*) and southern elephant seal (*Mirounga leonina*) also were decimated by hunting in the nineteenth century.

Laws (1985) estimated that this enormous reduction in the biomass of large whales might have released some 150 million tonnes of krill annually to predators (the remaining whales, seals, seabirds and fish). Although many of the estimates of population food consumption are necessarily tentative, Laws (1985) calculated that much of this krill was redistributed to seals (the crabeater seal and the Antarctic fur seal) and birds, both of which have become considerably more abundant over the past three decades. Chinstrap (*Pygoscelis antarctica*), Adelie (*P. adeliae*) and macaroni (*Eudyptes chrysolophus*) penguins, which together account for about 90% of the Antarctic avian biomass, have also increased in numbers. It is also possible that the present minke whale population is larger than it was before whaling, perhaps a response to the great reduction in blue whale numbers. It seems clear that the relative abundance of krill consumers and other prey in the Southern Ocean are quite different today from what they were prior to the exploitation of the large whales. Although we cannot be certain, it seems reasonable that the observed population increases of some species are due to the increase in food availability brought about by the removal of large whale predation (Laws 1985).

8.7.3 Bottom-up effects

Changes in primary and secondary productivity affecting resource availability to consumers at upper trophic levels can have significant and at times dramatic effects on the abundance of consumers. As discussed previously (Section 8.5), Antarctic fur seal pups suffer reduced growth and increased mortality during years of low krill abundance.

On average, El Niño events occur approximately every 4 years in the eastern tropical Pacific, although the strength and exact timing of these events varies greatly. During these events, the southeast trade winds decrease significantly, the flows of the Peru, South Equatorial and Cromwell countercurrents diminish and warm waters accumulate in the eastern tropical Pacific resulting in reduced upwelling and primary and secondary productivity. Galápagos fur seals have an unusually long lactation period of about 2 years that is thought to have evolved as a hedge against minor El Niño events (Trillmich & Limberger 1985). However, during a severe El Niño, the effects of reduced food availability on seabirds (Schreiber & Schreiber 1984) and marine mammals (Trillmich & Limberger 1985; Trillmich & Ono 1991) are dramatic. For example, during the severe El Niño between August 1982 and July 1983 pup production of Galápagos fur seals was only 11% of what it had been and none of these pups survived the first 5 months after birth. The foraging trips of females were also longer than had been previously measured. The effect of the El Niño on adult male fur seals was also dramatic. Prior to the El Niño about 30 large males held territories during at least part of the season, but in 1983, the season after the El Niño, none of these large males was seen and only five smaller than average adult males held territories (Trillmich & Limberger 1985). These oceanographic conditions sweep northward and southward from the equator in the eastern Pacific during strong El Niño events, impacting on the foraging ecology, reproduction and survival of numerous marine birds and mammals at high latitudes. Although these demographic effects are dramatic, their acute nature relative to the longevity of most marine birds and mammals makes any longer term impact on population size uncertain.

Although the effects of El Niño are dramatic and well documented, large-scale, low-frequency (i.e. interdecadal) changes in aquatic ecosystems also occur (McGowan *et al.* 1998; see National Research Council 1996 for references) and these changes can also have significant impacts on foraging success. Changes in body growth may reflect density-dependent changes or changes in the ecosystem affecting food availability or energy expenditure by individuals. Trites and Bigg (1992) analysed morphometric measurements from a large sample of northern fur seals shot at sea over the period 1958–74. Although fur seals attained larger body sizes as the population at the Pribilof Islands declined, fluctuations in the growth rate and condition of immature fur seals suggest underlying long-term changes in the marine environment. Anderson and Piatt (1999) found remarkable shifts in the epibenthic fish and crustacean faunas in the Gulf of Alaska over a 30-year period, providing compelling evidence for the effects of changing ocean temperature. Populations of piscivorous marine birds and mammals in the western Gulf of Alaska and southern Bering Sea ecosystems are probably impacted by these shifts. Unfortunately, a concurrent increase in ground fish landings during roughly the same period makes it difficult to evaluate the extent to which these various changes are responsible for recent population declines of harbour seals and Steller's sea lions (National Research Council 1996).

8.8 CONCLUSIONS

Foraging success determines an individual's intake of nutrients which in turn affects fitness. Despite its importance, the foraging ecology of most marine mammals is poorly understood and the consequences of variation in feeding success on fitness have been investigated in only a few species.

The feeding behaviour of marine mammals is clearly constrained by phylogeny, as illustrated by the extreme modification of the feeding apparatus of the baleen whales and sirenians. Foraging behaviour is also greatly influenced by body size through direct effects on diving physiology and fasting ability. Large-bodied marine mammals, such as baleen whales and elephant seals, are adapted to exploit distant, rich and patchy food sources and are able to store energy in the form of blubber for later use. Smaller species, such as the sea otter, must feed almost continually on predicable local prey.

Body size not only affects the temporal scale of feeding, but strongly influences the spatial scales of foraging. As large-bodied, long-lived species, many marine mammals forage at multiple scales, thereby reducing the effects of local variability in prey abundance. There is growing evidence that marine mammals respond to changes in prey abundance by altering their foraging behaviour in ways that are consistent with predictions from optimal foraging theory. Such changes in foraging behaviour can have consequences for both survival and reproductive success.

Although the role of sea otters in structuring nearshore marine communities is reasonably well understood, much less is known about the ecological roles of other marine mammals. Nevertheless, we expect reasonable advances in our understanding of marine mammal foraging ecology with the continuing development of smaller, more accurate and multichannel telemetry and data-logging devices, coupled with new approaches to the study of diets and energy expenditure. However, to understand the consequences of foraging success on population biology will require programmes of long-term, integrated research on marine mammals.

REFERENCES

Aarefjord, H., Bjorge, A.J., Kinze, C.C. & Lindstedt, I. (1995) Diet of harbour porpoise (*Phocoena phocoena*) in Scandinavian waters. In: *Biology of the Phocoenids* (A. Bjorge & G.P. Donovan, eds), pp. 211–222. International Whaling Commission, Cambridge, UK.

Ackman, R.G. (1980) Fish lipids, part 1. In: *Advances in Fish Science and Technology* (J.J. Connell, ed.), pp. 86–103. Fishing News Books, Oxford.

Alcock, J. (1972) The evolution of the use of tools by feeding animals. *Evolution* 26, 464–473.

Ames, A.L., Van Veet, E.S. & Sackett, W.M. (1996) The use of stable carbon isotope analysis for determining the dietary habits of the Florida manatee, *Trichechus manatus latirostris*. *Marine Mammal Science* 12, 555–563.

Anderson, P.J. & Piatt, J.F. (1999) Community reorganization in the Gulf of Alaska following ocean climate regime shift. *Marine Ecology Progress Series* 189, 117–123.

Bailey, K.M. & Ainley, D.G. (1982) The dynamics of California sea lion predation on Pacific hake. *Fisheries Research* 1, 163–176.

Baird, R.W. & Dill, L.M. (1995) Occurrence and behaviour of transient killer whales: seasonal and pod-specific variability, foraging behaviour, and prey handling. *Canadian Journal of Zoology* 73, 1300–1311.

Baird, R.W. & Stacey, P.J. (1988) Variation in saddle patch pigmentation in populations of killer whales (*Orcinus orca*) from British Columbia, Alaska, and Washington State. *Canadian Journal of Zoology* 66, 2582–2585.

Barrett-Lennard, L.G., Ford, J.K.B. & Heise, K.A. (1996) The mixed blessing of ecolocation: differences in sonar use by fish eating and mammal-eating killer whales. *Animal Behaviour* 51, 553–565.

Barros, N.B. & Odell, D.K. (1990) Food habits of bottlenose dolphins in the southeastern United States. In: *The Bottlense Dolphin* (S. Leatherwood & R.R. Reeves, eds), pp. 309–328. Academic Press, San Diego, CA.

Barros, N.B. & Wells, R.S. (1998) Prey and feeding patterns of resident bottlenose dolphins (*Tursiops truncatus*) in Sarasota Bay, Florida. *Journal of Mammalogy* 79, 1045–1059.

Benoit, D. & Bowen, W.D. (1990) Seasonal and geographic variation in the diet of grey seals (*Halicheorus grypus*) in eastern Canada. In: *Population Biology of Sealworm (Pseudoterranova decipiens) in Relation to its Intermediate and Seal Hosts* (W.D. Bowen, ed.), pp. 215–226. Canadian Bulletin of Fisheries and Aquatic Science, Ottawa.

Berger, J. (1998) Future prey: some consequences of the loss and restoration of large carnivores. In: *Behavioral Ecology and Conservation Biology* (T. Caro, ed.), pp. 80–103. Oxford University Press, Oxford.

Best, P.B. (1979) Social organization of sperm whales, *Physeter macrocephalus*. In: *Behavior of Marine Animals. Current Perspectives in Research, Vol. 3: Cetaceans* (H.E. Winn & B.L. Olla, eds), pp. 227–289. Plenum Press, New York.

Bigg, M. (1982) An assessment of killer whale (*Orcinus orca*) stocks off Vancouver Island, British Columbia. *Report of the International Whaling Commission* 32, 655–666.

Bigg, M.A., Ellis, G.M., Ford, J.K.B. & Balcomb, K.C. (1987) *Killer Whales: a Study of their Identification, Genealogy, and Natural History in British Columbia and Washington State*. Phantom Press, Nanaimo, BC.

Boggs, C.L. (1992) Resource allocation: exploring connections between foraging and life history. *Functional Ecology* 6, 508–518.

Boness, D.J. & Bowen, W.D. (1996) The evolution of maternal care in pinnipeds. *Bioscience* 46, 645–654.

Boness, D.J., Bowen, W.D. & Oftedal, O.T. (1994) Evidence of a maternal foraging cycle resembling that of otariid seals in a small phocid, the harbor seal. *Behavioural Ecology and Sociobiology* 34, 95–104.

Bowen, W.D. (1997) Role of marine mammals in aquatic ecosystems. *Marine Ecology Progress Series* 158, 267–274.

Bowen, W.D. & Harrison, G.D. (1994) Offshore diet of grey seals *Halichoerus grypus* near Sable Island, Canada. *Marine Ecology Progress Series* 112, 1–11.

Bowen, W.D. & Siniff, D.B. (1999) Distribution, population biology, and feeding ecology of marine mammals. In: *Biology of Marine Mammals* (J.E.I. Reynolds & S.A. Rommel, eds), pp. 423–484. Smithsonian Press, Washington, DC.

Bowen, W.D., Oftedal, O.T. & Boness, D.J. (1992) Mass and energy transfer during lactation in a small phocid, the harbor seal (*Phoca vitulina*). *Physiological Zoology* **65**, 844–866.

Boyd, I.L. (1996) Temporal scales of foraging in a marine predator. *Ecology* **77**, 426–434.

Boyd, I.L. (1998) Time and energy constraints in pinniped lactation. *American Naturalist* **152**, 717–728.

Boyd, I.L. & Croxall, J.P. (1996) Dive durations in pinnipeds and seabirds. *Canadian Journal of Zoology* **74**, 1696–1705.

Boyd, I.L., Arnould, J.P.Y., Barton, T. & Croxall, J.P. (1994) Foraging behaviour of Antarctic fur seals during periods of contrasting prey abundance. *Journal of Animal Ecology* **63**, 703–713.

Boyd, I.L., McCaffery, D.J. & Walker, T.R. (1997) Variation in foraging effort by lactating Antarctic fur seals: response to simulated foraging costs. *Behavioural Ecology and Sociobiology* **40**, 135–144.

Breen, P.A., Carson, T.A., Foster, J.B. & Stewart, E.A. (1982) Changes in subtidal community structure associated with British Columbia transplants. *Marine Ecology Progress Series* **7**, 13–20.

Brodie, P.F. (1975) Cetacean energetics: an overview of intraspecific size vairation. *Ecology* **56**, 152–161.

Burgess, W., Tyack, P., Le Boeuf, B.J. & Costa, D.P. (1998) An intelligent acoustic recording tag, first results from free-ranging northern elephant seals. *Deep Sea Research II* **45**, 1327–1351.

Burton, R.K. & Koch, P.L. (1999) Isotopic tracking of foraging and long-distance migration in northeastern Pacific pinnipeds. *Oecologia* **119**, 578–585.

Carpenter, S.R. & Kitchell, J.F. (1993) *The Trophic Cascade in Lakes*. Cambridge University Press, Cambridge, UK.

Charnov, E.L. (1976) Optimal foraging: the marginal value theorem. *Theoretical Population Biology* **9**, 129–136.

Clarke, M.R. (1980) Cephalopods in the diet of sperm whales in the southern hemisphere and their bearing on sperm whale biology. *Discovery Report* **37**, 1–324.

Clarke, M.R. (1986) Cephalopods in the diet of odontocetes. In: *Research on Dolphins* (M.M. Bryden & R. Harrison, eds), pp. 281–321. Clarendon Press, Oxford.

Cockcroft, V.G. & Ross, G.J.B. (1990) Food and feeding in the Indian ocean bottlenose dolphin off Southern Natal, South Africa. In: *The Bottlenose Dolphin* (S. Leatherwood & R.R. Reeves, eds), pp. 295–308. Academic Press, San Diego, CA.

Coltman, D.W., Bowen, W.D., Boness, D.J. & Iverson, S.J. (1997) Balancing foraging and reproduction in male harbour seals: an aquatically mating pinniped. *Animal Behaviour* **54**, 663–678.

Coltman, D.W., Bowen, W.D., Iverson, S.J. & Boness, D.J. (1998) The energetics of male reproduction in an aquatically mating pinniped, the harbour seal. *Physiological Zoology* **71**, 387–399.

Costa, D.P. (1982) Energy, nitrogen, and electrolyte flux and sea water drinking in the sea otter *Enhydra lutris*. *Physiological Zoology* **55**, 35–44.

Costa, D.P. (1988) Methods for studying the energetics of freely diving animals. *Canadian Journal of Zoology* **66**, 45–52.

Costa, D.P. & Gentry, R.L. (1986) Free-ranging energetics of northern fur seals. In: *Fur Seals: Maternal Strategies on Land and at Sea* (R.J. Gentry & G.L. Kooyman, eds), pp. 79–101. Princeton University Press, Princeton, NJ.

Costa, D.P., Croxall, J.P. & Duck, C.D. (1989) Foraging energetics of Antarctic fur seals in relation to changes in prey availability. *Ecology* **70**, 596–606.

Croxall, J.P. (1995) Remote-recording of foraging patterns in seabirds and seals for studies of predator–prey interactions in marine systems. In: *Ecology of Fjords and Coastal Waters* (H.R. Skjoldal, C. Hopkins, K.E. Erikstad & H.P. Leinaas, eds), pp. 429–442. Elsevier Science, Berlin.

Darling, J. (1977) *Population biology and behaviour of the grey whale (Eschrichutis robustus) in Pacific Rim National Park, British Columbia*. MSc Thesis, University of Victoria, Victoria, BC.

Davis, R.W., Fuiman, L.A., Williams, T.M. *et al.* (1999) Hunting behavior of a marine mammal beneath the Antarctic fast ice. *Science* **283**, 993–996.

DeNiro, M.J. & Epstein, S. (1978) Influence of diet on the distribution of carbon isotopes in animals. *Geochimica et Cosmochimica Acta* **42**, 495–506.

DeNiro, M.J. & Epstein, S. (1981) Influence of diet on the distribution of nitrogen isotopes in animals. *Geochimica et Cosmochimica Acta* **45**, 341–351.

Desportes, G. & Mouritsen, R. (1993) Preliminary results on the diet of long-finned pilot whales off the Faroe Islands. In: *Biology of Northern Hemisphere Pilot Whales* (G.P. Donovan, C.H. Lockyer & A.R. Martin, eds), pp. 305–324. Report of the International Whaling Commission, Special Issue 14. International Whaling, Commission, Cambridge, UK.

Doidge, D.W. & Croxall, J.P. (1985) Diet and energy budget of the Antarctic fur seal *Arctocephalus gazella* at South Georgia. In: *Antarctic Nutrient Cycles and Food Webs* (W.R. Siegfried, P.R. Condy & R.M. Laws, eds), pp. 543–550. Springer-Verlag, Berlin.

Domning, D.P. (1978) Sirenian evolution in the North Pacific Ocean. *University of California Publications of Geological Science* **118**, 1–176.

Domning, D.P. (1980) Feeding position preference in manatees (*Trichechus*). *Journal of Mammalogy* **61**, 544–547.

Duggins, D.O. (1980) Kelp beds and sea otters: an experimental approach. *Ecology* **61**, 447–453.

Emlen, J.M. (1973) *Ecology: an Evolutionary Approach*. Addison-Wesley Publishing Co., Reading, MA.

Estes, J.A. (1989) Adaptations for aquatic living by carnivores. In: *Carnivore Behavior Ecology and Evolution* (J.L. Gittleman, ed.), pp. 242–282. Cornell University Press, Ithaca, NY.

Estes, J.A. (1996) The influence of large, mobile predators in aquatic food webs: examples from sea otters and kelp forests. In: *Aquatic Predators and Their Prey* (S.P.R. Greenstreet & M.L. Tasker, eds), pp. 65–72. Fishing News Books, Oxford.

Estes, J.A. & Duggins, D.O. (1995) Sea otters and kelp forests in Alaska: generality and variation in a community ecological paradigm. *Ecological Monograph* **65**, 75–100.

Estes, J.A. & Palmisano, J.F. (1974) Sea otters: their role in structuring nearshore communities. *Science* **185**, 1058–1060.

Estes, J.A. & Steinberg, P.D. (1988) Predation, herbivory, and kelp evolution. *Paleobiology* **14**, 19–36.

Estes, J.A. & VanBlaricom, G.R. (1985) Sea otters and shellfisheries. In: *Marine Mammals and Fisheries* (J.R. Beddington, R.J.H. Beverton & D.M. Lavigne, eds), pp. 187–235. Allen & Unwin, London.

Estes, J.A., Jameson, R.J. & Johnson, A.M. (1981) Food selection and some foraging tactics of sea otters. In: *Worldwide Furbearer Conference Proceedings* (J.A. Chapman & D. Pursley, eds), pp. 606–641. University of Maryland Press, Washington, DC.

Estes, J.A., Jameson, R.J. & Rhode, E.B. (1982) Activity and prey selection in the sea otter: influence of population status on community structure. *American Naturalist* **120**, 242–258.

Estes, J.A., Tinker, M.T., Williams, T.M. & Doak, D.F. (1998) Killer whale predation on sea otters linking coastal with oceanic ecosystems. *Science* **282**, 473–476.

Fay, F.H. & Stocker, S.W. (1982) *Reproductive Success and Feeding Habits of Walruses Taken in 1982 Spring Harvest, with Comparisons from Previous Years*. Final Report Eskimo Walrus Commission. Eskimo Walrus Commission, Nome, Alaska.

Fay, F.H., Feder, H.M. & Stoker, S.W. (1977) *An Estimation of the Impact of the Pacific Walrus Population on its Food Resources in the Bering Sea*. Final Report US Marine Mammal Commission No. PB-273–505. National Technical Information Service, Springfield, VA.

Fitch, J.E. & Brownell, R.L. (1968) Fish otoliths in cetacean stomachs and their importance in interpreting feeding habits. *Journal of the Fisheries Research Board of Canada* **25**, 2561–2575.

Ford, J.K.B., Ellis, G.M. & Balcomb, K.C. (1994) *The Natural History and Geneology of Orcinus orca in British Columbia and Washington State*. UBC Press, Vancouver, BC.

Ford, J.K.B., Ellis, G.M., Barrett-Lennard, L.G. *et al.* (1998) Dietary specialization in two sympatric populations of killer whales (*Orcinus orca*) in coastal British Columbia and adjacent waters. *Canadian Journal of Zoology* **76**, 1456–1471.

Fretwell, S.D. (1977) The regulation of plant communities by food chains exploiting them. *Perspectives in Biological Medicine* **20**, 169–185.

Fretwell, S.D. (1987) Food chain dynamics: the central theory of ecology? *Oikos* **50**, 291–301.

Frost, K.J., Lowry, L.F., Version Hoef, J.M., Iverson, S.J. & Gotthardt, T. (1998) *Monitoring Habitat Use and Trophic Interactions of Harbor Seals in Prince William Sound, Alaska*. Exxon Valdez Oil Spill Restoration Project Annual Report. Alaska Department of Fish and Game, Division of Wildlife Conservation, Fairbanks, Alaska.

Fry, B. (1988) Food web structure on Georges Bank from stable C, N, and S isotopic compositions. *Limnology and Oceanography* **33**, 1182–1190.

Gannon, D.P., Craddock, J.E. & Read, A.J. (1997) Autumn food habits of harbor porpoises, *Phocoena phocoena*, in the Gulf of Maine. *Fishery Bulletin* **96**, 428–437.

Gannon, D.P., Read, A.J., Craddock, J.E., Fristrup, K.M. & Nicolas, J.R. (1998) Feeding ecology of long-finned pilot whales *Globicephalus melas* in the western North Atlantic. *Marine Ecology Progress Series* **148**, 1–10.

Gaskin, D.E. (1982) *The Ecology of Whales and Dolphins*. Heinemann Educational Books, London.

Gentry, R.L., Kooyman, G.L. & Goebel, M.E. (1986) Feeding and diving behavior of northern fur seals. In: *Fur Seals: Maternal Strategies on Land and at Sea* (R.L. Gentry & G.L. Kooyman, eds), pp. 61–78. Princeton University Press, Princeton, NJ.

Goebel, M.E., Bengtson, J.L., DeLong, R.L., Gentry, R.L. & Loughlin, T.R. (1991) Diving patterns and foraging locations of female northern fur seals. *Fishery Bulletin* **89**, 171–179.

Gould, S.J. & Lewontin, R.L. (1979) The spandrels of San Marcos and the Panglossian paradigm: a critique of the adaptionist programme. *Proceedings of the Royal Society of London, Series B* **205**, 581–598.

Hain, J.H.W., Carter, G.R., Kraus, S.D., Mayo, C.A. & Winn, H.E. (1982) Feeding behaviour of the humpback whale, *Megaptera novaeangliae*, in the western North Atlantic. *Fishery Bulletin* **80**, 259–268.

Hammill, M.O. & Smith, T.G. (1991) The role of predation in the ecology of the ringed seal in Barrow Strait, Northwest Territories. *Marine Mammal Science* **7**, 123–135.

Hammond, P.S. & Prime, J.H. (1990) The diet of British grey seals (*Halichoerus grypus*). In: *Population Biology of Sealworm (Psuedoterranova decipiens) in Relation to its Intermediate and Seal Hosts* (W.D. Bowen, ed.), pp. 243–254. Canadian Bulletin of Fisheries and Aquatic Science, Ottawa.

Hammond, P.S., Hall, A.J. & Prime, J.H. (1994) The diet of grey seals around Orkney and other island and mainland sites in north-eastern Scotland. *Journal of Animal Ecology* **31**, 340–350.

Hariston, H., Smith, F.E. & Slobodkin, L. (1960) Community structure, population control and competition. *American Naturalist* **94**, 421–425.

Hartman, D.S. (1979) *Ecology and Behavior of the Manatee (Trichechus manus) in Florida*. American Society of Mammalogists Special Publication No. 5. American Society of Mammalogists, Lawrence, KS.

Hatfield, B.B., Marks, D., Tinker, M.T. & Nolan, K. (1998) Attacks on sea otters by killer whales. *Marine Mammal Science* **14**, 888–894.

Haug, T., Gjosaeter, H., Lindstrom, U. & Nilssen, K.T. (1995) Diet and food availability for north-east Atlantic minke whales (*Balaenoptera acutorostrata*), during the summer of 1992. *ICES Journal of Marine Science* **52**, 77–86.

Highsmith, R.C. & Coyle, K.O. (1992) Productivity of arctic amphipods relative to gray whale energy requirements. *Marine Ecology Progress Series* **83**, 141–150.

Hobson, K.A. & Welch, H.E. (1992) Determination of trophic relationships within a high Arctic marine food web using del13C and del15N analysis. *Marine Ecology Progress Series* **84**, 9–18.

Hoelzel, A.R. (1991) Killer whale predation on marine mammals at Punte Norte, Argentina; food sharing, provisioning and foraging strategy. *Behavioural Ecology and Sociobiology* **29**, 197–204.

Hoelzel, A.R. (1993) Foraging behaviour and social group dynamics in Puget Sound killer whales. *Animal Behaviour* **45**, 581–591.

Hoelzel, A.R., Dorsey, E.M. & Stern, J. (1989) The foraging specializations of individual minke whales. *Animal Behaviour* **38**, 786–794.

Hoelzel, A.R., Dahlheim, M. & Stern, S.J. (1998) Low genetic variation among killer whales (*Orcinus orca*) in the eastern North Pacific, and genetic differentiation between foraging specialists. *Journal of Heredity* **89**, 121–128.

Hoese, H.D. (1971) Dolphins feeding out of water in a salt marsh. *Journal of Mammalogy* **52**, 222–223.

Houston, A.I. (1995) Energetic constraints and foraging efficiency. *Behaviour Ecology* **6**, 393–396.

Hunt, G.L., Heinemann, D. & Everson, I. (1992) Distributions and predator–prey interactions of macaroni penguins, Antarctic fur seals, and Antarctic krill near Bird Island, South Georgia. *Marine Ecology Progress Series* **86**, 15–30.

Hunter, M.D. & Price, P.W. (1992) Playing chuts and ladders: heterogeneity and the relative roles of bottom-up and top-down forces in natural communities. *Ecology* **73**, 724–732.

Iverson, S.J. (1988) *Composition, intake and gastric digestion of milk lipids in pinnipeds*. PhD Thesis, University of Maryland, College Park, MD.

Iverson, S.J. (1993) Milk secretion in marine mammals in relation to foraging: can milk fatty acids predict diet? *Symposium, Zoological Society of London* **66**, 263–291.

Iverson, S.J., Oftedal, O.T., Bowen, W.D., Boness, D.J. & Sampugna, J. (1995) Prenatal and postnatal transfer of fatty acids from mother to pup in the hooded seal. *Journal of Comparative Physiology B* **165**, 1–12.

Iverson, S.J., Arnould, J.P.Y. & Boyd, I.L. (1997a) Milk fatty acid signatures indicate both minor and major shifts in the diet of lactating Antarctic fur seals. *Canadian Journal of Zoology* **75**, 188–198.

Iverson, S.J., Frost, K.J. & Lowry, L.F. (1997b) Fatty acid signatures reveal fine scale structure of foraging distribution of harbor seals and their prey in Prince William Sound. *Marine Ecology Progress Series* **151**, 255–271.

Janzen, D.H. & Martin, P.S. (1982) Neotropical anachronisms: the fruits the gomphotheres ate. *Science* **215**, 19–27.

Jefferson, T.A., Stacey, P.J. & Baird, R.W. (1991) A review of killer whale interactions with other marine mammals: predation to co-existence. *Mammal Review* **21**, 151–180.

Jurasz, C.M. & Jurasz, V.P. (1979) Feeding modes of the humpback whale, *Megaptera novaeangliae*, in southeast Alaska. *Scientific Reports of the Whales Research Instititue, Tokyo* **31**, 69–83.

Kajimura, H. (1980) Distribution and migration of northern fur seals (*Callorhinus ursinus*) in the eastern Pacific. In: *Further Analysis of Pelagic Fur Seal Data Collected by the United States and Canada During 1958–74. Proceedings of the 23nd Annual Meeting of the Standing Scientific Committee, North Pacific Fur Seal Committee 7–11 April, Moscow, USSR* (H. Kajimura, R.H. Lander, M.A. Perez, A.E. York & M.A. Bigg, eds), pp. 6–43. NMFS, NOAA National Marine Mammal Laboratory, Seattle, WA.

Kajimura, H. (1984) *Opportunistic Feeding of the Northern Fur Seal, Callorhinus ursinus, in the Eastern North Pacific Ocean and Eastern Bering Sea*. NOAA Technical Report No. NMFS SSRF-779. US Department of Commerce, Washington, DC.

Katona, S. & Whitehead, H. (1988) Are Cetacea ecologically important? *Oceanography and Marine Biology Annual Review* **26**, 553–568.

Kenyon, K.W. (1969) The sea otter in the eastern Pacific Ocean. *North American Fauna* **68**, 1–352.

Kirsch, P.E., Iverson, S.J. & Bowen, W.D. (2000) Effect of a low fat diet on body composition and blubber fatty acids of captive juvenile harp seals (*Phoca groenlandica*). *Physiological and Biochemical Zoology* **73**, 45–59.

Koopman, H.N., Iverson, S.J. & Gaskin, D.E. (1996) Stratification and age-related differences in blubber fatty acids of the male harbour porpoise (*Phocoena phocoena*). *Journal of Comparative Physiology* **165**, 628–639.

Krebs, J.R. & Davies, N.B. (1979) *Behavioural Ecology. An Evolutionary Approach*. Blackwell Scientific Publications, Oxford.

Kvitek, R.G. & Oliver, J.S. (1988) Sea otter foraging habits and effects on prey population in soft-bottom communities. In: *The Community Ecology of Sea Otters* (G.R. VanBlaricom & J.A. Estes, eds), pp. 22–47. Springer-Verlag, New York.

Kvitek, R.G. & Oliver, J.S. (1992) Influence of sea otters on soft-bottom prey communities in southeast Alaska. *Marine Ecology Progress Series* **82**, 103–113.

Kvitek, R.G., Bowlby, C.E. & Staedler, M. (1993) Diet and foraging behavior of sea otters in southeast Alaska. *Marine Mammal Science* **9**, 168–181.

Lambertsen, R.H. (1983) Internal mechanism of rorqual feeding. *Journal of Mammalogy* **64**, 76–88.

Laws, R.M. (1977) The significance of vertebrates in the Antarctic marine ecosystem. In: *Adaptation Within Antarctic Ecosystems. Third Symposium on Antarctic Biology* (G.A. Llano, ed.), pp. 411–438. Smithsonian Institution, Washington, DC.

Laws, R.M. (1984) Seals. In: *Antarctic Ecology* (R.M. Laws, ed.), pp. 621–716. Academic Press, London.

Laws, R.M. (1985) The ecology of the Southern Ocean. *American Scientist* **73**, 26–40.

Lawson, J.W., Anderson, J.T., Dalley, E.L. & Stenson, G.B. (1998) Selective foraging by harp seals *Phoca groenlandica* in nearshore and offshore waters of Newfoundland, 1993 and 1994. *Marine Ecology Progress Series* **163**, 1–10.

Levin, S.A. (1992) The problem of pattern and scale in ecology. *Ecology* **73**, 1943–1967.

Lindberg, D.L., Estes, J.A. & Warheit, K.A. (1998) Human influences on trophic cascades in rocky intertidal communities. *Ecological Applications* **8**, 880–890.

Lowry, M.S., Oliver, C.W. & Macky, C. (1990) Food habits of California sea lions, *Zalophus californianus*, at San Clemente Island, California, 1981–86. *Fishery Bulletin* **88**, 509–521.

Lunn, N.J. & Boyd, I.L. (1993) Effects of maternal age and condition on parturition and the perinatal period of female Antarctic fur seals. *Journal of Zoology, London* **229**, 55–67.

Lunn, N.J., Boyd, I.L. & Croxall, J.P. (1994) Reproductive performance of female Antarctic fur seals: the influence of age, breeding experience, environmental variation and individual quality. *Journal of Animal Ecology* **63**, 827–840.

Lyons, K.J. (1991) *Variation in feeding behavior of the sea otter, between individuals and with reproductive condition.* PhD Thesis, University of California, Santa Cruz.

McCafferty, D.J., Boyd, I.L., Walker, T.R. & Taylor, R.I. (1998) Foraging responses of Antarctic fur seals to changes in the marine environment. *Marine Ecology Progress Series* **166**, 285–299.

McGowan, J.A., Cayan, D.R. & Dorman, L.M. (1998) Climate–ocean variability and ecosystem response in the Northeast Pacific. *Science* **281**, 210–217.

McKinnon, J. (1993) Feeding habits of the dusky dolphin, *Lagenorhynchus obscurus*, in the coastal waters of central Peru. *Fishery Bulletin* **92**, 569–578.

Marshall, C.D., Huth, C.D., Edmonds, V.M., Halin, D.L. & Reep, R.L. (1998) Prehensile use of perioral bristles during feeding and associated behaviors of the Florida manatee (*Trichechus manatus latirosttris*). *Marine Mammal Science* **14**, 274–289.

Marshall, G.J. (1998) Crittercam: an animal-borne imaging and data logging system. *Marine Technology Society Journal* **32**, 11–17.

Mayo, C.A. & Marx, M.K. (1990) Surface foraging behaviour of the North Atlantic right whale, *Eubalaena glacialis*, and associated zooplankton characteristics. *Canadian Journal of Zoology* **68**, 2214–2220.

Mead, G.J.D. (1989) Beaked whales of the genus—Mesoplodon. In: *Handbook of Marine Mammals, Vol. 4: River Dolphins and the Larger Toothed Whales* (S.H. Ridgway & R. Harrison, eds), pp. 349–430. Academic Press, New York.

Minagawa, M. & Wada, E. (1984) Stepwise enrichment of 15N along food chains. Further evidence and the relation between delta15N and animal age. *Geochimica et Cosmochimica Acta* **48**, 1135–1140.

Miyazaki, N., Kusaka, T. & Nishiwali, M. (1973) Food of *Stenella caeruleoalba*. *Scientific Reports of the Whales Research Institute* **14**, 265–275.

Monson, D., Estes, J.A., Siniff, D.B. & Bodkin, J.B. (2000) Life history plasticity and population regulation in sea otters. *Oikos* **90**, 457–468.

National Research Council (1996) *The Bering Sea Ecosystem.* National Research Council, Washington, DC.

Nerini, M. (1984) A review of gray whale feeding ecology. In: *The Gray Whale, Eschrichtius robustus* (M.L. Jones, S.L. Swartz & S. Leatherwood, eds), pp. 423–450. Academic Press, Olando, FL.

Nichol, L.M. & Shackleton, D.M. (1995) Seasonal movements and foraging behaviour of northern resident killer whales (*Orcinus orca*) in relation to the inshore distribution of salmon (*Oncorhynchus* spp.) in British Columbia. *Canadian Journal of Zoology* **74**, 983–991.

Nowacek, D.P. (1999) *Sound use, sequential behavior and ecolgy of foraging bottlenose dolphins, Tursiops truncatus.* PhD Thesis, MIT and Woods Hole Oceanographic Institution, Woods Hole.

Olesiuk, P.F., Bigg, M.A., Ellis, G.M., Crockford, S.J. & Wigen, R.J. (1990) *An Assessment of the Feeding Habits of Harbour seals (Phoca vitulina) in the Strait of Georgia, British Columbia, Based on Scat Analysis.* Canadian Technical Report, Fisheries and Aquatic Science, No. 1730. Government of Canada, Ottawa.

Oliver, J.S. & Slattery, P.N. (1985) Destruction and opportunity on the sea floor: effects of gray whale feeding. *Ecology* **66**, 1965–1975.

Oliver, J.S., Slattery, P.N., O'Conner, E.F. & Lowry, L.F. (1983) Walrus, *Odobenus rosmarus*, feeding in the Bering Sea: a benthic perspective. *Fishery Bulletin* **81**, 501–512.

Oliver, J.S., Kvitek, R.G. & Slattery, P.N. (1985) Walrus feeding disturbance: scavenging habits and recolonization of the Bering Sea benthos. *Journal of Experimental Marine Biology and Ecology* **91**, 233–246.

Orians, G.H. & Pearson, N.E. (1979) On the theory of central place foraging. In: *Analysis of Ecological Systems* (D.J. Horn, R.D. Mitchell & G.R. Stairs, eds), pp. 155–157. Ohio State University, Columbus, OH.

Orton, L.S. & Brodie, P.F. (1987) Engulfing mechanics of fin whales. *Canadian Journal of Zoology* **65**, 2898–2907.

Ostrom, P.H., Lien, J. & Macko, S.A. (1993) Evaluation of the diet of Sowerby's beaked whale, *Mesoplodon bidens*, based on isotopic comparisons among northwestern Atlantic cetaceans. *Canadian Journal of Zoology* **71**, 858–861.

Packer, C. (1983) Sexual dimorphism: the horns of African antelopes. *Science* **221**, 1191–1193.

Perrin, W.F., Warner, R.R., Ficus, C.H. & Holts, D.B. (1973) Stomach contents of porpoise, *Stenella* sp. and yellowfin tuna, *Thunnus albacares*, in mixed species aggregations. *Fishery Bulletin* **70**, 1077–1092.

Peters, R.H. (1983) *The Ecological Implications of Body Size.* Cambridge University Press, Cambridge, UK.

Peterson, B.J. & Fry, B. (1987) Stable isotopes in ecosystem studies. *Annual Review of Ecology and Systematics* **181**, 293–320.

Pierce, G.J. & Boyle, P.R. (1991) A review of methods for diet analysis in piscivorous marine mammals. *Oceanography and Marine Biology* **29**, 409–486.

Pierce, G.J., Boyle, P.R., Watt, J. & Grisley, M. (1993) Recent advances in diet analysis of marine mammals. *Symposium, Zoological Society of London* **66**, 241–261.

Pivorunas, A. (1979) The feeding mechanisms of baleen whales. *American Scientist* **67**, 432–440.

Power, M.E. (1990) Effects of fish in river food webs. *Science* **250**, 411–415.

Power, M.E., Tilman, D., Estes, J.A. *et al.* (1996) Challenges in the quest for keystones. *Bioscience* **46**, 609–620.

Preen, A. (1995) Impacts of dugong foraging on seagrass habitats: observational and experimental evidence for cultivation grazing. *Marine Ecology Progress Series* **124**, 210–213.

Prothero, J. (1995) Bone and fat as a function of body weight in adult mammals. *Comparative Biochemistry and Physiology* **111A**, 633–639.

Radinsky, L.B. (1968) Evolution of somatic sensory specialization in otter brains. *Journal of Comparative Neurology* **134**, 495–506.

Ramsay, M.A. & Hobson, K.A. (1991) Polar bears make little use of terrestrial food webs: evidence from stable-carbon isotope analysis. *Oecologia* **86**, 598–600.

Rau, G.H., Ainley, D.G., Bengtson, J.L., Torres, J.J. & Hopkins, T.L. (1992) 15N/14N and 13C/12C in Weddell Sea birds, seals, and fish: implications for diet and trophic structure. *Marine Ecology Progress Series* **84**, 1–8.

Recchia, C.A. & Read, A.J. (1989) Stomach contents of harbour porpoises, *Phocoena phocoena* (L.), from the Bay of Fundy. *Canadian Journal of Zoology* **67**, 2140–2146.

Repenning, C.A. (1976) Adaptive evolution of the sea lions and walruses. *Systematic Zoology* **25**, 375–390.

Rice, D.W. (1989) Sperm whale—*Physeter macrocephalus*. In: *Handbook of Marine Mammals, Vol. 4. River Dolphins and the Larger Toothed Whales* (S.H. Ridgway & R. Harrison, eds), pp. 177–233. Academic Press, New York.

Riedman, M.L. & Estes, J.A. (1990) *The Sea Otter (Enhydra lutris): Behavior, Ecology, and Natural History*. Biological Report No. 90(14). US Fish and Wildlife Service, Washington, DC.

Rossbach, K.A. & Herzing, D.L. (1997) Underwater observations of benthic feeding bottlenose dolphins (*Tursiops truncatus*) near Grand Bahama Island, Bahamas. *Marine Mammal Science* **13**, 498–504.

Schell, D.M., Saupe, S.M. & Haubenstock, N. (1989) Bowhead whale (*Balaena mysticetus*) growth and feeding as estimated by del13C techniques. *Marine Biology* **103**, 433–443.

Schreiber, R.W. & Schreiber, E.A. (1984) Pacific seabirds and the El Niño Southern Oscillation. 1892 to 1983 perspectives. *Science* **225**, 713–716.

Sharpe, F.A. & Dill, L.M. (1997) The behavior of Pacific herring schools in response to artificial humpback whale bubbles. *Canadian Journal of Zoology* **75**, 725–730.

Sibly, R.M. & Calow, P. (1986) *Physiological Ecology of Animals*. Blackwell Scientific Publications, Oxford.

Silber, G.K., Newcomer, M.W. & Perez-Cortes, H. (1990) Killer whales (*Orcinus orca*) attack and kill a Bryde's whale (*Balaenopteri edeni*). *Canadian Journal of Zoology* **68**, 1603–1606.

Simenstad, C.A., Estes, J.A. & Kenyon, K.W. (1978) Aleuts, sea otters, and alternate stable-state communities. *Science* **200**, 403–411.

Simila, T. & Ugarte, F. (1993) Surface and underwater observations of cooperatively feeding killer whales in northern Norway. *Canadian Journal of Zoology* **71**, 1494–1499.

Sinclair, E., Loughlin, T. & Pearcy, W. (1994) Prey selection by northern fur seals (*Callorhinus ursinus*) in the eastern Bering Sea. *Fishery Bulletin* **92**, 144–156.

Smith, G.J.D. & Gaskin, D.E. (1974) The diet of harbor porpoises (*Phocoena phocoena* (L.)) in coastal waters of eastern Canada, with special reference to the Bay of Fundy. *Canadian Journal of Zoology* **52**, 777–782.

Smith, R.J. & Read, A.J. (1992) Consumption of euphausiids by harbour porpoise (*Phocoena phocoena*) calves in the Bay of Fundy. *Canadian Journal of Zoology* **70**, 1629–1632.

Smith, S.C. & Whitehead, H. (1993) Variations in the feeding success and behaviour of Galapagos sperm whales (*Physeter macrocephalus*) as they relate to oceanographic conditions. *Canadian Journal of Zoology* **71**, 1991–1996.

Smith, T.G. (1980) Polar bear predation of ringed and bearded seals in the land-fast sea ice habitat. *Canadian Journal of Zoology* **58**, 2201–2209.

Smolker, R.A., Richards, A.F., Connor, R.C., Mann, J. & Berggren, P. (1997) Sponge-carrying by Indian Ocean bottlenose dolphins: possible tool use by a delphinid. *Ethology* **103**, 454–465.

Steinberg, P.D. (1989) Biogeographical variation in brown algal polyphenolics and other secondary metabolites: comparison between Australasia and North America. *Oecologia* **78**, 374–383.

Steinberg, P.D. & van Altna, I. (1992) Tolerance of marine invertebrate herbivores to brown algal phlorotannins in temperate Australasia. *Ecological Monograph* **62**, 189–222.

Steinberg, P.D., Estes, J.A. & Winter, F.C. (1995) Evolutionary consequences of food chain length in kelp forest communities. *Proceedings of the National Academy of Sciences, USA* **92**, 8145–8148.

Stephens, D.W. & Krebs, J.R. (1986) *Foraging Theory*. Princeton University Press, Princeton, NJ.

Stevens, T.A., Duffield, D.A., Asper, E.D. *et al.* (1989) Preliminary findings of restriction fragment differences in mitochondrial DNA among killer whales (*Orcinus orca*). *Canadian Journal of Zoology* **67**, 2592–2595.

Stewart, B.S. & Delong, R.L. (1993) Seasonal dispersal and habitat use of foraging northern elephant seals. *Symposium, Zoological Society of London* **66**, 179–194.

Stewart, B.S. & DeLong, R.L. (1995) Double migrations of the northern elephant seal, *Mirounga angustirostris*. *Journal of Mammalogy* **76**, 196–205.

Stirling, I. & Archibald, W.R. (1977) Aspects of predation of seals by polar bears. *Journal of the Fisheries Research Board of Canada* **34**, 1126–1129.

Terborgh, J., Estes, J.A., Paquet, P. *et al.* (1999) The role of top carnivores in regulating terrestrial ecosystems. In: *Continental Conservation* (M.E. Soulé & J. Terborgh, eds), pp. 39–64. Island Press, Washington, DC.

Thompson, P.M., Pierce, G.J., Hislop, J.R.G., Miller, D. & Diack, J.S.W. (1991) Winter foraging by common seals (*Phoca vitulina*) in relation to food availability in the inner Moray Firth, N.E. Scotland. *Journal of Animal Ecology* **60**, 283–294.

Thompson, P.M., Tollit, D.J., Greenstreet, S.P.R., Mackay, A. & Corpe, H.M. (1996) Between-year variations in the diet and behaviour of harbour seals, *Phoca vitulina*, in the Moray Firth; causes and consequences. In: *Aquatic Predators and Their Prey* (S.P.R. Greenstreet & M.L. Tasker, eds), pp. 44–52. Blackwell Scientific Publications, Oxford.

Thompson, P.M., Tollit, D.J., Corpe, H.M., Reid, R.J. & Ross, H.M. (1997) Changes in haematological parameters in relation to prey switching in a wild population of harbour seals. *Functional Ecology* **11**, 743–750.

Thompson, P.M., Mackay, A., Tollit, D.J., Enderby, S. & Hammond, P.S. (1998) The influence of body size and sex on the characteristics of harbour seal foraging trips. *Canadian Journal of Zoology* **76**, 1044–1053.

Tollit, D.J., Greenstreet, S.P.R. & Thompson, P.M. (1997) Prey selection by harbour seals, *Phoca vitulina*, in relation to variations in prey abundance. *Canadian Journal of Zoology* **75**, 1508–1518.

Trillmich, F. & Limberger, D. (1985) Drastic effects of El Nino on Galapagos pinnipeds. *Oecologia* **67**, 19–22.

Trillmich, F. & Ono, K.A. (1991) Pinnipeds and El Niño. *Responses to Environmental Stress*. Springer-Verlag, Berlin.

Trites, A.W. & Bigg, M.A. (1992) Changes in body growth of northern fur seals from 1958 to 1974: density effects or changes in the ecosystem? *Fisheries Oceanography* **1**, 127–136.

VanBlaricom, G.R. (1988) Effects of foraging by sea otters on mussel-dominated intertidal communities. In: *The Community Ecology of Sea Otters* (G.R. VanBlaricom & J.A. Estes, eds), Springer-Verlag, Heidelberg, Germany.

Wada, E., Terazaki, M., Kabaya, Y. & Nemoto, T. (1987) 15N and 13C abundances in the Antarctic Ocean with emphasis on the biogeochemical structure of the food web. *Deep Sea Research* **34**, 829–841.

Walker, B.G. & Bowen, W.D. (1993) Changes in body mass and feeding behaviour in male harbour seals, *Phoca vitulina*, in relation to female reproductive status. *Journal of Zoology, London* **231**, 423–436.

Wallace, S.D. & Lavigne, D.M. (1992) *A Review of Stomach Contents of Harp Seals (Phoca groenlandica) from the Northwest Atlantic*. Technical Report No. 92–03. International Marine Mammal Association, Guelph, ON.

Watt, J., Siniff, D.B. & Estes, J.A. (2001) Interdecadal change in diet and population of sea otters at Amchitka Island, Alaska. *Oecologia* **124**, 289–298.

Weinrich, M.T., Schilling, M.R. & Belt, C.R. (1992) Evidence for acquisition of a novel feeding behavior: lobtail feeding in humpback whales, *Megaptera novaeangliae*. *Animal Behaviour* **44**, 1059–1072.

Wells, R.S., Scott, M.D. & Irvine, A.B. (1987) The social structure of free-ranging bottlenose dolphins. In: *Current Mammalogy* (H.H. Genoways, ed.), pp. 247–305. Plenum Press, New York.

Whitehead, H. (1996) Variation in the feeding success of sperm whales: temporal scale, spatial scale and relationship to migrations. *Journal of Animal Ecology* **65**, 429–438.

Würsig, B. (1986) Delphinid foraging strategies. In: *Dolphin Cognition and Behavior: a Comparative Approach* (R.J. Schusterman, J.A. Thomas & F.G. Wood, eds), pp. 347–359. Lawrence Erlbaum, Hillsdale, NJ.

Würsig, B. & Clark, C. (1993) Behavior. In: *The Bowhead Whale* (J.J. Burns, J.J. Montague & C.J. Cowles, eds), pp. 157–199. Society of Marine Mammalogy Special Publication No. 2. Allen Press, Lawrence, KS.

Würsig, B. & Würsig, M. (1980) Behaviour and ecology of the dusky dolphin, *Lagenorhynchus obscurus*, in the south Atlantic. *Fisheries Bulletin (Dublin)* **77**, 871–890.

Ydenberg, R.C., Welham, C.V.J., Schmid-Hempel, R., Schmid-Hempel, P. & Beauchamp, G. (1994) Time and energy constraints and the relationship between currencies in foraging theory. *Behavioural Ecology* **5**, 28–34.

Yoshida, H. & Yamaguchi, H. (1985) Distribution and feeding habits of the pelagic smooth lumpsucker, *Aptocyclus ventricosus* (Pallas), in the Aleutian Basin. *Bulletin Faculty of Fisheries, Hokaido University* **36**, 200–209.

Young, D.D. & Cockcroft, V.G. (1994) Diet of common dolphins (*Delphinus delphis*) off the south-east coast of southern Africa: opportunism or specialization? *Journal of Zoology, London* **234**, 41–53.

CHAPTER 9

Energetics: Consequences for Fitness

Ian L. Boyd

9.1 INTRODUCTION

This chapter is about how energy is acquired and used to maximize fitness in marine mammals. In simple terms, energetics includes the processes used to change fuel in the form of chemical energy into work and heat, but it is more than the documentation of a physical process. The use that animals make of energy, and particularly the efficiency with which they use energy, can be critical to their genetic fitness. The way in which energy is obtained and used, within the context of the natural cycles and variability in its availability as food within the environment, can have profound implications for the phenology of reproduction and the evolution of life histories. Animals take in energy in the form of food as a fuel, a proportion of which is burned to produce work and heat together with various by-products. Like engines, animals can use different forms of fuel and they also have different ways of using the fuel to produce work and heat. However, unlike engines, animals can store energy as tissue and they can employ strategies, including adjustments of behaviour and investment in reproduction, to maximize the fitness value of the energy they obtain.

Energy flux and its dynamics in ecosystems is also central to our understanding of how ecosystems function (Lindeman 1942), the role of populations within ecosystems and the contribution that individuals make to populations. In the past, the objective of studying energetics in marine mammals has often been to provide a framework for management of marine resources and has been firmly directed towards understanding energetics at the level of populations (Lavigne *et al.* 1982, 1985; Markussen & Øritsland 1991; Perez & McAlister 1993; Shelton

et al. 1995). One of the objectives in this chapter is to provide a foundation for the development of understanding energetics at the level of individuals and to examine how this can be used to understand population ecology.

Previous reviews by Costa (1991, 1993) have provided an insight into some of the behavioural and physiological correlates with energetics amongst the pinnipeds. These reviews emphasized a close interaction between body mass in pinnipeds, reproductive strategies and phylogeny. This chapter attempts to take these principles forward but not to review all the information available about marine mammal energetics. Lavigne *et al.* (1982) have provided a comprehensive assessment of bioenergetics in pinnipeds, as has Lockyer (1981) for cetaceans. This chapter will highlight the information about energetics that may determine the distribution and abundances of marine mammals and that may have constrained the evolution of marine mammal life histories.

The chapter is divided into five further sections. The first of these sets the scene by examining the context within which the energetics of marine mammals has evolved. The evolutionary implications of our understanding of energetics are returned to at several stages through the remainder of the chapter. Section 9.3 deals with the tricky issue of definitions so that it is clear to the reader what all the terminology means. The concept of metabolic rate is then examined—how it is measured, the implications and importance of considering body size by comparing metabolic rates among and within species, and how the surface area laws may apply to marine mammals in particular. Section 9.5 examines the use and conservation of energy by marine mammals including the most reasonable maximum sustainable rate of energy use, the way in which marine mammals apportion energy use to different activities, and the overarching problem of thermoregulation. The penultimate section looks at how marine mammals may balance their energy budgets through the development of optimal life histories and body sizes that match the distribution and abundance of their food. Finally, Section 9.7 examines briefly the importance of understanding energetics for the management of marine resources and I end with a plea for the research community to take a strategic approach to building energy-based models of marine mammal ecology and dynamics and, in particular, to ensure that research focuses on the most important variables (i.e. those which explain the greatest proportion of the variance in the results) in these types of models.

The study of energetics is central to our understanding of many, perhaps all, aspects of the ecology of marine mammals. Natural selection will operate forcefully upon individuals that do not find efficient ways of using energy and balancing their energy budgets. This is likely to be the main driving force behind the evolution of traits as diverse as lactation strategies, diving-time budgets, growth and reproductive rates, phenology of the reproductive cycle and life histories.

9.2 MARINE MAMMAL ENERGETICS IN AN EVOLUTIONARY CONTEXT

9.2.1 Allocating energy: interpretations and implication for life histories

Like all animals, marine mammals have to make choices about how they use the energy they obtain from foraging. They also have to make choices about how, and on what, to feed in order to maximize their genetic fitness. We should not underestimate the complexity of these decisions and make simple interpretations of our observations based upon the notion that marine mammals are simply slaves to the physical properties of their bodies and their environment. Let us examine a simple example.

If we measure the body condition (say by weighing) of a number of individuals and we find

several are in poorer condition than the rest, then we might conclude that those individuals are the weakest of the bunch and are at greater risk of dying. Taken at face value, this is a reasonable conclusion. However, it ignores what can be broadly termed the objective function of those individuals. If the objective of an individual is to maximize fitness (defined perhaps as the number of grandchildren it eventually produces) then we can obtain a different interpretation of the range in mass of the group of individuals we have just measured. If, for example, there is a cost to carrying a lot of fat or to being large (e.g. reduced escape speed from predators; more attractive to predators; reduced ability to catch certain prey or reduced ability to dive because of buoyancy constraints) then it may be disadvantageous to carry a lot of fat. In addition, those individuals from our sample that are in 'good' condition may only be that way because they are less certain of being able to satisfy their energy demands from their environment than those that are in 'poor' condition so, therefore, they carry more insurance in the form of more fat. In other words, those in 'poor' condition may simply be more efficient foragers and they may have no need to carry the additional, and potentially costly, body mass. If this is true then our final interpretation of the measurements we have made might be exactly opposite to our initial, simplistic view. For marine mammals, there is no empirical example of such reverse logic but, where the allocation of energy is concerned, this illustrates that interpretations are unlikely to be simple. In such a case, one has to consider both the costs and the benefits of a particular state before being able to interpret its importance.

Marine mammals can allocate energy across three different uses: (i) it can be burned in metabolism, (ii) allocated to growth of body tissues, or (iii) used in reproduction. The partitioning of energy between metabolism and growth is a highly dynamic process that can take place across timescales of minutes to months (in the case of seasonal fattening). However, the allocation process is best illustrated over longer timescales (years) when examining the trade-off between growth and reproduction.

This was recognized first in marine mammals by Laws (1956) who saw a relationship between body mass and the reproductive maturity state of individuals. Unfortunately, until relatively recently, the whole subject of when marine mammals become sexually mature got hung up on age as the driving force, mainly because this is what demographics and population models demanded. Only with the recovery of state-dependent approaches to population dynamics (Caswell 1989; McNamara & Houston 1996) has it been possible to return to a more biologically meaningful approach to the subject. In a recent analysis of pinnipeds (Boyd 2000a), I was able to examine the relative influence that several different state variables (such as body mass and age) had upon the decision that individuals made to allocate resources to reproduction. Not withstanding the arguments about simplistic interpretations of body mass (see above), this showed that, in general, body mass was more important than age for determining whether or not an animal became pregnant and it established a link between the allocation of energy to growth as opposed to reproduction. It suggests that small, young animals do not reproduce because they are below the critical mass (presumably determined through natural selection) at which energy is allocated mainly to reproduction rather than to growth. The distinction between age and mass as a driving force in this relationship is important because, in environments where food is abundant, we would expect a greater availability of energy for growth in young individuals causing them to reach puberty (critical mass for reproduction) in a shorter time than when food is less abundant. In addition, we would expect individuals with abundant food to remain above this critical mass and they would therefore reproduce continuously, whereas those that fall below the critical mass would allocate more of their energy to growth and delay reproduction. In these circumstances age *per se* has little or nothing to do with the energy allocation process.

9.2.2 Living in a cold and conductive environment

Energetics in marine mammals has evolved to cope with the joint challenges of being warm-blooded while living in a highly conductive and relatively cold medium and of having to submerge to feed or escape predators. The structure of the marine environment, involving a more or less unpredictable and patchy distribution of food over large spatial and temporal scales (Steele 1985), has almost certainly also contributed to the evolution of marine mammal energetics, especially through its effect upon energy storage and expenditure strategies. But the hypothesis being pursued in this chapter is that such structural features of the environment are secondary in importance to the over-riding influence of heat management. This is because water is 25 times more conductive to heat than air ($X_{air} = 2.41 \times 10^2$ W/m/K; $\lambda_{water} = 0.591$ W/m/K) which means that, for any warm-blooded vertebrate, keeping warm is likely to be an overwhelming priority. This has profound implications for how animals manage energy flux.

For example, the reproductive pattern in almost all pinnipeds is constrained by the need to give birth out of the water and offspring are more or less terrestrial for a period of a few days to several months. Although many authors have investigated the implications of this constraint for social structure (Bartholomew 1970, Stirling 1983) and foraging strategies (Costa 1991, Boyd 1998) there has been relatively little attention given to why the constraint exists at all. In contrast to cetaceans, which generally occupy tropical to temperate regions of the globe, pinnipeds are most abundant in the polar to temperate regions (see Chapter 1). Since cetaceans and sirenians have managed to overcome the problem of aquatic lactation, it is likely that pinnipeds could also have done so had this been necessary. The thermal challenges of living in cold climates may, at some stage, have meant that it is energetically impractical for small neonatal endotherms to maintain thermal equilibrium in cold water. Consequently, pinnipeds have opted for a semiterrestrial reproductive cycle in which the neonate is protected to some degree from the necessity to be insulated against the thermal challenge of being born into water at, or close to, freezing. Although this is not currently supported by experimental evidence, other studies of the metabolic responses of grey, harbour and northern fur seal pups to changes in air temperature and weather (Blix *et al.* 1979; Trites 1990; Hansen *et al.* 1995; Hansen & Lavigne 1997) support the view that pinniped distributions may be limited to some extent by the thermal environment. It therefore seems probable that their behaviour is limited in a similar way (Hind & Gurney 1998).

Marine mammals have specific morphological and anatomical adaptations to manage heat and many of these are described more fully in Chapter 3. Heat is a product of energy metabolism and, to retain a constant body temperature that exceeds the ambient environmental temperature, marine mammals must balance heat input from metabolism with heat loss to the environment. Poor heat management due to lack of insulation, for example, would lead to an increased requirement for heat production through metabolism and, in turn, an increased requirement for fuel in the form of food. Similarly, small body size would lead to greater potential for heat loss because of the increased surface area of small individuals in proportion to their volume. Consequently, we might expect small marine mammals to use energy at a higher rate than large marine mammals. The rate at which energy is used by marine mammals, possibly because of issues relating to heat management, is also an important feature that affects their foraging ecology (see Chapter 8). Energy demand determines which habitats are suitable for occupation because, to survive, marine mammals must balance their energy gains from foraging with their energy expenditures, although, as we shall see, the time period over which energy gains and expenditures must be balanced depends upon body size and this, in turn, influences the suitability of habitat for different species.

9.2.3 Metabolic machinery

As far as we know the metabolic machinery, including the basic biochemical pathways used for metabolism in marine mammals do not differ from those of other mammalian homeotherms. Fat, protein and usually relatively small amounts of carbohydrate, from the diet form the bulk of the metabolic fuel that is broken down to glucose for input to glycolysis. Like other mammals, marine mammals can terminate the metabolic pathway after glycolysis with the net gain of 2 molecules of adenosine triphosphate (ATP) plus lactic acid. However, in the presence of oxygen, aerobic metabolism will continue through the Krebs cycle to produce 18 molecules of ATP from each molecule of glucose. Although physiologically and biochemically similar in marine mammals and other species, the degree to which glycolysis is used compared with oxidative metabolism, and how marine mammals then recycle lactic acid (the main by-product of glycolysis), is a subject of particular interest. This stems from the need for parsimonious use of oxygen during diving (see Chapter 3). During diving, when marine mammals have access to strictly limited oxygen, there may be increased use of anaerobic metabolism by some tissues. Overall metabolic rate in individuals is the sum of the metabolic rates of all the tissues and organs, some of which may be more or less in a state of hypometabolism (reduced metabolic rate) depending on the availability of oxygen (Guppy et al. 1986; Hochachka 1986).

9.3 DEFINITIONS

Although energetics is about how animals balance their intake of energy with their use of energy, in effect the measurement of metabolic rate (rate of energy use in metabolism) is pivotal to the study of energetics. The definition of metabolic rate is sometimes confused because the units used to express metabolic rate are inconsistent within the literature. Therefore, I have provided a definition for metabolic rate and have explained the underlying rationale for the use of different units in Box 9.1.

9.3.1 Measuring metabolic rate

There are many ways in which it is possible to measure the metabolic rate of an animal. In practice, though, only a few of these can be applied to marine mammals (Box 9.2). Metabolic rate is equivalent to the steady-state heat production by an animal (Box 9.1) and this is measured most directly using calorimetry (C, Box 9.2). In effect, calorimetry is the only truly direct method of measuring metabolic rate and all other methods are approximations of this. The minimum heat loss method (MHL, Box 9.2) is the only alternative method that has attempted to carry out a form of calorimetry on these animals (Folkow & Blix 1992). It assumes that marine mammals are like passive bodies that transfer heat

BOX 9.1 DEFINITION AND UNITS OF METABOLIC RATE

Metabolic rate is the rate at which animals use energy. Its units are watts (W), which are joules per second. Joules (J) are units of work, where 1 J is the work done to move 1 kg through 1 m. In contrast, calories (cal) are units of heat, where 1 cal is the energy required to raise 1 ml of water through 1°K at standard temperature and pressure. This defines the difference between heat and temperature. Whereas temperature is an absolute measure relative to a known standard, heat is the energy contained by a body due to the kinetic energy of its molecules. Heat does not include the potential energy contained in atomic bonds within chemical compounds. The total kinetic energy in the form of heat contained in a body is determined by a combination of the temperature and the heat capacity of the body. The basic constituents of animal tissues—water, fat, protein and carbohydrate—have different heat capacities. Therefore the total kinetic energy contained in the body of animals will not only depend on its temperature but also on its body composition.

Joules are the standard international (SI) units used in animal energetics in preference to calories. However, some studies have expressed energy in terms of calories and the distinction is particularly difficult for endothermic (warm-blooded) animals. For most species, the objective of metabolism is to do work so joules are an appropriate unit but, for endotherms, a proportion of metabolism may take place specifically to produce heat for temperature regulation. In such cases there may be some justification for using calories as the units. One joule is 4.185 cal.

The literature about marine mammals also contains information about metabolic rates that are expressed in dimensions other than watts. These expressions take two forms: (i) those that are in units of joules per unit time, which can, if necessary, be readily translated into watts, and (ii) those that are in other units that have no direct relationship to watts. Examples of each of these are expressions of average daily metabolic rate (ADMR) which may be given in units such as megajoules per day (MJ/day), and the rate of oxygen consumption in units of ml O_2/s, respectively. Technically, the latter is only a surrogate measure of metabolic rate because it measures neither the work done nor the heat produced. However, in practice, the rate of oxygen consumption is often assumed to approximate closely to metabolic rate. Rate of oxygen consumption per unit body mass (kg) is also abbreviated to the expression $\dot{V}o_2$ (alliterated as vee-dot-O-two).

BOX 9.2 METHODS USED TO MEASURE METABOLIC RATE IN MARINE MAMMALS

Method	Acronym	Description
Calorimetry	C	Direct measurement of heat production
Minimum heat loss	MHL	Uses the physical characteristics of blubber, measurements of surface area and blubber depth, as well as estimates of heat loss across the lungs, to estimate total heat loss (Folkow & Blix 1992)
Oxygen respirometry	R-O_2	Measurement of oxygen consumption and also often carbon dioxide in expired gases
Lung capacity	LC	Based on a relationship between lung capacity, respiration rate and metabolic rate (Scholander 1940)
Total energy budget	TEB	Calculation of the total energy used during a period of time by estimating the sum of the costs of all activities
Doubly labelled water	IR-DLW	Indirect respirometry using doubly labelled water to measure water and carbon dioxide flux (Costa 1987)
Single labelled water	IR-SLW	Indirect respirometry using single labelled water to measure water flux (Costa 1987)
Heart rate	IR-f_H	Indirect respirometry using heart rate as a measure of the rate of delivery of oxygen to the tissues. Based on the Fick equation (Butler 1993)
Proximate body composition	PBC	Measurement of the change in total body gross energy during a period of fasting
Mass balance	MB	The use of the difference between gross energy input in food and gross energy output as waste products to estimate metabolic rate, assuming the animal is in a stable state

produced to their environment across an insulating layer, the blubber. This principle has formed the basis of several models of marine mammal energetics (Øritsland & Markussen 1990; Hind & Gurney 1997) but some recent studies of the skin temperatures (Williams *et al.* 1999; Boyd 2000b) have shown that both pinnipeds and cetaceans appear to have a high degree of control over the rate at which heat is lost across the skin. These studies call into question approaches to the measurement and modelling of metabolic rate in marine mammals using the physics of convective and conductive heat transfer.

In the laboratory, the rate of use of oxygen ($R-O_2$, Box 9.2) is the most common form of measurement. This method is known as respirometry or indirect calorimetry. Assuming animals are metabolizing mainly lipids, there is a direct conversion from the flux of oxygen into units of energy. Most of the remaining methods (LC, TEB, IR-DLW, IR-SLW, IR-f_H, Box 9.2) are even less direct measures of metabolic rate and, as a group, they can be classified as indirect respirometry. This is because they are usually calibrated against oxygen flux which is itself only an indirect measure of metabolic rate. One method, IR-DLW, which measures the rates of flux of carbon dioxide isotopically, has been used extensively to measure the metabolic rate of marine mammals in the field and I will return to a discussion of this method and its implications for our understanding of marine mammal metabolic rates later in this chapter. The change in body composition of an individual while fasting or the mass balance can provide a more direct method of examining the rate of energy use by individuals, but it requires that one can estimate accurately body composition at the beginning and the end of a period of fast (or a period of time when the energy assimilated by the individual is known), and this is only possible in special circumstances in the field, such as during lactation in some pinnipeds.

By virtue of their aquatic existence and difficulties with capture, marine mammals as a group do not lend themselves to measurement of their metabolic rates although the range of methods available (Box 9.2) can be used to estimate metabolic rate in just about any species. However, all these methods have inaccuracies associated with them and it is essential when examining patterns of metabolic rate

in marine mammals that any interpretations are made in the light of the methods used for measurement.

9.3.2 Metabolic fuels

The substrates used in metabolism can be viewed as metabolic fuels. This includes fat, protein and carbohydrate, or some mixture of these pure substrates. When these are burned individually, or in combination, in the presence of excess oxygen, the ratio of carbon dioxide produced to oxygen consumed during the combustion is known as the respiratory quotient ($RQ = CO_2/O_2$ where quantities are in moles).

The fuel used by marine mammals for metabolism is important in several ways. First, most methods for measuring metabolic rate depend for their accuracy on being able to establish which substrates an animal is using to run its metabolic machinery. Second, the substrates being used in metabolism provide an insight into the way animals partition their use of different fuels, especially when these have multiple roles, as in the case of fat which can be used as a basic fuel for metabolism, for energy storage and as insulation. And, third, different fuels may be burned in particular circumstances, such as during diving or fasting, and this may provide an insight into the metabolic physiology of an individual.

The respiratory exchange ratio R_E ($\dot{V}CO_2/\dot{V}O_2$) provides the only practical measure of which fuels are being burned. Each substrate has a theoretical respiratory quotient (RQ, see above) and if the fuel that an animal is using is formed of only one substrate the $R_E = RQ_{substrate}$. However, it is more common to find that R_E lies between the RQ for the different substrates, suggesting that the animal may be using a fuel that is a mixture of different substrates.

Relatively few studies of marine mammal metabolic rate have reported R_E values (Table 9.1). Although R_E varies between 0.66 and 0.85 the most common range is 0.71–0.76. This suggests that metabolism is normally fuelled by a mixture of fat ($RQ = 0.71$) and protein ($RQ = 0.79$) with a very small amount of carbohydrate ($RQ = 1$). The RQ of marine mammal diets usually lies in the range 0.74–0.77 (Costa 1987; Boyd *et al.* 1995b), so, based on the range of R_E measured in marine

Table 9.1 Measurements of steady-state (mostly averaged over a minimum of 5–10 min in each trial) metabolic rates in marine mammals.

Species	Metabolic rate[1] (W)	R_E ($\dot{V}_{CO_2}/\dot{V}_{O_2}$)	Sample size	Mass[2] (kg)	Type of measurement[3]	Predicted BMR (W)	Multiple of predicted BMR	Method of measurement[4]	Source[5]
Cetacea									
Tursiops truncatus (adult)	223.7	–	4	171	RMR	160.9	2.1	R-O₂	Irving *et al.* 1941
Tursiops truncatus (adult)	332.8	–	2	145	RMR	141.5	2.4	R-O₂	Williams *et al.* 1993
Phocoena phocoena (juvenile)	149.5	–	1	19	RMR	30.9	4.8	R-O₂	Scholander 1940
Phocoena phocoena (juvenile)	155.7	–	3	35.2	AMR	48.9	3.2	R-O₂	Worthy *et al.* 1987
Balaenoptera musculus	51319	–	–	122000	RMR	22209	2.3	LC	Lockyer 1981
Balaenoptera physalus	6200	–	–	48000	FMR	11032	0.6	PBC	Brodie 1975
Balaenoptera physalus	10559	–	–	30000	RMR	7755	1.4	LC	Lockyer 1981
Balaenoptera physalus	25866	–	–	70000	RMR	14641	1.8	LC	Lockyer 1981
Physeter catodon	4325	–	–	11380	RMR	3749	1.5	LC	Lockyer 1981
Balaenoptera acutorostrata	3367	–	8	3636	FMR	1593	2.8	MHL	Blix & Folkow 1995
Hyperoodon rostratus	1095	–	–	1400	RMR	779	1.4	LC	Lockyer 1981
Pinnipedia									
Pusa hispida (adult)	52.5	–	2	47.5	BMR	61.5	0.8	R-O₂	Innes & Ronald 1981
Pusa hispida (nursing pup)	31.8	–	3	–	FMR	–	3.8	IR-DLW	Lydersen & Hammill 1993
Phoca vitulina (juvenile)	65.6	0.7	3	22	RMR	34.5	1.9	R-O₂	Irving *et al.* 1935
Phoca vitulina (juvenile)	92.6	0.71–0.75	5	30.8	RMR	44.2	2.2	R-O₂	Irving & Hart 1957
Phoca vitulina (juvenile)	71.1	0.69–0.76	6	27.4	RMR	40.7	1.8	R-O₂	Hart & Irving 1959
Phoca vitulina (adult)	130	–	1	98.0	BMR	105.9	1.2	R-O₂	Matsuura & Whittow 1973
Phoca vitulina (juvenile)	57.8	0.78	5	29.1	SMR	42.4	1.6	R-O₂	Hansen *et al.* 1995
Phoca vitulina (juvenile)	55.2	0.71–0.76	2	33.0	RMR	46.6	1.2	R-O₂	Davis *et al.* 1985
Phoca vitulina (juvenile)	175.2	0.71–0.76	2	33.0	AMR	46.6	3.8	R-O₂	Davis *et al.* 1985
Phoca vitulina (juvenile)	119.6	–	4	38.1	ADMR	52.2	2.3	MB	Markussen *et al.* 1990
Phoca vitulina (juvenile)	152.7	–	4	32.0	RMR	45.5	3.4	R-O₂	Williams *et al.* 1991
Phoca vitulina (adult)	95.0	0.71–0.76	1	63.0	RMR	75.7	1.2	R-O₂	Davis *et al.* 1985
Phoca vitulina (adult)	208.9	0.71–0.76	1	63.0	AMR	75.7	2.8	R-O₂	Davis *et al.* 1985
Phoca vitulina (adult F)	223.4	–	17	87.5	FMR	97.3	2.3	IR-SLW, PBC	Bowen *et al.* 1992
Phoca vitulina (adult M)	607.6	–	1	93.0	ADMR	101.3	6.0	IR-DLW	Reilly & Fedak 1991
Phoca vitulina (adult M)	356.5	–	17	107.5	FMR	113.6	3.1	IR-SLW, PBC	Coltman *et al.* 1998

(*continued on p. 254*)

Table 9.1 (cont'd)

Species	Metabolic rate[1] (W)	R_E ($\dot{V}_{CO_2}/\dot{V}_{O_2}$)	Sample size	Mass[2] (kg)	Type of measurement[3]	Predicted BMR (W)	Multiple of predicted BMR	Method of measurement[4]	Source[5]
Pagophilus groenlandicus (juvenile)	80.8	0.70–0.72	2	38.5	RMR	52.6	1.5	R-O₂	Irving & Hart 1957
Pagophilus groenlandicus (adult F)	201.4	0.70–0.75	1	105	AMR	111.0	1.8	R-O₂	Øritsland & Ronald 1975
Pagophilus groenlandicus (adult)	149.4	–	3	141	BMR	139.0	1.1	R-O₂	Gallivan 1981
Pagophilus groenlandicus (adult)	99.1	–	11	15.4	FMR	26.8	4.0	IR-DLW	Lydersen & Kovacs 1996
Pagophilus groenlandicus (nursing pup)	60.7	–	5	c. 95	SMR	103.5	0.7	R-O₂	Hedd et al. 1997
Pagophilus groenlandicus (adult F, winter)	86.3	–	5	c. 95	SMR	103.5	1.0	R-O₂	Hedd et al. 1997
Pagophilus groenlandicus (adult F, summer)	85.3	–	5	c. 95	ADMR	103.5	1.0	R-O₂	Hedd et al. 1997
Pagophilus groenlandicus (adult F, winter)	117.6	–	5	c. 95	ADMR	103.5	1.3	R-O₂	Hedd et al. 1997
Erignathus barbatus (nursing pup)	346.6	–	3	46.7	FMR	60.7	6.0	IR-DLW	Lydersen et al. 1996
Halichoerus grypus (adult F)	232.5	–	2	160	RMR	153.1	1.5	R-O₂	Fedak & Anderson 1982
Halichoerus grypus (adult and juvenile)	204.2	–	3	111	ADMR	116.3	1.8	MB	Ronald et al. 1984
Halichoerus grypus (juvenile)	131–262	0.66–0.85	4	91.8	RMR	100.4	1.3–2.6	R-O₂	Boily & Lavigne 1995
Halichoerus grypus (adult)	141–236	0.66–0.85	3	190.7	RMR	173.7	0.8–1.4	R-O₂	Boily & Lavigne 1995
Halichoerus grypus (nursing pup)	63.7	–	8	–	FMR	–	4.5	IR-DLW	Lydersen et al. 1995
Halichoerus grypus (adult F, lactating)	347.2	–	7	158.1	FMR	151.0	2.3	IR-SLW	Reilly et al. 1996
Halichoerus grypus (pup, fasting)	156.2	–	4	28.9	FMR	42.2	3.7	IR-DLW	Reilly et al. 1996
Cystophora cristata (nursing pup)	200.8	–	6	24.3	FMR	37.2	5.8	IR-DLW	Lydersen et al. 1997
Cystophora cristata (post-weaned pup)	184.5	–	6	42.5	FMR	56.6	3.2	IR-DLW	Lydersen et al. 1997

Species									
Mirounga angustirostris (juvenile)	148.6	–	6	136.5	RMR	135.4	1.1	R-O_2	Webb *et al.* 1998
Mirounga angustirostris (adult F)	469.8	–	1	265	FMR	222.4	2.1	TEB	Sakamoto *et al.* 1989
Mirounga angustirostris (adult F, lactating)	766.2	–	6	425	FMR	318.4	2.4	IR-SLW	Costa *et al.* 1986
Mirounga angustirostris (adult F, moulting)	538.2	–	8	393	FMR	300.3	1.8	IR-SLW	Worthy *et al.* 1992
Mirounga leonina (adult F, lactating)	963.7	–	27	419	FMR	315.1	3.1	IR-SLW	Fedak *et al.* 1996
Mirounga leonina (adult F, moulting)	961.2	–	20	483	FMR	350.5	2.8	IR-SLW	Boyd *et al.* 1993
Leptonychotes weddellii (adult)	717.5	0.69	–	425	BMR	318.5	2.2	R-O_2	Kooyman *et al.* 1973
Leptonychotes weddellii (adult)	589.8	0.69	–	425	AMR	318.5	1.8	R-O_2	Kooyman *et al.* 1973
Leptonychotes weddellii (adult)	429.2	–	–	355	AMR	276.9	1.6	R-O_2	Kooyman *et al.* 1980
Leptonychotes weddellii (adult)	226.1	–	3	263	AMR	222.2	1.0	R-O_2	Kooyman *et al.* 1983
Leptonychotes weddellii (adult)	517.6	–	5	350	AMR	275.3	1.9	R-O_2	Ponganis *et al.* 1993a
Leptonychotes weddellii (juvenile)	314.7	–	1	150	RMR	145.8	2.2	R-O_2	Ponganis *et al.* 1993a
Leptonychotes weddellii (juvenile)	418.0	–	1	150	AMR	145.8	2.8	R-O_2	Ponganis *et al.* 1993b
Zalophus californianus (juvenile)	85.2	–	2	34.9	RMR	48.7	1.7	C, R-O_2	South *et al.* 1976
Zalophus californianus (juvenile)[6]	107.6	0.76	4	25.5	AMR	38.4	2.8	R-O_2	Feldkamp 1987
Zalophus californianus (juvenile)[6]	39.9	–	3	21.3	AMR	33.6	2.7	R-O_2	Williams *et al.* 1991
Zalophus californianus (juvenile)[6]	125.0	0.76	6	32.3	AMR	45.9	2.7	R-O_2	Boyd *et al.* 1995b
Zalophus californianus (juvenile)	126.5	–	6	32.3	AMR	45.9	2.8	IR-f_H	Boyd *et al.* 1995b
Zalophus californianus (juvenile)	158.9	–	6	32.3	AMR	45.9	3.5	IR-DLW	Boyd *et al.* 1995b
Zalophus californianus (juvenile)	105.4	0.74	3	54–69	BMR	74.7	1.4	R-O_2	Matsuura & Whittow 1973
Zalophus californianus (adult F, foraging)	438.9	–	9	81.7	FMR	92.5	4.7	IR-DLW	Costa *et al.* 1991

(*continued on p. 256*)

Table 9.1 (cont'd)

Species	Metabolic rate[1] (W)	R_E ($\dot{V}CO_2/\dot{V}O_2$)	Sample size	Mass[2] (kg)	Type of measurement[3]	Predicted BMR (W)	Multiple of predicted BMR	Method of measurement[4]	Source[5]
Callorhinus ursinus (adult F, fasting)	199.4	–	7	42.7	FMR	56.5	3.5	IR-SLW	Costa & Gentry 1986
Callorhinus ursinus (adult F, foraging)	353.4	–	11	43.2	FMR	57.0	6.2	IR-DLW	Costa & Gentry 1986
Callorhinus ursinus (juvenile M)	158.8	–	1	29.3	RMR	42.6	3.7	IR-DLW, MB	Costa 1987
Arctocephalus gazella (adult F, foraging)	305.6	–	23	32.1	FMR	45.7	6.7	IR-DLW	Costa et al. 1989
Arctocephalus gazella (adult F, foraging)	576.2	–	4	188	FMR	172.1	3.3	PBC, IR-SLW	Boyd & Duck 1991
Arctocephalus gazella (adult M, fasting)	233.3	–	9	36.8	FMR	50.6	4.6	IR-DLW	Arnould et al. 1996
Arctocephalus gazella (adult F, foraging)	193.4	–	8	39.4	FMR	52.8	3.7	IR-SLW	Costa & Trillmich 1988
Arctocephalus gazella (juvenile M, fasting)	189.1	–	9	26.6	FMR	39.7	4.7	IR-SLW	Costa & Trillmich 1988
Arctocephalus galapagoensis (adult F, fasting)	55.4	–	3	37.4	FMR	51.2	1.1	IR-DLW	Costa & Trillmich 1988
Sirenia									
Trichechus latirostris	354.1	–	2	250	RMR	211.8	1.5	R-O$_2$	Scholander & Irving 1941

[1] Where metabolic rates were given at source in terms of oxygen consumption, these were expressed in terms of power (W) assuming that fat was the dominant metabolic fuel and that the energy yield from fat metabolism was 19.67 J/ml O$_2$ (Schmidt-Nielsen 1997). Metabolic rates are expressed as means.
[2] Mean of values are quoted in the source.
[3] See Box 9.3 for definitions of acronyms.
[4] See Box 9.2 for descriptions of the different methods and their acronyms.
[5] Only sources from which the relevant data coud be extracted in numerical form were used. Studies where data were only presented graphically or where species were combined have not been included.
[6] Metabolic rate at 1.5 m/s, i.e. minimum cost of transport speed.

mammals, it would appear that marine mammals burn fuels roughly in proportion to their presence in the diet. Davis *et al.* (1993) calculated for harbour seals that catabolism of lipids provided 87% and 95% of metabolic energy for resting and exercising, respectively. They also suggested that marine mammals probably use lipids as a greater proportion of metabolic fuel than terrestrial mammals, possibly because of the relatively greater presence of lipids in the diet of marine mammals than in those of terrestrial carnivores.

Most of the studies involving the measurement of R_E were made on postabsorptive animals (i.e. they had not fed recently) that may have been metabolizing fuels in the manner that is more normal during fasting. In these circumstances, we might expect most of the fuel that is burned to be fat because of its importance as a storage tissue and, therefore, that the R_E would be close to 0.71. As expected, Boily and Lavigne (1995) found that R_E declined with time after feeding even in postabsorptive individuals from a mean of about 0.8 to a mean of about 0.72. This type of progression in R_E has been observed in pinniped pups at the beginning of the postweaning fast (Nordøy & Blix 1991; Castellini & Rea 1992; Nordøy *et al.* 1993). It may be attributed to the selection of metabolic substrates, which includes a progression from larger amounts of carbohydrates soon after feeding to mainly fat as the animals enter a fasting state after abandonment by the mother at weaning.

Extreme low values of R_E (< 0.71) indicate that animals have moved from a fasting state into a highly conservative starvation state, and they may be attributable to the burning of ketone bodies, which are derived from protein breakdown, particularly by the central nervous system. Ketones tend to replace carbohydrate as a metabolic substrate when animals enter a starvation state (Owen *et al.* 1967).

9.4 METABOLIC RATE

9.4.1 Comparisons of basal metabolic rate between species

Amongst marine mammals, adult body size varies from 28 kg in the Galápagos fur seal to > 105 kg in the blue whale, the largest cetacean. This is a size range of about four orders of magnitude and is greater than for any other group of endothermic vertebrates. Considerations of scale mean that the problems of heat management in a blue whale will be different from those in a Galápagos fur seal. Kleiber (1975) introduced the concept of scaling of metabolic rate in mammals by showing, in an interspecific comparison, that the basal metabolic rate (BMR, Box 9.3) increased with body mass raised to the power of 0.75.

Early measurements of BMR in marine mammals (reviewed by Lavigne *et al.* 1986) suggested that they departed from the expected relationship by having BMRs that were greater than those expected from the Kleiber relationship for terrestrial mammals. Back in 1942, Scholander *et al.* recognized that the apparently high metabolic rates of marine mammals were likely to have resulted from stress induced

---

BOX 9.3 DEFINITIONS OF ACRONYMS

MR	Metabolic rate
AMR	Active metabolic rate. Metabolic rate measured during an activity such as swimming
BMR	Basal metabolic rate (after Kleiber 1975). Metabolic rate of a sexually mature, resting, thermoneutral, postabsorptive, non-growing individual. Often it is assumed to be the lowest stable measurement of metabolic rate from an individual
Predicted BMR	BMR calculated from the Kleiber equation
RMR	Resting metabolic rate. Metabolic rate while at rest but not fulfilling all the requirements of Kleiber (1975)
FMR	Field metabolic rate. Metabolic rate measured in free-ranging animals in the field
SMR	Standard metabolic rate. Defined as for BMR
ADMR	Average daily metabolic rate. This is normally the AMR expressed on a daily basis

during experiments and this was reiterated by Irving (1972). These considerations were often overlooked and it was not until Lavigne *et al.* (1986) showed that when the strict criteria required to determine BMR were applied to the data, the metabolic rate–body mass relationship for marine mammals tended to conform with the general Kleiber relationship in which BMR = 3.39 $M^{0.75}$, where M is the mass in kilograms and where BMR is expressed in watts. In determinations for phocid seals, which conformed to the criteria for measuring BMR, Lavigne *et al.* (1986) found the relationship BMR = 1.93 $M^{0.87}$. Since this was not significantly different from the Kleiber equation, the Kleiber relationship has usually been used to predict the BMR of marine mammals of different body size. Nevertheless, even the analysis by Lavigne *et al.* (1986) is not without its difficulties. Unfortunately, they did not provide information about the sources they used to examine the metabolic rate–body mass relationship and some of those also given by Schmitz and Lavigne (1984) did not conform to the criteria required to determine BMR. I suspect that, as in the case of the analyses presented here, many of the data points used by Lavigne *et al.* (1986) were also not statistically independent. Therefore, the relationship between BMR and mass given by Lavigne *et al.* (1986) should be treated with caution.

The fact is that measurement of BMR is extremely difficult in a marine mammal, but in an attempt to try to put the record straight, I have compiled most of the measurements that are available from the literature (Table 9.1) and the metabolic rate–body mass relationship is illustrated in Fig. 9.1.

According to this analysis, only five of the 76 studies used here conform to the type of measurement that could strictly be defined as a measure of BMR, and these reveal measured BMR in the range of 0.8–2.2 times the predicted BMR from Kleiber's equation based on terrestrial mammals. When the conditions for the inclusion of a study are relaxed and resting metabolic rate (RMR) is also included, a wider range of multiples of predicted BMR is evident (Fig. 9.1).

Consequently, since there is no apparent independent confirmation that the Kleiber relationship applies to marine mammals, the use of the predicted BMR is slightly controversial. Strictly speaking, it

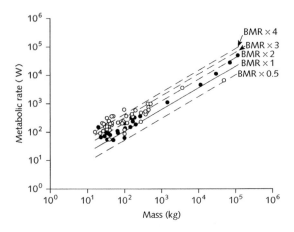

Fig. 9.1 The relationship between body mass and metabolic rate in 76 studies of marine mammals (Table 9.1). Symbols distinguish between measurements of BMR/RMR (●) and other measurements including ADMR, FMR and AMR (○). The solid line shows the predicted BMR for terrestrial mammals (Kleiber 1975) and the dashed lines show the relationship for different multiples of the predicted BMR.

is also only valid for interspecific comparisons of metabolic rate.

There are several important points of interpretation to note about Fig. 9.1. For example, the relatively few points available for the large cetaceans will have a large influence upon the overall relationship and these are some of the least confident measurements because they rely upon the measurement of lung capacity as an indication of resting metabolic rate (Scholander 1940), a highly indirect measurement of metabolic rate. Also, not all of the points are completely independent and, statistically, it is difficult to justify placing a linear regression through these points. This problem arises because some species are represented by many studies whereas others are represented by only one. The additional presence in some cases of two determinations, such as resting and active metabolic rates, for the same individuals also confounds the independence of the data points. Nevertheless, despite all these problems the evidence from Fig. 9.1 suggests that a relationship similar to that of the Kleiber relationship for terrestrial mammals does exist in interspecific comparisons of marine mammals even though it is difficult to quantify this from the data that are currently available.

9.4.2 Comparisons of metabolic rate within species

There is contradictory evidence of a relationship between body mass and metabolic rate within species. Measurements of BMR in harbour seals of different body size suggested that the Kleiber relationship may apply in intraspecific comparisons (Hansen *et al.* 1995). The exponent of 0.71 found by Hansen *et al.* (1995) was not significantly different from the Kleiber exponent of 0.75, but it also does not differ significantly from 0.66, the exponent expected if metabolic rate scales to the surface area available for heat loss. In another study examining field metabolic rate (FMR) in Antarctic fur seals, Costa and Trillmich (1988) found an exponent of metabolic power on body mass of 0.58 and amongst harbour seals the exponent appears to be about 0.49 (Rosen & Renouf 1998). Conversely, there was no significant relationship between body mass, expressed either as lean body mass or total body mass, and FMR amongst moulting female southern elephant seals (Boyd *et al.* 1993).

9.4.3 Effect of fat deposition on measurement of metabolic rate

Lavigne *et al.* (1982) and other authors highlighted the potential problem of the relatively large amount of body fat carried by marine mammals when making inter- and intraspecific comparisons of metabolic rate in relation to body mass. Since blubber has a relatively low metabolic rate and can account for > 40% of body mass but can be reduced by > 50% over a period of a few weeks, especially during lactation, clearly such changes in mass will cause large changes in the apparent BMR when expressed on a mass-specific basis. A large layer of relatively metabolically inactive blubber will have the effect of reducing mass-specific metabolic rate.

In a highly controlled study of harp seal metabolism, Aarseth *et al.* (1999) showed that the BMR in harp seals was largely independent of the amount of fat carried by individuals. This result supports the hypothesis that fat deposition should be discounted when adjusting metabolic rate for the body mass of individuals. This careful study contradicts Lavigne *et al.* (1986) who, with the support of the allometric

studies of McNab (1968) and Kleiber (1975), suggested that marine mammals did not have unusually large fat reserves by mammalian standards and that it was justifiable to include fat content in calculations of metabolic rate. I suggest that the evidence to support this is now less clear because there is a high degree of uncertainty about which value of mass (fat excluded or fat included) should be used in the allometric relationship involving metabolic rate. Such problems are most acute when intraspecific comparisons are being made because these usually involve relatively small ranges of body mass, many of which can be caused by changes in fat content.

It will become apparent in subsequent sections why we should be interested in reconciling the issue of what predicted BMR based upon the allometric equation derived by Lavigne *et al.* (1986) actually means in marine mammals. It becomes central to how we view the energy expenditures of free-ranging individuals and how they are likely to respond to environmental variation.

9.4.4 Field metabolic rates

Despite all these problems, predicted BMR based on the Kleiber relationship has provided a useful benchmark against which to compare the FMRs (Box 9.3) of different species with widely varying body mass (Table 9.1). In order to make such comparisons, FMR is usually expressed as a multiple of the predicted BMR. This is necessary because it is vital to understand how FMR varies between species and individuals of widely varying body mass in order to examine relationships between energetics and ecology.

As illustrated by Table 9.1, the metabolic rate measured for different species depends to a degree upon the circumstances in which measurements are made. Nagy (1987) provided an interspecific comparison of FMR in mammals which included measurements from three pinnipeds. For non-herbivorous eutherian mammals, his comparison showed that FMR was related to body mass to the power of 0.86. However, metabolic rate has now been measured most frequently across a wider size range of fasting and foraging marine mammals and it is possible to look again at Nagy's result for marine mammals alone. Based on the data in Table 9.1, this

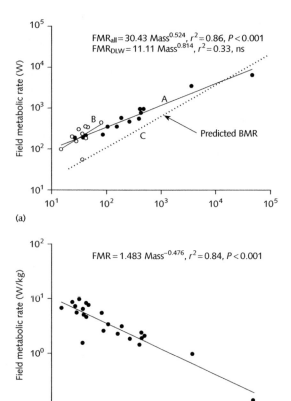

$FMR_{all} = 30.43\ Mass^{0.524}$, $r^2 = 0.86$, $P < 0.001$
$FMR_{DLW} = 11.11\ Mass^{0.814}$, $r^2 = 0.33$, ns

(a)

$FMR = 1.483\ Mass^{-0.476}$, $r^2 = 0.84$, $P < 0.001$

(b)

Fig. 9.2 The relationship of body mass with (a) FMR, and (b) FMR expressed as a proportion of body mass for marine mammals. Data and sources are given in Table 9.1. Three lines are shown in (a): line A is the least-squares regression through all of the data points; line B is the least-squares regression through the points marked by open circles which represent measurements made using the IR-DLW method (Box 9.3); and line C is the relationship of predicted BMR for terrestrial mammals (Kleiber 1975).

gave an allometric relationship in which FMR varied with body mass to the power of 0.52 (Fig. 9.2a). Expressing metabolic rate in proportion to body mass (Fig. 9.2b) showed more clearly how metabolic costs tend to decline with increasing body mass.

Considering the relationship in Fig. 9.2a, the upper 95% confidence limit on the estimate of the exponent (0.52) was 0.61. Comparing this exponent with Nagy's, the relationship suggests that FMR in marine mammals is significantly less sensitive to increases in body mass than it is in other mammals.

However, it is important to moderate this conclusion based on the possible effects of the same set of problems that existed for the allometry of BMR. These include different methods of measurement and potential bias because of outlying points which may have had a strong influence upon the relationship. Removal from Fig. 9.2a of the two species with the largest body mass (fin and minke whales), which could have had undue influence upon the allometric regression, did not result in a significant change in the slope of the relationship (slope = 0.54). This suggests that these two outliers did not bias the overall relationship. Figure 9.2a also shows the effects of considering only FMR measurements made using doubly labelled water. This method has been used mainly on marine mammals of small body mass. Removal of these values also did not produce a significant difference in the slope of the regression (slope = 0.52). Nevertheless, if one uses only FMR measurements obtained using doubly labelled water, the slope of the regression (slope = 0.81) was not significantly different from that obtained by Nagy (1987), primarily for terrestrial mammals, even though the regression in this case was not significant due mainly to the rather low FMR for Galápagos fur seals (Table 9.1 and Fig. 9.2a).

9.4.5 Metabolic rate: the consequences of surface area

It appears that the FMR of marine mammals converges with the predicted BMR for terrestrial mammals as body mass increases (Fig. 9.2). This supports the results of calculated heat losses in marine mammals based upon surface area (Ryg *et al.* 1993), which suggest that the larger whales do not need to raise their metabolic rates to keep warm. Therefore, there appear to be thermal advantages to being large in the marine environment.

The lack of energetic costs associated with supporting tissues against gravity and the lack of gravitational effects on locomotion are likely to be a partial explanation for this effect and it is also probable that the necessity to thermoregulate has an important part to play. For example, if surface area was the dominant factor affecting heat loss amongst marine mammals, we would expect an exponent for the relationship between FMR and mass of approximately 0.66, but the exponent in Fig. 9.2 is significantly less than 0.66. Ryg *et al.* (1993) also calculated that the rate of heat loss from marine mammals varied with mass to the power of 0.59 (a result similar to that found in my analysis) which was significantly less than 0.66. However, as pointed out by Boyd and Croxall (1996), the effective areas of heat loss in highly insulated marine mammals are likely to be across the appendages (although see Williams *et al.* 1999; Boyd 2000b). Although, for a variety of marine endotherms, including some marine mammals, surface area scales to volume or mass to the power of 0.64–0.68 as expected (Ryg *et al.* 1993; Boyd & Croxall 1996), the surface area of the flippers scales to volume to the power of 0.49 (Ryg *et al.* 1993; Boyd & Croxall 1996) which is not significantly different from the exponent of the relationship between FMR and body mass (Fig. 9.2a). Therefore, a potential explanation for the interspecific relationship between FMR and body mass in marine mammals could be that it is driven by the effective surface area for heat exchange, i.e. the surface area of the flippers.

9.4.6 Summary of interactions between metabolic rate and body mass

Overall, there remains uncertainty about whether BMR in marine mammals conforms to the pattern found in terrestrial mammals. This is partly due to a lack of suitable data across a representative size range of species, but it is also because of difficulties with measuring BMR in marine mammals and with defining mass. Measurements of FMR in marine

mammals provide a tighter and more convincing allometric relationship with body mass than those of BMR. They suggest that mass-specific metabolic rate (Fig. 9.2b) declines more rapidly with increasing body mass than is the case for terrestrial mammals and they also support the view that active metabolic rates in the rorquals are unlikely to be greater than the predicted BMR for a terrestrial mammal of equivalent mass (however, note that this comparison is not entirely valid because there are no terrestrial mammals with body mass as great as those of the rorquals). The reasons for this are probably related to the management of heat in the highly conductive environment of the oceans. Small body size appears to incur additional costs, possibly because a greater surface area relative to body mass results in a greater elevation of FMR relative to predicted BMR than for marine mammals with a large body mass.

9.4.7 Evolutionary implications of the allometry of metabolic rate

The fact that metabolic rate does not increase in direct proportion to body size has important implications for marine mammal biology. Mass-specific metabolic rate declines with increasing body size (Fig. 9.2b) whereas the storage capacity for energy as fat increases in approximate proportion to increasing body mass (Table 9.2) (Prothero 1995). Consequently, the mass-specific rate at which energy stores are used declines with increasing body mass. The theoretical time taken to exhaust 50% of the fat reserves, when based on the relationship between

Table 9.2 Mass, fat mass and percentage of body mass composed of fat for species of phocid pinnipeds at the beginning of lactation.

Species	Mass	Fat mass	% fat mass	Source
Mirounga leonina	513	163	32	Fedak *et al.* 1996
Mirounga angustirostris	425	195	46	Costa *et al.* 1986
Halichoerus grypus	178	37	21	Fedak & Anderson 1982; Reilly *et al.* 1996
Pagophilus groenlandicus	120	37	31	Stewart & Lavigne 1984*
Phoca vitulina	90	28	31	Bowen *et al.* 1992*
Cystophora cristata	180	55	30	Bowen *et al.* 1987*

* Based on sculp mass × 0.69 (Reilly & Fedak 1990).

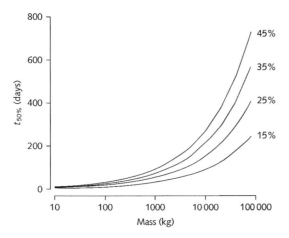

Fig. 9.3 The theoretical relationship between body mass and the time taken to expend 50% of the adipose energy stores in marine mammals when fat reserves represent 15%, 25%, 35% and 45% of body mass.

FMR and body mass given in Fig. 9.2b, and at different levels of fat reserves, is shown in Fig. 9.3. Of course, this ignores additional expenditures such as lactation or moult and also increasing costs of thermoregulation as the blubber layer declines in thickness (Øritsland & Markussen 1990), but it illustrates how increasing body mass can have a disproportionately large effect on the time that individuals may be able to survive on adipose reserves alone. This could vary from a few days to more than 2 years, depending on body size and the initial level of fat reserves. Inevitably, this will affect the frequency with which individuals must feed and has implications for patterns of migration in relation to body mass. It may be that many of the cetaceans with large body size that undergo long-distance migrations to high latitudes to feed can only use such a strategy because of their large size. As explained later, the interaction between body mass and metabolic rate may also be a principal driving force behind the evolution of pinniped reproductive behaviour.

9.4.8 Phylogeny or body size as a determinant of energetic strategies?

Although studies of phocid pinnipeds (Table 9.1) tend to indicate comparatively low metabolic rates, very few of these studies have examined metabolic rate over periods of more than a few hours. Those that have are generally from otariid pinnipeds, mainly because of the smaller cost of carrying out IR-DLW studies (Box 9.2) on these smaller pinnipeds in the field. However, as illustrated in Fig. 9.2, this practical difficulty with carrying out prolonged studies of metabolic rate in large marine mammals, mainly using IR-DLW methods, could have led to an erroneous interpretation concerning phylogenetic differences between the fundamental energetic strategies of otariid and phocid pinnipeds. Costa (1993) concluded that otariids attempt to maximize energy intake during foraging (energy-maximizing strategy) in order to sustain a high metabolic rate, whereas phocids, which were observed to have much lower metabolic rates, tended to maximize the efficiency of foraging (efficiency-maximizing strategy). Although the evidence for the two distinct strategies is weak, we can now see that the apparent otariid–phocid split is much more likely to have been the result of studies that concentrated on two groups of animals with highly contrasting body mass—studies of small otariids because it is practical to study them using IR-DLW and of large phocids (especially Weddell and elephant seals) because of their accessibility. This large–small split of species happened also to be a phocid–otariid split. It now seems more probable that phylogeny has relatively little to do with energetic strategies and that body size is the dominant force.

9.4.9 Maximal sustained metabolic rates

Based upon mass-corrected metabolic rates of marine mammals measured over durations of several days, mainly using doubly labelled water (IR-DLW), it appears that small marine mammals have metabolic rates that are at the high end of the range observed in mammals generally (Hammond & Diamond 1997). Experimental studies in the laboratory have shown that the maximum metabolic rate that can be sustained under aerobic metabolism (metabolic scope) is less than that commonly associated with the average metabolic rate of individuals in the field. For juvenile harbour seals, the maximum metabolic rate was 5.8–7.3 times the predicted BMR (Elsner 1987), and for juvenile California sea lions it was 6.2 times the predicted BMR (Butler *et al.*

1992). In bottlenose dolphins the range was greater at 6.6–9.9 times the predicted BMR (Williams *et al.* 1993). Such high rates of work are at the limit of what an animal can sustain under aerobic metabolism, possibly because of physiological limits to the rate at which it is possible for metabolites to be delivered to the metabolically active tissues and for animals to remain in energy balance (Hammond & Diamond 1997). Since many of the measurements of FMR in small marine mammals are in the region of 5–7 times the predicted BMR, it is possible that these small species normally operate close to their maximum sustainable metabolic rate. Although not impossible, it seems improbable that these marine mammals sustain such high rates of energy expenditure averaged across periods of days to weeks as suggested by some field studies. Putting this in a human context, the metabolic rate being measured in these marine mammals during the normal activities is greater than that sustained by human athletes in the Tour de France which is probably the most demanding endurance sport. Is it possible that some of these rates have been overestimated?

9.4.10 Possible biases in the measurement of field metabolic rate

Several methods are available for measuring metabolic rate in marine mammals (Box 9.2), but, of these, the doubly labelled water (IR-DLW) method has been used most often to measure FMR in free-ranging, foraging marine mammals. However, most of the very high values for FMR come from studies involving IR-DLW (Fig. 9.2). The IR-DLW method carries with it many assumptions (Costa 1987) and there have been few attempts to validate the method in marine mammals. Costa (1987) carried out a validation of the IR-DLW method on a single juvenile male northern fur seal. This showed good congruence between the IR-DLW estimate and an estimate using the mass balance (MB) method of measuring metabolic rate (Box 9.2). However, this type of MB study is complex to carry out and may be subject to biases if not carefully replicated. Subsequently, Boyd *et al.* (1995b) compared the metabolic rate of postabsorptive juvenile California sea lions while exercising and using different methods of measurement, including IR-DLW. This was cross-correlated with indirect respirometry using heart rate (IR-f_H, Box 9.2) and direct respirometry (R-O$_2$, Box 9.2). The study showed that, whereas IR-f_H provided an estimate of metabolic rate to within +3% of the estimate using respirometry, IR-DLW gave an overestimate of up to 36%. Although Speakman (1997) has criticized the methods used in this validation study, additional field data tend to support the conclusion that, on occasions, IR-DLW can overestimate FMR in pinnipeds. The most extensive datasets from Antarctic fur seals generally show FMRs measured using IR-f_H to be 30–40% below FMRs estimated from IR-DLW (Costa *et al.* 1989; Arnould *et al.* 1996; Boyd *et al.* 1999). Nevertheless, because of the possibility that the relationship between heart rate and metabolic rate has not been fully characterized, there is also uncertainty about the accuracy of some measurements of metabolic rate using heart rate in marine mammals. The bottom line is therefore not to overinterpret measurements of FMR, especially those using essentially unvalidated methods in which assumptions may not be upheld.

The assumptions underlying some of the alternatives to the IR-DLW method, such as the proximate body composition (PBC, Box 9.2) and the single labelled water (IR-SLW, Box 9.2) methods of measuring FMR in fasting pinnipeds are probably more robust and these show metabolic rates of 1.5–3.5 times the predicted BMR for pinnipeds (Table 9.1). Most of these measurements are for pinnipeds during terrestrial phases of their annual cycle, when we might expect reduced metabolic rates because the animals spend long periods inactive compared with periods spent actively swimming at sea.

9.5 ENERGY EXPENDITURE: THE USE AND CONSERVATION OF ENERGY

9.5.1 Metabolic rate in relation to activity

Activity has a metabolic cost and considerable efforts have been made to quantify these costs in marine mammals. If BMR is viewed as the lowest average metabolic rate at which it is possible for marine

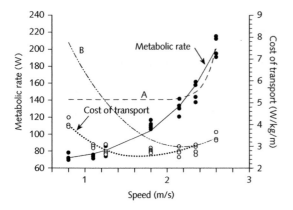

Fig. 9.4 The relationship between metabolic rate (●) and the cost of transport (○) with swimming speed for a 22.5 kg harbour seal (Feldkamp 1987). Also shown are the metabolic rate (curve A) and cost of transport relationship with speed (curve B) when metabolic costs of thermoregulation are present (see text for details).

mammals to survive, most energy expenditure can be accounted for by the overheads relating to other activities, such as locomotion, digestion and growth. Based on the results shown in Fig. 9.2, it would appear that these metabolic overheads become an increasingly important part of the energy budgets of individuals with declining body mass.

Locomotion has a potentially high cost. Like other swimming vertebrates, marine mammals exhibit a curvilinear increase in metabolic rate with increasing swimming speed (Fig. 9.4). This is mainly due to hydrodynamic constraints. As speed increases, drag also increases so that the power required to propel the animal increases in proportion to the cube of speed. This is in contrast to locomotion in terrestrial mammals in which drag forms a small component of the additional power requirement and in which power requirement increases roughly linearly with speed (see Chapter 3 for a further discussion of the effects of shape and size on swimming speed). However, marine mammals benefit from having few costs associated with supporting their posture because they are close to being neutrally buoyant (e.g. Feldkamp 1987). They also benefit, compared with terrestrial mammals, from greater efficiency associated with not having to raise and lower their centre of gravity as they move (Heglund et al. 1982; Feldkamp 1987). Consequently, swimming in marine

mammals can be less costly than locomotion in terrestrial mammals at equivalent speeds.

9.5.2 Implications of body size and swimming speed for the costs of locomotion

This is one possible reason why, with increasing body mass, the FMR of marine mammals tends to converge with the predicted BMR (Fig. 9.2). Passive drag increases in proportion to the frontal area of the animal which, in turn, is likely to increase in proportion to mass to the power of 0.67. Therefore, as mass increases, the frontal area per unit of mass declines as a proportion of mass causing a reduced rate of increase in drag with increasing mass. This will result in the energetic costs of swimming being a progressively smaller proportion of FMR as mass increases.

Calculations based on hydrodynamics suggested that locomotion costs were 40–60% of total energy expenditure in southern elephant seals (Boyd et al. 1994). However, while exercise at different swimming speeds in the laboratory has shown a smooth increase in metabolic rate with speed (e.g. Fig. 9.4), this is not as clearly defined in free-ranging conditions. In bottlenose dolphins, the metabolic cost of swimming at 2 m/s (2.6 times the predicted BMR) was not significantly different from the resting metabolic rate, although metabolic rate increased steeply at greater speeds (Williams et al. 1993). In Antarctic fur seals foraging during lactation, there was no relationship between metabolic rate and swimming speed (Boyd et al. 1999), possibly because animals that are free-swimming can employ behavioural strategies to reduce drag effects at different speeds. These could include swimming at depths that would be sufficient to eliminate surface drag waves or employing more efficient gaits, including porpoising.

The cost of transport can be expressed as the power required to move 1 kg of mass through 1 m. It is a useful means of comparing the locomotor efficiency of different swimming speeds. As indicated above, drag does not increase linearly with speed. This means that, at relatively low speeds, the increase in speed is not compensated for by increased drag and the cost of transport declines. However, as speed increases further there is a rapid increase in drag which demands greater power.

Consequently, the cost of transport then increases. The result of these processes is that marine mammals have a minimum cost of transport at some intermediate swimming speed (Fig. 9.4). For most marine mammals tested to date, this speed appears to be in the range of 1.5–2.2 m/s (Feldkamp 1987; Williams 1987; Williams *et al.* 1993). Observations of swimming speeds in free-ranging pinnipeds have tended to confirm that the preferred swimming speed for pinnipeds lies within this range (Ponganis *et al.* 1990, 1993b; Boyd *et al.* 1995a; Hindell & Lea 1998).

9.5.3 Effects of water temperature on swimming speeds

Hind and Gurney (1997) showed, using a theoretical hydrodynamic model of marine endotherms, that the cost of transport curve can be influenced by water temperature. They showed that, at low speeds, the heat generated as a by-product of locomotion is small. In these circumstances, there is a discrepancy between heat loss and the heat generated from basal metabolism plus that of locomotion and, in order to maintain a constant core temperature in a marine mammal the size of a small harbour seal (Fig. 9.4), metabolic rate would have to increase, although the magnitude of the increase would depend upon water temperature. However, the cost of thermoregulation decreases as the seal swims faster and the waste heat from locomotion offsets the heat loss. This principle is illustrated in Fig. 9.4 where, according to the empirical relationship in Fig. 9.2, it may be necessary for a small, foraging marine mammal to have an FMR which is about four times the predicted BMR. In Fig. 9.4, curve A shows the expected relationship between metabolic rate and swimming speed for a species in which it is necessary for metabolic rate to be increased in order to maintain core temperature. Curve B shows the corresponding cost of transport curve. The important feature of this relationship between cost of transport and speed is that the increased thermoregulatory costs result in an increased minimum cost of transport speed. In other words, as temperature declines, marine mammals can move with greater economy if they swim at progressively greater speeds.

9.5.4 Partitioning energy between activities

In common with other animals, marine mammals have a limited sustainable metabolic rate. Marine mammals may be restricted in the number of activities they can carry out simultaneously by the need to remain within their metabolic scope. For example, the metabolic costs of digestion can occur at the same time as swimming. However, size and composition differences of meals are likely to greatly affect the energy required for digestion. In harbour seals, feeding caused the metabolic rate to increase by up to 46% (Markussen *et al.* 1994). In Steller sea lions the increase was 10–12% depending upon meal size (Rosen & Trites 1997). Therefore, in some circumstances digestion and swimming could be mutually exclusive activities because, when they occur simultaneously, they could exceed an animal's aerobic capacity.

One of the practical problems with examining simple bivariate relationships between metabolic rate and an activity (such as swimming) in free-ranging marine mammals is that animals may be involved in several activities simultaneously. Unfortunately, it is usually impossible to measure all of the activities of a free-ranging marine mammal that contribute to overall metabolic rate. Using multivariate statistical methods to partition the variance in metabolic rate between factors such as the morphology of the animal, swimming speed, time of day and whether animals were mainly at the surface or mainly diving, it has been possible to explain up to 50% of the variation in metabolic rate (Boyd *et al.* 1999). Only by using this statistical method of partitioning between the variance caused by different factors was it possible to reveal an effect of swimming speed upon metabolic rate similar to that observed in controlled experiments.

9.5.5 Energetics of thermoregulation

Although the problems of thermoregulation are a recurring theme in this chapter, it is worth considering the evidence there is that marine mammals have a problem with managing heat, and also the mechanisms used by marine mammals for thermogenesis. In Antarctic fur seals, Butler *et al.* (1995) noted that animals resting in air at 8.1°C had a mean metabolic rate of 2.6 W/kg, but when they were resting in water at 6.8°C the metabolic rate more than doubled to 6.1 W/kg. The mechanisms used to generate this extra heat in adult marine mammals have not been investigated. Non-shivering thermogenesis occurs in some pinniped neonates (Grav *et al.* 1974; Blix *et al.* 1975) and a similar mechanism may exist in adults.

Surprisingly, most recent studies of thermoregulation in pinnipeds have concentrated upon thermoregulation in air (e.g. Matsuura & Whittow 1973; Hansen *et al.* 1995; Hansen & Lavigne 1997). Irving and Hart (1957) showed a lower critical temperature (defined as the temperature below which individuals must increase their metabolic rate in order to maintain their core body temperature) for harbour seals in water of 10°C, whereas in air it was less than −10°C.

Thermoregulation in air has been hypothesized by Lavigne (1982) as a potential factor that limits the distribution of some species of marine mammals (see Chapter 1). Studies of thermoregulation in juvenile harbour seals showed a broad thermoneutral zone (the range of environmental temperature in which an animal can maintain a constant core temperature without incurring additional metabolic costs) from −2.3°C to +25.1°C. Outside this zone there was a rapid increase in metabolic rate which doubled for each 10°C change in ambient air temperature above or below the thermoneutral zone (Hansen *et al.* 1995).

Phocid pinnipeds are generally most abundant in temperate or high latitudes; otariid pinnipeds are mainly found in temperate and subtropical latitudes; relatively few small cetaceans are found at high latitudes compared with the tropics; and large cetaceans are found at all latitudes (see Chapter 1). Hansen *et al.* (1995) noted that, globally, pinnipeds appear to be limited to areas where water temperatures do not exceed 20°C, but they also hypothesized that it was more likely to be ambient air temperatures than water temperatures that limited their distribution. They reasoned that relatively high water temperatures should not pose a thermoregulatory problem (Irving & Hart 1957; Hart & Irving 1959; Watts *et al.* 1993), citing the fact that cetaceans of similar size to pinnipeds must regulate body temperature in warm tropical waters. Instead,

the distribution may be influenced by thermal constraints on land (Watts 1992).

9.5.6 Energy conservation through hypometabolism (reduced metabolic rate)

Reduced metabolic rates have been measured or hypothesized to exist in marine mammals under specific circumstances, both as a strategy for saving energy and to reduce the rate of oxygen use during diving. During moulting, the RMR in harbour and spotted seals declined to 18.6% below normal (Ashwell-Erickson *et al.* 1986). Metabolic rates close to the predicted BMR have been measured during diving and foraging in Weddell seals (Table 9.1), and there is some evidence that hypometabolism may be a strategy employed by marine mammals to extend the duration of aerobic dives (Boyd 1997; Butler & Jones 1997; Hindell & Lea 1998; Kooyman & Ponganis 1998). The RMR was also observed to vary seasonally in harp seals (Hedd *et al.* 1997) suggesting the presence of seasonal hypometabolism, although a similar study on harbour seals showed no consistent seasonal change in RMR (Rosen & Renouf 1998). There is also evidence for diel variation in FMR in Antarctic fur seals after the effects of swimming and diving have been removed (Boyd *et al.* 1999), but no diel variation in RMR was observed in grey seals (Boily & Lavigne 1995). Metabolic rate has also been observed to decline during fasting (Worthy 1987; Nordøy *et al.* 1990; Markussen *et al.* 1992; Rea & Costa 1992).

During fasting, hypometabolism is usually preceded by a change from glucose as the main metabolic substrate to ketone bodies, suggesting a shift to protein metabolism (Nordøy & Blix 1991; Castellini & Rea 1992; Nordøy *et al.* 1993) but it is not clear if this is a cause or a consequence of hypometabolism. During diving, there is reduced peripheral blood circulation with reduced heart rate (Butler & Jones 1997) suggesting that some tissues may also have reduced perfusion with blood. Reduced blood flow to the splanchnic organs (liver, kidney, spleen, intestines and reproductive tract) could result in a substantial reduction in metabolic activity and, since these organs are a potentially significant proportion of the total mass of individuals, it is possible that a reduction of blood

BOX 9.4 Q_{10}

Q_{10} is an expression of the rate of change of metabolic rate with temperature. Metabolism is fundamentally an enzyme-mediated process within cells, and the enzymes tend to operate optimally at quite specific temperatures. In endotherms such as marine mammals this temperature is likely to be 37–38°C. Changes in temperature will lead to changes in enzyme function and this will result in a reduced metabolic rate.

Formally, $Q_{10} = [R_1/R_2]^{10/\Delta t}$, where R_1 and R_2 are the metabolic rates and Δt is the temperature difference between the two metabolic rates. A Q_{10} of 2 will represent a doubling of metabolic rate for each 10°C increase in temperature. In marine mammals, measurements of the whole body Q_{10} come from the studies of Matsuura and Whittow (1973) and

Hansen et al. (1995). They measured the rate of change in metabolic rate in animals exposed to greater than the upper critical temperature. The resulting Q_{10} was 4.14 in sea lions and 1.5 in harbour seals.

The Q_{10} given by Matsuura and Whittow (1973) represents the metabolic costs involved in maintaining a constant core temperature when the ambient temperature is outside the thermoneutral zone. Nevertheless, it is the only indication we have of how marine mammal tissues might respond to heterothermy. Taken on its face value, it might suggest that a reduction of temperature by 10°C could result in a reduction of metabolic rate to one-quarter of its level at normal body temperature.

flow to them could result in the reductions of metabolic rate that have been observed (Kooyman & Ponganis 1998). However, observations of elephant seal diving bouts, in which animals can make long dives with only short intervening surfacings, for periods of months, does not suggest that a temporary reduction of blood flow to the splanchnic organs could be solely responsible for hypometabolism. This is because the splanchnic organs are required for vital activities, such as digestion and reproduction, and it is therefore important for circulation to be maintained to these organs for substantial parts of an animal's life.

These studies suggest that pinnipeds have a high degree of control over their metabolic rate and that even basal metabolism may not be a constant. Pinnipeds are apparently able to reduce their metabolic rates either as a measure to conserve energy while fasting or as a means of conserving oxygen while diving.

9.5.7 Are marine mammals heterothermic?

One way in which marine mammals could reduce their metabolic costs is through reductions in body temperature. Marine mammals have a high degree of control over the circulation to different organs and tissues (Butler & Jones 1997) and this could be used as a mechanism for the redistribution of heat. This idea has developed mainly from studies of

penguins (e.g. Handrich et al. 1997), which appear to make use of regional heterothermy (defined as the ability to control body temperature in at least one region of the body and to control this in isolation from other regions) as a means of reducing their metabolic rate. Depending upon the Q_{10} of tissues (Box 9.4), this mechanism could result in substantial reductions in the overall metabolic rate. However, this concept of heat management as the basis of hypometabolism in marine mammals remains to be tested experimentally and is not currently strongly supported by the available evidence. The few studies that have examined body temperatures in diving marine mammals provide conflicting evidence in support of heterothermic behaviour. Aortic temperatures in Weddell seals decline by only 1–3°C during and after prolonged dives (Kooyman et al. 1980; Hill et al. 1987). Similarly, in multiple tissues, Scholander et al. (1942) found temperature reductions of only 1°C during long forced dives and Ponganis et al. (1993b) did not detect any change in muscle (an active tissue during diving) temperature in Weddell seals. Rectal temperature remained normal during diving in captive harp seals (Gallivan & Ronald 1979).

Therefore, there is currently little evidence for the existence of heterothermy as a means of regulating metabolic rate in marine mammals but it is possible that the critical experiments to test this hypothesis remain to be done.

9.6 BALANCING ENERGY BUDGETS

9.6.1 Evolutionary implications

All animals must balance their energy budgets if they are to survive and reproduce. The timescale over which they have to do this depends upon body size (Section 9.4) and, as shown in Fig. 9.3, body size is an important determinant of the time over which animals must balance their energy budgets. The larger the animal is, the longer the timescale will be. Many baleen whales that migrate into high latitudes in the summer do so to feed on the seasonally high food densities in these regions and they are likely to fast for the remainder of the year (Brodie 1975). Therefore, these animals probably only balance their energy budgets over a full annual cycle. Smaller species need to balance energy intake over expenditure on considerably shorter timescales (e.g. days to weeks).

An example of the close interaction that exists between body mass, energetics and behaviour is seen in the range of strategies used by female pinnipeds to feed their offspring. This probably results from the different potential fasting durations of pinnipeds with different body sizes and is described in greater detail by Boyd (1998), but the basic ideas are outlined here. Much of the recent literature concerning pinniped lactation has distinguished between two strategies, the foraging cycle strategy and the fasting strategy, that divide along phylogenetic lines between otariid and phocids, respectively (Bonner 1984; Oftedal et al. 1987; Boness & Bowen 1996). The problem faced by lactating pinnipeds is that they must provision their pups based on a food supply that could be from tens to thousands of kilometres from the pupping site. Large pinnipeds like elephant seals almost certainly still require a larger absolute rate of energy gain than small pinnipeds (Fig. 9.2a). Thus, to balance their energy budgets, large pinnipeds need to forage in locations which are likely to yield higher rates of energy gain. Therefore, large pinnipeds will be forced to forage more widely than small pinnipeds because they need to find richer patches of prey and these patches will be relatively uncommon and unpredictable in their occurrence.

This disadvantage for large pinnipeds is compensated for by their greater ability to fast (Fig. 9.3) and it is this ability that makes it possible for large pinnipeds to have a terrestrial phase within their reproductive cycles. Assuming that the mother's objective is to deliver as much energy to her offspring in the form of milk as quickly as possible and that other nutrients are not limiting, three main factors limit what pinnipeds are able to do in order to provision their pups during nursing. These are the potential fasting duration of the mother while still provisioning the pup, the potential fasting duration of the pup while the mother is absent foraging, and the maximum potential rate of delivery of milk. If the local environment has a sufficiently high energy yield for mothers to both balance their energy budgets and deliver enough food to their pups, then it is feasible for mothers to follow the foraging cycle strategy. This allows them to make foraging trips during lactation, and this is the strategy commonly used by small pinnipeds. However, due to the greater absolute energy requirements of large pinnipeds, a much greater energy yield would be required from the local environment to make the foraging cycle strategy feasible. Consequently, large pinnipeds in general do not forage during lactation. Only in some ice-breeding species which have food easily accessible is it economical to forage during lactation.

There appears to be a critical body mass in pinnipeds above which energetic constraints make it difficult for them to follow the foraging cycle strategy, but at which point pinnipeds are able to carry enough stored energy to sustain them through lactation. Harbour seals (80–100 kg), which forage during lactation (Boness et al. 1994), appear to be below this critical mass whereas grey seals (130–180 kg), which generally fast during lactation (Fedak & Anderson 1982), appear to be above the critical mass. There are, of course, other lactation strategies in pinnipeds but most of these, including foraging during lactation by pinnipeds that raise their pups on floating ice, can be explained by the type of energy balance model that explicitly includes all the energetic costs to mothers (including those involved in migration to and from the foraging grounds, while on the foraging grounds, and during nursing) (Boyd 1998).

The main object of this example is to illustrate, once again, that what appears at first to be a phylogenetic trait is more likely to be a consequence of adaptation within the energetic constraints imposed by both body size and environmental productivity. It is then possible to see how this is likely to limit the distribution of pinnipeds. Small species are generally constrained to forage close to the pupping grounds, at least during lactation, and they are constrained to raise their pups in regions where there is a local abundance of food. It is more difficult to extend the example to cetaceans because the phenology of their reproductive cycles is very different to that of pinnipeds. However, it is likely that similar principles exist which explain, for example, the differences in the foraging distributions of male and female sperm whales.

9.6.2 Optimization of energy use

The concept that animals may optimize their behavioural–time budgets to allow them to balance their energy budgets is useful but controversial (Stephens & Krebs 1986). The example given in the previous section was based upon a time–energy budget optimization model. In this case, natural selection was assumed to have been the driving force behind the optimization process. However, controversy arises when optimization processes are hypothesized to be the driving forces behind behaviour because the mechanisms by which animals can optimize their behaviour are unclear. For the most part, we have to assume that animals possess a set of decision rules that have been selected to allow animals to have a flexible response to different situations and that the resulting behaviour converges towards an optimal solution even though the animal has no means of knowing this. It may be rare for an animal to fully optimize its behaviour in order, for example, to maximize its rate of energy return on energy expended during foraging. Other factors can influence fitness, such as predation and, if behaviour is being adjusted to minimize the predation risk, then an animal may not achieve the optimum that might exist if energy was its only consideration.

It is also possible that marine mammals are highly sophisticated in the way that they manage their energy reserves. What we view as a probable objective for these animals, such as maximizing the rate of delivery of energy to offspring in the case of lactating pinnipeds, may be a simplistic view of the actual objectives of an individual. There is a large, complex and fairly inconclusive literature on parental investment patterns in pinnipeds (reviewed by Trillmich 1996) which generally fails to appreciate that female pinnipeds are likely to balance their energy investment in current offspring against the risk this has for lifetime fitness. Such a principle is likely to apply to all marine mammals and could result in mothers hedging their bets with respect to how much energy to deliver to their current offspring. Fedak *et al.* (1996) showed that the proportion of energy that female southern elephant seals apparently choose to invest in their offspring is often considerably less than the energy that is available. The remaining energy may be insurance against starvation before the animals are able to return to their foraging grounds (Boyd 1998). Therefore, the amount of hedging that goes on could, in turn, be influenced by the spatial and temporal heterogeneity of the environment and the perceived potential fitness value of a particular offspring. In an unpredictable environment we would expect mothers to hedge more than they might in a predictable environment.

9.6.3 Energetic currencies

Several energy currencies are recognized as being of potential importance to predators like marine mammals. These include the gross rate of gain (absolute rate of energy gain), energetic efficiency (gross rate of energy gained expressed as a proportion of energy used) and net rate of gain (gross rate of energy gained minus the rate energy expenditure) (Houston 1995). One can see immediately why these are important to animals. Take the case of a marine mammal that is diving for food. If there is a specific cost of feeding on a prey item at a particular depth and the energy gained from feeding on that item is less than or equal to the cost, there is no net gain from foraging on that food source. When the gain is greater than the cost there is a net gain from foraging on that food source. Moreover, for a standard food item, profitability will decline with depth (Carbone & Houston 1994), so a food item that may be fed upon at shallow depths may not be chosen when it is deeper because it will cost more to obtain. Individuals that make good decisions about the profitability of foraging will also be those that achieve greatest genetic fitness because they are more likely to be competitive for food when this is a limiting factor and therefore they are more likely to survive longer and produce higher quality offspring. However, current empirical observations of marine mammals are generally not sufficiently extensive to be able to distinguish between the energy currencies being used (Boyd 1999) even though different currencies can predict different behaviour (Thompson *et al.* 1993). It is also possible that animals switch between currencies in different circumstances which adds further complication.

These concepts can be influenced greatly by how we view the behaviour of marine mammals. For example, if, due to thermal considerations, it costs an animal the same to swim at its minimum cost of transport speed as it does to remain at rest (Fig. 9.4), then there is technically no energetic cost associated with searching for prey. In such circumstances, the only cost of a long versus a short dive for a standard prey item is that the long dive means that the average rate of energy gain across all dives will be reduced. This simple conceptual model would predict that marine mammals would always choose to dive for a standard prey item when it is shallow rather than deep. But it also suggests, based on the result in Fig. 9.4, that marine mammals with smaller body size should be more active (in the sense that they should have greater swimming speeds per unit body length and they should also be searching for prey for a greater proportion of available time) because the energetic overheads of searching for prey are likely to be relatively small, even though the mass-specific rate of energy expenditure will be high.

None of these ideas have been formally tested in marine mammals but they are of considerable importance to assessments of the potential effects that changes in food abundance and dispersion may have for marine mammal distribution and abundance and for the understanding of behaviour. The decisions made by individuals can be modelled and measured in relation to both their own energy budgets, and the distribution and profitabilities of the available prey. These individual-based behavioural models (Thompson *et al.* 1993; Boyd 1999) can be linked to energetics models that predict the demographic consequences of varying rates of energy gain (e.g. Øritsland & Markussen 1990). Thus, by examining the behaviour of individuals it may be possible to move towards a system for assessing how the dynamics of marine mammal populations are affected by changes in both the absolute abundance and the distribution of food (Thomson *et al.* 2000).

9.7 CONCLUSIONS

9.7.1 Energetics and the management of marine resources

Energy demand at the level of individuals can have profound importance for the management of marine resources. The energy requirements of individuals can be built, through the stratified sampling of different classes of individuals (e.g. males, females, adults, juveniles, reproductively active and non-reproductive), into a population energy budget (Lavigne *et al.* 1982, 1985; Markussen & Øritsland 1991; Perez & McAlister 1993; Shelton *et al.* 1995). The complexity of these approaches can vary considerably and there is a need for updated algorithms that assimilate data about behaviour, life history, demography and physiology to provide information about energy budgets that can be made spatially explicit, i.e. they also incorporate information about the dispersion of individuals to show where predators like marine mammals are making their greatest demands on resources. When tied to diet, this can provide an estimate of the contribution that a marine mammal population makes towards the mortality of a particular prey species that may also be the subject of commercial exploitation.

In a broader ecological context this information may be used to examine the flow of energy through ecosystems. Although these ideas have been in the literature for some time (e.g. Priddle *et al.* 1998), to date there are few examples of energetics applied in this context and there has been little recognition of the need to provide confidence intervals around estimates of prey consumption or of the need to integrate the results with other models of ecosystem dynamics. This approach can also be used, together with information about population dynamics, to examine the relative vulnerability of marine mammals to the consequences of changes in food availability.

9.7.2 A plea for a strategic approach

I reiterate the views of Lavigne *et al.* (1982) that an understanding of energetics in marine mammals is pivotal to the management of most large-scale marine ecosystems. This is because marine mammals are influential components of ecosystems, both ecologically and politically, and because we cannot begin to understand the role of marine mammals in relation to other biological components of an ecosystem without using a common currency—energy. In future, resources within ecosystems, including marine mammals and fisheries, will be managed as integrated, complex systems. Science has to provide the rationale and data to underpin such management and for this we require a strategic approach to the study of energetics.

This approach must be driven by hypotheses, mainly in the form of formal or semiformal models, that are challenged with data (Hilborn & Mangel 1997). In the complex world of marine mammal energetics, there is an absolute necessity for models because the practical difficulty and expense of obtain-

ing data in this field make it imperative that the few data that do emerge can fit into a wider context and have generality. Two examples from marine mammal energetics show how this process can operate. In this chapter, I have devoted much attention to the influence of body size on metabolic rate. Up to the mid 1980s the model that was being tested suggested that marine mammals have metabolic rates that were elevated above those of terrestrial mammals of equivalent size. This was questioned by Lavigne *et al.* (1986) who presented new data which supported the model of marine mammals as species that conformed to the relationship for terrestrial mammals. Now, I question the whole concept once more and, while still useful as a benchmark, I reject the notion that BMR is quantifiable in a practical sense in marine mammals and think that our attention should be turned instead to models based on measurements of FMR, which is quantifiable and is actually a much more useful measurement than BMR.

A second example comes from the ideas generated about optimal behaviour in air-breathing, diving homeotherms. Houston and Carbone (1992) proposed an energy balance model to explain diving behaviour which gained some support from empirical observations (Boyd *et al.* 1995a). Thompson *et al.* (1993) added complexity to this through the inclusion of a cost of transport function in the model and here, informally, I have suggested that

Fig. 9.5 Diagrammatic illustration of the structure of an energetics, scale-based approach to examining the processes that drive the responses of marine mammals to changes in their environment. It takes small-scale behaviour described by energy-based models of how animals dive for food and builds this into progressively larger scale behaviours involving bouts of dives, foraging trips and eventually to whole life histories. These influence the fitness of individuals in different environments that are defined by differing prey distributions, which are themselves affected by population size through competition for resources. See Box 9.5 for further explanation.

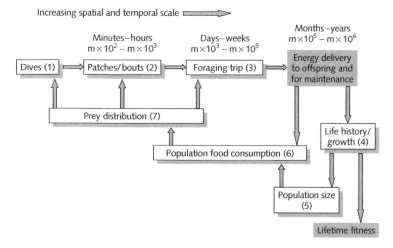

BOX 9.5 SCALE-DEPENDENT MODELLING OF POPULATION ENERGETICS

Figure 9.5 provides a diagrammatic illustration of the major modelling components required for a strategic and dynamic model of the population energetics of a marine mammal. It is implicit that the component models are energy-based and that they are driven by principles of thermodynamics. The arrows show influences and the flow of control.

Models 1, 2 and 3 (Fig. 9.5) are state-dependent dynamic models that operate at different spatial and temporal scales. Model 1 is equivalent to the types of models provided by Houston and Carbone (1992) and Thompson et al. (1993). In the case of model 3, the terminology used in this illustration may not be appropriate for a cetacean but it would still be necessary to provide a model of the dynamic nature of habitat use at the appropriate scale. The timescale involved in model 3 is defined by the period over which individuals must balance their energy budgets which is, in turn, dependent upon body mass. Boyd (1999) provides an example of such a model. The purpose of models 1, 2 and 3 is to examine the potential dynamics of energy flow to offspring and maintenance that results from changes in the prey distribution which is represented by model 7. This model could incorporate elements of the prey population structure, dynamics and patchiness that could be affected by competing predator species.

Model 4 represents the consequences of different energy flows for production and survival at all stages of the life cycle. This could be constructed in a similar way to the model described by Øritsland & Markussen (1990). The output of such a model will influence the number of animals in the population, which could be represented by standard Leslie matrix modelling. When combined with the energy flow, the results of model 5 provides a population energy consumption which can be translated into a food consumption (e.g. Lavigne et al. 1985). This will, in turn affect the prey distribution, thus completing the feedback loop to prey through potential competition.

A strategic model of this type has still not been constructed for marine mammals although many of the modelling components are already available in some form and substantial datasets exist to parameterize such a model and to test its predictions. Such a model would need to assess the level of certainty surrounding its predictions, possibly by using Bayesian analytical approaches (e.g. Thomson et al. 2000).

---

this complexity should be increased to include the effects of thermoregulation and body size (Fig. 9.4). Now, the challenge is to test these models empirically and to iterate the process.

I have said little about population energetics but, from the beginning, I recognize the need to integrate studies of energetics at the level of individuals so that they provide a dynamic picture of how marine mammals are likely to respond to change. This requires a strategic approach to the study of marine mammal energetics which is most likely to involve coupled models of different processes (Box 9.5 and Fig. 9.5). I have attempted in this chapter to present some of the general principles upon which such a strategy could be built. For example, the apparent over-riding effect of body size for time–energy budgets and for thermoregulation may permit interpolation of empirical data between species. But it must always be appreciated that such modelling exercises will always produce results that are couched in uncertainty. Two requirements of strategic modelling of energetics are to provide a realistic view of the uncertainty of the results and to identify the

critical experiments that are required in order to progress the models towards a more realistic and therefore a more certain representation of how marine mammals function in their environment.

REFERENCES

Aarseth, J.J., Nørdoy, E.E. & Blix, A.S. (1999) The effect of body fat on basal metabolic rate in adult harp seals (*Phoca groenlandica*). *Comparative Biochemistry and Physiology A* **124**, 69–72.

Arnould, J.P.Y., Boyd, I.L. & Speakman, J.R. (1996) The relationship between foraging behaviour and energy expenditure in Antarctic fur seals. *Journal of Zoology, London* **239**, 769–782.

Ashwell-Erickson, S.F.H., Fay, R., Elsner, R. & Wartzok, D. (1986) Metabolic and hormonal correlates of molting and regeneration of pelage in Alaskan harbour and spotted seals (*Phoca vitulina* and *Phoca largha*). *Canadian Journal of Zoology* **64**, 1086–1094.

Bartholomew, G.A. (1970) A model for the evolution of pinniped polygyny. *Evolution* **24**, 546–559.

Blix, A.S. & Folkow, L.P. (1995) Daily energy expenditure of free-living minke whales. *Acta Physiologica Scandinavica* **153**, 61–66.

Blix, A.S., Grav, H.J. & Ronald, K. (1975) Brown adipose tissue and the significance of the venous plexuses in pinnipeds. *Acta Physiologica Scandinavica* **94**, 133–135.

Blix, A.S., Miller, L.K., Keyes, M.C., Grav, H.J. & Elsner, R. (1979) Newborn northern fur seals (*Callorhinus ursinus*) —do they suffer from cold? *American Journal of Physiology* **236**, R322–R337.

Boily, P. & Lavigne, D.M. (1995) Resting metabolic rates and respiratory quotients of gray seals (*Halichoerus grypus*) in relation to time and duration of food deprivation. *Physiological Zoology* **68**, 1181–1193.

Boness, D.J. & Bowen, W.D. (1996) The evolution of maternal care in pinnipeds. *BioScience* **46**, 645–654.

Boness, D.J., Bowen, W.D. & Oftedal, O.T. (1994) Evidence of a maternal foraging cycle resembling that of otariid seals in a small phocid, the harbour seal. *Behavioral Ecology and Sociobiology* **34**, 95–104.

Bonner, W.N. (1984) Lactation strategies in pinnipeds: problem for a marine mammalian group. *Symposium of the Zoological Society of London* **51**, 253–272.

Bowen W.D., Boness, D.J. & Oftedal, O.T. (1987) Mass transfer from mother to pup and subsequent mass loss by the weaned pup in the hooded seal, *Cystophora cristata*. *Canadian Journal of Zoology* **65**, 1–8.

Bowen, W.D., Oftedal, O.T. & Boness, D.J. (1992) Mass and energy transfer during lactation in a small phocid, the harbor seal (*Phoca vitulina*). *Physiological Zoology* **65**, 844–866.

Boyd, I.L. (1997) The behavioural and physiological ecology of diving. *Trends in Ecology and Evolution* **12**, 213–217.

Boyd, I.L. (1998) Time and energy constraints in pinniped lactation. *American Naturalist* **152**, 717–728.

Boyd, I.L. (1999) Foraging and provisioning in Antarctic fur seals: inter-annual variability in time–energy budgets. *Behavioral Ecology* **10**, 198–208.

Boyd, I.L. (2000a) State-dependent fertility in pinnipeds: contrasting capital and income breeders. *Functional Ecology* **14**, 623–630.

Boyd, I.L. (2000b) Skin temperatures during free-ranging swimming and diving in Antarctic fur seals. *Journal of Experimental Biology* **203**, 1907–1914.

Boyd, I.L. & Croxall, J.P. (1996) Dive durations in pinnipeds and seabirds. *Canadian Journal of Zoology* **74**, 1696–1705.

Boyd, I.L. & Duck, C.D. (1991) Mass change and metabolism of territorial male Antarctic fur seals (*Arctocephalus gazella*). *Physiological Zoology* **64**, 375–392.

Boyd, I.L., Arnbom, T. & Fedak, M.A. (1993) Water flux, body composition, and metabolic rate during molt in female southern elephant seals (*Mirounga leonina*). *Physiological Zoology* **66**, 43–60.

Boyd, I.L., Arnbom, T. & Fedak, M.A. (1994) Biomass and energy consumption of the South Georgia population of southern elephant seals. In: *Elephant seals: Population Ecology, Behavior, and Physiology* (B.J. Le Boeuf & R.M. Laws, eds), pp. 98–117. University of California Press, Berkeley, CA.

Boyd, I.L., Reid, K. & Bevan, R.M. (1995a) Swimming speed and allocation of time during the dive cycle in Antarctic fur seals. *Animal Behaviour* **50**, 769–784.

Boyd, I.L., Woakes, A.J., Butler, P.J., Davis, R.W. & Williams, T.M. (1995b) Validation of heart rate and doubly labelled water as measures of metabolic rate during swimming in California sea lions. *Functional Ecology* **9**, 151–160.

Boyd, I.L., Bevan, R.M., Woakes, A.J. & Butler, P.J. (1999) Heart rate and behavior of fur seals: implications for the measurement of field energetics. *American Journal of Physiology* **276**, H844–H857.

Butler, P.J. (1993) To what extent can heart rate be used as an indicator of metabolic rate in free-living marine mammals? *Symposium of the Zoological Society of London* **66**, 317–332.

Butler, P.J. & Jones, D.R. (1997) Physiology of diving of birds and mammals. *Physiology Review* **77**, 837–899.

Butler, P.J., Woakes, A.J., Boyd, I.L. & Kanatous, S. (1992) Relationship between heart rate and oxygen consumption during steady-state swimming in California sea lions. *Journal of Experimental Biology* **170**, 35–42.

Butler, P.J., Bevan, R.M., Woakes, A.J., Croxall, J.P. & Boyd, I.L. (1995) The use of data loggers to determine the energetics and physiology of aquatic birds and mammals. *Brazilian Journal of Medical and Biological Research* **28**, 1307–1317.

Brodie, P.F. (1975) Cetacean energetics, an overview of intraspecific size variation. *Ecology* **56**, 152–161.

Carbone, C. & Houston, A.I. (1994) Patterns in the diving behaviour of the pochard, *Aythya ferina*: a test of an optimality model. *Animal Behaviour* **48**, 457–465.

Castellini, M.A. & Rea, L.D. (1992) The biochemistry of natural fasting and its limits. *Experimentia* **48**, 575–776.

Caswell, H. (1989) *Matrix Population Models*. Sinauer Associates, Sunderland, MA.

Coltman, D.W., Bowen, W.D., Iverson, S.J. & Boness, D.J. (1998) The energetics of male reproduction in an aquatically mating pinniped, the harbour seals. *Physiological Zoology* **71**, 387–399.

Costa, D.P. (1987) Isotopic methods for quantifying material and energy intake of free-ranging marine mammals. In: *Marine Mammal Energetics* (A.C. Huntley, D.P. Costa, G.A.J. Worthy & M.A. Castellini, eds), pp. 43–66. Society for Marine Mammalogy Special Publication No. 1. Allen Press, Lawrence, KS.

Costa, D.P. (1991) Reproductive and foraging energetics of pinnipeds: implications for life-history patterns. In: *Behavior of Pinnipeds* (D. Renouf, ed.), pp. 300–344. Chapman & Hall, London.

Costa, D.P. (1993) The relationship between reproductive and foraging energetics and the evolution of the Pinnipedia. *Symposium of the Zoological Society of London* **66**, 293–313.

Costa, D.P. & Gentry, R.L. (1986) Free-ranging energetics of northern fur seals. In: *Fur seals: Maternal Strategies on Land and at Sea* (R.L. Gentry & G.L. Kooyman, eds), pp. 79–101. Princeton University Press, Princeton, NJ.

Costa, D.P. & Trillmich, F. (1988) Mass changes and metabolism during the perinatal fast: a comparison between Antarctic (*Arctocephalus gazella*) and Galápagos fur seals (*Arctocephalus galapagoensis*). *Physiological Zoology* **61**, 160–169.

Costa, D.P., LeBoeuf, B.J., Huntley, A.C. & Ortiz, C.L. (1986) The energetics of lactation on the northern elephant seal, *Mirounga angustirostris*. *Journal of Zoology, London* **209**, 21–33.

Costa, D.P., Croxall, J.P. & Duck, C.D. (1989) Foraging energetics of Antarctic fur seals in relation to changes in prey availability. *Ecology* **70**, 596–606.

Costa, D.P., Antonelis, G.A. & DeLong, R.L. (1991) Effects of El Niño on the foraging energetics of the California sea lion. In: *Pinnipeds and El Niño* (F. Trillmich & K.A. Ono, eds), pp. 156–165. Springer-Verlag, Berlin.

Davis, R.W., Williams, T.M. & Kooyman, G.L. (1985) Swimming metabolism of yearling and adult harbour seals *Phoca vitulina*. *Physiology Zoology* **58**, 590–596.

Davis, R.W., Beltz, W.F., Peralta, F. & Witztum, J.L. (1993) Role of plasma and tissue lipids in the energy metabolism of the harbour seal. *Symposium of the Zoological Society of London* **66**, 369–382.

Elsner, R. (1987) The contribution of anaerobic metabolism to maximum exercise in seals. In: *Marine Mammal Energetics* (A.C. Huntley, D.P. Costa, G.A.J. Worthy & M.A. Castellini, eds), pp. 109–114. Society for Marine Mammalogy Special Publication No. 1. Allen Press, Lawrence, KS.

Fedak, M.A. & Anderson, S.S. (1982) The energetics of lactation: accurate measurements from a large wild mammal, the grey seal (*Halichoerus grypus*). *Journal of Zoology, London* **198**, 473–479.

Fedak, M.A., Arnbom, T. & Boyd, I.L. (1996) The relation between the size of southern elephant seal mothers, growth of their pups and the use of maternal energy, fat, and protein during lactation. *Physiological Zoology* **69**, 887–911.

Feldkamp, S.D. (1987) Swimming in the California sea lion: morphometrics, drag and energetics. *Journal of Experimental Biology* **131**, 117–135.

Folkow, L.P. & Blix, A.S. (1992) Metabolic rates of minke whales (*Balaenoptera acutorostrata*) in cold water. *Acta Physiologica Scandinavica* **146**, 141–150.

Gallivan, G.J. (1981) Ventilation and gas exchange in unrestrained harp seals (*Phoca groenlandica*). *Comparative Biochemistry and Physiology* **69A**, 809–813.

Gallivan, G.J. & Ronald, K. (1979) Temperature regulation in freely diving harp seals (*Phoca groenlandica*). *Canadian Journal of Zoology* **57**, 2256–2263.

Grav, H.J., Blix, A.S. & Pasche, A. (1974) How do seal pups survive birth in Arctic winter? *Acta Physiologica Scandinavica* **92**, 427–429.

Guppy, M., Hill, R.D., Schneider, R.C. *et al.* (1986) Microcomputer-assisted metabolic studies of voluntary diving of Weddell seals. *American Journal of Physiology* **250**, R175–R187.

Hammond, K.A. & Diamond, J. (1997) Maximal sustained energy budgets in humans and animals. *Nature* **386**, 457–460.

Handrich, Y., Bevan R.M., Charrassin, J.-B. *et al.* (1997) Hypothermia in foraging king penguins. *Nature* **388**, 64–67.

Hansen, S. & Lavigne, D.M. (1997) Temperature effects on the breeding distribution of grey seals (*Halichoerus grypus*). *Physiological Zoology* **70**, 436–443.

Hansen, S., Lavigne, D.M. & Innes, S. (1995) Energy metabolism and thermoregulation in juvenile harbour seals (*Phoca vitulina*) in air. *Physiological Zoology* **68**, 290–315.

Hart, J.S. & Irving, L. (1959) The energetics of harbour seals in air and in water with special consideration of seasonal changes. *Canadian Journal of Zoology* **37**, 447–457.

Hedd, A., Gales, R. & Renouf, D. (1997) Inter-annual consistency in the fluctuating energy requirements of captive harp seals *Phoca groenlandica*. *Polar Biology* **18**, 311–318.

Heglund, N.C., Cavagna, G.A. & Taylor, C.R. (1982) Energetics and mechanics of terrestrial locomotion. III. Energy changes of the centre of mass as a function of speed and body size in birds and mammals. *Journal of Experimental Biology* **97**, 1–21.

Hilborn, R. & Mangel, M. (1997) *The Ecological Detective.* Princeton University Press, Princeton, NJ.

Hill, R.D., Liggins, G.C., Schuette, A.H. *et al.* (1987) Heart rate and body temperature during free diving of Weddell seals. *American Journal of Physiology* **253**, R344–R351.

Hind, A.T. & Gurney, W.S.C. (1997) The metabolic cost of swimming in marine homeotherms. *Journal of Experimental Biology* **200**, 531–542.

Hind, A.T. & Gurney, W.S.C. (1998) Are there thermoregulatory constraints on the timing of pupping for harbour seals? *Canadian Journal of Zoology* **76**, 2245–2254.

Hindell, M.A. & Lea, M.-A. (1998) Heart rate, swimming speed, and estimated oxygen consumption of free-ranging southern elephant seal. *Physiological Zoology* **71**, 74–84.

Hochachka, P.W. (1986) Defense strategies against hypoxia and hypothermia. *Science* **231**, 234–241.

Houston, A.I. (1995) Energetic constraints and foraging efficiency. *Behavioral Ecology* **6**, 393–396.

Houston, A.I. & Carbone, C. (1992) The optimal allocation of time during the dive cycle. *Behavioral Ecology* **3**, 255–265.

Innes, S. & Ronald, K. (1981) Preliminary estimates on the cost of swimming in three species of phocid seal. In: *Fourth Biennial Conference on the Biology of Marine Mammals, San Francisco.* Society for Marine Mammalogy, Lawrence, KS.

Irving, L. (1972) *Arctic Life of Birds and Mammals Including Man.* Springer-Verlag, Berlin.

Irving, L. & Hart, J.S. (1957) The metabolism and insulation of seals as bare-skinned mammals in cold water. *Canadian Journal of Zoology* **37**, 447–457.

Irving, L., Scholander, P.F. & Grinnell, S.W. (1941) The respiration of the porpoise, *Truciops truncatus*. *Journal of Cellular and Comparative Physiology* **17**, 145–168.

Irving, L., Solandt, O.M., Solandt, D.Y. & Fisher, K.G. (1935) The respiratory metabolism of the seal and its adjustment to diving. *Journal of Cellular and Comparative Physiology* **7**, 137–151.

Kleiber, M. (1975) *The Fire of Life: an Introduction to Animal Energetics*. R.E. Kteiger Publishing Co., Huntington, NY.

Kooyman, G.L. & Ponganis, P.J. (1998) The physiological basis of diving to depth: birds and mammals. *Annual Reviews of Physiology* **60**, 19–32.

Kooyman, G.L., Kerem, D.H., Campbell, W.B. & Wright, J.J. (1973) Pulmonary gas exchange in freely diving Weddell seals (*Leptonychotes weddelli*). *Respiration Physiology* **17**, 283–290.

Kooyman, G.L., Wahrenbrock, E.A., Castellini, M.A., Davis, R.W. & Sinnett, E.E. (1980) Aerobic and anaerobic metabolism during voluntary diving in Weddell seals: evidence of preferred pathways from blood chemistry and behavior. *Journal of Comparative Physiology* **138**, 335–346.

Kooyman, G.L., Castellini, M.A., Davis, R.W. & Maue, R.A. (1983) Aerobic diving limits of immature Weddell seals. *Journal of Comparative Physiology* **151**, 171–174.

Lavigne, D.M. (1982) Pinniped thermoregulation: comments on the 'effects of cold on the evolution of pinniped breeding systems'. *Evolution* **36**, 409–414.

Lavigne, D.M., Barchard, W., Innes, S. & Øritsland, N.A. (1982) Pinniped bioenergetics. In: *Mammals of the Seas*, Vol. V, pp. 191–235. Food and Agriculture Organization, Rome.

Lavigne, D.M., Innes, S., Stewart, R.E.A. & Worthy, G.A.J. (1985) An annual energy budget for northwest Atlantic harp seals. In: *Marine Mammals and Fisheries* (J.R. Beddington, R.J.H. Beverton & D.M. Lavigne, eds), pp. 319–336. George Allen & Unwin, London.

Lavigne, D.M., Innes, S., Worthy, G.A.J. *et al.* (1986) Metabolic rates of seals and whales. *Canadian Journal of Zoology* **64**, 279–284.

Laws, R.M. (1956) Growth and sexual maturity in aquatic mammals. *Nature* **178**, 193–194.

Lindeman, R. (1942) The trophic–dynamic aspect of ecology. *Ecology* **23**, 399–418.

Lockyer, C. (1981) Growth and energy budgets of large baleen whales from the southern hemisphere. In: *Mammals of the Seas*, Vol. III, pp. 379–488. Food and Agriculture Organization, Rome.

Lydersen, C. & Hammill, M.O. (1993) Activity, milk intake and energy consumption in free-living ringed seal (*Phoca hispida*) pups. *Journal of Comparative Physiology B* **163**, 433–438.

Lydersen, C. & Kovacs, K.M. (1996) Energetics of lactation in harp seals (*Phoca groenlandica*) from the Gulf of St Lawrence, Canada. *Journal of Comparative Physiology B* **166**, 295–304.

Lydersen, C., Hammill, M.O. & Kovacs, K.M. (1995) Milk intake, growth and energy-consumption in pups of ice-breeding grey seals (*Halichoerus grypus*) from the Gulf of St Lawrence, Canada. *Journal of Comparative Physiology B* **164**, 585–592.

Lydersen, C., Kovacs, K.M., Hammill, M.O. & Gjertz, I. (1996) Energy intake and ultilisation by nursing bearded seal (*Erignathus barbatus*) pups from Svalbard, Norway. *Journal of Comparative Physiology B* **166**, 405–411.

Lydersen, C., Kovacs, K.M. & Hammill, M.O. (1997) Energetics during nursing and early post-weaning fasting in hooded seal (*Cystophora cristata*) pups from the Gulf of St Lawrence, Canada. *Journal of Comparative Physiology B* **167**, 81–88.

McNab, B.K. (1968) The influence of fat depositions on the basal rate of metabolism in desert homeotherms. *Comparative Biochemistry and Physiology* **26**, 337–343.

McNamara, J.M. & Houston, A.I. (1996) State-dependent life histories. *Nature* **380**, 215–221.

Markussen, N.H. & Øritsland, N.A. (1991) Food-energy requirements of the harp seal (*Phoca groenlandica*) population in the Barents and White seas. *Polar Research* **10**, 603–608.

Markussen, N.H., Ryg, M. & Øritsland, N.A. (1990) Energy requirements for maintenance and growth of captive harbour seals, *Phoca vitulina*. *Canadian Journal of Zoology* **68**, 423–426.

Markussen, N.H., Ryg, M. & Øritsland, N.A. (1992) Metabolic rate and body composition of harbour seals, *Phoca vitulina*, during starvation and refeeding. *Canadian Journal of Zoology* **70**, 220–224.

Markussen, N.H., Ryg, M. & Øritsland, N.A. (1994) The effect of feeding on the metabolic rate in harbour seals (*Phoca vitulina*). *Journal of Comparative Physiology B* **164**, 89–93.

Matsuura, D.T. & Whittow, C.G. (1973) Oxygen uptake of the California sea lion and harbour seal during exposure to heat. *American Journal of Physiology* **225**, 711–715.

Nagy, K.A. (1987) Field metabolic rate and food requirement scaling in mammals and birds. *Ecological Monographs* **57**, 111–128.

Nordøy, E.S. & Blix, A.S. (1991) Glucose and ketone body turnover in fasting grey seal pups. *Acta Physiologica Scandinavica* **141**, 565–571.

Nordøy, E.S., Ingebretsen, O.C. & Blix, A.S. (1990) Depressed metabolism and low protein catabolism in fasting grey seal pups. *Acta Physiologica Scandinavica* **139**: 361–369.

Nordøy, E.S., Aakvaag, A., & Larsen, T.S. (1993) Metabolic adaptations to fasting in harp seal pups. *Physiological Zoology* **66**, 926–945.

Oftedal, O.T., Boness, D.J. & Tedman, R.A. (1987) The behavior, physiology, and anatomy of lactation in the Pinnipedia. *Current Mammalogy* **1**, 175–245.

Øritsland, N.A. & Markussen, N.H. (1990) Outline of a physiologically based model for population energetics. *Ecological Modelling* **52**, 267–288.

Øritsland, N.A. & Ronald, K. (1975) Energetics of free diving harp seal (*Pagophilus groenlandicus*). *Rapports et Proces-Verbaux des Réunions, Conseil International pur L'Exploration de la Mer* **169**, 451–454.

Owen, O.E., Morgan, A.P., Kemp, H.G. *et al.* (1967) Brain metabolism during fasting. *Journal of Clinical Investigation* **46**, 1589–1595.

Perez, M.A. & McAlister, W.B. (1993) *Estimates of Food Consumption by Marine Mammals in the Eastern Bering Sea*. US Department of Commerce, NOAA Technical Memorandum No. NMFS-AFSC-14. US Department of Commerce, Washington, DC.

Ponganis, P.J., Ponganis, E.P., Ponganis, K. *et al.* (1990) Swimming velocities in otariids. *Canadian Journal of Zoology* **68**, 2105–2112.

Ponganis, P.J., Kooyman, G.L. & Castellini, M.A. (1993a) Determinants of the aerobic dive limit of Weddell seals: analysis of diving metabolic rates, post-dive end tidal Po_2's, and blood and muscle oxygen stores. *Physiological Zoology* **66**, 732–749.

Ponganis, P.J., Kooyman, G.L., Castellini, M.A., Ponganis, E.P. & Ponganis, K.V. (1993b) Muscle temperature and swim velocity profiles during diving in a Weddell seal, *Leptonychotes weddellii*. *Journal of Experimental Biology* **183**, 341–348.

Priddle, J., Boyd, I.L., Murphy, E.J., Whitehouse, M. & Croxall, J.P. (1998) Southern Ocean primary production —constraints from predator carbon demand and nutrient drawdown. *Journal of Marine Systems* **17**, 275–288.

Prothero, J. (1995) Bone and fat as a function of body weight in adult mammals. *Comparative Biochemistry and Physiology* **111A**, 633–639.

Rea, L.D. & Costa, D.P. (1992) Changes in the standard metabolism during long-term fasting in northern elephant seal pups (*Mirounga angustrostris*). *Physiological Zoology* **65**, 97–111.

Reilly, J.J. & Fedak, M.A. (1990) Measurement of the body composition of living gray seals by hydrogen isotope dilution. *Journal of Applied Physiology* **69**, 885–891.

Reilly, J.J. & Fedak, M.A. (1991) Rates of water turnover and energy expenditure of free-living male common seals (*Phoca vitulina*). *Journal of Zoology, London* **223**, 461–468.

Reilly, J.J., Fedak, M.A., Thomas, D.H., Coward, W.A.A. & Anderson, S.S. (1996) Water balance and the energetics of lactation in grey seals (*Halichoerus grypus*) as studied by isotopically labelled water methods. *Journal Zoology, London* **238**, 157–165.

Ronald, K., Keiver, K.M., Beamish, F.W.H. & Frank, R. (1984) Energy requirements for maintenance and faecal urinary loss of the grey seal (*Halichoerus grypus*). *Canadian Journal of Zoology* **62**, 1101–1105.

Rosen, D.A.S. & Renouf, D. (1998) Correlates of seasonal changes in metabolism in Atlantic harbour seals (*Phoca vitulina concolor*). *Canadian Journal of Zoology* **76**, 1520–1528.

Rosen, D.A.S. & Trites, A.W. (1997) Heat increment of feeding in Steller sea lions, *Eumetopias jubatus*. *Comparative Biochemistry and Physiology* **118A**, 877–881.

Ryg, M., Lydersen, C., Knutsen, L.Ø. *et al.* (1993) Scaling of insulation in seals and whales. *Journal of Zoology, London* **230**, 193–206.

Sakamoto, W., Naito, Y., Huntley, A.C. & Le Boeuf, B.J. (1989) Daily gross energy requirements of a female northern elephant seal *Mirounga angustirostris* at sea. *Nippon Suisan Gakkaishi* **55**, 2057–2063.

Schmidt-Nielsen, K. (1997) *Animal Physiology*, 5th edn. Cambridge University Press, Cambridge, UK.

Scholander, P.F. (1940) Experimental investigations on the respiratory function in diving mammals and birds. *Hvalradets Skrifter* **22**, 1–131.

Scholander, P.F. & Irving, L. (1941) Experimental investigations of the respiration and diving of the Florida manatee. *Journal of Cellular and Comparative Physiology* **17**, 169–175.

Scholander, P.F., Irving, L. & Grinnell, S.W. (1942) On the temperature and metabolism of the seal during diving. *Journal of Cellular and Comparative Physiology* **19**, 67–78.

Schmitz, O.J. & Lavigne, D.M. (1984) Intrinsic rate of increase, body size, and specific metabolic rate in marine mammals. *Oecologia* **62**, 305–309.

Shelton, P.A., Warren, W.G., Stenson, G.B. & Lawson, J.W. (1995) *Quantifying Some of the Major Sources of Uncertainty Associated with Estimates of Harp Seal Prey Consumption. Part II: Uncertainty in Consumption Estimates Associated with Population Size, Residency, Energy Requirement and Diet*. NAFO SCR Document No. 95/93, N2615. Northwest Atlantic Fisheries Organization, St Johns, Newfoundland.

South, F.E., Luecke, R.H., Zatzman, M.L. & Shanklin, M.D. (1976) Air temperature and direct patitional calorimetry of the California sea lion (*Zalophus californianus*). *Comparative Biochemistry and Physiology* **54**, 269–294.

Speakman, J.R. (1997) *Doubly Labelled Water: Theory and Practice*. Chapman & Hall, London.

Steele, J.H. (1985) A comparison of terrestrial and marine ecological systems. *Nature* **313**, 355–358.

Stephens, D.W. & Krebs, J.R. (1986). *Foraging Theory*. Princeton University Press, Princeton, NJ.

Stewart, R.E.A. & Lavigne, D.M. (1984) Energy transfer and female condition in nursing harp seals, *Phoca groenlandica*. *Holarctic Ecology* **7**, 182–194.

Stirling, I. (1983) The evolution of mating systems in pinnipeds. In: *Recent Advances in the Study of Mammalian Behavior* (J.F. Eisenberg & D.G. Kleinman, eds), pp. 489–527. American Society of Mammalogists, Pittsburgh.

Thompson, D., Hiby, A.R. & Fedak, M.A. (1993) How fast should I swim? Behavioural implications of diving physiology. *Symposium of the Zoological Society of London* **66**, 349–368.

Thomson, R.B., Butterworth, D.S., Boyd, I.L. & Croxall, J.P. (2000) Modelling the consequences of Antarctic krill harvesting on Antarctic fur seals. *Ecological Applications* **10**, 1806–1819.

Trillmich, F. (1996) Parental investment in pinnipeds. *Advances in the Study of Animal Behavior* **25**, 533–577.

Trites, A.W. (1990) Thermal budgets and climate spaces: the impact of weather on the survival of Galapagos (*Arctocephalus galapagoensis* Heller) and northern fur seal pups (*Callorhinus ursinus* L.). *Functional Ecology* **4**, 753–768.

Watts, P. (1992) Thermal constraints on hauling out by harbour seals (*Phoca vitulina*). *Canadian Journal of Zoology* **70**, 553–560.

Watts, P., Hansen, S.E. & Lavigne, D.M. (1993) Models of heat loss by marine mammals: thermoregulation below the zone of irrelevance. *Journal of Theoretical Biology* **163**, 505–525.

Webb, P.M., Andews, R.D., Costa, D.P. & Le Boeuf, B.J. (1998) Heart rate and oxygen consumption of northern elephant seals during diving in the laboraory. *Physiological Zoology* **71**, 116–125.

Williams, T.M. (1987) Approaches for the study of exercise physiology and hydrodynamics in marine mammals. In: *Marine Mammal Energetics* (A.C. Huntley, D.P. Costa, G.A.J. Worthy & M.A. Castellini, eds), pp. 127–145. Society for Marine Mammalogy Special Publication No. 1. Allen Press, Lawrence, KS.

Williams, T.M., Kooyman, G.L. & Croll, D.A. (1991) The effect of submergence on heart rate and oxygen consumption of swimming seals and sea lions. *Journal of Comparative Physiology B* **160**, 637–644.

Williams, T.M., Friedl, W.A. & Haun, J.E. (1993) The physiology of bottlenose dolphins (*Tursiops truncatus*): heart rate; metabolic rate and plasma lactate concentration during exercise. *Journal of Experimental Biology* **179**, 31–46.

Williams, T.M., Noren, D., Berry, P. *et al.* (1999) The diving physiology of the bottlenose dolphin (*Turciops truncatus*) —III. Thermoregulation at depth. *Journal of Experimental Biology* **202**, 2763–2769.

Worthy, G.A.J. (1987) Metabolism and growth of young harp and grey seals. *Canadian Journal of Zoology* **65**, 1377–1382.

Worthy, G.A.J., Innes, S., Braune, B.M. & Stewart, R.E.A. (1987) Rapid acclimation of cetaceans to an open-system respirometer. In: *Marine Mammal Energetics* (A.C. Huntley, D.P. Costa, G.A.J. Worthy & M.A. Castellini, eds), pp. 115–126. Society for Marine Mammalogy Special Publication No. 1. Allen Press, Lawrence, KS.

Worthy, G.A.J., Morris, P.A., Costa, D.P. & Le Boeuf, B.J. (1992) Moult energetics of the northern elephant seal (*Mirounga angustirostris*). *Journal of Zoology, London* **227**, 257–265.

Life History and Reproductive Strategies

Daryl J. Boness, Phillip J. Clapham and Sarah L. Mesnick

10.1 INTRODUCTION

Life history theory has emerged to explain why individual animals behave the way they do over their lifetime. It concerns decisions individuals make to maximize their fitness and reflects phenotypic variation, adaptation and constraints (Stearns 1992; Charnov 1993). An integral component of maximizing fitness involves reproductive strategies. This chapter will describe the variation in life history and reproductive strategies among marine mammal species and discuss how this relates to constraints, ecological variation and major selective pressures.

For mammals, which are air-breathers and warm-blooded, living in a marine environment poses some challenges that are different from living on land. Indeed, some marine mammals are amphibious, which may present conflicting challenges. There are similarities with terrestrial mammals in some characteristics of marine mammal life history and reproductive patterns, but also differences that are related to the marine or amphibious existence. For example, most marine mammals are large, with the baleen whales reaching phenomenal sizes. This is probably driven by thermoregulatory needs together with a requirement for large lipid stores to enable the animal to survive long periods without food (Brodie 1975, 1977) (see Chapter 9). Another example is the extremely short lactation strategies of many phocid seals, probably driven by unstable ice conditions, long distances to food resources and potential predation pressure. We will review what is known about the major life history and reproductive strategies of marine mammals, highlighting those taxa that have been studied most intensively.

10.2 MAJOR LIFE HISTORY TRAITS

Marine mammals evolved from terrestrial ancestors to varying degrees to cope with the marine environment (see Chapter 2). All marine mammals forage at sea (depending on how polar bears are classified). Some also reproduce there too and spend their entire life in water. However, for pinnipeds most or all of reproduction occurs on land (or ice) as does moulting. Both sea otters and pinnipeds are known to come ashore to rest and bask in the sun. The amphibious existence of pinnipeds has a profound influence on reproductive patterns and cycles.

Despite being large for mammals, the range in body size is substantial, with the sea otter being the smallest and the blue whale being the largest (Table 10.1). There is probably a multitude of reasons behind the large size, but one that is of primary importance, is the high thermal conductivity of water (see Chapter 9). Boyd explains the reasons for this in Chapter 9 and of particular importance to this chapter is that larger animals are able to build up larger blubber stores than smaller animals. The ability to store large amounts of blubber becomes important for many species because their reproductive strategies require fasting. Bartholomew (1970) and others (Stirling 1983; Boness 1991) have argued that large size and insulative blubber were preadaptations that permitted fasting and the opportunity for high levels of polygyny to evolve. The largest marine mammals, the mysticete whales, undergo extensive migrations to largely unproductive tropical waters where they may fast for as long as several months. The selective advantage of these extraordinary movements remains unclear

Table 10.1 Selected life history traits of marine mammals.

Species	Age at sexual maturity (years)[1]		Adult length (cm; m for mysticetes)[2]		Adult mass (kg; t for mysticetes and sperm whales)[2]		Gestation (months)[3]	Interbirth interval (years)	Source[7]
	Males	Females	Males	Females	Males	Females			
PINNIPEDIA									
Phocidae									
Cystophora cristata, hooded seal	4–6	2–9 (3)	228	200	300	160–179	11.5/7.8	1	1, 2, 3, 4
Erignathus barbatus, bearded seal	6–7	3–6 (6)	216–235	223–233	244–275	229–275	11.5/9.5	1	1, 3, 5
Halichoerus grypus, grey seal	6	3–5 (5)	210–230	184–201	240–298	174–207	11.3/7.9	1	1, 2, 3, 6
Histriophoca fasciata, ribbon seal	3–5 (5)	2–4 (3)	170	168	77	88		1	1, 2, 3
Hydrurga leptonyx, leopard seal	4–6 (5)	2–7 (4)	285	315			c. 11/9.4	1	1, 2
Leptonychotes weddellii, Weddell seal	3–6	2–6	240–247	246–262	340	379–447	10.3/8.7	1	1, 2, 3
Lobodon carcinophaga, crabeater seal	2–6 (4)	2–6 (4)	225	224	221	224	11/5	1	1, 2, 3
Mirounga angustirostris, northern elephant seal	5	2–6 (4)	402	322	1704	504–513	11/7	1	1, 2, 3, 6, 7
Mirounga leonina, southern elephant seal	4–6	2–7	420–500	260–297	c. 3250	c. 390–790	11.3/7.3	1	1, 2, 3, 6
Ommatophoca rossii, Ross seal	2–7 (5)	3–5	227	229	129–216	159–204	c. 11/8.2	1	1, 8
Pagophilus groenlandicus, harp seal	4–5 (5)	3–7 (6)	171–173	169–171	135	109	11.8/7	1	1, 2, 3
Phoca hispida, ringed seal	5–7	3–7 (6)	130–138	130–135	60–68	50–62	c. 11.8/9.1	1	1, 2, 3
Phoca largha, largha seal	3–6	2–5	153–170	149–162	85–110	65–115	c. 11/9	1	1, 2, 8
Phoca sibirica, Baikal seal	4	3–6	c. 130	c. 130	63–70			1	1, 8
Phoca vitulina, harbour seal	3–7 (5)	2–7 (5)	156–191	148–171	88–142	65–107	11.1/9.1	1	1, 2, 3, 9
Otariidae									
Arctocephalus australis, South American fur seal	7	2–3 (3)	189*	143	159*	49*	11.8/7.8	1	1, 10
Arctocephalus forsteri, New Zealand fur seal	10–12	2–6	142–250	125–150	120–185	25–50	11.8/7.8	1	1, 8
Arctocephalus galapagoensis, Galápagos fur seal	9	4	152	120	64	27	11.8	1	11, 12
Arctocephalus gazella, Antarctic fur seal	3–4	3–4	195–212	130–140	186	39	11.8/7.6	1	1, 2, 3, 13

(*continued on p. 280*)

Table 10.1 (cont'd)

Species	Age at sexual maturity (years)[1]		Adult length (cm; m for mysticetes)[2]		Adult mass (kg; t for mysticetes and sperm whales)[2]		Gestation (months)[3]	Interbirth interval (years)	Source[7]
	Males	Females	Males	Females	Males	Females			
Arctocephalus philippii, Juan Fernández fur seal			210	150	140	50	11.8	1	8, 14
Arctocephalus pusillus doriferus, Australian fur seal	4–5	3–6	200–225	125–170	279	76	11.8/8.8	1	1, 8, 15
Arctocephalus pusillus pusillus, South African fur seal	3–4	3–4	220	160	247	57	11.8/7.8	1	1, 3, 16
Arctocephalus townsendi, Guadalupe fur seal			193	137	165	50	11.8	1	8, 17
Arctocephalus tropicalis, subantarctic fur seal	3–4	4–6	180	145	131	36	11.8/7.5	1	1, 3, 18
Callorhinus ursinus, northern fur seal	5	3–7 (6)	314	129	155–220	40–45	11.8/8.3	1	1, 2, 3
Eumetopias jubatus, northern sea lion	3–8	2–7 (5)	282	228	566	273	11.5/8.0	1	1, 2, 3, 8
Neophoca cinerea, Australian sea lion			335	274	300	77	11.8	1.4	1, 8, 19
Otaria flavescens, southern sea lion	5–6	3–4	250	190	233–313	121–140	11.8	1	1, 8, 20
Phocarctos hookeri, Hooker's sea lion			240–290	180–183	400	230	11.8	1	1, 8, 21
Zalophus californianus, California sea lion	4–5	4–5	229	223	200–289	78–88	11/8	1	1, 2, 3
Odobenidae *Odobenus rosmarus,* walrus	7–10 (10)	4–12 (6)	320	272	1200	830	15/10.5	2	1, 2, 22
SIRENIA **Trichechidae** *Trichechus manatus,* manatee	2–11	3	315	280	*c.* 685	*c.* 500	12–14	3	23, 24
Dugongidae *Dugong dugon,* dugong	9–15 (10)	9–15	260	270			13.9	3–6	25, 26
Mustelidae *Enhydra lutris,* sea otter	5–6	2–5 (4)	129	120	29	20	11/7.5	1	27

CETACEA, ODONTOCETI

Physeteridae									
Physeter macrocephalus, sperm whale	18–21	7–13 (9)	1520–1610	1040–1100	17–29	5–12	15–16	3–6	28, 29
Kogiidae									
Kogia breviceps, pygmy sperm whale			270–330	266–330	318–408		9–11	2	30, 31
Kogia sima, dwarf sperm whale			234*	218–234	136–276			2	30, 32
Ziphiidae									
Berardius bairdii, Baird's beaked whale	8–10		1000	1050	8000–10 000		17		33, 34
Berardius arnuxii, Arnoux's beaked whale	8–10		c. 900*	885*			17		33, 34
Iniidae									
Inia geoffrensis, boto, Amazon river dolphin			209–255	183–228	71–159.5	67–97	8.5	c. 1.5	35, 36, 37
Pontoporiidae									
Pontoporia blainvillei, franciscana	2–3		121–158	137–174	25–43	30–52	10.5	2	37, 38
Platanistidae									
Platanista spp.	10?		185–212	<199–252	84*	51–83	11–12?	2?	37, 39
Phocoenidae									
Neophocaena phocaenoides, finless porpoise	3–6		132–227	132–206	30–80		10–11		40, 41
Phocoenoides dalli dalli, Dall's porpoise, NW Pacific	3.6–4		175–180	174–177	170		10.7	1	42
Phocoenoides dalli true, True's porpoise, Japan	7.9		196	187	170		11.4	3	42
Phocoena phocoena, harbour porpoise, Bay of Fundy	3–4		143	158	50	65	11	1	43, 44
Phocoena sinus, vaquita	3–6		135	141	42	44	11	>1	45, 46
Phocoena spinipinnis, Burmeister's porpoise, Peru			170	166	72*	79*	11–12		47, 48
Monodontidae									
Delphinapterus leucas, beluga	8–9	4–7	350–470	310–390	1352	956	14.5	3	49, 50, 51
Monodon monoceros, narwhal	11–13	5–8	410–470	340–415	1600*	1000*	14–15.3	2–3	49, 52, 53
Delphinidae									
Lagenorhynchus acutus, Atlantic white-sided dolphin	6–12		244–275	194–243	234*	182*	10–12		54, 55, 56
Sotalia fluviatilis spp., tucuxi			140–187	139–183	32–46	40–53	10.2–12		54, 57

(continued on p. 282)

Table 10.1 (cont'd)

Species	Age at sexual maturity (years)[1]		Adult length (cm; m for mysticetes)[2]		Adult mass (kg; t for mysticetes and sperm whales)[2]		Gestation (months)[3]	Interbirth interval (years)	Source[7]
	Males	Females	Males	Females	Males	Females			
Steno bredanensis, rough-toothed dolphin	14	10–17	209–265	212–255	90–155	90–155			54, 58
Lagenorhynchus obscurus, dusky dolphin	4–8	4–8	167–211	167–204.5	70–85	69–78		2.4	54, 59, 60
Lagenorhynchus obliquidens, Pacific white-sided dolphin	7–9	7–9	170–234	170–236	198*	148*	10–12		54, 61
Cephalorhynchus hectori, Hector's dolphin	6–9	7–9	138*	153*	53*	57*		2–3+	62, 63
Cephalorhynchus commersonii, Commerson's dolphin	5–8	5–8	130–167	139–174	78*	86*	12		62, 64, 65
Globicephala macrorhynchus, short-finned pilot whale	14–23	7–12 (9)	413.7	316	1200	570	14.9	6.9	66
Globicephala melas, long-finned pilot whale	11–16	6–7	*c.* 550	*c.* 460	2750*	*c.* 1600	15	3.3	66, 67
Orcinus orca, killer whale	*c.* 16	*c.* 10	520–975	457–853	10 488*	5708*	15	5.2	54, 68, 69
Stenella attenuata spp., spotted dolphin	12–15	9–11	166–257	163–240	46–162	41–114	11.2–11.5	3	54, 70, 71, 72
Stenella longirostris spp., spinner dolphin	6–10	4–7	136–235	129–204	23–78	23–78	*c.* 10.5	*c.* 3	54, 73, 74, 75
Lagenodelphis hosei, Fraser's dolphin	*c.* 7	*c.* 7	234	241		209*	10–12?		54, 76
Delphinus delphis, short-beaked common dolphin	3–12	5–12	171–260	167–244		136*	10–11	1.3–2.6	54, 77, 78
Stenella coeruleoalba, striped dolphin, western Pacific	7–15	9	236	225.3	116–165	110–153	12	4	41, 54, 79
Tursiops truncatus, common bottlenose dolphin, Sarasota Bay, Florida	8–10	5–10	263	250	282*	263*	12	3	80, 81
Tursiops aduncus, Indian Ocean bottlenose dolphin, Natal South Africa	10–12	9.5–15	243	238	176	160	12.3	3	54, 82
CETACEA, MYSTICETI									
Balaenoptera musculus, blue whale[4]	5–10	5–10	28*, 22.6#	31*, 24#	110 at 25 m	190*	11–12	2–3	83, 84, 85
Balaenoptera physalus, fin whale[4]	5–8	5–8	24*, 19#	26*, 20#	49 at 20 m	70 at 23 m	11–12	2–3	83, 84, 86, 87

Species									
Balaenoptera borealis, sei whale	8	8	12.9#	13.6#	16 at 14 m	22 at 15 m	c. 11	2–3	83, 84
Balaenoptera edeni, Bryde's whale[5] South Africa offshore, Japan	9–10	10	13.2#	12.5#	14* at 13 m	16* at 14 m	c. 12	2–3	83, 88
Balaenoptera edeni, Bryde's whale, South Africa inshore	10–11	9–10	13.7#	13.1#			c. 12	2–3	88
Balaenoptera acutorostrata, minke whale[4,6]	6–8	6–8	9.8*, 7.2#	10.7*, 18.0#	9 at 8 m	9 at 9 m	10	2–3	31, 83, 84
Megaptera novaeangliae, humpback whale	5	5	14.8*, 13#	15.5*, 13.9#	28 at 13 m	41 at 14 m	11–12	2–3	83, 89, 90
Eubalaena glacialis, northern right whale	8–9	8–9	17.1*	17.4*	67 at 17 m	107 at 17 m	12?	3–5	91, 92, 93, 94
Eubalaena australis, southern right whale	9–10	9–10	c. 14	16.8*, 13#			12–13	3–5	95, 96
Balaena mysticetus, bowhead whale	12–20	12–20	16.2*, 12#	18.0*, 12.5#			13–14	3–4?	97
Eschrichtius robustus, gray whale	8	8	11.1#	11.7#			13–14	2–3	84, 98

[1] Age at sexual maturity reflects the age of onset of sexual maturity. The precise measure may vary (e.g. 50% vs 100%) since papers often do not mention how the onset was determined. We believe the imprecision which results from this yields only small differences that should not affect major patterns. For some we give ranges where information is less precise and give in parentheses the age at which most seals of that sex are thought to become sexually mature.

[2] Lengths and weights are asymptotic measurements if available, or are means. When neither is available the range of adult size is given unless there is only a maximum value for the species (designated by an *) or a value for length at age of sexual maturity (designated by #). Cells were merged when separated values were not recorded for each sex.

[3] The number following the slash reflects that part of gestation after implantation has occurred.

[4] Length data are for the southern hemisphere animals; northern hemisphere animals are somewhat smaller.

[5] Taxonomy of Bryde's whales is unclear; neither this nor the inshore population is likely to be similar to the dwarf form. South African populations are given as examples of the range in 'regular' form.

[6] Minke whales may comprise three species worldwide.

[7] 1, Riedman 1990; 2, McLaren 1993; 3, Costa 1991; 4, Kovacs & Lavigne 1986; 5, Anderson & Fedak 1985; 6, Boness & Bowen 1996; 7, Deutsch et al. 1990; 8, Reijnders et al. 1993; 9, Walker & Bowen 1993b; 10, Vaz Ferreira & Ponce de Leon 1987; 11, Croxall & Gentry 1987; 12, Trillmich 1987; 13, Boyd & Duck 1991; 14, Torres 1987; 15, Shaughnessy & Warneke 1987; 16, David 1987; 17, Flescher 1987; 18, Bester 1987; 19, Walker & Ling 1981a; 20, Hamilton 1934; 21, Walker & Ling 1981b; 22, Fay 1982; 23, Marmontel 1995; 24, Hernandez et al. 1995; 25, Marsh 1980; 26, Marsh 1995; 27, Riedman & Estes 1990; 28, Rice 1989; 29, Best et al. 1984; 30, Caldwell & Caldwell 1989; 31, Leatherwood & Reeves 1983; 32, Leatherwood et al. 1988; 33, Mead 1984; 34, Balcomb 1989; 35, Best & da Silva 1989; 36, Best & da Silva 1993; 37, Brownell 1984; 38, Brownell 1989; 39, Reeves & Brownell 1989; 40, Goa & Zhou 1993; 41, Kasuya 1999; 42, Houck & Jefferson 1999; 43, Gaskin et al. 1984; 44, Read 1999; 45, Hohn et al. 1996; 46, Vidal et al. 1999; 47, Brownell & Clapham 1999; 48, Reyes & Van Waerebeek 1995; 49, Braham 1984; 50, Brodie 1989; 51, Sergeant & Brodie 1969; 52, Hay & Mansfield 1989; 53, Silverman 1979; 54, Perrin & Reilly 1984; 55, Perrin et al. 1999; 56, Sergeant et al. 1980; 57, da Silva & Best 1996; 58, Miyazaki & Perrin 1994; 59, Brownell & Cipriano 1999; 60, Van Waerebeek & Read 1994; 61, Brownell et al. 1999; 62, Goodall 1994; 63, Slooten & Dawson 1994; 64, Goodall et al. 1988; 65, Lockyer 1988; 66, Kasuya & Marsh 1984; 67, Sergeant 1962; 68, Dahlheim & Heyning 1999; 69, Olesiuk et al. 1990; 70, Myrick et al. 1986; 71, Perrin & Hohn 1994; 72, Perrin et al. 1976; 73, Perrin & Gilpatrick 1994; 74, Perrin et al. 1989; 75, Perrin 1998; 76, Perrin et al. 1994a; 77, Evans 1994; 78, Heyning & Perrin 1994; 79, Perrin et al. 1994b; 80, Wells & Scott 1999; 81, Wells et al. 1987; 82, Cockcroft & Ross 1990; 83, Lockyer 1976; 84, Lockyer 1984; 85, Mizroch et al. 1984; 86, Aguilar & Lockyer 1987; 87, Agler et al. 1993; 88, Best 1977; 89, Clapham & Mayo 1990; 90, Clapham & Mead 1999; 91, Klumov 1962; 92, Omura et al. 1969; 93, Hamilton et al. 1998; 94, Kraus et al., in press; 95, Payne et al. 1990; 96, Best 1994; 97, Koski et al. 1993; 98, Rice 1983.

(see Chapter 7), but may relate to exploitating pulses of productivity in high latitudes in summer, combined with a reduced risk of killer whale predation on calves (Corkeron & Connor 1999; but see rebuttal by Clapham 2001) and thermo-regulatory benefits of overwintering in warm water (Brodie 1977).

Physiological requirements associated with foraging under water for an air-breather are also likely to have been important factors promoting large size in marine mammals. Larger animals have a greater capacity to dive deeper and for longer periods (Kooyman 1989; Boyd 1997) (see Chapter 3).

There is a suite of characteristics in the life cycle of animals that tend to be associated with large-sized animals (Calder 1984; Reiss 1989; Stearns 1992). Large animals tend to be relatively long lived, grow slowly, delay sexual maturation and produce few offspring per cycle but invest heavily in each. This suite of characters ultimately relates to trade-offs between the energetic costs of growth, maintenance and reproduction. Fine-tuning is influenced by ecological variables such as the quality and distribution of food resources and predation pressures.

For all marine mammals, the normal number of offspring produced during a given breeding event is one. Low rates of multiple fetuses occur in stranded cetaceans and sirenians, but there are no reliable records of living twins or multiplets among members of these two orders. Similarly, rare observations of twinning in pinnipeds suggest that both offspring cannot be reared successfully (Riedman 1990).

Gestation length is relatively long, ranging from 7 to 17 months (Table 10.1). In pinnipeds and sea otters a phenomenon known as embryonic diapause exists. This is a delay in development of the blastocyst after fertilization (Daniel 1981; Boyd 1991; Boyd et al. 1999). The evolution of a diapause may be related to a need to synchronize the timing of breeding; it is common in highly seasonal breeders where males and females are spatially separated outside the breeding season. Thus, with regard to energy use, one must distinguish between the period from fertilization to birth and that period when resources are being devoted to development of the fetus. The longest gestation occurs in the beaked whale (Table 10.1). This is consistent with life history theory in that larger animals tend to have relatively higher costs of maintenance and growth and devote a lower proportion of their resources to reproduction. The largest animals of all, the mysticetes, have gestations lasting typically about 1 year and the highest rates of fetal growth in the animal kingdom 20 times that of some primates (Frazer & Huggett 1973). This telescoped gestation, and the attendant highly seasonal breeding of most mysticetes, may result from the need for reproductive timing to coincide with annual patterns of prey availability.

The fact that a long gestation has evolved among all marine mammals is probably associated with offspring being born into a harsh environment in which young must be well developed to survive (a phenomenon known as precocial in contrast to altricial—where offspring are less well developed at birth). Marine mammals are not alone among mammals in having precocial young. Ungulates, which are often under strong predation pressure, also tend to produce precocial young (Stearns 1992). Among cetaceans, sirenians and sea otters, the young must be capable of swimming at birth in order to keep up with their mothers. Mobility is essential to avoid predators. In pinnipeds, which give birth on land or ice, mobility at birth may be less important because females are relatively sedentary at this time. Also, for most species, predation pressure on land or ice is low.

Among phocid seals, one example of variation in degree of precociality is seen in the different degrees to which a blubber layer is developed and whether a natal coat is present (Oftedal et al. 1991). Most pups are born with a natal coat and small amounts of subcutaneous blubber. They rely on the natal coat and brown fat to provide warmth until building a blubber layer from high-fat milk. In harbour and hooded seals, where pups enter the water shortly after birth (1–4 days), all or a substantial proportion of pups are born having shed their natal coat in utero and weigh 12–13% of maternal mass compared to 5–10% in other phocids. They are composed of 10–14% body fat at birth (cf. 5–8% for phocids not entering the water early; Bowen et al. 1992; Oftedal et al. 1993).

Sexual maturation (defined as the age of capability of being fertilized or fertilizing) is delayed in all marine mammals (Table 10.1) as would be expected from life history theory. Underlying this relationship is a trade-off between investing energy in growth versus reproduction. Individuals that reproduce when they are too small are likely to produce offspring that have a low probability of survival (this is true especially for females) and may jeopardize their own survival or future breeding success (Reiter & Le Boeuf 1991; Sydeman *et al.* 1991).

The time to reach sexual maturity in marine mammals is variable but generally is several years (Table 10.1). Pinnipeds typically reach sexual maturity earlier than cetaceans, with maturity occurring between 3 and 7 years in the former. Among the cetaceans, age at sexual maturity is more likely to be greater than 7 years and may even occur between 15 and 20 years (e.g. bowhead whale; Koski *et al.* 1993). The extreme delay in bowheads may be related to the thermodynamic constraints of a life spent largely in very high latitudes among ice. Among odontocetes, the onset of sexual maturity ranges widely from 2 years to about 20 years (Table 10.1). To a point, the variation in age at onset of sexual maturity in marine mammals follows the expectation that larger species, which tend to have relatively higher costs of maintenance and growth and devote a lower proportion of their resources to reproduction, mature later. However, the relatively early maturity of balaenopterids (e.g. humpback whales at 5 years) does not fit.

The age of sexual maturity is not necessarily expected to be the same for males and females (see Table 6.4 in Stearns 1992). Wiley (1974) coined the term bimaturism to describe the condition in polygynous species where males compete with one another for females and tend to delay maturity and grow larger before beginning to compete. Sexual dimorphism in body size and secondary sexual characteristics are common in such species as well. Because females gain fecundity with size at a faster rate than males, in species where larger size in males is not sexually selected (i.e. males do not control access to females), females are expected to mature later than males.

As polygyny is common among marine mammals, and may be extreme in some pinnipeds (Boness *et al.* 1993), we should expect and do see a tend-

ency towards males deferring maturity (Table 10.1 and Fig. 10.1). In some mysticetes (e.g. humpback whales; Clapham 1992, 1996) both sexes become sexually mature at similar ages, but the inability of young males to compete successfully with fully grown males may result in a delay in the age at which they actively participate in breeding. A similar delay in the age of 'social maturity' (distinguished from physiological sexual maturity), is seen in some odontocetes and most pinnipeds. It is not unusual for female delphinids to attain sexual maturity sometime before, and at a smaller size than, males (Cockcroft & Ross 1990), yet sexual bimaturation is most extreme in species such as sperm whales, pilot whales and killer whales. In these species, and in many pinnipeds, males continue to grow after reaching sexual maturity and it is unlikely that they will mate successfully for several years after becoming sexually mature (Best *et al.* 1984; Kasuya & Marsh 1984).

There are a few species in each taxa in which females appear to mature later than males or in which maturation is the same for both sexes (Fig. 10.1). Presumably in these species polygyny is at a low level or size is not an important factor in male–male competition.

Age of sexual maturity is not necessarily constant for a given species or even in a given population. This life history trait appears to be density-dependent, with low-density populations exhibiting earlier maturation than higher density ones. An excellent discussion of this for pinnipeds is presented in Riedman (1990) and data for a number of cetacean species can be found in Perrin *et al.* (1984). Much of our knowledge about variation in sexual maturity among and within marine mammal populations comes from exploited populations now being protected in much of the world (Carrick *et al.* 1962; Laws 1962; Gamble 1973; Perrin *et al.* 1984; Kasuya 1999). Density dependence has often been claimed for some southern hemisphere mysticetes, for which whaling removed the majority of many populations. However, this phenomenon is difficult to disentangle from sampling and other methodological problems and its occurrence remains in dispute (Mizroch & York 1984). Similarly, density-dependence effects on reproductive parameters are difficult to test in odontocetes because there are

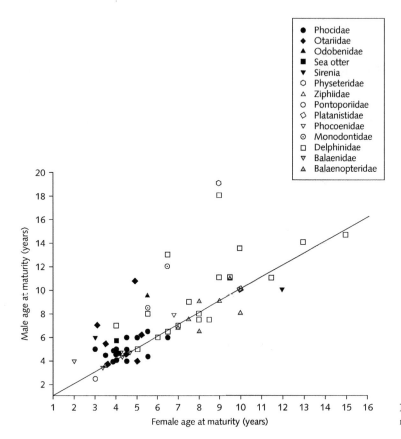

Fig. 10.1 The relationship between male and female age at sexual maturity.

few species for which large amounts of data exist through time, and there are potentially confounding factors acting in a population under exploitation (Perrin & Reilly 1984). Striped dolphins, studied extensively because of a long-standing fishery in Japan, show a decline in the age at onset of sexual maturity from 9.4 years to 7.5 years in schools sampled since 1968 (Kasuya 1985). This decline is thought to be a response to increased per capita food availability resulting from decreased density caused by the fishery. However, confusion about stock identification and possible changes in food availability may affect this interpretation (Kasuya 1999).

Reproductive cycles in marine mammals are relatively long (Table 10.1), as is the case with those of similar-sized terrestrial mammals. The interval between births ranges from 1 to 7 years. As the interbirth interval is particularly important to under-

standing male mating strategies, we will elaborate more on temporal aspects of breeding and variation within and among species later in our discussion of reproductive strategies.

10.3 REPRODUCTIVE STRATEGIES

Mating system theory provides a general framework from which we can understand the reproductive strategies of males and females, and has been formulated within the same evolutionary principles of life history theory. The framework has emerged over the past two or three decades and comes from several seminal works over this period (Williams 1966; Orians 1969; Trivers 1972; Jarman 1974; Wilson 1975; Bradbury & Vehrencamp 1977; Emlen & Oring 1977; Rubenstein & Wrangham 1986; Clutton-Brock 1988, 1989; Westneat *et al.*

1990; Davies 1991). At the heart of understanding variation in reproductive strategies is the fundamental idea that individual animals behave so as to maximize their reproductive success over their lifetime. Furthermore, females and males need not act in cooperation, and in fact, they may be in direct conflict or try to deceive one another. Generally, females have evolved to invest more heavily in rearing offspring and males more in maximizing the number of mates. When offspring survival will fail without the aid of males to assist in rearing then males contribute to offspring care as well as females. The potential for males to mate with many females depends on the temporal and spatial distribution of receptive females, which in turn depends on ecological and social factors (resource distribution, predation pressure and costs and benefits of group living). Lastly, phylogenetic constraints may predispose taxa toward particular reproductive strategies.

10.3.1 Mating strategies

As is true of most mammals, with maternal lactation providing the majority of nutrient requirements for dependent offspring, marine mammal males contribute little or nothing to parental care. Consequently, males devote their efforts to competing for females, and polygyny (i.e. successful competitors mate and fertilize more than one female) should predominate. Mating strategies of marine mammal males, driven primarily by sexual selection, will involve different forms of competition (Boness 1991; Le Boeuf 1991; Wells *et al.* 1999), including: (i) direct contests over females (contest competition); (ii) attempts to endure longer than other males (endurance competition); (iii) efforts to search and locate females more efficiently (scramble competition); (iv) attempts to entice and attract females (mate choice competition); or (v) attempts to outcompete with higher quality or greater quantities of sperm or methods for removing other male's sperm (sperm competition) (Andersson 1994). These forms of competition need not be mutually exclusive and perhaps more often than not several will be operating in a given population.

Although slightly different schemes have been used to categorize variation in mating systems (e.g. Bradbury & Vehrencamp 1977; Emlen & Oring 1977; Clutton-Brock 1989; Davies 1991), male and female mating strategies fall into a few types. Among the basic male strategies are: (i) defend territories containing resources used by females; (ii) defend non-resource-based territories that are clustered and behave to attract females (also known as leks); (iii) follow or defend one or more females directly (unimale or multimale groups, or female clusters); and (iv) search for receptive females, spending little time with them other than to mate. Among the basic female strategies are: (i) accept males that attempt to mate with them; (ii) investigate and choose males based on direct or indirect benefits provided by males; (iii) mate promiscuously, which promotes sperm competition; and (iv) incite male–male aggressive encounters and mate with the winner. In Table 10.2 we summarize what is known about the mating systems of marine mammals.

10.3.2 Female dispersion

10.3.2.1 Temporal distribution

The potential for males to fertilize multiple mates is highest when receptive females are moderately asynchronous and spatially clumped (Emlen & Oring 1977; Boness 1991; Nunn 1999). Considerable data are available on the temporal distribution of receptive females in pinnipeds but much less information is available for cetaceans and sirenians. Furthermore, the data are of different types and quality for the various taxa. For pinnipeds the data consist primarily of the temporal spread of births or mating. Among mammals a postpartum oestrus is common, allowing the spread of births or other characteristics linked to birth to be substituted for oestrus, since oestrus may be difficult to observe or detect. For cetaceans, the data are mostly from the observed spread of fetuses, corpora lutea or births. Data from fetuses and corpora lutea are often what is available because they can be obtained from animals that are taken in hunts or washed up on beaches. Observing births or mating at sea is extremely difficult and rare.

Table 10.2 Major aspects of mating systems of marine mammals, including male and female mating tactics and possible mechanisms of competition.

Species	Female spatial, social and movement patterns during receptivity	Mating location	Female mating (males)	Male mating pattern	Male mating tactics	Mechanisms of male competition[1]	Source[2]
Phocids							
Cystophora cristata	Mildly clustered and sedentary on ice; dispersed and mobile at sea	At sea	Single?	Polygynous—low skew	Defence of receptive females or clusters? Roving?	C, E, Sc?	1, 2
Erignathus barbatus	Mildly clustered and sedentary on ice; dispersed and mobile at sea	At sea	Single?	Polygynous—low skew?	Defence of clustered positions (lek)? Singing displays?	C?, M?	1
Halichoerus grypus	Moderately clustered; small movements on land	On land or ice	Multiple	Polygynous—moderate skew	Defence of female cluster	C, E, M, Sp?	1, 3, 4, 5
	Mildly clustered and more mobile at sea	At sea	Multiple?	Polygynous—low skew?	Defence of female cluster or of receptive females	C?, E, Sc, Sp?	1, 6
Leptonychotes weddellii	Moderately clustered and sedentary on ice; mildly dispersed and mobile at sea	At sea	Single?	Polygynous—moderate skew?	Defence of clustered territories (Lek?/resource areas?)	C, E, M?	1, 7
Lobodon carcinophaga	Dispersed on ice	On ice/at sea?	Single?	Polygynous—low skew	Defence of single females	C?, E?	1
Mirounga angustirostris	Highly clustered and relatively sedentary until departure	On land	Multiple	Polygynous—high skew	Defence of female clusters	C, E, M, Sp?	1
Mirounga leonina	Highly clustered and relatively sedentary until departure	On land	Multiple	Polygynous—high skew	Defence of female clusters	C, E, M?, Sp?	1
Monachus schauinslandi	Mildly clustered and sedentary on land; dispersed and mobile at sea	At sea	Multiple?	Polygynous—low skew?	Roving; defence of receptive single females	Sc	1
Pagophilus groenlandicus	Moderately clustered and mobile on ice; mildly dispersed and mobile at sea	At sea	Single? Multiple?	Polygynous—moderate skew?	Clustered defence of positions (lek)? Roving? Serial defence of single females?	C?, E?, M?, Sc?	8
Phoca hispida	Dispersed in ice dens and mobile; dispersed and mobile at sea	At sea; in ice dens	Single?	Polygynous—low skew?	Defence of spaced dens or territories encompassing multiple dens? Roving? Serial defence of single females?	C?, Sc?, E?	9, 10
Phoca vitulina	Moderately clustered and sedentary on land; dispersed and mobile at sea	At sea	Single?	Polygynous—low skew	Defence of clustered territories (lek?/resource?); roving? Serial defence of single females?	C, E, M	1, 11, 12, 13
Otariids							
Arctocephalus australis	Highly clustered and small thermoregulatory movements and investigation of males?	On land	Single? Multiple?	Polygynous—high skew	Defence of clustered resource territories (thermoregulatory sites)	C, E, M?	1, 14

Species	Ranging pattern	Parturition site	Litter size	Mating system	Mating strategy	Classification	References
Arctocephalus forsteri	Highly clustered and small thermoregulatory movements	On land	Single?	Polygynous—high skew	Defence of clustered resource territories (birth and thermoregulatory sites)	C, E	1, 15
Arctocephalus galapagoensis	Moderately clustered and foraging trips	On land	Single?	Polygynous—moderate skew	Defence of clustered resource territories (birth and thermoregulatory sites)?	C, E	1
Arctocephalus gazella	Highly clustered and no movement	On land	Single	Polygynous—high skew	Defence of clustered resource territories (birth and thermoregulatory sites)	C, E	1, 16
Arctocephalus philippii	Highly clustered and small thermoregulatory movements	On land or at sea	Single?	Polygynous—moderate skew	Defence of clustered resource territories (birth and thermoregulatory sites)	C, E	1
Arctocephalus pusillus doriferus	Highly clustered and minimal movements	On land	Single?	Polygynous—high skew?	Defence of clustered resource territories (birth sites)?	C, E?	1
Arctocephalus pusillus pusillus	Highly clustered and foraging trips	On land or at sea	Single?	Polygynous—high skew	Defence of clustered resource territories (birth sites)?	C, E?	1
Arctocephalus townsendi	Highly clustered and small thermoregulatory movements	On land	Single	Polygynous—moderate skew	Defence of clustered resource territories (birth and thermoregulatory sites)	C, E	1, 17
Arctocephalus tropicalis	Highly clustered and minimal movements	On land	Single?	Polygynous—high skew?	Defence of clustered resource territories (birth sites)?	C, E?	1
Callorhinus ursinus	Highly clustered and no movements	On land	Single?	Polygynous—high skew	Defence of clustered resource territories (birth sites)	C, E	1, 18
	Highly clustered and small thermoregulatory movements	On land	Single?	Polygynous—high skew?	Defence of female clusters	C, E?	1, 19
Eumetopias jubatus	Highly clustered and foraging trips and thermoregulatory movements	On land	Single? Multiple?	Polygynous—high skew	Defence of clustered resource territories (birth and thermoregulatory sites)	C, E	1
Neophoca cinerea	Moderately clustered and thermoregulatory movements	On land	Single?	Polygynous—moderate skew	Defence of clustered resource territories (birth and thermoregulatory sites)? or defence of female clusters?	C, E?	1
Otaria flavescens	Highly clustered and minimal movements	On land	Single?	Polygynous—high skew	Defence of clustered resource territories (birth and thermoregulatory sites)	C, E?	1
	Highly clustered and thermoregulatory movements	On land	Single? Multiple?	Polygynous—moderate skew	Defence of female clusters	C, E?	1
Phocarctos hookeri	Highly clustered and roaming at oestrus	On land	Multiple?	Polygynous—high skew?	Defence of clustered territories (lek?/resource areas?)	C, E?, M?	1
Zalophus californianus	Highly clustered and foraging trips and thermoregulatory movements	On land or at sea	Single?	Polygynous—high skew	Defence of clustered resource territories (birth and thermoregulatory sites)	C, E	1
	Moderately clustered in rafts and thermoregulatory movements	On land	Single?	Polygynous—moderate skew?	Defence of clustered positions (lek)?	C, E?, M?, Sc?	1

(continued on p. 290)

Table 10.2 (*cont'd*)

Species	Female spatial, social and movement patterns during receptivity	Mating location	Female mating (males)	Male mating pattern	Male mating tactics	Mechanisms of male competition[1]	Source[2]
Odobenid							
Odobenus rosmarus	Patchily clustered on ice and mobile at sea	At sea	Multiple	Polygynous—moderate skew?	Defence of clustered positions (lek)? whistling displays?	C, E?, M?	1
	Small clusters at sea and mobile	At sea	Single?	Polygynous—moderate skew?	Defence of female clusters	C, E?, M?	20
Sea otter							
Enhydra lutris	Moderately clustered and relatively sedentary	At sea	Single?	Polygynous—moderate skew?	Defence of clustered resource territories (food, kelp beds)	C, E	21, 22
Sirenian							
Trichechus manatus	Highly dispersed and mobile	At sea	Multiple	Polygynous—low skew?	Defence of receptive females; roving	C?, M?, Sc?	22, 23
Dugong dugon	Highly dispersed and mobile	At sea	Multiple	Polygynous—low skew?	Defence of receptive females; roving	C, M?, Sc?	22, 24
	Moderately dispersed and mobile	At sea	Multiple	Polygynous—moderate skew?	Defence of clustered territories (lek?)	C, E?, M?	25
Odontocetes[3]							
Physeter macrocephalus	Highly dispersed (pelagic); historical areas of high abundance; mobile (slow). Diffusely seasonal reproduction. Stable units of (related and unrelated) females and dependent young merge to form medium-sized groups	At sea	Not known	Polygynous—high skew?	Roving; males observed head butting, fighting and heavily scarred	C, Sc	26, 27, 28, 29, 30
Phocoenoides dalli	Highly dispersed (pelagic) and mobile (fast). Seasonal reproduction. Small groups	At sea	Not known	Polygynous—moderate/high skew?	Roving; possible serial defence of single females during mating season	Csd?, Sc?	31, 32
Phocoena phocoena	Dispersed (coastal) and mobile (slow). Seasonal reproduction. Small groups	At sea	Multiple?	Polygynous—low skew?	Roving?	Sc?, Sp?	33, 34
Tursiops truncatus, Sarasota Bay, Florida	Resident in coastal waters and mobile (fast). Diffusely seasonal reproduction. Small, unstable, sex- and age-segregated groups—fission-fusion	At sea	Multiple	Polygynous—moderate skew?	Roving; individual or coalition defence of oestrous females	Csd?, E?, Sc, Sp?	22, 35, 36, 37

Species		Location	Multiple paternity	Mating system	Male strategy		References
Tursiops aduncus, Shark Bay, Australia	Resident in coastal waters and mobile (fast). No distinct mating season. Small, unstable, sex- and age-segregated groups—fission–fusion	At sea	Multiple	Polygynous—moderate skew?	Roving; aggressive herding and defence of females by male coalitions	C, E?, Sc, Sp?	37, 38, 39, 40
Orcinus orca, 'resident'	Resident in coastal waters and mobile (slow). Seasonal reproduction, individual oestrous period about 40 days. Stable life-long bonds in purely matrilineal medium-sized groups	At sea	Multiple?	Polygynous—skew?	Roving (mating in multipod associations)	M?, Sc?	41, 42, 43, 44
Globicephala macrorhynchus, Canary Islands	Highly dispersed (pelagic) and mobile (slow). Diffusely seasonal reproduction. Stable medium-sized groups—avunculate	At sea	Not known	Polygynous—skew?	Roving (mating in multipod associations)?	Csd?, M?, Sc?	41, 45, 46
Delphinapterus leucas	Migratory between pelagic feeding and coastal breeding areas (mating occurs along the way). Mobile (slow). Seasonal reproduction. Small to medium-sized groups nested in larger sex- and age-segregated aggregations	At sea	Not known	Polygynous—skew?	Not known	Csd?, M?, Sc?	47
Monodon monoceros	Migratory between pelagic feeding and coastal breeding areas (mating occurs along the way). Mobile (slow). Seasonal reproduction. Small to medium-sized groups nested in larger sex- and age-segregated aggregations	At sea	Multiple?	Polygynous—high skew?	Roving? Visual display as sexual advertisement? Head scarring and broken tusks indicate violent fights	C, M?	48, 49
Hyperoodon ampullatus, Nova Scotia	Dispersed, pelagic but localized population. Mobile (slow). Diffusely seasonal reproduction with occasional mating year-round. Small sex- and age-segregated groups	At sea	Not known	Polygynous—high skew?	Defence of preferred spatial locations? Coalition defence of single females?	C?, E?, Sc?	50, 51, 52
Mesoplodon spp.	Highly dispersed (pelagic) and mobile (slow). Reproductive period unknown. Small groups	At sea	Not known	Polygynous—high skew?	Not known	C?	53, 54, 76

(continued on p. 292)

Table 10.2 (*cont'd*)

Species	Female spatial, social and movement patterns during receptivity	Mating location	Female mating (males)	Male mating pattern	Male mating tactics	Mechanisms of male competition[1]	Source[2]
Sotalia fluviatilis fluviatilis	Follow river fluctuations? Mobile (slow). Seasonal reproduction. Small groups, composition unknown	In rivers	Multiple	Polygynous—skew?	Roving?	Sc?, Sp?	55, 56, 57
Lagenorhynchus obscurus	Dispersed (coastal/shelf) and mobile (fast). Seasonal reproduction. Small stable subgroups within groups that vary in size seasonally	At sea	Multiple?	Polygynous—low skew?	Roving?	Sc?, Sp?	58, 59
Stenella attenuata, eastern tropical Pacific	Highly dispersed (pelagic) and mobile (fast). Diffusely seasonal reproduction, multiple peaks. Small age- and sex-segregated subgroups in segregated large-sized schools	At sea	Multiple?	Polygynous—moderate skew?	Roving?	M?, Sc?, Sp?	60, 61
Stenella longirostris, Hawaii inshore, eastern tropical Pacific	Dispersed to highly dispersed (pelagic and inshore) and mobile (fast). Seasonal reproduction. Small age- and sex-segregated subgroups in medium–large schools	At sea	Mutliple	Polygynous—moderate/high skew?	Roving?	M?, Sc?, Sp?	62, 63
Lagenodelphis hosei	Highly dispersed (pelagic) and mobile (fast swimming). Diffusely seasonal reproduction. Large mixed age and sex schools	At sea	Not known	Polygynous—moderate skew?	Roving?	M?, Sc?, Sp?	64
Delphinus delphis	Highly dispersed (shelf) and mobile (fast). Seasonal reproduction, multiple peaks. Sex- and age-segregated large–very large schools	At sea	Multiple mates?	Polygynous—moderate skew?	Roving?	M?, Sc?, Sp?	65, 66, 67
Mysticetes *Megaptera novaeangliae*	Dispersed throughout breeding range. Not clustered (occasional brief dyads with other females) or tied to resources; parturient females may prefer shallow, sheltered water	At sea	Unknown multiple across years	Polygynous—low skew?	Defence of females (precopulatory or mate guarding?) Singing as advertisment and perhaps to mediate intrasexual interactions. May be novel form of lek	C, M?	68, 69, 70, 71, 72

Species		At sea	Multiple	Polygynous—low skew?		Sc, Sp	
Eubalaena glacialis	Dispersed and mobile. Reproduction year-round; unclear where breeding occurs	At sea	Multiple	Polygynous—low skew?	Usually non-agonistic defence of single female	Sc, Sp	73
Eubalaena australis	Dispersed and mobile. Reproduction year-round; unclear where breeding occurs	At sea	Multiple	Polygynous—low skew?	Usually non-agonistic defence of single female	Sc, Sp	73
Balaena mysticetus	Dispersed and mobile. Seasonal reproduction, multiple peaks. Alone or in unstable groups	At sea or in sea ice	Multiple?	Polygynous—low skew?	Individual defence of female?	Sc, Sp	73, 74
Eschrichtius robustus	Clustered in coastal lagoons and probably also dispersed offshore. Seasonal reproduction. Alone or short-term associations	At sea	Multiple mates	Polygynous—low skew?	Individual and usually non-agonistic defence of single female?	C, Sc	73, 75

[1] C, contest competition; Csd, contest competition based on sexual dimorphism; E, endurance competition; M, mate choice competition based on displaying; Sc, scramble competiton; Sp, sperm competition.

[2] 1, Boness *et al.* 1993; 2. Kovacs *et al.* 1996; 3, Amos *et al.* 1995; 4, Ambs *et al.* 1999; 5, Worthington Wilmer *et al.* 2000; 6, Tinker *et al.* 1995; 7, Hoelzel *et al.* 1999; 8, Bartsh *et al.* 1992; 9, Kovacs 1995; 10, Kelly & Quakenbush 1990; 11, Kelly & Wartzok 1996; 12, Perry 1993; 13, Van Parijs *et al.* 1997; 14, P. Majluf unpublished data; 15, Troy 1997; 16, Arnould & Duck 1997; 17, Gallo 1994; 18, Gentry 1998; 19, DeLong 1982; 20, Sjare & Stirling 1996; 21, Riedman & Estes 1990; 22, Wells *et al.* 1999; 23, Rathbun *et al.* 1995; 24, Preen 1989; 25, Anderson 1998; 26, Rice 1989; 27, Whitehead & Arnbom 1987; 28, MacLeod 1998; 29, Richard *et al.* 1996; 30, Christal 1999; 31, Jefferson 1990; 32, Houck & Jefferson 1999; 33, Read 1999; 34, Fontaine & Barrette 1997; 35, Wells 1987; 36, Moors 1997; 37, Tolley *et al.* 1995; 38, Ross 1977; 39, Cockcroft & Ross 1990; 40, Connor *et al.* 1992a; 41, Heimlich-Boran 1993; 42, Dahlheim & Heyning 1999; 43, Clark & Odell 1999; 44, Baird 2000; 45, Kasuya & Marsh 1984; 46, Barnard & Reilly 1999; 47, Brodie 1989; 48, Hay & Mansfield 1989; 49, Mansfield 1975; 50, Mead 1989b; 51, Gowans 1999; 52, Hooker 1999; 53, Mead 1984; 54, McCann 1974; 55, da Silva & Best 1996; 57, Best & da Silva 1984; 58, Brownell & Cipriano 1999; 59, Van Waerebeek & Read 1994; 60, Perrin & Hohn 1994; 61, Pryor & Shallenberger 1991; 62, Perrin & Gilpatrick 1994; 63, Norris *et al.* 1994; 64, Jefferson *et al.* 1997; 65, Evans 1994; 66, Madsen & Herman 1980; 67, Harrison *et al.* 1972; 68, Payne & McVay 1971; 69, Herman & Tavolga 1980; 70, Tyack & Whitehead 1983; 71, Clapham & Palsbøll 1997; 72, Clapham 1996; 73, Brownell & Ralls 1986; 74, Würsig & Clark 1993; 75, Swartz 1986; 76, Mead 1989a.

[3] For odontocetes: Female spatial, social and movement pattern during receptivity: group size: small = < 10; medium = 10–50; large = > 50. Female mating: females are categorized as having 'multiple' mates if there are observations of multiple matings or they are polyoestrous, and 'single' mate if females were observed to mate only with one male during the breeding season. Male mating pattern: males are considered to pursue multiple mating opportunities and therefore to be 'polygynous' although there is little direct evidence of this. Males were subjectively categorized as having 'high skew' if there is extreme sexual dimorphism in body size or shape or teeth which would indicate that only the largest and oldest males are likely to obtain access to males and, therefore have greater mating success (e.g. sperm whales). Males were categorized as having 'moderate skew' if there is moderate sexual dimorphism. Males were categorized as 'low skew' if there is no sexual dimorphism or reverse sexual dimorphism (females larger than males). Certain males may be disproportionately successful in sperm competition, enabling them to achieve a disproportionate share of fertilizations. However, we assume here that unless males have morphological adaptations (as evidenced by sexual dimorphism) to obtaining more mates, they are equally likely to mate. Whether these matings lead to fertilizations remains to be determined. Mechanisms of competition: males were categorized as competing via particular mechanisms based on the pattern of sexual dimorphism and relative testes size; contest (fighting morphologies such as teeth, scars, larger body size or shape, larger flukes), scramble (males that rove; dimorphism associated with swimming and manoeuverability such as enlarged flippers and caudal peduncles), and sperm competition (if testes to body size ratio was subjectively chosen at greater than 1.0%).

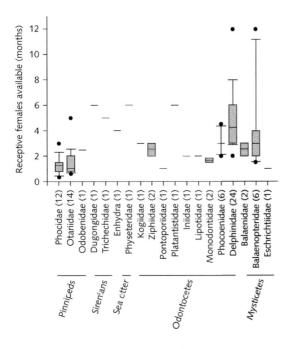

Fig. 10.2 Estimates of the synchrony in availablility of receptive females for the various marine mammal taxa. The boxes indicate the 25th and 75th percentiles; the error bars indicate the 10th and 90th percentiles; the black circles are outliers indicating 5% and 95%; and the solid horizontal lines are the means. The numbers in brackets refer to the number of species represented in the data.

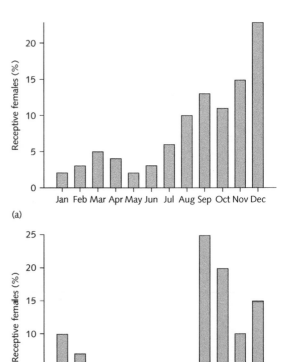

Fig. 10.3 Examples of the temporal spread of receptive females for two cetacean species: (a) killer whales, and (b) sperm whales.

Of marine mammals, pinniped females tend to be the most synchronous in timing of receptivity (Fig. 10.2). Both otariids and phocids tend to have a spread in the peak availability of receptive females of a month or less. The most important factor underlying the high synchrony is the spatial separation between aquatic food resources and the limited suitable land or ice habitats for giving birth (Bartholomew 1970). Other factors may be important as well (e.g. male harassment: Boness *et al.* 1995; habitat stability: Stirling 1983). The most synchronous species are two phocids—the harp and hooded seals—in which all females in a breeding colony are estimated to become receptive during a 10–15-day period (based on spread of births) (Oftedal *et al.* 1987). This level of reproductive synchrony, however, should be sufficient to allow considerable opportunity for individual males to acquire multiple females providing they are not too dispersed spatially. As a group, sirenians are the

most asynchronous with the peak in availability of females averaging about 6 months for the various species. In many of the odontocetes, sirenians and the sea otter, despite a restricted peak in availability of receptive females, there is a small proportion of females in a population that become receptive throughout all or much of the year. This can be seen for sperm and killer whales in Fig. 10.3.

The mysticetes have breeding periods strongly linked to the climatic seasons as in many terrestrial mammals, with mating and calving occurring over several months in winter. This lack of synchrony, together with the general lack of stable group structure and widespread distribution of females in space and time, makes it impossible for males to monopolize multiple females simultaneously; hence, extreme polygyny is unlikely to be found in any baleen whale.

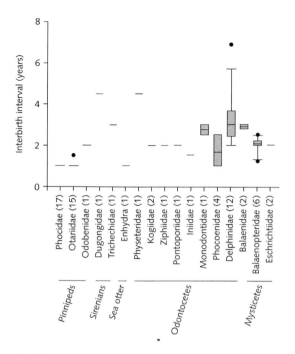

Fig. 10.4 Interbirth intervals for the various marine mammal taxa. The boxes indicate the 25th and 75th percentiles; the error bars indicate the 10th and 90th percentiles; the black circles are outliers indicating 5% and 95%; and the solid horizontal lines are the means. The numbers in brackets refer to the number of species represented in the data.

Another characteristic of temporal dispersion of receptive females that must be considered in terms of the potential for polygyny or promiscuity is the length of the reproductive cycle of individual females. As a measure of this we can use the average interval between births for individual females (Table 10.1 and Fig. 10.4). The multiyear cycles of sirenians and most cetaceans means that only a portion of the female population is receptive in any given year, effectively reducing the availability of potential mates in comparison to the annual cycle of pinnipeds and sea otters.

10.3.2.2 Spatial distribution

Bartholomew (1970) was among the first to establish the importance of retention of terrestrial birthing as a major factor in contributing to the spatial clustering of breeding female pinnipeds. With increased adaptations for aquatic locomotion, pinnipeds have

reduced terrestrial mobility (Tarasoff *et al.* 1972; English 1977; Beentjes 1990), making them more vulnerable to terrestrial predators. Oceanic islands or isolated mainland beaches undoubtedly afford the greatest protection from predators but these habitats are relatively rare. Terrestrial predators still exist for some extant species; for example polar bears prey on northern ice-breeding seals (Stirling 1988) and jackals on Cape fur seals (David 1987) (see Riedman 1990 for a review). While high density may be explained by the lack of suitable beach area for South American fur seals in Peru (Fig. 10.5a), there appears to be plenty of unoccupied space for Juan Fernández fur seals at a colony in Chile (Fig. 10.5b) yet females are still clustered.

At least four factors might help account for female clustering at higher levels than space alone would predict. These include marginal male, selfish herd, male harassment, and thermoregulatory effects. The marginal male effect hypothesizes that sexual selection should favour aggregation of females in so far as peripheral females are fertilized by males that are less competitive and have lower fitness (McLaren 1967). Evidence of more competitive males occupying central areas of breeding sites is available for numerous species of birds and mammals (e.g. Kruijt & de Vos 1988; Pemberton *et al.* 1992; Höglund & Alatalo 1995), including some pinnipeds (Le Boeuf 1974; Boness & James 1979; Anderson & Fedak 1985; Gisiner 1985; Haley *et al.* 1994).

The selfish herd hypothesis, proposed by Hamilton (1971), argues that animals become gregarious because it reduces the probability of a given individual becoming the target of a predator. There is considerable evidence for this explanation of herding in ungulates. A strong case is presented for pronghorn antelopes (Byers 1997). While there is presently little clear evidence for this effect being important in pinnipeds (but see Harcourt 1992), it is likely to have been a significant factor in the evolution of sociality in odontocetes (see Chapter 12).

Another antipredator explanation to grouping behavior is increased vigilance of groups to warn of pending predation. Some evidence suggests this is a factor in clustering in harbour seals (Terhune 1985; da Silva & Terhune 1988). Evidence of alarm signals, as seen in many terrestrial animals that form social groups (see Bradbury & Vehrencamp 1998), are not

(a)

(b)

Fig. 10.5 Examples of clustering on the breeding grounds in two fur seal species regardless of space constraints: (a) South American fur seals at Punta San Juan, Peru (courtesy of Patricia Majluf); (b) Juan Fernández fur seals at Robinson Crusoe Island, Chile (courtesy of Daryl Boness).

well known in marine mammals, except in sperm whales. Sperm whales are reported to communicate alarm over distances of up to 11 km in response to attacks by killer whales (Pitman *et al.* 2001) or whalers (Caldwell *et al.* 1966). The whales receiving the alarm signals may either flee or come to the assistance of the individuals under attack.

Intraspecific male harassment may act in similar ways to interspecific predation to select for clustering behaviour. Solitary southern sea lion females are more prone to attacks by young conspecific males than are females that give birth to young in groups (Campagna *et al.* 1992). Similar attacks occur in the Australian sea lion (Higgins & Tedman 1990), although differential mortality has not been related to clustering in this species. In the grey seal, a temporal clustering has been shown to reduce the impact of male harassment on maternal performance and pup growth (Boness *et al.* 1995). This effect undoubtedly works in combination with spatial clustering since at the end of the breeding season female density is markedly reduced as many females have left land because they have completed lactation. Not only is there safety in numbers but, perhaps most importantly, clusters of females are defended by dominant or territorial males, which chase marginal males away. Le Boeuf (1972) suggested that elephant seal females choose to associate with dominant males because they minimize disruptions from peripheral males and this may con-

tribute to the evolution of gregariousness in this species. As for the grey seals, disruptions to elephant seal females are costly. Female northern elephant seals compete for central positions within aggregations, a location that affords the greatest protection from disruption by marginal males (Cox & Le Boeuf 1977). The avoidance of intraspecific male harassment through clustering around protective males may be an important factor in the evolution of female gregariousness in a wide variety of taxa (reviewed in Mesnick 1997).

A final factor that may contribute to enhanced clustering in pinnipeds is thermoregulatory needs. This is mostly concerned with keeping cool since pinnipeds have evolved a subcutaneous blubber layer to cope with retaining body heat in the high thermal conductivity of water. Thus, females may cluster to make use of patchily distributed or limited thermoregulatory sites (e.g. shade-producing boulders: Trillmich 1984; Higgins 1990; tide pools: Campagna & Le Boeuf 1988b; Carey 1992; wet shoreline: Whittow 1978; Trillmich & Majluf 1981).

Thermoregulatory needs might also result in increased movement and dispersion of females. For example, females of many otariid species make daily thermoregulatory movements to the water or water's edge during midday when the sun is hottest (reviewed in Boness *et al.* 1993). Such movements are likely to reduce the defensibility of females and lower the potential for polygyny.

Among many phocids, mating takes place at sea but is not likely to be related to thermoregulation since many of these species breed in polar or subpolar habitats. Females of these species are moderately clustered on land or ice while they nurse their pups, but become receptive about the time they wean their pups, or shortly after, and disperse from land (Ray *et al.* 1969; Boness *et al.* 1988; Walker & Bowen 1993b; McRae & Kovacs 1994). In some species, such as harbour seals (Boness *et al.* 1994), harp seals (Lydersen & Kovacs 1993) and Weddell seals (Bartsh *et al.* 1992), females may spend time in the water even before weaning. The ease of mobility of females in an aquatic medium and the lack of dependent offspring, which tie females to a particular location, increase female dispersion and reduces the ability of males to monopolize females.

The other marine mammals that live and mate at sea will not have the constraints on female spatial patterns that exist for the land-breeding pinnipeds, with the exception of the sea otter, which relys on kelp beds to minimize the energy required to keep from drifting. Of the various marine mammals sea otters are the least well adapted to marine travelling and deep diving. Most sea otters are found in relatively small coastal home ranges that contain their food resources and kelp patches, which provide shelter from rough seas (Garshelis *et al.* 1984; Ralls *et al.* 1996). The patchiness of kelp beds and shallow water suitable for sea otter foraging tends to lead to females forming some level of clustering in which their home ranges overlap (see Riedman & Estes 1990; Wells *et al.* 1999). Regardless, sea otters are much less clustered than most pinniped females.

Few of the factors that determine clustering behaviour in pinnipeds appear to apply to baleen whales, although our knowledge of their breeding distribution and behaviour is limited to a few species. While predation (by killer whales or sharks) almost certainly occurs on some mysticete calves, fatal attacks appear to be comparatively rare, and there is no evidence that the social organization or spatial distribution of any baleen whale is influenced by antipredator considerations (Clapham 1999; but see Corkeron & Connor 1999). It is possible that male harassment may affect the behaviour of females (calves are often in the way of aggressive interactions between male humpback whales, for example) (Glockner-Ferrari & Ferrari 1990), but this does not seem to promote clustering in any species for which there are good data. In all mysticetes, the apparent lack of stable groups (of either sex), and the absence of prey resources on winter calving grounds, result in wide distribution of receptive females in space and time. Although humpback whale mothers and calves may loosely cluster in areas of sheltered water (e.g. behind reefs), this distribution is not static or severely restricted, as it is on pinniped pupping beaches, and monopolization by males of groups of females does not occur. This is the case even in the most extreme example of mysticete clustering, in which large numbers of female gray whales use coastal lagoons for calving (Jones *et al.* 1984).

During the summer feeding season, the unpredictable and patchy nature of mysticete prey resources (schooling fish, euphausiids or smaller zooplankton) also functions to limit clustering behaviour. Aggregations of whales often occur in productive areas, but groups within these aggregations are small and generally unstable; this is presumably a function of small and variable patch size (Whitehead 1983; Clapham 1993). Among the balaenopterids, singletons and pairs are most common, although larger groups are frequently recorded in association with large patches of food.

Similarly, few of the factors that determine clustering behaviour in pinnipeds appear to apply to odontocetes, although our knowledge of the distribution and behaviour of females during the breeding season is limited to a few well-studied species. Odontocete groups vary in size (from solitary individuals to schools of thousands), the duration of social bonds (ephemeral to lifelong) and the degree of relatedness among group members. Typically, odontocete social groups are mobile and dispersed, sometimes over many nautical miles. These characteristics result in a wide and unpredictable distribution of receptive females in space and time and limit the ability of males to monopolize females.

Variation in group size among odontocetes is substantial, ranging from one or two (mothers and dependent calves being the only permanent social groups) in river dolphins and some coastal porpoises to 10 000 in some pelagic delphinids. Several factors influence the formation of odontocete groups (see Chapter 12). The primary factors are related to foraging mode, habitat, predator avoidance and communal care of the young (Wells *et al.* 1980). A minimum number necessary to protect against predation (the selfish herd hypothesis) may determine the lower end of group size while the upper end may be limited only by resource availability. Communal care of the young (communal nursing, babysitting and defence) may be especially important for deep divers such as sperm whales. In many odontocete species, group structure is hierarchical: there appears to be a primary group, comprised of long-term associates and/or kin, and secondary groups formed by primary groups joining together, for feeding, socializing or mating. Odontocete social groups may also be segregated by sex, age, kinship and reproductive state (Best 1979; Perrin & Reilly 1984; Wells *et al.* 1987; Kasuya 1999) (see Chapter 12). Spatial separation of adults and juveniles, especially during a breeding period, is known to occur in many pinniped species (e.g. Davies 1949; Le Boeuf 1972; Vaz Ferreira 1975; Miller 1975) as well as in other mammals. For example, male bachelor herds are common among ungulate and equid species (Leuthold 1970; David 1973; Clutton-Brock *et al.* 1982; Rubenstein 1986).

Among the sirenians, manatees and dugongs are similar in their range of spatial patterns and movements. Females are often observed alone or with their calves, although larger aggregations are observed in relation to food and warm-water resources. Aerial surveys of dugongs in Australia produced between 60% and 90% of sightings of lone animals (Marsh & Saalfeld 1989; Marsh *et al.* 1994). Slightly lower percentages (40–55%) of aerial sightings of manatees were of single animals (Hartman 1979; Irvine *et al.* 1982). The group size of manatees and dugongs observed is highly variable, ranging from two to over 600 (Anderson & Birtles 1978; Preen 1989; Reynolds & Wilcox 1994; Marsh 1995; Anderson 1998). There is no evidence of social relationships or group cohesion other than cow–calf pairs (see Chapter 12). The larger groups in the case of manatees tend to be winter aggregations at power plants that produce warm-water outflows; among dugongs they appear to be related to large patches of sea grass (Preen 1995).

Sirenians, like sea otters are constrained by their feeding habits to relatively shallow waters.

These species, however, are vegetarians rather than carnivores. The relatively poor-quality food and its occurrence in shallow water are extremely important factors in the spatial patterns and movement of sirenians (Wells *et al.* 1999). Daily foraging movements may take manatees and dugongs over ranges of 20–40 km (Rathbun *et al.* 1990; Preen 1992; Reid *et al.* 1995).

In summary, the temporal–spatial pattern of receptive females is a product of ecological and social factors. Three major factors are resource distribution, predation pressure and the costs and benefits of group living. Pinniped females that mate on land tend to be reproductive annually, relatively synchronous in receptivity and spatially aggregated. As a consequence, males of these species will probably compete to monopolize access to females and competition among males may be the most intense. Pinniped species or populations in which individuals mate at sea may be similarly synchronized but are more mobile and thus more dispersed than females of terrestrially mating species. Contest competition and control of resources or aggregations of females through territoriality is less likely in these species. Sirenians and mysticete females tend to be solitary and have multiyear reproductive cycles, which may make females of these species the least able to be monopolized by males. Although some mysticetes (notably humpback and gray whales) aggregate at traditional breeding sites, the absence of food resources and predation pressure cause females to be scattered and this precludes monopolization by males. Most odontocete females form social groups and males may compete to monopolize these groups, or individual females within them. The multiyear reproductive cycle of females, the extended period of breeding, and relatively large home ranges of groups probably reduces the ability of males to monopolize large numbers of females effectively.

10.3.3 Male mating strategies

10.3.3.1 *Terrestrially mating pinnipeds*

Among the terrestrial mating pinnipeds, as noted above, an important factor in male mating strategies is the reduced spatial scale of reproductive activity produced by the limited mobility and probable high costs of movements on land. Indeed the reduced movement may make it difficult to distinguish between strategies involving defence of females directly versus territorial defence of resources. Virtually all otariids are terrestrial breeders, although evidence of some mating at sea, or the potential for it (because females begin foraging trips before becoming receptive), is available for a few species (Gentry 1970; David & Rand 1986; Higgins *et al.* 1988; Heath 1989; Francis & Boness 1991). All otariid populations that have been studied exhibit polygyny. In most cases the primary male strategy is individual defence of territories, which are resource-based and the resources are suitable birth sites or thermoregulatory sites (Table 10.2) (Boness 1991; Boness *et al.* 1993). Clear indication that males employ territorial

defence (cf. defence of females) is based on several lines of evidence, most clearly seen in studies of Juan Fernández fur seals (Francis & Boness 1991), northern fur seals (Peterson 1965; Gentry 1998) and Antarctic fur seals (McCann 1980; Arnould & Duck 1997). For example, males arrive at the traditional mating grounds before females. They display and fight with one another (contest competition) before and during the presence of females, and the displays and fights occur at the periphery of areas used by males, establishing boundaries (Fig. 10.6a). Males subsequently have exclusive access to mating with females that are in their territories (i.e. no overlap in space use; Fig. 10.6a), which females use to give birth and care for their young or to cool off (or both in some populations).

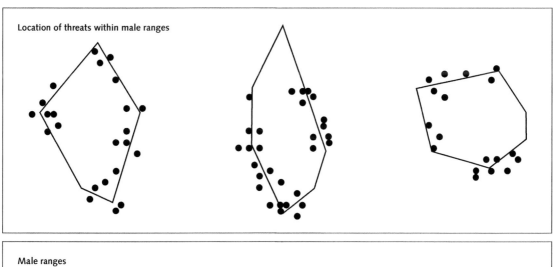

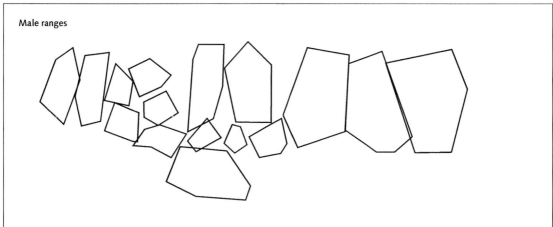

(a)

Fig. 10.6 Examples of data in: (a) the Juan Fernández fur seal. (*continued on p. 300*)

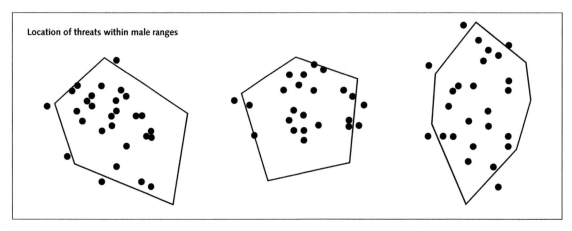

Location of threats within male ranges

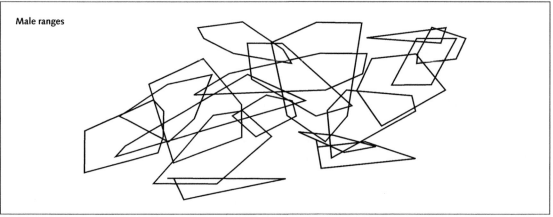

Male ranges

(b)

Fig. 10.6 (*cont'd*) (b) The grey seal. These allow the differentiation between defence of an area (territoriality, location of threats at boundaries and no overlap in male ranges) and defence of females directly (location of threats within male ranges and overlapping ranges).

In at least one otariid species, the southern sea lion (Campagna & Le Boeuf 1988a), and possibly others (DeLong 1982; Higgins 1990), there is evidence that at some colonies males may exhibit defence of female clusters rather than territorial defence. Female defence also appears to be the primary strategy for males of the three terrestrially mating phocid species (grey seal: Anderson *et al.* 1975; Boness & James 1979; Godsell 1991; Tinker *et al.* 1995; northern elephant seal: Le Boeuf & Peterson 1969; Le Boeuf 1974; Haley *et al.* 1994; southern elephant seal: Carrick *et al.* 1962; McCann 1981). In these species, male arrival may not precede the females on the mating grounds. Displays or threats occur throughout the area within

which a male moves, not just along the periphery (Fig. 10.6b). There is considerable overlap in male ranges (Fig. 10.6b), and males may change their locations when females move. Within this strategy there is variation in male tactics. For example, among northern and southern elephant seals an alpha male maintains control over a tight cluster of females using vocal threats over relatively long distances and will fight when necessary (Le Boeuf 1974; McCann 1981). In contrast, grey seal males defend smaller and loosely clustered groups of females using face-to-face threats, focus attention on females in the cluster closest to oestrus, and shift their locations as necessary to obtain proximity to females that will become receptive (Boness & James 1979; Godsell

1991). In all three species, some males that are unable to maintain control over access to females attempt to 'sneak' copulations or capture females as they leave the colony (Cox & Le Boeuf 1977; Boness et al. 1982). Boness et al. (1993) reviewed what we know about alternative strategies in terrestrially breeding pinnipeds (see also Twiss et al. 1998; Lidgard et al., in press).

Relatively recent studies of two otariid species and the walrus suggest the potential that males in some populations may employ clustered mating territories or a lek-type strategy (Fay et al. 1984; Heath 1989; P. Majluf, unpublished data). Distinguishing this from territorial defence of birth and thermoregulatory sites can be difficult because the aggregated nature of breeding colonies of seals on land necessarily results in a clustering of male territories. Some might consider this characteristic alone enough to classify all pinnipeds as lekking species (Höglund & Alatalo 1995), however, others require further evidence (e.g. Bradbury & Gibson 1983; Gosling & Petrie 1990). One way to distinguish between leks and resource territories, based on other species such as ungulates, where both resource defence territories and clustered territories exist, is that territory size should be smaller when lekking occurs (Clutton-Brock et al. 1988; Gosling 1991; Balmford et al. 1993a). Territory size is relatively small in the South American fur seal (P. Majluf, unpublished data), but not in the California sea lion (Heath 1989). Another possibility is that there might be phenotypic enhancements (morphological or behavioural), known as secondary sexual characters, which are used by lekking males to attract females, and by which females might choose males (e.g. Møller & Pomiankowski 1993; Jennions & Petrie 1997). Male California sea lions have a sagittal crest not found in females and they 'bark' incessantly. Yet it is unclear whether these traits are used as displays for females or in male–male competition. It is also unclear whether aspects of these traits correlate with reproductive success.

The reproductive success of individual males ranges widely for both female defence and territorial defence strategies and appears to produce comparable levels of polygyny. Measures of success in pinniped studies are of varying types and qualities (see Boness et al. 1993). In some species, the most successful male

may mate with as many as 50–100 females in a given season (Boness et al. 1993). However, we must be cautious about the use of estimates based on behavioural observations. Recent studies of marine and terrestrial mammals using genetic techniques to assess paternity have found that the level of success of the most successful males in polygynous species has been substantially overestimated using behavioural estimates (Pemberton et al. 1992; Amos et al. 1993; Morin et al. 1994; Craighead et al. 1995; Coltman et al. 1998a; Ambs et al. 1999; Hoelzel et al. 1999). Hoelzel et al. (1999), however, did find that relative success based on paternity and behavioural observation yielded comparable results for southern elephant seals.

Why do land-mating otariid males tend to employ a strategy of defending territories encompassing female resources, whereas the few land-mating phocids tend to defend females directly? One possible explanation for this difference is a higher energetic cost to terrestrial locomotion in phocids. Phocids have lost features for efficient movement on land to a greater extent than otariids (Tarasoff et al. 1972; English 1977; Beentjes 1990). Maintenance of territorial boundaries is likely to require more movement than direct defence of females (e.g. boundary displays described in many species: Peterson & Bartholomew 1967; Peterson 1968; Miller 1991). The fact that territorial defence appears to occur in populations of some phocids that mate at sea (Hewer 1957; Bartsh et al. 1992; Perry 1993), where mobility is much less costly energetically, lends some support to this hypothesis. In New Zealand fur seals, the lack of a relationship between mating success and intrasexual activity but a strong positive relationship between success and intersexual activity suggests that herding of females may be a more important factor than maintaining boundaries for some otariids (Troy 1997).

Two other possible factors explain why territorial defence occurs in some situations and female defence in others—the degree to which females are patchily or evenly distributed and the intensity of male competition (Emlen & Oring 1977; Stamps & Buechner 1985; Lott 1991). Examples of this are seen in the southern sea lion in Patagonia (Campagna & Le Boeuf 1988a, 1988b) and in the northern fur seal in California (DeLong 1982). Territorial defence seems to occur when females are patchily distributed in uneven terrain, but female defence occurs when females are continuous in uniform habitats. It may be difficult to separate the effects of male–male competition from female distribution, however, since large continuous groups of females also tend to yield higher densities of males and more frequent male–male aggression (Campagna & Le Boeuf 1988a, 1988b; Gentry 1998).

10.3.3.2 Aquatically mating pinnipeds

Nearly half of the pinniped species mate at sea, including the majority of phocids. Although an increasing number of studies of these species are occurring, for most we still have minimal information on male mating strategies. As mentioned above, for a few species that mate at sea, there is reasonable evidence to indicate that individual males are defending territories off the beach or ice where females reside. In these cases, the nearshore areas used by females appear to be restricted and predictable (e.g. Kaufman et al. 1975; Francis & Boness 1991; Perry 1993). However, the greater mobility and dispersion of females at sea and the lack of a connection to land because offspring have been weaned, theoretically should lead to different primary male strategies than those used by species where females are clustered and sedentary on land. In species such as the harp, hooded, ringed and bearded seals (Boness et al. 1988; Cleator et al. 1989; Kelly & Quakenbush 1990; Kovacs 1990, 1995), there is a small amount of quantitative and anecdotal data that allow speculation on male strategies. The evidence suggests that these species' strategies more closely resemble roving (i.e. scramble competition), sequential defence of single females, or clustered males advertising vocally to females (Table 10.2).

Male reproductive behaviour of one aquatically mating species, the harbour seal, has been given considerable attention (Sullivan 1981; Davis & Renouf 1987; Godsell 1988; Thompson et al. 1989; Perry 1993; Walker & Bowen 1993a, 1993b; Hanggi & Schusterman 1994; Coltman et al. 1997, 1998b; Van Parijs et al. 1997). Although females gather on land to give birth and nurse their offspring, they begin foraging trips to sea before weaning and disperse from land after weaning, becoming receptive just about the time of weaning (Bigg 1969; Boulva & McLaren 1979; Perry 1993; Boness et al. 1994; Van Parijs et al. 1997). Copulations have not been observed on land at any colony even though males spend time ashore among females and pups.

Based on radio tracking and time–depth recorders, harbour seal males reduce their ranges and spend less time in deeper water about the time receptive females become available (D.J. Boness, unpublished data; Coltman et al. 1997; Van Parijs et al. 1997). They also lose more mass after this time, suggesting they begin fasting (Walker & Bowen 1993a; Coltman et al. 1998b). With underwater video cameras and transmitters attached to males, D.J. Boness, (unpublished data) found that males in Nova Scotia continue to forage substantially up to the time females begin weaning pups and are often beyond 4 km from the beaches where females care for their pups. Subsequent to the onset of receptive females, males spend considerable time giving visual and vocal displays within 1.5 km of shore and in restricted 'individual home ranges' (1–2 km²). Similar nearshore behaviour was observed from cliff tops at a colony in California (Hanggi & Schusterman 1994) and underwater recordings near breeding beaches and foraging areas in Scotland revealed male vocalizations (Van Parijs et al. 1999).

Genetic paternity analyses have shown that harbour seal male success is moderate to low, with most males fertilizing one or no females and the maximum number of females fertilized for any given male being five (Coltman et al. 1998a). For a given female, the male siring her offspring was usually not holding a position adjacent to her postion on land as speculated by Sullivan (1981), but averaged being several kilometres away (D.J. Boness, unpublished data). The combination of these various data suggest that the primary harbour seal male mating strategy at Sable Island, and other locations, is something akin to lekking as reported for many birds and mammals (Wiley 1991; Balmford et al. 1993b; Höglund & Alatalo 1995; Widemo 1998). Perry (1993), on the other hand, observed males clearly defending discrete aquatic territories immediately adjacent to females on the beach. However, based on paternity analyses of a small sample of seals, none of the females was fertilized by any of the territorial males.

10.3.3.3 Other marine mammals

Sea otters, cetaceans and sirenians mate at sea and indeed spend virtually all of their time there. Sea otters in some areas may come ashore to bask in the sun for short periods of time. Outside the breeding season sea otter males and females are usually segregated (Jameson 1989) and males form large rafting groups. The probable reason for segregation outside the breeding season is resource limitation since females and young remain in the same home ranges year-round (Wells *et al.* 1999). During the breeding season, male sea otters migrate to where females have home ranges. They establish ranges that encompass one or more female home ranges, which contain food resources and sheltered resting places. The male areas are often exclusive of other males and appear to be territories (Garshelis *et al.* 1984; Jameson 1989), although aggressive competition is less common than subtle threats. These spatial and behavioural patterns led Wells *et al.* (1999) to conclude that the primary male mating strategy in sea otters is territorial defence of resources. The importance of food resources in the territories of males has been clearly demonstrated by a strong correlation between an estimate of food quality within a male's territory and his mating success (Fig. 10.7) (Garshelis *et al.* 1984).

Data on male mating strategies in cetaceans are much more limited than in pinnipeds and the sea otter. Evidence of territorial defence of resources is almost completely lacking among cetaceans (except maybe the northern bottlenose whale: Hooker 1999) and there is limited evidence for a form of lekking behaviour in one mysticete (the humpback whale: Mobley & Herman 1985; Clapham 1996). Among the mysticetes, substantial data on the breeding behaviour of males exist only for the humpback, right and gray whales. Of these three species, the humpback has been the most intensively studied. Male humpbacks adopt one or more of three primary strategies: they display by singing long, complex songs (Payne & McVay 1971) (see Chapter 6); they compete directly with other males for females in competitive groups (Tyack & Whitehead 1983; Clapham *et al.* 1992); and they escort females, including those with newborn calves (Glockner-Ferrari & Ferrari 1990). Movement by males within and among breeding areas (presumably to search for receptive females) is also a strategy akin to roving.

Singing by humpback males presumably acts to attract females, although whether songs contain cues to mate quality remains in dispute (Chu 1988; Clapham 1996). Singing may also function (perhaps secondarily) to space males in a breeding area or to aid in establishment of dominance hierachies (Darling 1983; Darling & Bérubé 2001). Songs change over time yet all members of a population sing essentially the same song at any one time (see Chapter 6). How this coordination is accomplished, and what selective factors drive the changes, are unclear. Although female choice based upon song characteristics would presumably represent a powerful selective force for change, there is no direct evidence that this is the primary factor involved. Mobley and Herman (1985) have suggested that the aggregation and displaying of humpback whales at specific sites constitutes lekking. However, this ignores the requirement in traditional definitions of leks that lek territories be relatively stable in space, which is not the case with humpbacks; Clapham (1996) suggests that a novel classification is required.

Direct competition often involves two or more males engaging in agonistic interactions over a central female. While these competitive groups occasionally contain more than 20 males, actual challenges to the principal escort of the female are usually confined to a few animals (Clapham *et al.* 1992). Other

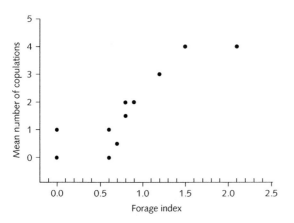

Fig. 10.7 The relationship between male mating success and the amount of food resources in a sea otter territory. (Adapted from Garshelis *et al.* 1984.)

males remain on the periphery, perhaps waiting to engage in 'sneaking' behaviour. Principal escorts are occasionally displaced by challenging males. Eventually, all competitive groups break up, leaving the central female with the winning principal escort. Whether mating follows is unknown; it is possible that an escort's defence of a female may sometimes reflect mate guarding following earlier copulation (Clapham 1996).

The relative importance and success of each male strategy is unclear, primarily because observations of copulation are almost non-existent, and because molecular testing of paternity is hampered by the wide distribution of animals and the difficulty of sampling a large pool of potential fathers. In addition, we know nothing about a key variable, the duration of oestrus, in this or any other mysticete. Given the variation in size and fitness of individual males, it is very likely that (as in better studied taxa) different males will utilize different strategies depending on age, size or number and quality of available females. Given the relatively low rate of annual calving in this species, many observers believe that escorting mother–calf pairs is a low-return strategy adopted by inferior males, but direct evidence of this is lacking (Clapham 1996).

Despite strongly seasonal calving (in winter), right whales engage in sexual activity year-round (Payne 1986; Stone et al. 1988). Since the gestation period is 1 year (Best 1994) and there is no evidence for delayed implantation, mating leading to conception presumably occurs primarily in winter, and the function of sexual activity at other times is not clear. Observations of multiple males mating with single females, together with the huge (1 t) size of the testes, strongly suggest that sperm competition is a principal mating strategy in this species, and probably also in bowhead and gray whales (Brownell & Ralls 1986). In these species, the level of aggression in male–male interactions is low compared to that observed in humpback whales, which is also consistent with the predominance of sperm competition as a strategy.

With the exception of humpbacks, virtually nothing is known of the mating systems of balaenopterid whales. Both blue and fin whales appeared to be widely dispersed during the winter breeding season (Watkins et al. 2000). Male fin whales make a pat-terned 20 Hz call, which Watkins et al. (2000) have termed a breeding display, but observations of courtship or competitive interactions at this time are sparse.

The mating strategies of sirenian males appear to overlap the range of strategies seen among the mysticetes. Male manatees and dugongs tend to be solitary, roaming over large areas that encompass the home ranges of several females, and perhaps are searching for oestrous females (Bengtson 1981). Periodically they are seen in 'mating herds', where several males follow and try to mate with a single female (Hartman 1979; Preen 1989; Rathbun et al. 1995), somewhat similar to the mating groups observed in mysticete whales. We know too little about these groups to fully understand the behaviour or success of individuals in them. In manatees, an oestrous herd of several males and a single female may last for several weeks. Although herd composition changes, it is unclear whether any male remains with the female for the whole time (Hartman 1979; Rathbun et al. 1995), nor who fertilizes the female.

Mating herds in dugongs appear to be of much shorter duration, lasting no more than a day. Male fights and threats occur in both the manatee and dugong herds. As Rathbun notes in Wells et al. (1999), we do not know whether the male strategy associated with mating herds in sirenians is a form of scramble competition or more akin to a type of lekking similar to that suggested for the humpback whale (Mobley & Herman 1985; Clapham 1996). In at least one location, a more classic kind of lekking has been suggested for the dugong (Anderson 1997). In a small cove in Shark Bay, Australia 20 dugongs patrolled exclusive areas and engaged in activities that appeared to be related to both male competition and mate attraction, including vocalizations by the occupants of the areas. Females were seen with a single male rather than mating herds.

Among odontocetes, the high mobility and spatial dispersion of receptive females (Table 10.2) has two important consequences for male mating strategies: there is decreased opportunity to control access to aggregated receptive females and diminished paternity assurance. It is not surprising, therefore, that the most common odonotcete male mating strategy appears to be searching for receptive females and spending little time with them except to mate

(Table 10.2). It is likely that mate guarding, or monopolization of an individual female long enough to ensure conception, occurs in odontocetes, although this has been well documented only in bottlenose dolphins (Connor *et al.* 1992b; Moors 1997).

Our knowledge of male mating strategies among the odontocetes is sparse or non-existent for most species. Moreover, sexual behaviour in many species is an important component of non-reproductive social interactions that have little to do with fertilization (Kasuya *et al.* 1993; Norris *et al.* 1994; Wells *et al.* 1999) (see Chapter 12), making it difficult to infer mating strategy from incomplete observations. Therefore, we provide examples of odontocete male mating strategies for a few species that have been studied more thoroughly. In some locations, bottlenose dolphins form stable alliances of a few males that may work alone or with other closely associated alliances (Connor *et al.* 1992a, 1992b) (see Chapter 12) and either 'capture' individual females through aggressive herding (Australia: Connor *et al.* 1996) or form temporary consortships without obvious aggressive herding (Florida: Wells *et al.* 1987, 1999). In another location males apparently do not form alliances nor aggressively herd females, although single males may accompany groups of females throughout the breeding season (Scotland: Wilson 1995, cited in Wells *et al.* 1999). The extent to which this sequential female defence strategy is successful is uncertain, however, since individual females can cycle multiply and associate with several males during the season in which they conceived, both of which may encourage mate choice and sperm competition (Connor *et al.* 1996). The difference in use of aggressive herding by males at different locations may be inversely correlated to male body size and the degree of sexual dimorphism (Tolley *et al.* 1995). This behaviour may vary between the *truncatus* and *aduncus* forms of bottlenose dolphins, which are now thought to represent different species (Curry 1997; Le Duc *et al.* 1999).

Male sperm whales near the Galápagos Islands rove between groups of females, searching for potential mates (Whitehead & Arnbom 1987). One or more large males attend groups of females (sometimes simultaneously) for periods of time varying from 5 min to 19 h. Rather than males herding female groups, females will alter course and speed so they can join a large male hundreds of metres away. Males are not aggressive toward one another within groups of females, despite several accounts in the literature of males fighting outside of groups (Caldwell *et al.* 1966). In modelling roving sperm whale behaviour, Whitehead (1990) suggests that males should rove among groups of females if the travel time between encountering groups of females is less than the oestrous period of a female. Moreover, males should delay breeding until attaining a large size so that they are competitively dominant (Whitehead 1994). The conclusions of this model also appear to hold for other large, sexually dimorphic mammals such as elephants and polar bears. One thing not yet clear for sperm whales, given an apparent roving strategy, is the role of the relatively large nose of the male and its possible use as a sound-generating organ. Auditory clicks may function in male–male competition or advertisement to attract females (Weilgart & Whitehead 1988; Cranford 1999).

The striking sexual dimorphism of killer whales and pilot whales is generally associated with polygyny. Although confirmed observations of mating are rare, pods of resident killer whales in the Pacific Northwest are frequently observed associating with one another, particularly during the summer months when prey abundance is high (Baird, 2000). In these multipod groups there is much sexual activity amongst all members of the pods, young and old alike. Since no dispersal of either males or females occurs from resident pods (Bigg *et al.* 1990), it is thought that mating occurs during these encounters. Similarly, in long-finned pilot whales captured in the Faroese fishery, genetic evidence suggests that males remain in their natal groups but do not mate within them (Amos *et al.* 1991). Paternity analyses indicate that the young were sired by related males not captured with the group, implying that pilot whales must mate when two or more groups meet or when adult males pay brief visits to other groups (Amos *et al.* 1993).

For the remaining odontocetes, very little is known about mating. The best we can do is make predictions about the mechanisms of male competition that might be operating based on the pattern of sexual dimorphism and characters such as relative testis size and bodily scarring. These characters also provide clues to the intensity of sexual selection (the skew in male mating success in a species). Based on studies of terrestrial mammals, a positive correlation is generally assumed between the amount of sexual dimorphism in a species and the degree of polygyny. Caution must be taken in making such inferences, however, because dimorphic traits may also reflect ecological differences between the sexes (e.g. beak length between the sexes in the platanistids: Reeves & Brownell 1989) or may be important for females and their young (e.g. genital pigment patterns in Dall's porpoises: Morejohn *et al.* 1973). Also the connection between exaggerated male traits and male behaviour or mating success remains untested among odontocetes.

Sexual dimorphism in size, weaponry (big teeth, large heads, strong flukes) and bodily scarring, together with observations of males fighting might suggest the importance of contest competition for access to mates. Males are larger than females in many odontocete species, with the most pronounced sexual size dimorphism found in sperm whales (Table 10.1). Tooth eruption or enlargement and unpigmented scarring occur only in adult males in several species (McCann 1974; Best 1981; Heyning 1984; MacLeod 1998). On the other hand sexual dimorphism in propulsion structures (e.g. flippers, caudal penduncle flukes) might suggest scramble competition at work. In three-dimensional habitats, agility, rather than size or strength may be selected for (Wells *et al.* 1999). Lastly, sexual dimorphism in signalling structures or acoustic signals might suggest the importance of mate choice competition and attempts by males to entice and attract females. Among odontocetes, sexually dimorphic acoustic signals are known only in sperm whales (Caldwell *et al.* 1966). However, because odontocetes produce a wide range of sounds, acoustic displays are likely to occur in several other species as well.

Among mammals, the relationship between relative testes size and mating system is so strong that relative testes size can be used as a good indicator of mating system and relatively large testes are indicative of potential sperm competition (Brownell & Ralls 1986; Kenagy & Trombulak 1986; Gomiendo *et al.* 1998). Relative testis size ranges dramatically among odontocete species, from less than 0.05% of body mass (several *Mesoplodon* species, the franciscana, baiji

and sperm whale: Mead 1984; Brownell 1984, 1989; Rice 1989) to 5% or greater (harbour porpoise, finless porpoise, tucuxi and dusky dolphin: Goa & Zhou 1993; Best & da Silva 1984; Fontaine & Barrette 1997; Brownell & Cipriano 1999). Human males, for comparison, have testes of about 0.08% of body mass (Kenagy & Trombulak 1986).

Using the characters and pattern of sexual dimorphism described above to make inferences about mechanisms of competition in odontocetes for which direct behavioural observations are not available, we find that all mechanisms of male competition for access to females are likely to occur (Table 10.2). Scramble and contest competition are likely to be common mechanisms in odontocetes and sperm competition may be particularly important in some of the delphinids and porpoises.

10.3.4 Female mating strategies

Although female mammals maximize their reproductive success by being good mothers, they may also employ mating strategies to enhance their fitness by choosing males with immunologically compatible genes or with high-quality genes that will give their offspring a competitive edge in reproduction (Bateson 1993; Andersson 1994; Eberhard 1996). This choice may occur either pre- or postcopulation, the latter of which may lead to sperm competition. Virtually nothing is known about postcopulatory choice by female marine mammals and only minimal effort has gone into precopulation mate choice.

In pinnipeds there are at least two otariid species for which some evidence of females choosing males exists (Heath 1989; P. Majluf, unpublished data). In both species, multiyear studies show that females changed pupping locations from one year to the next to remain within the territory of a male who changed location.

Indirect choice of males by females inciting male–male competition has been demonstrated fairly convincingly for northern elephant seals (Cox & Le Boeuf 1977) and grey seals (Boness *et al.* 1982). In both species, females respond aggressively to the approach and attempted mounts by males, being more likely to respond vigorously if the male is a younger or transient male. The net result is that the alpha male in the case of elephant seals or a dominant male in the case of grey seals is likely to challenge the younger male and prevent them from copulating. Female competition for access to preferred mates may be intense. In the northern elephant seal, females compete for central positions in female aggregations (Cox & Le Boeuf 1977) but whether this enhances the likelihood of being fertilized by the alpha male is not certain (see Hoelzel *et al.* 1999).

Female promiscuity may result in selection for sperm competition among males and has been reported in all three land-breeding phocids (Carrick *et al.* 1962; Le Boeuf 1972; Boness & James 1979; Amos *et al.* 1993; Ambs *et al.* 1999). Despite a common belief that it is rare among otariids, female promiscuity may occur in as many as 30% of females for half the otariids studied (Boness *et al.* 1993).

Little is known regarding female choice or female mating strategies in mysticetes. Female humpback whales are known to sometimes aggressively reject subadult males (Clapham *et al.* 1992), and anecdotal evidence suggests they may incite competition among males (Clapham 1996). Molecular analysis of paternity has shown that female humpbacks are fertilized by different males in successive breeding seasons (Clapham & Palsbøll 1997), but whether they engage in multiple matings within a season remains unknown. As noted above, it is possible that the songs of males contain cues that serve as a basis for female choice, but no such cue has been positively identified.

Female mating strategies in odontocetes are also little understood. It is thought that females are able to reject males by outmanoeuvering them or rolling belly up. Female bottlenose dolphins at Shark Bay, Australia, and Sarasota Bay, Florida, have been observed successfully eluding males who were attempting to mate with them (Connor *et al.* 1992a; Moors 1997). Whether female odontocetes advertise receptivity, incite male–male competition or evaluate potential mates by testing their ability to follow them has not as yet been demonstrated for any species. As noted above, it is possible that the numerous visual and acoustic signals of males contain information that females might use to assess male quality, strength, endurance or competitive ability, but no such cues have been identified.

Observational studies of multiple copulations by individual females within a breeding season suggest female promiscuity in several species (bottlenose dolphins: Wells *et al.* 1987; Connor *et al.* 1996; bouto and tucuxi: Best & da Silva 1984; harbour porpoise: Fontaine & Barrette 1997). Additionally, hormonal studies with captive animals of several species found females to be polyoestrous (bottlenose dolphins and common dolphins: Kirby & Ridgway 1984; narwhals: Hay 1984; false killer whales: Odell & McClune 1999; killer whales: Dahlheim & Heyning 1999). In combination, these findings suggest sperm competition may be important. As our understanding of the physiology of female receptivity grows, we will be better able to understand female mating strategies and how they influence male behaviour.

10.3.5 Maternal strategies

Individual females can most directly influence their lifetime reproductive success and fitness by successfully rearing the single offspring produced during each reproductive event. We refer to the behavioural and physiological patterns of maternal care and lactation exhibited by individual females as their maternal strategies. Although the strategies of aquatic and amphibious species may be expected to differ, ecological and phylogenetic conditions have led to both convergent and divergent patterns among and between the various marine mammal taxa. One important factor is body size (Costa 1991, 1993; Boness & Bowen 1996) (see Chapter 9). While much research has been conducted and numerous recent reviews written on maternal strategies in pinnipeds (Bonner 1984; Gentry & Kooyman 1986; Oftedal *et al.* 1987; Trillmich 1996; Boyd 1998), little systematic study and analysis of this topic has been done on cetaceans or sirenians (but see Lockyer 1984; Perrin & Reilly 1984; Wells *et al.* 1999). We will rely heavily on an excellent review of cetacean lactation by Oftedal (1997). Our discussion will focus on length of lactation, pattern of maternal foraging to sustain lactation, and milk composition.

10.3.5.1 Length of lactation

The length of lactation among all marine mammals varies over a phenomenal range from 4 days to 3 or more years (Table 10.3). Within the cetaceans and pinnipeds we see what appears to be a dichotomy of

Table 10.3 Major aspects of maternal strategies of marine mammals, including maternal mass, duration of lactation, maternal feeding pattern and milk composition.

Species	Maternal parturition mass (kg)	Maternal length (cm)	Lactation duration (days)	Maternal feeding pattern	Trip[1] duration (days)	Milk fat (%)	Source[2]
PHOCIDS							
Cystophora cristata	179	200	4	Fasting		61	1
Erignathus barbatus	250	228	12–18	Fasting?/incidental feeding?		50	2, 3, 4,
Halichoerus grypus	207	192	16	Fasting (land)/incidental feeding? (ice)		60	1, 4
Leptonychotes weddellii	447	254	53	Incidental feeding?		48	1
Mirounga angustirostris	504	322	27	Fasting		54	1
Mirounga leonina	515	278	24	Fasting		47	1
Pagophilus groenlandicus	129	170	12	Incidental feeding?		57	1
Phoca hispida	62	132	48	Regular feeding bouts	? Hours		3, 5
Phoca vitulina	84	159	24	Feeding–nursing cycle, mid–late lactation	0.3	50	1
OTARIIDS							
Arctocephalus australis	45	137	365	Feeding–nursing cycle, early–late lactation	4.6	55	6
Arctocephalus galapagoensis	37	120	540	Feeding–nursing cycle, mid–late lactation	1.1	29	1
Arctocephalus gazella	39	135	117	Feeding–nursing cycle, mid–late lactation	2.7–5.9	42	1, 7
Arctocephalus philippii	48	150	210	Feeding–nursing cycle, mid–late lactation	12.3	41	8
Callorhinus ursinus	37	129	125	Feeding–nursing cycle, mid–late lactation	5.9–9.4	42	1, 7
Eumetopias jubatus	273	235	330	Feeding–nursing cycle, mid–late lactation	1.2	24	1, 7
Neophoca cinerea	84	274	540	Feeding–nursing cycle, mid–late lactation	1.7–2.1	26	3, 9
Zalophus californianus	88	180	300	Feeding–nursing cycle, mid–late lactation	1.4–2.5	44	1, 7
Zalophus wollebaeki	80		365	Feeding–nursing cycle, mid–late lactation	0.8	19	3, 6
ODOBENID							
Odobenus rosmarus	830	284	730–900	Regular feeding, young in attendance		30[1]	10
SEA OTTER							
Enhydra lutris	20	120	180	Regular feeding, young in attendance		23[1]	11, 12

ODONTOCETES

				Feeding		References
Delphinids						
Delphinus delphis	112	170	420–570	Regular feeding, young in attendance	30	7, 13
Sousa chinensis		280		Regular feeding, young in attendance	10	14
Stenella attenuata		187	390–810	Regular feeding, young in attendance	22[1]	7, 13
Tursiops truncatus	251	251	540–600	Regular feeding, young in attendance	24	7, 13
Physeterids						
Kogia breviceps		300		Regular feeding, young in attendance	15	7, 13
Kogia sima	155	270		Regular feeding, young in attendance	19	7, 13
Physeter macrocephalus	13 500	1090	600–1200	Regular feeding, young in attendance	26	7, 13
Monodontid						
Delphinapterus leucas		430	600–720	Regular feeding, young in attendance	27[3]	7, 13
Phocoenid						
Phocoena phocoena	53	180	240–360	Regular feeding, young in attendance	46[3]	7, 13
MYSTICETES						
Balaenopterids						
Balaenoptera musculus	100 000	3100	180–210	Fasting/minimal feeding in early lactation; feeding likely in late lactation; young in attendance	39	7, 13
Balaenoptera physalus	65 000	2600	180–210	Fasting/minimal feeding in early lactation; feeding likely in late lactation; young in attendance	33	7, 13
Balaenoptera acutorostrata	8000	1070	150–180	Fasting/minimal feeding in early lactation; feeding likely in late lactation; young in attendance	30	7, 13
Megaptera novaeangliae	30 000	1520	300–330	Fasting/minimal feeding in early lactation; extensive feeding in late lactation; young in attendance	44	7, 13
Eschrichtiids						
Eschrichtius robustus	25 000	1500	210–240	Fasting/minimal feeding in early lactation; feeding likely in late lactation; young in attendance	53	7, 13

[1] Only applicable to some pinnipeds.

[2] 1, Boness & Bowen 1996; 2, Bowen 1991; 3, Costa 1991; 4, Lydersen *et al.* 1994; 5, Kelly & Wartzok 1996; 6, Oftedal *et al.* 1987; 7, Wells *et al.* 1999; 8, Francis *et al.* 1998; 9, Higgins 1990; 10, Fay 1982; 11, Estes 1989; 12, Riedman & Estes 1990; 13, Oftedal 1997; 14, Gaskin 1982. Also see Van Lennep & Van Utrecht 1953.

[3] Lactation stage is not known or is not mid-lactation (see Oftedal *et al.* 1987 for definition and discussion).

lactation lengths, this being somewhat clearer in the pinnipeds. Furthermore, the dichotomy within these suborders generally follows family lines. The phocids have short lactation lengths of 4–50 days and the otariids and odobenids nurse their young for 4–36 months. None of the cetaceans exhibit the extremely short lactation of the phocids, but the mysticetes tend to have lactation lengths of 5–8 months with some evidence of 10–12-month long lactation in a few species such as humpback, bowhead and right whales. The odontocetes, by contrast, tend to nurse their young in excess of 1 year and in many species for 2 or more years; the harbour porpoise may be unusual, with lactation being as short as 8 months.

Both dugongs and manatees have lactation lengths of more than 1 year and manatees in some situations may nurse for as long as 4 years (Marsh 1995; Oshea & Hartley 1995; Rathbun et al. 1995). Sea otters nurse their young for an average of 6 months (Riedman & Estes 1990).

Among the short-lactation pinnipeds, the phocids, the short-lactation cetaceans and the mysticetes (note that short means different things in these different taxa), young generally do not appear to consume solid food before weaning. In the case of many phocids, pups typically do not leave land or ice until they are weaned, so there is virtually no opportunity to forage. However, even in those species such as harbour and Weddell seals, where pups spend substantial periods of time in the water before weaning, there is little evidence that the pups feed (Bowen 1991; Muelbert & Bowen 1993). There is some indication that right and gray whale young may consume solid food near the middle of lactation in the case of right whales and at the end of lactation in the gray whales, but in both cases at around 6 months of age. By contrast, in the long-lactating pinnipeds, cetaceans, sirenians and sea otters, milk provisioning diminishes over time as the young increasingly consume solid food (Riedman & Estes 1990; Bowen 1991; Boness & Bowen 1996; Oftedal 1997). Odontocete calves may begin to chase small prey when they are a few months old (Oftedal 1997).

In many of the long-lactating species, the precise time of weaning is often difficult to determine because the weaning process tends to be prolonged and because suckling may continue beyond the time when milk is a primary source of nutrition. Protracted nursing is likely to perform a social bonding function as well as a nutritive one (Wells et al. 1999). In several odontocete species (e.g. short-finned pilot whales: Kasuya & Marsh 1984; Marsh & Kasuya 1984; bottlenose dolphins: Cockcroft & Ross 1990; and sperm whales: Best 1979; Best et al. 1984) some individuals may suckle several months to years beyond the typical weaning age for the species. For example, Indian Ocean bottlenose dolphins usually wean between 12 and 18 months but non-nutritional suckling may continue for as long as 3 years for some mothers and calves.

10.3.5.2 Maternal foraging pattern

For the most part, the short lactation length of the phocid seals is associated with maternal fasting throughout lactation (Table 10.3) (fasting strategy: Bonner 1984; Oftedal et al. 1987; Boness & Bowen 1996) (see Chapter 9). However, there is a growing body of evidence to suggest there may be interplay between body size, lactation length and maternal foraging in phocids such that smaller individuals of a given species must supplement body fat stores with food obtained from foraging to sustain lactation (Testa et al. 1989; Costa 1991; Hammill et al. 1991; Lydersen & Kovacs 1993; Boness et al. 1994; Boness & Bowen 1996; Boyd 1998). The most clearly documented case of this is the harbour seal (Bowen et al. 1992; Boness et al. 1994; Bowen et al. 1994, 2001). At the end of lactation most phocid females make extensive migratory movements to forage intensively (see Chapter 7).

The relatively long lactation period and small body size of otariids have led to maternal foraging in all species (Table 10.3) (foraging cycle strategy: Bonner 1984; Oftedal et al. 1987; Boness & Bowen 1996) (see Chapter 9). The typical pattern is for females to give birth at traditional pupping grounds and to fast for 1–2 weeks while remaining on land continuously with their pups. Females then begin to alternate between foraging excursions (leaving pups ashore) and nursing visits ashore. There is considerable variation both among and within species in the length of the foraging trips but much less variation in the length of shore visits. This variation appears to be most closely associated with the local abundance and distribution of food resources (Boness & Bowen 1996; Gentry & Kooyman 1986), which may vary with short- and long-term climatic changes (Ono *et al.* 1987; Trillmich & Ono 1991; Alverson 1992; Polovina *et al.* 1994; Francis *et al.* 1998).

Among the mysticetes, maternal fasting during the first few months of lactation occurs in all species studied (Table 10.3). For most species, minimal to moderate maternal feeding may occur in late lactation, and in at least one species, the humpback whale (with close to 1-year-long lactation), feeding is extensive. For many mysticetes, long migrations occur between the calving grounds in warmer water and foraging areas in colder polar or high-latitude temperate waters (see Chapter 7). Females build up large fat stores before migrating to the calving grounds, as the phocid females do (see Chapter 9). The requirement for intensive lactation together with winter fasting is presumably inextricably tied to the large body size of these animals and the attendant ability to store large reserves of fat (Oftedal 1997) (see Chapter 9).

The long-lactating odontocetes do not make large migratory movements between foraging and calving areas (see Chapter 7), nor do they fast or reduce foraging effort during lactation (Table 10.3). Unlike the otariid seals, they are not constrained in space by having to have land for their young, so odontocete females simply continue to forage with their young in tow rather than making foraging trips to and from a centralized location.

One pinniped species, the walrus, has evolved aquatic nursing (Loughrey 1959; Miller & Boness 1983), and consequently behaves somewhat similarly to odontocetes, taking their young with them when they forage. However, unlike most odontocetes (exceptions being the Arctic beluga and narwhal), female walruses make large migratory movements with their calves, following the recession or expansion of the ice edge (Fay 1982).

10.3.5.3 Milk composition

Milk composition must be intimately linked to maternal behavioural strategies since milk is the primary or only source of nutrients for young mammals prior to becoming independent. Rates of energy provisioning need not be as high in species that have extended lactation compared to those that have abbreviated lactation because, in the former, the young have a longer period in which to develop and learn the skills needed to forage. As fat is the primary source of energy in milks, we might expect the phocids, which have the shortest lactation periods, to have the highest fat milks. One must be cautious in making comparisons across species since milk composition does vary depending on the stage of lactation (Oftedal *et al.* 1987). The data summarized here are the best available data, but one must consider the generalizations tentative, especially for the cetacean data.

Milk fat content in phocids ranges between about 45% and 60%, considerably higher, on average, than any of the other marine mammal groups (Table 10.3). The milk fats of mysticetes and otariid seals range from about 25% to 50%. Odontocetes, as a group, have the lowest fat milks ranging from 10% to 30%.

The high fat content of phocid milks is clearly related to the short, intensive lactation (Oftedal *et al.* 1987). However, given the long lactation for most otariids, these fat contents may seem unusually high. The reason is most likely the 'feast and famine' nature of nursing pups. The foraging cycle strategy results in periodic fasting for pups, which requires energy-dense milk to sustain both fasting and growth. Short-term cyclical mass gain followed by loss have been shown in association with maternal foraging, but are superimposed on longer term mass gain (Goldsworthy 1992, 1995). Additionally, milk fat content correlates positively with the length of maternal foraging trips (conversely with pup fasting) (Trillmich & Lechner 1986; Boness & Bowen 1996).

The relatively low fat content in odontocete and sea otter milks (Table 10.3) is consistent with calves having continuous access to their mothers over a long lactation. These circumstances should reduce pressure for rapid calf growth and a high rate of energy transfer fuelled by extremely high-fat milk. Milk composition is unknown in sirenians, but we would expect it to be low given the similar circumstance between them and odontocetes and sea otters.

At the beginning of this section we alluded to the importance of body size in determining maternal strategies. As noted by Oftedal (1993) and Boyd (see Chapter 9), the allometric relationship between energy storage and energy expenditure suggests that a fasting strategy, as in the phocids and mysticetes (also seen in bears), require being large. The general correlation between the body size of different taxa of marine mammals and maternal strategy described in this section is consistent with this hypothesis. However, one surprising case that does not fit the trend as one might expect is the small harbour porpoise, which has one of the shortest lactation periods of the odontocetes. Upon close inspection we see that there is an array of traits that set this species apart from the norm. There seems to be an acceleration of reproductive and life history traits, including an annual calving by individuals, beginning calving at 4 years of age, a relatively short lactation of only 8 months, and relatively short lifespan of 10 years (Read & Hohn 1995).

10.4 CONCLUSIONS

Marine mammals are large and long-lived, which undoubtedly relates to being adapted to all or part of their life in an aquatic environment. Thermoregulatory needs are an important component underlying their large size. Large variation in body size and degree of aquatic living have resulted in different selective pressures favouring different temporal and spatial patterns of females, and hence, different reproductive strategies by males. At one end of a continuum of spatial clustering and temporal synchrony of reproductive females are the pinnipeds in which the retention of terrestrial reproduction, in total or in part, and the decoupling of the ability to forage and care for offspring simultaneously, has led to relatively high levels. At the other end of the continuum are sirenians and some cetaceans (mostly mysticetes) in which there are large interbirth intervals leading to considerable asynchrony in reproductive females and a tendency for a dispersed spatial pattern. Intermediate are those cetaceans (mostly odontocetes) that form social groups but for which females do not have annual reproductive cycles, and pinnipeds that mate aquatically.

Male mating strategies that appear to be favoured by high levels of temporal and spatial clustering of reproductive females are territorial defence of resources used by females or defence of females in clusters or of individuals sequentially, and involve primarily contest and endurance competition. In contrast, high temporal asynchrony, mobility of females in the water and dispersed spacing appear to preclude either territorial or female defence tactics and favour roving in search of receptive females. Alternatively, males may attempt to attract females, which will probably involve scramble competition, mate choice or sperm competition (see Chapter 12). Among species with intermediate spatial and temporal clustering of females we are still unlikely to see territorial defence tactics but may see the other tactics, with roving possibly being the most common. Having suggested these generalizations we must caution that at present the only group of species for which we have solid data on a reasonable portion of the species are the pinnipeds that mate on land. Thus, these ideas must remain speculative and serve as a basis for further work.

Considerable variation also exists in the maternal strategies of marine mammals. Again large size and ability to store energy is an important factor, as is the decoupling of food resources and breeding location. The most intensive lactation (short lactation and high-fat milk resulting in a high rate of energy transfer) is seen in those species that have extended fasting (phocids and some mysticetes) and that are relatively large. In species that have substantial maternal foraging (otariids, odobenids, sirenians, sea otters and odontocetes) and are relatively small, lactation is much less intense. The social nature of odontocetes is likely to be an important factor in the more extended lactation and developmental period of these species and, in turn, is probably driven by more intense predation than occurs in other marine mammals.

While substantial strides have been made in our understanding of the life history and reproductive strategies of marine mammals, there is much that we do not yet know. As our greatest needs are in a better understanding of those species that are mostly or entirely aquatic in their reproductive activity, progress will probably be slow. There are a number of technological advances, such as the miniaturization of video cameras that can be attached to animals (see Box 8.3), that will help in obtaining new information, but the costs of research at sea is likely to be the biggest constraint on progress.

REFERENCES

Agler, B., Schooley, R.L., Frohock, S.E., Katona, S.K. & Seipt, I.E. (1993) Reproduction of photographically identified fin whales, *Balaenoptera physalus*, from the Gulf of Maine. *Journal of Mammalogy* **74**, 577–587.

Aguilar, A. & Lockyer, C.H. (1987) Growth, physical maturity and mortality of fin whales *Balaenoptera physalus* inhabiting the temperate waters of the Northeast Atlantic. *Canadian Journal of Zoology* **65**, 253–264.

Alverson, D.L. (1992) A review of commercial fisheries and the Steller sea lion *(Eumetopias jubatus):* the conflict arena. *Review of Aquatic Science* **6**, 203–256.

Ambs, S.M., Boness, D.J., Bowen, W.D., Perry, E.A. & Fleischer, R.C. (1999) Proximate factors associated with high levels of extraconsort fertilization in polygynous grey seals. *Animal Behaviour* **58**, 527–535.

Amos, B., Barrett, J. & Dover, G.A. (1991) Breeding behaviour of pilot whales revealed by DNA fingerprinting. *Heredity* **67**, 49–55.

Amos, W., Twiss, S., Pomeroy, P. & Anderson, S.S. (1993) Male mating success and paternity in the grey seal, *Halichoerus grypus*: a study using DNA fingerprinting. *Proceedings of the Royal Society of London* **252**, 199–207.

Amos, W., Twiss, S., Pomeroy, P. & Anderson, S.S. (1995) Evidence for mate fidelity in the gray seal. *Science* **268**, 1897–1898.

Anderson, P.K. (1997) Shark Bay dugongs in summer. I: Lek mating. *Behaviour* **134**, 433–462.

Anderson, P.K. (1998) Shark Bay dugongs *(Dugong dugong)* in summer. II: Foragers in a *Halodule*-dominated community. *Extrait de Mammalia* **62** (3), 409–425.

Anderson, P.K. & Birtles, A. (1978) Behaviour and ecology of the dugong, *Dugong dugong* (Sirenia): observations in Shoalwater and Cleveland Bays, Queensland. *Australian Wildlife Research* **5**, 1–23.

Anderson, S.S. & Fedak, M.A. (1985) Grey seal males: energetic and behavioral links between size and sexual success. *Animal Behaviour* **33**, 829–838.

Anderson, S.S., Burton, R.W. & Summers, C.F. (1975) Behaviour of grey seals (*Halichoerus grypus*) during a breeding season at North Rona. *Journal of Zoology, London* **177**, 179–195.

Andersson, M. (1994) *Sexual Selection*. Princeton University Press, Princeton, NJ.

Arnould, J.P.Y. & Duck, C.D. (1997) The costs and benefits of territorial tenure, and factors affecting mating success in male Antarctic fur seals. *Journal of Zoology, London* **241**, 649–664.

Baird, R.W. (2000) The killer whale—foraging specializations and group hunting. In: *Cetacean Societies: Field Studies of Dolphins and Whales* (J. Mann, R.C. Connor, P.L. Tyack & H. Whitehead, eds), pp. 127–153. University of Chicago Press, Chicago.

Balcomb III, K.C. (1989) Barid's beaked whale *Berardius bairdii* (Stejneger, 1833): Arnoux's beaked whale *Berardius arnuxii* (Duvernoy, 1851). In: *Handbook of Marine Mammals, Vol. 4: River Dolphins and the Larger Toothed Whales*

(S.H. Ridgway & R. Harrison, eds), pp. 261–288. Academic Press, London.

Balmford, A., Bartos, L., Brotherton, P. *et al.* (1993a) When to stop lekking: density-related variation in the rutting behaviour of sika deer. *Journal of Zoology, London* **231**, 652–656.

Balmford, A.P., Deutsch, J.C., Nefdt, R.J.C. & Clutton-Brock, T. (1993b) Hotspot models of lek evolution: testing the predictions in three ungulate species. *Behavioral Ecology and Sociobiology* **33**, 57–65.

Barnard, H.J. & Reilly, S.B. (1999) Pilot whales—*lobicephala* Lesson, 1828. In: *Handbook of Marine Mammals, Vol. 6: The Second Book of Dolphins and Porpoises* (S.H. Ridgway & R. Harrison, eds), pp. 245–279. Academic Press, London.

Bartholomew, G.A. (1970) A model for the evolution of pinniped polygyny. *Evolution* **24**, 546–559.

Bartsh, S.S., Johnston, S.D. & Siniff, D.B. (1992) Territorial behavior and breeding frequency of male Weddell seals (*Leptonychotes weddelli*) in relation to age, size and concentrations of serum testosterone and cortisol. *Canadian Journal of Zoology* **70**, 680–692.

Bateson, P. (1993) *Mate Choice*. Cambridge University Press, Cambridge, UK.

Beentjes, M.P. (1990) Comparative terrestrial locomotion of the Hooker's seal lion *Phocarctos hookeri* and the New Zealand fur seal *Arctocephalus forsteri*: evolutionary and ecological implications. *Zoological Journal of the Linnean Society* **98**, 307–325.

Bengtson, J.L. (1981) *Ecology of manatees in the St Johns River, Florida*. PhD Thesis, University of Minnesota, Minneapolis.

Best, P.B. (1977) Two allopatric forms of Bryde's whale off South Africa. In: *Sei and Bryde's Whales*, pp. 10–35. Special Issue No. 1. International Whaling Commission, Cambridge, UK.

Best, P.B. (1979) Social organization in sperm whales, *Physeter macrosephalus*. In: *Behavior of Marine Animals, Vol. 3: Cetaceans* (H.E. Winn & B.L. Olla, eds), pp. 227–289. Plenum Press, New York.

Best, P.B. (1994) Seasonality of reproduction and the length of gestation in southern right whales, *Eubalaena australis*. *Journal of Zoology* **232**, 175–189.

Best, P.B., Canham, P.A.S. & MacLeod, N. (1984) Patterns of reproduction in sperm whales, *Physeter macrocephalus*. In: *Reproduction in Whales, Dolphins, and Porpoises* (W.F. Perrin, R.L.J. Brownell & D.P. DeMaster, eds), pp. 51–80. Special Issue No. 6. International Whaling Commission, Cambridge, UK.

Best, R.C. (1981) The tusk of the narwhal (*Monodon monoceros* L.): interpretation of its function (Mammalia: Cetacea). *Canadian Journal of Zoology* **59**, 2386–2393.

Best, R.C. & da Silva, V.M.F. (1984) Preliminary analysis of reproductive parameters of the boutu, *Inia geoffrensis*, and the tucuxi, *Sotalia fluviatilis*, in the Amazon River system. In: *Reproduction in Whales, Dolphins, and Porpoises* (W.F. Perrin, R.L.J. Brownell & D.P. DeMaster, eds), pp. 361–369. Special Issue No. 6. International Whaling Commission, Cambridge, UK.

Best, R.C. & da Silva, V.M.F. (1989) Amazon River dolphin, boto *Inia geoffrensis* (de Blainville, 1817). In: *Handbook of Marine Mammals, Vol. 4: River Dolphins and the Larger Toothed Whales* (S.H. Ridgway & R. Harrison, eds), pp. 1–23. Academic Press, London.

Best, R.C. & da Silva, V.M.F. (1993) Inia geoffrensis. *Mammalian Species* **426**, 1–8.

Bester, M.N. (1987) Subantarctic fur seal, *Arctocephalus tropicalis*, at Gough Island (Tristan Da Cunna Group). In: *Status, Biology, and Ecology of Fur Seals. Proceedings of an International Symposium and Workshop* (J.P. Croxall & R.L. Gentry, eds), pp. 57–60. NOAA Technical Report NMFS No. 51. National Technical Information Service, Springfield, VA).

Bigg, M.A. (1969) The harbour seal in British Columbia. *Bulletin of the Fisheries Research Board of Canada* **172**, 1–33.

Bigg, M.A., Olesiuk, P.F., Ellis, G.M., Ford, J.K.B. & Balcomb, K.C. (1990) Social organization and genealogy of resident killer whales (*Orcinus orca*) in the coastal waters of British Columbia and Washington State. In: *Individual Recognition of Cetaceans: Use of Photo-Identification and Other Techniques to Estimate Population Parameters* (P.S. Hammond, S.A. Mizroch & G.P. Donovan, eds), pp. 383–406. Special Issue No. 12. International Whaling Commission, Cambridge, UK.

Boness, D.J. (1991) Determinants of mating systems in the Otariidae (Pinnipedia). In: *Behaviour of Pinnipeds* (D. Renouf, ed.), pp. 1–44. Chapman & Hall, London.

Boness, D.J. & Bowen, W.D. (1996) The evolution of maternal care in pinnipeds. *Bioscience* **46**, 645–654.

Boness, D.J. & James, H. (1979) Reproductive behaviour of the grey seal (*Halichoerus grypus*) on Sable Island, Nova Scotia. *Journal of Zoology, London* **188**, 477–500.

Boness, D.J., Anderson, S.S. & Cox, C.R. (1982) Functions of female aggression during the pupping and mating season of grey seals, *Halichoerus grypus* (Fabricius). *Canadian Journal of Zoology* **60**, 2270–2278.

Boness, D.J., Bowen, W.D. & Oftedal, O.T. (1988) Evidence of polygyny from spatial patterns of hooded seals (*Cystophora cristata*). *Canadian Journal of Zoology* **66**, 703–706.

Boness, D.J., Bowen, W.D. & Francis, J.M. (1993) Implications of DNA fingerprinting for mating systems and reproductive strategies of pinnipeds. *Symposium of the Zoological Society of London* **66**, 61–93.

Boness, D.J., Bowen, W.D. & Oftedal, O.T. (1994) Evidence of a maternal foraging cycle resembling that of otariid seals in a small phocid, the harbor seal. *Behavioral Ecology and Sociobiology* **34**, 95–104.

Boness, D.J., Bowen, W.D. & Iverson, S.J. (1995) Does male harassment of females contribute to reproductive synchrony in the grey seal by affecting maternal performance? *Behavioral Ecology and Sociobiology* **36**, 1–10.

Bonner, W.N. (1984) Lactation strategies in pinnipeds: problems for a marine mammalian group. *Symposia of the Zoological Society of London* **51**, 253–272.

Boulva, J. & McLaren, I.A. (1979) Biology of the harbour seal, *Phoca vitulina*, in eastern Canada. *Bulletin of the Fisheries Research Board of Canada* **200**, 1–24.

Bowen, W.D. (1991) Behavioural ecology of pinniped neonates. In: *Behaviour of Pinnipeds* (D. Renouf, ed.), pp. 66–127. Chapman & Hall, London.

Bowen, W.D., Oftedal, O.T. & Boness, D.J. (1992) Mass and energy transfer during lactation in a small phocid, the harbor seal (*Phoca vitulina*). *Physiological Zoology* **65**, 844–866.

Bowen, W.D., Oftedal, O.T., Boness, D.J. & Iverson, S.J. (1994) The effect of maternal age and other factors on birth mass in the harbour seal. *Canadian Journal of Zoology* **72**, 8–14.

Bowen, W.D., Iverson, S.J., Boness, D.J. & Oftedal, O.T. (2001) Energetics of lactation in harbour seals: effect of body mass on sources and level of energy allocated to offspring. *Functional Ecology* **15**, 325–334.

Boyd, I.L. (1991) Environmental and physiological factors controlling the reproductive cycle of pinnipeds. *Canadian Journal of Zoology* **69**, 1135–1148.

Boyd, I.L. (1997) The behavioral and physiological ecology of diving. *Trends in Evolution and Ecology* **12** (6), 213–217.

Boyd, I.L. (1998) Time and energy constraints in pinniped lactation. *American Naturalist* **12**, 213–217.

Boyd, I.L. & Duck, C. (1991) Mass changes and metabolism in territorial male Antarctic fur seals (*Arctocephalus gazella*). *Physiological Zoology* **64**, 375–392.

Boyd, I.L., Lockyer, C. & Marsh, H.D. (1999) Reproduction in marine mammals. In: *Biology of Marine Mammals* (J.E. Reynolds & S.A. Rommel, eds), pp. 218–286. Smithsonian Institution Press, Washington, DC.

Bradbury, J.W. & Gibson, R. (1983) Leks and mate choice. In: *Mate Choice* (P. Bateson, ed.), pp. 109–138. Cambridge University Press, Cambridge, UK.

Bradbury, J.W. & Vehrencamp, S.L. (1977) Social organization and foraging in emballonurid bats. III. Mating systems. *Behavioral Ecology and Sociobiology* **2**, 1–17.

Bradbury, J.W. & Vehrencamp, S.L. (1998) *Principles of Animal Communication.* Sinauer Associates, Inc., Sunderland, MA.

Braham, H.W. (1984) Review of reproduction in the white whale, *Delphinapterus leucas*, narwhal *Monodon monoceros*, and Irrawaddy dolphin, *Orcaella brevirostris*, with comments on stock assessment. In: *Reproduction in Whales, Dolphins and Porpoises* (W.F. Perrin, R.L.J. Brownell & D.P. DeMaster, eds), pp. 81–90. Special Issue No. 6. International Whaling Commission, Cambridge, UK.

Brodie, P.F. (1975) Cetacean energetics, an overview of intraspecific size variation. *Ecology* **56**, 152–161.

Brodie, P.F. (1977) Form, function and energetics of Cetacea: a discussion. In: *Functional Anatomy of Marine Mammals* (R.J. Harrison ed.), Vol. 3, pp. 45–58. Academic Press, New York.

Brodie, P.F. (1989) The white whale- *Delphinapterus leucas* (Pallas, 1776). In: *Handbook of Marine Mammals, Vol. 4: River Dolphins and the Larger Toothed Whales* (S.H. Ridgway & R. Harrison, eds), pp. 119–144. Academic Press, London.

Brownell, R.L.J. (1984) Review of the reproduction in platanistid dolphins. In: *Reproduction in Whales, Dolphins and Porpoises* (W.F. Perrin, R.L.J. Brownell & D.P. DeMaster, eds), pp. 149–158. Special Issue No. 6. International Whaling Commission, Cambridge, UK.

Brownell, R.L.J. (1989) Franciscan *Pontoporia blainvillei* (Gervais and d'Orbigny, 1984). In: *Handbook of Marine Mammals, Vol. 4: River Dolphins and the Larger Toothed Whales* (S.H. Ridgway & R. Harrison, eds), pp. 45–67. Academic Press, London.

Brownell, R.L.J. & Cipriano, F. (1999) Dusky dolphin *Lagenorhynchus obscurus*. In: *Handbook of Marine Mammals, Vol. 6: The Second Book of Dolphins and Porpoises* (S.H. Ridgway & R. Harrison eds), pp. 85–104. Academic Press, London.

Brownell, R.L.J. & Clapham, P.J. (1999) Burmeister's porpoise *Phocaena spinipinnis* Burmeister, 1865. In: *Handbook of Marine Mammals, Vol. 6: The Second Book of Dolphins and Porpoises* (S.H. Ridgway & R. Harrison, eds), pp. 393–410. Academic Press, London.

Brownell, R.L.J. & Ralls, K. (1986) Potential for sperm competition in baleen whales. In: *Behaviour of Whales in Relation to Management* (G.P. Donovan, ed.), pp. 97–112. Special Issue No. 8. International Whaling Commission, Cambridge, UK.

Brownell, R.L.J., Walker, W.A. & Forney, K.A. (1999) Pacific white-sided dolphin *Lagenorhynchus obliquidens*, 1865. In: *Handbook of Marine Mammals, Vol. 6: The Second Book of Dolphins and Porpoises* (S.H. Ridgway & R. Harrison, eds), pp. 57–84. Academic Press, London.

Byers, J.A. (1997) *American Pronghorn: Social Adaptations and the Ghosts of Predators Past*. University of Chicago Press, Chicago.

Calder, W.A. (1984) *Size, Function and Life History*. Harvard University Press, Cambridge, MA.

Caldwell, D.K. & Caldwell, M.C. (1989) Pygmy sperm whale *Kogia breviceps* (de Blaiville, 1838); dwarf sperm whale *Kogia simus* (Owen, 1866). In: *Handbook of Marine Mammals, Vol. 4: River Dolphins and the Larger Toothed Whales* (S.H. Ridgway & R. Harrison, eds), pp. 235–260. Academic Press, London.

Caldwell, D.K., Caldwell, M.C. & Rice, D.W. (1966) Behavior of the sperm whale. In: *Whales, Dolphins and Porpoises* (K.S. Norris, ed.), pp. 617–717. University of California Press, Berkeley.

Campagna, C. & Le Boeuf, B.J. (1988a) Reproductive behaviour of southern sea lions. *Behaviour* 104, 233–262.

Campagna, C. & Le Boeuf, B.J. (1988b) Thermoregulatory behaviour of southern sea lions and its effect on mating strategies. *Behaviour* 107, 72–90.

Campagna, C., Bisioli, C., Quintana, F., Perez, F. & Vila, A. (1992) Group breeding in sea lions: pups survive better in colonies. *Animal Behaviour* 43, 541–548.

Carey, P.W. (1992) Agonistic behaviour in female New Zealand fur seals, *Arctocephalus forsteri*. *Ethology* 92, 70–80.

Carrick, R., Csordas, S.E. & Ingham, S.E. (1962) Studies on the southern elephant seal, *Mirounga leonina* (L.) IV. Breeding and development. *CSIRO Wildlife Research* 7, 161–197.

Charnov, E.L. (1993) *Life History Invariance*. Oxford University Press, Oxford.

Christal, J. (1999) *An analysis of sperm whale social structure; patterns of association and genetic relatedness*. PhD Thesis, Dalhousie University, Halifax, Nova Scotia.

Chu, K.C. (1988) Dive times and ventilation patterns of singing humpback whales (*Megaptera Novaeangliae*). *Canadian Journal of Zoology* 66, 1322–1327.

Clapham, P.J. (1992) Age of attainment at sexual maturity in humpback whales, *Megaptera novaeangliae*. *Canadian Journal of Zoology* 70, 1470–1472.

Clapham, P.J. (1993) Social organization of humpback whales on a North Atlantic feeding ground. *Symposium Zoological Society of London* 66, 131–145.

Clapham, P.J. (1996) The social and reproductive biology of humpback whales: an ecological perspective. *Mammal Review* 26, 27–49.

Clapham, P.J. (1999) The humpback whale: seasonal feeding and breeding in baleen whale. In: *Cetacean Societies* (J. Mann, P.L. Tyack, R. Connor & H. Whitehead, eds), pp. 173–196. University of Chicago Press, Chicago.

Clapham, P.J. (2001) Why do ballen whales migrate? A response to Corkeron and Connor. *Marine Mammal Science* 17, 432–436.

Clapham, P.J. & Mayo, C.A. (1990) Reproduction of humpback whales (*Megaptera novaeangliae*) observed in the Gulf of Maine. In: *Individual Recognition of Cetaceans: Use of Photo-Identification and Other Techniques to Estimation Population Parameters* (P.S. Hammond, S.A. Mizroch & G.P. Donovan, eds), pp. 171–175. Special Issue No. 12. International Whaling Commission, Cambridge, UK.

Clapham, P.J. & Mead, J.G. (1999) *Megaptera novaeangliae*. *Mammalian Species* 604, 1–9.

Clapham, P.J. & Palsbøll, P.J. (1997) Molecular analysis of paternity shows promiscuous mating in female humpback whales (*Megaptera novaeangliae*, Borowski). *Proceedings of the Royal Society of London* B264, 95–98.

Clapham, P.J., Palsbøll, P.J., Mattila, D.K. & Vasquez, O. (1992) Composition and dynamics of humpback whale competitive groups in the West Indies. *Behaviour* 122, 182–194.

Clark, S.T. & Odell, D.K. (1999) Allometric relationships and sexual dimorphism in captive killer whales. (*Orcinus orca*). *Journal of Mammalogy* 80, 777–785.

Cleator, H.J., Stirling, I. & Smith, T.G. (1989) Underwater vocalizations of the bearded seal (*Erignathus barbatus*). *Canadian Journal of Zoology* 67, 1900–1910.

Clutton-Brock, T.H. (1988) *Reproductive Success*. University of Chicago Press, Chicago.

Clutton-Brock, T.H. (1989) Mammalian mating systems. *Proceedings of the Royal Society of London* B236, 339–372.

Clutton-Brock, T.H., Guinness, F.E. & Albon, S.D. (1982) *Red Deer*. University of Chicago Press, Chicago.

Clutton-Brock, T.H., Green, D., Hiraiwa-Hasegawa, M. & Albon, S.D. (1988) Passing the buck: resource defence, lek breeding and mate choice in fallow deer. *Behavioral Ecology and Sociobiology* **23**, 281–296.

Cockcroft, V.G. & Ross, G.J.B. (1990) Age, growth, and reproduction of bottlenose dolphins *Tursiops truncatus* from the east coast of Southern Africa. *Fishery Bulletin* **88**, 289–302.

Coltman, D.W., Bowen, W.D., Boness, D.J. & Iverson, S.J. (1997) Balancing foraging and reproduction in the male harbour seal, an aquatically mating pinniped. *Animal Behaviour* **54**, 663–678.

Coltman, D.W., Bowen, W.D., Iverson, S.J. & Boness, D.J. (1998b) The energetics of male reproduction in an aquatically mating pinniped: the harbour seal. *Physiological Zoology* **71**, 387–399.

Coltman, D.W., Bowen, W.D. & Wright, J.M. (1998a) Male mating success in an aquatically mating pinniped, the harbor seal (*Phoca vitulina*), assessed by microsatellite DNA markers. *Molecular Ecology* **7**, 627–638.

Connor, R.C., Smolker, R.A. & Richards, A.F. (1992a) Dolphin alliances and coalitions. In: *Coalitions and Alliances in Humans and Other Animals* (A.H. Harcourt & F.B.M. DeWall, eds), pp. 415–443. Oxford University Press, Oxford.

Connor, R.C., Smolker, R.A. & Richards, A.F. (1992b) Two levels of alliance formation among male bottlenose dolphins (*Tursiops* sp.). *Proceedings of the National Academy of Science* **89**, 987–990.

Connor, R.C., Richards, A.F., Smolker, R.A. & Mann, J. (1996) Patterns of female attractiveness in Indian Ocean bottlenose dolphins. *Behavior* **133**, 37–69.

Corkeron, P.J. & Connor, R.C. (1999) Why do baleen whales migrate? *Marine Mammal Science* **15**, 1228–1245.

Costa, D.P. (1991) Reproductive and foraging energetics of pinnipeds: implications for life history patterns. In: *The Behaviour of Pinnipeds* (D. Renouf, ed.), pp. 300–344. Chapman & Hall, London.

Costa, D.P. (1993) The relationship between reproductive and foraging energetics and the evolution of the Pinnipedia. *Symposia of the Zoological Society of London* **66**, 293–314.

Cox, C.R. & Le Boeuf, B.J. (1977) Female incitation of male competition: a mechanism in sexual selection. *American Naturalist* **111**, 317–335.

Craighead, L., Paetkau, D., Reynolds, H.V., Vyse, E.R. & Strobeck, C. (1995) Microsatellite analysis of paternity and reproduction in Arctic grizzly bears. *Journal of Heredity* **86**, 255–261.

Cranford, T.W. (1999) The sperm whale's nose: sexual selection on a grand scale? *Marine Mammal Science* **15**, 1133–1157.

Croxall, J.P. & Gentry, R.L. (1987) *Status, Biology and Ecology of Fur Seals*. NOAA Technical Report NMFS No. 51. US Department of Commerce, Washington, DC.

Curry, B.E. (1997) *Phylogenetic relationships among bottlenose dolphins (genus Tursiops) in a world-wide context*. PhD Thesis, Texas A & M University, Galveston.

da Silva, V.M.F. & Best, R.C. (1994) Tucuxi *Sotalia fluviatilis* (Gervais, 1853). In: *Handbook of Marine Mammals,*

Vol. 5: The First Book of Dolphins (S.H. Ridgway & R. Harrison, eds), pp. 43–69. Academic Press, London.

da Silva, V.M.F. & Best, R.C. (1996) *Sotalia fluviatilis. Mammalian Species* **527**, 1–7.

da Silva, J. & Terhune, J.M. (1988) Harbour seal grouping as an anti-predator strategy. *Animal Behaviour* **36**, 1309–1316.

Dahlheim, M.E. & Heyning, J.E. (1999) Killer whale *Orcinus orca* (Linnaeus, 1758). In: *Handbook of Marine Mammals, Vol. 6: The Second Book of Dolphins and Porpoises* (S.H. Ridgway & R. Harrison, eds), pp. 281–322. Academic Press, London.

Daniel, J.C.J. (1981) Delayed implantation in the northern fur seal (*Callorhinus ursinus*) and other pinnipeds. *Journal of Reproductive Fertilization and Fertility* **29**, 35–50.

Darling, J.D. (1983) *Migrations, abundance and behavior of Hawaiian humpback whales, Megaptera novaeangliae (Borowski)*. PhD Thesis, University of California, Santa Cruz.

Darling, J.D. & Bérubé, M. (2001) Interactions of singing humpback whales with other males. *Marine Mammal Science* **17**, 570–584.

David, J.H.M. (1973) The behaviour of the bontebok, *Damaliscus dorcas dorcas*, (Pallas 1766), with special reference to territorial behaviour. *Zeitschrift für Tierpsychologie* **107**, 38–107.

David, J.H.M. (1987) South African fur seal, *Arctocephalus pusillus pusillus*. In: *Status, Biology, and Ecology of Fur Seals* (J.P. Croxall & R.L. Gentry, eds), pp. 65–73. US Department of Commerce, Washington, DC.

David, J.H.M. & Rand, R.W. (1986) Attendance behaviour of South African fur seals. In: *Fur Seals: Maternal Strategies on Land and at Sea* (R.L. Gentry & G.L. Kooyman, eds), pp. 126–141. Princeton University Press, Princeton, NJ.

Davies, J.L. (1949) Observations of the grey seal (*Halichoerus grypus*) at Ramsey Island, Pembrokeshire. *Proceedings Zoological Society of London* **119**, 673–692.

Davies, N.B. (1991) Mating systems. In: *Behavioural Ecology* (J.R. Krebs & N.B. Davies, eds), 3rd edn, pp. 263–299. Blackwell Scientific Publications, London.

Davis, M.B. & Renouf, D. (1987) Social behaviour of harbour seals, *Phoca vitulina*, on haulout grounds at Miquelon. *Canadian Field-Naturalist* **101**, 1–5.

DeLong, R.L. (1982) *Population biology of northern fur seals at San Miguel Island, California*. PhD Thesis, University of California, Berkeley.

Deutsch, C.J., Haley, M.P. & Le Boeuf, B.J. (1990) Reproductive effort of male northern elephant seals: estimates from mass loss. *Canadian Journal of Zoology* **68**, 2580–2593.

Eberhard, W.G. (1996) *Female Control: Sexual Selection by Cryptic Female Choice*. Princeton University Press, Princeton, NJ.

Emlen, S.T. & Oring, L.W. (1977) Ecology, sexual selection and the evolution of mating systems. *Science* **197**, 215–223.

English, A.W. (1977) Functional anatomy of the hands of the fur seals and sea lions. *American Journal of Anatomy* **147**, 1–18.

Estes, J. (1989) Adaptations for aquatic living by carnivores. In: *Carnivore Behavior, Ecology, and Evolution* (J.L. Gittleman, ed.), pp. 242–282. Cornell University Press, Ithaca.

Evans, W.E. (1994) Common dolphin, white-bellied porpoise —*Delphinus delphis* Linnaeus, 1758. In: *Handbook of Marine Mammals, Vol. 5: The First Book of Dolphins* (S.H. Ridgway & R. Harrison, eds), pp. 191–224. Academic Press, London.

Fay, F.H. (1982) Ecology and biology of the Pacific walrus, *Odobenus rosmarus divergens* Illiger. *North American Fauna* **74**, 279 pp.

Fay, F.H., Ray, G.C. & Kibalichich, A.A. (1984) Time and location of mating and associated behavior of the Pacific walrus, *Odobenus rosmarus divergens*, Illiger. In: *Soviet-American Cooperative Research on Marine Mammals. Vol. 1. Pinnipeds* (F.H. Fay & G.A. Fedoseev, eds), pp. 89–100. NOAA Technical Reports NMFS No. 12 National Technical Information Service, Springfield, VA.

Fleischer, L.A. (1987) Guadalupe fur seals, *Arctocephalus townsendi*. In: *Status, Biology, and Ecology of Fur Seals. Proceedings of an International Symposium and Workshop* (J.P. Croxall & R.L. Gentry, eds), pp. 43–48. NOAA Technical Report NMFS No. 51. National Technical Information Service, Springfield, VA.

Fontaine, P.M. & Barrette, C. (1997) Megatestes: anatomical evidence for sperm competition in the harbor porpoise. *Extrait de Mammalia* **61**, 65–71.

Francis, J.M. & Boness, D.J. (1991) The effect of thermoregulatory behavior on the mating system of the Juan Fernández fur seal, *Arctocephalus philippii*. *Behaviour* **119**, 104–127.

Francis, J.M., Boness, D.J. & Ochoa-Acuña, H. (1998) A protracted foraging and attendance cycle in female Juan Fernández fur seals. *Marine Mammal Science* **14**, 552–574.

Francis, R.C., Hare, S.E., Hollowed, A.B. & Wooster, W.S. (1998) Effect of interdecadal climate variability on the oceanic ecosystems of the NE Pacific. *Fisheries Oceanography* **7**, 1–21.

Frazer, J.F.D. & Huggett, St.G.A. (1973) Specific fetal growth rates of cetaceans. *Journal of Zoology* **169**, 111–126.

Gallo, J.P. (1994) *Factors affecting the population status of Guadalupe fur seal, Arctocephalus townsendi (Merriam, 1897) at Isla de Gualupe, Baja California, Mexico.* PhD Thesis, University of California, Santa Cruz.

Gamble, R. (1973) Some effects of exploration in reproduction of whales. *Journal of Reproduction and Fertility* **19**, 533–553.

Garshelis, D.L., Johnson, A.M. & Garshelis, J.A. (1984) Social organization of sea otters in Prince William Sound, Alaska. *Canadian Journal of Zoology* **62**, 2648–2658.

Gaskin, D.E. (1982) *The Ecology of Whales and Dolphins.* Heinemann, London.

Gaskin, D.E., Smith, G.J.D., Watson, P., Yasui, W.Y. & Yurick, D.B. (1984) Reproduction in the porpoises (*Phocoenidae*): implications for management. In: *Reproduction in Whales, Dolphins and Porpoises* (W.F. Perrin, R.L. Brownell & D.P. DeMaster, eds), pp. 135–148. Special Issue No. 6. International Whaling Commission, Cambridge, UK.

Gentry, R.L. (1970) *Social behavior of the Steller sea lion.* PhD Thesis, University of California, Santa Cruz.

Gentry, R.L. (1998) *Behavior and Ecology of the Northern Fur Seal.* Princeton University Press, Princeton, NJ.

Gentry, R.L. & Kooyman, G.L. (1986) *Fur Seals: Maternal Strategies on Land and at Sea.* Princeton University Press, Princeton, NJ.

Gisiner, R.C. (1985) *Male territorial and reproductive behavior in the Steller sea lion, Eumetopias jubatus.* PhD Thesis, University of California, Santa Cruz.

Glockner-Ferrari, D.A. & Ferrari, M.J. (1990) Reproduction in the humpback whale (*Megaptera novaeangliae*) in Hawaiian waters, 1975–1988: the life history, reproductive rates and behavior of known individuals identified through surface and underwater photography. In: *Individual Recognition of Cetaceans: Use of Photo-Identification and Other Techniques to Estimation Population Parameters* (P.S. Hammond, S.A. Mizroch & G.P. Donovan, eds), pp. 161–169. Special Issue No. 12. International Whaling Commission, Cambridge, UK.

Goa, A. & Zhou, K. (1993) Growth and reproduction of three populations of finless porpoise, *Noephocaena phocaenoides*, in Chinese waters. *Aquatic Mammals* **19**, 3–12.

Godsell, J. (1988) Herd formation and haul-out behaviour in harbour seals (*Phoca vitulina*). *Journal of Zoology, London* **215**, 83–98.

Godsell, J. (1991) The relative influence of age and weight on the reproductive behaviour of male grey seals *Halichoerus grypus*. *Journal of Zoology, London* **224**, 537–551.

Goldsworthy, S.D. (1992) *Maternal care in three species of southern fur seal (Arctocephalus spp.).* PhD Thesis, Monash University, Australia.

Goldsworthy, S.D. (1995) Differential expenditure of maternal resources in Antarctic fur seals, *Arctocephalus gazella*, at Heard Island, southern Indian Ocean. *Behavioral Ecology* **6**, 218–228.

Gomiendo, M., Harcourt, A.H. & Roldan, E.R.S. (1998) Sperm competition in mammals. In: *Sperm Competition and Sexual Selection* (T.H. Birkhead & A.P. Möller, eds), pp. 467–755. Academic Press, London.

Goodall, R.N.P. (1994) Commerson's dolphin—*Cephalorhynchus commersonii* (Lacépe 1804). In: *Handbook of Marine Mammals, Vol. 5: The First Book of Dolphins* (S.H. Ridgway & R. Harrison, eds), pp. 269–287. Academic Press, London.

Goodall, R.N.P., Norris, K., Galeazzi, A.R., Oporto, J.A. & Cameron, I.S. (1988) On the Chilean dolphin, *Cephalorhynchus eutropia* (Gray, 1846). In: *Biology of the Genus Cephalorhynchus* (R.L.J. Brownell & G.P. Donovan, eds), pp. 197–257. Special Issue No. 9. International Whaling Commission, Cambridge, UK.

Gosling, L.M. (1991) The alternative mating strategies of male topi, *Damaliscus lunatus*. *Applied Animal Behaviour Science* **29**, 107–119.

Gosling, L.M. & Petrie, M. (1990) Lekking in topi: a consequence of satellite behaviour by small males at hotspots. *Animal Behaviour* **40**, 272–287.

Gowans, S. (1999) *Social organization and population structure of nothern bottlenose whales in the Gully*. PhD Thesis,
Dalhousie University, Nova Scotia.

Haley, M.P., Deutsch, C.J. & Le Boeuf, B.J. (1994) Size,
dominance and copulatory success in male northern elephant seals, *Mirounga angustirostris*. *Animal Behaviour*
48, 1249–1260.

Hamilton, J.E. (1934) The southern sea lion, *Otaria byronia*.
Discovery Reports **8**, 269–318.

Hamilton, P.K., Knowlton, A.R., Marx, M.K. & Kraus, S.D.
(1998) Age structure and longevity in North Atlantic right
whales (*Eubalaena glacialis*) *and* their relation to reproduction. *Marine Ecology Progress Series* **171**, 285–292.

Hamilton, W.D. (1971) Geometry of the selfish herd. *Journal
of Theoretical Biology* **31**, 295–311.

Hammill, M.O., Lydersen, C., Ryg, M. & Smith, T.G.
(1991) Lactation in the ringed seal (*Phoca hispida*) based
on cross-sectional sampling. *Canadian Journal of Fisheries
and Aquatic Sciences* **48**, 2417–2476.

Hanggi, E.B. & Schusterman, R.J. (1994) Underwater
acoustic displays and individual variation in male harbour
seals, *Phoca vitulina*. *Animal Behaviour* **48**, 1275–1283.

Harcourt, R. (1992) Factors affecting early mortality in the
South American fur seal (*Arctocephalus australis*) in Peru:
density-related and predation. *Journal of Zoology, London*
226, 259–270.

Harrison, R.J., Brownell, R.L.J. & Boice, R.C. (1972) Reproduction and gonadal appearances in some odontocetes. In:
Functional Anatomy of Marine Mammals (R.J. Harrison,
ed.), Vol. 1, pp. 361–429. Academic Press, London.

Hartman, D.S. (1979) *Ecology and Behavior of the Manatee
(Trichechus manatus) in Florida*. Special Publication No. 5.
American Society of Mammalogists, Lawrence, KS.

Hay, K.A. (1984) *The life history of the narwhal (Monodon
monoceros L.) in the eastern Canadian Arctic*. PhD Thesis,
McGill University, Montreal.

Hay, K.A. & Mansfield, A.W. (1989) Narwhal—*Monodon
monoceros* Linnaeus, 1758. In: *Handbook of Marine
Mammals, Vol. 4: River Dolphins and the Larger Toothed
Whales* (R. Ridgway & R. Harrison, eds), pp. 145–176.
Academic Press, London.

Heath, C.B. (1989) *The behavioral ecology of the California
sea lion*. PhD Thesis. University of California, Santa Cruz.

Heimlich-Boran, J.R. (1993) *Social organization of the short-
finned pilot whale, Globicephala macrorhynchus, with special
reference to the comparative social ecology of delphinids*. PhD
Thesis, University of Cambridge, Cambridge, UK.

Herman, L.M. & Tavolga, W.N. (1980) The communication
systems of cetaceans. In: *Cetacean Behavior: Mechanisms
and Functions* (L.M. Herman, ed.), pp. 149–207. Wiley
Interscience, New York.

Hernandez, P., Reynolds III, J.E., Marsh, H. & Marmontel,
M. (1995) Age and seasonality in spermatogenesis of Florida
manatees. In: *Population Biology of the Florida Manatee*
(T.J. O'Shea, B.B. Ackerman & H.F. Percival, eds),
pp. 84–97. US Department of Interior, Washington, DC.

Hewer, H.R. (1957) A Hebridean breeding colony of grey
seals, *Halichoerus grypus* (Fab.), with comparative notes on

the grey seals of Ramsey Island, Pembrokeshire. *Proceedings
of the Royal Society of London* **128**, 23–66.

Heyning, J.E. (1984) Functional morphology involved in
intraspecific fighting of the beaked whales. *Mesoplodon
carlhubbsi*. *Canadian Journal of Zoology* **62**, 1645–1654.

Heyning, J.E. & Perrin, W.F. (1994) Evidence for two
species of common dolphins (genus *Delphinus*) from the
eastern North Pacific. *Contributions in Science* **442**, 1–35.

Higgins, L.V. (1990) *Reproductive behavior and maternal
investment of Australian sea lions*. PhD Thesis, University
of California, Santa Cruz.

Higgins, L.V. & Tedman, R.A. (1990) Effect of attacks by
male Australian sea lions, *Neophoca cinerea*, on mortality of
pups. *Journal of Mammalogy* **71**, 617–619.

Higgins, L.V., Costa, D.P., Huntley, A.C. & Le Boeuf, B.J.
(1988) Behavioral and physiological measurements of
maternal investment in the Steller sea lion, *Eumetopias
jubatus*. *Marine Mammal Science* **4**, 44–58.

Hoelzel, A.R., Le Boeuf, B.J., Reiter, J. & Campagna, C.
(1999) Alpha male paternity in elephant seals. *Behavioral
Ecology and Sociobiology* **46**, 298–306.

Höglund, J. & Alatalo, R.V. (1995) *Leks*. Princeton University Press, Princeton, NJ.

Hohn, A.A., Read, A.J., Fernandez, S., Vidal, O. & Findley,
L.T. (1996) Life history of the vaquita, *Phocoena sinus*
(Phocoenidae, Cetacea). *Journal of Zoology, London* **239**,
235–251.

Hooker, S.K. (1999) *Resource and habitat use of northern
bottlenose whales in the Gully: ecology, diving and ranging
behaviour*. PhD Thesis, Dalhousie University, Nova Scotia.

Houck, W.J. & Jefferson, T.A. (1999) Dall's porpoise
Phocoenoides dalli (True, 1885). In: *Handbook of Marine
Mammals, Vol. 6: River Dolphins and the Larger Toothed
Whales* (S.H. Ridgway & R. Harrison, eds), pp. 443–472.
Academic Press, London.

Irvine, A.B., Caffin, J.E. & Kochman, H.I. (1982) Aerial
surveys of manatees and dolphins in western peninsular
Florida. *Fishery Bulletin* **80**, 621–630.

Jameson, R.J. (1989) Movements, home range, and territories
of male sea otters off central California. *Marine Mammal
Science* **5**, 159–172.

Jarman, P.J. (1974) The social organization of antelope in
relation to their ecology. *Behaviour* **48**, 215–267.

Jefferson, T.A. (1990) Sexual dimorphism and development
of external features in Dall's porpoise *Phocoenoides dalli*.
Fishery Bulletin **88**, 119–132.

Jefferson, T.A., Pitman. R.L., Leatherwood, S. & Dollar,
M.L.L. (1997) Development and sexual variation in the
external appearance of Fraser's dolphins (*Lagenodelphis
hosei*). *Aquatic Mammals* **23**, 145–153.

Jennions, M.D. & Petrie, M. (1997) Variation in mate choice
and mating preferences: a review of causes and consequences.
Biological Review **72**, 283–327.

Jones, M.L., Swartz, S.L. & Leatherwood, S. (1984) *The
Gray Whale*. Academic Press, New York.

Kasuya, T. (1985) Review of the biology and exploitation of
striped dolphins in Japan. *Journal of Cetacean Research
Management* **1**, 81–100.

Kasuya, T. (1999) Finless porpoise *Neophocaena phocaenoides* (G. Cuvier, 1829). In: *Handbook of Marine Mammals, Vol. 6: River Dolphins and the Larger Toothed Whales* (S.H. Ridgway & R. Harrison, eds), pp. 411–442. Academic Press, London.

Kasuya, T. & Marsh, H. (1984) Life history and reproductive biology of the short-finned pilot whale, *Globicephala macrorhynchus*, of the Pacific coast of Japan. In: *Reproduction in Whales, Dolphins and Porpoises* (W.F. Perrin, R.L.J. Brownell & D.P. DeMaster, eds), pp. 259–310. Special Issue No. 6. International Whaling Commission, Cambridge, UK.

Kasuya, T., Marsh, H. & Amino, A. (1993) Non-reproductive mating in short-finned pilot whales. In: *Biology of Northern Hemisphere Pilot Whales* (G.P. Donovan, C.H. Lockyer & A.R. Martin, eds), pp. 425–437. Special Issue No. 14. International Whaling Commission, Cambridge, UK.

Kaufman, G.W., Siniff, D.B. & Reichle, R. (1975) Colony behavior of Weddell seals, *Leptonychotes weddelli*, at Hutton Cliffs, Antarctica. *Rapport et Proces-Verbaux Des Reunions, Consiel International Exploration de Mer* **169**, 228–246.

Kelly, B.P. & Quakenbush, L.T. (1990) Spatiotemporal use of lairs by ringed seals (*Phoca hispida*). *Canadian Journal of Zoology* **68**, 2503–2512.

Kelly, B.P. & Wartzok, D. (1996) Ringed seal diving behavior in the breeding season. *Canadian Journal of Zoology* **74**, 1547–1555.

Kenagy, G.J. & Trombulak, S.C. (1986) Size and function of mammalian testis in relation to body size. *Journal of Mammalogy* **67**, 1–22.

Kirby, V.L. & Ridgway, S.H. (1984) Hormonal evidence of spontaneous ovulation in captive dolphins, *Tursiops truncatus* and *Delphinus delphis*. In: *Reproduction in Whales, Dolphins, and Porpoises* (W.P. Perrin, R.L. Brownell & D.P. DeMaster, eds), pp. 459–464. Special Issue No. 6. International Whaling Commission, Cambridge, UK.

Klumov, S.K. (1962) The right whales in the Pacific Ocean. *Trudy Inst Okeanol* **58**, 202–297 (in Russian).

Kooyman, G.L. (1989) *Diverse Divers*. Springer-Verlag, Berlin.

Koski, W.R., Davis, R.A., Miller, G.W. & Withrow, D.E. (1993) Reproduction. In: *The Bowhead Whale* (J.J. Burns, J.J. Montague & C.J. Cowless, eds), pp. 239–274. Special Publication No. 2. Society for Marine Mammalogy, Lawrence, KS.

Kovacs, K.M. (1990) Mating strategies in male hooded seals (*Cystophora cristata*). *Canadian Journal of Zoology* **68**, 2499–2502.

Kovacs, K.M. (1995) Harp and hooded seals—a case study in the determinants of mating systems in pinnipeds. In: *Whales, Seals, Fish and Man* (A.S. Blix, L. Walloe & O. Ultang, eds), pp. 329–335. Elsevier, Amsterdam.

Kovacs, K.M. & Lavigne, D.M. (1986) Maternal investment and neonatal growth in phocid seals. *Journal of Animal Ecology* **55**, 1035–1051.

Kovacs, K.N., Lydersen, C., Hammill, M. & Lavigne, D.M. (1996) Reproductive effort of male hooded seals (*Cystophora cristata*): estimates from mass loss. *Canadian Journal of Zoology* **74**, 1521–1530.

Kraus, S.D., Hamilton, P.K., Kenney, R.D., Knowlton, A.R. & Slay, C.K. (in press) Status and trends in reproduction of the North Atlantic right whale. *Journal of Cetacean Research and Management*, in press.

Kruijt, J.P. & de Vos, G.J. (1988) Individual variation in reproductive success in male black grouse, *Tetrao tetrix* L. In: *Reproductive Success Studies of Individual Variation in Contrasting Breeding Systems* (T.H. Clutton-Brock, ed.), pp. 279–290. University of Chicago Press, Chicago.

Laws, R.M. (1962) Some effects of whaling on the southern stocks of baleen whales. In: *The Exploitation of Natural Animal Populations* (E.D. LeCren & M.W. Holdgate, eds), pp. 137–158. Oxford University Press, Oxford.

Le Boeuf, B.J. (1972) Sexual behaviour in the northern elephant seal, *Mirounga angustirostris*. *Behaviour* **41**, 1–25.

Le Boeuf, B.J. (1974) Male–male competition and reproductive success in elephant seals. *American Zoologist* **14**, 163–176.

Le Boeuf, B.J. (1991) Pinniped mating system on land, ice, and in the water; emphasis on the Phocidae. In: *Behaviour of Pinnipeds* (D. Renouf, ed.), pp. 45–65. Chapman & Hall, London.

Le Boeuf, B.J. & Peterson, R.S. (1969) Social status and mating activity in elephant seals. *Science* **163**, 91–93.

Leatherwood, S. & Reeves, R.R. (1983) *The Sierra Club Handbook of Whales and Dolphins*. Sierra Club Books, San Francisco.

Leatherwood, S., Reeves, R.R., Perrin, W.F. & Evans, W.E. (1988) *Whales, Dolphins, and Porpoises of the Eastern North Pacific and Adjacent Arctic Waters. A Guide to Their Identification*. Dover Publications, New York.

LeDuc, R.G., Perrin, W.F. & Dizon, A.E. (1999) Phylogenetic relationships among the delphinid cetaceans based on full cytochrome B sequences. *Marine Mammal Science* **15**, 619–648.

Leuthold, W. (1970) Observations on the social organization of impala (*Aepyceros melampus*). *Zeitschrift für Tierpsychologie* **27**, 693–719.

Lidgard, D.C., Boness, D.J. & Bowen, W.D. (in press). A novel mobile approach to investigating grey seal male mating tactics. *Journal of Zoology* (in press).

Lockyer, C. (1976) Body weights of some species of large whales. *Journal Des Conseil International pour L'Exploration de la Mer* **36**, 259–273.

Lockyer, C. (1984) Review of baleen whale (Mysticeti) reproduction and implications for management. In: *Reproduction of Whales, Dolphins and Porpoises* (W.F. Perrin, R.L. Brownell & D.P. Demaster, eds), pp. 27–50. Special Issue No. 6. International Whaling Commission, Cambridge, UK.

Lockyer, C., Goodall, R.N.P. & Galeazzi, A.R. (1988) Age and body-length characteristics of *Cephalorhynchus commersonii* from incidentally-caught specimens off Tierra Del Fuego. In: *The Biology of the Genus Cephalorhynchus* (R.L.J. Brownell & G.P. Donovan, eds), pp. 103–118. Special Issue No. 9. International Whaling Commission, Cambridge, UK.

Lott, D.F. (1991) *Intraspecific Variation in the Social Systems of Wild Vertebrates*. Cambridge University Press, Cambridge, UK.

Loughrey, A.G. (1959) Preliminary investigation of the Atlantic walrus, *Odobenus rosmarus rosmarus* (Linnaeus). *Wildlife Management Bulletin* **14**, 1–119.

Lydersen, C., Hammill, M.O. & Kovacs, K.M. (1994) Activity of lactating ice-breeding grey seals, *Halichoerus grypus*, from the Gulf of St Lawrence, Canada. *Animal Behaviour* **48**, 1417–1425.

Lydersen, C. & Kovacs, K.M. (1993) Diving behaviour of lactating harp seal, *Phoca groenlandica*, females from the Gulf of St Lawrence, Canada. *Animal Behaviour* **46**, 1213–1221.

McCann, C. (1974) Body scarring on cetacea–odontocetes. *Scientific Reports of the Whale Research Institute* **26**, 145–155.

McCann, T.S. (1980) Territoriality and breeding behaviour of adult male Antarctic fur seal, *Arctocephalus gazella*. *Journal of Zoology, London* **192**, 295–310.

McCann, T.S. (1981) Aggression and sexual activity of male southern elephant seals, *Mirounga leonina*. *Journal of Zoology, London* **195**, 295–310.

McLaren, I.A. (1967) Seals and group selection. *Ecology* **48**, 104–110.

McLaren, I.A. (1993) Growth in pinnipeds. *Biological Review* **68**, 1–79.

MacLeod, C.D. (1998) Intraspecific scarring in odontocete cetaceans: an indicator of male 'quality' in aggressive social interactions. *Journal of Zoology, London* **244**, 71–77.

McRae, S.B. & Kovacs, K.M. (1994) Paternity exclusion by DNA fingerprinting, and mate guarding in the hooded seal *Cystophora cristata*. *Molecular Ecology* **3**, 101–107.

Madsen, C.J. & Herman, L.M. (1980) The social ecology of inshore odontocetes. In: *Cetacean Behavior: Mechanisms and Functions* (L.M. Herman, ed.), pp. 101–147. Wiley & Sons, New York.

Mansfield, A.W., Smith, T.G. & Beck, B. (1975) The narwhal, *Monodon monoceros*, in eastern Canadian waters. *Journal of the Fisheries Research Board of Canada* **32**, 1041–1046.

Marmontel, M. (1995) Age and reproduction in female Florida manatees. In: *Population Biology of the Florida Manatee* (T.J. O'Shea, B.B. Ackerman & H.F. Percival, eds), pp. 98–119. US Department of Interior, Washington, DC.

Marsh, H. (1980) Age determination of the dugong (*Dugong dugong*) (Muller) in northern Australia and its biological implications. In: *Age Determination of Toothed Whales and Sirenians* (W.F. Perrin & A.C. Myrick, eds), pp. 81–201. Special Issue No. 3. International Whaling Commission, Cambridge, UK.

Marsh, H. (1995) Life history and patterns of breeding and population dynamics of the dugong. In: *Population Biology of the Florida Manatee* (T.J. O'Shea, B.B. Ackerman & H.F. Percival, eds), Vol. 1, pp. 75–83. US Department of the Interior, Washington, DC.

Marsh, H. & Kasuya, T. (1984) Ovarian changes in the short-finned pilot whale, *Globicephala macrorhynchus*, off the Pacific coast of Japan. In: *Reproduction of Whales, Dolphins and Porpoises* (W.F. Perrin, R.L. Brownell & D.P. Demaster, eds), pp. 311–335. Special Issue No. 6. International Whaling Commission, Cambridge, UK.

Marsh, H., Prince, R.I.T., Saalfeld, W.K. & Shepherd, R. (1994) The distribution and abundance of dugong in Shark Bay, Western Australia. *Australian Wildlife Research* **21**, 149–161.

Marsh, H. & Saalfeld, W.K. (1989) Distribution and abundance of dugong in the northern Great Barrier Reef Marine Park. *Australian Wildlife Research* **16**, 429–440.

Mead, J.G. (1984) Survey of reproductive data for the beaked whales (Ziphiidae). In: *Reproduction of Whales, Dolphins and Porpoises* (W.F. Perrin, R.L. Brownell & D.P. Demaster, eds), pp. 91–97. Special Issue No. 6. International Whaling Commission, Cambridge, UK.

Mead, J.G. (1989a) Beaked whales of the genus *Mesoplodon*. In: *Handbook of Marine Mammals, Vol. 4: River Dolphins and the Larger Toothed Whales* (S.H. Ridgway & R. Harrison, eds), pp. 349–430. Academic Press, London.

Mead, J.G. (1989b) Bottlenose whales *Hyperoodon ampullatus* (Forster, 1770) and *Hyperoodon planifrons* (Flower, 1882). In: *Handbook of Marine Mammals, Vol. 4: River Dolphins and the Larger Toothed Whales* (S.H. Ridgway & R. Harrison, eds), pp. 321–348. Academic Press, London.

Mesnick, S.L. (1997) Sexual alliances: evidence and evolutionary implications. In: *Feminism and Evolutionary Biology* (P.A. Gowaty, ed.), pp. 207–160. Chapman & Hall, New York.

Miller, E.H. (1975) Annual cycle of fur seals, *Arctocephalus forsteri* (Lesson), on the Open Bay Islands, New Zealand. *Pacific Science* **29**, 139–152.

Miller, E.H. (1991) Communication in pinnipeds, with special reference to non-acoustic signaling. In: *The Behaviour of Pinnipeds* (D. Renouf ed.), pp. 129–235. Chapman & Hall, London.

Miller, E.H. & Boness, D.J. (1983) Summer behaviour of the Atlantic walrus (*Odobenus rosmarus rosmarus*) on Coats Island, NWT. *Zeitschrift für Säugetierkunde* **48**, 298–313.

Miyazaki, N. & Perrin, W.F. (1994) Rough-toothed dolphin— *Steno bredanensis* (Lesson, 1828). In: *Handbook of Marine Mammals, Vol. 6: The Second Book of Dolphins and Porpoises* (S.H. Ridgway & R. Harrison, eds), pp. 1–22. Academic Press, London.

Mizroch, S.A. & York, A.E. (1984) Have pregnancy rates of southern hemisphere fin whales (*Balaenoptyra physalus*) increased? In: *Reproduction of Whales, Dolphins and Porpoises* (W.F. Perrin, R.L. Brownell & D.P. Demaster, eds), pp. 401–410. Special Issue No. 6. International Whaling Commission, Cambridge, UK.

Mizroch, S.A., Rice, D.W. & Breiwick, J.M. (1984) The blue whale, *Balaenoptera musculus*. *United States National Marine Fisheries Service Marine Fisheries Review* **46**, 15–19.

Mobley, J.R. & Herman, L.M. (1985) Transience of social affiliations among humpback whales (*Megaptera novaeangliae*) on the Hawaiian wintering grounds. *Canadian Journal of Zoology* **63**, 763–772.

Møller, A.P. & Pomiankowski, A. (1993) Why have birds got multiple sexual ornaments? *Behavioral Ecology and Sociobiology* **32**, 167–176.

Moors, T.L. (1997) *Is 'menage a trois' important in dolphin mating systems? Behavioral patterns of breeding female bottlenose dolphins*. MS Thesis, University of California, Santa Cruz.

Morejohn, G.V., Loeb, V. & Baltz, D.M. (1973) Coloration and sexual dimorphism in the Dall porpoise. *Journal of Mammalogy* **54**, 977–982.

Morin, P.A., Wallis, J., Moore, J.J. & Woodruff, D.S. (1994) Paternity exclusion in a community of wild chimpanzees using hypervariable simple sequence repeats. *Molecular Ecology* **3**, 469–478.

Muelbert, M.M. & Bowen, W.D. (1993) Duration of lactation and postweaning changes in mass and body composition of harbour seal, *Phoca vitulina*, pups. *Canadian Journal of Zoology* **71**, 1405–1414.

Myrick, A.C., Hohn, A.A., Barlow, J. & Sloan, P.A. (1986) Reproductive biology of female spotted dolphins, *Stenella attenuata*, from the eastern tropical Pacific. *Fishery Bulletin* **84**, 247–259.

Norris, K.S., Wursig, B., Wells, R. & Wursig, M. (1994) *The Hawaiian Spinner Dolphin*. University of California Press, Berkeley, CA.

Nunn, C.L. (1999) The number of males in primate social groups: a comparative test of the socioecological model. *Behavioral Ecology and Sociobiology* **46**, 1–13.

Odell, D.K. & McClune, K.M. (1999) False killer whales—*Psuedorca crassidens* (Owen, 1846). In: *Handbook of Marine Mammals, Vol. 6: The Second Book of Dolphins and Porpoises* (D.K. Ridgway & R. Harrison, eds), pp. 213–244. Academic Press, London.

Oftedal, O.T. (1993) The adaptation of milk secretion to the constraints of fasting in bears, seals, and baleen whales. *Journal of Dairy Science* **76**, 3234–3246.

Oftedal, O.T. (1997) Lactation in whales and dolphins: evidence of divergence between baleen and toothed species. *Journal of Mammary Gland Biology and Neoplasia* **2**, 205–230.

Oftedal, O.T., Boness, D.J. & Tedman, R.A. (1987) The behavior, physiology, and anatomy of lactation in the Pinnipedia. *Current Mammalogy* **1**, 175–245.

Oftedal, O.T., Bowen, W.D. & Boness, D.J. (1993) Energy transfer by lactating seals and nutrient deposition in their pups during the four days from birth to weaning. *Physiological Zoology* **66**, 412–436.

Oftedal, O.T., Bowen, W.D., Widdowson, E. & Boness, D.J. (1991) The prenatal molt and its ecological significance in hooded and harbor seals. *Canadian Journal of Zoology* **69**, 2489–2493.

Olesiuk, P.K., Bigg, M.A. & Ellis, G.M. (1990) Life history and population dynamics of resident killer whales (*Orcinus orca*) in the coastal waters of British Columbia and Washington State. In: *Individual Recognition of Cetaceans: Use of Photo-Identification and Other Techniques to Estimate Population Parameters* (P.S. Hammond, S.A. Mizroch & G.P. Donovan, eds), pp. 209–244. Special Issue No. 12. International Whaling Commission, Cambridge, UK.

Omura, H., Ohsumi, S., Nemoto, T., Nasu, K. & Kasuya, T. (1969) Black right whales in the North Pacific. *Scientific Reports of the Whale Research Institute* **21**, 1–77.

Ono, K.A., Boness, D.J. & Oftedal, O.T. (1987) The effect of a natural environmental disturbance on maternal investment and pup behavior in the California sea lion. *Behavioral Ecology and Sociobiology* **21**, 109–118.

Orians, G.H. (1969) On the evolution of mating systems in birds and mammals. *American Naturalist* **103**, 589–603.

Oshea, T.J. & Hartley, W.C. (1995) Reproduction and early-age survival of manatees at Blue Spring, Upper St Johns River, Florida. In: *Population Biology of the Florida Manatee* (T.J. O'Shea, B.B. Ackerman & H.F. Percival, eds), Vol. 1, pp. 157–170. US Department of the Interior, Washington, DC.

Payne, R. (1986) Long-term behavioral studies of the southern right whale (*Eubalaena australis*). In: *Right Whales: Past and Present Status* (R.L.J. Brownell, P.B. Best & J.H. Prescott, eds), pp. 161–167. Special Issue No. 10. International Whaling Commission, Cambridge, UK.

Payne, R. & McVay, S. (1971) Songs of humpback whales. *Science* **173**, 585–597.

Payne, R., Rowntree, V., Perkins, J.S., Cooke, J.G. & Lankester, K. (1990) Population size, trends and reproductive parameters of right whales (*Eubalaena australis*) off Peninsula Valdes, Argentina. In: *Individual Recognition of Cetaceans: Use of Photo-Identification and Other Techniques to Estimate Population Parameters* (P.S. Hammond, S.A. Mizroch & G.P. Donovan, eds), pp. 271–278. Special Issue No. 12. International Whaling Commission, Cambridge, UK.

Pemberton, J.M., Albon, S.D., Guinness, F.E., Clutton-Brock, T.H. & Dover, G.A. (1992) Behavioral estimates of male mating success tested by DNA fingerprinting in a polygynous mammal. *Behavioral Ecology* **3**, 66–75.

Perrin, W.F. (1998) *Stenella longirostris. Mammalian Species* **599**, 1–7.

Perrin, W.F. & Gilpatrick, J.W.J. (1994) Spinner dolphin *Stenella longirostris* (Gray, 1828). In: *Handbook of Marine Mammals, Vol. 5: The First Book of Dolphins* (S.H. Ridgway & R. Harrison, eds), pp. 99–128. Academic Press, London.

Perrin, W.F. & Hohn, A.A. (1994) Pantropical spotted dolphin *Stenella attenuata*. In: *Handbook of Marine Mammals, Vol. 5: The First Book of Dolphins* (S.H. Ridgway & R. Harrison, eds), pp. 71–98. Academic Press, London.

Perrin, W.F. & Reilly, S.B. (1984) Reproductive parameters of dolphins and small whales of the family Delphinidae. In: *Reproduction of Whales, Dolphins and Porpoises* (W.F. Perrin, R.L.J. Brownell & D.P. DeMaster, eds), pp. 97–133. Special Issue No. 6. International Whaling Commission, Cambridge, UK.

Perrin, W.F., Coe, J.M. & Zweifel, J.R. (1976) Growth and reproduction of the spotted porpoise, *Stenella attenuata*, in the offshore easter tropical Pacific. *Fisheries Bulletin* **74**, 229–269.

Perrin, W.F., Brownell, R.L. & DeMaster, D.P., eds (1984) *Reproduction of Whales, Dolphins and Porpoises* Special Issue No. 6. International Whaling Commission, Cambridge, UK.

Perrin, W.F., Miyazaki, N. & Kasuya, T. (1989) A dwarf form of the spinner dolphin (*Stenella longirostris*) from Thailand. *Marine Mammal Science* **5**, 213–227.

Perrin, W.F., Caldwell, D.K. & Caldwell, M.C. (1994a) Atlantic spotted dolphin—*Stenella frontalis* (G. Cuvier,

1829). In: *Handbook of Marine Mammals, Vol. 5: The First Book of Dolphins* (S.H. Ridgway & R. Harrison, eds), pp. 173–190. Academic Press, London.

Perrin, W.F.C., Wilson, C. & Archer III, F. (1994b) Striped dolphin *Stenella coeruleoalba* (Meyen, 1833). In: *Handbook of Marine Mammals, Vol. 5: The First Book of Dolphins* (S.H. Ridgway & R. Harrison, eds), pp. 129–159. Academic Press, London.

Perry, E.A. (1993) *Aquatic territory defense by male harbour seals (Phoca vitulina) at Miquelon: relationship between active defense and male reproductive success.* PhD Thesis, Memorial University of Newfoundland, St Johns.

Peterson, R.S. (1965) *Behavior of the northern fur seal.* DSc Thesis, Johns Hopkins University, Baltimore.

Peterson, R.S. (1968) Social behavior in pinnipeds. In: *Behavior and Physiology of Pinnipeds* (R.J. Harrison, R.C. Hubbard, R.S. Peterson, D.E. Rice & R.J. Schusterman, eds), pp. 3–53. Appleton-Century Crofts, New York.

Peterson, R.S. & Bartholomew, G.A. (1967) *The Natural History of the California Sea Lion.* Special Publication No. 1. American Society of Mammalogy, Lawrence, KS.

Pitman, R.L., Balance, L.T., Mesnick, S.L. & Chivers, S. (2001) Killer whale predation on sperm whales: observations and implications. *Marine Mammal Science* **17**, 494–507.

Polovina, J.J., Mitchum, G.T., Graham, N.E. *et al.* (1994) Physical and biological consequences of climate event in the central North Pacific. *Fisheries Oceanography* **3**, 15–21.

Preen, A. (1989) Observations of mating behavior in dugongs (*Dugong dugon*). *Marine Mammal Science* **5**, 382–387.

Preen, A. (1992) *Interaction between dugongs and seagrasses in a subtropical environment.* PhD Thesis, James Cook University of North Queensland, Townsville.

Preen, A. (1995) Impacts of dugong foraging on seagrass habitats: observational and experimental evidence for cultivation grazing. *Marine Ecology Progress Series* **1224**, 201–213.

Pryor, K. & Shallenberger, I.K. (1991) Social structure in the spotted dolphins (*Stenella attenuata*) in the tuna purse seine fishery in the eastern Tropical Pacific. In: *Dolphin Societies* (E. Pryor & K.S. Norris, eds). University of California Press, Berkeley.

Ralls, K., Eagle, T. & Siniff, D.B. (1996) Movements and spatial use patterns of California sea otters. *Canadian Journal of Zoology* **74**, 1841–1849.

Rathbun, G.B., Reid, J.P., Bonde, R.K. & Powell, J.A. (1995) Reproduction in free-ranging Florida manatees. In: *Population Biology of the Florida Manatee* (T.J. O'Shea, B.B. Ackerman & H.F. Percival eds), Vol. 1, pp. 135–157. US Department of the Interior, Washington, DC.

Rathbun, G.B., Reid, J.P. & Carowan, G. (1990) Distribution and Movement Patterns of Manatees (*Trichechus manatus*) in northwestern peninsular Florida. *Florida Marine Research Publications* **48**, 1–33.

Ray, C., Watkins, W.A. & Burns, J.J. (1969) The underwater song of *Erignathus* (bearded seal). *Zoologica* **54**, 79–86.

Read, A.J. (1999) Harbour porpoise, *Phocoena phocoena* (Linnaeus, 1758). In: *Handbook of Marine Mammals, Vol. 6: The Second Book of Dolphins and Porpoises* (S.H. Ridgway & R. Harrison, eds), pp. 323–355. Academic Press, London.

Read, A.J. & Hohn, A.A. (1995) Life in the fast lane: the life history of harbor porpoises from the Gulf of Maine. *Marine Mammal Science* **11**, 423–440.

Reeves, R.R. & Brownell, R.L.J. (1989) Susu *Platanista gangetica* (Roxburgh, 1801) and *Platanista minor* (Owen, 1853). In: *Handbook of Marine Mammals, Vol. 4: River Dolphins and the Larger Toothed Whales* (S.H. Ridgway & R. Harrison, eds), pp. 69–99. Academic Press, London.

Reeves, R.R., Smeenk, C., Kinze, C.C., Brownell, R.L.J. & Lien, J. (1999) White-beaked dolphin *Lagenorhynchus albirostris* (Gray, 1846). In: *Handbook of Marine Mammals, Vol. 6: The Second Book of Dolphins and Porpoises* (S.H. Ridgway & R. Harrison, eds), pp. 1–30. Academic Press, London.

Reid, J.P., Bonde, R.K. & O'Shea, T.J. (1995) Reproduction and mortality of radio-tagged and recognizable manatees on the Atlantic coast of Florida. In: *Population Biology of the Florida Manatee* (T.J. O'Shea, B.B. Ackerman & H.F. Percival, eds), pp. 171–191. US Department of Interior, Washington, DC.

Reijnders, P., Brasseur, van der Toorn, S. *et al.* (1993) *Seals, Fur Seals, Sea Lions, and Walrus.* IUCN, Gland, Switzerland.

Reiss, M.J. (1989) *The Allometry of Growth and Reproduction.* University of Cambridge Press, Cambridge, UK.

Reiter, J. & Le Boeuf, B.J. (1991) Life history consequences of variation in age at primiparity in northern elephant seals. *Behavioral Ecology and Sociobiology* **28**, 153–160.

Reyes, J.C. & Van Waerebeek, K. (1995) Aspects of the biology of Burmeister's porpoise from Peru. In: *Biology of Phocoenids* (A. Bjørge & G.P. Donovan, eds), pp. 349–364. Special Issue No. 16. International Whaling Commission, Cambridge, UK.

Reynolds III, J.E. & Wilcox, J.R. (1994) Observations of Florida manatees (*Trichechus manatus latirostris*) around selected power plants in winter. *Marine Mammal Science* **10**, 163–177.

Rice, D.W. (1983) Gestation period and fetal growth of the grey whale. *Reports of the International Whaling Commission* **33**, 549–554.

Rice, D.W. (1989) Sperm whale *Physeter macrocephalus* Linnaeus, 1758. In: *Handbook of Marine Mammals, Vol. 4: River Dolphins and the Larger Toothed Whales* (S.H. Ridgway & R. Harrison, eds), pp. 177–234. Academic Press, London.

Richard, K.R., Dillon, M.C., Whitehead, H. & Wright, J.M. (1996) Patterns of kinship in groups of free-living sperm whales (*Physeter macrocephalus*) revealed by mulitple molecular genetic analyses. *Proceedings of the National Academy of Sciences* **93**, 8792–8795.

Riedman, M. (1990) *The Pinnipeds.* University of California Press, Berkeley.

Riedman, M.L. & Estes, J.A. (1990) *The Sea Otter (Enhydra lutris): Behavior, Ecology, and Natural History.* US Fish and Wildlife Service, Washington, DC.

Ross, G.J.B. (1977) The taxonomy of bottlenose dolphins *Tursiops* species in South African waters, with notes on their biology. *Annals Cape Providence Museum (Natural History)* **11**, 135–194.

Rubenstein, D.I. (1986) Ecology and sociality in horses and zebras. In: *Ecological Aspects of Social Evolution* (D.I. Rubenstein & R.W. Wrangham, eds), pp. 282–302. Princeton University Press, Princeton, NJ.

Rubenstein, D.I. & Wrangham, R.W. (1986) *Ecological Aspects of Social Evolution*. Princeton University Press, Princeton, NJ.

Sergeant, D.E. (1962) The biology of the pilot or pothead whale *Globicephala melaena* (Traill) in Newfoundland waters. *Bulletin of the Fisheries Research Board of Canada* **132**, 1–84.

Sergeant, D.E. & Brodie, P.F. (1969) Body size in white whales, *Delphinapterus leucas*. *Journal of the Fisheries Research Board of Canada* **26**, 2561–2580.

Sergeant, D.E., St Aubin, D.J. & Geraci, J.R. (1980) Life history and northwest Atlantic status of the Atlantic white-sided dolphin, *Lagenorhynchus acutus*. *Cetology* **37**, 1–12.

Shaughnessy, P.D. & Warneke, R.M. (1987) Australian fur seal, *Arctocephalus pusillus doriferus*. In: *Status, Biology, and Ecology of Fur Seals. Proceedings of an International Symposium and Workshop* (J.P. Croxall & R.L. Gentry, eds), pp. 73–78. NOAA Technical Report NMFS No. 51. National Technical Information Service, Springfield, VA.

Silverman, H.B. (1979) *Social organization and behavior of the narwhal, Monodon monoceros in Lancaster Sound, Pond Inlet, and Tremblay Sound, Northwest Territories*. MS Thesis, McGill University, Montreal.

Sjare, B. & Stirling, I. (1996) The breeding behavior of Atlantic walruses, *Odobenus rosmarus rosmarus*, in the Canadian High Arctic. *Canadian Journal of Zoology* **74**, 897–911.

Slooten, E. & Dawson, S.M. (1994) Hecto's dolphin—*Cephalorhynchus hectori* (Van Beneden, 1881). In: *Handbook of Marine Mammals, Vol. 5: The First Book of Dolphins* (S.H. Ridgway & R. Harrison, eds), pp. 311–333. Academic Press, London.

Stamps, J.A. & Buechner, M. (1985) The territorial defense hypothesis and the ecology of insular vertebrates. *Quarterly Review of Biology* **60**, 155–181.

Stearns, S.C. (1992) *The Evolution of Life Histories*. Oxford University Press, Oxford.

Stirling, I. (1983) The evolution of mating systems in Pinnipeds. In: *Recent Advances in the Study of Behavior* (J.F. Eisenberg & D.G. Kleiman, eds), pp. 489–527. Special Publication No. 7. American Society of Mammalogists, Lawrence, KS.

Stirling, I. (1988) *Polar Bears*. University of Michigan Press, Ann Arbor, MI.

Stone, G.S., Kraus, S.D., Prescott, J.H. & Hazard, K.W. (1988) Significant aggregations of the endangered right whale (*Eubalaena glacialis*), on the continental shelf of Nova Scotia. *Canadian Field-Naturalist* **102**, 471–474.

Sullivan, R.M. (1981) Aquatic displays and interactions in harbor seals (*Phoca vitulina*), with comments on mating systems. *Journal of Mammalogy* **62**, 825–831.

Swartz, S. (1986) Grey whale migratory, social and breeding behavior. In: *Behaviour of Whales in Relation to Management* (G.P. Donovan, ed.), pp. 207–229. Special Issue

No. 8. International Whaling Commission, Cambridge, UK.

Sydeman, W.J., Huber, H.R., Emslie, S.D., Ribic, C.A. & Nur, N. (1991) Age-specific weaning success of northern elephant seals in relation to previous breeding experience. *Ecology* **72**, 2204–2217.

Tarasoff, F.J., Bisaillon, A., Pierard, J. & Whitt, A.P. (1972) Locomotory patterns and external morphology of the river otter, sea otter, and harp seal (Mammalia). *Canadian Journal of Zoology* **50**, 915–929.

Terhune, J. (1985) Scanning behavior of harbor seals on haul-out sites. *Journal of Mammalogy* **66**, 392–395.

Testa, J.W., Hill, S.E.B. & Siniff, D.B. (1989) Diving behavior and maternal investment in Weddell seals (*Leptonychotes weddelli*). *Marine Mammal Science* **5**, 399–406.

Thompson, P.M., Fedak, M.A., McConnell, B.J. & Nicholas, K.S. (1989) Seasonal and sex-related variation in the activity patterns of common seals (*Phoca vitulina*). *Journal of Applied Ecology* **26**, 521–535.

Tinker, T.M., Kovacs, K.M. & Hammill, M.O. (1995) The reproductive behavior and energetics of male gray seals (*Halichoerus grypus*) breeding on land-fast ice substrate. *Behavioral Ecology and Sociobiology* **36**, 159–170.

Tolley, K.A., Read, A.J., Wells, R.S. *et al.* (1995) Sexual dimorphism in wild bottlenose dolphins (*Trusiops truncatus*) from Sarasota, Florida. *Journal of Mammalogy* **74** (4), 1190–1198.

Torres, D.T. (1987) Juan Fernández fur seal, *Arctocephalus philippii*. In: *Status, Biology, and Ecology of Fur Seals; Proceedings of an International Symposium and Workshop* (J.P. Croxall & R.L. Gentry, eds), pp. 37–41. NOAA Technical Report NMFS No. 51. National Technical Information Service, Springfield, VA.

Trillmich, F. (1984) The natural history of the Galapagos fur seal (*Arctocephalus galapagoensis*, Heller 1904). In: *Key Environment Series: Galapagos* (R. Perry, ed.), pp. 215–223. Pergamon Press, Oxford.

Trillmich, F. (1987) Galapagos fur seal, *Arctocephalus galapagoensis*. In: *Status, Biology, and Ecology of Fur Seals* (J.P. Croxall & R.L. Gentry, eds), pp. 23–28. US Department of Commerce, Washington, DC.

Trillmich, F. (1996) Parental investment in pinnipeds. *Advances in the Study of Behavior* **25**, 533–577.

Trillmich, F. & Lechner, E. (1986) Milk of the Galapagos fur seal and sea lion, with a comparison of the milk of eared seals (Otariidae). *Journal of Zoology* **209**, 271–277.

Trillmich, F. & Majluf, P. (1981) First observations on colony structure, behavior, and vocal repertoire of the South American fur seal (*Arctocephalus australis*, Zimmerman, 1783) in Peru. *Zeitschrift für Säugetierkunde* **46**, 310–322.

Trillmich, F. & Ono, K.A. (1991) *Pinnipeds and El Niño: Responses to Environmental Stress*. Springer-Verlag, Heidelberg.

Trivers, R.L. (1972) Parental investment and sexual selection. In: *Sexual Selection and the Descent of Man, 1871–1971* (B. Campbell, ed.), pp. 136–179. Heinemann, London.

Troy, S.K. (1997) *Territorial behaviour and mating success in male New Zealand fur seals, Arctocephalus forsteri.* PhD Thesis, University of Melbourne, Melbourne.

Twiss, S.D., Anderson, S.S. & Monogham, P. (1998) Limited intra-specific variation in male grey seal (*Halichoerus grypus*) dominance relationships in relation to variation in male mating success and female availability. *Journal of Zoology, London* **246**, 259–267.

Tyack, P.L. & Whitehead, H. (1983) Male competition in large groups of wintering humpback whales. *Behaviour* **83**, 1–23.

Van Lennep, E.W. & Van Utrecht, W.L. (1953) Preliminary report on the study of the mammary glands of whales. *Norsk Hvalangsttidende* **42**, 249–258.

Van Parijs, S.M., Thompson, P.M., Tollit, D.J. & Mackay, A. (1997) Changes in the distribution and activity of male harbour seals during the mating season. *Animal Behaviour* **54**, 35–43.

Van Parijs, S.M., Hastie, G.D. & Thompson, P.M. (1999) Geographical variation in temporal and spatial vocalization patterns of male harbour seals in the mating season. *Animal Behaviour* **58**, 1–9.

Van Waerebeek, K. & Read, A.J. (1994) Reproduction of dusky dolphins, *Lagenorhynchus obscurus*, from coastal Peru. *Journal of Mammalogy* **75**, 1054–1062.

Vaz Ferreira, R. (1975) Behaviour of the southern sea lion, *Otaria flavescens* (Shaw) in the Uruguayan Islands. *Rapport et Proces-Verbaux Des Reunions, Consiel International Exploration de Mer* **169**, 219–227.

Vaz Ferreira, R. & de Ponce Leon (1987) South American fur seal, *Arctocephalus australis* in Uruguay. In: *Status, Biology, and Ecology of Fur Seals. Proceedings of an International Symposium and Workshop* (J.P. Croxall & R.L. Gentry, eds), pp. 29–32. NOAA Technical Report NMFS No. 51. National Technical Information Service, Springfield, VA.

Vidal, O., Brownell, R.L. & Findley, L.T. (1999) Vaquita *Phocoena sinus* Norris and McFarland, 1958. In: *Handbook of Marine Mammals, Vol. 6: The Second Book of Dolphins and Porpoises* (S.H. Ridgway & R. Harrison, eds), pp. 357–378. Academic Press, London.

Walker, B.G. & Bowen, W.D. (1993a) Behavioral differences among adult male harbour seals during the breeding season may provide evidence of reproductive strategies. *Canadian Journal of Zoology* **71**, 1585–1591.

Walker, B.G. & Bowen, W.D. (1993b) Changes in body mass and feeding behaviour in male harbour seals, *Phoca vitulina* in relation to female reproductive status. *Journal of Zoology, London* **231**, 423–436.

Walker, G.E. & Ling, J.K. (1981a) Australian sea lion *Neophoca cinerea* (Peron, 1816). In: *Handbook of Marine Mammals, Vo. 1: The Walrus, Sea Lions, Fur Seals and Sea Otter* (S.H. Ridgway & R.J. Harrison, eds), pp. 99–118. Academic Press, London.

Walker, G.E. & Ling, J.K. (1981b) New Zealand sea lion—*Phocartos hookeri.* In: *Handbook of Marine Mammals, Vol. 1: The Walrus, Sea Lions, Fur Seals and Sea Otter* (S.H. Ridgeway & R.J. Harrison, eds), pp. 25–38. Academic Press, London.

Watkins, W.A., Daher, M.A., Reppucci, G.M. *et al.* (2000) Seasonality and distribution of whale calls in the North Pacific. *Oceanography* **13**, 62–67.

Weilgart, L.S. & Whitehead, H. (1988) Distinctive vocalizations from mature sperm whales (*Physeter macrocephalus*). *Canadian Journal of Zoology* **66**, 1931–1937.

Wells, R.S. & Scott, M.D. (1999) Bottlenose dolphin *Tursiops truncatus* (Montagu, 1821). In: *Handbook of Marine Mammals, Vol. 6: The Second Book of Dolphins and the Porpoises* (S.H. Ridgway & R. Harrison, eds), pp. 137–182. Academic Press, London.

Wells, R.S., Irvine, A.B. & Scott, M.D. (1980) The social ecology of inshore odontocetes. In: *Cetacean Behavior: Mechanisms and Functions* (L.M. Herman, ed.), pp. 263–317. John Wiley & Sons, New York.

Wells, R.S., Scott, M.D. & Irvine, A.B. (1987) The social structure of free-ranging bottlenose dolphins. *Current Mammalogy* **1**, 247–305.

Wells, R.S., Boness, D.J. & Rathbun, G.B. (1999) Behavior. In: *The Biology of Marine Mammals* (J.E. Reynolds III & S. Rommell, eds), pp. 324–422. Smithsonian Institution Press, Washington, DC.

Westneat, D.F., Sherman, P.W. & Morton, M.L. (1990) The ecology and evolution of extra-pair copulations in birds. *Current Ornithology* **7**, 331–369.

Whitehead, H. (1983) Structure and stability of humpback whale groups off Newfoundland. *Canadian Journal of Zoology* **61**, 1391–1397.

Whitehead, H. (1990) Rules for roving males. *Journal of Theoretical Biology* **145**, 355–368.

Whitehead, H. (1994) Delayed competitive breeding in roving males. *Journal of Theoretical Biology* **166**, 127–133.

Whitehead, H. & Arnbom, T. (1987) Social organization of sperm whales off the Galapagos Islands, February–April, 1985. *Canadian Journal of Zoology* **65**, 913–919.

Whittow, G.C. (1978) Thermoregulatory behavior of Hawaiian monk seal, *Monachus schauinslandi. Pacific Science* **32**, 47–60.

Widemo, F. (1998) Competition for females on leks when male competitive abilities differ: empirical test of a model. *Behavioral Ecology* **9**, 427–431.

Wiley, R.H. (1974) Evolution of social organization and life-history patterns among grouse. *Quarterly Review of Biology* **49**, 227.

Wiley, R.H. (1991) Lekking in birds and mammals: behavioral and evolutionary issues. *Advances in the Study of Behavior* **20**, 201–291.

Williams, G.C. (1966) *Adaptation and Natural Selection.* Princeton University Press, Princeton, NJ.

Wilson, E.O. (1975) *Sociobiology.* Belknap Press, Cambridge, MA.

Worthington Wilmer, J., Overall, A.J., Pomeroy, P.P., Twiss, S.D. & Amos, W. (2000) Patterns of paternal relatedness in British grey seal colonies. *Molecular Ecology* **9**, 283–292.

Würsig, B. & Clark, C. (1993) Behavior. In: *The Bowhead Whale* (J.J. urns, J.J. Montague & C.J. Cowles, eds), pp. 157–199. Society for Marine Mammalogy Special Publication No. 2. Allen Press, Lawrence, KS.

CHAPTER 11

Population Genetic Structure

A. Rus Hoelzel, Simon D. Goldsworthy and Robert C. Fleischer

11.1 INTRODUCTION

Marine mammals have adapted to marine habitats in ways that to varying degrees reflect their earlier evolutionary history in terrestrial environments. This is especially relevant to the evolution of population genetic structure when reproductive constraints affect patterns of dispersal, movement and demography. A detailed review of reproductive biology and behaviour is given in Chapter 10, and patterns of distribution are reviewed in Chapter 1. Here we consider some of the implications of adaptation to the marine habitat for the spatial and temporal patterns of population genetic structure in marine mammals.

Some marine mammal taxa give birth on land or ice (all pinniped species and the polar bear), others give birth at sea (all cetaceans and sirenians), and one reproduces near the shore, but is not dependent on land for breeding (the sea otter). Those species that mate and give birth on land or ice are tied to breeding sites, and suitable habitat is limiting. Therefore, geographical range, philopatry and mating system are likely to be especially important in shaping the genetic population structure for these species. Species that give birth at sea are typically less restricted, but still require suitable habitat with respect to resource and possibly thermal requirements. Even so, population boundaries are often less predictable for the fully aquatic marine mammals.

Site fidelity is generally high in otariids where the same animals may breed within the same small part of a colony from one year to the next (see Chapter 10). Some pack-ice breeding seals, which tend to be monogamous, breed over a broad area and

are less site faithful. Among the fully aquatic species, one group of sirenians (the manatees) and some species of cetaceans (in the genus *Plantinestidae*) inhabit rivers and estuaries, while most species have oceanic distributions (see Chapter 1). The extent and pattern of dispersal varies greatly, with some species having a nearly global distribution (such as sperm whales) while others are more restricted (such as the Florida manatee and Commerson's dolphin). A further complexity involves the regular migrations undertaken by some species, in some cases over vast distances.

For many marine mammals, another important factor is their history of extensive and indiscriminate harvesting for various products including fur and oil (see Chapter 14). In some species this resulted in genetic population bottlenecks, reductions in range and local extinctions. For several species, including the Stellar sea cow (*Hydrodamalis gigas*), the Caribbean monk seal (*Monachos tropicalis*) and the Japanese sea lion (*Zalophus japonicus*), it resulted in extinction. Some species were depleted to the extent that their populations have remained small long after the hunting ceased. For example, the northern right whale, once abundant in the North Atlantic, now numbers only about 500 whales, and the population appears to be in decline (Caswell *et al.* 1999).

The development during the past decade of powerful techniques that can assay the variability of molecular genetic markers has greatly facilitated studies about the evolution of behaviour and population structure in these difficult study subjects. The methods include the enzymatic amplification of DNA by the polymerase chain reaction (PCR), automated DNA sequencing (in particular of highly variable mitochondrial DNA) and assays of size variation in variable number of tandem repeat (VNTR) markers (in particular microsatellites). The results of these studies are discussed below, while a brief review of the genetic markers and some of the methods used to analyse them are presented in Box 11.1. For more detailed discussions of the methods see Hillis *et al.* (1996) and Hoelzel (1998a). The molecular approach is an essential component

of our understanding and conservation of natural populations, as it allows the determination of otherwise intractable information, from paternity and kin associations to the pattern and level of genetic diversity within and among populations. We need to conserve diversity at this level, as this is the raw material that permits a species to adapt to a changing environment over time.

BOX 11.1 MOLECULAR MARKERS AND LABORATORY METHODS

A wide range of molecular markers are currently in use for the analyses of population genetic structure of marine mammals. Allozymes (enzymes with different mobilities in electrophoretic gels, reflecting different genotypes; see Lewonton & Hubby 1966) were used extensively in the past, but here we will concentrate on more recently developed and powerful DNA methods. Ideally, such methods should have high resolving power, but also be easy and inexpensive. This is especially true for population genetic analyses, because relatively large sample sizes are usually needed.

Laboratory methods

The polymerase chain reaction (PCR): PCR is an *in vitro* chemical reaction that results in the copying (or 'amplification') of specific sequences of DNA using a thermally stable DNA polymerase and synthesized primers. PCR consists of three primary steps repeated 25–50 times: (i) denaturation of double-stranded DNA (the template) to single strands by heating to nearly boiling temperature (typically 94°C); (ii) annealing of synthetic primers that are complementary to sequences that flank the region of interest (usually at temperatures in the range of 40–60°C); and (iii) extension of a newly synthesized, complementary sequence (at 72°C, the optimal temperature for the polymerase). During extension, the thermally stable polymerase (usually *Taq*) removes deoxynucleotides from solution and covalently binds them into the new, complementary strand. With each PCR repeat or cycle, the number of copies of the sequence increases (in theory as much as doubling). Amplified, double-stranded sequences, called PCR products, can be analysed for variation via digestion with restriction enzymes, DNA sequencing or other methods such as SSCP (see below).

DNA sequencing: This requires labelled synthetic primers or nucleotides (e.g. with fluorescent dyes in most types of automated sequencing). It differs from PCR in that it also requires nucleotides that lack the hydroxyl group necessary for complementary strand

synthesis (i.e. dideoxynucleotides or ddNTPs). The ddNTPs 'terminate' a polymerase-generated sequence extension at random; this results in a nested series of labelled, single-stranded products in size increments of one base. Run side by side through a high-resolution gel, these products allow the sequence of base pairs in the DNA molecule to be 'read'.

Single-stranded conformational polymorphisms (SSCPs): These are detected by denaturing PCR products and separating them in a non-denaturing polyacrylamide gel. PCR products or alleles separate because of differences in their secondary structures caused by differences in their sequences. SSCP is most useful for rapid diagnosis of different genotypes, with resolution down to one base pair (1 bp) for short (less than 300 bp) sequences.

Restriction fragment length polymorphism (RFLP): 'Restriction' enzymes which cleave DNA always at the same short (usually 4–6 bp) sequence can be used to detect variation at the cleavage sites. This is sometimes used in conjunction with PCR.

Random amplified polymorphic DNA (RAPD) and amplified fragment length polymorphism (AFLP): Both of these involve the use of PCR primers that anneal 'randomly' to sequences within the genome. Thus, while they often can quickly resolve a great number of variable markers, their sequences and functions are usually unknown, and the resulting patterns can be difficult to interpret, especially at the population level.

DNA markers

Mitochondrial DNA (mtDNA): A circular DNA molecule found in the mitochondria of most eukaryotes. Animal mtDNA is haploid and in most cases inherited only along maternal lineages. It shows little evidence of recombination, so forces that affect variation in one part of the molecule will have an impact on variation in the entire molecule. It is relatively variable compared to the

(continued)

BOX 11.1 (*cont'd*)

nuclear genome, in part a result of higher mutation rates, and this constitutes a major advantage for its use in population genetics. Different regions or genes of the molecule have different substitution rates. For example, rRNAs, tRNAs and some proteins are relatively slow, while other proteins, and most of the non-coding 'control region', evolve more rapidly. Variation in mtDNA is now mostly assayed by PCR followed by direct sequencing or SSCP.

Variable number of tandem repeat loci (VNTRs): These include minisatellites ('DNA fingerprints') and microsatellites (also known as simple tandem repeats or STRs). Structurally, VNTRs consist of a repeated motif of DNA (e.g. 'C A' repeated as in the microsatellite: C A C A C A C A). Allelic variation is due mostly to differences in the number of tandem repeats within an array, and is generated by very high mutation rates (10^{-3}–10^{-5}/gene/generation). VNTR variation at microsatellite loci is assayed by PCR amplification, followed by electrophoretic separation in high-resolution, polyacrylamide gels. VNTRs are highly variable markers in marine mammals, and are very useful for determining relatedness, assessing mating systems, and elucidating population variation and structure.

Nuclear introns: These are non-coding regions within genes. Thus they tend to be more variable (less constrained by selection) than coding regions. Primers for PCR and sequencing are initially designed from more conserved exon (protein-coding) sequences such that they amplify across the intron. This gives them wide applicability. Substitution rates tend to be slower and variability lower in nuclear introns in comparison even to conserved regions of mtDNA.

Sex-linked genes: Genes that are found on the X and Y chromosomes are useful for determining the sex of individuals. However, variable Y chromosome markers can also be used to assess structure and evolution within male lineages.

11.2 POPULATION GENETICS OF REDUCED OR RECOVERING POPULATIONS

The impacts of overhunting during the recent past are especially well documented for some pinniped species (e.g. Bonnell & Selander 1974; Busch 1985; Hoelzel *et al.* 1993). Highly polygynous pinnipeds (such as the northern elephant seal) were most severely impacted by bottlenecks as the extreme skew in mating success (Boness 1991) reduces effective population size and thus the ability of a population to retain genetic variation (see simulations in Hoelzel 1999a). With worldwide protection established early in the twentieth century, many of the impacted pinniped populations are now recovering (Bester 1987; Shaughnessy & Goldsworthy 1990; Boyd 1993; Guinet *et al.* 1994; Hofmeyr *et al.* 1997). However, in many cases populations have remained small for long periods because of continued hunting pressure, or as a consequence of their life history. Factors such as reproduction rate and generation time will impact on the period of demographic recovery. Hoelzel (1999a) modelled this effect for the northern elephant seal, and showed

that a 30% decrease in reproductive output approximately doubled the length of time it took a bottle-necked population to recover (Fig. 11.1).

Heterozygosity decays more quickly in small populations, and in many cases demographic and genetic data are consistent, with depleted populations showing low levels of diversity. However, some species reported to have experienced historic population bottlenecks retained genetic variation (e.g. the Guadalupe fur seal, *Arctocephalus townsendii*, Juan Fernández fur seal, *A. philippii*, Antarctic fur seal, *A. gazella*, and the sea otter, *Enhydra lutris*) (Cronin *et al.* 1996; Bernardi *et al.* 1998; Wynen *et al.* 2000; Goldsworthy *et al.*, in press), while others (e.g. the polar bear) not known to have gone through human-induced population bottlenecks, show low levels (Cronin *et al.* 1991; Shields & Kocher 1991; but see Paetkau *et al.* 1995). For example, the bowhead whale (*Balaena mysticetus*) population in the Bering, Chukchi and Beaufort Seas, though proposed by some to have undergone a bottleneck, has retained high levels of microsatellite DNA variation. Tests for a 'bottleneck footprint' in the allele frequencies (Luikart *et al.* 1998) for this species gave no indication of a recent bottleneck (Rooney *et al.* 1999). Note that various

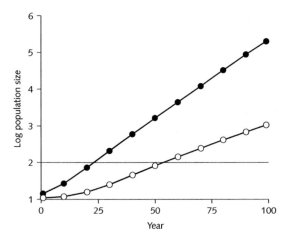

Fig. 11.1 Simulation of population expansion based on life history data for the northern elephant seal. Each simulation was repeated 500 times, and started with a bottleneck population size of 10 seals. Only the means are shown (but note that standard deviations are large, reflecting demographic stochasticity). One simulation represents a reproductive output of one pup per year (closed circles), while the other represents two pups every 3 years (open circles). A horizontal line indicates the point at which each population surpasses 100 individuals. (Data from Hoelzel 1999a).

factors can contribute to the level of variation at different markers, including genetic drift, selection and effective population size, so comparisons between species and among markers should be treated with caution. A number of case studies are useful in illustrating the above points, and these are reviewed in more detail below.

The northern elephant seal was heavily exploited during the nineteenth century and was reduced to a bottleneck population size estimated to be only 10–30 individuals (Bartholomew & Hubbs 1960; Hoelzel et al. 1993). They were originally distributed from Point Reyes near San Francisco to Magdalena Bay, Baja California. The hunt began around 1810 and became commercially inviable by the 1860s due to overexploitation. Hunters and collectors continued to take northern elephant seals until 1884 when over 150 were taken. After this they were not found again until 1892, when seven of eight animals found on Guadalupe Island were killed. These were thought to be the last of the species at the time. Fortunately this was not the case

and in 1922 the population was estimated at 350, at which time they received legal protection in the USA and Mexico (Bartholomew & Hubbs 1960). In recent years they have recolonized much of their former range, and the population is currently estimated at over 150 000 (Stewart et al. 1994).

Molecular genetic variation (Box 11.1) in the northern elephant seal is low at mtDNA (Table 11.1) (Hoelzel et al. 1993), allozyme (Bonnell & Selander 1974; Hoelzel et al. 1993), major histocompatibility complex (MHC) (Hoelzel et al. 1999a), microsatellite and minisatellite DNA loci (Hoelzel et al. 1999b), consistent with predictions, given the severity of the bottleneck (Hoelzel et al. 1993; Hoelzel 1999a). Analysis of samples collected prior to the bottleneck at five loci showed unique genotypes at all loci compared to the postbottleneck population (Hoelzel 1999b; Hoelzel et al., in review). By comparing postpopulation bottleneck genetic diversity with demographic simulation models and historical data, Hoelzel et al. (1993) estimated the severity of the population bottleneck to be less than 30 seals over a 20-year period, or a single-year bottleneck of less than 20 seals.

Similarly, Hawaiian monk seals were hunted to near extinction by the end of the nineteenth century (Kenyon & Rice 1959), although the extent and duration of the population bottleneck are not well understood. The species then underwent a partial recovery to approximately 3000 individuals, and then suffered a 50% decline in numbers between the late 1950s and the 1970s (Kenyon 1972; Johnson et al. 1982; Ragen 1993). This decline has been partially attributed to human disturbance at breeding colonies (Kenyon 1972; Gerrodette & Gilmartin 1990). Data from mtDNA sequencing of the control region has revealed levels of genetic diversity that appear very low in comparison with other species (Table 11.1) and no evidence of subpopulation differentiation (Kretzmann et al. 1997). In fact, only three mtDNA haplotypes have been detected in 50 seals, with the majority of seals (86%) sharing one haplotype (Kretzmann et al. 1997). These data have been supported by multilocus DNA fingerprinting (Box 11.1), where putative unrelated individuals show high levels of band sharing ranging from 49% to 73% (Kretzmann et al. 1997).

Table 11.1 Measures of mtDNA control region genetic variation in marine mammals.

Species	Number of populations	Haplotypes/individuals	Variable sites/total sites (%)[1]	Pairwise uncorrected distance	Percentage nucleotide diversity (π)	Population diversity	Source[2]
Juan Fernández fur seal (*Arctocephalus philippii*)	1	13/28	39/313 (12.5)	0.034	3.0		1
Guadalupe fur seal (*Arctocephalus townsendi*)	1	7/25	18/313 (5.7)	0.016–0.036	2.0		2
Antarctic fur seal (*Arctocephalus gazella*)	1	25/47	58/291 (19.9)	0.038		ϕ_{ST} = 0.074	1, 3
Subantarctic fur seal (*Arctocephalus tropicalis*)	8	26/145	45/268 (16.8)		3.2		4
	1	7/8	36/291 (12.4)	0.048	3.4		1
South American fur seal (*Arctocephalus australis*)	5	33/103	46/273 (16.8)		4.8	ϕ_{ST} = 0.190	5
	1	4/5	39/297 (13.3)		7.8		
Australian fur seal (*Arctocephalus pusillus doriferus*)	4	3/4	2/287 (0.7)		0.4		5
New Zealand fur seal (*Arctocephalus forsteri*)	8	16/17	40/276 (14.5)		5.1		5, 6
Northern fur seal (*Callorhinus ursinus*)	1	4/5	17/287 (5.90)		2.7		5
Steller's sea lion (*Eumetopias jubatus*)	2	52/224	29/238 (12.2)	0.017			7
California sea lion (*Zalophus californianus*)	4	11/40	29/319 (9.2)	0.011–0.053			8
Australian sea lion (*Neophoca cinerea*)	1	1/5	0/289 (–)		0.0		5
Southern sea lion (*Otaria bryonia*)	1	4/5	5/288 (1.7)		0.8		5
Hooker's sea lion (*Phocarctos hookeri*)	2	4/6	3/286 (1.0)		0.4		5
Southern elephant seal (*Mirounga leonina*)	2	26/48	26/300 (8.7)	0.023	1.95	ϕ_{ST} = 0.57	9
	5	19/37	28/444 (6.3)		0.43		10
Northern elephant seal (*Mirounga angustirostris*)	1	2/40	3/300 (1.0)	0.010	0.7		9
Hawaiian monk seal (*Monachus schauinslandi*)	5	3/50	2/359 (0.6)	0.0007			11
Harbour seal (*Phoca vitulina*)	24	34/277	40/453 (8.8)	0.002–0.075	3.28		12
	4	47/86	30/320 (9.4)				13
Manatee (*Trichechus manatus*)	8	15/86	52/410 (12.7)		4.0	ϕ_{ST} = 0.8	14
Sperm whale (*Physeter macrocephalus*)	3	13/37	12/954 (1.25)	0.001–0.008	0.2		15
Southern right whale (*Eubalaena australis*)	2	7/45	20/289 (6.9)	0.007–0.052	2.1	ϕ_{ST} = 0.157	16
Northern right whale (*Eubalaena glacialis*)	2	5/180	7/500 (1.4)	–			17
Fin whale (*Balaenoptera physalus*)	6	48/295	30/288 (10.4)	0.003–0.020	1.1	F_{ST} = 0.002–0.601	18
Minke whale (*Balaenoptera acutorostrata*)	4	25/87	20/346 (5.6)	0.009–0.035	0.64	γ_{ST} = 0.024	19
Antarctic minke whale (*Balaenoptera bonaerensis*)	2	23/23	26/346 (7.5)	0.004–0.063	1.6	γ_{ST} = 0.070	19
Humpback whale (*Megaptera novaeangliae*)	6	31/136	36/288 (12.5)	0.004–0.011	2.6	K_{ST} = 0.04–0.33	20
Narwhal (*Monodon monoceros*)	6	5/74	4/287 (1.4)	0.003–0.070	0.17	H_{ST} = 0.04–0.73	21
Bottlenose dolphin (*Tursiops truncatus*)	2	18/55	45/300 (15.0)	0.001–0.009	0.6, 2.7	ϕ_{ST} = 0.604	22
Killer whale (*Orcinus orca*)	8	12/65	18/960 (1.9)	–	0.52	–	23
Harbour porpoise (*Phocoena phocoena*)	4	75/253	61/342 (17.8)		1.1	ϕ_{ST} = 0.011	24

[1] Calculated excluding sequence alignment gaps.

[2] 1, Goldsworthy *et al.* 2000; 2, Bernardi *et al.* 1998; 3, Goldsworthy *et al.* 1999; 4, Wynen *et al.* 2000; 5, Wynen *et al.* 2001; 6, Lento *et al.* 1997; 7, Bickham *et al.* 1996; 8, Maldonado *et al.* 1995; 9, Hoelzel 1993; 10, Slade 1997; 11, Kretzmann *et al.* 1997; 12, Stanley *et al.* 1996; 13, Lamont *et al.* 1996; 14, Garcia-Rodriguez *et al.* 1998; 15, Lyrholm *et al.* 1996; 16, Baker *et al.* 1999; 17, Malik *et al.* 1999; 18, Berube *et al.* 1998; 19, Bakke *et al.* 1996; 20, Palsboll *et al.* 1995; 21, Palsboll *et al.* 1997a; 22, Hoelzel *et al.* 1998b; 23, A.R. Hoelzel *et al.*, unpublished data; 24, Rosel *et al.* 1999.

Both Hawaiian monk seals and northern elephant seals share a similar history of exploitation, leading to a population bottleneck. However, despite sharing similarly low genetic diversity, they have shown markedly different capacities for recovery (Stewart *et al.* 1994). Possible explanations include different sensitivities to inbreeding depression (Kretzmann *et al.* 1997). One of the major threats to the recovery of the Hawaiian monk seal is the generally low reproductive success of females, and Kretzmann *et al.* (1997) hypothesize that this may be related to low genetic diversity.

The Juan Fernández fur seal (*Arctocephalus philippii*) is limited in range to the islands that make up the Juan Fernández Archipelago (Alejandro Selkirk, Robinson Crusoe and Santa Clara) and San Félix group (San Félix and San Ambrosio) (Torres 1987). The Juan Fernández Archipelago is about 650 km west of Valparaiso, Chile, and the San Félix group lies about 900 km further north. Early accounts of the species on both Alejandro Selkirk and Robinson Crusoe Islands during the late 1600s and early 1700s estimated the numbers of seals in the millions, perhaps exceeding 4 million in number at the end of the seventeenth century (see Hubbs & Norris 1971 for a review). Intensive exploitation of fur seals that began in the late 1600s resulted in near extinction by the late 1800s (Hubbs & Norris 1971; Torres 1987). For nearly 100 years the species was thought to be extinct, until 1965 when about 200 seals were found on Alejandro Selkirk Island. Since then, the population has grown steadily at 15–20% per year, with recent estimates of greater than 6000 animals (Torres 1987).

Goldsworthy *et al.* (in press) compared the mtDNA control region sequence of seals from two islands in the Juan Fernández Archipelago and, contrary to expectation, found relatively high levels of genetic variation (Table 11.1). The higher than expected levels were attributed to two main factors. First, although Juan Fernández fur seals were hunted in large numbers from the late 1600s through to the early 1800s, demand declined after that following the collapse of the fur seal markets in China (Richards 1994). Second, some sites where fur seals bred, such as caves or narrow rock ledges at the base of cliffs, would have been mostly inaccessible to sealers and thus provided some refuge (in contrast to more accessible species like elephant seals). The 'refugia' hypothesis (Goldsworthy *et al.*, in press) may also hold for other fur seal populations that were thought to have undergone bottlenecks.

The Antarctic fur seal is another similar example. Their commercial exploitation was intense, with an estimated 250 000 taken from the South Shetland Islands in a single season, and 1.2 million seals taken at South Georgia by 1822 (Bonner 1958; Bonner & Laws 1964; Headland 1984). Other populations throughout the Southern Ocean were also systematically sought and harvested, until just one seal was found and killed at South Georgia in 1915. Since then, the population has recovered and has an estimated annual pup production of approximately 378 000 (estimated in 1990–91; Boyd 1993). This recent history certainly suggests a population bottleneck. However, like the Juan Fernández fur seal, molecular genetic data indicate that the extent and duration of this bottleneck may not have been enough to result in a significant reduction in genetic variation (Wynen *et al.* 2000; Goldsworthy *et al.*, in press). Genetic variation at both nuclear and mtDNA loci (Table 11.1) is relatively high. Multilocus DNA fingerprinting revealed very low ($S = 0.28$) band-sharing coefficients (indicating high diversity) compared to Hawaiian monk seals ($S = 0.73$) and Guadalupe fur seals ($S = 0.59$) (Goldsworthy *et al.* 1999).

In common with fur seals, populations of sea otter were sought in the 1700s and 1800s for their valuable skins. Pre-exploitation populations of sea otters may have numbered in the order of several hundred thousands, but commercial harvesting resulted in local extinctions or drastic reductions in the size of populations across the range of the species. Eleven populations survived with total numbers reduced to perhaps less than 2000 (Cronin *et al.* 1996). Since legal protection, most populations have recovered, and in some locations the species has been reintroduced (Cronin *et al.* 1996). Studies examining the genetic variation in sea otters have identified low allozyme variation in three populations that were monomorphic at 24 loci (Cronin & May in Cronin *et al.* 1996). However, polymorphism at mtDNA and allozyme loci have been found within most other populations (Sanchez 1992; Cronin *et al.* 1996; Lidicker & McCollum 1997).

BOX 11.2 BIOPSY SAMPLING

The most common method for sampling free-ranging cetacean species has been the use of a small cylindrical dart fired from a bow or air rifle. Various dart designs have been proposed over the years (e.g. see review in Hoelzel 1991), however, most are now similar in design. A 4–8 mm diameter cylinder, sharpened at the leading edge, and providing some mechanism for the retention of the sample (such as inward-facing barbs) and a stop-plate to limit penetration, is attached to a dart body, arrow or bolt to be fired from a rifle or bow. A typical design for an air rifle is illustrated in Fig. 1.

The dart is either tethered or designed to float independently. Once retrieved it must be processed quickly, as the initial degradation of DNA by nucleases will start immediately. Skin tissue for subsequent DNA extraction should be immersed in a 20% solution of DMSO (dimethylsulphoxide) in water, saturated with NaCl (Hoelzel & Dover 1989). The salt inhibits the activity of the nucleases, while the DMSO facilitates

permeation of the preservative into the cells. Trials have shown this to be effective for 2 years at room temperature (Amos & Hoelzel 1991). An alternative is immersion in 100% ethanol. Biopsy samples collected as above provide ample DNA for PCR-based studies. The blubber should be stored separately in glass and kept frozen for use in subsequent stable isotope, fatty acid and biotoxin analyses (see Chapters 8 and 14). An alternative for sampling only skin, especially suited to bow-riding dolphins, is the 'scrub-pad' method. A strip of Teflon pan scrubber (e.g. the 3M™ product) is attached to the end of a pole and pushed along the back of the dolphin or whale. The entire strip is then immersed in the salt/DMSO preservative, and DNA is extracted from the sloughed skin caught in the scrub-pad. Another alternative for some species is to simply follow the whale in the water and collect naturally sloughing skin (most common for sperm whales). Tissue sampling from other marine mammals is typically more consistent with standard methods for restraining and sampling mammals for blood or skin biopsies. However, pinniped species can often be sampled without restraint using 'ear-clip' pliers (used to mark the ears of cattle) and collecting a small clip of skin from the webbing on the hind-flippers. The utility of biopsy samples ranges over a spectrum of applications from the identification of individuals (Amos & Hoelzel 1990; Palsbøll *et al.* 1997b) to population genetic studies and various other analyses such as stable isotope (see Chapter 8) and biotoxin analyses (see Chapter 14).

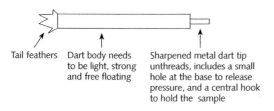

Tail feathers | Dart body needs to be light, strong and free floating | Sharpened metal dart tip unthreads, includes a small hole at the base to release pressure, and a central hook to hold the sample

Fig. 1 Air rifle biopsy dart design.

These results are consistent with theoretical predictions that population bottlenecks in sea otters did not result in a major loss of genetic variation among all populations during the period of population reduction and recovery (Ralls *et al.* 1983). Consistent with these results, Bodkin *et al.* (1999) found a positive correlation between genetic diversity and the size of remnant sea otter populations. The duration of the minimum population size was also an important factor.

Whaling records document the depletion of numerous whale stocks, but details of population size and patterns of recovery are even more difficult to assess than for amphibious species. The logistics of sample collection are also more complicated, and in the past were dependent on strandings and commercial catches. More recently, methods for the remote sampling of small tissue samples have been

developed (Box 11.2) and this has greatly facilitated population genetic studies for these species.

One species known to have been reduced to very low numbers is the northern right whale. The commercial hunt for this species began in the sixteenth century and continued into the early twentieth century. Even as populations became clearly depleted, hunting continued until their protection in the 1930s. Recent growth rates are estimated at 2.5% per annum (Knowlton *et al.* 1994). This rate and the current population estimates are consistent with a sustained bottleneck size of less than 100 animals at the population nadir. The southern right whale, on the other hand, is unlikely to have suffered a severe population bottleneck in spite of hunting pressure from the eighteenth into the twentieth centuries. Preliminary analyses of minisatellite (Shaeff *et al.* 1997) and mtDNA variation (Shaeff *et al.*

1991) suggested less diversity in the northern right whale, though the sample sizes were small. Recently, Malik *et al.* (1999) investigated variation at the mtDNA control region from 180 presumably unrelated northern right whales. These whales were sampled from two regions in the Bay of Fundy (North Atlantic Ocean). They used the single-stranded conformational polymorphism (SSCP) method (Box 11.1) together with sequence analysis and found only seven polymorphic sites defining five haplotypes (Table 11.1).

While, as noted above, direct comparisons must be made with caution, this appears to be considerably less variation than that found in some other baleen whale species (Table 11.1). For example, Bérubé *et al.* (1998) sequenced mtDNA control regions for fin whales from six putative populations in the North Atlantic and found greater variation (Table 11.1). If we consider just the population from the region where the right whales were sampled, there were 14 haplotypes among 29 fin whales, while in the Gulf of St Lawrence there were 22 haplotypes among 98 fin whales. There is also considerably more diversity at this locus in the southern right whale where there were 32 haplotypes among 99 individuals (see Rosenbaum *et al.* 2000), although variation within some local populations was lower. Baker *et al.* (1999) describe four haplotypes among 20 southern right whales sampled off the Auckland Islands, and five haplotypes among 20 animals sampled off southwest Australia.

The vaquita (*Phocoena sinus*) is another cetacean known to be rare. It also has the most restricted range of any cetacean species (see Gerrodette *et al.* 1995). Found only in the northern Gulf of California, it has a current population estimate of less than 600 animals (Jaramillo-Legorreta *et al.* 1999). Rosel and Rojas-Bracho (1999) investigated mtDNA control region variation (for 322 bp) from 43 vaquitas and found no variation at all. Limited data suggest that the vaquita population has been small since its discovery in 1958, and a persistent small effective population size or a founder event may explain the lack of variation.

In a number of cases a population bottleneck has been proposed for cetacean species based on low genetic diversity alone. For example, Palsbøll

et al. (1997a) investigated diversity at 287 bp from the mtDNA control region among 427 narwhals from the northwest Atlantic. They combined DNA sequencing (for 74 samples) with restriction fragment length polymorphism (RFLP) screening (353 samples) to reveal only four polymorphic sites defining five haplotypes. Furthermore, over 90% of sequenced individuals had one of two haplotypes. Their estimate for nucleotide diversity was 0.0017 for either the 74 DNA sequences or the entire sample of 427 genotypes. By all measures, this is a very low level of genetic diversity. According to reports from the International Whaling Commission there are approximately 30 000–40 000 narwhals in the waters west of Greenland, and probably far fewer east of Greenland. Essentially nothing is known about past levels of abundance. Pod structure in narwhals is highly matrifocal, and this can limit mtDNA diversity if haplotypes become fixed within social matrilines (see below). However, Palsbøll *et al.* (1997a) concluded that this mechanism alone is insufficient to explain the low levels of variation observed, and instead proposed a small founder population followed by a rapid, recent population expansion.

Various other matrifocal cetacean species have also been found to have low levels of mtDNA diversity, including the killer whale (Hoelzel & Dover 1991; Hoelzel *et al.* 1998a) and the sperm whale (Lyrholm *et al.* 1996; Lyrholm & Gyllensten 1998). Hoelzel *et al.* (1998a) sequenced 520 bp from the mtDNA control region from 66 killer whales from the eastern North Pacific. They found 11 polymorphic sites defining five haplotypes. Later work encompassing samples from around the world (including the eastern North Pacific, the western South Pacific, the western and eastern North Atlantic, and the western South Atlantic) and based on the entire control region (920 bp) shows only 12 haplotypes among 81 samples (A.R. Hoelzel *et al.* unpublished data). Lyrholm *et al.* (1996) sequenced the whole mtDNA control region (954 bp) from 37 sperm whales and found 12 variable sites defining 13 haplotypes. They calculated a worldwide average nucleotide diversity (including samples from the North Atlantic, North Pacific and southern hemisphere) of $\pi = 0.002$, comparable to that seen for the narwhal and northern elephant seal

1870–1987, *n* = 36

1988–1998, *n* = 46

(a)

(b)

Fig. 11.2 Parsimony network of mtDNA haplotypes for New Zealand Hector's dolphins from the east coast of the South Island collected from (a) 1870 to 1987, and (b) 1988 to 1998. The relative frequency of the different haplotypes is indicated by the size of the circles, and the difference among them by the number of cross-marks. The black circles show which haplotypes are present in the given population. (Adapted from Pichler & Baker 2000).

(Table 11.1). Assuming a rate of evolutionary change at this locus comparable to that seen in other cetacean species, the authors estimated the time when all modern sperm whales shared the same mtDNA haplotype, and based on this calculation proposed a population bottleneck in this species approximately 6000–25 000 years ago.

Whitehead (1998) suggested that low mitochondrial DNA diversity in cetacean species such as the sperm whale and killer whale may be a consequence of their matrilineal social structure. He proposes that variation could be reduced at neutral mtDNA loci by 'hitchhiking', through selection on maternally transmitted cultural traits, a concept borrowed from the idea that variation can be reduced through selection at a linked locus. This theory depends on the assumption that selection for cultural traits occurs in cetaceans, and on the condition that the lateral transmission of relevant behaviour to unrelated females will be at a rate of less than 0.5%. Such a theory could explain why a number of matrifocal species (killer whale, sperm whale, short-finned pilot whale, long-finned pilot whale) all seem to have low levels of mtDNA diversity compared to other cetacean species. However, it is not clear that it fits the data better than the alternative that these species have low effective population sizes. As described above, Lyrholm *et al.* (1996) suggest that the pattern of low variation seen in the sperm whale can be best explained by a bottleneck event. Some also doubt that the condition of strict matrilineal transmission of cultural traits is likely for these species (Mesnik *et al.* 1999). Tiedemann and Milinkovitch (1999) further suggest that given

matrilineal social structure and heterogeneity in reproductive success in space and time (perhaps associated with environmental factors), mtDNA variation will decrease for the population independent of any hitchhiking effects.

An interesting example of ongoing fisheries-related mortality correlating to the loss of variation over time, has been described by Pichler and Baker (2000) for Hector's dolphin. This species has a slow rate of reproduction (<5% per year; Slooten & Lad 1991) and is typically found in small, local populations. Hector's dolphins are caught as a bycatch in nets used for a fishery established off New Zealand in the 1920s. Pichler and Baker (2000) acquired samples from specimens dating from the 1870s through to the present day. They compared a 206 bp fragment from the mtDNA control region between 108 contemporary (collected from 1988 to 1998) and 55 historical (collected from 1870 to 1987) samples. A significant decrease in haplotype diversity and the overall number of matrilines was shown (Fig. 11.2).

11.3 PATTERNS OF GENETIC STRUCTURE

Population differentiation by genetic drift, selection or both can generate intraspecific structure that can be estimated and interpreted using molecular genetic methods (Box 11.3). In this section we review some of the patterns that have been observed among marine mammal species, illustrating trends with specific examples.

BOX 11.3 ANALYTICAL METHODS

Analysis of data

This short review is divided into analytical methods concerned with the assessment of within-population variation, and those concerned with structure and patterns of relatedness among populations (although some apply to both).

Assessment of within-population variation

The primary measures of genetic variability in populations are average heterozygosity (or gene or nucleotide diversity for DNA sequences), and the mean number of alleles (or number of segregating sites). Population genetics is largely concerned with the conditions that affect these two measures. Heterozygosity (H) and allelic diversity reflect long-term effective population size and inbreeding level ('effective' population size is the size of an idealized population that would show the same rate of decay of heterozygosity as the actual population).

Mutation is the ultimate source of genetic variability, and its rate (μ) varies among different types of genetic markers (e.g. point mutations occur less frequently in mtDNA than slippage mutations do in microsatellites). Genetic variation in a population or species is a balance between the rates of mutation, the random 'fixation' or loss of variants (genetic drift), selection and gene flow. Fixation refers to the condition when the allele is present in all individuals within the population. Genetic drift and natural selection are the two primary factors that affect gene frequency. In smaller populations, heterozygosity is lost more rapidly, and allele frequencies are more subject to change by genetic drift. When mutation restores variation at the same rate as genetic drift removes it, a population is in mutation–drift equilibrium.

A population that declines rapidly to a small size undergoes a genetic bottleneck. Bottlenecks result in losses of heterozygosity and allelic diversity, but the latter is proportionately more impacted than the former (Nei *et al.* 1975; Nei 1987). Therefore, bottlenecks can leave 'signatures' detected by assessing the levels of nucleotide diversity relative to allelic diversity (see Luikart *et al.* 1998).

Assessment of among-population variation and gene flow

Among-population genetic variation is assessed and interpreted by three primary approaches: calculation of genetic distances, estimation of inbreeding or fixation indices, and by visualizing population relationships using hierarchical structures or 'trees'.

Genetic distances: Genetic distance measures have been developed for different types of molecular markers. For allelic frequency data (e.g. for allozymes and microsatellites), a variety of distance measures have been developed that are based on the proportion of shared alleles (e.g. 'D' and 'D_{xy}'; for a review see Nei 1987). Some measures developed for microsatellites are based on a different evolutionary model, more appropriate for the 'stepwise' mutation at these loci (e.g. '$\delta\mu^2$', see Goldstein & Pollock 1997). For DNA sequence data, distance is often measured simply by the proportion of nucleotide sites that differ between sequences. This measure is often 'corrected' by a particular model of evolution, for example, the 'Jukes–Cantor' model corrects for multiple substitutions at a site (which will accumulate over time, since there are only four nucleic acid bases that constitute the DNA sequence), while the 'Kimura 2-parameter' method corrects for both multiple substitutions and different classes of base substitution: transitions (A–G or C–T) and transversions (all other types). Measures reflecting more complex models of base substitution are also sometimes used, depending on the type of locus, and the expected extent of the genetic distance.

Inbreeding coefficients or fixation indices: Wright (1951) developed a system for assessing the probability of fixation of variants (see above) by inbreeding within and among subpopulations of a larger population. This system can be likened to an analysis of variance, with the standardized partitioning of genetic variation into within- and among-subpopulation variation. Within-subpopulation inbreeding coefficients (F_{IS}) are estimated from the deviation of observed from predicted heterozygosity, and give the probability that alleles are identical by descent (IBD; identical because they share a common ancestor) within subpopulations. Positive values of F_{IS} indicate a deficiency of heterozygotes, most probably caused by inbreeding. The among-subpopulation equivalent, F_{ST} and its analogues (e.g. G_{ST}, R_{ST}, etc.), ranges from 0 (panmixia—free mixing among all subpopulations) to 1 (complete isolation between subpopulations). F_{ST} is the standardized variance in allele frequencies across subpopulations. Put another way, it represents the proportion of the total variance that explains variance among subpopulations ($F_{ST} = (H_T - H_S)/H_T$, where H_S is the mean of within-subpopulation heterozygosity, and H_T is the heterozygosity of the total population). F_{ST} assumes an 'infinite alleles' model of evolution (all new mutations are novel) and provides a measure of the amount of genetic structure among populations. An analogue, R_{ST} (e.g. Slatkin 1995), assumes a 'stepwise' model of evolution (new alleles derived in steps, which can occur again by the same process), essentially correcting for 'saturation' caused by stepwise back-mutations (and usually applied to

(*continued*)

BOX 11.3 (*cont'd*)

microsatellite data). Another F_{ST} analogue for which a stepwise mutation model can be applied is ϕ_{ST}, which can be estimated by partitioning the genetic variation among different levels of a hierarchy in an analysis of molecular variance (AMOVA) (Excoffier *et al.* 1992).

Phylogeography: Genetic variation and structure can also be assessed visually in the form of networks or trees. Networks indicate the relationships of genotypes or populations based on the mutational steps among sequences or distances (see Fig. 11.2). Networks are usually not 'rooted' by ancestral or 'outgroup' genotypes. Trees can be constructed by a number of phylogenetic or clustering methods. Some methods, such as 'neighbour joining' use matrices of among-individual or population genetic distances. 'Maximum parsimony' identifies related taxa by means of nested

sets of shared, derived characters (mutational steps), and constructs phylogenetic trees from these data. This method searches for the tree that requires the fewest steps (most parsimonious). 'Minimum spanning networks' can be obtained by parsimony criteria, but differ from phylogenetic trees in that existing genotypes can be placed in the network as ancestral genotypes (a more realistic approach for population data). 'Maximum likelihood' estimation is an increasingly applied method that generates the most likely tree under a variety of complex models of evolution. Sampling error in phylogenetic reconstructions can be estimated by generating pseudoreplicates in a 'bootstrapping' procedure. This gives an estimate for the relative support for different lineages in the reconstruction. For further discussion on phylogenetic methods see Chapter 2, Swofford *et al.* (1996) and Page and Holmes (1998).

11.3.1 Amphibious species

In both the southern and northern hemispheres, several species of pinniped have large distributions that are essentially circumpolar (e.g. pack-ice breeding seals and seals breeding on subantarctic islands) or wide ranging (e.g. walrus and harbour seals occur in both the Pacific and Atlantic Oceans). Polar bears also have a large distribution across the Arctic and subarctic regions. There are few data on the population structure of any of the pack-ice seals; however, the population structure of several species of pinniped occurring in the subarctic, subantarctic and temperate regions have been investigated. Interspecific comparisons suggest some differences in the extent of population genetic structure among northern and southern hemisphere species (see below).

In the walrus (*Odobenus rosmarus*), restriction enzyme analysis of the three mtDNA NADH-dehydrogenase (ND) segments ND1, ND2 and ND3/4 has been used to support the designation of two subspecies, *O. r. rosmarus* (the Atlantic walrus) and *O. r. divergens* (the Pacific walrus) (Cronin *et al.* 1994). That study further suggested population differentiation in the North Atlantic. This was confirmed in a more recent study investigating population structure among samples from northwest and eastern Greenland, Svalbard and Franz Joseph Land in the eastern North Atlantic at mtDNA and nuclear microsatellite loci (Andersen *et al.* 1998).

RFLP analysis of mtDNA ND segments (the same method and locus as for Cronin *et al.* 1994) revealed a clear distinction between northwest Greenland and each of the other three populations (Fig. 11.3). Pairwise comparisons of ϕ_{ST} (Box 11.3) indicated very high levels of differentiation between northwest Greenland and each of the other populations (with between-population differences explaining 74–80% of the variance), and no significant differentiation between the other three populations. Although the sample size from northwest Greenland was relatively small ($n = 19$), the genetic distance is large enough to suggest a real difference. The microsatellite DNA data, based on 11 loci, showed significant differentiation between all populations, though not as pronounced as for the mtDNA data for northwest Greenland.

The harbour seal (*Phoca vitulina*) has one of the broadest geographical distributions of any seal, extending from the eastern Baltic Sea westward across the Atlantic and Pacific Oceans to southern Japan. The worldwide genetic structure of this species has been investigated by Stanley *et al.* (1996) who examined 435 bp of control region mtDNA sequence in 227 harbour seals from 24 localities across the species range. Similar to walrus populations, phylogenetic analysis of harbour seal data has revealed that Atlantic and Pacific Ocean populations, and east and west coast populations in these oceans are highly differentiated. This is consistent

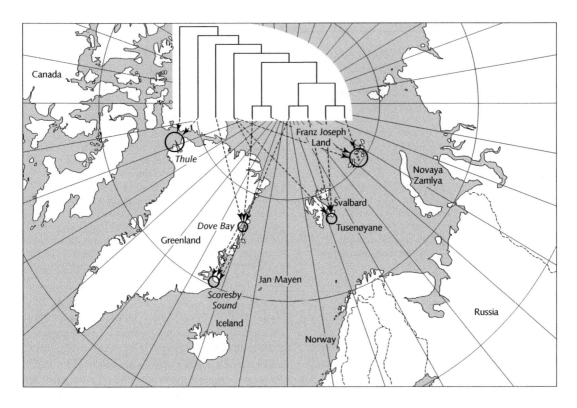

Fig. 11.3 Walrus population differentiation in the North Atlantic based on an RFLP analysis of the mtDNA ND region. The tree was constructed using the Wagner parsimony method and is based on 2000 bootstrap replications. The arrows indicate the geographical source for each haplotype. (Adapted from Andersen *et al.* 1998).

with the current taxonomic subdivision of this species into four subspecies (see also Arnason *et al.* 1995; Kappe *et al.* 1997). Such a broad pattern of population genetic structure is thought to reflect large-scale historic events (Stanley *et al.* 1996). Results are consistent with the ancient isolation of populations in the Pacific and Atlantic Ocean brought about by ice ages with extensive continental ice sheets and sea ice. In both oceans, populations appear to have been established from east to west, with the most recent established populations being those in Europe. These may reflect recolonization from ice age refugia after the last glaciation (Stanely *et al.* 1996).

Genetic population structure has been examined in detail in two of the subspecies of the harbour seal, *P. v. vitualina* in Europe, and *P. v. richardsi* on the east Pacific Ocean coast (Lamont *et al.* 1996; Goodman 1998). Both studies found significant levels of genetic population structure that were consistent with geographical separation. The study of

Goodman (1998) examined seven microsatellite loci in a large sample of seals (1029) from 12 localities in Europe. He identified six distinct populations throughout Europe. Where there were continuous distributions of breeding animals along coastlines there was little genetic substructuring, but differentiation increased with geographical distance.

In grey seals (*Halichoerus grypus*), Allen *et al.* (1995) examined microsatellite DNA variation at two breeding colonies in Scotland, separated by about 500 km, to determine the level of philopatry and migration between them. As previous tracking and tagging studies had shown that grey seals can travel thousands of kilometres (McConnell *et al.* 1992), Allen *et al.* (1995) were surprised to find highly significant differences in allele frequencies between the breeding sites for all eight loci they examined, indicating extensive genetic differentiation. They concluded that despite the large distances travelled by seals outside the breeding season, the

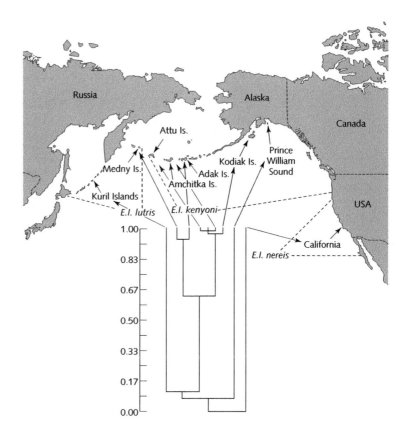

Fig. 11.4 Sea otter (*Enhydra lutris*) population differentiation in the North Pacific based on a mtDNA RFLP analysis. A UPGMA tree is shown and the scale indicates the Nei genetic identity. The arrows indicate the geographical source for each haplotype, and the designated subspecies boundaries are indicated. (Adapted from Cronin *et al.* 1996).

philopatry of pups and adults were important aspects of the breeding biology of these seals.

In sea otters, three subspecies have been described based on cranial morphology: *Enhydra lutris lutris* from the northwestern Pacific Ocean, *E. l. kenyoni* from the Aleutian Islands east and south to Oregon, and *E. l. nereis* along the Californian coast (Wilson *et al.* 1991). Cronin *et al.* (1996) examined population structure in these putative subspecies using restriction enzyme analysis of four segments of mtDNA (NADH ND1, ND3/4, ND5/6 and 12S–16S ribosomal RNA). They found that the Californian subspecies (*E. l. nereis*) was monophyletic while the other two subspecies were polyphyletic (Fig. 11.4). For example, the population of sea otters at Medny Island, which is considered to be *E. l. lutris* based on morphology, has haplotype frequencies much more similar to populations of *E. l. kenyoni* than to another putative population of *E. l. lutris* (Cronin *et al.* 1996). Cronin *et al.* (1996) propose four genetically distinct populations

of sea otters, not consistent with subspecies status: California, Prince William Sound, the Kodiak–Adak–Amchitka–Attu–Medny Islands and the Kuril Islands (Fig. 11.4). They also suggest that the haplotype frequencies among these populations are distinct enough to indicate differentiation prior to human exploitation.

In polar bears, early studies of genetic population structure had been hampered by the very low levels of variation found in this species at allozyme (Allendorf *et al.* 1979; Larsen *et al.* 1983) and mtDNA loci (Cronin *et al.* 1991; Shields & Kocher 1991). However, Paetkau *et al.* (1995) used eight hypervariable microsatellite loci to compare four northeastern Canadian polar bear populations: the northern Beaufort Sea, southern Beaufort Sea, western Hudson Bay and Davis Strait in the Labrador Sea. Using this method they found considerable genetic variation with high levels of heterozygosity (ranging from 25% to 86% per locus) within each population. Allele frequency differences were tested using a G-test, and all population comparisons showed significantly different allele distributions. This suggested restricted gene flow between local populations despite the long-distance seasonal movements typical of polar bears (Paetkau *et al.* 1995). The genetic relatedness between North American populations and other circumpolar Arctic populations of polar bears has yet to be established.

In contrast to polar bears, examination of Steller's sea lion (*Eumetopias jubatus*) control region mtDNA sequences has revealed very high levels of genetic variability, with no common haplotype or haplotypes predominating throughout the species range (Bickham *et al.* 1996). Instead, Bickham *et al.* (1996) found 52 haplotypes, most occurring at relatively low frequency among 224 individuals (Table 11.1). Many of the haplotype lineages were specific to either an eastern (southeastern Alaska and Oregon) or western population (Commander Islands to the Gulf of Alaska), consistent with two genetically divergent populations of Steller sea lions (Bickham *et al.* 1996). The very high level of sequence divergence observed between the most divergent haplotypes in the Steller sea lions, relative to that observed between Californian and Steller sea lions, suggest that the haplotypes in Steller sea lions have persisted for an estimated 380 000 years (Bickham *et al.* 1996).

The identification of two genetically differentiated stocks of Steller's sea lions may have important management implications. The species has suffered major declines in numbers from around 240 000–300 000 in the 1960s to about 116 000 in 1989 (Loughlin *et al.* 1992), representing a 40–50% decline in the worldwide population. The species is currently listed as endangered in the United States. However, the two genetically divergent stocks of sea lions have shown markedly different population trends in recent times, with western stocks experiencing drastic reductions (some by up to 81% since the 1960s), while eastern stocks have remained stable or have increased slightly in number (Merrick *et al.* 1991; Loughlin *et al.* 1992).

As for the Steller's sea lion, two genetically distinct populations of Californian sea lion have been identified. Maldonado *et al.* (1995) compared three colonies along the Pacific coast (southern and Baja California and the Gulf of California). Although

they found no variation for 368 bp of mtDNA cytochrome *b* sequences, they found 11 haplotypes in a 360 bp region of the mtDNA control region. Four haplotypes were restricted to the Gulf of California population and this population was phylogenetically distinct from populations along the Pacific coast. Given the low level of within-population variation, the average sequence divergence between them was substantial at 4.3%, suggesting long-term isolation. However, the relationship between Pacific coast and Gulf of California populations to the Galápagos Islands subspecies has yet to be established.

In the southern hemisphere, genetic population structure has been examined in detail in three species with large geographical ranges, the southern elephant seal, the Antarctic fur seal (*Arctocephalus gazella*) and the subantarctic fur seal (*A. tropicalis*). Southern elephant seals breed on subantarctic islands and have a circumpolar distribution. Examination of population structure using mtDNA control region sequence and nuclear DNA markers has revealed four main populations: Heard Island, South Georgia, Peninsula Valdez (Argentina) and Macquarie Island (Fig. 11.5) (Hoelzel *et al.* 1993; Slade 1997; Slade *et al.* 1998; Hoelzel *et al.* 2001). Much of this population structure is due to the divergent Peninsula Valdez and Macquarie Island populations, the lineages of which form monophyletic groups (Slade *et al.* 1998). In contrast, the South Georgia and Heard Island populations share mtDNA haplotype lineages (Slade *et al.* 1998). Slade *et al.* (1998) suggest that such population structure may reflect historical associations of populations related to habitat availability during the last ice age. However, the low variation found on the mainland population in Argentina, in contrast to greater diversity on the islands, may reflect a founding event establishing the mainland rookery, and continuing gene flow at some level among the island sites (Hoelzel *et al.* 2001).

While there was strong geographical structure for mtDNA, there was less for nuclear DNA markers (Slade *et al.* 1998; Hoelzel *et al.* 2001). These differences may be attributable to differences in the rate of gene flow for males and females, or differences in the characteristics of the mtDNA and nuclear markers, including rates of mutation

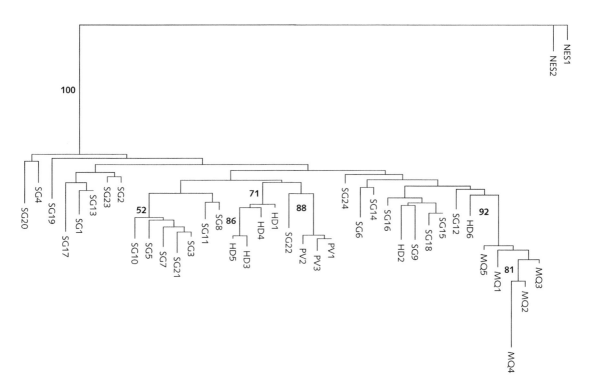

Fig. 11.5 Southern elephant seal neighbour-joining phylogeny based on mtDNA control region sequence. The geographical source of the haplotypes is indicated by the haplotype names: HD, Heard Island; MQ, Macquarie Island; NES, northern elephant seal; PV, Peninsula Valdez, Argentina; SG, South Georgia. Bootstrap support is given in bold numbers. (Data from Hoelzel *et al.* 2001).

and/or genetic drift. The former interpretation would be consistent with ecological studies on male and female dispersal in this species (Hindell *et al.* 1991; Slade 1997).

Recently, Wynen *et al.* (2000) examined the worldwide population structure of the two Southern Ocean fur seals with circumpolar distributions (Antarctic and subantarctic fur seals). The extent to which the current genetic differentiation in these species reflects past population subdivision is difficult to determine, as both species were subject to intensive sealing during the eighteenth and nineteenth centuries resulting in significant population reductions and local extinctions (see above). For subantarctic fur seals, Wynen *et al.* (2000) examined a 316 bp region of the mtDNA control region from 103 individuals (plus 89 screened for RFLPs) from five major breeding populations. They found 33 haplotypes, 13 of which were represented in more than one individual (Table 11.1). The geographical distribution of these haplotypes is shown in Fig. 11.6. They found a high degree of lineage structure, with three distinct lineages apparent, and some geographical structure in the distribution of lineages, particularly among the three populations that survived commercial sealing (Gough, Marion and Amsterdam Islands). The distribution of lineages within *Arctocephalus tropicalis* suggests that the Marion Island population was the major source for immigrants for recently recolonized sites at Macquarie Island and Iles Crozet, with some input from Ile Amsterdam (Wynen *et al.* 2000).

For Antarctic fur seals, Wynen *et al.* (2000) examined mtDNA control region sequences from 145 individuals from eight populations, finding 26 haplotypes, 16 of which were represented by more than one individual (Table 11.1). In comparison with the subantarctic fur seal, the relationship between

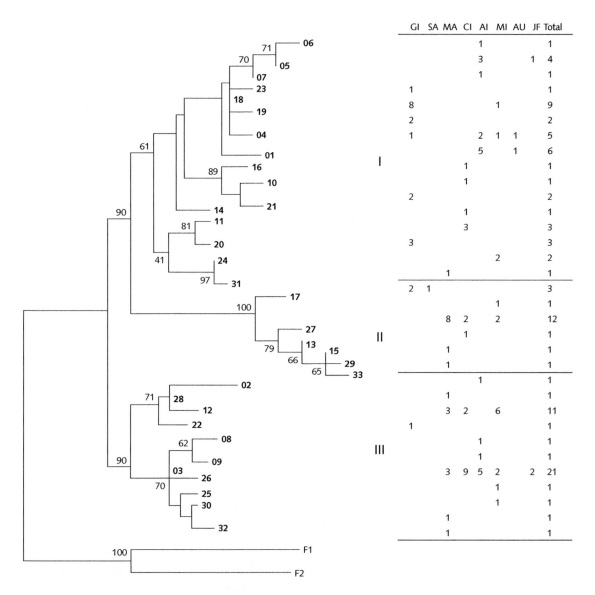

Fig. 11.6 Neighbour-joining tree of 33 sequence haplotypes of mtDNA control regions observed in five populations of *Arctocephalus tropicalis* and vagrants, the variable sites for each haplotype and its geographical distribution. Labels are arbitrarily assigned to haplotypes from 1 to 33 (bold numerals). Bootstrap values are shown only at nodes which were supported in over 60% of the 1000 replications. The three major clades are labelled as I, II and III. Variable sites are numbered according to their position within the 316 bp aligned sequence. The locations where samples were taken from are Gough Island (GI), South Africa (SA), Marion Island (MA), Iles Crozet (CI), Ile Amsterdam (AI), Macquarie Island (MI), Australia (AU) and the Juan Fernández Archipleago (JF). The outgroup is *A. forsteri* (F1 and F2). (Adapted from Wynen *et al.* 2000.)

haplotypes relevealed little lineage structure, although the geographical distribution of haplotypes suggest two genetically differentiated regions. An eastern region was centred on Iles Kerguelen and Macquarie Island, and a western region included South Georgia, the South Shetland Islands, Bouvetøya and Marion Islands. The populations at Iles Crozet and Heard Island show a mixture of haplotypes from both regions. Wynen *et al.*'s (2000) results suggest that postsealing populations survived at

South Georgia, Bouvetøya and Iles Kergulen, with the former two populations probably being the source of immigrants to the recently recolonized Marion and South Shetland Islands, and the latter providing the source for Macquarie Island. The recolonized populations at Iles Crozet and Heard Island may have received immigrants from several sources. The possible existence of three remnant populations of Antarctic fur seal surviving sealing is at odds with the regularly reported idea that the species only survived at Bird Island (next to South Georgia).

Hybrids between Antarctic and subantarctic fur seals have been inferred using a combination of mtDNA sequencing of the control region and DNA fingerprinting (Goldsworthy *et al.* 1999). The paternity of pups was determined from DNA fingerprinting, and species-specific mtDNA control region haplotypes of each parent were compared. Using this method, Goldsworthy *et al.* (1999) determined that 33% (13 of 39 cases where paternity was determined) of the pups examined in their study were the product of subantarctic male/Antarctic female parents, indicating high levels of hybridization and/or backcrossing in this population. Importantly, only three (23%) of the 13 known cases of hybridization/backcrossing were also classed as phenotypic hybrids on the basis of external characteristics, while three of the four putative hybrids based on phenotype had parents of differing mtDNA haplotype (Goldsworthy *et al.* 1999). These results indicate that estimates of the extent of hybridization in the population that are based on phenotypic traits may significantly underestimate the real extent of hybridization in the population. It is unknown whether hybrid fur seals have significantly reduced fitness compared with non-hybrids, although this seems likely given the phylogenetically distinct lineages seen for these two species based on mtDNA loci (Wynen *et al.* 2000, 2001).

The worldwide distribution of Cape and Australian fur seal subspecies (*Arctocephalus pusillus*) are restricted to the southern coasts of Africa and southeastern Australia, respectively, and are geographically isolated by the southern Indian Ocean. Lento *et al.* (1997) found relatively low levels of variation, but only one common haplotype between the two subspecies. A phylogenetic reconstruction suggests that the Australian population may have resulted from a recent historical founder from South Africa.

For the New Zealand fur seal (*A. forsteri*), mtDNA cytochrome *b* sequence data from 56 animals from across the species range revealed two deeply divergent lineages (Lento *et al.* 1997). These lineages differ by an average divergence of 3.4%, an order of magnitude greater than that found between Australian and Cape fur seals. The maximum divergence found among New Zealand fur seals was 4.2%, which Lento *et al.* (1997) argue is close to the approximate threshold of divergence (5.0%) found among closely related mammalian taxa using cytochrome *b*. A considerable proportion of the genetic variation found among New Zealand fur seals was attributable to geographical structuring, with the majority of New Zealand samples coming from one lineage, and those from Western Australia coming from the other. Samples from Tasmania were intermediate between the two regions, with six of the nine haplotypes from each lineage being present in the 12 samples (Lento *et al.* 1997).

In addition to the highly divergent lineage in New Zealand fur seals, Wynen *et al.* (2001) found two differentiated lineages in South American fur seals (*A. australis*) based on cytochrome *b* sequence data. Surprisingly, the relationships among these four lineages were poorly resolved (Fig. 11.7), which reinforces the need for further work on the classification of otariid species (see Chapter 2).

From the limited number of species where genetic population structure has been examined, the extent of population subdivision present among species is highly variable. This may be due to interspecific differences in the extent of site fidelity and philopatry, differences in the duration of isolation among populations, and historical effects on habitat availability and species distribution. However, geographical barriers to gene flow appear to be the most important determinants of population subdivision in the northern hemisphere. The major difference in the distribution pattern of northern and southern hemisphere pinniped populations is that in the northern hemisphere, the Pacific and Atlantic Oceans are separated by two extensive land masses to the east and west, and by polar ice

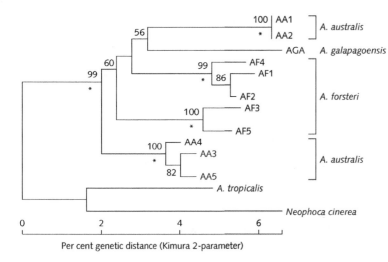

Fig. 11.7 Neighbour-joining tree reconstructed from combined cytochrome *b* and control region data for clade 3 from Fig. 2 in Wynen *et al.* (2001), with *Neophoca cinerea* and *Arctocephalus tropicalis* as outgroups. Figures at the nodes indicate bootstrap values of 50% or greater obtained after 2000 replications. * indicates where branch lengths are significant.

in the north. For example, the coupling of these geographical barriers with periodic ice ages has probably been largely responsible for much of the stock subdivisions in harbour seals and the walrus. In the southern hemisphere, although ice ages may have been important in the genetic subdivision of some species, there are fewer obvious barriers to gene flow. Many species have large circumpolar distributions, and their distributions are not typically broken by continental land masses. However, population structure exists for many of these species, which in some cases may simply reflect the large distances between breeding habitats (see above).

11.3.2 Aquatic species

In recent reviews of cetacean population genetic structure, Hoelzel (1994, 1998b) highlights the observation that geographical barriers, or the apparent lack of them, are not necessarily good indicators of population genetic structure in these species. The lack of genetic concordance to geographical distribution is particularly striking in species that show differentiation among putative populations in sympatry. This is rare among animal species in general, but may be more common among delphinid cetaceans. A complicating factor is our relatively poor understanding of taxonomy among these species (see Chapter 2), and this will be discussed further in the context of specific examples.

The most striking example is probably that of the killer whale (*Orcinus orca*). Killer whales are distributed worldwide, especially in temperate to polar regions (see Chapter 1), but the best studied populations are in the eastern North Pacific. Here a clear distinction between resource specialists has been described, based on prey choice (see Bigg *et al.* 1990). One putative population, known as 'transients' prey primarily on other marine mammals, while the other, known as 'residents', prey primarily on fish. Pod size and other aspects of social structure differ (see review in Bigg *et al.* 1987), but their geographical ranges overlap extensively. These putative populations have been compared at mitochondrial and nuclear DNA markers, and found to be genetically isolated (Hoelzel & Dover 1991; Hoelzel *et al.* 1998a; A.R. Hoelzel *et al.*, unpublished data). A comparison of 1812 bp from the mtDNA control region and adenosine triphosphatase (ATPase) region showed no shared haplotypes and a measure of population differentiation (F_{ST}) indicated that 92% of the genetic variance was due to differences between the two populations. There was no variation within two regional populations of 'resident' whales (off Washington State and off Alaska), and little variation among 'transient' pods. Variation at five microsatellite DNA loci showed a similar pattern, though the degree of differentiation between transients and Alaskan residents was greater than that between transients and the Washington State residents (Hoelzel *et al.* 1998a; A.R. Hoelzel *et al.*, unpublished data). Male-mediated gene flow remains a possibility.

Taken in isolation, the level and pattern of differentiation is enough to suggest the possibility of a species-level distinction. However, this is less clear when populations from around the world are compared (Hoelzel & Dover 1991; A.R. Hoelzel *et al.*, unpublished data). The eastern North Pacific transient population stands out as distinct from all other geographical populations (including Iceland, New Zealand and Argentina) regardless of feeding strategy, and shows about the same genetic distance from each (including the eastern North Pacific residents). The maximum level of divergence (0.9%) is also low for a mammalian species-level difference. The data are more consistent with the retention of at least these two distinct matrilines within the species over evolutionary time, and may reflect the impact of a historical bottleneck (A.R. Hoelzel *et al.*, unpublished data). However, the current level of gene flow between sympatric residents and transients is low enough that incipient speciation is a possibility.

Various other dolphin species show differential habitat use among groups of sympatric or parapatric individuals, especially with respect to the utilization of nearshore and offshore habitat. Nearshore and offshore animals often differ in some aspect of morphology, which could reflect either genetic or environmental factors (the latter through an influence on development). For example, the two morphs of spotted dolphins (*Stenella attenuata*) differ in tooth and jaw structure (Douglas *et al.* 1984). Two morphs of common dolphins (*Delphinus delphis*) differ in beak length, and are sometimes found in sympatry in nearshore habitats. Rosel *et al.* (1994) compared eight short-beaked animals from California, four from the Black Sea and six from the eastern tropical Pacific with 11 long-beaked animals from California. Although the sample size was small, their data suggest a significant difference between the two morphs. Further, the sympatric short- and long-beaked forms in California were more differentiated (1.09%) than the allopatric short-beaked populations (0.02%).

The best known example is that of the bottlenose dolphin (*Tursiops truncatus*), which occurs in coastal and offshore populations throughout its range. However, the taxonomy of *Tursiops* popula-

tions has a history of being difficult to resolve. In the eastern North Pacific, nearshore and offshore forms were originally classified as two different species: *T. gilli* (the nearshore form) and *T. nuuanu*, though a reappraisal recognizing extensive overlap in morphotypes later reclassified both as *T. truncatus* (Walker 1981). In the South Indian and South Pacific Oceans, a smaller morphotype has been described in nearshore habitats, known as the 'aduncus' type (Ross 1977). Recent population genetic comparisons between 'aduncus' and 'truncatus' forms in China based on mtDNA control region sequence (Wang *et al.* 1999) support the current consensus that these represent different species (cf. LeDuc *et al.* 1999; Rice 1999). However, in some locations differentiation is more consistent with intraspecific differences.

In the western North Atlantic, the nearshore and offshore forms have been described in some detail and show both morphometric (especially cranial characteristics, but also body size; Mead & Potter 1995) and genetic differentiation (Hoelzel *et al.* 1998b). There were also consistent differences between the two types in feeding behaviour with the nearshore form feeding primarily on coastal fish, while the offshore form concentrated on deep-water squid (Mead & Potter 1995). The genetic distinction, based on both mtDNA ($\phi_{ST} = 0.604$) and nuclear DNA (five microsatellite loci; $R_{ST} = 0.373$), indicated low levels of gene flow (or no gene flow in the recent past) between the two populations (Fig. 11.8) (Hoelzel *et al.* 1998b). However, not all nearshore and offshore forms of this species show substantial genetic differentiation (Hoelzel *et al.* 1998b; A. Natoli *et al.*, unpublished data).

A number of species show significant differentiation between geographical populations, though it is not always clear what represents the likely geographical boundary (see Hoelzel 1994). The best studied species in this respect is the humpback whale (*Megaptera novaeangliae*). From over a dozen published studies, thousands of humpback whales have been genotyped from across their worldwide range. Initial studies in the North Pacific based on mtDNA variation showed significant differentiation between putative stocks (Baker *et al.* 1993). This work was later extended to include both mtDNA and nuclear DNA markers and a broader range of

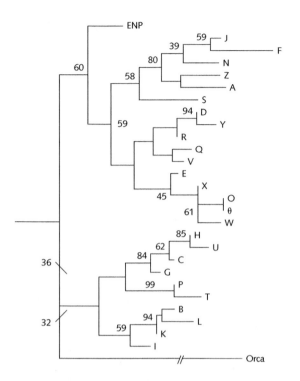

Fig. 11.8 Bottlenose dolphin differentiation in nearshore and offshore habitats based on mtDNA control region sequences. Figures at the nodes indicate bootstrap support. Haplotypes are represented by letters. BAH, Bahamas; ENA, eastern North Atlantic; ESA, eastern South Atlantic; WNAP, western North Atlantic pelagic population; WNAC, western North Atlantic coastal population. (Adapted from Hoelzel *et al.* 1998b.)

populations (Palumbi & Baker 1994; Baker *et al.* 1998a). The mtDNA data confirmed earlier studies indicating the influence of maternal fidelity to migratory destinations (Baker *et al.* 1990, 1994). The authors identified two primary eastern North Pacific stocks—one that breeds in Hawaiian waters and migrates to feeding grounds off Alaska, and another that breeds off Mexico and migrates to feeding grounds off California (Fig. 11.9) (cf. Fig. 7.8). Significant geographical structure was also found for the nuclear markers, but these loci showed lower levels of differentiation and suggested the possibility of greater male-mediated gene flow among the two stocks (Baker *et al.* 1998a).

In the North Atlantic, genetic (Palsbøll *et al.* 1997b) and photographic (e.g. Clapham *et al.* 1993) mark-recapture studies suggest a single breeding population in the West Indies and provided matches to feeding aggregations from the Gulf of Maine to the Barents Sea (Fig. 11.9). Population genetic studies have tended to support the idea of a single breeding population for most of the North Atlantic (Palsbøll *et al.* 1995; Larsen *et al.* 1996), though it remains possible that at least some

whales that feed in the eastern North Atlantic, breed off West Africa (Larsen *et al.* 1996). Two primary mtDNA lineages were described, one shared across all North Atlantic populations, and another found only in the western North Atlantic (Palsbøll *et al.* 1995). A study based on variation at four microsatellite DNA loci also suggested differentiation between the eastern and western North Atlantic, though the sample size for the eastern population was low (11 samples from Iceland) (Valsecchi *et al.* 1997).

Global perspectives based on mtDNA (Fig. 11.9) have shown shared lineages and suggested trans-equatorial migration at some time in the past (Baker *et al.* 1990, 1993). Lineages found in the southern hemisphere are represented both in the North Pacific and North Atlantic (Medrano-Gonzalez *et al.* 1995; Baker *et al.* 1998a, 1998b). A study of mtDNA control region variation among 84 humpback whales from four putative stocks in the Southern Ocean identified three differentiated populations (western Australia, eastern Australia and Tonga, and Gorgona Island off western Africa). This was consistent with female migratory fidelity

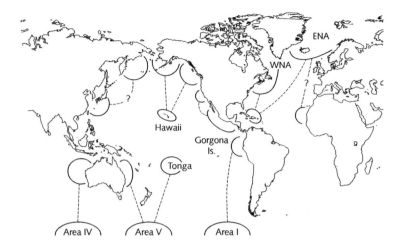

Fig. 11.9 Links between humpback whale seasonal populations (low latitudes for breeding and high latitudes for feeding) based on DNA analyses. Two links marked '?' are only hypothetical. See text for further discussion.

to feeding grounds in management areas IV, V and I (see Donovan 1991), respectively, in the Antarctic (Baker *et al.* 1998b).

Most of the data for humpback whales suggests at least female fidelity between feeding and breeding grounds, which would allow the designation of broad geographical regions for different stocks. However, some species are more likely to converge in mixed assemblages of breeding stocks on seasonal feeding grounds (see Hoelzel 1998b). One example is the minke whale (*Balaenoptera acutorostrata*). Minke whale stocks on either side of Japan (the Korean vs. the western North Pacific stocks) are genetically differentiated at both allozyme (Wada 1991) and mtDNA loci (Goto & Pastene 1996). Both studies found evidence for the seasonal mixing of these stocks on feeding grounds in the Okhotsk Sea. In another example, a temporal mixing of two genetic stocks, primarily in the early part of the feeding season, was reported for minke whale samples from the Antarctic (management areas IV and V), based on mtDNA variation (Pastene *et al.* 1996). Genetic studies (based on mtDNA) of minke whales in the North Atlantic and North Sea show no clear distinction between stocks off Norway and off Iceland, but in each location there are two distinct mtDNA lineages (Palsbøll 1990; Bakke *et al.* 1996), which may represent different breeding stocks.

In an extensive study of mtDNA (288 bp from the control region) and six microsatellite loci, Bérubé *et al.* (1998) investigated fin whale (*Balaenoptera*

physalus) variation across the North Atlantic from six putative populations. All 309 North Atlantic whales were sampled on summer feeding grounds. Little is known about breeding behaviour in this and other species in the genus *Balaenoptera*, but earlier studies had suggested the possibility of mixing breeding stocks on feeding grounds off Iceland (Danielsdottir *et al.* 1991, 1992). Results reported by Bérubé *et al.* (1998) indicated non-significant F_{ST} values for some transatlantic comparisons, such as the Gulf of St Lawrence and West Greenland compared to Spain. They also showed reduced heterozygosity for combined samples across the North Atlantic. Together these data suggest multiple stocks in the North Atlantic, and the mixing of some stocks on feeding grounds. However, they also found evidence of differentiation, especially between the Mediterranean and other populations, and between a sample from the Sea of Cortez in the North Pacific ($n = 61$), and all North Atlantic populations. Further, given that the estimated divergence time (based on mtDNA) between Pacific and Atlantic samples was less than the time since the rise of the Panama Isthmus, they suggest occasional gene flow between the two oceans.

Among the odontocetes, the sperm whale is known to have a very broad geographical range, and this is especially true of the adult males. Lyrholm and Gyllensten (1998) describe differentiation between ocean basins based on mtDNA control region sequences. However they found no genetic structure at nine microsatellite loci, and interpret this to indicate greater male-mediated gene flow among oceans (Lyrholm *et al.* 1999). In contrast, beluga (*Delphinapterus leucas*) had been divided into numerous putative stocks based on local concentrations and migration routes in the Arctic. Sequence data from the mtDNA control region comparing 628 beluga from 25 sites in the Canadian Arctic indicated two lineages, suggesting postglacial recolonization of the Arctic from two different refugia (Brown-Gladden *et al.* 1997). One lineage was represented primarily in the west from the Chukchi Sea to Baffin Bay, while the other was found primarily in the Hudson Bay. A third lineage was represented in both the Hudson Bay and the St Lawrence River. O'Corry-Crowe *et al.* (1997) studied populations representing five summer concentrations around the coast of Alaska from Cook Inlet to the Beaufort Sea. Their mtDNA control region data also supported the idea of a rapid radiation of beluga into the western Nearctic, and indicated significant differentiation among the five putative populations ($\phi_{ST} = 0.33$).

Harbour porpoises also show fine-scale population genetic structure with respect to the movement of females. In a study based on mtDNA control region sequence and six microsatellite DNA loci, Rosel *et al.* (1999) found significant differentiation for mtDNA comparing four putative summer-breeding populations ranging from the Gulf of Maine to West Greenland. For example, the proximate populations in the Gulf of St Lawrence and Newfoundland showed an F_{ST} of 0.024 ($P = 0.02$). However, none of the pairwise comparisons for the combined microsatellite loci showed a significant difference among populations. The authors suggest that this may imply greater male-mediated gene flow. In winter, the stocks disperse, and it is thought that they mainly migrate south. Samples collected off the US mid-Atlantic states were compared with the samples collected in summer, and the pattern of variation suggested the seasonal mixing of stocks in this region (Rosel *et al.* 1999). The level of mtDNA nucleotide diversity for all western North Atlantic populations was nearly two times higher than for populations studied at the same locus in the northeastern Atlantic (Walton 1997).

In a study of mtDNA RFLP variation in the North Pacific, Dall's porpoise (*Phocoenoides dalli*)

populations in the Bering Sea and Aleutians were found to be significantly differentiated from a parapatric region in the western North Pacific ($G_{ST} = 0.14$, $P < 0.01$) (McMillan & Bermingham 1996). This was consistent with an earlier study based on allozymes (Winans & Jones 1988).

Sirenian species, while fully aquatic, are less mobile than cetaceans and have not been observed in water far from land. There are two species of manatee in the genus *Trichechus*, the West Indian manatee (*T. manatus*) and the Amazonian manatee (*T. inunguis*) (see Chapter 1). Two subspecies have been proposed for the West Indian manatee based on variation in cranial characters (Domning & Hayek 1986). Of these, the Florida manatee (*T. m. latirostris*) is restricted to the Florida peninsula, while the Antillian manatee (*T. m. manatus*) is distributed throughout the Caribbean and Central and South America. Their distribution is patchy, and recent habitat degradation and epizootics associated with the 'red tide' have threatened local populations (O'Shea *et al.* 1991, 1995).

A recent study investigated mtDNA control region variation among eight putative populations of *T. manatus* and included 86 individuals (Table 11.1) (Garcia-Rodriguez *et al.* 1998). Sampled populations were from Florida, Mexico, the Dominican Republic, Puerto Rico and sites along the South American coastline from Columbia to Brazil. Florida ($n = 23$), Mexico ($n = 6$) and Brazil ($n = 6$) were each fixed for a single, different haplotype. The other four populations were variable with 2–7 haplotypes ($n = 4$–22). Only the Dominican Republic and Puerto Rico were very similar to each other ($\chi^2 = 0.12$, $P = 0.8$), while all other comparisons showed significantly different haplotype frequencies ($P = 0.01$–0.001).

A neighbour-joining phylogeny identified three geographically consistent clusters with one dominated by haplotypes from Florida and the West Indies, another by Mexico, Columbia and Venezuela, and the third by Guyana and Brazil. These lineages are not concordant with previous subspecific designations. The authors conclude that open waters and unsuitable coastal habitat may have constituted significant barriers to gene flow, resulting in differentiation among these three lineages. An additional 16 samples from *T. inunguis* were analysed and compared with the 86 *T. manatus* samples. Surprisingly, pairwise differences between the two species were of the same level as pairwise differences among the three principal clusters of *T. manatus* haplotypes (d = 0.04–0.08).

11.4 CONCLUSIONS

Marine mammals are large, highly mobile animals that live in an environment with few obvious physical boundaries. All else being equal, we may expect these species to show rather little structure among populations, as the potential for gene flow seems high. There are certainly some marine fish species that meet this expectation (see review by Graves 1995). However, many marine mammals in fact show relatively extensive structure among populations, and while some differentiation may be expected on the basis of allopatry and isolation by distance or restricted gene flow due to physical boundaries, many examples cannot be easily explained in this way. Instead, many of the observed patterns are probably due to a complex interaction between historical changes in marine environments (e.g. the impact of ice ages), resource requirements and specializations, and aspects of life history and demographics.

REFERENCES

Allen, P.J., Amos, W., Pomeroy, P.P. & Twiss, S.D. (1995) Microsatellite variation in grey seals (*Halichoerus grypus*) shows evidence of genetic differentiation between two British breeding colonies. *Molecular Ecology* **4**, 653–662.

Allendorf, F.W., Christiansen, F.B., Donson, T., Eanes, W.F. & Frydenberg, O. (1979) Electrophoretic variation in large mammals. 1. The polar bear, *Thalarctos maritimus*. *Hereditas* **91**, 19–22.

Amos, W. & Hoelzel, A.R. (1990) DNA fingerprinting cetacean biopsy samples for individual identification. *International Whaling Commission Special Issue* **12**, 79–86.

Amos, W. & Hoelzel, A.R. (1991) Long-term preservation of whale skin for DNA analysis. *International Whaling Commission Special Issue* **13**, 99–104.

Andersen, L.W., Born, E.W., Gjertz, I. *et al.* (1998) Population structure and gene flow of the Atlantic walrus (*Odobeus rosmarus rosmarus*) in the eastern Atlantic Arctic based on mitochondrial DNA and microsatellite variation. *Molecular Ecology* **7**, 1323–1336.

Arnason, U., Bodin, K., Gullberg, A., Ledje, C. & Mouchaty, S. (1995) A molecular view of pinniped relationships with particular emphasis on the true seals. *Journal of Molecular Evolution* **40**, 78–85.

Baker, C.S., Palumbi, S.R., Lambertsen, R.H. *et al.* (1990) Influence of seasonal migration on geographic distribution of mitochondrial DNA haplotypes in humpback whales. *Nature* **344**, 238–240.

Baker, C.S., Perry, A., Bannister, J.L. *et al.* (1993) Abundant mitochonrial DNA variation and worldwide population structure of humpback whales. *Proceedings of the National Academy of Sciences of the USA* **90**, 8239–8243.

Baker, C.S., Slade, R.B., Bannister, J.L. *et al.* (1994) Hierarchical structure of mtDNA gene flow among humpback whales, world-wide. *Molecular Ecology* **3**, 313–327.

Baker, C.S., Medrano-Gonzalez, L., Calambokidis, J. *et al.* (1998a) Population structure of nuclear and mitochondrial DNA variation among humpback whales in the North Pacific. *Molecular Ecology* **7**, 695–707.

Baker, C.S., Florez-Gonzalez, L., Abernethy, B. *et al.* (1998b) Mitochondrial DNA variation and maternal gene flow among humpback whales of the southern hemisphere. *Marine Mammal Science* **14**, 721–737.

Baker, C.S., Patenaude, N.J., Bannister, J.L., Robins, J. & Kato, H. (1999) Distribution and diversity of mtDNA lineages among southern right whales (*Eubalaena australis*) from Australia and New Zealand. *Marine Biology* **134**, 1–7.

Bakke, I., Johansen, S., Bakke, O. & El-Gewely, M.R. (1996) Lack of population subdivision among minke whales (*Balaenoptera acutorostrata*) from Iceland and Norwegian waters based on mtDNA sequences. *Marine Biology* **125**, 1–9.

Bartholomew, G.A. & Hubbs, C.L. (1960) Population growth and seasonal movements of the northern elephant seal, *Mirounga angustirostris*. *Mammalia* **24**, 313–324.

Bernardi, G., Fain, S.R., Gallo-Reynoso, J.P., Figueroa-Carranza, A.L. & Le Boeuf, B.J. (1998) Genetic variability in Guadalupe fur seals. *Journal of Hereditary* **89**, 301–305.

Bérubé, M., Aguilar, A., Dendanto, D. *et al.* (1998) Population genetic structure of North Atlantic, Mediterranean Sea and Sea of Cortez fin whales, *Balaenoptera physalus*: analysis of mitochondrial and nuclear loci. *Molecular Ecology* **7**, 575–584.

Bester, M.N. (1987) Subantarctic fur seal, *Arctocephalus tropicalis*, at Gough Island (Tristan Da Cunha Group). In: *Status, Biology, and Ecology of Fur Seals* (J.P. Croxall & R.L. Gentry, eds), pp. 57–60. NOAA Technical Report NMFS No. 51.

Bickham, J.W., Patton, J.C. & Loughlin, T.R. (1996) High variability for control-region sequences in a marine mammal: implications for conservation and biogeography of Steller sea lions (*Eumetopias jubatus*). *Journal of Mammalogy* **77**, 95–108.

Bigg, M.A., Ellis, G.M., Ford, J.K.B. & Balcomb, K.C. (1987) *Killer Whales: a Study of their Identification, Genealogy and Natural History in British Columbia and Washington State*. Phantom Press, Nanaimo, BC.

Bigg, M.A., Olesiuk, P.K., Ellis, G.M., Ford, J.K.B. & Balcomb, K.C. (1990) Social organization and genealogy of resident killer whales (*Orcinus orca*) in the coastal waters of British Columbia and Washington State. *International Whaling Commission Species Issue* 12, 383–406.

Bodkin, J.L., Ballachey, B.E., Cronin, M.A. & Schribner, K.T. (1999) Population demographics and genetic diversity in remnant and translocated populations of sea otters. *Conservation Biology* 13, 1378–1385.

Boness, D.J. (1991) Determinants of mating systems in the Otariidae (Pinnipedia). In: *The Behaviour of Pinnipeds* (D. Renouf, ed.), pp. 1–36. Chapman & Hall, London.

Bonnell, M. & Selander, R.K. (1974) Elephant seals: genetic variation and near extinction. *Science* 184, 908–909.

Bonner, W.N. (1958) Notes on the southern fur seal in South Georgia. *Proceedings of the Zoological Society of London* 130, 241–252.

Bonner, W.N. & Laws, R.M. (1964) Seals and sealing. In: *Antarctic Research* (R. Priestley, R.J. Adie & G.D.Q. Robin, eds), pp. 163–190. Butterworths, London.

Boyd, I.L. (1993) Pup production and distribution of breeding Antarctic fur seals (*Arctocephalus gazella*) at South Georgia. *Antarctic Science* 5, 17–24.

Brown-Gladden, J.G., Ferguson, M.M. & Clayton, J.W. (1997) Matriarchal genetic population structure of North American beluga whales *Delphinapterus leucas* (Cetacea: Monodontidae). *Molecular Ecology* 6, 1033–1046.

Busch, B.C. (1985) *The War Against the Seals*. McGill–Queen's University Press, Montreal, Canada.

Caswell, H., Fujiwara, M. & Brault, S. (1999) Declining survival probability threatens the North Atlantic right whale. *Proceedings of the National Academy of Sciences of the USA* 96, 3308–3313.

Clapham, P.J., Baraff, L.S., Carlson, C.A. *et al.* (1993) Seasonal occurrence and annual return of humpback whales in the southern Gulf of Maine. *Canadian Journal of Zoology* 71, 440–443.

Cronin, M.A., Amstrup, S.C., Garner, G.W. & Vyse, E.R. (1991) Interspecific and intraspecific mitochondrial DNA variation in North American bears (*Ursus*). *Canadian Journal of Zoology* 69, 2985–2992.

Cronin, M.A., Hills, S., Born, E.W. & Patton, J.C. (1994) Mitochondrial DNA variation in Atlantic and Pacific walruses. *Canadian Journal of Zoology* 72, 1035–1043.

Cronin, M.A., Bodkin, J., Ballachey, B., Estes, J. & Patton, J.C. (1996) Mitochondrial-DNA variation among subspecies and populations of sea otters (*Enhydra lutris*). *Journal of Mammalogy* 77, 546–557.

Danielsdottir, A.K., Duke, E.J. & Arnason, A. (1992) *Mitochondrial DNA Analysis of North Atlantic Fin Whales and Comparison With Four Species of Whales: Sei, Minke, Pilot and Sperm Whales*. Report of the International Whaling Commission No. SC/F91/F17. International Whaling Commission, Cambridge, UK.

Danielsdottir, A.K., Duke, E.J., Joyce, P. & Arnason, A. (1991) Preliminary studies on the genetic variation at enzyme loci in fin whales and sei whales from the North Atlantic. *International Whaling Commssion Special Issue* 13, 115–124.

Domning, D.P. & Hayek, L.C. (1986) Interspecific and intraspecific morphological variation in manatees (Sirenia: *Trichechus*). *Marine Mammal Science* 2, 87–144.

Donovan, G.P. (1991) A review of IWC stock boundaries. *International Whaling Commission Special Issue* 13, 39–70.

Douglas, M.E., Schnell, G.D. & Hough, D.J. (1984) Differentiation between inshore and offshore spotted dolphins in the eastern tropical Pacific Ocean. *Journal of Mammalogy* 65, 375–387.

Excoffier, L., Smouse, P. & Quattro, J. (1992) Analysis of molecular variance inferred from metric distances among DNA haplotypes: applications to human mitochondrial DNA restriction data. *Genetics* 131, 479–491.

Garcia-Rodriguez, A.I., Bowen, B.W., Domning, D. *et al.* (1998) Phylogeography of the West Indian manatee (*Trichechus manatus*): how many populations and how many taxa? *Molecular Ecology* 7, 1137–1149.

Gerrodette, T. & Gilmartin, W.G. (1990) Demographic consequences of changed pupping and hauling sites of the Hawaiian monk seal. *Conservation Biology* 4, 423–430.

Gerrodette, T., Fleischer, L.A., Perez-Cortes, H. & Ramirez, B.V. (1995) Distribution of the vaquita, *Phocoena sinus*, based on sightings from systematic surveys. *International Whaling Commission Special Issue* 16, 271–281.

Goldstein, D.B. & Pollock, D.D. (1997) Launching microsatellites: a review of mutation processes and methods of phylogenetic inference. *Journal of Heredity* 88, 335–342.

Goldsworthy, S.D., Boness, D.J. & Fleischer, R.C. (1999) Mate choice among sympatric fur seals: female preference for conphenotypic males. *Behavioral Ecology and Sociobiology* 45, 253–267.

Goldsworthy, S.D., Francis, J., Boness, D. & Fleischer, R. (in press) Variation in the mitochondrial control region in the Juan Fernández fur seal (*Arctochepalus philippii*). *Heredity*, in press.

Goodman, S.J. (1998) Patterns of extensive genetic differentiation and variation among European harbor seals (*Phoca vitulina vitulina*) revealed using microsatellite DNA polymorphisms. *Molecular Biology and Evolution* 15, 104–118.

Goto, M. & Pastene, L.A. (1996) Population genetic structure in the western North Pacific minke whale examined by two independent RFLP analyses of mitochondrial DNA. Report of the International Whaling Commission No. SC/48/NP5. International Whaling Commission, Cambridge, UK.

Graves, J.E. (1995) Conservation genetics of fishes in the pelagic marine realm. In: *Conservation Genetics; Case Histories from Nature* (J.C. Avise & J.L. Hamrick, eds), pp. 335–366. Chapman & Hall, New York.

Guinet, C., Jouventin, P. & Georges, J.-Y. (1994) Long term changes of fur seals *Arctocephalus gazella* and *Arctocephalus tropicalis* on subantarctic (Crozet) and subtropical (St Paul and Amsterdam) islands and their possible relationship to El Nino Southern Oscillation. *Antarctic Science* 6, 473–478.

Headland, R. (1984) *The Island of South Georgia.* Cambridge University Press, Cambridge, UK.

Hillis, D.M., Moritz, C. & Mable, B.K., eds (1996) *Molecular Systematics,* 2nd edn. Sinauer Associates, Sunderland, MA

Hindell, M.A., Burton, H.R. & Slip, D.J. (1991) Foraging areas of southern elephant seals, *Mirounga leonina,* as inferred from water temperature data. *Australian Journal of Marine and Freshwater Research* 42, 115–128.

Hoelzel, A.R. (1991) *Genetic Ecology of Whales and Dolphins.* International Whaling Commission Special Issue No. 13. International Whaling Commission, Cambridge, UK.

Hoelzel, A.R. (1994) Ecology and genetics of whales and dolphins. *Annual Review of Ecology and Systematics* 25, 377–399.

Hoelzel, A.R. (1998a) *Molecular Genetic Analysis of Populations; a Practical Approach,* 2nd edn. Oxford University Press, Oxford.

Hoelzel, A.R. (1998b) Genetic structure of cetacean populations in sympatry, parapatry and mixed assemblages; implications for conservation policy. *Journal of Heredity* 98, 451–458.

Hoelzel, A.R. (1999a) Impact of a population bottleneck on genetic variation and the importance of life history; a case study of the northern elephant seal. *Biological Journal of the Linnean Society* 68, 23–39.

Hoelzel, A.R. (1999b) Assessing the impact of a population bottleneck in the northern elephant seal. In: *Abstracts from the 13th Annual Meeting of the Society for Conservation Biology.*

Hoelzel, A.R. & Dover, G.A. (1989) Molecular techniques for examining genetic variation and stock identity in cetacean species. *International Whaling Commission Special Issue* 11, 81–120.

Hoelzel, A.R. & Dover, G.A. (1991) Genetic differentiation between sympatric killer whale populations. *Heredity* 66, 191–196.

Hoelzel, A.R., Halley, J., O'Brien, S.J. *et al.* (1993) Elephant seal genetic variation and the use of simulation models to investigate historical bottlenecks. *Journal of Heredity* 84, 443–449.

Hoelzel, A.R., Dahlheim. M. & Stern, S.J. (1998a) Low genetic variation among killer whales (*Orcinus orca*) in the eastern North Pacific, and genetic differentiation between foraging specialists. *Journal of Heredity* 89, 121–128.

Hoelzel, A.R., Potter, C.W. & Best, P. (1998b) Genetic differentiation between parapatric 'nearshore' and 'offshore' populations of the bottlenose dolphin. *Proceedings of the Royal Society of London B* 265, 1–7.

Hoelzel, A.R., Stephens, J.C. & O'Brien, S.J. (1999a) Molecular genetic diversity and evolution at the MHC DQB locus in four species of pinnipeds. *Molecular Biology and Evolution* 16, 611–618.

Hoelzel, A.R., LeBoeuf, B.J., Reiter, J. & Campagna, C. (1999b) Alpha male paternity in elephant seals. *Behavioral Ecology and Sociobiology* 46, 298–306.

Hoelzel, A.R., Campagna, C. & Arnbom, T. (2001) Genetic and morphometric differentiation between island and mainland populations of the southern elephant seal. *Proceedings of the Royal Society of London, Series B* 268, 325–332.

Hoelzel, A.R., Natoli, A., Dahlheim, M., Baird, R.W. & Black, N. (in review) World-wide variation and demographic history of the killer whale.

Hofmeyr, G.J.G., Bester, M.N. & Jonker, F.C. (1997) Changes in population sizes and distribution of fur seals at Marion Island. *Polar Biology* 17, 50–158.

Hubbs, C.L. & Norris, K.S. (1971) Original teeming abundance, supposed extinction, and survival of the Juan Fernández fur seal. *Antarctic Research Series* 18, 35–52.

Jaramillo-Legorreta, A.M., Rojas-Bracho, L. & Gerrodette, T. (1999) A new abundance estimate for vaquitas: first step for recovery. *Marine Mammal Science* 15, 957–973.

Johnson, A.M., DeLong, R.L., Fiscus, C.H. & Kenyon, K. (1982) Population status of the Hawaiian monk seal (*Monachus schauinslandi*). *Journal of Mammalogy* 63, 415–421.

Kappe, A.L., Bijlsma, R., Osterhaus, A.D.M.E., Van Delden, W. & Van de Zande, L. (1997) Structure and amount of genetic variation at minisatellite loci within the subspecies complex of *Phoca vitulina* (the harbour seal). *Heredity* 78, 457–463.

Kenyon, K.W. (1972) Man versus monk seal. *Journal of Mammalogy* 53, 687–696.

Kenyon, K.W. & Rice, D.W. (1959) Life history of the Hawaiian monk seal. *Pacific Science* 13, 359–367.

Knowlton, A.R., Kraus, S.D. & Kenney, R.D. (1994) Reproduction in North Atlantic right whales (*Eubalaena glacialis*). *Canadian Journal of Zoology* 72, 1297–1305.

Kretzmann, M.B., Gilmartin, W.G., Meyer, A. *et al.* (1997) Low genetic variability in the Hawaiian monk seal. *Conservation Biology* 11, 482–490.

Lamont, M.M., Vida, J.T., Jeffries, S. *et al.* (1996) Genetic substructure of the Pacific harbor seal (*Phoca vitulina richardsi*) off Washington, Oregon, and California. *Marine Mammal Science* 12, 402–413.

Larsen, A.H., Sigurejonsson, J., Gisli Vikingsson, N.O. & Palsbøll, P.J. (1996) Population genetic analysis of nuclear and mitochondrial loci in skin biopsies collected from central and northeastern North Atlantic humpback whales (*Megaptera novaeangliae*): population identity and migratory destinations. *Proceedings of the Royal Society of London B* 263, 1611–1618.

Larsen, T., Tegelström, H., Kumar Juneja, R. & Taylor, M.K. (1983) Low protein variability and genetic similarity between populations of the polar bear (*Ursus maritimus*). *Polar Research* **1**, 97–105.

LeDuc, R.G., Perrin, W.F. & Dizon, A.E. (1999) Phylogenetic relationships among the delphinid cetaceans based on full cytochrome b sequences. *Marine Mammal Science* **15**, 619–648.

Lento, G.M., Haddon, M., Chambers, G.K. & Baker, C.S. (1997) Genetic variation of southern hemisphere fur seals (*Arctocephalus* spp.): investigation of population structure and species identity. *Journal of Heredity* **88**, 202–208.

Lewonton, R.C. & Hubby, J.L. (1966) A molecular approach to the study of genetic heterozygosity in natural populations. II. Amounts of variation and degree of heterozygosity in natural populations of *Drosophila pseudoobscura*. *Genetics* **54**, 595–609.

Lidicker, W.Z. & McCollum, F.C. (1997) Allozymic variation in California sea otters. *Journal of Mammalogy* **78**, 417–425.

Loughlin, T.R., Perlov, A.S. & Vladimirov, V.A. (1992) Rangewide survey and estimation of total number of Steller sea lions in 1989. *Marine Mammal Science* **8**, 220–239.

Luikart, G., Allendorf, F.W., Cornuet, J.M. & Sherwin, W.B. (1998) Distortion of allele frequency distributions provides a test for recent population bottlenecks. *Journal of Heredity* **89**, 238–247.

Lyrholm, T. & Gyllensten, U. (1998) Global matrilineal population structure in sperm whales as indicated by mtDNA sequences. *Proceedings of the Royal Society of London B* **265**, 1679–1684.

Lyrholm, T., Leimar, O. & Gyllensten, U. (1996) Low diversity and biased substitution patterns in the mtDNA control region of sperm whales: implications for estimates of time since common ancestry. *Molecular Biology and Evolution* **13**, 1318–1326.

Lyrholm, T., Leimar, O., Johannson, B. & Gyllensten, U. (1999) Sex-biased dispersal in sperm whales: contrasting mitochondrial and nuclear genetic structure of global populations. *Proceedings of the Royal Society of London B* **266**, 347–354.

McConnell, B.J., Chambers, C., Nicholas, K.S. & Fedak, M.A. (1992) Satellite tracking of gray seals (*Halichoerus gryopus*). *Journal of Zoology* **226**, 271–282.

McMillan, W.O. & Bermingham, E. (1996) The phylogeographic pattern of mitochondrial DNA variation in the Dall's porpoise *Phocoenoides dalli*. *Molecular Ecology* **5**, 47–61.

Maldonado, J.E., Davila, F.O., Stewart, B.S., Geefen, E. & Wayne, R.K. (1995) Intraspecific genetic differentiation in California sea lions (*Zalophus californianus*) from southern California and the Gulf of California. *Marine Mammal Science* **11**, 46–58.

Malik, S., Brown, M.W., Kraus, S.D. *et al.* (1999) Assessment of mitochondrial DNA structuring and nursery use in the North Atlantic right whale (*Eubalaena glacialis*). *Canadian Journal of Zoology* **77**, 1217–1222.

Mead, J.G. & Potter, C.W. (1995) Recognising two populations of the bottlenose dolphin (*Tursiops truncatus*) off the Atlantic coast of North America: morphologic and ecological considerations. *Index of Biological Integrity Report* **5**, 31–44.

Medrano-Gonzalez, L., Aguayo-Lobo, A., Urban-Ramirez, J. & Baker, C.S. (1995) Diversity and distribution of mitochondrial DNA lineages among humpback whales, *Megaptera novaeangliae*, in the Mexican Pacific Ocean. *Canadian Journal of Zoology* **73**, 1735–1743.

Merrick, R.L., Rerm, L.M., Everitt, R.D., Ream, R.R. & Lessard, L.A. (1991) *Aerial and Ship-based Surveys of Northern Sea Lions (Eumetopias jubatus) in the Gulf of Alska and Aleutian Islands During June and July 1990*. NOAA Technical Memorandum NMFS No. F/NWC-196.

Mesnik, S.L., Taylor, B.L., LeDuc, R.G. *et al.* (1999) Culture and genetic evolution in whales. *Science* **284** (5432), 2055a.

Nei, M. (1987) *Molecular Evolutionary Genetics*. Columbia University Press, New York.

Nei, M., Maruyama, T. & Chakraborty, R. (1975) The bottleneck effect and genetic variability in populations. *Evolution* **29**, 1–10.

O'Corry-Crowe, G.M., Suydam, R.S., Rosenberg, A., Frost, K.J. & Dizon, A.E. (1997) Phylogeography, population structure and dispersal patterns of the beluga whale *Delphinapterus leucas* in the western Nearctic revealed by mitochondrial DNA. *Molecular Ecology* **6**, 955–970.

O'Shea, T.J., Rathbun, G.B., Bonde, R.K., Buergelt, C.D. & Odell, D.K. (1991) An epizootic of Florida manatees associated with a dinoflagellate bloom. *Marine Mammal Science* **7**, 165–179.

O'Shea, T.J., Ackerman, B.B. & Percival, H.F. (1995) Introduction. In: *Population Biology of the Florida Manatee* (T.J. O'Shea, B.B. Ackerman & H.F. Percival, eds), pp. 1–5. National Biological Service Technical Report No. 1. US Department of the Interior, Washington, DC.

Paetkau, D., Calvert, W., Stirling, I. & Strobeck, C. (1995) Microsatellite analysis of population structure in Canadian polar bears. *Molecular Ecology* **4**, 347–354.

Page, R.D.M. & Holmes, E.C. (1998) *Molecular Evolution: a Phylogenetic Approach*. Blackwell Science, Oxford.

Palsbøll, P.J. (1990) *Preliminary Results of Restriction Fragment Length Analysis of Mitochondrial DNA in Minke Whales from the Davis Strait, Northwest and Central Atlantic*. Report of the International Whaling Commission No. SC/42/NHMi35. International Whaling Commission, Cambridge, UK.

Palsbøll, P.J., Clapham, P.J., Mattila, D.K. *et al.* (1995) Distribution of mtDNA haplotypes in North Atlantic humpback whales: the influence of behavior on population structure. *Marine Ecology Progress Series* **116**, 1–10.

Palsbøll, P.J., Heide-Jorgensen, M.P. & Dietz, R. (1997a) Population structure and seasonal movements of narwhals, *Monodon monoceros*, determined from mtDNA analysis. *Heredity* **78**, 284–292.

Palsbøll, P.J., Allen, J., Bérubé, M. *et al.* (1997b) Genetic tagging of humpback whales. *Nature* **388**, 767–769.

Palumbi, S.R. & Baker, C.S. (1994) Contasting population structure from nuclear intron sequences and mtDNA of humpback whales. *Molecular Biology and Evolution* **11**, 426–435.

Pastene, L.A., Goto, M., Itoh, S. & Numachi, K.I. (1996) Spatial and temporal patterns of mitochondrial DNA variation in minke whales from Antarctic areas IV and V. *Report of the International Whaling Commission* **46**, 305–314.

Pichler, F.B. & Baker, C.S. (2000) Loss of genetic diversity in the endemic Hector's dolphin due to fisheries-related mortality. *Proceedings of the Royal Society of London B* **267**, 97–102.

Ragen, T.J. (1993) *Status of the Hawaiian Monk Seal in 1992.* Report No. H-93-05. Southwest Fisheries Science Centre, Honolulu Laboratory, Honolulu.

Ralls, K., Ballou, J. & Brownwell, R.L. (1983) Genetic diversity in Californian sea otters: theoretical considerations and management implications. *Biological Conservation* **25**, 209–232.

Rice, D.W. (1999) *Marine Mammals of the World; Systematics and Distribution.* Special Publication No. 4. Society for Marine Mammalogy, Lawrence, KA.

Richards, R. (1994) 'The upland seal' of the Antipodes and Macquarie Islands: a historian's perspective. *Journal of the Royal Society of New Zealand* **24**, 289–295.

Rooney, A.P., Honeycutt, R.L., Davis, S.K. & Derr, J.N. (1999) Evaluating a putative bottleneck in a population of bowhead whales from patterns of microsatellite diversity and genetic disequilibria. *Journal of Molecular Evolution* **49**, 682–690.

Rosel, P.E., Dizon, A.E. & Heyning, J.E. (1994) Genetic analysis of sympatric populations of common dolphins (genus *Delphinus*). *Marine Biology* **119**, 159–167.

Rosel, P.E., France, S.C., Wang, J.Y. & Kocher, T.D. (1999) Genetic structure of harbour porpoise *Phocoena phocoena* populations in the northwest Atlantic based on mitochondrial and nuclear markers. *Molecular Ecology* **8**, S41–S54.

Rosel, P.E. & Rojas-Bracho, L. (1999) Mitochondrial DNA variation in the critically endangered vaquita *Phocoena sinus* Norris and Macfarland, 1958. *Marine Mammal Science* **15**, 990–1003.

Rosenbaum, H.C., Brownell, R.L., Brown, M.W. *et al.* (2000) World-wide genetic differentiation of *Eubalaena*: questioning the number of right whale species. *Molecular Ecology* **9**, 1793–1802.

Ross, G.J.B. (1977) The taxonomy of bottlenose dolphins *Tursiops* species in South African waters, with notes on their biology. *Annals of Cape Provincial Museum* **11**, 135–194.

Sanchez, M.S. (1992) *Differentiation and variability of mitichondrial DNA in three sea otter, Enhydra lutris, populations.* MSc Thesis, University of California, Santa Cruz.

Shaeff, C.M., Kraus, S.D., Brown, M.W. *et al.* (1991) Preliminary analysis of mtDNA variation in and between right whale species *Eubalaena glacialis* and *E. australis*. *International Whaling Commission Special Issue* **13**, 217–224.

Shaeff, C.M., Kraus, S.D., Brown, M.W. *et al.* (1997) Comparison of genetic variability of North and South Atlantic right whales (*Eubalaena*), using DNA fingerprinting. *Canadian Journal of Zoology* **75**, 1073–1080.

Shaughnessy, P.D. & Goldsworthy, S.D. (1990) Population size and breeding season of the Antarctic fur seal *Arctocephalus gazella* at Heard Island—1987/88. *Marine Mammal Science* **6**, 292–304.

Shields, G.F. & Kocher, T.D. (1991) Phylogenetic relationships of North American ursids based on analysis of mitichondrial DNA. *Evolution* **45**, 218–221.

Slade, R.W. (1997) Genetic studies of the southern elephant seal. *Mirounga leonina*. In: *Marine Mammal Research in the Southern Hemisphere, 1: Status, Ecology and Medicine* (M. Hindell & C. Kemper, eds), pp. 11–29. Surrey Beatty & Sons, Chipping Norton, UK.

Slade, R.W., Moritz, C., Hoelzel, A.R. & Burton, H.R. (1998) Molecular population genetics of the southern elephant seal *Mirounga leonina*. *Genetics* **149**, 1945–1957.

Slatkin, M. (1995) A measure of population subdivision based on microsatellite allele frequencies. *Genetics* **139**, 457–462.

Slooten, E. & Lad, F. (1991) Population biology and conservation of Hector's dolphin. *Canadian Journal of Zoology* **69**, 1701–1707.

Stanley, H.F., Casey, S., Carnahan, J.M. *et al.* (1996) Worldwide patterns of mitochondrial DNA differentiation in the harbor seal (*Phoca vitulina*). *Molecular Biology and Evolution* **13**, 368–382.

Stewart, B.S., Yochem, P.K., Huber, H.R. *et al.* (1994) History and present status of the northern elephant seal. In: *Elephant Seals: Population Ecology, Behavior and Physiology* (B.J. LeBoeuf & R.M. Laws, eds), pp. 29–48. University of California Press, Berkely, CA.

Swofford, D.L., Olsen, D.G., Waddell, P.J. & Hillis, D.M. (1996) Phylogenetic inference. In: *Molecular Systematics*, 2nd edn (D.M. Hillis, C. Moritz & B.K. Mable, eds), pp. 407–514. Sinauer Associates, Sunderland, MA.

Tiedemann, R. & Milinkovitch, M.C. (1999) Cultural and genetic evolution in whales. *Science* **284** (5432), 2055a.

Torres, D.N. (1987) Juan Fernández fur seal, *Arctocephalus philippii*. In: *Status, Biology, and Ecology of Fur Seals* (J.P. Croxall & R.L. Gentry, eds), pp. 37–41. NOAA Technical Report NMFS No. 51.

Valsecchi, E., Palsbøll, P.J., Hale, P. *et al.* (1997) Microsatellite genetic distances between oceanic populations of the humpback whale (*Megaptera novaeangliae*). *Molecular Biology and Evolution* **14**, 355–362.

Wada, S. (1991) *Genetic Distinction Between Two Minke Whale Stocks in the Okhotsk Sea Coast of Japan.* Report of the International Whaling Commission No. SC/43/Mi32. International Whaling Commission, Cambridge, UK.

Walker, W.A. (1981) Geographic variation in morphology and biology of bottlenose dolphins (*Tursiops*) in the eastern North Pacific. NOAA/NMFS Southwest Fisheries Science Center Administrative Report No. LJ-81-3c.

Walton, M.J. (1997) Population structure of harbour porpoises *Phocoena phocoena* in the seas around the UK and adjacent waters. *Proceedings of the Royal Society of London B* **264**, 89–94.

Wang, J.Y., Chou, L.S. & White, B.N. (1999) Mitochondrial DNA analysis of sympatric morphotypes of bottlenose dolphins (genus: *Tursiops*) in Chinese waters. *Molecular Ecology* **8**, 1603–1612.

Whitehead, H. (1998) Cultural selection and genetic diversity in matrilineal whales. *Science* **282**, 1708–1711.

Wilson, D.E., Bogan, M.A., Brownwell Jr, R.L., Burdin, A.M. & Maminov, M.K. (1991) Geographic variation in sea otters, *Enhydra lutris*. *Journal of Mammalogy* **72**, 22–36.

Winans, G.A. & Jones, L.J. (1988) Electrophoretic variability in the Dall's porpoise (*Phocoenoides dalli*) in the North Pacific Ocean and Bering Sea. *Journal of Mammalogy* **69**, 4–21.

Wright, S. (1951) The genetical structure of populations. *Annals of Eugenics* **15**, 323–354.

Wynen, L.P., Goldsworthy, S.D., Guinet, C. *et al.* (2000) Post-sealing genetic variation and population structure of two species of fur seal (*Arctocephalus gazella* and *A. tropicalis*). *Molecular Ecology* **9** (3), 299–314.

Wynen, L.P., Goldsworthy, S.D., Insley, S.J. *et al.* (2001) Phylogenetic relationships within the eared seals (Otariidae: Carnivora), with implications for the historical biogeography of the family. *Molecular Phylogenetics and Evolution* **21**, 270–284.

CHAPTER 12

Ecology of Group Living and Social Behaviour

Richard C. Connor

12.1 INTRODUCTION

The evolution of social behaviour in marine mammals has been shaped fundamentally by three main ecological characteristics: where they give birth, where they forage and what they eat. Cetaceans, pinnipeds, polar bears and sea otters consume a diet of animal tissue, mostly fish, squid and crustaceans, that yields a much higher rate of energy intake than the mostly vegetarian diet of sirenians (see Chapter 8). Diet also affects the cost of group formation, thus exerting a profound influence on their social lives. Individuals can join groups if their food is abundant and clumped but must disperse if their prey is dispersed. For example, fin whales may form larger groups than minke whales in the Gulf of California because the fins' euphausiid prey occurs in larger patches than the fish favoured by minkes (Tershy 1992).

Pinnipeds (and occasionally sea otters) give birth on land or ice, removing their vulnerable newborns from the threat of predation by sharks and killer whales. The vulnerability of newborn mammals in the three-dimensional aquatic habitat may have played a fundamental role in shaping the complex social lives we find in some of the toothed whales (Connor & Norris 1982). The need to give birth in a relatively safe environment but to feed in a productive habitat may be a factor in the extensive migration by baleen whales between warm tropical and cold polar waters (Corkeron & Connor 1999). Some marine mammal mothers must leave their offspring while they forage. Otariid females leave their pups in the relative safety of the colony on land during foraging excursions. Sperm whales leave vulnerable offspring at the surface in the company of

'babysitters' while they forage at depth (Whitehead 1996). Baleen whales and phocids both travel to relatively safe havens to give birth (land or ice and tropical waters) and do not leave their offspring to forage during lactation. The energy for lactation in both groups comes from fat stores they have acquired since their previous effort to reproduce (see Chapter 9).

As with other chapters in this volume, the selection of topics and species is naturally limited to the minority of marine mammals that have been studied. Field studies of pinnipeds are largely limited to breeding colonies, and only a handful of cetaceans have been studied in any depth in the wild. This research will provide the backbone of this chapter, which will be sprinkled with observations from other, less-well studied species. Following the discussion on group living, the chapter focuses primarily on social relationships and social structure, topics of import for odontocetes but much less so for other marine mammal taxa. The last section considers why this is so. Further discussion about social behaviours in other marine mammals can be found in Chapter 10.

12.2 GROUP LIVING IN MARINE MAMMALS

The groups formed by marine mammals can be spectacular in their size and density. Schools of thousands of oceanic dolphins may fill the view of fortunate shipboard observers. The beaches of southern Namibia are transformed into a sea of fur seals during the breeding season (Fig. 12.1). Some groups of marine mammals are much less obvious to observers. Bowhead whales travelling through

Fig. 12.1 A sea of cape fur seals (*Arctocephalus pusillus*) summering on the shores of Namibia. (Courtesy of A.R. Hoelzel.)

Bering Sea pack ice may travel in groups of 10–15 individuals spread out over 4–8 km². Clark (1991) suggests that migrating bowheads may learn about the location of ice obstacles from the echoes of their distant neighbours' vocalizations.

12.2.1 Costs and benefits of living in groups

With no trees to climb, burrows to scamper into or rocks to hide behind, the open sea appears to be a habitat devoid of refuge from predators (Norris & Dohl 1980). Marine mammals that do not have the option to flee from the water onto land or ice may seek shelter where they are more difficult to find or capture. Elephant seals may find refuge from killer whales at depth (Le Boeuf *et al.* 1989), and nearshore dolphins and baleen whales may hide in the shallows in river mouths, the surf zone or in kelp beds (Saayman & Taylor 1979; Jefferson *et al.* 1991). When disturbed, dwarf sperm whales (*Kogia simus*) can disappear, like squid behind ink, in a large cloud of reddish excreta (Scott & Cordaro 1987). Cover may also be sought in groups, which dilute the predation risk to individuals, providing that the ability of predators to detect groups does not increase in proportion to group size, and that predators can only consume one or a few prey at a time (Inman & Krebs 1987, and see further discussion in Chapter 8). Predation risk probably is the most common reason animals join groups. Groups need not be limited to members of your

own species, however. Mixed groups of spinner and spotted dolphins in the eastern tropical Pacific might form because of benefits from reduced predation risk, much like mixed groups of forest-living primates (Würsig *et al.* 1994; Nöe & Bshary 1997). Non-socially transmitted parasites may provide a similar motive for group formation. When horses from small groups joined larger groups they enjoyed a reduction in number of blood-sucking flies (Duncan & Vigne 1979). The 'cookie cutter' shark (*Isistius brasiliensis*) is essentially a monstrous biting-fly to dolphins in tropical and subtropical waters, removing chunks of flesh up to 7 cm across (Jones 1971). Cookie-cutters are thought to attack while dolphins are feeding (Jones 1971), suggesting a dilution benefit to group-hunting dolphins. The behaviour of group members, whether active (e.g. alarm calls) or passive (fleeing), may alert group members to predators. The vigilance of other group members may allow individuals to reduce their own vigilance in favour of foraging (Pulliam & Caraco 1984; but see Lima 1995a, 1995b). While overall vigilance increased with group size in harbour seals, individual scanning rate decreased (Terhune 1985; Da Silva & Terhune 1988). Group-living animals may also jointly defend against predators. Sperm whales sometimes form protective circles around calves when threatened by killer whales, pilot whales and false killer whales (e.g. Arnbom *et al.* 1987; Weller *et al.* 1996). Humpback dolphins (*Sousa* spp.), Galápagos fur seals and sea lions have been observed mobbing sharks (Saayman & Taylor 1979; Trillmich 1996).

Competition for resources such as food or space with members of your own or other species may also favour group defence of resources (Wrangham 1980; Buss 1981) though there are few observations of this in marine mammals. One pod of killer whales was twice observed to displace a smaller pod from a preferred sea lion hunting area along the shore in Argentina (Hoelzel 1991). In Norway, a group of killer whales feeding on herring left the area when approached rapidly by another group (Bisther & Vongraven 1995).

The benefits of cooperative feeding (see Chapter 8) may also favour the formation of groups. Dusky dolphins (*Lagenorhyncus obscurus*) in Argentina cooperatively herd schools of fish to the surface. Most of the dolphins circle outside to maintain the integrity of the fish school while individuals dart through snatching a mouthful of fish (Würsig & Würsig 1979). Atlantic spotted dolphins (*Stenella frontalis*) in the Gulf of Mexico and killer whales in Norway use bursts of bubbles while feeding cooperatively on schooling fish (Similä & Ugarte 1993; Fertl & Würsig 1995; Similä 1997). Group-feeding humpback whales in the Northeast Pacific wrap herring schools in a cylindrical bubble net and herd them toward the surface (Jurasz & Jurasz 1979; Hain *et al*. 1982; Sharpe & Dill 1995; see also Fig. 8.11). Bottlenose dolphins may cooperate to trap fish between dolphin groups or against or on the shore (Morosov 1970; Würsig 1986; see also Fig. 8.10). This latter behaviour has given rise to remarkable cooperative fishing associations between bottlenose dolphins (and in one case humpback dolphins) and traditional net fishers working along the shore in several parts of the world (Busnel 1973; Hall 1985; Pryor *et al*. 1990). By hunting in groups, killer whales are able to attack the largest baleen and toothed whales (blue whale: Tarpy 1979; sperm whale: Arnbom *et al*. 1987). The size of prey schools or individual prey may also restrict the size of cooperative groups. 'Transient' killer whales in British Columbia most commonly hunt for harbour seals in groups of three, which yields a higher per individual energy intake than larger or smaller groups (Fig. 12.2)

(Baird & Dill 1996; Baird 2000). In many cases, killer whale social group size apparently exceeds the optimal size for foraging (Hoelzel 1991), a common finding in other mammals.

Females at risk from male harassment may group around a male 'hired gun' that protects them from other males (Wrangham 1979). Female southern sea lions (*Otaria byronia*) hauled out on the beach apart from the protective males in the main colony suffered much greater pup mortality from harassing males (Campagna *et al*. 1992).

Animals may also form groups in response to challenges from their habitat. They may huddle to stay warm, or travel near others to reduce the cost of swimming or flight (Lissaman & Schollenberger 1970; Partridge *et al*. 1983; Hainsworth 1987; Abrahams & Colgan 1985; but see Pitcher & Parrish 1993).

Finally, individuals may be forced to aggregate because needed resources are highly clumped. Belugas gather in the relatively warm tidal creeks during the annual moult (Sergeant & Brodie 1969). Many pinniped species may be forced to aggregate in limited haul-out areas. Cases of forced aggregation differ importantly from the other cases mentioned because the individuals do not necessarily benefit from the presence of others.

Group living is never a free ride. The price of mingling with conspecifics includes a risk from socially transmitted parasites and competition for limiting resources (Alexander 1974). The benefits of group formation must exceed these ubiquitous costs for animals to seek each other's company.

12.3 WHO'S WHO IN MARINE MAMMAL GROUPS

In the majority of mammals, females remain in or closer to the area where they were born than males (Greenwood 1980), increasing the opportunity to interact with same-sex relatives. Such interactions range from occasional encounters with neighbours in mammals that are basically solitary to highly developed social relationships in stable groups. In a few species, such as chimpanzees (*Pan troglogytes*), males bond with kin and remain in their natal range and females disperse. In some

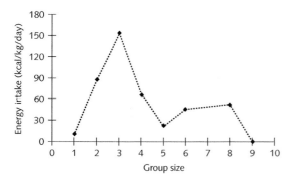

Fig. 12.2 Energy intake (kcal/kg/day) versus group size for transient killer whales, showing a peak in energy intake rates for individuals in groups of three. (From Baird & Dill 1996, courtesy of Oxford University Press.)

species males disperse with kin, and join an unrelated group of related females (e.g. lions: Packer *et al.* 1988, 1991).

The most generally accepted explanation for sexually dimorphic patterns of philopatry in terrestrial mammals is inbreeding avoidance (Greenwood 1980; Pusey & Wolf 1996). One sex or the other must disperse to avoid mating with close relatives. However, the same rules may not apply to many marine mammals, which have home ranges that dwarf those found on terra firma. For blue whales, an entire ocean basin might be considered 'home'. Such large ranges are not defensible (even if food patches occasionally are) so the ranges of individuals will typically overlap many conspecifics, including unrelated, potential mates. Thus, when discussing philopatry and dispersal in many marine mammals, it becomes especially important to distinguish geographic from social philopatry (Isbell & van Vuren 1996; Connor 2000). Male bottlenose dolphins may maintain their natal range in their adult home range but interact very little with their maternal relatives (Connor *et al.* 2000b). Thus male bottlenose dolphins exhibit a degree of geographic but not social philopatry. Resident killer whales, on the other hand, exhibit both; males and females are thought to remain in their natal group for life. It is likely that killer whales mate with members of different groups when they join together (Baird 2000).

In pinnipeds, the potential for philopatry is much greater for land-breeding species compared to those that breed on ephemeral ice (Trillmich 1996). Because individuals must be followed from birth to document natal philopatry, it has been reported in few species. Adult fidelity to breeding sites occurs in a number of land-breeding pinnipeds. For example, in northern fur seals, most females and some males are thought to return to their birth sites to breed (Gentry 1998). The territories of male northern fur seals rarely shift more than 10 m throughout their life (Gentry 1998). However, high breeding colony fidelity does not automatically lead to preferential interactions with kin. Female harbour seals (*Phoca vitulina*) on Sable Island exhibit high breeding colony fidelity but poor within-colony fidelity and possibly as a result, do not preferentially foster related pups (Schaeff *et al.* 1999).

12.4 SOCIAL MARINE MAMMALS: INTERACTIONS, RELATIONSHIPS AND SOCIAL STRUCTURE

Social relationships emerge from the pattern of social interactions between individuals over time (Hinde 1976). Individuals have relationships with a number of others in a population forming a network of relationships or social structure. Social relationships and structure are not exclusive attributes of group-living animals. A so-called 'solitary' mammal defending his or her territory against intruders has a relationship with its neighbours, even if their interactions are mostly hostile and based largely on long-distance communication. Social bonds are social relationships that include a consistent affiliative component. I favour the term 'bonding strategy' because it promotes an understanding of bonds as essentially social tools by which individuals attempt to increase their reproductive success, and which should vary with ecological conditions, age, sex, social position and the strategies employed by others.

12.4.1 Affiliative behaviour

Everyone is familiar with the picture of one monkey or ape carefully picking through the fur of another. Such 'grooming' is an example of affiliative or 'friendly' behaviour. Affiliative behaviour can be used to strengthen bonds, mend a damaged bond, reduce tension in stressful circumstances or simply to obtain a needed service from another, potentially anonymous, individual. Impalla (*Aepcyceros melampus*) use specially adapted teeth to rid themselves of ticks, but hard-to-reach places on the head or neck require the services of another. Impala communicate their need by giving a few nibbles to a similarly infested neighbour, thereby initiating a series of grooming exchanges that benefit both individuals. (Hart & Hart 1992). Female bonobos (pygmy chimpanzees, *Pan paniscus*) in potential conflict over a fruit tree rub their genitals together to ease the tension (Kano 1992). Stress reduction can be measured as a reduced heart rate in a rhesus monkey being gently stroked by another. The more utilitarian form of grooming employed by rhesus monkeys, to pick off parasites, does not affect the heart rate (Boccia *et al.* 1989).

Outside mating relationships and mother–calf bonds (see Chapter 10) and such potentially anonymous encounters illustrated by the impalla allogrooming, the frequency and diversity of affiliative behaviour in marine mammals is expected to accord with the importance of social bonds and the frequency with which such bonds are initiated and maintained. Bottlenose dolphins in Shark Bay pet each other with their flippers, rub their bodies against the extremities of others (flukes, flippers and fins) and often engage in more intense body-to-body contact (Connor *et al.* 2000b). Beyond its social value, such contact behaviour may help remove sloughing skin and perhaps parasites in some cases. Males often engage in mutual stroking bouts with alliance partners (Connor *et al.* 1992a), and females being harassed by males may rest their pectoral fin against the side of another female (Richards 1996). Similar petting and rubbing is described from a number of odontocetes, including sperm whales (Whitehead & Weilgart 2000), spinner dolphins (Johnson & Norris 1994), killer whales (Osborne 1986) and Commerson's dolphins (Johnson & Moewe 1999). Remarkably, the leading edge of the left pectoral fin is serrated in the majority of male but only a minority of female Commerson's dolphins (Goodall *et al.* 1988). In their study of two captive male and four female Commerson's dolphins, Johnson and Moewe (1999) found that the males performed the vast majority of pectoral fin 'touches' (96%) and predominately with their serrated left pectoral fins (94%). The left pectoral bias was regardless of whether the recipient was male or female, or whether the socializing included genitally directed behaviours. A need to maintain or reaffirm bonds occurs when individuals that have been separate for a time reunite. Intense socializing may occur when resident killer whale pods come together in 'greeting ceremonies' (Osborne 1986) that invite comparison to similar behaviour in elephants (Moss & Poole 1983; Moss 1988). Synchrony may also reflect the strength of social bonds in some marine mammals. Allied male bottlenose dolphins in Shark Bay often surface side by side synchronously (Fig. 12.3) (Connor *et al.* 2000b). Synchronous surfacing by mothers and

Fig. 12.3 An alliance of three male bottlenose dolphins in Shark Bay, Western Australia dive in precise synchrony. Such synchrony is thought to be a measure of affiliation.

infants has been reported in marine mammals as diverse as bottlenose dolphins and manatees, and may be important for infant protection (Hartman 1979; Mann & Smuts 1999). In some marine mammals, probable affiliative interactions among adults have been described, but what, if any, role the behaviours play in bond formation remains obscure. For example, bonds have not been described among adult manatees which, upon meeting at 'rendezvous sites' (often areas where waterways intersect), may nuzzle, rub and 'kiss' each other (Hartman 1979). In a study of captive California sea lions (Schusterman *et al.* 1992b) measured a variety of behaviours they considered affiliative, including sleeping together, maintaining physical contact, extended periods of side by side swimming or one individual following another.

12.4.2 Aggression, agonistic displays and dominance

Many of the vocal and physical threats employed by marine mammals are typical mammalian behaviours such as growling, opening the mouth, jerking the head and lunging at or charging an opponent (e.g. sea otters: Riedman & Estes 1990; California sea lions: Peterson & Bartholomew 1967; bottlenose dolphins: Connor *et al.* 2000b). A few of the more unusual behaviours are worth mentioning. Male walruses threaten others by throwing their heads back so their saber-like tusks point at their rival

(Miller 1975). Male hooded seals may threaten an opponent with an inflated proboscus, as well as a 'red baloon' of skin that is inflated from the nostril (Miller & Boness 1979). Differences between related species are illustrated by the territorial display of California and Steller's sea lions; the rapid lateral head shake ('no, no, no') of the former is matched by rapid nodding ('yes, yes, yes') in the latter (Schusterman *et al.* 1992b). Male bottlenose dolphins in Shark Bay, Western Australia threaten female consorts with a low-frequency click train called 'pops' (Connor & Smolker 1996). Competing male humpback whales may increase their apparent size by inflating the area of their throat pleats (Tyack & Whitehead 1983; Baker & Herman 1984) or bending into an 'S' posture (Helweg *et al.* 1992). Johnson and Norris (1994) suggest that the 'S' posture in spinner dolphins may be an imitation of the aggressive swimming patterns of the grey reef shark (*Carcharhinus amblyrhinchos*) or it may simply highlight their ventral 'postanal hump' (Norris *et al.* 1994).

If interactions escalate beyond the threat stage, again, many marine mammals do the normal mammalian thing—they bite each other. The use of teeth during conflicts often leaves 'toothrake scars' on victims (e.g. cetaceans: Norris 1967; dugongs: Preen 1989). Teeth have become modified into more elaborate weapons in some species. Male beaked whales sport a pair of teeth which project outside the mouth from the lower mandible and, judging from scarring patterns, are used in a tusk-like fashion in male–male conflicts (McCann 1974; Heyning 1984; MacLeod 1998). The 3 m tusk of the male narwhal (and, rarely, females!) may be used in fencing contests but observations are limited (Silverman & Dunbar 1980). We can predict that occasional lethal use of tusks may occur when two males are evenly matched and the resource they are contesting is valuable (most likely a fertile female) (Enquist & Leimar 1990). We can make a similar prediction of the tusks of walruses, which are found in both sexes but are larger and may grow up to 1 m in males (Sjare & Stirling 1996).

The use of tusks during fighting by male elephant seals is on vivid display for several months annually for scientists and the public on Ano Nuevo Island off central California (Le Boeuf 1974). Although impressive, the 6 cm tusks of male elephant seals are small compared to narwhale and walrus tusks, and males are well defended by thickened skin in the most accessible 'strike zone' on their chests. Further, the stakes can be very high for male elephant seals, who may be fighting for access to dozens of receptive females on the beach (Le Boeuf & Reiter 1988) (see Chapter 10). The small tusks of male dugongs are also used in fighting (Preen 1989).

A variety of non-dental structures may be modified for combat in cetaceans. Male northern bottlenose whales (*Hyperoodon ampullatus*) may use their large bony foreheads in headbutting interactions (Gowans & Rendell 1999). Southern right whales bump and scrape each other with their callosities, which are larger and more numerous on males (Payne & Dorsey 1983). In many cetaceans, parts of the body that are not obviously modified for combat, such as the peduncle or flukes, may be used to hit opponents. Male humpback whales competing for the 'principle escort' position next to a female may incur bloody dorsal fins and rostral tubercles from hitting each other (Tyack & Whitehead 1983; Baker & Herman 1984; Clapham 2000) but such interactions are rarely lethal (Pack *et al.* 1998). Hitting may be used to commit infanticide in bottlenose dolphins and to kill harbour porpoises (Ross & Wilson 1996; Patterson *et al.* 1998). In one study, half of the 10 apparent victims of bottlenose dolphin infanticide showed no external signs of trauma but all showed extensive internal injury (Dunn *et al.* 1999). It is notable that the only lethal fight observed among killer whales was decided by a blow rather than a bite (Connor *et al.* 2000a). For biologists describing cetacean social relationships, the propensity for such potentially lethal behaviour to leave individuals without external wounds means that the importance and severity of aggression may often be underestimated (Connor *et al.* 2000a).

An agonistic interaction is one that mediates dominance/submissive relationships among individuals and may not involve overt aggression at all. For example, an agonistic interaction between two bottlenose dolphins may be signalled by a flinch on the part of the submissive individual in response to a mere approach (considered a neutral behaviour) by the dominant individual (Samuels & Gifford 1997). In bottlenose dolphins and primates, such submissive gestures are more reliable indicators of dominance than are aggressive behaviours (Hausfater 1975; Samuels & Gifford 1997). Male walruses exhibit submission by erecting their mystacial vibrissae while rapidly repeating a monosyllabic vocalization and stretching their neck up or up and away from the opponent (Miller 1975).

Studies of agonistic interactions can reveal much about social relationships in an animal society. Samuels and Gifford (1997) found the same pattern of dominance relationships in captive bottlenose dolphins as is found in primates such as baboons and chimpanzees: males were dominant over females even when some females were larger and dominance relations between females were much more stable and contested less than relations between the two males in the study.

12.4.3 Sociosexual behaviour

Non-conceptive sexual behaviour occurs in a wide range of birds and mammals. Some cases of non-conceptive sex may be related directly to mating strategies (e.g. promiscuous mating by females to confuse paternity, see Wrangham 1993) but other examples are clearly involved in the mediation of social relationships. 'Communication' or 'social' sex (Wickler 1967) is defined as sexual interactions in which neither partner obtains conceptive benefits and excludes homosexual behaviour that is a 'substitute' for absent females and the abnormal behaviour of captive reared animals (de Waal 1990, 1992). Among non-human terrestrial mammals, the bonobo may exceed all others in both the variety of acts performed and the age/sex class of participants (Kuroda 1984; de Waal 1987, 1990; Kitamura 1989; Kano 1992). Diversity also characterizes the function of bonobo social sex; for example, it can be used to develop bonds, mend relationships or reduce tension (Wrangham 1993).

Non-conceptive sexual behaviour has been reported in a diverse range of marine mammals, including northern fur seals (Gentry 1998), baleen whales (right whales: Payne 1995), toothed whales (e.g. Saayman & Taylor 1979; Norris & Dohl 1980; Wells 1984) and manatees (Hartman 1979). Sociosexual behaviour in some delphinids may rival that found in bonobos in the range of participants,

if not variety, of sexual acts. Adult homosexual behaviour and sexual behaviour by immature and even infant dolphins toward adults was first described in captivity, but occurs commonly in the wild (Saayman & Taylor 1979; Norris & Dohl 1980). In Shark Bay, homosexual behaviour that includes mounting and rostrogenital contact is observed more often between adult males than females, but females have been observed to mount each other and even, on one occasion, an adult male that was herding her (Connor & Smolker 1996). Sex is more commonly observed among immature dolphins, and again, following earlier captive observations (Caldwell & Caldwell 1972), infant male dolphins may even mount their mothers or adult males, with erections (Connor et al. 2000b). Sexual behaviour of immatures may often be 'play' or 'practice.' Juvenile bottlenose dolphins in Shark Bay have been observed in 'play herding' groups in which the composition of 'alliances' and the 'herded' individual changed frequently and both males and females took both roles (Connor et al. 2000b). In this case, much more than the sex act itself was being 'practised'. Homosexual behaviour between adult males in Shark Bay may play a role in dominance interactions (cf. Östman 1991). For, example, during one 65 min interaction, members of one alliance directed extensive sexual and aggressive behaviour toward members of another alliance (Connor & Smolker 1996). In contrast, males have also been observed mounting their alliance partners in a relaxed manner indicative of affiliative interactions. Role changes in sexual behaviour among bottlenose dolphins, humpback dolphins and immature male killer whales suggests a communication or practise function rather than dominance (Saayman & Taylor 1979; Rose 1992; Connor et al. 2000b). As in bonobos, the function of dolphin social sex that occurs in affiliative contexts may be to 'test the bond' (Zahavi 1977). 'Testing the bond' is an important and useful concept: a social behaviour may be used to test the strength of a social bond if the behaviour is, in the absence of a bond, stressful and not tolerated (e.g. open mouth kissing in bonobos and humans). A stressful signal will be a reliable one; it should only be tolerated by the recipient if the recipient values the bond (see also Wrangham 1993).

12.5 BONDING STRATEGIES IN ODONTOCETES

In this section we examine the patterns of bond formation that give rise to descriptions of social structure in odontocetes. The bonding strategies of females should be closely linked to calf protection and access to resources, so we expect to find a close link between the factors favouring group formation (see above) and female bonding strategies. Male bonding strategies, on the other hand, should be based on access to receptive females, and will therefore link with discussions of mating strategies (see Chapter 10). Whitehead (1990) found that if the receptive period of individual females exceeds the time it takes for males to travel between groups then males should rove rather than remain with one group of females during a breeding season or permanently. Behavioural and/or genetic studies have elucidated the basic framework of social bonds among a half-dozen genera of odontocetes and in every case males are rovers (Connor *et al.* 2000a). Given that males rove, the possible bonding strategies available to males are simplified to two questions: 'To rove with others?' And, if answered in the affirmative, 'With whom to rove?' The few odontocetes studied to date have provided a range of answers. Sperm whales rove alone, bottlenose dolphins rove with males and killer and pilot whales rove with their natal kin.

12.5.1 Females bonded, males rove alone: sperm whales

Female sperm whales live in stable matrilineal groups of about 10 related individuals that often form temporary associations with other groups (Richard *et al.* 1996; Whitehead & Weilgart 2000). There is some evidence of occasional group fusions or splits over timescales of 5–10 years, which accords with recent observations that some stable female groups contain unrelated matrilines (Dillon 1996; Richard *et al.* 1996; Christal *et al.* 1998). Bonds between female sperm whales are thought to be based on cooperative care of offspring who do not dive as deep as their mothers (Gordon 1987; Papastavrou *et al.* 1989) and are highly vulnerable to predation, especially by killer whales and large sharks (Best *et al.*

1984; Arnbom *et al.* 1987; Brennan & Rodriguez 1994). Whitehead (1996) found evidence of baby-sitting in female groups. Females and juveniles stagger their dives to a greater extent when a calf is in the group than when there is no calf. The majority diet of female sperm whales consists of relatively small (< 1200 g) squid that are thought to occur singly or in small groups (Clarke 1980; Whitehead & Weilgart 2000). A role for cooperative feeding or defence of resources in the evolution of female–female bonds in sperm whales is contraindicted by the enormous, indefensible ranges of female groups (one to several thousand kilometres) and the observation that females spread out when foraging (Whitehead 1989; Whitehead & Weilgart 2000).

Male sperm whales disperse and join 'bachelor groups'. As they slowly grow to be much larger than females they become increasingly solitary and roam to much higher latitudes where they may find better feeding opportunities needed to support their large mass (Best 1979; Rice 1989). Males mature much later than females and appear to fit a model where they delay breeding in order to grow larger and to be more competitive (Whitehead 1994). Males visiting the tropics may form loose aggregations (Christal & Whitehead 1997) but do not form bonds when roving among female groups (Whitehead & Weilgart 2000). However, occasional strandings of groups of adult male sperm whales (Rice 1989; Reeves & Whitehead 1997) suggest that male–male bonds may sometimes occur.

Many features of sperm whale life history and social structure are remarkably similar to elephants (Weilgart *et al.* 1996), including their large size, extreme size dimorphism, relatively large brains, extensive home ranges and general ecological success. Sperm whales and elephants also share a hierarchical social organization based upon cooperative groups of related females and males that initially form bachelor groups but rove singly between groups of females as they mature.

12.5.2 Roving male alliances, weaker female–female bonds: bottlenose dolphins and bottlenose whales

Bottlenose dolphins (*Tursiops* spp.) live in classic fission–fusion societies in which individuals associate in small groups that often change in size and composition (Würsig & Würsig 1979; Wells *et al.* 1987; Smolker *et al.* 1992; Connor *et al.* 2000b). Strong male–male bonds have been described from two sites: in Shark Bay, Western Australia (*T. aduncus*) and Sarasota Bay, Florida (*T. truncatus*). In both cases bonds between some males are as strong as those between females and their dependent offspring and may last for one or two decades (Connor *et al.* 2000b). At both sites, female–female bonds are weaker but variable.

In Shark Bay, males in pairs and trios cooperate to form consortships with individual females that are often initiated and maintained by aggressive herding (Connor *et al.* 1992a, 1992b, 1996, 2000b). Some alliances are very stable and may last over a decade; the alliances of other males are quite labile (Fig. 12.4) (Connor *et al.* 1999). Moderately strong bonds between alliances are based upon cooperative attacks or defence in interactions against other alliances over female consorts (Connor *et al.* 1992b). In Sarasota, stable male pairs are common but trios are unknown and some males do not form alliances (Wells *et al.* 1987; Wells 1991; Connor *et al.* 2000b). Sarasota *Tursiops* are much larger and possibly exhibit more size dimorphism than in Shark Bay where adults are the size of Sarasota 2–3 year olds. Bottlenose dolphins are even larger in the Moray Firth, Scotland, where strong bonds have not been found among any adults (Wilson *et al.* 1992, 1993; Connor *et al.* 2000b).

Female bonds in both Shark Bay and Sarasota are variable with some females sharing moderately strong bonds with several individuals while others are largely solitary. Moderately social females are clustered in 'bands', which in Sarasota are often composed of female relatives but may contain unrelated matrilines (Duffield & Wells 1991). In neither site do any females form the strong bonds that characterize stable male alliances; rather social females appear to maintain a large network of weak or moderate bonds. A female's associates at a given time may depend on her reproductive state (Wells *et al.* 1987; Connor *et al.* 2000b).

Recently, stronger male–male bonds and weaker female–female bonds have been described in a member of the beaked whale family, the bottlenose whale (*Hyperoodon ampulatus*) (Gowans 1999).

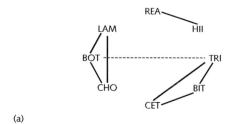

(a)

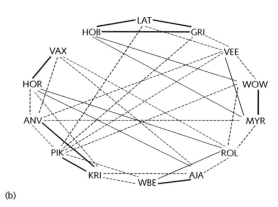

(b)

Fig. 12.4 Contrasting patterns of male alliance formation in Shark Bay, Western Australia. Sociograms depicting alliance formation between males in: (a) three stable alliances (one pair and two trios) that associated on a regular basis, and (b) a large 'superalliance' of 14 males (lines connect males that shared association coefficients of at least 10, on a scale of 1–100; thicker bars reflect stronger associations). Males in the 14-member superalliance formed trios and pairs with many other group members to consort females and joined forces against teams of stable alliances. (From Connor *et al.* 1999, courtesy of Macmillan Magazines Ltd.)

This is somewhat surprising, as bottlenose whales are large (8–10 m), deep divers like sperm whales (Hooker & Baird 1999) and might, therefore, be expected to exhibit more similarity with sperm whale rather than bottlenose dolphin society.

12.5.3 Roving with natal kin: bisexual bonds in killer whales

Two sympatric populations of killer whales in the Northeast Pacific differ in feeding habits, ranging, group size and other characteristics (reviewed by Baird 2000). In the fish-eating 'resident' killer whales, individuals associate in pods and pods associate only with other pods from their community, although the ranges of the known communities overlap.

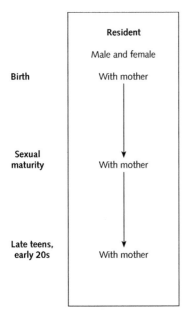
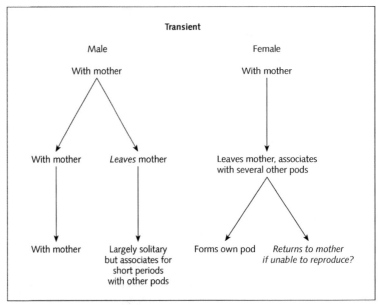

Fig. 12.5 Resident and transient killer whale association patterns. Both sexes are thought to remain in their natal group in residents, while some dispersal occurs in transients. Links labelled in italics are speculative at this juncture, emphasizing the point that while killer whales have been studied for many years in the Canadian southwest, much remains to be discovered. (From Baird 2000, courtesy of Chicago University Press.)

The most stable associations within pods are matrilineal units, which range from two to nine individuals and include from one to four generations (Bigg *et al.* 1990; Baird 2000). A typical matrilineal unit contains a grandmother, her adult son, her adult daughter and the offspring of her daughter. Individuals rarely spend more than a few hours apart from other members of their matrilineal unit. One to 11 matrilineal units associate in subpods, which temporarily travel apart from other members of the pod. Pods contain one to three subpods. A process of gradual fissioning along maternal lines is thought to be responsible for the formation of new matrilineal groups, subpods and pods.

Mammal-eating transients generally live in much smaller groups than residents and some dispersal occurs but again, some adult males may remain in their natal pods (Fig. 12.5).

Most hypotheses for male philopatry in killer whales focus on benefits that males provide to the offspring of related females, including observations of assistance in hunting and teaching (Lopez & Lopez 1985; Guinet 1991; Baird 2000). Cooperative

feeding has been reported in some populations that feed on fish (Norway: Similä & Ugarte 1993) but may be less important in others (British Columbia: Hoelzel 1993; Baird 2000). Killer whales that feed on marine mammals often hunt cooperatively (Baird 2000). In Argentina, male and female killer whales may help 'teach' juveniles the technique of strand feeding, in which the whales beach themselves temporarily in pursuit of southern elephant seal and southern sea lion pups. Stranding adults have been observed capturing and tossing prey to juveniles that stranded with them but failed to capture a pup (Lopez & Lopez 1985). Connor (2000) suggested that mothers may provide assistance to their adult sons, noting observations that older mothers and adult sons tend to travel independently of other pod members, including dependent offspring (Bigg *et al.* 1990).

Sons and daughters may remain in their mother's group in long-finned pilot whales (*Globicephala melas*). Amos *et al.* (1993) employed genetic analyses on pilot whale schools captured in drive fisheries to demonstrate that mature males neither mate within, nor disperse from, their natal pods. It remains unclear whether the large pilot whale schools examined by Amos *et al.* (1993) represent a social unit more like killer whale pods or communities (Connor *et al.* 2000a).

12.6 ECOLOGICAL FACTORS IN THE EVOLUTION OF ODONTOCETE BONDING STRATEGIES

Because they bear the cost of parental care, female reproductive and bonding strategies are determined by access to food and safety from predators and, in some cases, males (Bradbury & Vehrencamp 1977; Emlen & Oring 1977; Wrangham 1980) (see Chapter 10). Male mating and bonding strategies will be determined primarily by the distribution of females but may, in turn, impact on female strategies, especially those male behaviours that impose a cost on females, such as infanticide (Sterck *et al.* 1997). Thus, females are usually given 'priority' in models, such that if philopatry is favoured for females, then males are forced to emigrate because of the cost of inbreeding. Female philopatry may be favoured because of non-social factors such as knowledge of resource distribution, or social factors such as benefits conferred from cooperative defence of resources or against predators. Where selection for female philopatry is weak, the door is left open for bonds among male kin if they are favoured and females may be forced to emigrate to avoid the costs of inbreeding. The priority given females appears justified for terrestrial mammals: only a small minority of mammals exhibit male philopatry (Greenwood 1980).

Several prominent ecological influences on odontocete societies, including a possible explanation for why cetaceans do not follow terrestrial rules of philopatry, are reviewed briefly below. This discussion is not intended to be complete, but rather to serve as an introduction to thinking about ecological influences on odontocete social bonds (for a more exhaustive coverage, see Mann *et al.* 2000).

Cetaceans are designed to move efficiently through a medium that supports them against the pull of gravity. Although their total cost of transport is similar to that seen in terrestrial species (see Chapter 3), this total is made up of separate components associated with the costs of 'maintenance' and 'locomotion'. Cetaceans have high maintenance and low locomotion costs compared to terrestrial mammals. Connor and colleagues (Connor *et al.* 1998; Connor 2000) suggested that a key factor allowing bisexual philopatry is this low cost of locomotion enjoyed by cetaceans, which allows them to pursue widely scattered resources over large home ranges. Home and day ranges for many dolphins are an order of magnitude greater than those found in terrestrial mammals, in many cases divorcing the issue of philopatry from inbreeding (e.g. Connor 2000). A male resident killer whale, travelling with his maternal relatives, will range over a large area that contains other groups with unrelated, potential mates. It is also important that cetacean offspring are 'followers' and thus not tied to a breeding site (Connor *et al.* 2000a).

Predators have already been mentioned as an important reason for cetaceans to form groups, and in some species may be a primary determinant of social bonds as well. The function of bonds among female sperm whales may be to protect their highly vulnerable offspring from predators (Whitehead & Weilgart 2000). It is expected that predators will influence female social bonds in many species and perhaps even male–male bonds. Wilson *et al.* (1993) suggested that low shark predation risk might explain the lack of male alliances among bottlenose dolphins in the Moray Firth. In this scenario, being solitary is the optimal mating strategy, but when survival is factored into the mix, sharks tip the balance toward alliance formation (see Connor *et al.* 2000b). Wells (1991) found that shark bite scars were more numerous on paired males than on singletons in Sarasota, but males in pairs survived longer.

Connor *et al.* (1996) suggested that female Indian Ocean bottlenose dolphins might endure extended periods of harassment by males as part of a strategy to confuse paternity and thus reduce the risk of infanticide. Evidence for infanticide was subsequently discovered in several other populations (Patterson *et al.* 1998; Dunn *et al.* 1999). Many odontocetes exhibit characteristics that are thought to favour infanticide in terrestrial mammals (Connor *et al.* 2000a). These include: (i) broadly seasonal breeding that allows females that have lost offspring time to conceive again in the same season (Hrdy 1979), and (ii) a longer lactation than gestation period so that postpartum mating cannot be employed to eliminate the advantage of infanticide (van Schaik & Kappeler 1997).

Bonds with their maternal kin are valuable to females when they confer protection from predators or allow cooperative defence of resources. Food patches that can support only one individual do not select for cooperative defence or bond formation. If those resources are distributed in small patches that cannot feed more than an individual, then females that forage continuously cannot afford to travel together. If foraging occurs in distinct bouts, however, individuals may form groups between bouts if the benefits outweigh the travel costs (Connor 2000). Here again, the relatively low cost of locomotion may favour social living in cetaceans compared to terrestrial species. For many cetacean species, we have to consider vertical travel in addition to horizontal travel. Female-bonded sperm whales make round trips to feed at depth where they spread out while feeding on small (1 kg) squid. Whitehead (1989) suggested that the females spread out to avoid resource competition. Greater resource competition may also limit group size in transient relative to resident killer whales.

The wide ranging habits of female odontocetes and the mobility of the resources they pursue preclude anything approaching the spatial territoriality found in many terrestrial mammals but not group defence of resources. In the Moray Firth, Scotland, spatial relations among bottlenose dolphins are generally maintained in spite of seasonal shifts in habitat use by the whole population. Individuals moving further into the firth in summer are 'replaced' by others moving in from outside areas, with little social mixing. Wilson *et al.* (1997) suggest that this stratified pattern of habitat use can be understood if groups defend the area they are in at a given time (see also Connor 2000). Their concept of area defence might have wide application among group-living odontocetes that pursue prey that is found in large but fluctuating concentrations in space and time. Actual observations of displacement occurring at a resource are limited to killer whales in Argentina (Hoelzel 1991) and Norway (Bisther & Vongraven 1995).

Another kind of resource odontocetes might potentially compete for is alloparental care, which has been reported in several odontocetes (Whitehead & Mann 2000). Reproductive suppression is currently unknown in odontocetes, but is expected where alloparental care is depreciable (i.e. benefits of the care decline with the number of offspring, see Clutton-Brock 1991). Two kinds of possible alloparental care have been described in sperm whales, communal nursing and babysitting. Calves have been observed to attempt suckling from several different females (Gordon 1987), and Best *et al.* (1984) suggested that communal suckling may explain why there are almost always more lactating females than calves in sperm whale groups. Communal nursing has not been conclusively established (see Whitehead & Mann 2000) but would be just the kind of depreciable care that should favour reproductive suppression. The more clearly established kind of allocare in sperm whales, babysitting, is less clear-cut. While vigilance is non-depreciable, active defence against predators might be depreciable if a babysitter has a more difficult time defending more than one calf. The 'teaching' described in killer whales sometimes involves food sharing (e.g. Lopez & Lopez 1985; Hoelzel 1991; Baird 2000) which is depreciable.

Resource abundance and distribution can also impact odontocete bonds indirectly, by changing the rate at which conspecifics encounter each other in competitive circumstances (e.g. in conflict over a resource). Connor *et al.* (2000a,b) suggested that population differences in alliance formation among male bottlenose dolphins might be explained by differences in the rate males come into conflict over oestrous females. The 'rate of interaction' might be primarily affected by population density, but also by other ecological factors such as detection distance (Smolker *et al.* 1992) or day range. The three-dimensional underwater environment may also impact on the optimal strategy for winning contests, placing a premium on manoeuvrability (Stirling 1975; Connor *et al.* 2000a). Selection for manoeuvrability may reduce selection for sexual size dimorphism, at least for smaller species, and possibly, in turn, impact on selection for male alliance formation (see Connor *et al.* 2000a, 2000b).

12.7 THE FUTURE: RESEARCH ON THE SOCIAL LIVES OF MARINE MAMMALS

If the past gives us any indication, then the study of odontocete societies has a most promising future. Each of the handful of species that has been studied in any detail has yielded remarkable discoveries (see Connor *et al.* 1998). Other species whose social lives have been exposed to a lesser extent hint at exciting new social patterns. For example, what kind of society do Baird's beaked whales (*Berardius bairdii*) live in, given that males may outlive females by up to 30 years (Kasuya *et al.* 1997)?

Why is there not a science of social bonds in other marine mammals: the baleen whales, pinnipeds and sirenians? One is hard pressed to find evidence of social bonds outside mother–offspring pairs for any marine mammal other than odontocetes. Reviews of social interactions in these taxa are justly found in discussions of parental care and reproductive strategies (see Chapter 10).

A partial explanation for the relative complexity of odontocete societies may be found in the predation risk suffered by newborns. Born in water and relatively small, odontocete newborns are probably at greater risk than the majority of pinniped and baleen whale calves. Feeding competition may also disfavour groups in baleen whales relative to odontocetes (Gaskin 1982; Connor 2000).

On the other hand, it is becoming clear that the potential for interactions among related female pinnipeds on rookeries and baleen whales throughout their range has been overlooked (Trillmich 1996; Connor 2000). Any tendency toward natal philopatry in pinnipeds would permit bonds among maternal kin on breeding grounds. California sea lions imprint strongly on their mothers and, at least in captivity, for life (natural or human surrogates) (Schusterman *et al.* 1992a). Northern fur seals can recognize their mothers for at least 4 years after weaning (Insley 2000). In a captive study, Schusterman *et al.* (1992b) found that mothers, daughters and sisters engaged in more affiliative and less aggressive behaviour with each other than non-relatives. Most pinniped species have not been studied at sea where matrilineal kin might continue to interact.

Schusterman *et al.* (1992a) describe qualitatively different interactions among sea lions at sea where real estate is not at a premium: 'The transition from land to water brings about a very rapid change in behaviour from intense social competition to non-aggression and play'. Potentially, therefore, a science of pinniped social bonds that focuses on reduced aggression among relatives in breeding colonies, might focus on affiliative interactions offshore.

There are also hints of persistent social bonds in some baleen whales. Adult gray whales have been observed to assist a mother in freeing her stranded calf (Swartz 1986). Female humpback whales usually return to the same tropical breeding areas but individuals have moved between breeding areas as far apart as Hawaii and Japan (Clapham 1996, 2000). Potential interactions among philopatric (and possibly related) humpback whales are not restricted to breeding sites but may extend to migration and high-latitude feeding areas as well (D'Vincent *et al.* 1985; Perry *et al.* 1988).

REFERENCES

Abrahams, M.V. & Colgan, P.W. (1985) Risk of predation, hydrodynamic efficiency, and their influence on school structure. *Environmental Biology of Fishes* **13**, 195–202.

Alexander, R.D. (1974) The evolution of social behavior. *Annual Review of Ecology and Systematics* **5**, 325–383.

Amos, B., Schloetterer, C. & Tautz, D. (1993) Social structure of pilot whales revealed by analytical DNA profiling. *Science* **260**, 670–672.

Arnbom, T., Papastavrou, V., Weilgart, L.S. & Whitehead, H. (1987) Sperm whales react to an attack by killer whales. *Journal of Mammalogy* **68**, 450–453.

Baird, R.W. (2000) The killer whale—foraging specializations and group hunting. In: *Cetacean Societies: Field Studies of Dolphins and Whales* (J. Mann, R.C. Connor, P.L. Tyack & H. Whitehead, eds), pp. 127–153. University of Chigaco Press, Chicago.

Baird, R.W. & Dill, L.M. (1996) Ecological and social determinants of group size in transient killer whales. *Behavioral Ecology* **7**, 408–416.

Baker, C.S. & Herman, L.M. (1984) Aggressive behavior between humpback whales (*Megaptera novaeangliae*) wintering in Hawaiian waters. *Canadian Journal of Zoology* **62**, 1922–1937.

Best, P.B. (1979) Social organization in sperm whales, *Physeter macrocephalus*. In: *Behavior of Marine Mammals: Current Perspectives in Research, Vol. 3. Cetaceans* (H.E. Winn & B.L. Olla, eds), pp. 227–289. Plenum Press, New York.

Best, P.B., Canham, P.A.S. & Macleod, N. (1984) Patterns of reproduction in sperm whales, *Physeter macrocephalus*. *Reports of the International Whaling Commission Special Issue* **6**, 51–79.

Bigg, M.A., Olesiuk, P.F., Ellis, G.M., Ford, J.K.B. & Balcomb, K.C. (1990) Social organization and genealogy of *resident* killer whales (*Orcinus orca*) in the coastal waters of British Columbia and Washington State. *Reports of the International Whaling Commission Special Issue* **12**, 383–405.

Bisther, A. & Vongraven, D. (1995) Studies of the social ecology of Norwegian killer whales (*Orcinus orca*). In: *Whales, Seals, Fish and Man* (A.S. Blix & L. Walloe, eds), pp. 169–176. Elsevier, Amsterdam.

Boccia, M.L., Reite, M. & Laudenslager, M. (1989) On the physiology of grooming in a pigtail macaque. *Physiology and Behavior* **45**, 667–670.

Bradbury, J.W. & Vehrencamp, S.L. (1977) Social organization and foraging in emballonurid bats. III. Mating systems. *Behavioral. Ecology and Sociobiology* **2**, 1–17.

Brennan, B. & Rodriguez, P. (1994) Report of two orca attacks on cetaceans in Galápagos. *Noticias de Galápagos* **54**, 28–29.

Busnel, R.G. (1973) Symbiotic relationship between man and dolphins. *New York Academy Scientific Transactions* **35**, 112–131.

Buss, L.W. (1981) Group living, competition, and the evolution of cooperation in a sessile invertebrate. *Science* **213**, 1012–1014.

Caldwell, D.K. & Caldwell, M.C. (1972) *The World of the Bottlenose Dolphin*. Lippincott, Philadelphia.

Campagna, C., Bisioli, C., Quintana, F., Perez, F. & Vila, A. (1992) Group breeding in sea lions: pups survive better in colonies. *Animal Behavior* **43**, 541–548.

Christal, J. & Whitehead, H. (1997) Aggregations of mature sperm whales on the Galapagos Islands breeding ground. *Marine Mammal Science* **13**, 59–69.

Christal, J., Whitehead, H. & Lettevall, E. (1998) Sperm whale social units: variation and change. *Canadian Journal of Zoology* **76**, 1431–1440.

Clapham, P.J. (1996) The social and reproductive biology of humpback whales: an ecological perspective. *Mammal Review* **26**, 27–49.

Clapham, P.J. (2000) The humpback whale: seasonal feeding and breeding in a baleen whale. In: *Cetacean Societies: Field Studies of Dolphins and Whales* (J. Mann, R.C. Connor, P.L. Tyack & H. Whitehead, eds), pp. 173–196. University of Chicago Press, Chicago.

Clark, C.W. (1991) Moving with the herd. *Natural History* **3**, 38.

Clarke, M.R. (1980) Cephalopoda in the diet of sperm whales of the southern hemisphere and their bearing on sperm whale biology. *Discovery Reports* **37**, 1–324.

Clutton-Brock, T.H. (1991) *The Evolution of Parental Care*. Princeton University Press, Princeton, NJ.

Connor, R.C. (2000) Group living in whales and dolphins. In: *Cetacean Societies: Field Studies of Dolphins and Whales* (J. Mann, R.C. Connor, P.L. Tyack & H. Whitehead, eds), pp. 199–218. University of Chicago Press, Chicago.

Connor, R.C. & Norris, K.S. (1982) Are dolphins reciprocal altruists? *American Naturalist* **119**, 358–374.

Connor, R.C. & Smolker, R.A. (1996) 'Pop' goes the dolphin: a vocalization male bottlenose dolphins produce during consortships. *Behaviour* **133**, 643–662.

Connor, R.C., Smolker, R.A. & Richards, A.F. (1992a) Dolphin alliances and coalitions. In: *Coalitions and Alliances in Humans and Other Animals* (A.H. Harcourt & F.B.M. de Waal, eds), pp. 415–443. Oxford University Press, Oxford.

Connor, R.C., Smolker, R.A. & Richards, A.F. (1992b) Two levels of alliance formation among bottlenose dolphins (*Tursiops* sp.). *Proceedings of the National Academy of Sciences* **89**, 987–990.

Connor, R.C., Richards, A.F., Smolker, R.A. & Mann, J. (1996) Patterns of female attractiveness in Indian Ocean bottlenose dolphins. *Behaviour* **133**, 37–69.

Connor, R.C., Mann, J., Tyack, P.L. & Whitehead, H. (1998) Social evolution in toothed whales. *Trends in Ecology and Evolution* **13**, 228–232.

Connor, R.C., Heithaus, R.M. & Barre, L.M. (1999) Super-alliance of bottlenose dolphins. *Nature* **371**, 571–572.

Connor, R.C., Read, A.J. & Wrangham, R.W. (2000a) Male reproductive strategies and social bonds. In: *Cetacean Societies: Field Studies of Dolphins and Whales* (J. Mann, R.C. Connor, P.L. Tyack & H. Whitehead, eds), pp. 247–269. University of Chicago Press, Chicago.

Connor, R.C., Wells, R.S., Mann, J. & Read, A.J. (2000b) The bottlenose dolphin: social relationships in a fission–fusion society. In: *Cetacean Societies: Field Studies of Dolphins and Whales* (J. Mann, R.C. Connor, P.L. Tyack & H. Whitehead, eds), pp. 91–126. University of Chicago Press, Chicago.

Corkeron, P. & Connor, R.C. (1999) Why do baleen whales migrate? *Marine Mammal Science* **15**, 1228–1245.

D'Vincent, C.G., Nilson, R.M. & Hanna, R.E. (1985) Vocalization and coordinated feeding behavior of the humpback whale in southeastern Alaska. *Scientific Reports of the Whales Research Institute, Tokyo* **36**, 41–47.

Da Silva, J. & Terhune, J.M. (1988) Harbour seal grouping as an anti-predator strategy. *Animal Behaviour* **36**, 1309–1316.

de Waal, F.B.M. (1987) Tension regulation and nonreproductive functions of sex in captive bonobos (*Pan paniscus*). *National Geographic Research* **3**, 318–335.

de Waal, F.B.M. (1990) Sociosexual behavior used for tension regulation in all age and sex combinations among bonobos. In: *Pedophilia: Biosocial Dimensions* (J.R. Feierman, ed.), pp. 378–393. Springer-Verlag, New York.

de Waal, F.B.M. (1992) Appeasement, celebration and food sharing in the two *Pan* species. In: *Human Origins* (T. Nishida, W.C. McGrew, P. Marler, M. Pickford & F. de Waal, eds), pp. 37–50. University of Tokyo Press, Tokyo.

Dillon, M.C. (1996) *Genetic structure of sperm whale populations assessed by mitochondrial DNA sequence variation*. PhD Thesis, Dalhousie University, Halifax, Nova Scotia.

Duffield, D.A. & Wells, R.S. (1991) The combined application of chromosome protein and molecular data for the investigation of social unit structure and dynamics in *Tursiops truncatus*. *Reports of the International Whaling Commission Special Issue* **13**, 155–170.

Duncan, P. & Vigne, N. (1979) The effect of group size in horses on the rate of attacks by blood-sucking flies. *Animal Behaviour* **27**, 623–625.

Dunn, D.G., Barco, S., McLellan, W.A. & Pabst, D.A. (1999) Evidence for infanticide in bottlenose dolphins (*Tursiops truncatus*) of the western north Atlantic (abstract). In: *Abstracts, 13th Biennial Conference on the Biology of Marine Mammals, Maui, Hawaii.*

Emlen, S.T. & Oring, L.W. (1977) Ecology, sexual selection, and the evolution of mating systems. *Science* **197**, 215–223.

Enquist, M. & Leimar, O. (1990) The evolution of fatal fighting. *Animal Behaviour* **39**, 1–9.

Fertl, D. & Würsig, B. (1995) Coordinated feeding by Atlantic spotted dolphins (*Stenella frontalis*) in the Gulf of Mexico. *Aquatic Mammals* **21**, 3–5.

Gaskin, D.E. (1982) *The Ecology of Whales and Dolphins.* Heinemann, London.

Gentry, R.L. (1998) *Behavior and Ecology of the Northern Fur Seal.* Princeton University Press, Princeton, NJ.

Goodall, R.N.P., Galeazzi, A.R. & Sobral, A.P. (1988) Flipper serration in *Cephalorhynchus commersonii*. *Reports of the International Whaling Commission Special Issue* **9**, 161–171.

Gordon, J.C.D. (1987) Sperm whale groups and social behaviour observed off Sri Lanka. *Reports of the International Whaling Commission* **37**, 205–217.

Gowans, S. (1999) *Social organization and population structure of northern bottlenosewhales in the Gully.* PhD Thesis, Dalhousie University, Halifax, Nova Scotia.

Gowans, S. & Rendell, L. (1999) Head-butting in northern bottlenose whales (*Hyperoodon ampullatus*): a possible function for big heads? *Marine Mammal Science* **15**, 1342–1350.

Greenwood, P.J. (1980) Mating systems, philopatry and dispersal in birds and mammals. *Animal Behaviour* **28**, 1140–1162.

Guinet, C. (1991) L'orque (*Orcinus orca*) autour de l'archipel Crozet—comparaison avec d'autres localites. *Rev. Ecol. (Terre Vie)* **46**, 321–337.

Hain, J.H.W., Carter, G.R., Kraus, S.D., Mayo, C.A. & Winn, H.E. (1982) Feeding behaviour of the humpback whale, *Megaptera novaeangliae*, in the western North Atlantic. *Fishery Bulletin* **80**, 259–268.

Hainsworth, F.R. (1987) Precision and dynamics of positioning by Canada geese flying in formation. *Journal of Experimental Biology* **128**, 445–462.

Hall, H.J. (1985) Fishing with dolphins?: affirming a traditional aboriginal fishing story in Moreton Bay, SE Queensland. In: *Focus on Stradbroke: New Information on North Stradbroke Island and Surrounding Areas, 1974–1984.*

Hart, B.L. & Hart, L.A. (1992) Reciprocal allogrooming in impala, *Aepyceros melampus*. *Animal Behaviour* **44**, 1073–1083.

Hartman, D.S. (1979) *Ecology and Behavior of the Manatee (Trichechus manuatus) in Florida.* Special Publication No. 5. American Society of Mammalogists.

Hausfater, G. (1975) Dominance and reproduction in baboons (*Papio cynocephalus*): a quantitative analysis. In: *Contributions to Primatology*, Vol. 7. Karger, Basel.

Helweg, D.A., Bauer, G.B. & Herman, L.M. (1992) Observations of an S-shaped posture in humpback whales (*Megaptera novaeangliae*). *Aquatic Mammals* **18**, 74–78.

Heyning, J.E. (1984) Functional morphology involved in intraspecific fighting of the beaked whale. *Mesoplodon carlhubbsi*. *Canadian Journal of Zoology* **62**, 1645–1654.

Hinde, R.A. (1976) Interactions, relationships and social structure. *Man* **1**, 1–17.

Hoelzel, A.R. (1991) Killer whale predation on marine mammals at Punta Norte, Argentina; food sharing, provisioning and foraging strategy. *Behavioural Ecology and Sociobiology* **29**, 197–204.

Hoelzel, A.R. (1993) Foraging behavior and social group dynamics in Puget sound killer whales. *Animal Behavior* **45**, 581–591.

Hooker, S.K. & Baird, R. (1999) Deep-diving behaviour of the northern bottlenose whale *Hyperoodon ampulatus* (Cetacea: Ziphiidae). *Proceedings of the Royal Society of London, Series B* **266**, 671–676.

Hrdy, S.B. (1979) Infanticide among mammals: a review, classification, and examination of the implications for the reproductive strategies of females. *Ethology and Sociobiology* **1**, 13–40.

Inman, A.J. & Krebs, J. (1987) Predation and group living. *Trends in Ecology and Evolution* **2**, 31–32.

Insley, S. (2000) Long-term vocal recognition in a non-human mammal, the northern fur seal. *Nature* **406**, 404–405.

Isbell, L.A. & van Vuren, D. (1996) Differential costs of locational and social dispersal and their consequences for female group-living primates. *Behaviour* **133**, 1–36.

Jefferson, T.A., Stacey, P.J. & Baird, R.W. (1991) A review of killer whale interactions with other marine mammals: predation to co-existence. *Mammal Review* **21**, 151–180.

Johnson, C.M. & Moewe, K. (1999) Pectoral fin preference during contact in Commerson's dolphins (*Cephalorhynchus commersonii*). *Aquatic Mammals* **25**, 73–77.

Johnson, C.M. & Norris, K.S. (1994) Social behavior. In: *The Hawaiian Spinner Dolphin* (K.S. Norris, B. Würsig, R.S. Wells & M. Würsig, eds), pp. 243–286. University of California Press, Berkeley.

Jones, E.C. (1971) *Isistius brasiliensis*, a squaloid shark, the probable cause of crater wounds on fishes and cetaceans. *Fishery Bulletin* **69**, 791–798.

Jurasz, C.M. & Jurasz. V.P. (1979) Feeding modes of the humpback whale, *Megaptera novaeangliae*, in southeast Alaska. *Scientific Reports of the Whales Research Institute, Tokyo* **31**, 69–83.

Kano, T. (1992) *The Last Ape: Pygmy Chimpanzee Behavior and Ecology.* Stanford University Press, Stanford, CA.

Kasuya, T., Balcomb, K. & Brownell Jr, R.L. (1997) Life history of Baird's beaked whales off the Pacific coast of Japan. *Reports of the International Whaling Commission* **47**, 969–979.

Kitamura, K. (1989) Genito-genital contacts in the pygmy chimpanzees (*Pan paniscus*). *African Study Monographs* **10**, 49–67.

Kuroda, S. (1984) Interaction over food among pygmy chimpanzees. In: *The Pygmy Chimpanzee: Evolutionary Biology and Behavior* (R. Susman, ed.), pp. 301–324. Plenum Press, New York.

Le Boeuf, B.J. (1974) Male–male competition and reproductive success in elephant seals. *American Zoologist* **14**, 163–176.

Le Boeuf, B.J., Naito, Y., Huntley, A.C. & Asaga, T. (1989) Prolonged, continuous, deep diving in female northern elephant seals, *Mirounga angustrostris. Canadian Journal of Zoology* **67**, 2514–2519.

Le Boeuf, B.J. & Reiter, J. (1988) Lifetime reproductive success in northern elephant seals. In: *Lifetime Reproductive Success* (T.H. Clutton-Brock, ed.), pp. 344–362. University of Chicago Press, Chicago.

Lima, S.L. (1995a) Back to the basics of anti-predator vigilance: the group-size effect. *Animal Behavior* **49**, 11–20.

Lima, S.L. (1995b) Collective detection of predatory attack by social foragers: fraught with ambiguity? *Animal Behavior* **50**, 1097–1108.

Lissaman, P.B.S. & Schollenberger, C.A. (1970) Formation flight of birds. *Science* **18**, 1003–1005.

Lopez, J.C. & Lopez, D. (1985) Killer whales of Patagonia (*Orcinus orca*) and their behavior of intentional stranding while hunting nearshore. *Journal of Mammalogy* **66**, 181–183.

MacLeod, C.K. (1998) Intraspecific scarring in odontocete cetaceans: an indicator of male 'quality' in aggressive social interactions? *Journal of Zoology* **244**, 71–77.

McCann, C. (1974) Body scarring on Cetacea—Odontocetes. *Scientific Reports of the Whales Research Institute, Tokyo* **1974**, 145–155.

Mann, J. & Smuts, B.B. (1999) Behavioral development of wild bottlenose dolphin newborns. *Behavior* **136**, 529–566.

Mann, J., Connor, R., Tyack, P. & Whitehead, H. (2000) *Cetacean Societies: Field Studies of Whales and Dolphins.* University of Chicago Press, Chicago.

Miller, E.H. (1975) Walrus ethology, I. The social role of tusks and applications of multidimensional scaling. *Canadian Journal of Zoology* **53**, 590–613.

Miller, E.H. & Boness, D.J. (1979) Remarks on display functions of the snout of the grey seal, *Halichoerus grypsus* (Fab.), with comparative notes. *Canadian Journal of Zoology* **57**, 140–148.

Morosov, D.A. (1970) Dolphins hunting. *Rybnoe Khoziaistvo* **46**, 16–17.

Moss, C. (1988) *Elephant Memories.* William Morrow, Inc., New York.

Moss, C.J. & Poole, J.H. (1983) Relationships and social structure of African elephants. In: *Primate Social Relationships: an Integrated Approach* (R.H. Hinde, ed.), pp. 315–325. Sinauer Associates, Inc., Sunderland, MA.

Nöe, R. & Bshary, R. (1997) The formation of red colobus–diana monkey associations under predation pressure from chimpanzees. *Proceedings of the Royal Society of London, Series B* **264**, 253–259.

Norris, K.S. (1967) Aggressive behavior in Cetacea. In: *Aggression and Defenses: Neural Mechanisms and Social Patterns* (C.D. Clemente & D.B. Lindsley, eds), pp. 225–224. University of California Press, Berkeley.

Norris, K.S. & Dohl, T.P. (1980) The structure and function of cetacean schools. In: *Cetacean Behavior* (L.M. Herman, ed.), pp. 211–261. John Wiley & Sons, New York.

Norris, K.S., Würsig, B., Wells, R.S. and Würsig, M. (1994) *The Hawaiian Spinner Dolphin.* University of California Press, Berkeley.

Osborne, R.W. (1986) A behavioral budget of Puget Sound killer whales. In: *Behavioral Biology of Killer Whales* (B.C. Kirkevold & J.S. Lockard, eds), pp. 211–249. Alan R. Liss, Inc., New York.

Östman, J. (1991) Changes in aggressive and sexual behavior between two male bottlenose dolphins (*Tursiops truncatus*) in a captive colony. In: *Dolphin Societies* (K. Pryor & K.S. Norris, eds), pp. 305–317. University of California, Berkeley.

Pack, A.A., Salden, D.R., Ferrari, M.J. *et al.* (1998) Male humpback whale dies in competitive group. *Marine Mammal Science* **14**, 861–873.

Packer, C., Herbst, L., Pusey, A.E. *et al.* (1988) Reproductive success in lions. In: *Reproductive Success* (T.H. Clutton-Brock, ed.), pp. 363–383. University of Chicago Press, Chicago.

Packer, C., Gilbert, D.A., Pusey, A.E. & O'Brien, S.J. (1991) A molecular genetic analysis of kinship and cooperation in African lions. *Nature* **351**, 562–565.

Papastavrou, V., Smith, S.C. & Whitehead, H. (1989) Diving behaviour of the sperm whale, *Physeter macrocephalus*, off the Galápagos Islands. *Canadian Journal of Zoology* **67**, 839–846.

Partridge, B.L., Johansson, J. & Kalish, J. (1983) The structure of schools of giant bluefin tuna in Cape Cod Bay. *Environmental Biology of Fishes* **9**, 253–262.

Patterson, I.A.P., Reid, R.J., Wilson, B. *et al.* (1998) Evidence for infanticide in bottlenose dolphins: an explanation for violent interactions with harbour porpoises? *Proceedings of the Royal Society of London, Series B* **265**, 1–4.

Payne, R. (1995) *Among Whales.* Simon & Schuster, New York.

Payne, R.S. & Dorsey, E. (1983) Sexual dimorphism and aggressive use of callosities in right whales (Eubalaena australis). In: Communication and Behavior of Whales (R. Payne, ed.), pp. 295–329. Westview Press, Boulder, CO.

Perry, A., Baker, C.S. & Herman, L.M. (1988) Population characteristics of individually identified humpback whales in the central and eastern North Pacific: a summary and critique. *Reports of the International Whaling Commission, Special Issue* **13**, 307–317.

Peterson, R.S. & Bartholomew, G.A. (1967) *The Natural History and Behavior of the California Sea Lion*. Special Publication No. 1. American Society of Mammologists,

Pitcher, T.J. & Parrish, J.K. (1993) Functions of shoaling behaviour in teleosts. In: *Behaviour of Teleost Fishes*, 2nd edn (T.J. Pitcher, ed.), pp. 363–439. Chapman & Hall, London.

Preen, A. (1989) Observations of mating behavior in dugongs (*Dugong dugon*). *Marine Mammal Science* 5, 382–387.

Pryor, K., Lindbergh, J., Lindbergh, S. & Milano, R. (1990) A human–dolphin fishing cooperative in Brazil. *Marine Mammal Science* 6, 77–82.

Pulliam, H.R. & Caraco, T. (1984) Living in groups: is there an optimal group size?. In: *Behavioral Ecology: an Evolutionary Approach*, 2nd edn (J.R. Krebs & N.B. Davies, eds), pp. 122–147. Sinauer, Sunderland, MA.

Pusey, A.E. & Wolf, M. (1996) Inbreeding avoidance in animals. *Trends in Ecology and Evolution* 11, 201–206.

Reeves, R.R. & Whitehead, H. (1997) Status of the sperm whale (*Physeter macrocephalus*) in Canada. *Canadian Field Naturalist* 111, 293–307.

Rice, D.W. (1989) Sperm whale. *Physeter macrocephalus* Linnaeus, 1758. In: *Handbook of Marine Mammals, Vol. 4. River Dolphins and the Larger Toothed Whales* (S.H. Ridgway & R. Harrison, eds), pp. 177–233. Academic Press, London.

Richard, K.R., Dillon, M.C., Whitehead, H. & Wright, J.M. (1996) Patterns of kinship in groups of free-living sperm whales (*Physeter macrocephalus*) revealed by multiple molecular genetic analyses. *Proceedings of the National Academy of Sciences, USA* 93, 8792–8795.

Richards, A.F. (1996) *Life history and behavior of female dolphins in Shark Bay, Western Australia*. PhD Thesis, University of Michigan, Ann Arbor, MI.

Riedman, M.L. & Estes, J.A. (1990) *The Sea Otter (Enhydra lutris): Behavior, Ecology, and Natural History*. Biological Report No. 90 (14). US Fish and Wildlife Service, Washington, DC.

Rose, N.A. (1992) *The social dynamics of male killer whale, Orcinus orca, in Johnstone Strait, British Columbia*. PhD Thesis, University of California, Santa Cruz.

Ross, H.M. & Wilson, B. (1996) Violent interactions between bottlenose dolphins and harbour porpoises. *Proceedings of the Royal Society of London, Series B* 263, 283–286.

Saayman, G.S. & Taylor, C.K. (1979) The socioecology of humpback dolphins (*Sousa* sp.). In: *Behavior of Marine Animals, Vol. 3* (H.E. Winn & B.L. Olla, eds), pp. 165–226. Plenum Press, New York.

Samuels, A. & Gifford, T. (1997) A quantitative assessment of dominance relations among bottlenose dolphins. *Marine Mammal Science* 13, 70–99.

Schaeff, C.M., Boness, D.J. & Bowen, W.D. (1999) Female distribution, genetic relatedness, and fostering behaviour in harbour seals, *Phoca vitulina*. *Animal Behaviour* 57, 427–434.

Schusterman, R.J., Gisiner, R. & Hanggi, E.B. (1992a) Imprinting and other aspects of pinniped–human interactions. In: *The Inevitable Bond: Examining Scientist–Animal Interactions* (H. Davis & D. Balfour, eds), pp. 334–356. Cambridge University Press, New York.

Schusterman, R.J., Hanggi, E.B. & Gisiner, R. (1992b) Acoustic signalling in mother–pup reunions, interspecies bonding, and affiliation by kinship in California sea lions (*Zalophus californianus*). In: *Marine Mammal Sensory Systems* (J. Thomas, ed.), pp. 533–551. Plenum Press, New York.

Scott, M.D. & Cordaro, J.G. (1987) Behavioral observations of the dwarf sperm whale, *Kogia simus*. *Marine Mammal Science* 3, 353–354.

Sergeant, D.E. & Brodie, P.F. (1969) Body size in white whales, *Delphinapterus leucas*. *Journal of the Fisheries Research Board of Canada* 26, 2561–2580.

Sharpe, F.A. & Dill, L.M. (1995) The bubble helix: sonar studies of feeding humpback whales (abstract). *Abstracts, Eleventh Biennial Conference on the Biology of Marine Mammals, Orlando, Florida*.

Silverman, H.B. & Dunbar, M.J. (1980) Aggressive tusk use by the narwhal (*Monodon monocerus* L.). *Nature* 284, 57–58.

Similä, T. (1997) Sonar observations of killer whales (*Orcinus orca*) feeding on herring schools. *Aquatic Mammals* 23, 119–126.

Similä, T. & Ugarte, F. (1993) Surface and underwater observations of cooperatively feeding killer whales in northern Norway. *Canadian Journal of Zoology* 71, 1494–1499.

Sjare, B. & Stirling, I. (1996) The breeding behavior of Atlantic walruses, *Odobenus romarus rosmarus*, in the Canadian high arctic. *Canadian Journal of Zoology* 74, 897–911.

Smolker, R.A., Richards, A.F., Connor, R.C. & Pepper, J.W. (1992) Sex differences in patterns of association among Indian Ocean bottlenose dolphins. *Behaviour* 123, 38–69.

Sterck, E.H.M., Watts, D.P. & van Schaik, C.P. (1997) The evolution of female social relationships in nonhuman primates. *Behavioral Ecology and Sociobiology* 41, 291–309.

Stirling, I. (1975) Factors affecting the evolution of social behaviour in the Pinnipedia. *Rapports et Proces-Verbaux des Renunions, Conseil International pur L'Exploration de la Mer* 169, 205–212.

Swartz, S. (1986) Gray whale migratory, social and breeding behavior. *Reports of the International Whaling Commission, Special Issue* 8, 207–229.

Tarpy, C. (1979) Killer whale attack. *National Geographic* 155, 542–545.

Terhune, J. (1985) Scanning behavior of harbor seals on haul-out sites. *Journal of Mammalogy* 66, 392–395.

Tershy, B.R. (1992) Body size, diet, habitat use, and social behavior of *Balaenoptera* whales in the Gulf of California. *Journal of Mammalogy* 73, 477–486.

Trillmich, F. (1996) Parental investment in pinnipeds. *Advances in the Study of Behavior* 25, 533–577.

Tyack, P.L. & Whitehead, H. (1983) Male competition in large groups of wintering humpback whales. *Behaviour* 83, 132–154.

van Schaik, C.P. & Kappeler, D.M. (1997) Infanticide risk and the evolution of male–female association in primates. *Proceedings of the Royal Society of London, Series B* **264**, 1687–1694.

Weilgart, L.S., Whitehead, H. & Payne, K. (1996) A colossal convergence. *American Scientist* **84**, 278–287.

Weller, D.W., Würsig, B., Whitehead, H. *et al.* (1996) Observations of an interaction between sperm whales and short-finned pilot whales in the Gulf of Mexico. *Marine Mammal Science* **12**, 588–593.

Wells, R.S. (1984) Reproductive behavior and hormonal correlates in Hawaiian spinner dolphins, *Stenella longirostris*. *Reports of the International Whaling Commission, Special Issue* **6**, 465–472.

Wells, R.S. (1991) The role of long-term study in understanding the social structure of a bottlenose dolphin community. In: *Dolphin Societies: Discoveries and Puzzles* (K. Pryor & K.S. Norris, eds), pp. 199–225. University of California Press, Berkeley, CA.

Wells, R.S., Scott, M.D. & Irvine, A.B. (1987) The social structure of free-ranging bottlenose dolphins. In: *Current Mammalogy* (H. Genoways, ed.), Vol. 1, pp. 263–317. Plenum Press, New York.

Whitehead, H. (1989) Formations of foraging sperm whales, *Physeter macrocephalus*, off the Galápagos Islands. *Canadian Journal of Zoology* **67**, 2131–2139.

Whitehead, H. (1990) Rules for roving males. *Journal of Theoretical Biology* **145**, 355–368.

Whitehead, H. (1994) Delayed competitive breeding in roving males. *Journal of Theoretical Biology* **166**, 127–133.

Whitehead, H. (1996) Babysitting, dive synchrony, and indications of alloparental care in sperm whales. *Behavioral Ecology and Sociobiology* **38**, 237–244.

Whitehead, H. & Mann, J. (2000) Female reproductive strategies of cetaceans: life histories and calf care. In: *Cetacean Societies: Field Studies of Dolphins and Whales* (J. Mann, R.C. Connor, P.L. Tyack & H. Whitehead, eds), pp. 219–246. University of Chicago Press, Chicago.

Whitehead, H. & Weilgart, L. (2000) The sperm whale: social females and roving males. In: *Cetacean Societies: Field Studies of Dolphins and Whales* (J. Mann, R.C. Connor,

P.L. Tyack & H. Whitehead, eds), pp. 154–172. University of Chicago Press, Chicago.

Wickler, W. (1967) Sociosexual signals and their intraspecific imitation among primates. In: *Primate Ethology* (D. Morris, ed.), pp. 69–147. Doubleday, New York.

Wilson, B., Thompson, P. & Hammond, P. (1992) The ecology of bottle-nosed dolphins, *Tursiops truncatus*, in the Moray Firth. In: *European Research on Cetaceans, Proceedings of the Sixth Annual Conference of the European Cetacean Society* (P.G.H. Evans, ed.). European Cetacean Society, Cambridge, UK.

Wilson, B., Thompson, P. & Hammond, P. (1993) An examination of the social structure of a resident group of bottle-nosed dolphins (*Tursiops truncatus*) in the Moray Firth, NE Scotland. In: *European Research on Cetaceans, Proceedings of the Seventh Annual Conference of the European Cetacean Society* (P.G.H. Evans, ed.). European Cetacean Society, Cambridge, UK.

Wilson, B., Thompson, P.M. & Hammond, P.S. (1997) Habitat use by bottlenose dolphins: seasonal distribution and stratified movement patterns in the Moray Firth, Scotland. *Journal of Applied Ecology* **34**, 1365–1374.

Wrangham, R.W. (1979) On the evolution of ape social systems. *Social Science Information* **18**, 335–368.

Wrangham, R.W. (1980) An ecological model of female-bonded primate groups. *Behaviour* **75**, 262–292.

Wrangham, R.W. (1993) The evolution of sexuality in chimpanzees and bonobos. *Human Nature* **4**, 47–79.

Würsig, B. (1986) Delphinid foraging stategies. In: *Dolphin Cognition and Behavior: a Comparative Approach* (R.J. Schusterman, J.A. Thomas & F.G. Wood, eds), pp. 347–359. Lawrence Erlbaum Associates, Hillsdale, NJ.

Würsig, B., Wells, R.S. & Norris, K.S. (1994) Food and feeding. In: *The Hawaiian Spinner Dolphin* (K.S. Norris, B. Würsig, R.S. Wells & M. Würsig, eds), pp. 216–231. University of California Press, Berkeley, CA.

Würsig, B. & Würsig, M. (1979) Behavior and ecology of the bottlenose dolphin, *Tursiops truncatus*, in the South Atlantic. *Fishery Bulletin* **77**, 399–412.

Zahavi, A. (1977) The testing of a bond. *Animal Behaviour* **25**, 246–247.

CHAPTER 13

Problem Solving and Memory

Ronald J. Schusterman and David Kastak

13.1 INTRODUCTION

In this chapter we will review marine mammal cognition, defined loosely as information processing, or the intervening neural stages between perception and action. We will contrast cognition with simpler models of behaviour (stimulus–response or reflex models), in asserting that the external environment of the marine mammal is coded through peripheral (sensory) and central mechanisms into imaginal representations. We will not speculate further on the topography of these representations, since that is beyond the scope of most of the existing work on the behaviour of marine mammals. We will emphasize an approach that unites experimental psychology and behavioural ecology: that is, based on both laboratory and field studies of behaviour. By synthesizing data and theory from these two approaches, we can begin to determine the cognitive mechanisms involved in natural behaviour. Most of the data in this area are limited to two species, the California sea lion (*Zalophus californianus*) and the bottlenose dolphin (*Tursiops truncatus*), and this is particularly true for the laboratory approach. It is for that reason that our discussion will focus with few exceptions on these animals.

13.2 CLASSIFICATION, ABSTRACTION AND MEMORY

13.2.1 Immediate sensory experience

An organism obtains information about its environment through sensory perception (see Chapter 5); therefore, accurate perception is crucial in performing even the most basic tasks related to foraging, navigation, social behaviour and predator avoidance. Sensory channels are, to a certain extent, domain specific, that is, they operate on a restricted set of ecologically appropriate inputs. This domain specificity can be illustrated by examining vision in pinnipeds. For example, the northern elephant seal's visual system is designed not only to respond rapidly to changes in light level (i.e. dark adapt) but to take advantage of very low light levels of shorter wavelengths of light that predominate in deep water. In contrast, shallower diving pinnipeds like harbour seals and California sea lions dark adapt more slowly and are less sensitive to light than are elephant seals (Levenson & Schusterman 1997, 1999) (see Chapter 5). This trend appears to hold for other sensory domains, such as sound pressure detection, as well (Kastak & Schusterman 1998, 1999).

The ability of a marine mammal, for instance, to learn simple discriminative behaviours will be fundamental to our understanding of how it forms a concept. Since the memory of the animal or the orientation of its receptors may change from one encounter to another, it is likely that most animals, including marine mammals, will have a tendency to generalize, classify or in some way respond in a similar way to objects and events previously encountered. The initial experience of catching and eating prey of a particular type (i.e. having a particular shape, style of movement, location, taste, etc.) should be a good predictor of the animal's behaviour the next time around.

Such problem solving relies on more general, less domain specific cognitive approaches. There are findings from both the laboratory and the field demonstrating that odontocete cetaceans, pinnipeds, otters and sirenians can spontaneously classify and

BOX 13.1 STIMULUS CONTROL

Laboratory experiments in animal cognition are typically variants of the following common procedures. The **simple discrimination** is a task in which a response to one of two or more stimuli is reinforced while responses to any other stimulus is not reinforced. The stimulus discriminative for reinforcement is often termed the S+, and the stimulus discriminative for non-reinforcement, or extinction, is often termed the S–. Stimulus presentation can be simultaneous (the S+ and S– are presented together) or sequential (the S+ and S– are presented in series). In the sequential case, the subject is trained to respond to the S+ when it appears, and to withhold response to the S– when it appears. This is often called a **go/no-go** procedure. Occasionally, an experiment is designed such that there are two separate responses (e.g. point right or left) or the subject is required to withhold the response until the end of a sampling interval during which two stimuli (usually acoustic) are presented. This type of task is referred to as a **two-alternative forced choice** procedure. The simplest discriminations, in terms of procedural ease, involve visual displays of three-dimensional objects or planometric stimuli (patterns on a flat background). Acoustic discriminations are somewhat more challenging, because the stimuli must be presented successively, and the subject must remember the salient characteristics of the first stimulus in order to respond accurately.

In a variation of the simple discrimination, responses to the S+ are reinforced until the subject responds at a criterional level (e.g. 90% correct in a block of trials). Following acquisition of the criterion, the reinforcement contingencies are reversed such that the previous S– becomes the S+. The subject must now learn to inhibit responding to the previous S+ and respond only to the previous S–. The cycle is repeated following acquisition of criterion during each phase of the experiment. Repeated reversals often result in the development of a strategy such as 'win–stay/lose–shift', that enables the subject to respond accurately following only one trial of a

reversal. This 'learning to learn' or to understand the nature of the task is called **learning set**.

In **conditional** or **contextual discriminations**, responding is controlled not by the S+ and/or S– but by an additional stimulus, often called the **sample**. The sample is discriminative for responding to one of two or more comparisons, thus the task is sometimes called **matching-to-sample** (MTS) (Fig. 13.1). There are three distinct categories of MTS: identity, oddity and symbolic. In identity matching, one of the comparisons is a duplicate of the sample (i.e. regardless of the characteristics of the sample, choose the comparison that is the same). Correct responding in identity matching often, but not always, generates an identity rule or concept. Oddity matching is the direct opposite of identity matching, and may also generate a rule (i.e. choose the stimulus that is not the same as the sample). In symbolic or arbitrary MTS, the sample and S+ do not share any features in common, but are related arbitrarily, thus no rule can facilitate the solution of completely novel problems with novel stimuli. It is important to distinguish between a poorly controlled MTS test, which allows the animal to choose an alternative stimulus merely by 'exclusion' or the process of elimination (i.e. a rule of thumb), and a MTS test that rigorously controls for exclusion effects (Table 13.1) (Kastak & Schusterman 1992; Schusterman *et al.* 1993a).

Artificial language procedures used with bottlenose dolphins and California sea lions are variants of symbolic MTS. The discriminative stimuli consist of sequences of gestural cues or acoustic signals that become related to specific objects, object properties (location, size or brightness), or actions. When strung together, the various signs convey instructions to the subject (e.g. fetch the black ball). Although seemingly complicated, performance on artificial language tasks can be explained using concepts no more complicated than sequential conditional discriminations (for discussion of alternate interpretations, see Herman 1988, 1989; Schusterman & Gisner 1988, 1989, 1997).

---

remember their experiences with individuals, objects, events, time and space, according to distinctive (perceptually based) stimulus attributes, or, at least, can be readily trained to do so (Box 13.1).

13.2.2 Conceptual abstractions

Marine mammals, in addition to solving problems based on concrete stimulus properties, are able to learn more general or abstract strategies—'select

whatever object paid off on the last trial and avoid the object that did not pay off the last time'. Thus, they develop a 'win–stay/lose–shift' or 'learning set' strategy (Box 13.1) (Schusterman 1962, 1964). Table 13.1 shows a break down of performance on laboratory-based cognitive tasks in pinnipeds and cetaceans. In addition to a learning set, both bottlenose dolphins and California sea lions, following limited experience, can perform generalized identity matching (i.e. form a 'sameness' concept that can be

Table 13.1 Summary of major rule-based learning and concept formation studies in marine mammals. The table is broken down into species, sensory modality, tasks and stimulus dimensions. Tasks involving simple psychophysics (e.g. auditory thresholds) have not been included. Rule-based studies whose primary emphasis was the study of working memory are also not included.

Species	Sensory modality	Task	Stimuli or stimulus dimensions	Source
Odontocetes Delphinidae *Tursiops truncatus* (bottlenose dolphin)	Vision	Two-choice discrimination	Shape	Kellogg & Rice 1966
		Conditional discrimination (artificial language)	Size, shape, brightness, movement, constancy, transposition, presence, absence	Herman et al. 1984; Herman & Forestell 1985
		Identity MTS (exclusion and reflexivity)	Shape	Herman et al. 1989
		Two-choice discrimination (symmetry vs asymmetry)	Shape	von Fersen et al. 1993
		Same/different	Shape	Mercado et al. 2000
	Audition	Two-choice discrimination, learning set	Complex auditory characteristics	Herman & Arbeit 1973
		MTS (oddity vs identity)	Complex auditory characteristics	Herman & Gordon 1974
		Arbitrary (symbolic) MTS	Complex auditory characteristics	Herman & Thompson 1982
		Identity MTS (exclusion)	Complex auditory characteristics	Herman & Thompson 1982
		Spatial two-alternative forced-choice procedure (functional relations)	Pure tones	von Fersen & Delius 2000
	Audition/vision	Conditional discrimination (artificial language)	Size, shape, brightness, movement, constancy, transposition, presence, absence	Herman et al. 1984
	Audition (echolocation)	Identity MTS	Target shape, structure	Roitblat et al. 1990; Pack & Herman 1995; Harley et al. 1996; Herman et al. 1998
	Audition (echolocation) and vision	Conditional discrimination	Target shape, structure	Pack & Herman 1995; Harley et al. 1996; Herman et al. 1998

(continued on p. 374)

Table 13.1 (*cont'd*)

Species	Sensory modality	Task	Stimuli or stimulus dimensions	Source
Pinnipeds				
Otariidae				
Zalophus californianus (California sea lion)	Vision	Two-choice discrimination or go/no-go conditioned vocalization	Shape	Schusterman 1966
		Attention shift	Size to shape	Schusterman 1966, 1967
		Two-choice discrimination	Shape constancy	Schusterman & Thomas 1967
		Conditioned vocalization	Form/reversal	Schusterman 1967
		Two-choice discrimination	Form/reversal learning set	Schusterman 1968
		Conditional discrimination (artificial language)	Size, shape, brightness, movement, constancy, transposition, presence, absence	Schusterman & Krieger 1984, 1986; Schusterman & Gisiner 1988; Gisiner & Schusterman 1992
		Arbitrary (symbolic) MTS (exclusion)	Shape	Schusterman *et al.* 1993a
		Arbitrary (symbolic) MTS and two-choice discrimination (equivalence and functional relations)	Shape	Schusterman & Kastak 1993, 1998
		Identity MTS (reflexive relations)	Shape	Kastak & Schusterman 1994
		Identity MTS (exclusion)	Shape	Pack *et al.* 1991
		Functional relations and differential outcome	Shape	Reichmuth Kastak *et al.* 2001
		Identity MTS (rotational configurations or mental rotation)	Shape	Mauck & Denhardt 1997
Phocidae				
Phoca vitulina (harbour seal)	Vision	Two-choice discrimination, learning set	Shape	Schusterman 1968
		Two-choice discrimination/ identity MTS	Shape	Constantine 1981*
		Identity MTS (reflexivity)	Brightness	Renouf & Gaborko 1988*
		Arbitrary (symbolic) MTS	Spatial location	Renouf & Gaborko 1988, 1989
		Arbitrary (symbolic) MTS	Shape	Hanggi & Schusterman 1995*
	Audition	Identity MTS (reflexivity)	Frequency; complex acoustic characteristics	Renouf & Gaborko 1988*

MTS, matching-to-sample.
* Denotes failure by the subject to learn or generalize rule.

BOX 13.2 STIMULUS EQUIVALENCE

In the late 1960s, Murray Sidman, then at Northeastern University, was involved in a series of studies using a matching-to-sample procedure to assess reading comprehension in mentally retarded children (Sidman 1994). Sidman and colleagues noted that following explicit training of: (i) an associative relation between the spoken name for an object and the object's picture, and (ii) an associative relation between the spoken name for an object and the printed word of the object, a novel relation emerged (i.e. became apparent without explicit training) between the object and the object's printed word. Sidman referred to the sets of objects and referents (stimuli that 'referred' to the objects, like names) as equivalence classes, and also showed that an equivalence relation between any two members of a class would satisfy the mathematical properties of reflexivity, symmetry and transitivity. Thus, explicit training of relations between stimuli A and B followed by explicit training of relations between stimuli B and C, may result in a three-member equivalence class, confirmed by demonstrating the emergence of the following relations—reflexivity (AA, BB, CC), symmetry (CB and BA) and transitivity (CA). In this way, stimuli can mediate relations between all other stimuli in the same category (the same way that the words 'dog' and 'perro' and the picture of a dog may each evoke the same representation), allowing the solution of a variety of novel problems.

Although it was once thought that stimulus equivalence could only be demonstrated in language-proficient subjects, recent research has shown that some marine mammals, in addition to some terrestrial animals, can form equivalence relations. The ramifications of these recent experiments are clear—rather than being a manifestation of linguistic skills, equivalence, having been demonstrated in non-verbal animals, must be a more basic function that probably subserves many cognitive capabilities, especially those involved in social behaviour and communication. Equivalence relations (unlike many sensory processes) are not domain specific or modular. That is, the selective pressures acting on the brain's ability to categorize objects and events resulted from an array of non-social as well as social problems. Thus, equivalence theory provides a simple yet powerful model to explain many facets of complex behaviour without resorting to invocations of domain-specific 'social intelligence' or 'ecological intelligence' that require complicated learning theories to account for what are probably associative mechanisms. This is not to say that all aspects of cognitive function are likewise non-domain specific; for instance, there has been recent research on the modularity of some aspects of human social knowledge, imitation, etc., which seem to be uniquely human cognitive traits and may be related to the development of language, surely a cognitive module in its own right.

A possible example of specialization or modularity in marine mammal cognition is the singing of male humpback whales (Tyack 1981). In terms of both structure and function, humpback whale songs have been compared to the songs of birds (Tyack 1999) (see Chapter 6). However, unlike bird song, we presently have little knowledge of the neurological and behavioural mechanisms involved in the learning and memory of humpback whale song. It is quite possible that like birds' memory for songs, humpback whales' memory for song is highly species specific. However, laboratory research (see text) has revealed only small quantitative differences between marine mammal species in the capacity and durability of memory. And thus with the limited studies available, the captive studies of short-term memory in bottlenose dolphins and sea lions measured in similar ways have shown similar abilities which are comparable to terrestrial mammals and birds.

---

applied to novel stimuli). Acquisition of such a concept is a prerequisite for many other 'higher order' tasks such as classification of stimuli that are not perceptually similar, yet may share some common function (Box 13.2). This non-similarity based classification is a prerequisite for the emergence of meaning or object representations in social interactions and communication.

13.2.3 Memory

In many aspects of conspecific communication, foraging, predator avoidance and navigation, both learning and memory are important. Forming a search image for a particular prey type, recognizing the signature whistle or vocalization of one's kin, friends or foes, or acquiring a conditioned response are all examples of learning, but they are also examples of memory. In contrast to research on learning, which usually deals with associative processes, i.e. how information about relationships between events is acquired, research on memory deals with how the information is coded, stored, retained and retrieved. Psychology and the neurosciences generally distinguish between two types of memory—short-term memory (STM) or long-term memory (LTM). This section will focus on STM, or working memory, which is largely used to guide whatever tasks the individual is currently performing. For example, a male sea lion or fur seal occupying a breeding territory may have to keep track of all his opponents on adjoining territories. This focal male's working memory at any given moment would contain several pieces of information: what one opponent sounds like relative to the others, what a given opponent's motivational state might be (as reflected in territorial vocalizations), which opponent he most recently interacted with and what the outcome of that interaction was, topographical features delineating territorial boundaries, and other aspects of the physical and social environment. Notice that the focal animal's information must continually be updated; if he fails to notice one or more novel rivals on an adjoining territory and confuses the new opponent with an old one, he might lose all or part of his territory, which could have dramatic fitness consequences. In the foraging domain, a sea otter searching for fish, crustaceans or other prey items is required to remember which patches of prey it has already visited on a given day and must not confuse one day's visit with another. In most of these tasks involving foraging and reproductive behaviour, animals are required to remember significant details about the ongoing task and to ignore similar details from already completed tasks.

The most popular test of animal working memory is the delayed match-to-sampling (MTS) procedure shown in Fig. 13.1b. After the animal has learned to relate the conditional or sample stimulus to the comparison stimulus, memory trials are conducted. During these trials, the animal is required to choose the correct comparison stimulus in the absence of the sample. The delay between the removal of the sample and the presentation of the comparisons may vary, and this period of time is known as the 'retention interval'. Many studies of animal memory have focused on the retention interval with the idea that the memory for the sample–comparison relation either decays over time passively, or the memory is interfered with or retained by active processes (Domjan 1998). In memory tasks using delayed MTS, the stimuli are usually presented visually, but modifications have been made so the procedure can be adapted to different sensory modalities, including audition, taction, olfaction or some combination in a cross-modal test.

Herman and his associates (Herman & Gordon 1974; Thompson & Herman 1977; Herman et al. 1989) have done the most careful and systematic research measuring the durability and to some extent the capacity of STM in bottlenose dolphins. Originally, they showed that by gradually increasing the retention interval from 1 s to 2 min in an auditory delayed MTS, a female dolphin (Kea) could, after some practice, remember the same sample–comparison identity relation almost perfectly. Schusterman et al. (1991) attempted to replicate the dolphin's performance with two California sea lions (a 4-year-old female named Rio and a 13-year-old female named Rocky). The response task for the sea lions was a visual delayed symbolic MTS (Fig. 13.1). The results of the study demonstrate that the mature female Rocky could be trained to remember any given relationship for delays of up to 2 min with performance levels at 90% correct responses. The 4-year-old sea lion Rio achieved similar levels of performance at delays of up to 45 s.

On a California sea lion rookery, it would not be surprising if a very young pup (who had already been imprinted on its mother's signature vocalization) (Schusterman et al. 1992b), while listening to a cacophony of female pup attraction calls, would have its memory for its mother's vocalization interfered with by these similar sounding but irrelevant (and potentially dangerously distracting) signals. This kind of interference from new material with the to-be-remembered signal or event is called 'retroactive interference'. In California sea lions, and probably all otariid pinnipeds, the memory of both pups and their mothers for their respective signature vocalizations is highly resistant to retroactive interference. The same could be said about the mutual recognition of signature whistles by dolphin calves and their mothers (see Chapter 6). Nevertheless, since retroactive interference has been demonstrated in a variety of animals (for a review see Domjan 1998), it is worthwhile to determine whether the phenomenon is general enough to be demonstrated in marine mammals such as dolphins and sea lions. In another auditory identity-matching experiment, insertion of 13 s duration tones into a 15 s delay interval retroactively interfered with a

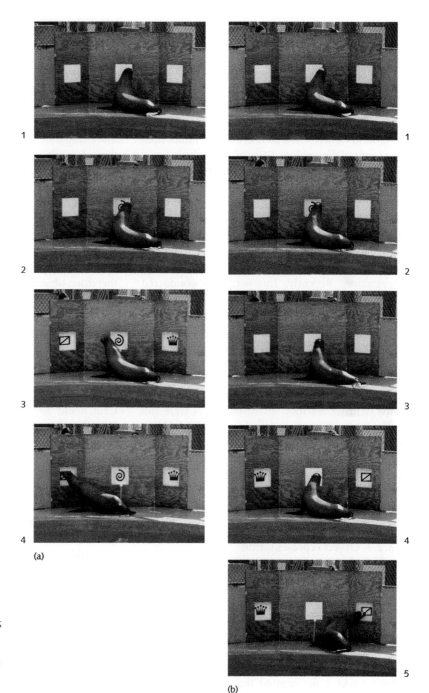

Fig. 13.1 Photograph of two sequences of matching-to-sample procedures. (a) A California sea lion in a simultaneous symbolic matching-to-sample procedure: 1, the animal waits for the sample presentation; 2, the sample is presented; 3, while the sample remains visible, the comparison stimuli are revealed; the sea lion readies itself to respond to the S+; 4, the sea lion responds by poking its head into the S+ box. (b) The same sea lion in a delayed symbolic matching-to-sample procedure: 1, the sea lion waits for the sample presentation; 2, the sample is presented; 3, a delay interval is inserted when no stimuli are visible; 4, comparison stimuli are shown; 5, the sea lion is released from the station and chooses the appropriate comparison stimulus by placing its head in the S+ box.

dolphin's representations of two different sample tones (Herman 1975). In a variety of memory experiments with bottlenose dolphins and California sea lions (Herman 1975; Schusterman *et al.* 1991)

retroactive interference of well-learned relations or associations has been readily demonstrated. Many types of surprising or unexpected stimulus material presented during the delay interval are likely to impair working memory. Results from experiments involving active disruption of memory, for instance the interposition of distracting events or stimuli (Schusterman *et al.* 1991, 1993b), indicate that both dolphins and California sea lions have to keep information in the working memory in an active state. Such an active state has been called 'rehearsal', and forgetting occurs when rehearsal is disrupted, rather than as a result of the gradual decay of a stimulus trace.

With respect to long-term retention of information, several studies have shown individual recognition in pinnipeds and dolphins, most frequently involving mutual recognition between mothers and their dependent offspring (e.g. Trillmich 1981; Sayigh *et al.* 1999). However, until recently, there has been no test of LTM in the context of individual recognition. Recently, Insley (2000) conducted a vocal recognition/playback study on LTM of mothers for the voices of their pups and visa versa in northern fur seals (*Callorhinus ursinus*). Despite the fact that mothers and their pups had not interacted for at least 8 months, Insley showed that females and their pups were still responsive to one another's vocal playback; 4 years later, a subset of former pups (females reaching sexual maturity) were still able to recognize the original recordings of their mother's voice. This finding in the field complements observations in captivity of hand-reared California sea lions, who show recognition for their original caregivers despite being separated for a month or more (Schusterman *et al.* 1992a).

13.2.4 Object representation and echolocation

Most objects have a salient colour, shape, texture, odour and taste; they appear, feel, smell and taste different from other objects. The congruence and apparently spontaneous integration of these different sensory inputs as features of just one and the same object is so obvious to us as verbal human beings that we take it for granted that a similar process occurs in non-verbal infants and other animals. For example, apes can accurately recognize by touch alone, objects that have only been seen, or conversely, can identify visually an object that they have previously experienced only by touch (Ettlinger & Wilson 1990). This kind of object recognition across

the senses is therefore not necessarily related to language, as it had once been believed. For an ape, as for us, there is a visual representation as well as a tactile representation and stimulus input from vision may evoke a tactile representation and vice versa.

The precise ways in which streams of sensory input or neuronal activation patterns related to two or more disparate modalities can give rise to perception of an object are at present unknown. A point of view based on nativist philosophy, which holds that some aspects of intellect (e.g. those related to object recognition) are formed before birth, would hold that infants are born with the capacity to immediately recognize stimuli across a variety of senses; that is, there is an isomorphy between visual, tactile and acoustic representations of an object and, therefore, object recognition occurs both within and between modalities without learning. In contrast, the empiricist philosophy, which states that recognizing contingencies between environmental stimuli (e.g. the sight of an object and the sound it makes when struck) is an ontogenetic learning process, would predict that such cross-modal relations have to be learned. In 1932 Senden (cited in Krech & Crutchfield 1959) reviewed 66 cases of recovery from blindness in adult humans. Senden found that these subjects, upon recovery of sight, were unable to visually identify stimuli with which they were acoustically and tactually familiar. Relations between the sight, sound and touch of objects had to be learned *de novo*. The inability to immediately recognize objects cross-modally in these subjects casts doubt upon the proposition that cross-modal recognition is immediate and spontaneous.

One of the most interesting questions in cross-modal perception is how echolocating animals represent echoic and visual aspects of stimuli. In order to examine the relationship between echoic and visual representations of single objects, Pack and Herman (1995) and Herman *et al.* (1998) tested a bottlenose dolphin in cross-modal identity MTS tasks. In these experiments, the dolphin was able to inspect sample objects echoically and then recognize them again when they were presented visually. These authors concluded that the echoic representations must in some way have given the dolphin access to the visual representations, and they suggested that echoic object representation was isomorphic with visual object representation. In similar experiments with another dolphin, Harley *et al.* (1996) concluded that the dolphin has an 'object-based representation system', in contrast to a representational system that identifies features of objects then relates them via previously learned cross-modal associations.

Despite claims of 'seeing through sound' (Herman *et al.* 1998) and 'echo-imaging' (Pack & Herman 1995), which imply that dolphin echo returns somehow directly translate to a visual representation, there is no solid behavioural or neurophysiological evidence that such isomorphy between sensory systems exists. We suspect that decomposition of an acoustic scene into recognizable elements should allow an individual dolphin to respond accurately to visual counterparts of echoic targets. Such performance does not necessitate a point-to-point transformation of the object representation from one modality (auditory) to another (visual). Instead, evidence suggests that cross-modal matching might be accomplished through categorization based on similarity of function or shared reinforcement history. In the case of the dolphin, specific visual and echoic object features common to many objects gain associative strengths through experience, and are grouped together because they share many visual–echoic links. Despite the physical dissimilarity between echo returns from an object and its light-reflecting characteristics, an experienced dolphin should be able to relate the two as members of an equivalence class (Box 13.2), with the mediating stimulus function being the relation between the visual and echoic features of the object. As has been pointed out, the intermodal equivalence of sonar cues and reflected light cues from common objects remains a rich and relatively untapped source for cognitive research with the highly encephalized brain of dolphins.

13.3 CONTINGENCY ANALYSIS, CONTEXTUAL STIMULUS CONTROL AND EQUIVALENCE

Recent work with some marine mammals has shown that these animals are capable of categorizing objects based on similarity of function. Classification not based on physical similarity is often called stimulus equivalence (Box 13.2). Behavioural ecologists and ethologists as well as natural historians by necessity must work with an animal's behaviour and therefore the basic units of cognition—representations, plans, rules and other mental processes—have to be eventually linked to overt behaviour (Sidman 2000). Because ultimately we deal with observable responses and not hypothetical mental constructs *per se*, it is helpful to discuss stimulus equivalence and complex behaviours in terms of contingencies—the events, responses and consequences linked to behaviour patterns. Relations between the stimulus (environmental variation), response and behavioural consequence or reinforcer are taken as a three-term contingency.

The initial event, or environmental variation, comprises what is known as a 'discriminative stimulus', an occurrence that sets the occasion for the reinforcement of a particular response (and the nonreinforcement of all other possible responses). This type of contingency is elegantly demonstrated whenever a captive marine mammal successfully performs a simple trick at the command of its trainer. The discriminative stimulus is a signal (usually vocal or gestural) produced by the trainer. This stimulus signals to the animal that a particular response (and only that particular response) will be reinforced. Following a correct response by the animal, a reward (usually a fish) is delivered, completing the simple three-term contingency of discriminative stimulus–response–reinforcer.

Box 13.1 describes how the three-term contingency can be expanded to a four-term contingency when it can be placed under 'conditional', 'instructional' or 'contextual' stimulus control (Sidman 2000). In this case, an equivalence class comprises not only the stimuli involved in explicitly trained and emergent relations, but also includes responses and reinforcers as well. This was shown quite convincingly by Reichmuth Kastak *et al.* (2001). They used class-specific fish reinforcers (herring and capelin) to facilitate the formation of equivalence relations in a simple discrimination procedure with two California sea lions. In this study there was powerful evidence that the reinforcers not only facilitated class formation but also became class members themselves. Therefore, equivalence classes are not immutable, but subject to expansion, merger and disintegration, based on controlling variables such as differential reinforcement or punishment, or differences in response topography. Stimuli may change or share class membership based on context, a phenomenon we will refer to when discussing the cognition of social behaviour. Finally, expanded models of equivalence explain cognitive performance in the laboratory ranging from artificial language comprehension to complex categorization and concept formation (Table 13.1).

13.3.1 Equivalence and animal language research

No single aspect of marine mammal cognition has received as much attention or stimulated as much discussion as animal language research (ALR). In experimental psychology, ALR is conducted on a variety of species ranging from birds and apes to sea lions and dolphins. One of the primary goals of ALR is to search for the rudiments of language in 'subhuman' species, using the doctrine of evolutionary continuity as the primary impetus. In particular, common chimpanzees, bonobos and gorillas are the best known subjects of such experiments; but despite the close degree of relatedness of these species to humans, no convincing evidence for a non-human communication system resembling human language has yet emerged. The rationale for ALR experimentation in the bottlenose dolphin is slightly different. The brain of *Tursiops* is extremely large relative to the size of its body. Furthermore, these animals are known to be highly social, and playback experiments suggest that individual recognition by vocalization occurs readily. Thus, a human–dolphin artificial language might be easily taught, and used as a tool to 'tap into' the dolphin brain. In the early 1980s, Herman and colleagues trained two dolphins to respond to artificial languages, one acoustic and one gestural, by combining sequences of cues that modified objects or their positions (e.g. left, right), that referred to object shape (e.g. ball, cone), and specified actions to be performed on the objects (e.g. jump over, tail touch). Sequences were delivered in a restricted order. For example, for the dolphin Ake, signals denoting modifiers always preceeded signals referring to objects and were followed by

signals informing the dolphin as to what action to take. Thus the artificial language had a relatively simple syntactic structure. Despite the apparent difficulty of the artificial language task, the performance of the dolphins showed that on a superficial level they understood the referential nature of the signals, and were sensitive to the sequence of commands (e.g. could differentiate between the instruction 'take the ball to the hoop' and 'take the hoop to the ball') (Gisiner & Schusterman 1992). However, in a series of critiques published in the *Psychological Record* in 1988 and 1989, Herman and Schusterman argued about the interpretation of the results of these ALR experiments (Herman 1988, 1989; Schusterman & Gisiner 1988, 1989). Did the dolphins, as proposed by Herman, understand elements of a 'true' language, or were they merely responding to complex sequences of instructions in a temporal modification of a MTS procedure (e.g. match the gesture for 'ball' with the object 'ball'), as proposed by Schusterman and colleagues (Box 13.1)?

Though arguments concerning the proper interpretation of data from ALR experiments may never be completely put to rest, it is our opinion that success in the types of experiments conducted by Herman with dolphins, as well as similar experiments conducted by Schusterman with California sea lions, does not require language or rudiments of language, even though such tasks can require significant cognitive skills. STM, keen perceptual systems, object and sequence recognition, and the ability to rapidly relate gestures and objects are among the many requisites for successful performance. However, just because a dolphin or a sea lion recognizes the functional equivalence of signals based on sequential position, it does not follow that the animal understands linguistic syntax *per se*. In the same way, though the ability to form equivalence relations is a necessary prerequisite for semantics, it is not sufficient for the emergence of 'meaning' in the linguistic sense. Such performance at best only resembles linguistic performance. Thus, with sufficient training a sea lion or dolphin may acquire a great deal of knowledge about sequential position as well as learning about the interchangeability of a symbol and its referent. However, this does not mean they possess a multimodal language that they use to communicate with one another about previous and future events, nor do such cognitive skills reflect a language analogous to natural human language that enable these animals to spontaneously think in symbols as well as images (Schusterman 1986).

Because new objects, actions and modifiers can be readily interchanged with familiar ones, the stimulus classes comprising actions, objects and modifiers in artificial languages appear (and indeed probably are) open ended, limited only by the experimenter's ability to contrive new signals and referents. Consequently, performance in these tasks appears linguistic but instead depends on prelinguistic cognitive skills like equivalence class formation and rule-based behaviours like conditional discriminations (Box 13.2). For example, in a sequence of stimuli, each signal (e.g. object or action) and its position relative to other signals constitutes an equivalence class. Thus all stimuli can be classified based on at least two cues: the position of the signal within the sequence (e.g. last slot = action) and the relation between a signal and all other signals of its class (e.g. all action signals form a class). When the experimenter adds contextual control (e.g. number or colour or size of the available objects; number and type of signs in a sequence), the subject can develop rules regarding which equivalence classes are appropriate to use for each particular sequence (Schusterman & Gisiner 1997). Based on these somewhat flexible rules governing sequence, novel objects, modifiers and actions can be added at the appropriate time in a sequence and be 'understood' immediately, based on a shared function with other signs of the same class. Apparently, equivalence of positions in a sequence is learned relatively rapidly, while learning the referentiality or meaning of a gesture takes longer (Schusterman & Gisiner 1997). Further, performance is disrupted when the subject's expectations about sequential position are contradicted by the experimenter (Gisiner & Schusterman 1992).

What these ALR studies demonstrate is that sea lions and dolphins, like rhesus monkeys (Chen *et al.* 1997), chunk information into sequential categories. In the case of marine mammals, the categories consisted of modifiers, objects and actions, usually in a relatively fixed order. By using two rules (Premack 1986) these animals were able to interpret sequences

containing as many as seven signs. Their working memories permitted these marine mammals to integrate arbitrary signals such as gestural cues and to maintain an instructional theme as if understanding a book passage that read, for example, 'take the large black ball to the white car' (Schusterman *et al.* 1993b; Schusterman & Gisiner 1997). As pointed out in the next section, such cognitive skills can be readily applied to more natural settings requiring either social, foraging or navigational/spatial knowledge.

We should caution that in the ALR experiments, mutual substitutability between gestures and the objects they signify has not been systematically tested. Nevertheless, the question remains: 'Is such "symbolic" performance merely an artefact of an artificial or arbitrary testing procedure?' We suspect that the answers are based in the evolution of complex social and ecological behaviour where 'meaning' is of utmost importance, regardless of linguistic competence.

13.3.2 Equivalence relations and behaviour in the field

Most previous laboratory studies of marine mammal cognition have failed to address what behavioural ecologists consider paramount: the role of cognition in the behaviour of free-ranging animals and the selective pressures acting on cognitive capacities. To biologists and psychologists interested in the comparative study of behaviour, the question becomes whether different types of intelligence and problem-solving abilities have evolved in response to species-typical environmental pressures.

The idea that behavioural traits, like morphological traits, are subject to the laws of natural selection is not a recent development. In fact, from its inception, the ethological tradition has relied on evolutionary explanations for species-specific behaviour patterns (Tinbergen 1963). However, laboratory and field studies investigating the evolution of complex or 'intelligent' behaviour in animals have only recently received much attention (this approach is reviewed in Shettleworth 1998).

Intelligence related to social behaviour has often been assumed to be under strong selective pressures in many species, such as some of the highly social cetaceans. This idea has received a great deal of attention in the literature on the behaviour of non-human primates. In highly social species, like chimpanzees and baboons, individual reproductive success is related to remembering not only the identities of other group members, but also kin and non-kin. These animals must recognize and remember family members, friends, foes and the alliances that form between them (sometimes involving many individuals), and use this information in novel ways in order to thrive in a complicated and often changing social environment. This argument has recently been applied to species such as *Tursiops truncatus*, which also exhibit highly complex social behaviour (Connor *et al.* 1998) (see Chapter 12). Indeed, dolphins appear to be particularly adept at recognizing individuals and manipulating social situations. Proficiency at vocal mimicry (some cetaceans are among the few mammals in which there is evidence of vocal learning; Richards *et al.* 1984) may be an adaptation enabling recognition and 'naming' other dolphins based on characteristic phonations that have been termed signature whistles (see Chapter 6). It has been hypothesized that signature whistles are a way of keeping track of who's who in the dolphin society (Tyack 1997).

Unfortunately, there is scant, if any, laboratory evidence for any marine mammal behaviours that appear to be useful only in social situations. Data from experiments on stimulus equivalence suggest that many of these animals are capable of cognitive strategies that allow them to solve problems of an arbitrary nature—strategies that are also likely to facilitate many types of social learning as well. Until there is evidence for a specific learning mechanism related to social cognition, more general associative theories must be invoked to account for social behaviour in marine mammals.

In the following examples of the role of cognition in the natural behaviour of free-ranging animals, we suggest that theories of stimulus equivalence offer a fairly simple way to explain many complex behavioural patterns, using nothing more than contingency analyses based on associative learning (Schusterman & Kastak 1993, 1998; Kastak & Schusterman 1994; Schusterman *et al.* 2000). Stimuli such as conspecifics, neighbours and kin must be recognized in highly social animals. By forming equivalence classes corresponding to these group distinctions,

an animal can gain an economy of cognitive processing by dealing with classes of stimuli that may not be perceptually similar. The essence of equivalence lies in the ability to treat members of a class as interchangeable, thus responses appropriate to one member of a class may be appropriate for all members of a class—trial and error learning gives way to conceptualization, saving time and energy.

Cognitive behaviours leading to individual and kin recognition appear to be particularly well developed in some cetaceans and a few pinnipeds. While most studies of marine mammal social behaviour have emphasized cetaceans (see Chapter 12), there are also several cases in which pinnipeds have been shown to engage in complex social interactions. For instance, after a feeding bout that may last several days, a female California sea lion will return to the rookery and become reunited with her pup following a vocal exchange between them. During the first several hours of the pup's life, associative cues relating the sight, sound and smell of its mother are especially powerful, resulting in a process termed imprinting (see Chapter 12). The imprinting phenomenon is usually described as a special form of perceptual learning that takes place only within a limited timeframe after birth. During this 'sensitive period' the pup gathers sensory information and forms representations of its mother based on hearing, vision, olfaction and touch. These cross-modal representations later allow the mature sea lion to recognize its sisters through their connection with their mother and each other (Schusterman *et al.* 1992b). The original cross-modal equivalence class may expand to include other relatives through primary stimulus generalization (related females might resemble the mother) or through other associative mechanisms (kin groups may share proximity on the rookery, or forage together in groups). Hanggi and Schusterman (1990) demonstrated this sort of kin recognition in captive California sea lions, showing that related half-sisters were less aggressive and more affiliative toward each other than non-relatives; a sea lion may determine degrees of kin relatedness by observing affiliative interactions between its mother and its sisters.

There is considerable experimental evidence that territorial male otariid pinnipeds recognize one another by their calls (see Chapters 6 and 10) and

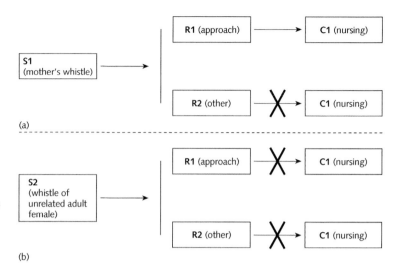

Fig. 13.2 Three-term contingency as applied to a dolphin calf that discriminates between: (a) its mother's whistle, and (b) the whistles of other individuals in its group (see Sayigh et al. 1999). See text for complete explanation.

subsequently group other territorial males as familiar or novel. For example, Steller sea lion territorial males become reproductively successful only after securing a territory for at least one full season (Gisiner 1985). The grouping of neighbouring males into classes of familiar and novel individuals appears to be critical in this process, which can take 3 or more years. Successful males do not expend energy aggressing against familiar males, but established territorial males will aggress against newcomers. There is strong evidence, spanning multiple breeding seasons, suggesting that this 'dear enemy' effect is not simply the result of habituation to neighbouring males but is indeed a classification of males into categories, and that members of each category are relatively interchangeable.

Although details regarding signal emission and reception are different in the dolphins and whales, scenarios involving individual recognition and conceptualization are likely to be verified among the highly and complexly organized societies of dolphins and whales (Connor et al. 1998) (see Chapter 12). Figure 13.2 illustrates a hypothetical three-term contingency drawn from actual playback experiments showing that dolphin mothers and their offspring respond preferentially to each other's signature whistles (Sayigh et al. 1999). Only in the presence of its mother's whistle, stimulus (S1), will a calf's approach response (R1) be followed by, or have as its consequence (C1), a nursing episode. Nursing does

not occur (crossed arrow) if the calf does anything else (R2) in the presence of its mother's whistle. Repeated pairings between these stimuli may ultimately result in an equivalence class comprising the calf's mother, the mother's whistle, the approach response and nursing. When other individuals frequently accompany the mother, their signature whistles and visual characteristics may become incorporated into this class (Box 13.2 and Fig. 13.3). However, the equivalence is context specific in that certain reinforcers (e.g. milk) may only be available from one member of the class (the mother), while other reinforcers (e.g. shelter, protection, affiliative behaviour) may be obtained from the other members of the class. Contextual cues including overt behavioural signals and motivational state may shift the importance of one attribute of the class (all adult females affiliative with the mother) to another attribute (the sight and sound of the only class member that is also a reliable source of milk).

Contextual control is also particularly important in describing both complex foraging situations and social relations in marine mammals. For example, Connor et al. (1998) reported that male bottlenose dolphins in Shark Bay, Australia form enduring pair or trio associations (see Chapter 12). These first-order teams form an association or ally themselves with one or two other teams in a second-order alliance. The principle function of these nested groups or 'superalliances' is to defend against or

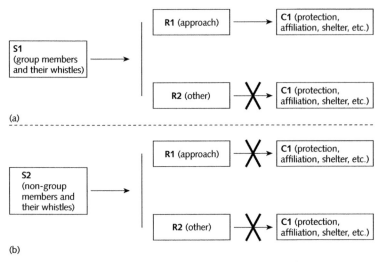

Fig. 13.3 Three-term contingency as applied to a dolphin calf that discriminates between: (a) group members (including its mother, aunts, companions, etc.) and their whistles, and (b) non-group members and their whistles.

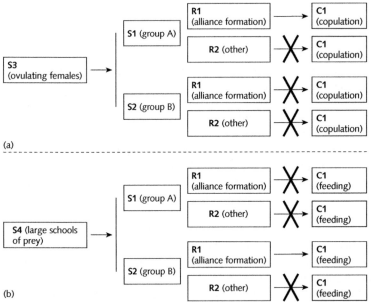

Fig. 13.4 Four-term contingency as applied to Connor *et al.*'s (1992) description of superalliances in bottlenose dolphins depending on different motivations: (a) reproduction; and (b) feeding. See text for complete explanation.

attack other alliances in competition over females. Equivalence relations and contextual control involved in these interactions are shown in Fig. 13.4, in that different superalliances are formed depending on the presence of different motivations for behaviour (reproduction or feeding). Further, within an alliance, individual dolphins that are good at sequestering females might be 'exchanged' for those that might be better at chasing fish into shoals when attention shifts to hunting, giving rise to the extreme fluidity of social behaviour that is characteristic of many dolphin species.

13.4 CONCLUSIONS

Despite the fact that general cognitive mechanisms explain much of the complex behaviour of marine mammals, ecological approaches to the evolution of intelligence in these animals may likewise account for such behaviour. California sea lions, for instance, are quite adept at forming rules and concepts that allow them to solve entire classes of problems efficiently. However, harbour seals have been shown on numerous occasions not to be able to form rules or concepts in laboratory settings using visual and auditory tasks (Table 13.1). However, some of the same seals readily learned tasks in which correct responses were based on spatial rule learning. Renouf and Gaborko (1988, 1989) suggested that these performance differences are related to harbour seal ecology—harbour seal mothers may depend on spatial cues rather than specific visual cues when searching for their offspring. Harbour seal pups, on the other hand, appear not to depend on acoustic signals for differentiating between their mothers and other females, because the females do not emit signature vocalizations to attract their pups. To a harbour seal female, all pups may in fact look alike, and to harbour seal pups all females may look and sound alike. In these animals, mothers infrequently leave their pups and thus selection pressures for mechanisms enabling visual and acoustic recognition might be weak. Spatial orientation and memory, which is important in philopatry and site fidelity, may substitute for individual recognition *per se*—the pup is recognized by the location it occupies rather than by its appearance.

Based on life history differences, then, we would predict that the phocids should perform less well on visual rule-based tasks than the otariids, in which mother–pup recognition has been shown to occur and to be mutual and multisensory, the latter aspect being important in the development of equivalence relations. The odobenids, and likewise many of the cetaceans, might be expected to perform well on such tasks in both visual and auditory modalities, in part because of the extended period of mother–calf interaction and presumed individual recognition.

The synergy of laboratory and field-based approaches is evident in the preceding examples, as it allows the development of testable hypotheses based on behaviour, life history and ecology of marine mammals. Such hypotheses can in turn be used to design laboratory tests for specific cognitive mechanisms such as equivalence, and the results used along with phylogenetic evidence to shed light on the evolution of marine mammal behaviour. However, within this framework for the study of marine mammal problem solving and memory, we have hardly begun to scratch the surface in the exploration of the cognitive abilities of these animals.

REFERENCES

Chen, S., Swartz, K.B. & Terrace, H.S. (1997) Knowledge of the ordinal position of list items in rhesus monkeys. *Psychological Science* 8, 86–86.

Connor, R.C., Mann, J., Tyack, P.L. & Whitehead, H. (1998) Social evolution in toothed whales. *Trends in Ecology and Evolution* 13, 228–232.

Connor, R.C., Smolker, R.A. & Richards, A.F. (1992) Dolphin alliances and coalitions. In: *Coalitions and Alliances in Humans and Other Animals* (A.H. Harcourt & F.B.M. De Waal, eds), pp. 415–443. Oxford University Press, Oxford.

Constantine, B.J. (1981) *An experimental analysis of stimulus control in simple conditional discriminations: a methodological study.* Unpublished PhD Thesis, Northeastern University, Boston.

Domjan, M. (1998) *The Principles of Learning and Behavior*, 4th edn. Brooks-Cole Publishing, Pacific Grove.

Ettlinger, G. & Wilson, W.A. (1990) Cross-modal performance: behavioural processes, phylogenetic considerations, and neural mechanisms. *Behavioural Brain Research* 40, 169–192.

Gisiner, R.C. (1985) *Male territoriality and reproductive behavior in the Steller sea lion, Eumetopias jubatus.* Unpublished PhD Thesis, University of California, Santa Cruz.

Gisiner, R.C. & Schusterman, R.J. (1992) Sequence, syntax and semantics: responses of a language trained sea lion (*Zalophus californianus*) to novel sign combinations. *Journal of Comparative Psychology* 106, 78–91.

Hanggi, E.B. & Schusterman, R.J. (1990) Kin recognition in captive California sea lions (*Zalophus californianus*). *Journal of Comparative Psychology* 104, 368–372.

Hanggi, E.B. & Schusterman, R.J. (1995) Conditional discrimination learning in a male harbor seal (*Phoca vitulina*). In: *Sensory Systems of Aquatic Mammals* (R.A. Kastelein, J.A. Thomas & P.E. Nachtigall, eds), pp. 543–559. DeSpil, Woerden, the Netherlands.

Harley, H.E., Roitblat, H.L. & Nachtigall, P.E. (1996) Object representation in the bottlenosed dolphin (*Tursiops truncatus*): integration of visual and echoic information.

Journal of Experimental Psychology: Animal Behavior Processes **22**, 164–174.

Herman, L.M. (1975) Interference and auditory short-term memory in the bottlenosed dolphin. *Animal Learning and Behavior* **3**, 43–48.

Herman, L.M. (1988) The language of animal research: reply to Schusterman and Gisiner. *Psychological Record* **38**, 349–362.

Herman, L.M. (1989) In which Procrustean bed does the sea lion sleep tonight? *Psychological Record* **39**, 19–42.

Herman, L.M. & Arbeit, W.R. (1973) Stimulus control and auditory discrimination learning sets in the bottlenose dolphin. *Journal of the Experimental Analysis of Behavior* **19**, 379–394.

Herman, L.M. & Forestell, P.H. (1985) Reporting presence or absence of named objects by a language-trained dolphin. *Neuroscience and Biobehavioral Reviews* **9**, 667–681.

Herman, L.M. & Gordon, J.A. (1974) Auditory delayed matching in the bottlenose dolphin. *Journal of the Experimental Analysis of Behavior* **21**, 19–26.

Herman, L.M. & Thompson, R.K.R. (1982) Symbolic, identity, and probe delayed matching of sounds by the bottlenosed dolphin. *Animal Learning and Behavior* **10**, 22–34.

Herman, L.M., Richards, D.G. & Wolz, J.P. (1984) Comprehension of sentences by bottlenosed dolphins. *Cognition* **16**, 129–219.

Herman, L.M., Hovancik, J.R., Gory, J.D. & Bradshaw, G.L. (1989) Generalization of visual matching by a bottlenose dolphin (*Tursiops truncatus*): evidence for invariance of cognitive performance with visual and auditory materials. *Journal of Experimental Psychology: Animal Behavior Processes* **15**, 124–136.

Herman, L.M., Pack, A.A. & Hoffman-Kuhnt, M. (1998) Seeing through sound: dolphins (*Tursiops truncatus*) perceive the spatial structure of objects through echolocation. *Journal of Comparative Psychology* **112**, 292–305.

Insley, S.J. (2000) Long-term vocal recognition in the northern fur seal. *Nature* **406**, 404–405.

Kastak, D. & Schusterman, R.J. (1992) Comparative cognition in marine mammals: a clarification on match-to-sample tests. *Marine Mammal Science* **8**, 414–417.

Kastak, D. & Schusterman, R.J. (1994) Transfer of visual identity matching-to-sample in two California sea lions (*Zalophus californianus*). *Animal Learning and Behavior* **22** (4), 427–435.

Kastak, D. & Schusterman, R.J. (1998) Low frequency amphibious hearing in pinnipeds: methods, measurements, noise, and ecology. *Journal of the Acoustical Society of America* **103**, 2216–2228.

Kastak, D. & Schusterman, R.J. (1999) In-air and underwater hearing sensitivity of a northern elephant seal (*Mirounga angustirostris*). *Canadian Journal of Zoology* **77**, 1751–1758.

Kellogg, W.N. & Rice, C.E. (1966) Visual discrimination and problem solving in a bottlenose dolphin. In: *Whales, Dolphins, and Porpoises* (K.S. Norris, ed.), pp. 731–754. University of California Press, Berkeley, CA.

Krech, D. & Crutchfield, R.S. (1959) *Elements of Psychology.* Knopf, New York.

Levenson, D.H. & Schusterman, R.J. (1997) Pupillometry in seals and sea lions: ecological implications. *Canadian Journal of Zoology* **75**, 2050–2057.

Levenson, D.H. & Schusterman, R.J. (1999) Dark adaptation and visual sensitivity in shallow and deep-diving pinnipeds. *Marine Mammal Science* **15**, 1303–1313.

Mauck, B. & Dehnhardt, G. (1997) Mental rotation in a California sea lion (*Zalophus californianus*). *Journal of Experimental Biology* **200**, 1309–1316.

Mercado III, E., Killebrew, D.A., Pack, A.A., Mácha, I.V.B. & Herman, L.M. (2000) Generalization of 'same–different' classification abilities in bottlenosed dolphins. *Behavioural Processes* **50**, 79–94.

Pack, A.A. & Herman, L.M. (1995) Sensory integration in the bottlenosed dolphin: immediate recognition of complex shapes across the senses of echolocation and vision. *Journal of the Acoustical Society of America* **98**, 722–733.

Pack, A.A., Herman, L.M. & Roitblat, H.L. (1991) Generalization of visual matching and delayed matching by a California sea lion (*Zalophus californianus*). *Animal Learning and Behavior* **19**, 37–48.

Premack, D. (1986) *'Gavagai' or the Future History of the Animal Language Controversy.* MIT Press, Cambridge, MA.

Reichmuth Kastak, C., Schusterman, R.J. & Kastak, D. (2001) Equivalence classification in California sea lions using class-specific reinforcers. *Journal of the Experimental Analysis of Behaviour* **76**, 131–158.

Renouf, D. & Gaborko, L. (1988) Spatial matching to sample in harbour seals (*Phoca vitulina*). *Biology of Behaviour* **13**, 73–81.

Renouf, D. & Gaborko, L. (1989) Spatial and visual rule use by harbour seals (*Phoca vitulina*). *Biology of Behaviour* **14** (2), 169–181.

Richards, D.G., Wolz, J.P. & Herman, L.M. (1984) Vocal mimicry of computer-generated sounds and vocal labeling of objects by a bottlenosed dolphin (*Tursiops truncatus*). *Journal of Comparative Psychology* **98**, 10–28.

Roitblat, H.L., Penner, R.H. & Nachtigall, P.E. (1990) Matching-to-sample by an echolocating dolphin (*Tursiops truncatus*). *Journal of Experimental Psychology: Animal Behavior Processes* **16**, 85–95.

Sayigh, L.S., Tyack, P.L., Wells, R.S. *et al.* (1999) Individual recognition in wild bottlenose dolphins: a field test using playback experiments. *Animal Behaviour* **57**, 41–50.

Schusterman, R.J. (1962) Transfer effects of successive discrimination-reversal training in chimpanzees. *Science* **137**, 422–423.

Schusterman, R.J. (1964) Successive discrimination-reversal training and multiple discrimination training in one-trial learning by chimpanzees. *Journal of Comparative Physiological Psychology* **58**, 153–156.

Schusterman, R.J. (1966) Serial discrimination-reversal learning with and without errors by the California sea lion. *Journal of the Experimental Analysis of Behavior* **9**, 593–600.

Schusterman, R.J. (1967) Attention shift and errorless reversal learning by the California sea lion. *Science* 156, 833–835.

Schusterman, R.J. (1968) Experimental laboratory studies of pinniped behavior. In: *The Behavior and Physiology of Pinnipeds* (R.J. Harrison, R.C. Hubbard, R.S. Peterson, C.E. Rice & R.J. Schusterman, eds), pp. 87–171. Appleton Century Crofts, New York.

Schusterman, R.J. (1986) Cognition and intelligence of dolphins. In: *Dolphin Cognition and Behavior: a Comparative Approach* (R.J. Schusterman, J.A. Thomas & F.G. Wood, eds), pp. 137–139. Lawrence Erlbaum Associates, Hillsdale, NJ.

Schusterman, R.J. & Gisiner, R. (1988) Artificial language comprehension in dolphins and sea lions: the essential cognitive skills. *Psychological Record* 38, 311–348.

Schusterman, R.J. & Gisiner, R. (1989) Please parse the sentence: animal cognition in the Procrustean bed of linguistics. *Psychological Record* 39, 3–18.

Schusterman, R.J. & Gisiner, R. (1997) Pinnipeds, porpoises and parsimony: animal language research viewed from a bottom-up perspective. In: *Anthropomorphism, Anecdotes and Animals: the Emperor's New Clothes?* (R.W. Mitchell, N.S. Thompson & H.L. Miles, eds), pp. 370–382. State University of New York Press, Albany, NY.

Schusterman, R.J. & Kastak, D. (1993) A California sea lion (*Zalophus californianus*) is capable of forming equivalence relations. *Psychological Record* 43, 823–839.

Schusterman, R.J. & Kastak, D. (1998) Functional equivalence in a California sea lion: relevance to social and communicative interactions. *Animal Behaviour* 55, 1087–1095.

Schusterman, R.J. & Krieger, K. (1984) California sea lions are capable of semantic comprehension. *Psychological Record* 34, 3–24.

Schusterman, R.J. & Krieger, K. (1986) Artificial language comprehension and size transposition by a California sea lion (*Zalophus californianus*). *Journal of Comparative Psychology* 100, 348–355.

Schusterman, R.J. & Thomas, T. (1967) Shape discrimination and transfer in the California sea lion. *Psychonomic Science* 5, 21–22.

Schusterman, R.J., Grimm, B.K., Gisiner, R. & Hanggi, E.B. (1991) Retroactive interference of delayed 'symbolic' matching-to-sample in California sea lions. Paper presented at the Annual Meeting of the Psychonomic Society, San Francisco, 1991.

Schusterman, R.J., Gisiner, R. & Hanggi, E. (1992a) Imprinting and other aspects of pinniped/human interactions. In: *The Inevitable Bond* (H. Davis & D. Balfour, eds), pp. 334–356. Cambridge University Press, New York.

Schusterman, R.J., Hanggi, E. & Gisiner, R. (1992b) Acoustic signaling in mother–pup reunions, interspecies bonding, and affiliation by kinship in California sea lions (*Zalophus californianus*). In: *Marine Mammal Sensory Systems* (J.A. Thomas, R.A. Kastelein & Y.A. Supin, eds), pp. 533–562. Plenum Press, New York.

Schusterman, R.J., Gisiner, R., Grimm, B.K. & Hanggi, E.B. (1993a) Behavior control by exclusion and attempts at establishing semanticity in marine mammals using match-to-sample paradigms. In: *Language and Communication: Comparative Perspectives* (H. Roitblat, L. Herman & P. Nachtigall, eds), pp. 249–274. Erlbaum, Hillsdale, NJ.

Schusterman, R.J., Hanggi, E.B. & Gisiner, R.C. (1993b) Remembering in California sea lions: using priming cues to facilitate language-like performance. *Animal Learning and Behavior* 21 (4), 377–383.

Schusterman, R.J., Reichmuth, C.J. & Kastak, D. (2000) How animals classify friends and foes. *Current Directions in Psychological Science* 9, 1–6.

Shettleworth, S.J. (1998) *Cognition, Evolution, and Behavior*. Oxford University Press, New York.

Sidman, M. (1994) *Equivalence Relations and Behavior: a Research Story*. Author's Cooperative, Boston.

Sidman, M. (2000) Equivalence relations and reinforcement contingency. *Journal of the Experimental Analysis of Behavior* 74, 127–146.

Thompson, R.K.R. & Herman, L.M. (1977) Auditory delayed discriminations by the dolphin: nonequivalence with delayed-matching performance. *Animal Learning and Behavior* 9, 9–15.

Tinbergen, N. (1963) On aims and methods of ethology. *Zeithschrift fur Tierpsychologie* 20, 410–433.

Trillmich, F. (1981) Mutual mother–pup recognition in Galapagos fur seals and sea lions: cues used and functional signitificance. *Behaviour* 78, 21–42.

Tyack, P.L. (1981) Interactions between singing Hawaiian humpback whales and conspecifics nearby. *Behavioral Ecology and Sociobiology* 8, 105–116.

Tyack, P.L. (1997) Development and social functions of signature whistles in bottlenose dolphins *Tursiops truncatus*. *Bioacoustics* 8, 21–46.

Tyack, P.L. (1999) Communication and cognition. In: *Biology of Marine Mammals* (J.E. Reynolds III & S.A. Rommel, eds), pp. 287–323. Smithsonian Institution Press, Washington, DC.

von Fersen, L., Manos, C.S., Goldowsky, B. & Roitblat, H. (1993) Dolphin detection and conceptualization of symmetry. In: *Marine Mammal Sensory Systems* (J.A. Thomas, R.A. Kastelein & A.Y. Supin, eds), pp. 733–762. Plenum Press, New York, NY.

von Fersen, L & Delius, J.D. (2000) Acquired equivalences between auditory stimuli in dolphins (*Tursiops truncatus*). *Animal Cognition* 3, 79–83.

CHAPTER 14

Conservation and Management

Randall R. Reeves and Peter J.H. Reijnders

14.1 INTRODUCTION

Although the exploitation of marine mammals began in ancient times and continues to the present, conservation began only in the early twentieth century. The conservation efforts were sporadic and narrowly focused until after the Second World War. Deliberate killing, whether to achieve subsistence, profit or predator control, was for centuries the principal concern. It alone determined the fates of some populations and species. During the past several decades, the emphasis has changed. Particularly as the scale of commercial hunting has declined, the perceived importance of other less obvious or less direct threats has grown. For many marine mammals, habitat deterioration, ship strikes, incidental mortality in fisheries, pollution and reduced prey resources have become at least as important as deliberate killing in deciding their fate.

A review of conservation issues can be species based, area based or threat based. In the present chapter we take mainly a threat-based approach. The range of threats facing marine mammals is vast, as is our ignorance about how to evaluate and manage those threats. On one hand, there are situations in which the needed actions are fairly obvious: close a fishery; stop a hunt; eliminate motor-vessel traffic in areas where the animals rest, socialize, feed or nurse their young. More often than not, the political, economic and cultural dimensions are more complex and contentious than the biological or ecological ones. On the other hand, there are situations in which the conservation imperatives are obscured by genuine scientific uncertainty. Why is a sea lion population collapsing even in the absence of overhunting or excessive bycatch? Why is a whale population not reproducing at the expected rate, judging by rates observed in other populations of closely related species? Is there a causal link between a large-scale die-off of seals and the high concentrations of certain contaminants in their body tissues? A necessary first step in determining how to prevent extinction or facilitate population recovery is to understand the cause or causes of population decline.

For each type of threat we attempt to describe its character, history, magnitude and trend. Is the threat long-standing or new? Is it increasing or decreasing? Is its geographical scale spreading? What is being done to combat the threat? One or several examples are given to illustrate each threat category. We also discuss the kinds of actions needed to address the threats more effectively.

14.1.1 Definitions and assumptions

The meaning of 'conservation' has become more and more ambiguous in recent decades. Depending upon the context and the source, conservation can mean managing wild animal populations for human benefit, completely protecting wild populations, or securing the welfare of individual animals. 'Animal rights', a concept rooted in philosophy and law, is sometimes invoked, but it often confounds the meaning of conservation. To many, the terms 'sustainable development' and 'sustainable use', with their human-centred emphases, have been adopted as substitutes for 'conservation', particularly since the early 1980s when intergovernmental organizations began to adopt them (Robinson 1993).

Our use of the term 'conservation' is intended to mean the preservation of biodiversity, i.e. the variety of organisms at all levels (e.g. genetic variants, arrays of species) and the ecosystems in which communities of organisms exist (Wilson 1992; Gaston 1998). It does not require, or exclude, the use of marine mammals by humans, but it emphasizes preservation of the self-renewing capacity of natural populations as the principal goal. Removal of animals from the wild for captive breeding and restocking or reintroduction can contribute to conservation in some instances. However, maintaining species or populations in artificial settings (e.g. zoos, oceanariums or 'seminatural' wildlife parks) does not, by itself, constitute conservation.

We assume that conservation is a good thing and that it need not be justified by economic or other benefits to humans. Conservation strategies may involve tangible rewards for people, or arguments claiming that human life will be improved as a result of conservation action. However, the value of conservation also exists a priori and need not be quantified or stated only in economic terms (see Norton 1987; Domning 1999).

We realize, of course, that extinction is a natural process and the ultimate fate of all forms of life. Ideally, conservation needs to be conceptualized and pursued in an evolutionary context, recognizing that human actions are superimposed on a dynamic, continuous process of natural selection. Intervention to prevent natural extinctions would be foolish and wrong. In practice, however, it is usually impossible to distinguish natural from man-made extinctions in a world where the effects of human presence are ubiquitous and frequently decisive.

14.1.2 What to conserve

The conservationist, having professed an interest in preserving biodiversity, must decide what needs to be conserved. What is the fundamental unit of biodiversity? For example, is it sufficient that a large population of white whales exists in the relatively pristine Barents Sea, little hunted and with access to abundant prey (cf. Gjertz & Wiig 1994)? If so, there may be less incentive to invest resources in preserving small isolated populations of white whales in the polluted, congested St Lawrence River (cf. Lesage & Kingsley 1998) or in Cook Inlet, Alaska, where removals by hunting are clearly unsustainable (cf. Marine Mammal Commission 1999). Should

extraordinary effort be made to conserve ringed seals in the Baltic Sea (cf. Härkönen et al. 1998) even though the same species is widespread and abundant (in the millions) in the circumpolar Arctic (cf. Reeves 1998)? Is it sufficient to preserve West Indian manatees in the southeastern United States or dugongs in Australia if, at the same time, they are hunted to extinction elsewhere in their range (cf. Marsh & Lefebvre 1994)? These questions strike at the heart of a fundamental dilemma in conservation biology: 'What is biodiversity and why should we care about preserving it?'

Traditionally, morphological differences and obvious discontinuities in distribution have been used to distinguish marine mammal populations as 'stocks' for management. In recent years, genetic analyses have come to play a prominent role in defining management units (e.g. Dizon et al. 1994; International Whaling Commission 1996; O'Corry-Crowe & Lowry 1997; Rosel 1997). Contaminant profiles, parasite faunas and acoustic dialects have been used to augment the other types of evidence for stock differences.

Dizon et al. (1992) used the minke whale, a species group subject to both commercial and 'aboriginal subsistence' whaling, as an example for elucidating a phylogeographic approach in which taxa are established by both mtDNA genetic distance and spatial distribution, i.e. gene flow and geographical localization (see Chapter 11). Separate stocks of minke whales (in fact different species) occur in the northern and southern hemispheres and the populations in the North Atlantic and North Pacific are also distinct. A dwarf form found off southern Africa, Australia, New Zealand and Brazil is well distinguished from the other southern hemisphere ('dark-shouldered') minke whale (Best 1985). Management units recognized by the International Whaling Commission (IWC) within the North Atlantic, North Pacific and Southern Ocean are generally validated by at least one of the available rationales—distribution, population response (demographic or behavioural), phenotypic information (morphological) and genotypic information (Dizon et al. 1992).

In a recent series of papers, Barbara Taylor and Andrew Dizon (Taylor & Dizon 1996, 1999; Taylor 1997) have developed the argument that the overarching goal of conservation should be to maintain the full historical range of wild species, including all of their feeding and breeding grounds. This goal requires that account be taken of more than just the genetic evidence of population structure. Selection of management units must also incorporate information on philopatry driven by behavioural differences that may or may not be reflected in genetic distinctiveness.

14.2 THREATS

14.2.1 Historical hunting, trapping and netting

Advocates of sustainable use argue that wild animal populations can be exploited indefinitely as long as removals are not excessive, recruitment is maintained and supportive ecosystems are preserved (cf. Prescott-Allen & Prescott-Allen 1996). A few marine mammal populations provide potentially illustrative examples, e.g. long-finned pilot whales at the Faroe Islands (Zachariassen 1993), South American fur seals in Uruguay (Vaz-Ferreira 1982) and white whales in Canada's Mackenzie Delta (McGhee 1974; Fraker 1980). Unfortunately, they are not typical. Deliberate killing has been responsible for the depletion of many marine mammal populations.

The literature on marine mammal exploitation is vast, and here we provide only a brief overview. Sealing began in the Stone Age when people attacked hauled out animals with clubs (Bonner 1982). Later methods involved the use of harpoons, traps, nets and gaff-like instruments for killing pups on ice or beaches. The introduction of firearms caused a marked increase in the proportion of animals killed but not retrieved (e.g. Fay *et al.* 1994; Born *et al.* 1995; Reeves *et al.* 1998). Markets for oil, hides and ivory (in the case of walruses) fuelled commercial hunting on a massive scale (Busch 1985). The origins of whaling are also ancient (Ellis 1991; McCartney 1995). From its beginnings in the Bay of Biscay, commercial whaling spread throughout the world's oceans and came to involve people of many nationalities. Modern whaling (Fig. 14.1), characterized by engine-driven catcher boats and

Fig. 14.1 A fin whale, one of 68 taken under a special scientific research permit in the 1988 season, on the flensing platform at the whaling station in Hvalfjördür, Iceland, 11 July 1988. Iceland, which withdrew from the International Whaling Commission in 1992, has announced its intention to resume commercial whaling. (Courtesy of Steve Leatherwood.)

deck-mounted harpoon cannons firing explosive grenades, began in Norway in the 1860s (Tønnessen & Johnsen 1982). Factory ships eventually operated in the Antarctic, the richest whaling ground on the planet.

Sirenians have also been subjected to intensive exploitation, mainly for their delectable meat and blubber and their strong hides. Steller's sea cow was hunted to extinction within about 25 years of its discovery by European sea otter and fur seal hunters. The other sirenians have survived only because of their more extensive ranges and greater aggregate abundance (Reynolds & Odell 1991; Marsh & Lefebvre 1994). Sea otters were hunted remorselessly to supply the Oriental fur market from the 1780s onwards (Busch 1985; Chanin 1985), and polar bears have been prime targets of both Eskimo 'subsistence' hunters and non-Eskimo sport hunters (Stirling 1986; Prestrud & Stirling 1994).

The legacy of overexploitation is responsible for many ongoing marine mammal conservation problems. Some populations of baleen whales exist at relict levels (i.e. only tens or a few hundreds of individuals) and their ability to achieve meaningful recovery is in doubt (Best 1993; Clapham *et al.* 1999). Small population size could have implications not only for genetic fitness to adapt to changing environmental conditions, but also for a population's social competence, foraging efficiency and ability to reproduce at optimal rates (see Whitehead *et al.* 2000). The selective removal of large male sperm whales by whalers may have reduced pregnancy rates and the disruption caused by whaling to matricentric groups of sperm whales may have left them more vulnerable to predation (Whitehead & Weilgart 2000). Sirenians have been eliminated in large parts of their range by over-exploitation and now exist in scattered, isolated populations, many of which may go extinct as a result of demographic or environmental stochasticity, or genetic effects, even if protected from hunting (cf. Caughley 1994).

14.2.2 Ongoing direct exploitation

Even though the scale of commercial exploitation has declined, hundreds of thousands of marine mammals are still killed every year for food, fish bait, fur or profit. There is not always an immediate conservation problem, but the legacy of unsustainable exploitation, coupled with increased killing power, growing human populations and accelerating demand for most products, means that the burden of proof should be on the exploiters to show that the animal populations are large and robust enough to support continued removals.

After a period of drastically reduced killing in the 1980s the Canadian commercial hunt for harp and hooded seals has been reinvigorated through a combination of governmental support and market demand (Lavigne 1999a; Lavigne *et al.* 1999). There are strong markets in Oriental communities for pinniped penises and bacula (Bräutigam & Thomsen 1993; Malik *et al.* 1997) as well as for bear gall bladders, including those from polar bears (Servheen *et al.* 1999). The sale of these items, along with polar bear hides, has helped offset the economic losses in some local hunting communities due to the decline in international markets for sealskins (cf. Wenzel 1991). The large 'subsistence' takes of walruses in Alaska, Canada, Greenland and Russia continue to supply raw ivory and carvings to an international market (Bräutigam & Thomsen 1993). Faroe Islanders continue to kill hundreds,

and in some years well over a thousand, long-finned pilot whales and Atlantic white-sided dolphins in a drive fishery that is centuries old (Zachariassen 1993). Aboriginal hunters in Russia, the United States (Alaska), Canada and Greenland kill at least several tens of bowhead and gray whales, at least many hundreds of belugas and narwhals, more than a hundred thousand seals (ringed, bearded, ribbon, harp, hooded and spotted) and at least 10 000 walruses every year. Most of the products circulate within the hunting communities, often on a cash basis (Reeves 1993; Heide-Jørgensen 1994), but narwhal ivory enters the international market mainly as a trophy item (Reeves 1992; Reeves & Heide-Jørgensen 1994). Norway and Japan continue their hunting of minke whales—the former in the North Atlantic and the latter in the Antarctic and North Pacific. Both countries are eager to reopen the international trade in whale meat and expand their whaling operations. Iceland, too, has stated its intention to resume commercial whaling (Sigurjónsson 1989).

It is impossible to make a reasonable guess at how many manatees are killed by villagers each year in West Africa and South America but the total (three species, combined) is probably in the thousands (cf. Reeves *et al.* 1988, 1996a; Rosas 1994). Hunting of dugongs continues in much of their range, including areas where they are almost extinct (Marsh *et al.* 1980–81, 1995, 1997; Marsh & Lefebvre 1994). The legal hunting of sea otters in Alaska has increased in recent years (Rotterman & Simon-Jackson 1988) as has the poaching of sea otters in Russia (Estes 1994). At least several hundred polar bears are killed each year (Prestrud & Stirling 1994; Servheen *et al.* 1999).

Marine mammals have been used extensively as bait for shark longlines (Romero *et al.* 1997) and crab traps (Lescrauwaet & Gibbons 1994) in South America. A continuing threat to river dolphins in India and Bangladesh is the use of their oil as fish attractant (Motwani & Srivastava 1961; Mohan *et al.* 1997; Smith *et al.* 1998).

The dichotomy between 'subsistence' and 'commercial' exploitation has been controversial, particularly in the case of whaling. Wenzel (1991), Freeman (1993) and most other anthropologists argue that 'subsistence' should be broadly defined and not exclude cash-based exchanges when these occur within a context that emphasizes local production and consumption. Indeed, the IWC recognizes Greenland whaling for minke and fin whales as 'aboriginal subsistence' whaling even though the meat and other products enter a country-wide cash-based exchange network (Caulfield 1993). In contrast 'small-type coastal' whaling in Japan, which also serves a domestic but cash-based market, is classified as commercial whaling. The basic concern of conservationists is that commercial incentives should not be allowed to override the capacity for enforcing regulations. Evidence of deliberate, large-scale misreporting of whaling data by the Soviet Union (Best 1988; Yablokov 1994; Zemsky *et al.* 1995; Tormosov *et al.* 1998) and Japan (Kasuya 1999) demonstrates the validity of this concern.

14.2.3 Control killing

Marine mammals, like other predators (e.g. canids, felids and ursids), have been viewed in certain contexts as a threat to people's livelihoods. In some instances the issue is perceived competition for fish or invertebrate resources (Fig. 14.2); in others, damage to fishing gear and the stealing of fish

Fig. 14.2 A California sea lion with a salmon in its jaws off Race Rocks, Vancouver Island, British Columbia. Predation by pinnipeds on commercially valuable fish makes them unpopular with fishermen. Bounties and other types of culling programmes have been widely used to 'control' pinniped populations and reduce their interference with fisheries, whether real or only perceived. (Courtesy of Robin Baird.)

from nets or lines (Northridge & Hofman 1999). Grey seals have been vilified because of their role as vectors for parasites that infect cod and other groundfish (Bonner 1982). Even manatees, properly regarded as inoffensive herbivores, are resented by farmers and fishermen in Africa who claim that they raid rice fields and remove fish from gillnets (Reeves *et al.* 1988). Polar bears are the only marine mammals that have a reputation of endangering human life and at least a few are killed each year in self-defence (Stirling 1986; Gjertz *et al.* 1993).

Pinnipeds have been subjected to control killing in many parts of the world. There were bounties on grey and harbour seals in Sweden from 1900 and in Canada from 1927 (Bonner 1982) and on harbour seals in the Wadden Sea from the late nineteenth century (Reijnders 1992a) and in Alaska between 1920 and 1967 (Hoover 1988). In addition to bounties, governments have sometimes taken a more active role in reducing marine mammal populations to protect fisheries (Bonner 1982; Earle 1996). In South Africa and Namibia, Cape fur seals have long been subjected to annual culling through a government-regulated commercial hunt. One major rationale for the cull is that seals scare fish away from gear or remove them once caught (Shaughnessy 1985). Another 'ecological' rationale is that the seals displace endangered seabirds from nesting colonies (Crawford *et al.* 1989).

For the most part, control programmes do not seem to have threatened the survival of the targeted species or populations. It is possible, however, that the bounty on St Lawrence River belugas earlier this century was partly responsible for endangering this whale population (Reeves & Mitchell 1984). Also, the large-scale killing of small cetaceans in Japan has been partially justified as a control measure to protect fisheries (Kasuya 1985) and some of those populations may be in trouble (cf. International Whaling Commission 1993). The killing of marine otters to protect prawn stocks may at least partly account for the present rarity of these mammals in Peruvian coastal waters (e.g. Grimwood 1969). Bonner (1982) suggested that an official control programme can be preferable, from a conservation perspective, to the *ad hoc* measures taken against marine mammals by fishermen.

14.3 NON-DELIBERATE REMOVALS

14.3.1 Bycatch

As humans have expanded their use of the marine environment, new risks to marine mammals have arisen and old risks have become more serious. With the decline and regulation of deliberate killing, the relative importance of non-deliberate killing, especially bycatch in fisheries, has increased (Brownell *et al.* 1989; Reijnders *et al.* 1993; Reeves & Leatherwood 1994a; Hofman 1995). For most marine mammal populations, accidental mortality from entanglement or entrapment in fishing gear now poses a much greater threat than deliberate exploitation. Increased bycatch is often linked to technological change in a fishery or an increase in local or regional fishing intensity.

Bycatch, i.e. the non-target portion of the catch, can include a multitude of species, from 'trash' fish and invertebrates to cetaceans, pinnipeds, birds and reptiles, taken using a variety of unselective fishing techniques. Passive gear, especially gillnets, generally kill more marine mammals than actively fished gear (Perrin *et al.* 1994). However, cetaceans and pinnipeds also die from encounters with active gear, including trawls (e.g. Shaughnessy & Payne 1979; Fertl & Leatherwood 1997) and purse seines (e.g. Dolar 1994). Baited longlines are hazardous for marine mammals not only because the animals can become hooked or entangled by them, but also because of the antipathy of fishermen that results from seeing killer whales, sperm whales, fur seals or monk seals stealing their catch (Section 14.2.3). A specific threat to the baiji is a fishing method called 'rolling hooks' used in the Yangtze River to snag bottom fish (Perrin 1999). The vaquita in the Gulf of California, Mexico, is endangered primarily by gillnets (Vidal 1995). Antishark nets used to protect bathing beaches threaten Indo-Pacific bottlenose and humpbacked dolphins in southern Africa (Cockcroft 1990) and dugongs in Australia (Heinsohn 1972). Entanglement in fishing gear is a major contributing factor in the mortality of right (International Whaling Commission 2000), humpback (Volgenau *et al.* 1995) and other whales, and

Fig. 14.3 A California sea lion hauled out on Los Islotes in the Gulf of California, Mexico (6 April 1995) exhibiting a serious entanglement injury—a deep open wound round the base of its neck. Derelict fishing gear and discarded packaging materials, as well as operational nets and lines, represent hazards to marine mammals. (Courtesy of Robin Baird.)

many pinnipeds (Reijnders *et al.* 1993) including the endangered monk seals (Lavigne 1999b; Ragen & Lavigne 1999). Gillnet entanglement is thought to be responsible for a recent decline in the population of California sea otters (Marine Mammal Commission 1999). The large purse seines used to catch tuna are an important cause of mortality for some pelagic dolphin populations, although it can be argued that because the dolphins killed in this fishery are deliberately chased and encircled, they are not really bycatch.

Marine debris, especially derelict fishing gear, is responsible for substantial incidental mortality of marine mammals (Laist *et al.* 1999). There is reason to believe that the scale of killing and maiming from entanglement in debris (Fig. 14.3) has been great enough to affect the populations of northern fur seals (Fowler 1982) and Hawaiian monk seals (Marine Mammal Commission 1999).

Marine mammals that are taken incidentally are sometimes retained and used as food or bait. When the markets for products change or the availability of target fish species declines, a directed fishery for marine mammals may develop. Peru (Read *et al.* 1988; Van Waerebeek & Reyes 1994), Sri Lanka (Leatherwood & Reeves 1989) and the Philippines (Dolar *et al.* 1994) offer particularly

Fig. 14.4 Fresh dusky dolphins on sale at a fish market in Pucusana, Peru. Very large catches of dusky dolphins have been made in Peruvian fisheries since at least the early 1970s. Because there is local demand for cetacean meat, fishermen have an incentive to bring ashore any dolphins and porpoises that become incidentally entangled along with sharks, rays and other large fish. They also sometimes set their nets with the deliberate intention of catching cetaceans, particularly dusky dolphins. (Courtesy of Koen Van Waerebeek, Centro Peruano de Estudios Cetológicos 1988.)

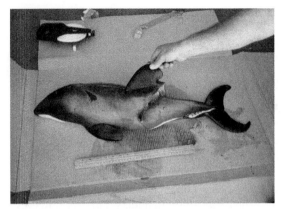

Fig. 14.5 A recently born Hector's dolphin killed by a propeller strike in Akaroa Harbour, New Zealand, January 1999 (see Stone & Yoshinaga 2000). With increased vessel traffic in the waters inhabited by marine mammals, the frequency of such mortality has undoubtedly increased. (Courtesy of Gregory S. Stone, New England Aquarium.)

clear examples (Fig. 14.4). In the Aru Islands (Indonesia), large-mesh shark nets introduced by Taiwanese and Japanese fishermen proved efficient at catching dugongs and thus were appropriated for that purpose by both foreign fishermen and local people who salvaged lost net fragments (Compost 1978). Thus the introduction of new technology, followed by the realization that it could be used to catch a traditionally prized species, led to intensified dugong exploitation. The incidental capture of dugongs in fishing nets also sustains a strong demand for dugong meat in villages along the Arabian Gulf (Baldwin & Cockcroft 1997).

14.3.2 Vessel collisions

Since the advent of the steam engine, marine and freshwater mammals have been at risk from vessel collisions. With their ever-increasing size, power and proliferation around the world, vessels have become a direct threat to marine mammals. Collisions were first recognized as a serious problem for the Florida manatee. Almost all adults in the population bear scars or other evidence of being struck. Since 1974 when observations began, collision with watercraft has consistently been the principal anthropogenic cause of death in this manatee population (Wright *et al.* 1995; Reynolds 1999). Although scattered records of collisions with whales had appeared in the literature, it was not until Kraus's (1990) study of right whale mortality that ship strikes became recognized as a serious conservation problem for whales (Katona & Kraus 1999). Even dolphins, especially young ones, can be vulnerable (Fig. 14.5) (Parsons & Jefferson 2000; Stone & Yoshinaga 2000; Laist *et al.* 2001).

14.3.3 River regulation devices

Manatees and freshwater cetaceans are at risk of being drowned or crushed in water regulation devices. Odell and Reynolds (1979) drew attention to the mortality of manatees in flood control dams and canal locks in Florida, speculating that animals are entrained by strong currents when the gates open and then become pinned or stuck in the opening. A calf may be drawn through the gate and washed downstream while its mother remains upstream, too large to go through the gate herself.

There is also evidence that endangered river dolphins die from the effects of river regulation devices. A baiji drowned after being entrained by the intake of a pumping station servicing ship locks on the Yangtze River (Zhou & Zhang 1991). Indus dolphins sometimes enter canals or seasonally inundated areas during the rainy season and are prevented by structures from returning to the main river channel (Reeves & Choudhry 1998).

14.3.4 Acute mortality events (e.g. toxic spills, explosions)

Die-offs of marine mammals have become a major conservation concern, especially since the mid to late 1980s. In 1987–88 some 5000–10 000 Baikal seals and about 17 000 European harbour seals died of morbillivirus infection, while more than 750 bottlenose dolphins died along the eastern United States, possibly from a disease syndrome involving a morbillivirus outbreak (Geraci *et al.* 1999). The aetiology of mass mortality is often difficult to establish even when the immediate cause of death is clear (Fig. 14.6). For example, although a die-off of about 150 Florida manatees in 1996 apparently was caused by exposure to a phytotoxin associated with red tide (Bossart *et al.* 1998), human activities may have increased the frequency and severity of red tides in Florida (Reynolds 1999). Therefore, even though the immediate cause of death was a naturally occurring organism, human influence on the environment may have contributed to the mass mortality (see Section 14.4.2 for more discussion).

Some die-offs have more straightforward causes. Among the most notorious marine mammal die-offs on record was that resulting from the 1989 *Exxon Valdez* oil spill in Prince William Sound, Alaska. Several thousand sea otters and about 300 harbour seals died from the direct effects of exposure to oil (Loughlin 1994; Monson *et al.* 2000; but see Hoover-Miller *et al.* 2001). Although all marine mammals could be negatively affected by an oil spill, sea otters, fur seals and polar bears are probably the most vulnerable because of their dependence on fur for insulation (Geraci & St Aubin 1990). Once the fur is badly oiled, the animal's fate is sealed.

Fig. 14.6 Dead Mediterranean monk seals on the beach in Mauritania during a major die-off in 1997. Toxic algae are thought to have caused the seal deaths, but no definitive aetiology was established. (Courtesy of Alex Aguilar.)

14.4 INDIRECT INFLUENCES

14.4.1 Environmental carrying capacity

Carrying capacity refers to the number of animals that the environment can support. When habitat is eliminated or degraded, or when the resources on which a species depends are depleted, the environmental carrying capacity for that species has been reduced. Although it is frequently mentioned as a potential problem (e.g. Evans 1990; Reijnders 1992b; Addink & Smeenk 1999) there is little conclusive evidence that prey depletion caused by human activities has reduced the carrying capacity for marine mammals. Major fishery-caused reductions in herring and mackerel resulted in a complex series of responses by cetacean populations in the Gulf of Maine (Kenney *et al.* 1996). The distribution of humpback whales changed as they switched from eating herring to eating mainly sand lance. When the sand lance population crashed a few years later, fin and humpback whales became uncommon and were 'replaced' by the more planktivorous right and sei whales. There was also a dramatic regional change in dolphin populations, with Atlantic white-sided dolphins becoming abundant and white-beaked dolphins rare. Harbour porpoises, which had depended heavily on Georges Bank herring before the stock crashed in 1980, apparently shifted their distribution inshore, where they became subject to a high rate of bycatch in bottom-set gillnets. It is uncertain whether any of these cetacean populations declined as a direct result of the fish stock depletion. However, changed distribution apparently led to increased incidental mortality for harbour porpoises. A similar situation occurred in Newfoundland during the late 1970s and early 1980s when the abundance of capelin was 'exceptionally low' offshore and the number of humpback whales inshore 'increased dramatically' (Whitehead & Carscadden 1985). It was during this time that humpback entrapment in inshore fishing gear emerged as a serious problem—for both the whales and the fishermen.

The northern (Steller) sea lion in the North Pacific provides another example of possible carrying-capacity reduction by fisheries. A major decline in the Alaskan population of sea lions has probably been driven by a combination of factors, including bycatch, shooting by fishermen, commercial hunting through 1972 and broad-scale changes in the ecosystem (Alverson 1992; National Marine Fisheries Service 1992; Marine Mammal Commission 1999). An ecosystem-wide change in fish populations caused primarily by commercial fishing is thought to have been particularly important. It is uncertain whether the effect comes more from reduced fish abundance or changed composition of the fish fauna. Between the 1960s and 1980s, walleye pollock replaced herring as the dominant pelagic species (in terms of biomass) in the Bering Sea and Gulf of Alaska. Sea lions may now eat more non-fatty fish (e.g. pollock) than

fatty fish (e.g. herring), resulting in a nutritional deficiency.

The carrying capacities of some Asian rivers have been reduced by the abstraction of water. For example, irrigation agriculture is the backbone of Pakistan's economy and it requires that most of the annual flow in the Indus River is diverted into canals before reaching the delta. As a result, Indus dolphins have probably lost at least half of their historical range (Reeves *et al.* 1991). Dams built for flood control or irrigation modify the flow and sediment regimes of rivers and block the migration and dispersal of many species, including dolphins and manatees (Reeves & Smith 1999; Smith & Reeves 2000).

Some dams may increase manatee habitat by promoting the growth of forage plants in reservoirs, and this could partly offset the impacts of population fragmentation (Rosas 1994). Moreover, the warm-water discharges at thermal power plants provide artificial winter refugia for manatees, so the number present in Florida waters during the winter months may now be larger than it was under natural conditions. However, these apparent 'benefits' to manatees from human activities have a downside. The clumping of animals near power plants can make them susceptible to die-offs caused by disease or point-source pollution. Operating lifespans of both dams and power plants are limited, so their associated 'benefits' are unreliable in the long term. Finally, power plant effluents have negative effects on seagrasses and may therefore reduce the amount of food locally available for manatees (Reynolds 1999).

The appropriation of bays for mariculture is another way in which human activity reduces the carrying capacity for marine mammals. Net enclosures take up space, and the waste products of fish farming contribute to eutrophication and chemical contamination of semienclosed waters. An emergent problem involves the operation of pinniped exclusion devices that produce loud, obnoxious sounds to prevent pinnipeds from raiding fish pens (Reeves *et al.* 1996b; Johnston & Woodley 1998). These 'seal scarers' may create more problems than they solve, as the noise is thought to drive inshore cetaceans (e.g. harbour porpoises) away from preferred feeding or resting areas (Morton 2000).

Fig. 14.7 Harbour seals housed at the Institute for Forestry and Nature Research at Texel, the Netherlands, during an experiment to investigate the effects of contaminants on reproductive performance. (Courtesy of Peter Reijnders.)

14.4.2 Effects of chemical and noise pollution

The effects of chemical pollution on marine mammals are a growing concern, but most research on the problem has consisted only of documenting tissue contaminant levels. Relatively little progress has been made towards understanding the nature of effects (O'Shea 1999; O'Shea *et al.* 1999; Reijnders *et al.* 1999).

Impaired reproductive performance associated with organochlorine contamination has been reported for California sea lions in the eastern North Pacific (DeLong *et al.* 1973), ringed and grey seals in the Baltic Sea (Helle 1980; Bergman & Olsson 1985), harbour seals in the Wadden Sea (Reijnders 1980) and belugas in the St Lawrence River (Béland *et al.* 1987; Martineau *et al.* 1994). In an experimental study, female harbour seals that were fed fish from the polluted Wadden Sea had poorer reproductive success than those fed less contaminated fish from the Atlantic (Reijnders 1986) (Fig. 14.7). The reproductive disorder occurred around the time of implantation and was accompanied by lowered levels of oestradiol 17β. The effects were attributed to polychlorinated biphenyls (PCBs) or PCB metabolites. The group of seals with the highest PCB intake had reduced blood levels of thyroid hormones and vitamin A (Brouwer *et al.* 1989). The biological significance may be that thyroid hormones play

an important role in spermatogenesis and vitamin A in reproductive success. Despite strongly suggestive findings in the other studies cited above, it has not been possible to confirm a causal link between specific chemicals and the observed effects on marine mammals (Reijnders & Brasseur 1992; O'Shea & Brownell 1994; Colborn & Smolen 1996); aetiology of the reproductive disorders remains uncertain (Reijnders 1999). Lower-than-expected blood testosterone levels in Dall's porpoises were associated with high PCB and dichloro-diphenyl-chloroethane (DDE) levels in their blubber (Subramanian *et al.* 1987). Again, however, the biological significance and mechanism of action are unclear.

A suite of pathological disorders including sterility, adrenocortical hyperplasia and osteoporosis has been observed in seals from the relatively contaminated Baltic and North Seas (Bergman & Olsson 1985; Stede & Stede 1990; Mortensen *et al.* 1992). Immunosuppression, thyroid hormone depletion and other non-reproductive disorders have also been associated with the presence of contaminants in tissues of marine mammals. It has been suggested that lowered immunocompetence induced by contaminants exacerbated die-offs of dolphins in the North Atlantic and Mediterranean Sea (Geraci 1989; Aguilar & Borrell 1994; Lahvis *et al.* 1995) and the seal die-offs mentioned earlier (Dietz *et al.* 1989; Grachev *et al.* 1989). Indeed, in an experiment with captive harbour seals, parameters related to immune function were compromised in a group fed contaminated fish from the Baltic Sea compared with a group that received a less contaminated diet from the North Sea (de Swart *et al.* 1994). Still, the significance of this finding remains uncertain. Focused studies are needed that combine immunology, toxicology and demography and that involve populations occurring over a gradient of environments, from highly polluted to relatively pristine (Reijnders *et al.* 1999).

The difficulty of establishing cause and effect relationships between pollution and the health of marine mammals is exemplified by the white whales in the St Lawrence River. Besides being exposed through their diet to a complex mixture of contaminants, these whales were depleted by past hunting, their habitat is infested by ships and recreational boats, and numerous St Lawrence tributaries have been dammed. One or more of these factors could help explain the population's small size and slowness of recovery (Reeves & Mitchell 1984). Although high levels of organochlorines, lead and mercury (Béland *et al.* 1993) and a range of pathological conditions (notably cancer and one case of true hermaphroditism; De Guise *et al.* 1994; Martineau *et al.* 1999) have been documented in this population, the lack of good population data has made it impossible to provide clear evidence of a primary effect of contaminants on vital parameters (Lesage & Kingsley 1998).

Very high tissue levels of trace elements, particularly cadmium and mercury, have been found in some marine mammals (Wagemann & Muir 1984). Given that no adverse effects were found in the sampled individuals, it has been suggested that at least some species may be adapted to such contamination as a result of natural exposure. Along these lines, an attractive hypothesis might be that the polar bear's ability to enzymatically break down even the more persistent PCBs has been acquired selectively. Polar bears prey regularly on seals, which accumulate considerable amounts of lipophilic compounds such as PCBs.

Stress from chronic or frequent exposure to aversive acoustic stimuli or other types of 'harassment' probably compromises the health of marine mammals. However, it has been extremely difficult to demonstrate a cause and effect relationship (Richardson *et al.* 1995). Female humpback whales and their calves traditionally congregated along the lee shore of Maui (Hawaii) in winter, but they have shifted their distribution farther offshore, apparently in response to the high volume of recreational boat traffic near shore (Tyack 1989). While some level of interactions between whales and vessels may be tolerable to the animals, there is probably a threshold of disturbance at which they move to less desirable areas.

The physiological effects of exposure to loud underwater noise can include temporary or permanent shifts in hearing thresholds, which degrade the animal's ability to forage and carry out other activities that depend on auditory acuity (Ketten 1995). Two humpback whales found dead in nets in Trinity Bay, Newfoundland, had severely damaged ear structures, leading researchers to consider whether poor hearing may have reduced the animals' ability to detect the netting (Todd *et al.* 1996). Whales in the area had recently been exposed to underwater explosions and drilling associated with offshore oil development. Recent mass strandings of beaked whales have been associated with military activities involving, *inter alia*, the use of underwater acoustic systems for submarine detection and tracking (Frantzis 1998; Rowles *et al.* 2000).

14.4.3 Climate change

Climate change is bound to affect the distribution, physiology and behaviour of marine mammals although it is difficult to predict how imminent and dramatic the impacts are or will be (International Whaling Commission 1997; Tynan & DeMaster 1997). A long-term global warming trend will particularly affect species that depend on sea ice, such as phocid seals, the walrus and the polar bear. Less stable ice in some parts of the Arctic already may have reduced the availability of ringed seals to polar bears, impairing the bears' reproductive success (Stirling & Derocher 1993; Stirling & Lunn 1997). There is evidence suggesting that a large-scale, long-term climatic event has altered deep-level oceanic mixing in the mid Pacific, perhaps contributing to a decline in survival of Hawaiian monk seal pups (Polovina *et al.* 1994). The consequences of climate change will differ according to whether a given species is able to follow its prey or find equivalent resources in alternative habitats. Some species may switch to prey that is less rich in energy. Some seal species that depend on land for whelping and lactation may encounter difficulties when the predicted sea level rise occurs.

14.5 MANAGEMENT OF THREATS (CONSERVATION MEASURES)

14.5.1 Regulation of killing and capture

The types of legal instruments and institutions needed to regulate the killing and removal of marine mammals from wild populations are widely agreed, at least in principle (Mangel *et al.* 1996). Management regimes need to have clearly expressed goals and objectives, whether to maximize production, minimize the risks of extinction or achieve a balance between human benefits and ecological risks. Ideally, people affected by management measures ('stakeholders') should perceive the regime as legitimate, i.e. believe that it is fair and reflects their values. Legitimacy also depends on credibility. If restrictions on human activity are to be accepted, people must be convinced that they are based on a genuine understanding of natural processes and limits. Management also needs to incorporate uncertainty so that the effects of miscalculation and error are reversible (Taylor & Gerrodette 1993).

Management regimes also need to be established at an appropriate scale. For some species, an international agreement or treaty is essential, e.g. the 1973 International Agreement on the Conservation of Polar Bears (Prestrud & Stirling 1994; Servheen et al. 1999) and the 1980 Convention on the Conservation of Antarctic Marine Living Resources (Kimball 1999). These instruments concern species whose distributions are circumpolar and include several national jurisdictions (the polar bear) and a large commons (the Antarctic). There is ongoing controversy over the scale of governance in the case of commercial whaling. Some insist that the migratory nature of whale stocks means that management authority should rest with an international body, namely the IWC (Holt 1985), others that a local or regional approach would be more equitable and effective (Sigurjónsson 1989). There is also disagreement within the IWC membership concerning the convention's applicability to small cetaceans (e.g. Birnie 1997).

Over the past 150 years the Pacific walrus population has, on at least three occasions, declined precipitously and then been allowed to recover. As Fay et al. (1989) put it, 'the Pacific walrus population is being made to fluctuate like an r-selected species, whereas it should be maintained at a high equilibrium with its environment, as befits a K-selected species . . .'. According to these authors, the problem resides in an insufficiently sensitive system for population monitoring (which may be responsible for delays in managerial response) and the separate, unilateral nature of management by the United States and the former Soviet Union.

Cultural checks to overexploitation have existed in many areas, but they are unlikely to be maintained in an increasingly 'globalized' world (Smith 1985). Ecuador's Siona Indians are said to adhere to a widespread taboo against killing and eating botos and to practice a self-imposed ban on manatee hunting because of low manatee numbers (Timm et al. 1989). In contrast, local people in Peru pay no heed to the manatee's legally protected status and hunt it regularly and openly (Reeves et al. 1996a). Most river people in Peru still recognize the taboos against killing freshwater dolphins, but some have adopted the view of commercial fishermen that dolphins should be killed to protect fish stocks and fishing gear (Reeves et al. 1999).

When local people have a long history of killing and consuming marine mammals to survive, special effort is required to accommodate both cultural and biological survival. Australia, the United States and Canada have engaged in protracted negotiations with aboriginal people to develop 'co-management' agreements (e.g. Smith & Marsh 1990; Huntington 1992; Marsh & Lefebvre 1994; Marsh et al. 1997; Baur et al. 1999). The goal is for hunting communities to manage their own affairs while at the same time ensuring the long-term conservation of marine mammal populations. 'Co-management' is always difficult because of the deep cultural (and often linguistic) differences, the reluctance of government agencies to cede authority and the amount of time and money needed for negotiations (Richard & Pike 1993). Having a large, productive animal population to begin with facilitates flexibility and compromise, but all too often the population at issue is already small and depleted.

When there is both a regulatory framework and a capability and willingness to enforce regulations, there is at least the potential to keep removals within sustainable bounds (i.e. 'sustainable use' may be feasible; see Section 14.2.1). Two approaches developed during the last few decades (discussed below, in Sections 14.5.2 and 14.5.3) are rooted in the principle that decisions need to incorporate uncertainty and provide some kind of buffer against errors in fact or judgement. This is frequently called the 'precautionary principle'.

14.5.2 The revised management procedure of the International Whaling Commission

The International Convention for the Regulation of Whaling (1946) is a utilitarian document and its wording is hard to misconstrue. The objective, plainly expressed in the preamble, was to provide for 'the proper conservation of whale stocks and thus make possible the orderly development of the whaling industry'(Committee for Whaling Statistics 1948). Reference is made to the 'common interest' of nations in bringing all whale stocks to 'the optimum level . . . as rapidly as possible without causing wide-spread economic and nutritional distress'. The IWC failed in its mandate until the 1970s, by which time many additional stocks had been depleted (Allen 1980). An important change was the implementation in 1972 of an international observer scheme. A new management procedure (NMP) adopted in 1975 provided for stocks to be classified in relation to their maximum sustainable yield (MSY) levels. Stocks judged to be below MSY were automatically protected while those above MSY were managed to remain so. Operationally, the lower bound of the MSY level was defined as 60% of the carrying-capacity population size. Although the NMP was a considerable improvement on the previous *ad hoc* system, it placed an unrealistic burden on the Scientific Committee to provide advice in the absence of adequate data. Amid widespread dissatisfaction with both the process and outcome of whaling management, the IWC agreed in 1982 to establish a worldwide 'pause' in commercial whaling. This measure, generally known as the 'moratorium', was seen by many activists as a permanent global ban rather than a period in which the status of stocks could be thoroughly reviewed and a new management scheme devised (Holt 1985).

In 1993 the IWC adopted a simple, elegant algorithm known as the revised management procedure, or RMP, with the goal of ensuring that depleted stocks recover and that no additional stocks fall below 54% of their initial abundance (Cooke 1994; Gambell 1999). The RMP is also intended to provide the highest possible continuing yield and ensure stable catch limits from year to year. At its most basic level, the RMP requires only a current estimate of population size and a catch history. Catch limits are 'tuned' according to what information is available. If population estimates are imprecise or dated, or the catch history is incomplete, then the allowable take is low. As surveys or other studies reveal more about the stock's condition, the take is adjusted to reflect reduced uncertainty. In this way, the procedure rewards effort at stock assessment and catch documentation. Usually catch limits are set at less than 2% of estimated abundance although in some circumstances they can be as high as 5%. When there is uncertainty about stock identity (see Chapter 11), the total catch limit is allocated to smaller areas in proportion to the estimated abundance.

The Achilles heel of the RMP is compliance. Clark (1976) elegantly demonstrated that when the monetary discount rate exceeds the rate of natural increase of a living resource, it may not be economically optimal to exploit it on a sustainable basis. Rather, there will be a strong incentive to 'mine' the resource to exhaustion and reinvest the profits. Thus the economic motivation may exist for the commercial whaling industry to hunt whales to extinction. Revelations that whaling catch data were falsified (see Section 14.2.2) make conservationists reluctant to embrace the RMP, which assumes that catch data and survey estimates of abundance are reliable. Not only is there a well-documented record of catch reporting fraud, but whalers are routinely engaged as spotters in surveys to estimate abundance. In today's atmosphere of distrust, implementation of the RMP has itself been stalled by internal IWC wrangling.

14.5.3 The 'potential biological removal' approach

The Marine Mammal Protection Act (MMPA) has shaped policies for conserving marine mammals in US waters since 1972. Some of the concepts and methods developed under the MMPA have been innovative and influential (e.g. Twiss & Reeves 1999). The management goal is for stocks to reach or remain at 'optimum sustainable population' (OSP) levels, which represent the number of animals providing 'maximum productivity' of the population or species, keeping in mind the carrying capacity of the habitat and the health of the ecosystem of which they are a part. Operationally, OSP is defined as falling somewhere between carrying capacity (historical or current) and the maximum net productivity level (MNPL) (Gerrodette & DeMaster 1990). Commercial exploitation is banned but aboriginal people are allowed to continue hunting marine mammals for 'subsistence'. It was not until the mid 1990s that a workable scheme was adopted for managing the bycatch in commercial fisheries.

This scheme seeks to estimate how many individuals can be 'safely' removed from a population annually while still meeting the goals of the MMPA (Barlow *et al.* 1995; Wade 1998). The mortality limit is called the potential biological removal, or PBR, calculated as the minimum population estimate (N_{min}) times one-half the maximum net productivity rate (R_{max}) times a recovery factor (*FR*). While N_{min} and R_{max} are supposed to be empirically derived, *FR* is set to reflect a stock's legal or recovery status, for example whether it is listed as endangered, threatened or depleted. Management measures are triggered by a finding that estimated total removals (including fishery bycatch, ship strikes and deliberate catch) exceed the PBR.

Even though the PBR approach provides a strong scientific basis for conservation action, it does not ensure that needed actions will be taken. For example, in the case of North Atlantic right whales, the elimination of whaling as a source of mortality was relatively easily accomplished. Reducing incidental deaths from ship strikes and entanglement in fishing gear is a much greater challenge. As Katona and Kraus (1999) state, 'the responsibility for killing North Atlantic right whales has broadened from a relatively small group of hunters seeking oil and baleen to a very diffuse group that includes most of us who eat seafood; purchase foreign autos, petroleum, appliances, or other products that arrive on ships; profit from the export of goods by sea; or benefit from the services of the Navy or Coast Guard'. Take-reduction teams, consisting of representatives of various stakeholder groups (e.g. the fishing and shipping industries, the military, government agencies, non-governmental conservation organizations, tourboat operators, etc.) carry on the hard work of finding ways to reduce marine mammal mortality without putting people out of work.

14.5.4 The role of technology

Technology or changes in techniques can contribute to the resolution of some conservation problems. For example, two innovations by fishermen—the 'backdown' procedure and the Medina panel—were instrumental in reducing dolphin mortality in the eastern tropical Pacific tuna fishery (Francis *et al.* 1992; Gosliner 1999). Attention had been drawn in the late 1960s to the fact that hundreds of thousands of dolphins were being killed each year in purse seines targeting yellowfin tuna. The fishermen had no wish to kill dolphins as they depended on the mammals to help them find and track fish schools. The strong tuna–dolphin bond is still not fully understood but may involve a trophic benefit, protection from predators or energy savings. After the tuna–dolphin complex has been trapped within the net, the fishermen attempt to release the dolphins by using the backdown procedure, invented and refined in the late 1960s to early 1970s. Once about half the net has been hauled aboard, the vessel is put into reverse, causing a long, narrow channel to form inside the net. The dolphins are herded towards the far end of this channel where they can escape over the submerged rim of the net. The Medina panel, named after the tunaboat captain who invented it, is a strip of small-mesh webbing in the backdown area of the net. The small mesh reduces the risk that a dolphin's beak, flipper or tail will become entangled while it awaits an opportunity to escape over the corkline. While it was technology that created the problem in the first place, these changes in operations and equipment have substantially reduced dolphin mortality.

Another example of a technical fix is the electronic pinger, a device attached to gillnets to warn or repel cetaceans (and pinnipeds) and thus prevent their entanglement. Experimental research with sonic devices and other types of gear modification began in the early 1980s (Todd & Nelson 1994) but it was not until 1994 that a properly designed field study demonstrated the efficacy of pingers for reducing harbour porpoise mortality in sink gillnets (Kraus *et al.* 1997). The use of pingers has since become part of an elaborate management scheme to keep porpoise mortality in the Gulf of Maine below the PBR level. While pingers are certainly not a panacea for all fishery conflicts and may have undesirable side-effects (such as displacement of animals away from important habitat, e.g. Reeves *et al.* 1996b; Dawson *et al.* 1998; Northridge & Hofman 1999; International Whaling Commission 2000), their experimental use in a drift gillnet fishery off California produced promising results for reducing mortality of several additional cetacean and pinniped species (Barlow & Cameron 1999).

14.5.5 Mitigation of indirect threats

There is an obvious need to mitigate or eliminate the indirect threats identified in Section 14.4. However, this can rarely be accomplished without a painstaking process of documentation and education, if not also innovation and invention. As a first step, it is necessary to establish that a perceived problem is real. This challenge is similar to ones facing the medical field. Without conducting experiments under realistic conditions using the subjects of concern (i.e. people), it may be impossible to establish the aetiology of a disease or the effectiveness of a cure. It is extremely difficult and expensive to demonstrate conclusively that noise or chemical pollution, for example, harms marine mammals. Experimental control is often infeasible when one is studying free-ranging animals in their natural environment. As a consequence, results are open to interpretive disagreement.

There has been hope that structures such as dams and flood-control devices could be modified to reduce their negative effects on marine mammals. For example, Reeves and Leatherwood (1994b) refer to the possible design of a dolphin 'swimway' that would function in a manner similar to that of a fish ladder, allowing river dolphins to move upstream and downstream past an irrigation barrage, and consideration has been given to devices and techniques for reducing manatee deaths in flood gates and navigation locks in Florida (Florida Manatee Recovery Team 1996).

14.5.6 Translocation

Translocation means taking wild animals from one part of their range and releasing them into another.

It can facilitate conservation by reintroducing a species where it has been extirpated, by reinforcing or supplementing an existing population with new animals, or (occasionally and as a last resort) by introducing a species outside its recorded distribution but within appropriate habitat (IUCN 1998). The best example of translocation for marine mammals is the sea otter, which had been functionally eliminated from most of its range by the end of the nineteenth century. The US government gave the species legal protection in 1911 and by the 1940s it was clear that a strong recovery was underway in the Aleutian Islands (Kenyon 1969). Thereafter, a series of translocation efforts re-established sea otters in much of their original range (Riedman & Estes 1990).

Although endowed with much charisma, sea otters are not universally admired. When reintroduced to an area, they quickly deplete local populations of certain shellfish, including ones that are valued by people. Conflict has been most acute in central California where today's population of about 2400 otters compares with 13 000–16 000 historically (DeMaster *et al.* 1996). The extreme historical reduction in sea otter numbers allowed dense populations of abalones, clams and crabs to become established, and these have supported important commercial and recreational fisheries (Estes & VanBlaricom 1985). The state's recent attempt to translocate otters to San Nicolas Island in order to establish a second population, and thus reduce the risk of an oil spill wiping out the entire California population, was controversial. It proved difficult not only to get the translocated otters to adapt to their new surroundings, but also to keep them from moving to sites along the mainland where fishermen had been assured that they would be excluded ('zonal management') (Gerber *et al.* 1999; Marine Mammal Commission 1999).

Translocation has also been used in efforts to increase survival and recruitment of Hawaiian monk seals (Lavigne 1999b; Ragen & Lavigne 1999). Weaned female pups at Kure Atoll were captured and kept in a shoreline enclosure for several weeks or months, then released. This 'headstart' programme was intended to protect the young seals from sharks, aggressive adult males and human disturbance during the critical period while they learned to forage independently. In another programme, undersized immature females were moved from French Frigate Shoals to Kure Atoll or Midway Atoll. Poor survivorship at French Frigate Shoals had been thought to be the result of inadequate food, so moving the seals was expected to improve their chances of survival and reduce the pressure on prey resources near their natal beaches. Also, it was hoped that the addition of young females would enhance the subpopulations at the two atolls. Monk seal recovery has been hampered by 'mobbing', in which several adult males severely injure or kill another seal while attempting to mount and mate with it. In the 1980s and 1990s adult males were translocated from Laysan Island to either Johnston Atoll or the main Hawaiian Islands with the goal of ameliorating the mobbing problem. While these efforts were carefully planned, conducted and monitored, they have not been sustained. Nor have they succeeded in halting the monk seal population's decline.

The salvage, captive rehabilitation and release of orphaned or stranded marine mammals is a form of translocation that, in some instances, might serve a conservation function. For example, the harbour seal population in the Dutch Wadden Sea had decreased from an estimated 17 000 to less than 500 animals within six to seven decades. Between 1974 and 1988, the percentage of pups in the total population was only 12% compared with a more 'normal' 20%, and first-year mortality was 60% instead of the normally observed 30%. Therefore, the rescue, rehabilitation and release of pups was considered a useful management tool to counter the low pup production and survival (Reijnders *et al.* 1996). In follow-up studies spanning 3 years it appeared that the percentage of tag recoveries, as well as the distance from the tag-and-release site to the place of recovery, did not differ between rehabilitated and released pups and pups of comparable age that were wild-caught, tagged and released. The conclusion that released animals were adapting well to their environment was confirmed in a short study in which released animals were instrumented with VHF-radiotransmitters in order to monitor their behaviour. Within 2 weeks the animals exhibited typical harbour seal haul-out behaviour.

Fig. 14.8 In Canada's lower Bay of Fundy, harbour porpoises frequently follow herring into weirs and then become trapped inside. In the past, fishermen usually shot such porpoises. Nowadays, many of the porpoises are 'rescued' by researchers and returned to the sea. This photo shows one of the authors (RR) releasing a porpoise taken from a weir in Passamaquoddy Bay, August 1980. (Courtesy of Randi Olsen.)

Rescue, rehabilitation and release has also been undertaken with a few young Mediterranean monk seals in Greece, Madeira and Mauritania, but whether these animals survived, mixed with wild groups and contributed to reproduction remains uncertain (Reijnders *et al.* 1996; Harwood *et al.* 1998).

Regardless of whether rehabilitation programmes contribute directly to the maintenance of wild populations, they certainly perform a public awareness function. The interest of the general public in caring for and returning young seals, manatees and other marine mammals to the wild (Fig. 14.8) can create opportunities to educate people about the status of populations and threats to the environment. However, some drawbacks connected to rescue, rehabilitation and release must be considered as well (cf. St Aubin *et al.* 1996). The principal ones are that the process is acting against natural selection and involves a risk of transferring pathogens from the captive group to the wild population. On balance, rescue, rehabilitation and release should be restricted mainly to endangered species and populations. Now that the harbour seal population in the entire Wadden Sea is recovering well, the responsible ministers for Denmark, Germany and the Netherlands have agreed that such programmes should be reduced to the lowest level possible (Reijnders 1996).

Fig. 14.9 There is only one baiji, or Yangtze River dolphin, in captivity and probably fewer than 100 in the wild, making this species the most endangered cetacean. Facilities for *ex situ* maintenance have been developed in China, but it has proven all but impossible to locate, capture and transport dolphins to these sites. (Courtesy of Steve Leatherwood.)

14.5.7 *Ex situ* approaches

Captive breeding has not been used as a conservation tool with marine mammals and it is unlikely to be used in the foreseeable future. The Yangtze River dolphin, or baiji, was little known outside China until the mid 1980s, by which time its population apparently had dwindled to no more than a few hundred individuals. An international panel of experts decided that a captive baiji population should be established in at least one 'seminatural reserve' within China (Perrin & Brownell 1989). The effective population in reserves would have had to be at least 20–25 to ensure long-term viability (Ralls 1989). However, more than a decade later, it is clear that such a target population size is completely unrealistic. A single female baiji has been caught and placed in the Shishou reserve and she died within 6 months (Liu *et al.* 1998). Meanwhile, the lone baiji in captivity, a male, has remained in its concrete tank (Fig. 14.9). Those responsible for implementing the baiji rescue effort appear to have been unwilling to risk losing the male by placing him in the reserve with the female. Under present circumstances, further efforts to capture baiji and place them in tanks or reserves would be very difficult to justify on conservation grounds.

14.5.8 Protected areas

The setting aside of areas where human activities are regulated to benefit wildlife is a standard approach to conservation. Perhaps the best-known examples of protected areas for marine mammals (Reeves 2000) are the very large IWC 'sanctuaries' in the Indian Ocean and the circumpolar Antarctic. These, however, are effectively nothing more than areas closed to commercial whaling. Other potential threats, such as those related to habitat degradation, pollution, fishing and artisanal (i.e. non-commercial) whaling, are unaffected by the IWC designations.

A number of pupping and mating areas for pinnipeds have been given special protection. For example, Seal Bay Conservation Park in South Australia is intensively managed to allow sea lion watching by tourists while at the same time ensuring that sea lions remain undisturbed (Robinson & Dennis 1988). Many of the areas in California where northern elephant seals and other pinnipeds come ashore to pup, nurse and mate are managed similarly.

The Banks Peninsula Marine Mammal Sanctuary in New Zealand was established specifically to protect Hector's dolphins from entanglement in gillnets (Dawson & Slooten 1993). Commercial gillnetting is not allowed and amateur gillnetting is subject to season and area closures within the sanctuary. In Baja California, Mexico, two of the most important gray whale 'nursery' lagoons (Ojo de Liebre and San Ignacio) are officially designated whale refuges, and access for tourism and research is managed through a permit system (Dedina & Young 1995). The lagoons are also included within the El Vizcaino Biosphere Reserve and a UNESCO World Heritage site (Reeves 2000). Mexico created a biosphere reserve in the upper Gulf of California partly to protect the highly endangered vaquita (Vidal 1995), but this seems not to have led to a reduction in gillnet fishing, the principal threat to this porpoise species (cf. International Whaling Commission 2000).

Marine protected areas can involve species-specific protection from harassment or take, a 'no take' policy, or seasonal or permanent prohibitions on certain human activities. Effectiveness, as always, depends on both the wisdom embodied in the design and the capacity for implementation and enforcement. Boundaries should take account of source–sink dynamics for the species of concern and be based on a solid understanding of ecological interactions (Mangel *et al.* 1996). To the extent possible, planning should include consideration of cyclic or other variability in distribution and movements. Ideally, the boundaries of a protected area would be set according to where the animals and their resource base are located at a given point in time rather than on fixed geographical positions (Hyrenbach *et al.* 2000).

14.5.9 Marine mammal watching as a conservation strategy

The idea that whales are worth more alive as tourist attractions than dead and rendered into useful products has become popular in some western nations (Fig. 14.10). Whale watching, and in some areas dolphin, seal, polar bear and sea otter watching, has become a profitable business. Whale- and dolphin-watching enterprises have even been established in the whaling countries of Japan, Norway and Iceland. Promotion of mammal watching can be regarded as a conservation strategy because it directs the

Fig. 14.10 The lagoons of Baja California, Mexico, are a whale-watching mecca. There, thousands of gray whales converge in the winter to mate and nurse their young calves. In Laguna San Ignacio, where this photo was taken in March 1985, whales often approach whale-watching vessels, occasionally allowing the tourists to reach out and touch them. Such 'friendly' behaviour has developed despite the fact that gray whales from this population are hunted on their summer feeding grounds in the Bering and Chukchi Seas. (Courtesy of Steve Leatherwood.)

emphasis away from 'consumptive' use and towards 'non-consumptive' or 'low consumptive' use. If it is decided, however, that the only 'consumption' of whales and other marine mammals should be by watching them, the implications need to be examined. These include not only the potential disturbance to the animals from relentless pursuit by tour boats, but also the broader ecological consequences of foregoing the use of marine mammal products and replacing them through farm, ranch and aquaculture production.

14.6 ECOSYSTEM CONSERVATION

The influence of ecologists has led to an increasing emphasis on the need to make ecosystems and communities, rather than individual species, the objects of conservation efforts. It has been argued that even though it is impossible to place boundaries around ecosystems (perhaps especially aquatic systems whose borders are dynamic and fluid), and our understanding of how they function is deplorably weak, we should at least make sure that our thinking is rooted in ecosystem level concerns (Mangel *et al.* 1996; Mangel & Hofman 1999). The tension between a focus on species or populations versus a focus on communities or ecosystems is not artificial. It affects everyday decisions about allocating human and financial resources.

The most realistic, and perhaps preferred, approach is to proceed on both fronts at once. In other words, efforts to enhance species or population survival should continue, while at the same time it is recognized that ecosystems need to be kept intact and functioning. We see particular value in maintaining the focus on species when the primary or most immediate threat is overkill. In the case of North Atlantic right whales, for example, various authors have raised the possibility that habitat-related factors, such as prey densities, exploitative competition (Clapham & Brownell 1996) and chemical or noise pollution, have retarded population recovery (Mitchell 1975; Reeves *et al.* 1978; Kenney *et al.* 1986). However, it has also been shown that the mortality from ship strikes and fishing gear entanglement is a serious problem (Kraus 1990; Caswell *et al.* 1999). While it is important to address ecosystem level threats that might be contributing to the right whales' low productivity, measures to reduce mortality are urgently needed. Once a population has been brought to such a low level (300–350 individuals), protecting each individual becomes a priority. Without enough potential mothers and fathers in the living population, no amount of tinkering with the ecosystem can bring about recovery. If for no other reason, mortality reduction is important because it is something that we humans are able to control. In contrast, repairing the ecosystem is a large-scale, long-term task requiring more knowledge and know-how than usually exist.

The ability of marine mammals to persist in badly degraded habitat as long as they are not killed outright can almost rival that of our own species. River dolphins, for example, survive in areas where the terrestrial landscape is given over entirely to peasant agriculture or sprawling urbanization (R.R. Reeves, personal observations in Bangladesh). They appear highly adaptable, managing to rest, socialize and forage in conditions that are in every sense marginal. Important though it is to protect, and where possible enhance, their habitat, the greatest immediate need is to reduce mortality, particularly that caused by entanglement in gillnets which are now almost ubiquitous. Until this blatant, obvious threat is removed or reduced to a negligible level, we should resist rushing to the simplistic conclusion that habitat deterioration is responsible for population declines or extirpations of all river dolphins. (It is, of course, equally simplistic to believe that mortality reduction alone is sufficient.)

A drawback of species-orientated conservation is that attention (and resources) tend to be committed only to taxa judged to be in trouble. In other words, we tend to set aside concerns about non-endangered species while devoting all of our energies to saving those most immediately at risk. In doing so, we often fail to prevent species from becoming endangered in the first place. Managing for abundance is usually not considered a legitimate pursuit of resource agencies unless, as is the case with some marine mammals, a large standing stock is required to sustain industrial-scale exploitation. Perrin (1999) points out that small cetaceans (and, by inference, sirenians) are not good candidates for 'sustainable development' projects because of their low rates of natural increase and the difficulty of assessing and monitoring their populations at the necessary spatial and temporal scales (cf. Taylor & Gerrodette 1993).

14.7 THE ROLE OF SCIENCE

Science plays a central role in marine mammal conservation (Twiss & Reeves 1999). At a minimum, it defines and characterizes problems. Ideally, it also provides the knowledge needed to address and solve those problems. In one sense, there can never be too much scientific understanding. Since all natural systems are dynamic, and no wild population is truly stationary except in theory, there will always be a need to verify and refine even the best (i.e. most certain) predictions.

In the realm of conservation, however, science is rarely an end in itself, and investments in scientific research need to be justified by their potential contribution to conservation objectives. This imperative may apply with special force to cases where *ex situ* conservation measures are essentially impractical, and knowing more about the mating system of the species, for example, is not necessarily relevant (cf. Berger 1996). Because research is what scientists are trained to do, they can easily fall into the trap of mistaking research for conservation. All too often, policymakers are happy to oblige. Sponsoring more surveys to improve the precision of estimates means that politically risky decisions can be postponed. This can result in overassessment and underprotection.

The vaquita is probably the world's most critically endangered marine cetacean. Most scientists have long been convinced that the population is too small (probably in the low to mid hundreds) to sustain the ongoing mortality in fishing gear (International Whaling Commission 1995; Vidal 1995; Perrin 1999; Rojas-Bracho & Taylor 1999). Substantial resources have nevertheless continued to be invested in abundance surveys (Barlow *et al.* 1997; Jaramillo-Legorreta *et al.* 1999) and recovery 'planning' (Vidal 1993; International Whaling Commission 2000) while few resources have been committed to addressing the threat of bycatch in fisheries. The human and financial resources available for conservation should logically be invested in bycatch reduction measures, possibly even if it means paying fishermen not to fish. Yet the political costs of more surveys and more planning meetings are slight compared with those of actions affecting employment and short-term human welfare. Thus political decisions are taken to promote more research rather than to implement unpopular management measures.

14.8 CONCLUSIONS

In recent years, the conservation 'establishment' has produced a plethora of plans—species recovery plans, action plans, biodiversity plans, habitat management plans, etc. In fact we, the authors of this chapter, have contributed our fair share to this outpouring of documentation and 'advice'. At the same time, in the wider political and economic arena, most nations have embraced the principle that planning is anathema and that the free market should be allowed to govern decision making. Far too often, conservation plans seem to be ends rather than means—token, feel-good gestures that serve as substitutes for actions that would limit free choice and protect the natural environment from the impacts of development. There is dissonance between all of the conservation planning by governments and non-governmental organizations and the accelerating global drive to deregulate and promote free enterprise. We agree with Meffe *et al.* (1999) that human population increases, and growth in per capita consumption, are at the root of the conservation crisis. Unless these processes are checked and even reversed, wild populations of many species, including marine mammals, are doomed.

We also believe that conservation needs to be buttressed by more than scientific arguments. Ultimately, for every conservation conflict, a rationale is required for choosing one action over another. There needs to be a convincing answer to the question: 'Why should the survival of a wild species take precedence over the satisfaction of a particular array of human needs and desires?' Without a strong philosophical underpinning, conservation efforts will, in most instances, fail in the long term. We suspect that the most satisfactory answers to the question of why conservation is important will come at least partly from a normative realm (religion, ethics, etc.) rather than from science alone (Domning 1999).

Finally, marine mammal conservation cannot be considered in isolation from the more general problem of environmental health at a planetary scale, as well as many local and regional scales. Just as a minimal number of reproductively active individuals is necessary to ensure a population's persistence, a minimal amount of suitable habitat is necessary to provide those individuals with sufficient resources to meet their nutritional, behavioural and social needs. Animals and their habitat are inseparable, and both need to be preserved.

REFERENCES

Addink, M.J. & Smeenk, C. (1999) The harbour porpoise (*Phocoena phocoena*) in Dutch coastal waters: analysis of stranding records for the period 1920–1994. *Lutra* **41**, 55–80.

Aguilar, A. & Borrell, A. (1994) Abnormally high polychlorinated biphenyl levels in striped dolphins (*Stenella coeruleoalba*) affected by the 1990–1992 Mediterranean epizootic. *Science of the Total Environment* **154**, 237–247.

Allen, K.R. (1980) *Conservation and Management of Whales.* Washington Sea Grant, University of Washington Press, Seattle.

Alverson, D.L. (1992) A review of commercial fisheries and the Steller sea lion (*Eumetopias jubatus*): the conflict arena. *Reviews in Aquatic Sciences* **6**, 203–256.

Baldwin, R. & Cockcroft, V.G. (1997) Are dugongs, *Dugong dugon*, in the Arabian Gulf safe? *Aquatic Mammals* **23**, 73–74.

Barlow, J. & Cameron, G.A. (1999) *Field Experiments Show that Acoustic Pingers Reduce Marine Mammal Bycatch in the California Drift Gillnet Fishery.* Document submitted to International Whaling Commission, Cambridge, No. SC/51/SM2.

Barlow, J., Swartz, S.L., Eagle, T.C. & Wade, P.R. (1995) *US Marine Mammal Stock Assessments: Guidelines for Preparation, Background, and a Summary of the 1995 Assessments.* NOAA Technical Memorandum No. NMFS-OPR-6. US Department of Commerce, Silver Spring, MD.

Barlow, J., Gerrodette, T. & Silber, G. (1997) First estimates of vaquita abundance. *Marine Mammal Science* **13**, 44–58.

Baur, D.C., Bean, M.J. & Gosliner, M.L. (1999) Laws governing marine mammal conservation in the United States. In: *Conservation and Management of Marine Mammals* (J.R. Twiss Jr & R.R. Reeves, eds), pp. 48–86. Smithsonian Institution Press, Washington, DC.

Béland, P.R., Michaud, R. & Martineau, D. (1987) Recensements de la population de bélugas (*Delphinapterus leucas*) du Saint-Laurent par embarcations en 1985. *Rapport*

Technique Canadien des Sciences Halieutiques et Aquatiques **1545**, 21 pp.

Béland, P.R., DeGuise, S., Girard, C. *et al.* (1993) Toxic compounds and health and reproductive effects in St Lawrence beluga whales. *Journal of Great Lakes Research* **19**, 766–775.

Berger, J. (1996) Animal behaviour and plundered mammals: is the study of mating systems a scientific luxury or a conservation necessity? *Oikos* **77**, 207–216.

Bergman, A. & Olsson, M. (1985) Pathology of Baltic grey seal and ringed seal females with special reference to adrenocortical hyperplasia: is environmental pollution the cause of a widely distributed disease syndrome? *Finnish Game Research* **44**, 47–62.

Best, P.B. (1985) External characters of southern minke whales and the existence of a diminutive form. *Scientific Reports of the Whales Research Institute (Tokyo)* **36**, 1–33.

Best, P.B. (1988) Right whales *Eubalaena australis* at Tristan da Cunha—a clue to the 'non-recovery' of depleted stocks? *Biological Conservation* **46**, 23–51.

Best, P.B. (1993) Increase rates of severely depleted stocks of baleen whales. *ICES Journal of Marine Science* **50**, 169–186.

Birnie, P.W. (1997) Small cetaceans and the International Whaling Commission. *Georgetown International Environmental Law Review* **10**, 1–27.

Bonner, W.N. (1982) *Seals and Man: a Study of Interactions.* Washington Sea Grant, University of Washington Press, Seattle.

Born, E.W., Gjertz, I. & Reeves, R.R. (1995) Population assessment of Atlantic walrus. *Norsk Polarinstituttt Meddelelser* **138**, 100 pp.

Bossart, G.D., Baden, D.G., Ewing, R.Y., Roberts, B. & Wright, S.D. (1998) Brevetoxicosis in manatees (*Trichechus manatus latirostris*) from the 1996 epizootic: gross, histologic, and imunohistochemical features. *Toxicologic Pathology* **26**, 276–282.

Bräutigam, A. & Thomsen, J. (1993) Appendix 2: harvest and international trade in seals and seal products. In: *Seals, Fur Seals, Sea Lions, and Walrus: Status Survey and Conservation Action Plan* (P. Reijnders, S. Brasseur, J. van der Toorn *et al.*, eds), pp. 84–87. International Union for Conservation of Nature and Natural Resources, Gland, Switzerland.

Brouwer, A., Reijnders, P.J.H. & Koeman, J.H. (1989) Polychlorinated biphenyl (PCB)-contaminated fish induces vitamin A and thyroid hormone deficiency in the common seal *Phoca vitulina. Aquatic Toxicology* **15**, 99–106.

Brownell Jr, R.L., Ralls, K. & Perrin, W.F. (1989) The plight of the 'forgotten' whales. *Oceanus* **32** (1), 5–13.

Busch, B.C. (1985) *The War Against the Seals: a History of the North American Seal Fishery.* McGill-Queen's University Press, Kingston.

Caswell, H., Fujiwara, M. & Brault, S. (1999) Declining survival probability threatens the North Atlantic right whale. *Proceedings of the National Academy of Sciences of the United States of America* **96**, 3308–3313.

Caughley, G. (1994) Directions in conservation biology. *Journal of Animal Ecology* **63**, 215–244.

Caulfield, R.A. (1993) Aboriginal subsistence whaling in Greenland: the case of Qeqertarsuaq municipality in West Greenland. *Arctic* **46**, 144–155.

Chanin, P. (1985) *The Natural History of Otters.* Facts on File, New York.

Clapham, P.J. & Brownell Jr, R.L. (1996) The potential for interspecific competition in baleen whales. *Report of the International Whaling Commission* **46**, 361–367.

Clapham, P.J., Young, S.B. & Brownell Jr, R.L. (1999) Baleen whales: conservation issues and the status of the most endangered populations. *Mammal Review* **29**, 35–60.

Clark, C.W. (1976) *Mathematical Bioeconomics: the Optimal Management of Renewable Resources.* Wiley-Interscience, New York.

Cockcroft, V.G. (1990) Dolphin catches in the Natal sharknets. *South African Journal of Wildlife Research* **20**, 44–51.

Colborn, T. & Smolen, M.J. (1996) Epidemiological analysis of persistent organochlorine contaminants in cetaceans. *Reviews in Environmental Contamination and Toxicology* **146**, 91–171.

Committee for Whaling Statistics, ed. (1948) *International Whaling Statistics XVIII.* Oslo, Norway.

Compost, A. (1978) *Pilot Survey of Exploitation of Dugong and Sea Turtle in the Aru Islands.* Report to National Parks Development Project, United Nations Development Programme, Jakarta.

Cooke, J.G. (1994) The management of whaling. *Aquatic Mammals* **20**, 129–135.

Crawford, R.J.M., David, J.H.M., Williams, A.J. & Dyer, B.M. (1989) Competition for space: recolonizing seals displace endangered, endemic seabirds off Namibia. *Biological Conservation* **48**, 59–72.

Dawson, S.M. & Slooten, E. (1993) Conservation of Hector's dolphins: the case and process which led to establishment of the Banks Peninsula Marine Mammal Sanctuary. *Aquatic Conservation: Marine and Freshwater Ecosystems* **3**, 207–221.

Dawson, S.M., Read, A. & Slooten, E. (1998) Pingers, porpoises and power: uncertainties with using pingers to reduce bycatch of small cetaceans. *Biological Conservation* **84**, 141–146.

De Guise, S., Lagacé, A. & Béland, P. (1994) True hermaphroditism in a St Lawrence beluga whale (*Delphinapterus leucas*). *Journal of Wildlife Diseases* **30**, 287–290.

de Swart, R.L., Ross, P.S., Vedder, L.J. *et al.* (1994) Impairment of immune function in harbour seals (*Phoca vitulina*) feeding on fish from polluted waters. *Ambio* **23**, 155–159.

Dedina, S. & Young, E. (1995) *Conservation and Development in the Gray Whale Lagoons of Baja California Sur, Mexico.* Report to US Marine Mammal Commission No. PB96-1113154. National Technical Information Service, Springfield, VA.

DeLong, R.L., Gilmartin, W.G. & Simpson, J.G. (1973) Premature births in California sea lions: association with high organochlorine pollutant residue levels. *Science* **181**, 1168–1170.

DeMaster, D.P., Marzin, C.M. & Jameson, R.L. (1996) Estimating the historical abundance of sea otters in California. *Endangered Species Update* **13**, 79–82.

Dietz, R., Heide-Jørgensen, M.P. & Härkönen, T. (1989) Mass deaths of harbour seals (*Phoca vitulina*) in Europe. *Ambio* **18**, 258–264.

Dizon, A.E., Lockyer, C., Perrin, W.F., DeMaster, D.P. & Sisson, J. (1992) Rethinking the stock concept: a phylogeographic approach. *Conservation Biology* **6**, 24–36.

Dizon, A.E., Perrin, W.F. & Akin, P.A. (1994) *Stocks of Dolphins (Stenella spp. and Delphinus delphis) in the Eastern Tropical Pacific: a Phylogeographic Classification*. NOAA Technical Report NMFS No. 119. US Department of Commerce, Seattle, WA.

Dolar, M.L.L. (1994) Incidental takes of small cetaceans in fisheries in Palawan, Central Visayas and northern Mindanao in the Philippines. *Report of the International Whaling Commission, Special Issue* **15**, 355–363.

Dolar, M.L.L., Leatherwood, S.J., Wood, C.J., Alava, M.N.R., Hill, C.L. & Aragones, L.V. (1994) Directed fisheries for cetaceans in the Philippines. *Report of the International Whaling Commission* **44**, 439–449.

Domning, D.P. (1999) Endangered species: the common denominator. In: *Conservation and Management of Marine Mammals* (J.R. Twiss Jr & R.R Reeves, eds), pp. 332–341. Smithsonian Institution Press, Washington, DC.

Earle, M. (1996) Ecological interactions between cetaceans and fisheries. In: *The Conservation of Whales and Dolphins: Science and Practice* (M.P. Simmonds & J. Hutchinson, eds), pp. 167–204. Wiley, Chichester, UK.

Ellis, R. (1991) *Men and Whales*. Alfred A. Knopf, New York.

Estes, J.A. (1994) Conservation of marine otters. *Aquatic Mammals* **20**, 125–128.

Estes, J.A. & VanBlaricom, G.R. (1985) Sea-otters and shellfisheries. In: *Marine Mammals and Fisheries* (J.R. Beddington, R.J. Beverton & D.M. Lavigne, eds), pp. 187–235. George Allen & Unwin, London.

Evans, P.G.H. (1990) European cetaceans and seabirds in an oceanographic context. *Lutra* **33**, 95–125.

Fay, F.H., Kelly, B.P. & Sease, J.L. (1989) Managing the exploitation of Pacific walruses: a tragedy of delayed response and poor communication. *Marine Mammal Science* **5**, 1–16.

Fay, F.H., Burns, J.J., Stoker, S.W. & Grundy, J.S. (1994) The struck-and-lost factor in Alaskan walrus harvests, 1952–72. *Arctic* **47**, 368–373.

Fertl, D. & Leatherwood, S. (1997) Cetacean interactions with trawls: a preliminary review. *Journal of Northwest Atlantic Fishery Science* **22**, 219–248.

Florida Manatee Recovery Team (1996) *Florida Manatee Recovery Plan*. US Fish and Wildlife Service, Southeast Region, Atlanta, GA.

Fowler, C.W. (1982) Interactions of northern fur seals and commercial fisheries. In: *Transactions of the 47th North American Wildlife and Natural Resources Conference*, pp. 278–292. Wildlife Management Institute, Washington, DC.

Fraker, M.A. (1980) Status and harvest of the Mackenzie stock of white whales (*Delphinapterus leucas*). *Report of the International Whaling Commission* **30**, 451–458.

Francis, R.C., Awbrey, F.T., Goudey, C.A. *et al.* (1992) *Dolphins and the Tuna Industry*. National Academy Press, Washington, DC.

Frantzis, A. (1998) Does acoustic testing strand whales? *Nature* **392**, 29.

Freeman, M.M.R. (1993) The International Whaling Commission, small-type whaling, and coming to terms with subsistence. *Human Organization* **52**, 243–251.

Gambell, R. (1999) The International Whaling Commission and the contemporary whaling debate. In: *Conservation and Management of Marine Mammals* (J.R. Twiss Jr & R.R. Reeves, eds), pp. 179–196. Smithsonian Institution Press, Washington, DC.

Gaston, K.J. (1998) Biodiversity. In: *Conservation Science and Action* (W.J. Sutherland ed.), pp. 1–19. Blackwell Science, Oxford.

Geraci, J.R. (1989) *Clinical Investigation of the 1987–88 Mass Mortality of Bottlenose Dolphins Along the United States Central and South Atlantic Coast*. Report to National Marine Fisheries Service, US Navy (Office of Naval Research) and Marine Mammal Commission, Guelph, Ontario, Canada.

Geraci, J.R. & St Aubin, D.J., eds (1990) *Sea Mammals and Oil: Confronting the Risks*. Academic Press, New York.

Geraci, J.R., Harwood, J. & Lounsbury, V.J. (1999) Marine mammal die-offs: causes, investigations, and issues. In: *Conservation and Management of Marine Mammals* (J.R. Twiss Jr & R.R. Reeves, eds), pp. 367–395. Smithsonian Institution Press, Washington, DC.

Gerber, L.R., Wooster, W.S., DeMaster, D.P. & VanBlaricom, G.R. (1999) Marine mammals: new objectives in US fishery management. *Reviews in Fisheries Science* **7**, 23–38.

Gerrodette, T. & DeMaster, D.P. (1990) Quantitative determination of optimum sustainable population level. *Marine Mammal Science* **6**, 1–16.

Gjertz, I., Aarvik, S. & Hindrum, R. (1993) Polar bears killed in Svalbard 1987–1992. *Polar Research* **12**, 107–109.

Gjertz, I. & Wiig, Ø. (1994) Distribution and catch of white whales (*Delphinapterus leucas*) at Svalbard. *Meddelelser om Grønland, Bioscience* **39**, 93–97.

Gosliner, M.L. (1999) The tuna–dolphin controversy. In: *Conservation and Management of Marine Mammals* (J.R. Twiss Jr & R.R. Reeves, eds), pp. 120–155. Smithsonian Institution Press, Washington, DC.

Grachev, M.A., Kumarev, V.P., Mamaev, L. *et al.* (1989) Distemper virus in Baikal seals. *Nature* **338**, 209–210.

Grimwood, I.R. (1969) *Notes on the Distribution and Status of some Peruvian Mammals 1968*. Special Publication No. 21. American Committee for International Wildlife Protection and New York Zoological Society, New York.

Härkönen, T., Stenman, O., Jüssi, M. *et al.* (1998) Population size and distribution of the Baltic ringed seal (*Phoca hispida botnica*). In: *Ringed Seals in the North Atlantic* (M.P. Heide-Jørgensen & C. Lydersen, eds), pp. 167–180.

NAMMCO Scientific Publications No. 1. North Atlantic Marine Mammal Commission, Tromsø, Norway.

Harwood, J., Lavigne, D. & Reijnders, P. (1998) *Workshop on the Causes and Consequences of the 1997 Mass Mortality of Mediterranean Monk Seals in the Western Sahara, Amsterdam, 11–14 December 1997.* IBN Scientific Contributions No. 11. Institute for Forestry and Nature Research, Wageningen, the Netherlands.

Heide-Jørgensen, M.P. (1994) Distribution, exploitation and population status of white whales (*Delphinapterus leucas*) and narwhals (*Monodon monceros*) in West Greenland. *Meddelelser om Grønland, Bioscience* **39**, 135–149.

Heinsohn, G.E. (1972) A study of dugongs (*Dugong dugon*) in northern Queensland, Australia. *Biological Conservation* **4**, 205–213.

Helle, E. (1980) Lowered reproductive capacity in female ringed seals (*Phoca hispida*) in the Bothnian Bay, northern Baltic Sea, with special reference to uterine occlusions. *Annals Zoologica Fennica* **17**, 147–158.

Hofman, R.J. (1995) The changing focus of marine mammal conservation. *Trends in Ecology and Evolution* **10**, 462–465.

Holt, S. (1985) Whale mining, whale saving. *Marine Policy* **9** (3), 192–213.

Hoover, A.A. (1988) Harbor seal *Phoca vitulina*. In: *Selected Marine Mammals of Alaska: Species Accounts with Research and Management Recommendations* (J.W. Lentfer, ed.), pp. 125–157. Marine Mammal Commission, Washington, DC.

Hoover-Miller, A., Parker, K.R. & Burns, J.J. (2001) A reassessment of the impact of the *Exxon Valdez* oil spill on harbor seals (*Phoca vitulina richardsi*) in Prince William Sound, Alaska. *Marine Mammal Science* **17**, 111–135.

Huntington, H.P. (1992) The Alaska Eskimo Whaling Commission and other cooperative marine mammal management organizations in northern Alaska. *Polar Record* **28**, 119–126.

Hyrenbach, K.D., Forney, K.A. & Dayton, P.K. (2000) Marine protected areas and ocean basin management. *Aquatic Conservation: Marine and Freshwater Ecosystems* **10**, 437–458.

International Whaling Commission (1993) Report of the Scientific Committee. *Report of the International Whaling Commission* **43**, 55–228.

International Whaling Commission (1995) Report of the Scientific Committee. *Report of the International Whaling Commission* **45**, 53–221.

International Whaling Commission (1996) Report of the Scientific Committee. *Report of the International Whaling Commission* **46**, 49–236.

International Whaling Commission (1997) Report of the IWC Workshop on Climate Change and Cetaceans. *Report of the International Whaling Commission* **47**, 291–320.

International Whaling Commission (2000) Report of the Scientific Committee. *Journal of Cetacean Research and Management* **2** (Suppl.).

IUCN (1998) *Guidelines for Re-introductions.* International Union for Conservation of Nature and Natural Resources, Species Survival Commission (IUCN/SSC), Gland, Switzerland.

Jaramillo-Legorreta, A.M., Rojas-Bracho, L. & Gerrodette, T. (1999) A new abundance estimate for vaquitas: first step for recovery. *Marine Mammal Science* **15**, 957–973.

Johnston, D.W. & Woodley, T.H. (1998) A survey of acoustic harassment device (AHD) use in the Bay of Fundy, NB, Canada. *Aquatic Mammals* **24**, 51–61.

Kasuya, T. (1985) Fishery–dolphin conflict in the Iki Island area of Japan. In: *Marine Mammals and Fisheries* (J.R. Beddington, R.J. Beverton & D.M. Lavigne, eds), pp. 253–272. George Allen & Unwin, London.

Kasuya, T. (1999) Examination of the reliability of catch statistics in the Japanese coastal sperm whale fishery. *Journal of Cetacean Research and Management* **1**, 109–122.

Katona, S.K. & Kraus, S.D. (1999) Efforts to conserve the North Atlantic right whale. In: *Conservation and Management of Marine Mammals* (J.R. Twiss Jr & R.R. Reeves, eds), pp. 311–331. Smithsonian Institution Press, Washington, DC.

Kenney, R.D., Hyman, M.A.M., Owen, R.E., Scott, G.P. & Winn, H.E. (1986) Estimation of prey densities required by western North Atlantic right whales. *Marine Mammal Science* **2**, 1–13.

Kenney, R.D., Payne, P.M., Heinemann, D.W. & Winn, H.E. (1996) Shifts in northeast shelf cetacean distributions relative to trends in Gulf of Maine/Georges Bank finfish abundance. In: *The Northeast Shelf Ecosystem: Assessment, Sustainability, and Management* (K. Sherman, N.A. Jaworski & T.J. Smayda, eds), pp. 169–196. Blackwell Science, Cambridge, MA.

Kenyon, K.W. (1969) *The Sea Otter in the Eastern Pacific Ocean.* North American Fauna No. 68. US Department of the Interior, Bureau of Sport Fisheries and Wildlife, Washington, DC.

Ketten, D.R. (1995) Estimates of blast injury and acoustic trauma zones for marine mammals from underwater explosions. In: *Sensory Systems of Aquatic Mammals* (R.A. Kastelein, J.A. Thomas & P.E. Nachtigall, eds), pp. 391–407. De Spil, Woerdwen, the Netherlands.

Kimball, L.A. (1999) The Antarctic treaty system. In: *Conservation and Management of Marine Mammals* (J.R. Twiss Jr & R.R. Reeves, eds), pp. 199–223. Smithsonian Institution Press, Washington, DC.

King, J.E. (1983) *Seals of the World.* Comstock Publishing Associates, Ithaca, NY.

Kraus, S.D. (1990) Rates and potential causes of mortality in North Atlantic right whales (*Eubalaena glacialis*). *Marine Mammal Science* **6**, 278–291.

Kraus, S.D., Read, A.J., Solow, A. *et al.* (1997) Acoustic alarms reduce porpoise mortality. *Nature* **388**, 525.

Lahvis, G.P., Wells, R.S., Kuehl, D.W. *et al.* (1995) Decreased lymphocyte responses in free-ranging bottlenose dolphins (*Tursiops truncatus*) are associated with increased concentrations of PCBs and DDT in peripheral blood. *Environmental Health Perspectives* **103**, 67–72.

Laist, D.W., Coe, J.M. & O'Hara, K.J. (1999) Marine debris pollution. In: *Conservation and Management of Marine*

Mammals (J.R. Twiss Jr & R.R. Reeves, eds), pp. 342–366. Smithsonian Institution Press, Washington, DC.

Laist, D.W., Knowlton, A.R., Mead, J.G., Collet, A.S. & Podesta, M. (2001) Collisions between ships and whales. *Marine Mammal Science* 17, 35–75.

Lavigne, D.M. (1999a) Estimating total kill of northwest Atlantic harp seals, 1994–1998. *Marine Mammal Science* 15, 871–878.

Lavigne, D.M. (1999b) The Hawaiian monk seal: management of an endangered species. In: *Conservation and Management of Marine Mammals* (J.R. Twiss Jr & R.R. Reeves, eds), pp. 246–266. Smithsonian Institution Press, Washington, DC.

Lavigne, D.M., Scheffer, V.B. & Kellert, S.R. (1999) The evolution of North American attitudes toward marine mammals. In: *Conservation and Management of Marine Mammals* (J.R. Twiss Jr & R.R. Reeves, eds), pp. 10–47. Smithsonian Institution Press, Washington, DC.

Leatherwood, S. & Reeves, R.R. (1989) *Marine Mammal Research 1985–1986.* Marine Mammal Technical Report No. 1. United Nations Environment Programme, Nairobi, Kenya.

Lesage, V. & Kingsley, M.C.S. (1998) Updated status of the St Lawrence River population of the beluga, *Delphinapterus leucas. Canadian Field-Naturalist* 112, 98–114.

Lescrauwaet, A.-C. & Gibbons, J. (1994) Mortality of small cetaceans and the crab bait fishery in the Magallanes area of Chile since 1980. *Report of the International Whaling Commission, Special Issue* 15, 485–494.

Liu, R., Yang, J., Wang Ding, Zhao, Q., Wei, Z. & Wang, X. (1998) Analysis on the capture, behavior monitoring and death of the baiji (*Lipotes vexillifer*) in the Shishou semi-natural reserve at the Yangtze River, China. *IBI Reports* 8, 11–21. International Marine Biological Research Institute, Kamogawa, Japan.

Loughlin, T.R., ed. (1994) *Marine Mammals and the Exxon Valdez.* Academic Press, San Diego.

McCartney, A.P., ed. (1995) *Hunting the Largest Animals: Native Whaling in the Western Arctic and Subarctic.* Studies in Whaling No. 3, Occasional Publication No. 36. Canadian Circumpolar Institute, University of Alberta, Edmonton.

McGhee, R. (1974) *Beluga Hunters. An Archaeological Reconstruction of the History and Culture of the Mackenzie Delta Kittegaryumiut.* Newfoundland Social and Economic Studies No. 13. Institute of Social and Economic Research, Memorial University of Newfoundland, St John's, Newfoundland.

Malik, S., Wilson, P.J., Smith, R.J., Lavigne, D.M. & White, B.N. (1997) Pinniped penises in trade: a molecular-genetic investigation. *Conservation Biology* 11, 1365–1374.

Mangel, M. & Hofman, R.J. (1999) Ecosystems: patterns, processes, and paradigms. In: *Conservation and Management of Marine Mammals* (J.R. Twiss Jr & R.R. Reeves, eds), pp. 87–98. Smithsonian Institution Press, Washington, DC.

Mangel, M., Talbot, L.M., Meffe, G.K. *et al.* (1996) Principles for the conservation of wild living resources. *Ecological Applications* 6, 338–362.

Marine Mammal Commission (1999) *Marine Mammal Commission Annual Report to Congress.* Marine Mammal Commission, Washington, DC.

Marsh, H. & Lefebvre, L.W. (1994) Sirenian status and conservation efforts. *Aquatic Mammals* 20, 155–170.

Marsh, H., Gardner, B.R. & Heinsohn, G.E. (1980–81) Present-day hunting and distribution of dugongs in the Wellesley Islands (Queensland): implications for conservation. *Biological Conservation* 19, 255–267.

Marsh, H., Rathbun, G.B., O'Shea, T.J. & Preen, A.R. (1995) Can dugongs survive in Palau? *Biological Conservation* 72, 85–89.

Marsh, H., Harris, A.N.M. & Lawler, I.R. (1997) The sustainability of the indigenous dugong fishery in Torres Strait, Australia/Papua New Guinea. *Conservation Biology* 11, 1375–1386.

Martineau, D., De Guise, S., Fournier, M. *et al.* (1994) Pathology and toxicology of beluga whales from the St Lawrence estuary, Québec, Canada. Past, present and future. *Science of the Total Environment* 154, 201–215.

Martineau, D., Lair, S., De Guise, S., Lipscomb, T.P. & Béland, P. (1999) Cancer in beluga whales from the St Lawrence estuary, Quebec, Canada: a potential biomarker of environmental contamination. *Journal of Cetacean Research and Management, Special Issue* 1, 249–265.

Meffe, G.K., Perrin, W.F. & Dayton, P.K. (1999) Marine mammal conservation: guiding principles and their implementation. In: *Conservation and Management of Marine Mammals* (J.R. Twiss Jr & R.R. Reeves, eds), pp. 437–459. Smithsonian Institution Press, Washington, DC.

Mitchell, E. (1975) Trophic relationships and competition for food in northwest Atlantic whales. In: *Proceedings of the Canadian Society of Zoologists Annual Meeting, June 2–5, 1974* (M.D.B. Burt, ed.), pp. 123–132.

Mohan, R.S.L., Dey, S.C., Bairagi, S.P. & Roy, S. (1997) On a survey of the Ganges River dolphin *Platanista gangetica* of Brahmaputra River, Assam. *Journal of the Bombay Natural History Society* 94, 483–495.

Monson, D.H., Doak, D.F., Ballachey, B.E., Johnson, A. & Bodkin, J.L. (2000) Long-term impacts of the *Exxon Valdez* oil spill on sea otters, assessed through age-dependent mortality patterns. *Proceedings of the National Academy of Sciences of the United States of America* 97, 6562–6567.

Mortensen, P., Bergman, A., Bignert, A. *et al.* (1992) Prevalence of skull lesions in harbour seals (*Phoca vitulina*) in Swedish and Danish museum collections: 1835–1988. *Ambio* 21, 520–524.

Morton, A. (2000) Occurrence, photo-identification and prey of Pacific white-sided dolphins (*Lagenorhynchus obliquidens*) in the Broughton archipelago, Canada 1984–1998. *Marine Mammal Science* 16, 80–93.

Motwani, M.P. & Srivastava, C.B. (1961) A special method of fishing for *Clupisoma garua* (Hamilton) in the Ganges River system. *Journal of the Bombay Natural History Society* 58, 285–286.

National Marine Fisheries Service (1992) *Final Recovery Plan for Steller Sea Lions Eumetopias jubatus.* National Marine Fisheries Service, Office of Protected Resources,

US Department of Commerce, National Oceanic and Atmospheric Administration, Silver Spring, MD.

Northridge, S.P. & Hofman, R.J. (1999) Marine mammal interactions with fisheries. In: *Conservation and Management of Marine Mammals* (J.R. Twiss Jr & R.R. Reeves, eds), pp. 99–119. Smithsonian Institution Press, Washington, DC.

Norton, B.G. (1987) *Why Preserve Natural Variety?* Princeton University Press, Princeton, NJ.

O'Corry-Crowe, G.M. & Lowry, L.F. (1997) Genetic ecology and management concerns for the beluga whale (*Delphinapterus leucas*). In: *Molecular Genetics of Marine Mammals* (A.E. Dizon, S.J. Chivers & W.F. Perrin, eds), pp. 249–274. Special Publication No. 3. Society for Marine Mammalogy, Lawrence, KA.

Odell, D.K. & Reynolds, J.E. (1979) Observations on manatee mortality in south Florida. *Journal of Wildlife Management* 43, 572–577.

O'Shea, T.J. (1999) Environmental contaminants and marine mammals. In: *Biology of Marine Mammals* (J.E. Reynolds III & S.A. Rommel, eds), pp. 485–563. Smithsonian Institution Press, Washington, DC.

O'Shea, T.J. & Brownell Jr, R.L. (1994) Organochlorine and metal contaminants in baleen whales: a review and evaluation of conservation implications. *Science of the Total Environment* 154, 179–200.

O'Shea, T.J., Reeves, R.R. & Long, A.K., eds (1999) *Marine Mammals and Persistent Ocean Contaminants: Proceedings of the Marine Mammal Commission Workshop, Keystone, Colorado, 12–15 October 1998.* Marine Mammal Commission, Washington, DC.

Parsons, E.C.M. & Jefferson, T.A. (2000) Post-mortem investigations on stranded dolphins and porpoises from Hong Kong waters. *Journal of Wildlife Diseases* 36, 342–456.

Perrin, W.F. (1999) Selected examples of small cetaceans at risk. In: *Conservation and Management of Marine Mammals* (J.R. Twiss Jr & R.R. Reeves, eds), pp. 296–310. Smithsonian Institution Press, Washington, DC.

Perrin, W.F. & Brownell Jr, R.L., eds (1989) Report of the workshop. In: *Biology and Conservation of the River Dolphins* (W.F. Perrin, R.L. Brownell Jr, K. Zhou & J. Liu, eds), pp. 1–22. Occasional Papers of the IUCN Species Survival Commission No. 3. IUCN, Gland, Switzerland.

Perrin, W.F., Donovan, G.P. & Barlow, J., eds (1994) *Gillnets and Cetaceans.* Report of the International Whaling Commission, Special Issue No. 15. International Whaling Commission, Cambridge, UK.

Polovina, J.J., Mitchum, G.T., Graham, N.E. *et al.* (1994) Physical and biological consequences of a climate event in the central North Pacific. *Fisheries Oceanography* 3, 15–21.

Prescott-Allen, R. & Prescott-Allen, C., eds (1996) *Assessing the Sustainability of Uses of Wild Species—Case Studies and Initial Assessment Procedure.* Occasional Papers of the IUCN Species Survival Commission No. 12. IUCN, Gland, Switzerland.

Prestrud, P. & Stirling, I. (1994) The International Polar Bear Agreement and the current status of polar bear conservation. *Aquatic Mammals* 20, 113–124.

Ragen, T.J. & Lavigne, D.M. (1999) The Hawaiian monk seal: biology of an endangered species. In: *Conservation and Management of Marine Mammals* (J.R. Twiss Jr & R.R. Reeves, eds), pp. 224–245. Smithsonian Institution Press, Washington, DC.

Ralls, K. (1989) A semi-captive breeding program for the baiji, *Lipotes vexillifer*: genetic and demographic considerations. In: *Biology and Conservation of the River Dolphins* (W.F. Perrin, R.L. Brownell Jr, K. Zhou & J. Liu, eds), pp. 150–156. Occasional Papers of the IUCN Species Survival Commission No. 3. IUCN, Gland, Switzerland.

Read, A.J., Van Waerebeek, K., Reyes, J.C., McKinnon, J.S. & Lehman, L.C. (1988) The exploitation of small cetaceans in coastal Peru. *Biological Conservation* 46, 53–70.

Reeves, R.R. (1992) Recent developments in the commerce in narwhal ivory from the Canadian Arctic. *Arctic and Alpine Research* 24, 179–187.

Reeves, R.R. (1993) The commerce in maktaq at Arctic Bay, northern Baffin Island, NWT. *Arctic Anthropology* 30 (1), 79–93.

Reeves, R.R. (1998) Distribution, abundance and biology of ringed seals (*Phoca hispida*): an overview. In: *Ringed Seals in the North Atlantic* (M.P. Heide-Jørgensen & C. Lydersen, eds), pp. 9–45. NAMMCO Scientific Publications No. 1. North Atlantic Marine Mammal Commission, Tromsø, Norway.

Reeves, R.R. (2000) *The Value of Sanctuaries, Parks, and Reserves (Protected Areas) as Tools for Conserving Marine Mammals.* Report to the Marine Mammal Commission, Bethesda, MD.

Reeves, R.R. & Chaudhry, A.A. (1998) Status of the Indus River dolphin *Platanista minor. Oryx* 32, 35–44.

Reeves, R.R. & Heide-Jørgensen, M.P. (1994) Commercial aspects of the exploitation of narwhals (*Monodon monoceros*) in Greenland, with emphasis on tusk exports. *Meddelelser om Grønland, Bioscience* 39, 119–134.

Reeves, R.R. & Leatherwood, S. (1994a) *Dolphins, Porpoises, and Whales: 1994–1998 Action Plan for the Conservation of Cetaceans.* IUCN/SSC Cetacean Specialist Group, IUCN, Gland, Switzerland.

Reeves, R.R. & Leatherwood, S. (1994b) Dams and river dolphins: can they co-exist? *Ambio* 23, 172–175.

Reeves, R.R. & Mitchell, E. (1984) Catch history and initial population of white whales (*Delphinapterus leucas*) in the River and Gulf of St Lawrence, eastern Canada. *Naturaliste Canadien* 111, 63–121.

Reeves, R.R. & Smith, B.D. (1999) Interrupted migrations and dispersal of river dolphins: some ecological effects of riverine development. In: *Proceedings of the CMS Symposium on Animal Migration, Gland, Switzerland, 13 April 1997* (UNEP/CMS, ed.), pp. 9–18. Convention on Migratory Species, Technical Series Publication No. 2. United Nations Environment Programme, Bonn/The Hague.

Reeves, R.R., Mead, J.G. & Katona, S. (1978) The right whale, *Eubalaena glacialis*, in the western North Atlantic. *Report of the International Whaling Commission* **28**, 303–312.

Reeves, R.R., Tuboku-Metzger, D. & Kapindi, R.A. (1988) Distribution and exploitation of manatees in Sierra Leone. *Oryx* **22**, 75–84.

Reeves, R.R., Chaudhry, A.A. & Khalid, U. (1991) Competing for water on the Indus plain: is there a future for Pakistan's river dolphins? *Environmental Conservation* **18**, 341–350.

Reeves, R.R., Leatherwood, S., Jefferson, T.A., Curry, B.E. & Henningsen, T. (1996a) Amazonian manatees, *Trichechus inunguis*, in Peru: distribution, exploitation, and conservation status. *Interciencia* **21**, 246–254.

Reeves, R.R., Hofman, R.J., Silber, G.K. & Wilkinson, D., eds (1996b) *Acoustic Deterrence of Harmful Marine Mammal–Fishery Interactions: Proceedings of a Workshop Held in Seattle, Washington, 20–22 March 1996.* NOAA Technical Memorandum No. NMFS-OPR-10. US Department of Commerce, National Oceanic and Atmospheric Administration, National Marine Fisheries Service, Silver Spring, MD.

Reeves, R.R., Wenzel, G.W. & Kingsley, M.C.S. (1998) Catch history of ringed seals (*Phoca hispida*) in Canada. In: *Ringed Seals in the North Atlantic* (M.P. Heide-Jørgensen & C. Lydersen, eds), pp. 100–129. NAMMCO Scientific Publications No. 1. North Atlantic Marine Mammal Commission, Tromsø, Norway.

Reeves, R.R., McGuire, T.L. & Zúñiga, E.L. (1999) Ecology and conservation of river dolphins in the Peruvian Amazon. *IBI Reports* **9**, 21–32. International Marine Biological Research Institute, Kamogawa, Japan.

Reijnders, P.J.H. (1980) Organochlorine and heavy metal residues in harbour seals from the Wadden Sea and their possible effects on reproduction. *Netherlands Journal of Sea Research* **14**, 30–65.

Reijnders, P.J.H. (1986) Reproductive failure in common seals feeding on fish from polluted coastal waters. *Nature* **324**, 456–457.

Reijnders, P.J.H. (1992a) Retrospective population analyses and related future management perspectives for the harbour seal *Phoca vitulina* in the Wadden Sea. In: *Proceedings of the 7th International Wadden Sea Symposium, Ameland, the Netherlands, 22–26 October 1990* (N. Dankers, C.J. Smit & M. Scholl, eds), pp. 193–197. Netherlands Institute of Sea Research, Publication Series No. 20. Netherlands Institute of Sea Research, Texel, the Netherlands.

Reijnders, P.J.H. (1992b) Harbour porpoises *Phocoena phocoena* in the North Sea: numerical responses to changes in environmental conditions. *Netherlands Journal of Aquatic Ecology* **26**, 5–86.

Reijnders, P.J.H. (1996) Developments of grey and harbour seal populations in the international Wadden Sea: reorientation on management and related research. *Wadden Sea Newsletter* **1996-2**, 12–16.

Reijnders, P.J.H. (1999) Reproductive and developmental effects of endocrine disrupting chemicals on marine mammals. In: *Marine Mammals and Persistent Ocean Contaminants: Proceedings of the Marine Mammal Commission Workshop, Keystone, Colorado, 12–15 October 1998* (T.J. O'Shea, R.R. Reeves & A.K. Long, eds), pp. 93–100. Marine Mammal Commission, Washington, DC.

Reijnders, P.J.H. & Brasseur, S.M.J.M. (1992) Xenobiotic induced hormonal and associated developmental disorders in marine organisms and related effects in humans. In: *Chemically Induced Alterations in Sexual and Functional Development: the Wildlife/Human Connection* (T. Colborn & C. Clement, eds), pp. 131–146. Princeton Scientific Publishers, Princeton, NJ.

Reijnders, P., Brasseur, S., van der Toorn, J. *et al.* (1993) *Seals, Fur Seals, Sea Lions, and Walrus: Status Survey and Conservation Action Plan.* IUCN/SSC Seal Specialist Group. International Union for Conservation of Nature and Natural Resources, Gland, Switzerland.

Reijnders, P., Brasseur, S.M.J.M. & Ries, E.H. (1996) The release of seals from captive breeding and rehabilitation programmes: a useful conservation management tool? In: *Rescue, Rehabilitation, and Release of Marine Mammals: an Analysis of Current Views and Practices* (D.J. St Aubin, J.R. Geraci & V.J. Lounsbury, eds), pp. 54–65. NOAA Technical Memorandum No. NMFS-OPR-8. US Department of Commerce, National Oceanic and Atmospheric Administration, National Marine Fisheries Service, Silver Spring, MD.

Reijnders, P.J.H., Aguilar, A. & Donovan, G.P., eds (1999) Chemical pollutants and cetaceans. *Journal of Cetacean Research and Management, Special Issue* **1**.

Reynolds III, J.E. (1999) Efforts to conserve the manatees. In: *Conservation and Management of Marine Mammals* (J.R. Twiss Jr & R.R. Reeves, eds), pp. 267–295. Smithsonian Institution Press, Washington, DC.

Reynolds III, J.E. & Odell, D.K. (1991) *Manatees and Dugongs.* Facts on File, New York.

Richard, P.R. & Pike, D.G. (1993) Small whale co-management in the eastern Canadian Arctic: a case history and analysis. *Arctic* **46**, 138–143.

Richardson, W.J., Greene Jr, C.R., Malme, C.I. & Thomson, D.H., eds (1995) *Marine Mammals and Noise.* Academic Press, San Diego.

Riedman, M.L. & Estes, J.A. (1990) *The Sea Otter (Enhydra lutris): Behavior, Ecology, and Natural History.* Biological Report No. 90(14). US Department of the Interior, Fish and Wildlife Service, Washington, DC.

Robinson, A.C. & Dennis, T.E. (1988) The status and management of seal populations in South Australia. In: *Marine Mammals of Australasia: Field Biology and Captive Management* (M.L. Augee, ed.), pp. 87–110. Royal Zoological Society of New South Wales, Mosman.

Robinson, J.G. (1993) The limits to caring: sustainable living and the loss of biodiversity. *Conservation Biology* **7**, 20–28.

Rojas-Bracho, L. & Taylor, B.L. (1999) Risk factors affecting the vaquita (*Phocoena sinus*). *Marine Mammal Science* **15**, 974–989.

Romero, A., Agudo, A.I. & Green, S.M. (1997) Exploitation of cetaceans in Venezuela. *Report of the International Whaling Commission* **47**, 735–746.

Rosas, F.C.W. (1994) Biology, conservation and status of the Amazonian manatee *Trichechus inunguis*. *Mammal Review* **24**, 49–59.

Rosel, P. (1997) A review and assessment of the status of the harbor porpoise (*Phocoena phocoena*) in the North Atlantic. In: *Molecular Genetics of Marine Mammals* (A.E. Dizon, S.J. Chivers & W.F. Perrin, eds), pp. 209–226. Society for Marine Mammalogy Special Publication No. 3. Society for Marine Mammalogy, Lawrence, KA.

Rotterman, L.M. & Simon-Jackson, T. (1988) Sea otter *Enhydra lutris*. In: *Selected Marine Mammals of Alaska: Species Accounts and Management Recommendations* (J. Lentfer, ed.), pp. 237–275. Marine Mammal Commission, Washington, DC.

Rowles, T., Ketten, D., Ewing, R. *et al.* (2000) *Mass stranding of multiple cetacean species in the Bahamas on March 15–17, 2000.* Unpublished report submitted to the International Whaling Commission, Document No. SC/52/E28.

Servheen, C., Herrero, S. & Peyton, B., eds (1999) *Bears: Status Survey and Conservation Action Plan.* IUCN/SSC Bear and Polar Bear Specialist Groups. International Union for Conservation of Nature and Natural Resources, Gland, Switzerland.

Shaughnessy, P.D. (1985) Interactions between fisheries and Cape fur seals in southern Africa. In: *Marine Mammals and Fisheries* (J.R. Beddington, R.J. Beverton & D.M. Lavigne, eds), pp. 119–134. George Allen & Unwin, London.

Shaughnessy, P.D. & Payne, A.I.L. (1979) Incidental mortality of Cape fur seals during trawl fishing activities in South African waters. *Fisheries Bulletin of South Africa* **12**, 20–25.

Sigurjónsson, J. (1989) To Icelanders, whaling is a godsend. *Oceanus* **32** (1), 29–36.

Smith, A. & Marsh, H. (1990) Management of traditional hunting of dugongs [*Dugong dugon* (Müller, 1776)] in the northern Great Barrier Reef, Australia. *Environmental Management* **14**, 47–55.

Smith, B.D., Aminul Haque, A.K.M., Hossain, M.S. & Khan, A. (1998) River dolphins in Bangladesh: conservation and the effects of water development. *Environmental Management* **22**, 323–335.

Smith, B.D. & Reeves, R.R., eds (2000) Report of the Workshop on the Effects of Water Development on River Cetaceans, 26–28 February 1997, Rajendrapur, Bangladesh. In: *Biology and Conservation of Freshwater Cetaceans in Asia* (R.R. Reeves, B.D. Smith & T. Kasuya, eds), pp. 15–22. Occasional Paper of the IUCN Species Survival Commission No. 23. International Union for Conservation of Nature and Natural Resources, Gland, Switzerland.

Smith, N.J.H. (1985) The impact of cultural and ecological change on Amazonian fisheries. *Biological Conservation* **32**, 355–373.

St Aubin, D.J., Geraci, J.R. & Lounsbury, V.J., eds (1996) *Rescue, Rehabilitation, and Release of Marine Mammals: an Analysis of Current Views and Practices. Proceedings of a Workshop Held in Des Plaines, Illinois, 3–5 December 1991.* NOAA Technical Memorandum No. NMFS-OPR-8. US Department of Commerce, National Oceanic and Atmospheric Administration, National Marine Fisheries Service, Silver Spring, MD.

Stede, G. & Stede, M. (1990) Orientierende Untersuchungen von Seehundschädeln auf pathalogische Knochenveränderungen. In: *Zoologische und Ethologische Untersuchungen Zum Robbensterben*, pp. 31–53. Institut für Haustierkunde, Kiel, Germany.

Stirling, I. (1986) Research and management of polar bears *Ursus maritimus*. *Polar Record* **23**, 167–176.

Stirling, I. & Derocher, A.E. (1993) Possible impacts of climatic warming on polar bears. *Arctic* **46**, 240–245.

Stirling, I. & Lunn, N.J. (1997) Environmental fluctuations in arctic marine ecosystems as reflected by variability in reproduction of polar bears and ringed seals. In: *Ecology of Arctic Environments* (S.J. Woodin & M. Marquiss, eds), pp. 167–181. Blackwell Science, Oxford.

Stone, G.S. & Yoshinaga, A. (2000) Hector's dolphin (*Cephalorhynchus hectori*) calf mortalities may indicate new risks from boat traffic and habituation. *Pacific Conservation Biology* **6**, 162–171.

Subramanian, A.N., Tanabe, S., Tatsukawa, R., Saito, S. & Miyazaki, N. (1987) Reductions in the testosterone levels by PCBs and DDE in Dall's porpoises of northwestern North Pacific. *Marine Pollution Bulletin* **18**, 643–646.

Taylor, B.L. (1997) Defining 'population' to meet management objectives for marine mammals. In: *Molecular Genetics of Marine Mammals* (A.E. Dizon, S.J. Chivers & W.F. Perrin, eds), pp. 49–66. Special Publication No. 3. Society for Marine Mammalogy, Lawrence, KA.

Taylor, B.L. & Dizon, A.E. (1996) The need to estimate power to link genetics and demography for conservation. *Conservation Biology* **10**, 661–664.

Taylor, B.L. & Dizon, A.E. (1999) First policy then science: why a management unit based solely on genetic criteria cannot work. *Molecular Ecology* **8**, S11–S16.

Taylor, B.L. & Gerrodette, T. (1993) The uses of statistical power in conservation biology: the vaquita and northern spotted owl. *Conservation Biology* **7**, 489–500.

Timm, R.M., Albuja, V.L. & Clauson, B.L. (1989) Siona hunting techniques for the larger aquatic vertebrates in Amazonian Ecuador. *Studies on Neotropical Fauna and Environment* **24**, 1–7.

Todd, S. & Nelson, D. (1994) Annex F. A review of modifications to the webbing and setting strategies of passive fishing gear to reduce incidental bycatch of cetaceans. *Report of the International Whaling Commission, Special Issue* **15**, 67–69.

Todd, S., Stevick, P., Lien, J., Marques, F. & Ketten, D. (1996) Behavioural effects of exposure to underwater explosions in humpback whales (*Megaptera novaeangliae*). *Canadian Journal of Zoology* **74**, 1661–1672.

Tønnessen, J.N. & Johnsen, A.O. (1982) *The History of Modern Whaling.* University of California Press, Berkeley, CA.

Tormosov, D.D., Mikhaliev, Y.A., Best, P.B. *et al.* (1998) Soviet catches of southern right whales *Eubalaena australis*, 1951–1971. Biological data and conservation implications. *Biological Conservation* **86**, 185–197.

Twiss Jr, J.R. & Reeves, R.R., eds (1999) *Conservation and Management of Marine Mammals.* Smithsonian Institution Press, Washington, DC.

Tyack, P. (1989) Let's have less public relations and more ecology. *Oceanus* **32** (1), 103–111.

Tynan, C.T. & DeMaster, D.P. (1997) Observations and predictions of arctic climate change: potential effects on marine mammals. *Arctic* **50**, 308–322.

Van Waerebeek, K. & Reyes, J.C. (1994) Interactions between small cetaceans and Peruvian fisheries in 1988/89 and analysis of trends. *Report of the International Whaling Commission, Special Issue* **15**, 495–502.

Vaz-Ferreira, R. (1982) *Arctocephalus australis* Zimmermann, South American fur seal. In: *Mammals in the Seas. Small Cetaceans, Seals, Sirenians and Otters. Selected Papers of the Scientific Consultation on the Conservation and Management of Marine Mammals and their Environment*, Vol. IV, pp. 497–508. FAO Advisory Committee on Marine Resources Research, Working Party on Marine Mammals, Food and Agriculture Organization, Rome.

Vidal, O. (1993) Aquatic mammal conservation in Latin America: problems and perspectives. *Conservation Biology* **7**, 788–795.

Vidal, O. (1995) Population biology and incidental mortality of the vaquita, *Phocoena sinus. Report of the International Whaling Commission, Special Issue* **16**, 247–272.

Volgenau, L., Kraus, S.D. & Lien, J. (1995) The impact of entanglements on two substocks of the western North Atlantic humpback whale, *Megaptera novaeangliae. Canadian Journal of Zoology* **73**, 1689–1698.

Wade, P. (1998) Calculating limits to the allowable human-caused mortality of cetaceans and pinnipeds. *Marine Mammal Science* **14**, 1–37.

Wagemann, R. & Muir, D.G.C. (1984) Concentrations of heavy metals and organochlorines in marine mammals of northern waters: overview and evaluation. *Canadian Technical Report of Fisheries and Aquatic Sciences* **1297**, 97 pp.

Wenzel, G. (1991) *Animal Rights Human Rights. Ecology, Economy and Ideology in the Canadian Arctic.* University of Toronto Press, Toronto.

Whitehead, H. & Carscadden, J.E. (1985) Predicting inshore whale abundance—whales and capelin off the Newfoundland coast. *Canadian Journal of Fisheries and Aquatic Sciences* **42**, 976–981.

Whitehead, H. & Weilgart, L. (2000) The sperm whale: social females and roving males. In: *Cetacean Societies: Field Studies of Dolphins and Whales* (J. Mann, R.C. Connor, P.L. Tyack & H. Whitehead, eds), pp. 154–172. University of Chicago Press, Chicago, IL.

Whitehead, H., Reeves, R.R. & Tyack, P.L. (2000) Science and the conservation, protection, and management of wild cetaceans. In: *Cetacean Societies: Field Studies of Dolphins and Whales* (J. Mann, R.C. Connor, P.L. Tyack & H. Whitehead, eds), pp. 308–332. University of Chicago Press, Chicago, IL.

Wilson, E.O. (1992) *The Diversity of Life.* Harvard University Press, Cambridge, MA.

Wright, S.D., Ackerman, B.B., Bonde, R.K., Beck, C.A. & Banowetz, D.J. (1995) Analysis of watercraft-related mortality of manatees in Florida, 1979–1991. In: *Population Biology of the Florida Manatee* (T.J. O'Shea, B.B. Ackerman & H.F. Percival, eds), pp. 259–268. National Biological Service, Information and Technology Report No. 1. US Department of the Interior, Washington, DC.

Yablokov, A.V. (1994) Validity of whaling data. *Nature* **367**, 108.

Zachariassen, P. (1993) Pilot whale catches in the Faroe Islands, 1709–1992. *Report of the International Whaling Commission, Special Issue* **14**, 69–88.

Zemsky, V.A., Berzin, A.A., Mikhaliev, Y.A. & Tormosov, D.D. (1995) Soviet Antarctic pelagic whaling after WWII: review of actual catch data. *Report of the International Whaling Commission* **45**, 131–135.

Zhou, K. & Zhang, X. (1991) *Baiji: the Yangtze River Dolphin and Other Endangered Animals of China.* Stone Wall Press, Washington, DC.

Index

Printed and bound by CPI Group (UK) Ltd, Croydon, CR0 4YY

Printed and bound by CPI Group (UK) Ltd, Croydon, CR0 4YY

27/10/2024

14580387-0004